Selected Titles in This Series

61 **T. N. Bailey and A. W. Knapp, Editors,** Representation theory and automorphic forms (International Centre for Mathematical Sciences, Edinburgh, Scotland, March 1996)

60 **David Jerison, I. M. Singer, and Daniel W. Stroock, Editors,** The legacy of Norbert Wiener: A centennial symposium (Massachusetts Institute of Technology, Cambridge, October 1994)

59 **William Arveson, Thomas Branson, and Irving Segal, Editors,** Quantization, nonlinear partial differential equations, and operator algebra (Massachusetts Institute of Technology, Cambridge, June 1994)

58 **Bill Jacob and Alex Rosenberg, Editors,** K-theory and algebraic geometry: Connections with quadratic forms and division algebras (University of California, Santa Barbara, July 1992)

57 **Michael C. Cranston and Mark A. Pinsky, Editors,** Stochastic analysis (Cornell University, Ithaca, July 1993)

56 **William J. Haboush and Brian J. Parshall, Editors,** Algebraic groups and their generalizations (Pennsylvania State University, University Park, July 1991)

55 **Uwe Jannsen, Steven L. Kleiman, and Jean-Pierre Serre, Editors,** Motives (University of Washington, Seattle, July/August 1991)

54 **Robert Greene and S. T. Yau, Editors,** Differential geometry (University of California, Los Angeles, July 1990)

53 **James A. Carlson, C. Herbert Clemens, and David R. Morrison, Editors,** Complex geometry and Lie theory (Sundance, Utah, May 1989)

52 **Eric Bedford, John P. D'Angelo, Robert E. Greene, and Steven G. Krantz, Editors,** Several complex variables and complex geometry (University of California, Santa Cruz, July 1989)

51 **William B. Arveson and Ronald G. Douglas, Editors,** Operator theory/operator algebras and applications (University of New Hampshire, July 1988)

50 **James Glimm, John Impagliazzo, and Isadore Singer, Editors,** The legacy of John von Neumann (Hofstra University, Hempstead, New York, May/June 1988)

49 **Robert C. Gunning and Leon Ehrenpreis, Editors,** Theta functions – Bowdoin 1987 (Bowdoin College, Brunswick, Maine, July 1987)

48 **R. O. Wells, Jr., Editor,** The mathematical heritage of Hermann Weyl (Duke University, Durham, May 1987)

47 **Paul Fong, Editor,** The Arcata conference on representations of finite groups (Humboldt State University, Arcata, California, July 1986)

46 **Spencer J. Bloch, Editor,** Algebraic geometry – Bowdoin 1985 (Bowdoin College, Brunswick, Maine, July 1985)

45 **Felix E. Browder, Editor,** Nonlinear functional analysis and its applications (University of California, Berkeley, July 1983)

44 **William K. Allard and Frederick J. Almgren, Jr., Editors,** Geometric measure theory and the calculus of variations (Humboldt State University, Arcata, California, July/August 1984)

43 **François Trèves, Editor,** Pseudodifferential operators and applications (University of Notre Dame, Notre Dame, Indiana, April 1984)

42 **Anil Nerode and Richard A. Shore, Editors,** Recursion theory (Cornell University, Ithaca, New York, June/July 1982)

41 **Yum-Tong Siu, Editor,** Complex analysis of several variables (Madison, Wisconsin, April 1982)

40 **Peter Orlik, Editor,** Singularities (Humboldt State University, Arcata, California, July/August 1981)

39 **Felix E. Browder, Editor,** The mathematical heritage of Henri Poincaré (Indiana University, Bloomington, April 1980)

38 **Richard V. Kadison, Editor,** Operator algebras and applications (Queens University, Kingston, Ontario, July/August 1980)

(Continued in the back of this publication)

Proceedings of Symposia in
PURE MATHEMATICS

Volume 61

Representation Theory and Automorphic Forms

Instructional Conference
International Centre for Mathematical Sciences
March 1996
Edinburgh, Scotland

T. N. Bailey
A. W. Knapp
Editors

American Mathematical Society
Providence, Rhode Island

International Centre for Mathematical Sciences
Edinburgh, Scotland

PROCEEDINGS OF AN INSTRUCTIONAL CONFERENCE
ON REPRESENTATION THEORY AND AUTOMORPHIC FORMS
EDINBURGH, SCOTLAND
MARCH 17–29, 1996

organized by the International Centre for Mathematical Sciences
with support from the European Commission and the EPSRC.

1991 *Mathematics Subject Classification.* Primary 11Rxx, 17Bxx, 22Exx, 43Axx;
Secondary 11Sxx.

Library of Congress Cataloging-in-Publication Data

Representation theory and automorphic forms : instructional conference, International Centre for Mathematical Sciences, March 1996, Edinburgh, Scotland / T. N. Bailey, A. W. Knapp, editors.
p. cm. — (Proceedings of symposia in pure mathematics, ISSN 0082-0717 ; v. 61)
Includes bibliographical references and index.
ISBN 0-8218-0609-2
1. Representation of groups—Congresses. 2. Semisimple Lie groups—Congresses. 3. Automorphic forms—Congresses. I. Bailey, T. N. II. Knapp, Anthony W. III. Series.
QA176.R455 1997
515'.7223—dc21 97-26278
CIP

---

10 9 8 7 6 5 4 3 2 1 02 01 00 99 98 97

Contents

Foreword

In 1997 the annual instructional conference of the International Centre for Mathematical Sciences in Edinburgh was devoted to the representation theory of semisimple groups, to automorphic forms, and to the relations between these subjects. It was organized by T. N. Bailey, L. Clozel, M. Duflo, and A. W. Knapp. The two-week meeting began with a rapid summary of basic theory and concluded with two lectures by Robert Langlands, returning from the award of the Wolf Prize. In between, fifteen other world experts gave courses of two to five lectures. There were close to one hundred participants, largely from Western Europe and North America, but also from Eastern Europe, Japan, and the Developing World. Funding for the conference was provided by the European Commission and the Engineering and Physical Sciences Research Council of the United Kingdom.

The papers in this volume consist of slightly expanded versions of the lectures, with some minor rearrangements. An exception is the paper by James Arthur, which is a version of a lecture given at a later conference. All papers were received before May 1, 1997, and were refereed. The papers are intended to provide overviews of the topics they address, and the authors have supplied extensive bibliographies to guide the reader who wants more detail. The editors hope that the papers will serve partly as guides to the literature and that readers at any level will be able to get an outline of new ideas that they will be able to fill in by following the references. As is true in the mathematical literature generally, different authors use slightly different definitions and notation. A global index at the end of the volume may help the reader reconcile the differences.

The aim of the conference was to provide an intensive treatment of representation theory for two purposes: One was to help analysts to make systematic use of Lie groups in work on harmonic analysis, differential equations, and mathematical physics, and the other was to treat for number theorists the representation-theoretic input to Wiles's proof of Fermat's Last Theorem.

It is tempting to think of the lectures and papers as consisting of a common core and two more advanced parts—one going in the direction of analysis on semisimple groups G and semisimple symmetric spaces G/H and the other going in the direction of properties of cusp and automorphic forms, their associated number theory, and properties of G/Γ for arithmetic subgroups Γ. But the editors have resisted the temptation to organize the proceedings in this fashion, because this would ignore the important historical interplay between the two subjects.

This interplay goes in both directions, as evidenced in many of the papers. The Langlands conjecture on discrete series of G, which is discussed in Schmid's paper, came about when Langlands took a known theorem about G/Γ, put $\Gamma = 1$, and made a heuristic calculation about what should happen. The standard intertwining operators for G, which are discussed in van den Ban's article, originally arose in the setting of G/Γ, but their beautiful properties are much clearer in the setting of G and lead to a better understanding of analytic continuation of Eisenstein series and L functions. Harish-Chandra's harmonic analysis on G, which is discussed in Helgason's paper, used Eisenstein integrals and cusp forms modeled on Eisenstein series and cusp forms for G/Γ. In turn Harish-Chandra's analysis on G is in part the model for analysis on the semisimple symmetric spaces G/H, discussed in the paper by van den Ban, Flensted-Jensen, and Schlichtkrull. Oddly, the analysis on G/H adapts two devices, truncation and the residual spectrum, that were first used for G/Γ but are not necessary in the analysis for G.

A great deal of the number-theoretic part of the representation theory in this volume is devoted to functoriality, a conjectural notion introduced by Langlands and applicable only in the setting of G/Γ. Rogawski's article shows how instances of functoriality lead to the Langlands proof of previously unsettled cases of Artin's conjecture; in turn, these cases of Artin's conjecture are what Wiles used from representation theory in his proof of Fermat's Last Theorem.

An important tool in addressing functoriality is the trace formula, which is discussed in several papers. One final instance of the interplay between G/Γ and G/H is that the notion of a semisimple symmetric space, which involves the fixed group of an involution, can be adapted from Lie groups to algebraic groups defined over number fields. In Jacquet's article this notion leads to a relative trace formula and to a conjecture characterizing the key ingredient, base change, in the work of Langlands on Artin's Conjecture. In his own article Langlands speculates that this formula of Jacquet is worth further examination by the coming generation.

The editors are grateful to David Vogan for his assistance with mathematical editing, to Lucy Young and Margaret Cook for making the arrangements for the conference, and to Sergei Gelfand, Christine Thivierge, and Thomas Costa at the American Mathematical Society for their work in publishing these proceedings.

Proceedings of Symposia in Pure Mathematics
Volume **61** (1997), pp. 1–27

Structure Theory of Semisimple Lie Groups

A. W. Knapp

This article provides a review of the elementary theory of semisimple Lie algebras and Lie groups. It is essentially a summary of much of [K3]. The four sections treat complex semisimple Lie algebras, finite-dimensional representations of complex semisimple Lie algebras, compact Lie groups and real forms of complex Lie algebras, and structure theory of noncompact semisimple groups.

1. Complex Semisimple Lie Algebras

This section deals with the structure theory of complex semisimple Lie algebras. Some references for this material are [He], [Hu], [J], [K1], [K3], and [V].

Let $\mathfrak{g}$ be a finite-dimensional Lie algebra. For the moment we shall allow the underlying field to be $\mathbb{R}$ or $\mathbb{C}$, but shortly we shall restrict to Lie algebras over $\mathbb{C}$.

Semisimple Lie algebras are defined as follows. Let $\operatorname{rad}\mathfrak{g}$ be the sum of all the solvable ideals in $\mathfrak{g}$. The sum of two solvable ideals is a solvable ideal [K3, §I.2], and the finite-dimensionality of $\mathfrak{g}$ makes $\operatorname{rad}\mathfrak{g}$ a solvable ideal. We say that $\mathfrak{g}$ is **semisimple** if $\operatorname{rad}\mathfrak{g} = 0$.

Within $\mathfrak{g}$, let $\operatorname{ad} X$ be the linear transformation given by $(\operatorname{ad} X)Z = [X, Z]$. The **Killing form** is the symmetric bilinear form on $\mathfrak{g}$ defined by $B(X, Y) = \operatorname{Tr}(\operatorname{ad} X \operatorname{ad} Y)$. It is **invariant** in the sense that $B([X, Y], Z) = B(X, [Y, Z])$ for all X, Y, Z in $\mathfrak{g}$.

Theorem 1.1 (Cartan's criterion for semisimplicity). *The Lie algebra $\mathfrak{g}$ is semisimple if and only if B is nondegenerate.*

Reference. [K3, Theorem 1.42].

The Lie algebra $\mathfrak{g}$ is said to be **simple** if $\mathfrak{g}$ is nonabelian and $\mathfrak{g}$ has no proper nonzero ideals. In this case, $[\mathfrak{g}, \mathfrak{g}] = \mathfrak{g}$. Semisimple Lie algebras and simple Lie algebras are related as in the following theorem.

Theorem 1.2. *The Lie algebra $\mathfrak{g}$ is semisimple if and only if $\mathfrak{g}$ is the direct sum of simple ideals. In this case there are no other simple ideals, the direct sum decomposition is unique up to the order of the summands, and every ideal is the sum of some subset of the simple ideals. Also in this case, $[\mathfrak{g}, \mathfrak{g}] = \mathfrak{g}$.*

1991 *Mathematics Subject Classification.* Primary 17B20, 20G05, 22E15.

REFERENCE. [K3, Theorem 1.51].

For the remainder of this section, $\mathfrak{g}$ will always denote a semisimple Lie algebra, and the underlying field will be $\mathbb{C}$. The dual of a vector space V will be denoted V^*.

We discuss root-space decompositions. For our semisimple Lie algebra $\mathfrak{g}$, these are decompositions of the form

$$\mathfrak{g} = \mathfrak{h} \oplus \bigoplus_{\alpha \in \Delta} \mathfrak{g}_\alpha .$$

Here $\mathfrak{h}$ is a **Cartan subalgebra**, defined in any of three equivalent ways [K3, §§II.2–3] as

(a) (usual definition) a nilpotent subalgebra $\mathfrak{h}$ whose normalizer satisfies $N_{\mathfrak{g}}(\mathfrak{h}) = \mathfrak{h}$,
(b) (constructive definition) the generalized eigenspace for 0 eigenvalue for $\operatorname{ad} X$ with X regular (i.e., characteristic polynomial $\det(\lambda 1 - \operatorname{ad} X)$ is such that the lowest-order nonzero coefficient is nonzero on X),
(c) (special definition for $\mathfrak{g}$ semisimple) a maximal abelian subspace of $\mathfrak{g}$ in which every $\operatorname{ad} H$, $H \in \mathfrak{h}$, is diagonable.

The elements $\alpha \in \mathfrak{h}^*$ are **roots**, and the $\mathfrak{g}_\alpha$'s are **root spaces**, the α's being defined as the nonzero elements of $\mathfrak{h}^*$ such that

$$\mathfrak{g}_\alpha = \{X \in \mathfrak{g} \mid [H, X] = \alpha(H)X \text{ for all } H \in \mathfrak{h}\}$$

is nonzero.

Let Δ be the set of all roots. This is a finite set. We recall the the classical examples of root-space decompositions [K3, §II.1].

Example 1. $\mathfrak{g} = \mathfrak{sl}(n, \mathbb{C}) = \{n\text{-by-}n \text{ complex matrices of trace } 0\}$. The Cartan subalgebra is

$$\mathfrak{h} = \{\text{diagonal matrices in } \mathfrak{g}\}.$$

Let

$$E_{ij} = \begin{cases} 1 & \text{in } (i,j)^{\text{th}} \text{ place} \\ 0 & \text{elsewhere.} \end{cases}$$

Let $e_i \in \mathfrak{h}^*$ be defined by

$$e_i \begin{pmatrix} h_1 & & \\ & \ddots & \\ & & h_n \end{pmatrix} = h_i .$$

Then each $H \in \mathfrak{h}$ satisfies

$$(\operatorname{ad} H)E_{ij} = [H, E_{ij}] = (e_i(H) - e_j(H))E_{ij}.$$

So E_{ij} is a simultaneous eigenvector for all $\operatorname{ad} H$, with eigenvalue $e_i(H) - e_j(H)$. We conclude that

(a) $\mathfrak{h}$ is a Cartan subalgebra,
(b) the roots are the $(e_i - e_j)$'s for $i \neq j$,
(c) $\mathfrak{g}_{e_i - e_j} = \mathbb{C}E_{ij}$.

Example 2. $\mathfrak{g} = \mathfrak{so}(2n+1, \mathbb{C}) = \{n\text{-by-}n \text{ skew-symmetric complex matrices}\}$. For this example one proceeds similarly. Let

$$\mathfrak{h} = \{H \in \mathfrak{so}(2n+1, \mathbb{C}) \mid H = \text{matrix below}\}.$$

Here

H is block diagonal with n 2-by-2 blocks and one 1-by-1 block,

the jth 2-by-2 block is $\begin{pmatrix} 0 & ih_j \\ -ih_j & 0 \end{pmatrix}$,

the 1-by-1 block is just (0).

Let $e_j(\text{above matrix } H) = h_j$ for $1 \le j \le n$. Then

$$\Delta = \{\pm e_i \pm e_j \text{ with } i \neq j\} \cup \{\pm e_k\}.$$

Formulas for the root vectors E_α may be found in [K3, §II.1].

Example 3. $\mathfrak{g} = \mathfrak{sp}(n, \mathbb{C})$.
This is the Lie algebra of all $2n$-by-$2n$ complex matrices X such that

$$X^t J + JX = 0, \qquad \text{where } J = \begin{pmatrix} 0 & I \\ -I & 0 \end{pmatrix}.$$

For this example the Cartan subalgebra $\mathfrak{h}$ is the set of all matrices H of the form

$$H = \begin{pmatrix} h_1 & & & & & \\ & \ddots & & & & \\ & & h_n & & & \\ & & & -h_1 & & \\ & & & & \ddots & \\ & & & & & -h_n \end{pmatrix}$$

Let $e_j(\text{above matrix } H) = h_j$ for $1 \le j \le n$. Then

$$\Delta = \{\pm e_i \pm e_j \text{ with } i \neq j\} \cup \{\pm 2e_k\}.$$

Formulas for the root vectors E_α may again be found in [K3, §II.1].

Example 4. $\mathfrak{g} = \mathfrak{so}(2n, \mathbb{C})$.
This example is similar to $\mathfrak{so}(2n+1, \mathbb{C})$ but without the $(2n+1)^{\text{st}}$ entry. The set of roots is

$$\Delta = \{\pm e_i \pm e_j \text{ with } i \neq j\}.$$

We return to the discussion of general semisimple Lie algebras $\mathfrak{g}$. The following are some elementary properties of root-space decompositions:

(a) $[\mathfrak{g}_\alpha, \mathfrak{g}_\beta] \subseteq \mathfrak{g}_{\alpha+\beta}$.
(b) If α and β are in $\Delta \cup \{0\}$ and $\alpha + \beta \neq 0$, then $B(\mathfrak{g}_\alpha, \mathfrak{g}_\beta) = 0$.
(c) If α is in Δ, then B is nonsingular on $\mathfrak{g}_\alpha \times \mathfrak{g}_{-\alpha}$.
(d) If α is in Δ, then so is $-\alpha$.
(e) $B|_{\mathfrak{h} \times \mathfrak{h}}$ is nondegenerate. Define H_α to be the element of $\mathfrak{h}$ paired with α.
(f) Δ spans $\mathfrak{h}^*$.

See [K3, §II.4]. We isolate some deeper properties of root-space decompositions as a theorem.

Theorem 1.3. *Root-space decompositions have the following properties:*

(a) *If* α *is in* Δ, *then* $\dim \mathfrak{g}_\alpha = 1$.
(b) *If* α *is in* Δ, *then* $n\alpha$ *is not in* Δ *for any integer* $n \geq 2$.
(c) $[\mathfrak{g}_\alpha, \mathfrak{g}_\beta] = \mathfrak{g}_{\alpha+\beta}$ *if* $\alpha + \beta \neq 0$.
(d) *The real subspace* $\mathfrak{h}_0$ *of* $\mathfrak{h}$ *on which all roots are real is a real form of* $\mathfrak{h}$, *and* $B|_{\mathfrak{h}_0 \times \mathfrak{h}_0}$ *is an inner product. Transfer* $B|_{\mathfrak{h}_0 \times \mathfrak{h}_0}$ *to the real span* $\mathfrak{h}_0^*$ *of the roots, obtaining* $\langle \cdot , \cdot \rangle$ *and* $| \cdot |^2$.

REFERENCE. [K3, §II.4].

Let us now consider root strings. By definition the α **string containing** β (for $\alpha \in \Delta$, $\beta \in \Delta \cup \{0\}$) consists of all members of $\Delta \cup \{0\}$ of the form $\beta + n\alpha$ with $n \in \mathbb{Z}$. The n's in question form an interval with $-p \leq n \leq q$ and $p - q = \dfrac{2\langle \beta, \alpha \rangle}{|\alpha|^2}$. Here $p - q$ is a measure of how centered β is in the root string. When $p - q$ is 0, β is exactly in the center. When $p - q$ is large and positive, β is close to the end $\beta + q\alpha$ of the root string. In any event, it follows that $\dfrac{2\langle \beta, \alpha \rangle}{|\alpha|^2}$ is always an integer. A consequence of the form of root strings is that if α is in Δ, then the orthogonal transformation of $\mathfrak{h}_0^*$ given by

$$s_\alpha(\varphi) = \varphi - \frac{2\langle \varphi, \alpha \rangle}{|\alpha|^2} \alpha$$

carries Δ into itself. The linear transformation s_α is called the **root reflection** in α.

An **abstract root system** is a finite set Δ of nonzero elements in a real inner product space V such that

(a) Δ spans V,
(b) all s_α for $\alpha \in \Delta$ carry Δ to itself,
(c) $\dfrac{2\langle \beta, \alpha \rangle}{|\alpha|^2}$ is an integer whenever α and β are in Δ.

We say that an abstract root system is **reduced** if $\alpha \in \Delta$ implies $2\alpha \notin \Delta$.

The relevance of these notions to semisimple Lie algebras is that the root system of a complex semisimple Lie algebra $\mathfrak{g}$ with respect to a Cartan subalgebra $\mathfrak{h}$ forms a reduced abstract root system in $\mathfrak{h}_0^*$. See [K3, Theorem 2.42].

There are four kinds of classical reduced root systems:

A_n has $V = \left\{ \sum_{i=1}^{n+1} e_i \right\}^\perp$ in $\mathbb{R}^{n+1}$ and $\Delta = \{e_i - e_j \mid i \neq j\}$. The system A_n arises from $\mathfrak{sl}(n+1, \mathbb{C})$.

B_n has $V = \mathbb{R}^n$ and $\Delta = \{\pm e_i \pm e_j \mid i \neq j\} \cup \{\pm e_k\}$. The system B_n arises from $\mathfrak{so}(2n+1, \mathbb{C})$.

C_n has $V = \mathbb{R}^n$ and $\Delta = \{\pm e_i \pm e_j \mid i \neq j\} \cup \{\pm 2e_k\}$. The system C_n arises from $\mathfrak{sp}(n, \mathbb{C})$.

D_n has $V = \mathbb{R}^n$ and $\Delta = \{\pm e_i \pm e_j \mid i \neq j\}$. . The system D_n arises from $\mathfrak{so}(2n, \mathbb{C})$.

We say that an abstract root system Δ is **reducible** if $\Delta = \Delta' \cup \Delta''$ with $\Delta' \perp \Delta''$. Otherwise Δ is **irreducible**.

Theorem 1.4. *A semisimple Lie algebra* $\mathfrak{g}$ *is simple if and only if the corresponding root system* Δ *is irreducible.*

REFERENCE. [K3, Proposition 2.44].

Now we introduce the notions of lexicographic ordering and positive roots for an abstract root system. The construction is as follows. Let $\varphi_1, \ldots, \varphi_m$ be a spanning set for V. Define φ to be **positive** (written $\varphi > 0$) if there exists an index k such that $\langle \varphi, \varphi_i \rangle = 0$ for $1 \leq i \leq k-1$ and $\langle \varphi, \varphi_k \rangle > 0$. The corresponding **lexicographic ordering** has $\varphi > \psi$ if $\varphi - \psi$ is positive. Fix such an ordering. Call the root α **simple** if $\alpha > 0$ and if α does not decompose as $\alpha = \beta_1 + \beta_2$ with β_1 and β_2 both positive roots.

Theorem 1.5. *If $l = \dim V$, then there are l simple roots $\alpha_1, \ldots, \alpha_l$, and they are linearly independent. If β is a root and is written as*

$$\beta = x_1\alpha_1 + \cdots + x_l\alpha_l,$$

then all the x_j have the same sign (if 0 is allowed to be positive or negative), and all the x_j are integers.

When standard choices are made, the following are the positive roots and simple roots for the classical reduced root systems:

A_n. The positive roots are the $e_i - e_j$ with $i < j$. The simple roots are all $e_i - e_{i+1}$ with $1 \leq i \leq n$.
B_n. The positive roots are the $e_i \pm e_j$ with $i < j$ and all e_k. The simple roots are e_n and all $e_i - e_{i+1}$ with $1 \leq i \leq n-1$.
C_n. The positive roots are the $e_i \pm e_j$ with $i < j$ and all $2e_k$. The simple roots are $2e_n$ and all $e_i - e_{i+1}$ with $1 \leq i \leq n-1$.
D_n. The positive roots are the $e_i \pm e_j$ with $i < j$. The simple roots are $e_{n-1} + e_n$ and all $e_i - e_{i+1}$ with $1 \leq i \leq n-1$.

A root α is called **reduced** if $\frac{1}{2}\alpha$ is not a root. Every simple root is reduced. By a **simple system** for Δ, we mean the set of simple roots for some ordering. By Theorem 1.5, a simple system $\{\alpha_1, \ldots, \alpha_l\}$ has the property that any root α, when expressed as $\sum_i x_i\alpha_i$, has all x_i of the same sign. Conversely any subset $\{\alpha_1, \ldots, \alpha_l\}$ of reduced roots with the property that any root α, when expressed as $\sum_i x_i\alpha_i$, has all x_i of the same sign is a simple system.

Let l be the dimension of the underlying space V of an abstract root system Δ. The number l is called the **rank**. If Δ is the root system of a semisimple Lie algebra $\mathfrak{g}$, we also refer to $l = \dim \mathfrak{h}$ as the **rank** of $\mathfrak{g}$. Relative to a given simple system $\{\alpha_1, \ldots, \alpha_l\}$, the **Cartan matrix** is the l-by-l matrix with entries

$$A_{ij} = \frac{2\langle \alpha_i, \alpha_j \rangle}{|\alpha_i|^2}.$$

It has the following properties:

(a) A_{ij} is in $\mathbb{Z}$ for all i and j,
(b) $A_{ii} = 2$ for all i,
(c) $A_{ij} \leq 0$ for $i \neq j$,
(d) $A_{ij} = 0$ if and only if $A_{ji} = 0$,
(e) there exists a diagonal matrix D with positive diagonal entries such that DAD^{-1} is symmetric positive definite.

An **abstract Cartan matrix** is a square matrix satisfying properties (a) through (e) as above. To such a matrix we can associate a **Dynkin diagram** in the standard way. See [K3, §II.5].

We come to the first principal result.

Theorem 1.6 (Isomorphism Theorem). *Let $\mathfrak{g}$ and $\mathfrak{g}'$ be complex semisimple Lie algebras with respective Cartan subalgebras $\mathfrak{h}$ and $\mathfrak{h}'$ and respective root systems Δ and Δ'. Suppose that a vector space isomorphism $\varphi : \mathfrak{h} \to \mathfrak{h}'$ is given with the property that φ carries Δ one-one onto Δ'. Let the mapping of Δ to Δ' be denoted $\alpha \mapsto \alpha'$. Fix a simple system Π for Δ. For each α in Π, select nonzero root vectors $E_\alpha \in \mathfrak{g}$ for α and $E_{\alpha'} \in \mathfrak{g}'$ for α'. Then there exists one and only one Lie algebra isomorphism $\tilde{\varphi} : \mathfrak{g} \to \mathfrak{g}'$ such that $\tilde{\varphi}|_{\mathfrak{h}} = \varphi$ and $\tilde{\varphi}(E_\alpha) = E_{\alpha'}$ for all $\alpha \in \Pi$.*

REFERENCE. [K3, Theorem 2.108].

Examples.

1) An automorphism of the Dynkin diagram yields an automorphism of the Lie algebra.

2) Let $\varphi = -1$ on $\mathfrak{h}$. This extends to $\tilde{\varphi} : \mathfrak{g} \to \mathfrak{g}$ and is used in constructing real forms of $\mathfrak{g}$. See Theorem 3.5 and the discussion that follows it.

The **Weyl group** $W(\Delta)$ of an abstract root system Δ is defined to be the finite group generated by all root reflections s_α for $\alpha \in \Delta$.

Theorem 1.7. *The Weyl group $W(\Delta)$ of the abstract root system Δ has the following properties:*

(a) *Fix a simple system $\Pi = \{\alpha_1, \ldots, \alpha_l\}$ for Δ. Then $W(\Delta)$ is generated by all s_{α_i}, $\alpha_i \in \Pi$. If α is any reduced root, then there exist $\alpha_j \in \Pi$ and $s \in W(\Delta)$ such that $s\alpha_j = \alpha$.*

(b) *If Π and Π' are two simple systems for Δ, then there exists one and only one element $s \in W(\Delta)$ such that $s\Pi = \Pi'$.*

REFERENCE. [K3, Proposition 2.62 and Theorem 2.63].

Briefly conclusion (b) says that $W(\Delta)$ acts simply transitively on the set of all simple systems. There is a geometric way of formulating this property. Regard V as the dual of its dual V^*, so that each root has a kernel in V^*. A **Weyl chamber** of V^* is a connected component of the subset of V^* on which every root is nonzero. Each Weyl chamber is an open convex cone, and each root has constant sign on each Weyl chamber. To each simple system corresponds exactly one Weyl chamber, namely the set where each simple root is positive. Conversely each Weyl chamber determines a simple system by this procedure. If the action of $W(\Delta)$ on V is transferred to an action on V^*, then (b) says that $W(\Delta)$ acts simply transitively on the set of Weyl chambers.

Dominance is a notion that plays a role with finite-dimensional representations and will be discussed in detail in §2. We call $\lambda \in V$ **dominant** if $\langle \lambda, \alpha \rangle \geq 0$ for all positive roots α. Equivalently $\langle \lambda, \alpha \rangle \geq 0$ is to hold for all simple roots α.

Theorem 1.8. *Fix an abstract root system Δ.*

(a) *If λ is in V, then there exists a simple system Π for which λ is dominant.*

(b) *If λ is in V and if a simple system is specified, then there is some element w of the Weyl group such that $w\lambda$ is dominant.*

REFERENCE. [K3, Proposition 2.67 and Corollary 2.68].

Here is a handy result that uses dominance in its proof.

Theorem 1.9 (Chevalley's Lemma). *Fix v in V, and let W_0 be the subgroup of $W(\Delta)$ fixing v. Then W_0 is generated by the root reflections s_α such that $\langle v, \alpha\rangle = 0$.*

REFERENCE. [K3, Proposition 2.72].

Examples.
1) The only reflections s_φ in $W(\Delta)$ are the root reflections.
2) If an element v of V is fixed by a nontrivial element of $W(\Delta)$, then some root is orthogonal to v.
3) Any element of order 2 in $W(\Delta)$ is the product of commuting root reflections.

The main correspondence involving complex semisimple Lie algebras relates three classes of objects and isomorphisms, identifying each one with the other two:

(1) complex semisimple Lie algebras and isomorphisms of Lie algebras,
(2) abstract reduced root systems and invertible linear maps carrying Δ to Δ' and respecting the integers $2\langle\beta, \alpha\rangle/|\alpha|^2$,
(3) abstract Cartan matrices and equality up to permutation of indices.

The passage from (1) to (2) is well defined because any two Cartan subalgebras of $\mathfrak{g}$ are conjugate via $\operatorname{Int}\mathfrak{g}$ (see [K3, Theorem 2.15]); here $\operatorname{Int}\mathfrak{g}$ is the analytic subgroup of $GL(\mathfrak{g})$ with Lie algebra $\operatorname{ad}\mathfrak{g}$. The passage from (1) to (2) is one-one by the Isomorphism Theorem (Theorem 1.6 above), and it is onto by a result known as the Existence Theorem (see [K3, Theorem 2.111]).

The passage (2) to (3) is well defined because any two simple systems are conjugate via the Weyl group (Theorem 1.7b above). It is one-one by Theorem 1.7a above, and it is onto by a case-by-case construction.

2. Finite-Dimensional Representations of Complex Semisimple Lie Algebras

This section deals with finite-dimensional representations of complex semisimple Lie algebras and with the tools needed in their study. Some references for this material are [Hu], [J], [K1], [K2], [K3], and [V].

Except for one segment about the universal enveloping algebra where $\mathfrak{g}$ will be allowed to be more general, the notation in this section will be as follows:

$$\begin{aligned}
\mathfrak{g} &= \text{complex semisimple Lie algebra}\\
\mathfrak{h} &= \text{Cartan subalgebra}\\
\Delta &= \Delta(\mathfrak{g},\mathfrak{h}) = \text{set of roots}\\
\mathfrak{h}_0 &= \text{real form of } \mathfrak{h} \text{ where roots are real-valued}\\
B &= \text{nondegenerate symmetric invariant bilinear form}\\
&\quad\ \text{on } \mathfrak{g} \text{ that is positive definite on } \mathfrak{h}_0\\
H_\lambda &= \text{member of } \mathfrak{h}_0 \text{ corresponding to } \lambda \in \mathfrak{h}_0^*
\end{aligned}$$

Here B can be the Killing form, but it does not need to be. In the definition of H_λ, it is understood that $(\,\cdot\,)^*$ refers to the vector space dual; the correspondence of λ to H_λ is the one induced by B.

A **representation** φ on a complex vector space V is a linear map $\varphi : \mathfrak{g} \to \operatorname{End} V$ with

$$\varphi[X,Y] = \varphi(X)\varphi(Y) - \varphi(Y)\varphi(X)$$

for all X and Y in $\mathfrak{g}$. Isomorphism of representations is called **equivalence**. An **irreducible representation** is a representation φ on a nonzero space V such that $\varphi(\mathfrak{g})U \subseteq U$ fails for all proper nonzero subspaces U.

Fix such a φ. For $\lambda \in \mathfrak{h}^*$, let V_λ be the set of all $v \in V$ with $(\varphi(H) - \lambda(H)1)^n v = 0$ for all $H \in \mathfrak{h}$ and some $n = n(H,V)$. If V_λ is nonzero, V_λ is called a **generalized weight space**, and λ is called a **weight**. If $\dim V$ is finite-dimensional, V is the direct sum of its generalized weight spaces. This is a generalization of the fact from linear algebra about a linear transformation L on a finite-dimensional V that V is the direct sum of the generalized eigenspaces of L. If λ is a weight, then the subspace

$$\{v \in V \mid \varphi(H)v = \lambda(H)v \text{ for all } H \in \mathfrak{h}\}$$

is nonzero and is called the **weight space** corresponding to λ.

A source of finite-dimensional representations of $\mathfrak{g}$ is group representations. Suppose that G is a compact connected Lie group whose Lie algebra $\mathfrak{g}_0$ has complexification $\mathfrak{g}$. A **representation** Φ of G on a complex vector space V is a continuous group homomorphism $\Phi : G \to GL(V)$. If V is finite-dimensional, then Φ is automatically smooth. We can differentiate to get a representation φ of $\mathfrak{g}_0$ on V, and then we can complexify, writing

$$\varphi(X + iY) = \varphi(X) + i\varphi(Y),$$

to obtain a representation φ of $\mathfrak{g}$ on V.

We can obtain some initial examples of this sort with $\mathfrak{g} = \mathfrak{sl}(n,\mathbb{C})$ and $\mathfrak{g} = \mathfrak{so}(n,\mathbb{C})$. We start with $G = SU(n)$ and $G = SO(n)$ in the two cases. Each of these has a standard representation on $\mathbb{C}^n$, given by the multiplication of matrices and column vectors. For each we can form a contragredient representation on the dual space $(\mathbb{C}^n)^*$. Then we can form tensor products of copies of the standard representation and its dual. Finally we can pass to skew-symmetric tensors, symmetric tensors, and similar such subspaces. Representations in polynomials arise as symmetric tensors in the tensor product of copies of $(\mathbb{C}^n)^*$.

More examples come by starting with the compact connected Lie group $G = U(2n) \cap Sp(n,\mathbb{C})$, whose complexified Lie algebra is $\mathfrak{sp}(n,\mathbb{C})$. In this case the standard representation has dimension $2n$.

In the examples below, we list some representations obtained in this way from $G = SU(n)$ and $G = SO(2n+1)$. In each case the weights are identified. Also the highest weight, i.e., the largest weight, is identified relative to the lexicographic ordering. The Cartan subalgebras and sets of positive roots for $\mathfrak{sl}(n,\mathbb{C})$ and $\mathfrak{so}(2n+1,\mathbb{C})$ are the ones in §1.

Examples. Let $\mathfrak{g} = \mathfrak{sl}(n,\mathbb{C})$. Here the Cartan subalgebra is the diagonal subalgebra.

1) Let V be the space of polynomials in $z_1, \ldots, z_n$ and their conjugates homogeneous of degree N. The action is

$$(\Phi(g)P)(z,\bar{z}) = P(g^{-1}z, \bar{g}^{-1}\bar{z}).$$

The weights are all expressions $\sum_{j=1}^{n}(l_j - k_j)e_j$ with all $k_j \geq 0$ and $l_j \geq 0$ and with $\sum_{j=1}^{n}(k_j + l_j) = N$. The highest weight is Ne_1.

2) Let V be the subspace of holomorphic polynomials in the preceding example. The action is $\Phi(g)(z) = P(g^{-1}z)$. The weights are all expressions $-\sum_{j=1}^{n} k_j e_j$ with all $k_j \geq 0$ and with $\sum_{j=1}^{n} k_j = N$. The highest weight is $-Ne_n$.

3) Let $V = \bigwedge^l \mathbb{C}^n$ with action

$$\Phi(g)(v_1 \wedge \cdots \wedge v_l) = gv_1 \wedge \cdots \wedge gv_l.$$

The weights are all expressions $\sum_{k=1}^{l} e_{j_k}$, and the highest weight is $\sum_{k=1}^{l} e_k$.

Examples. Let $\mathfrak{g} = \mathfrak{so}(2n+1, \mathbb{C})$. Here the Cartan subalgebra is block diagonal, containing n 2-by-2 skew-symmetric blocks and one 1-by-1 block whose entry is 0.

1) Let V be the space of all polynomials in $x_1, \ldots, x_{2n+1}$ that are homogeneous of degree N, the action being $\Phi(g)(x) = P(g^{-1}x)$. The weights are all expressions $\sum_{j=1}^{n}(l_j - k_j)e_j$ with all $k_j \geq 0$ and $l_j \geq 0$ and with $k_0 + \sum_{j=1}^{n}(k_j + l_j) = N$. The highest weight is Ne_1.

2) Let $V = \bigwedge^l \mathbb{C}^{2n+1}$ with $l \leq n$ and with action as in Example 3 for $\mathfrak{sl}(n, \mathbb{C})$. The weights are all expressions $\pm e_{j_1} \pm \cdots \pm e_{j_r}$ with $j_1 < \cdots < j_r$ and $r \leq l$. The highest weight is $\sum_{k=1}^{l} e_k$. When $V = \bigwedge^m \mathbb{C}^{2n+1}$ with $m > n$, we again get a representation, and it can be shown to be equivalent with the representation on $\bigwedge^{2n+1-m} \mathbb{C}^{2n+1}$.

A member λ of $\mathfrak{h}^*$ is said to be **algebraically integral** if $2\langle \lambda, \alpha \rangle / |\alpha|^2$ is in $\mathbb{Z}$ for each $\alpha \in \Delta$.

Some elementary properties of a finite-dimensional representation φ on a vector space V are as follows:

(a) $\varphi(\mathfrak{h})$ acts diagonably on V, so that every generalized weight vector is a weight vector and V is the direct sum of all the weight spaces,

(b) every weight is real-valued on $\mathfrak{h}_0$ and is algebraically integral,

(c) roots and weights are related by $\varphi(\mathfrak{g}_\alpha)V_\lambda \subseteq V_{\lambda+\alpha}$.

Properties (a) and (b) follow by restricting φ to copies of $\mathfrak{sl}(2, \mathbb{C})$ lying in $\mathfrak{g}$ and then using the representation theory of $\mathfrak{sl}(2, \mathbb{C})$, which we do not review. See [K3, §I.9].

Fix a lexicographic ordering, and let Δ^+ be the set of positive roots. Let $\Pi = \{\alpha_1, \ldots, \alpha_l\}$ be the corresponding simple system. There are three main theorems on representation theory in this section, and we come now to the first of the three.

Theorem 2.1 (Theorem of the Highest Weight). *Apart from equivalence the irreducible finite-dimensional representations φ of $\mathfrak{g}$ stand in one-one correspondence with the algebraically integral dominant linear functionals λ on $\mathfrak{h}$, the correspondence being that λ is the highest weight of φ_λ. The highest weight λ of φ_λ has these additional properties:*

(a) *λ depends only on the simple system Π and not on the ordering used to define Π.*

(b) *the weight space V_λ for λ is 1-dimensional.*

(c) *each root vector E_α for arbitrary $\alpha \in \Delta^+$ annihilates the members of V_λ, and the members of V_λ are the only vectors with this property.*

(d) *every weight of φ_λ is of the form $\lambda - \sum_{i=1}^{l} n_i\alpha_i$ with the integers ≥ 0 and the α_i in Π.*

(e) *each weight space* V_μ *for* φ_λ *has* $\dim V_{w\mu} = \dim V_\mu$ *for all* w *in the Weyl group* $W(\Delta)$, *and each weight* μ *has* $|\mu| \le |\lambda|$ *with equality only if* μ *is in the orbit* $W(\Delta)\lambda$.

REFERENCE. [K3, Theorem 5.5]. Later in this section we discuss tools used in the proof.

REMARK. As a consequence of (e), the Weyl group acts on the weights, preserving multiplicities. The extreme weights are those in the orbit of the highest weight.

We can immediately state the second main theorem of the section on representation theory. It concerns complete reducibility.

Theorem 2.2. *Let* φ *be a complex-linear representation of* $\mathfrak{g}$ *on a finite-dimensional complex vector space* V. *Then* V *is completely reducible in the sense that there exist invariant subspaces* $U_1, \ldots, U_r$ *of* V *such that* $V = U_1 \oplus \cdots \oplus U_r$ *and such that the restriction of the representation to each* U_i *is irreducible.*

REFERENCE. [K3, Theorem 5.29].

The proofs of Theorems 2.1 and 2.2 use three tools:

(a) universal enveloping algebra,
(b) Casimir element,
(c) Verma modules.

We review each of these in turn.

First we take up the universal enveloping algebra. In the discussion, we shall allow $\mathfrak{g}$ to be any complex Lie algebra. Let $T(\mathfrak{g})$ be the tensor algebra

$$T(\mathfrak{g}) = \mathbb{C} \oplus \mathfrak{g} \oplus (\mathfrak{g} \otimes \mathfrak{g}) \oplus (\mathfrak{g} \otimes \mathfrak{g} \otimes \mathfrak{g}) \oplus \cdots .$$

In $T(\mathfrak{g})$, let J be the two-sided ideal generated by all $X \otimes Y - Y \otimes X - [X,Y]$ with X and Y in the space $T^1(\mathfrak{g})$ of first-order tensors. The **universal enveloping algebra** of $\mathfrak{g}$ is the associative algebra (with identity) given by

$$U(\mathfrak{g}) = T(\mathfrak{g})/J.$$

Let $\iota : \mathfrak{g} \to U(\mathfrak{g})$ be the composition $\iota : \mathfrak{g} \cong T^1(\mathfrak{g}) \hookrightarrow T(\mathfrak{g}) \to U(\mathfrak{g})$, so that

$$\iota[X,Y] = \iota(X)\iota(Y) - \iota(Y)\iota(X).$$

The universal enveloping algebra is so named because of the following universal mapping property.

Theorem 2.3. *Whenever* A *is a complex associative algebra with identity and* $\pi : \mathfrak{g} \to A$ *is a linear mapping such that*

$$\pi(X)\pi(Y) - \pi(Y)\pi(X) = \pi[X,Y]$$

for all X, Y *in* $\mathfrak{g}$, *then there exists a unique algebra homomorphism* $\tilde{\pi} : U(\mathfrak{g}) \to A$ *such that* $\tilde{\pi}(1) = 1$ *and* $\pi = \tilde{\pi} \circ \iota$.

REFERENCE. [K3, Proposition 3.3].

REMARK. One thinks of $\tilde{\pi}$ in the theorem as an extension of π from $\mathfrak{g}$ to all of $U(\mathfrak{g})$. This attitude about $\tilde{\pi}$ implicitly assumes that ι is one-one, a fact that follows from Theorem 2.5 below.

Theorem 2.4. *Representations of $\mathfrak{g}$ on complex vector spaces stand in one-one correspondence with left $U(\mathfrak{g})$ modules in which* 1 *acts as* 1.

REFERENCE. [K3, Corollary 3.6].

REMARK. The one-one correspondence comes from $\pi \mapsto \tilde{\pi}$ in the notation of Theorem 2.3.

Theorem 2.5 (Poincaré-Birkhoff-Witt Theorem). *Let $\{X_i\}_{i\in A}$ be a basis of $\mathfrak{g}$, and suppose that a simple ordering has been imposed on the index set A. Then the set of all monomials*

$$(\iota X_{i_1})^{j_1} \cdots (iX_{i_n})^{j_n}$$

with $i_1 < \cdots < i_n$ and with all $j_k \geq 0$, is a basis of $U(\mathfrak{g})$. In particular the canonical map $\iota : \mathfrak{g} \to U(\mathfrak{g})$ is one-one.

REFERENCE. [K3, Theorem 3.8].

Let us now return to our assumption that $\mathfrak{g}$ is semisimple. We also return to the other notation listed at the start of this section. We shall apply the theorems about $U(\mathfrak{g})$ to a representation φ of $\mathfrak{g}$ on a finite-dimensional vector space V. We enumerate the positive roots as $\beta_1, \ldots, \beta_m$, and we let $H_1, \ldots, H_l$ be a basis of $\mathfrak{h}$. We use the ordered basis

$$E_{-\beta_1}, \ldots, E_{-\beta_m}, H_1, \ldots, H_l, E_{\beta_1}, \ldots, E_{\beta_m}.$$

in the Poincaré-Birkhoff-Witt Theorem. The theorem says that

$$E_{-\beta_1}^{p_1} \cdots E_{-\beta_m}^{p_m} H_1^{k_1} \cdots H_l^{k_l} E_{\beta_1}^{q_1} \cdots E_{\beta_m}^{q_m}$$

is a basis of $U(\mathfrak{g})$. If we apply members of this basis to a nonzero highest weight vector v of V, we get control of a general member of $U(\mathfrak{g})v$. In fact, $E_{\beta_1}^{q_1} \cdots E_{\beta_m}^{q_m}$ will act as 0 if $q_1 + \cdots + q_m > 0$, and $H_1^{k_1} \cdots H_l^{k_l}$ will act as a scalar. Thus we have only to sort out the effect of $E_{-\beta_1}^{p_1} \cdots E_{-\beta_m}^{p_m}$, and most of the conclusions in the Theorem of the Highest Weight (Theorem 2.1) follow readily.

This completes the discussion of the universal enveloping algebra. The second tool used in the proofs of Theorems 2.1 and 2.2 is the Casimir element. For our complex semisimple Lie algebra $\mathfrak{g}$, the **Casimir element** Ω is the member

$$\Omega = \sum_{i,j} B(X_i, X_j)\tilde{X}_i\tilde{X}_j$$

of $U(\mathfrak{g})$, where $\{X_i\}$ is a basis of $\mathfrak{g}$ and $\{\tilde{X}_i\}$ is the dual basis relative to B. One shows that Ω is defined independently of the basis $\{X_i\}$ and is a member of the center $Z(\mathfrak{g})$ of $U(\mathfrak{g})$. (See [K3, Proposition 5.24].)

Theorem 2.6. *Let Ω be the Casimir element. Let $\{H_i\}_{i=1}^l$ be an orthonormal basis of $\mathfrak{h}_0$ relative to B, and choose root vectors E_α so that $B(E_\alpha, E_{-\alpha}) = 1$ for all roots α. Then*

(a) $\Omega = \sum_{i=1}^l H_i^2 + \sum_{\alpha\in\Delta} E_\alpha E_{-\alpha}$.
(b) *Ω operates by the scalar $|\lambda|^2 + 2\langle\lambda, \delta\rangle = |\lambda + \delta|^2 - |\delta|^2$ in an irreducible finite-dimensional representation of $\mathfrak{g}$ of highest weight λ, where δ is half the sum of the positive roots.*
(c) *the scalar by which Ω operates in an irreducible finite-dimensional representation of $\mathfrak{g}$ is nonzero if the representation is not trivial.*

REFERENCE. [K3, Proposition 5.28].

The Casimir element is used in the proof of complete reducibility (Theorem 2.2). The key special case is that V has an irreducible invariant subspace of codimension 1 and dimension > 1. Then $\ker\Omega$ is the required invariant complement.

This completes the discussion of the Casimir element. The third tool used in the proofs of Theorems 2.1 and 2.2 is the theory of Verma modules. Fix a lexicographic ordering, and introduce $\mathfrak{b} = \mathfrak{h} \oplus \bigoplus_{\alpha>0} \mathfrak{g}_\alpha$. For $\nu \in \mathfrak{h}^*$, make $\mathbb{C}$ into a 1-dimensional $U(\mathfrak{h})$ module $\mathbb{C}_\nu$ by defining an action of $\mathfrak{h}$ by $H(z) = \nu(H)z$ for $z \in \mathbb{C}$. Make $\mathbb{C}_\nu$ into a $U(\mathfrak{b})$ module by having $\bigoplus_{\alpha>0} \mathfrak{g}_\alpha$ act by 0. For $\mu \in \mathfrak{h}^*$, define the **Verma module** $V(\mu)$ by

$$V(\mu) = U(\mathfrak{g}) \otimes_{U(\mathfrak{b})} \mathbb{C}_{\mu-\delta},$$

where δ is half the sum of the positive roots. (The term "$-\delta$" in the definition is the usual convention and has the effect of simplifying calculations with the Weyl group.)

Verma modules have the following elementary properties:

(a) $V(\mu) \neq 0$,
(b) $V(\mu)$ is a universal highest weight module for highest weight modules of $U(\mathfrak{g})$ with highest weight $\mu - \delta$,
(c) each weight space of $V(\mu)$ is finite-dimensional,
(d) $V(\mu)$ has a unique irreducible quotient $L(\mu)$.

(See [K3, §V.3].)

The use of Verma modules allows one to prove the hard step of the Theorem of Highest Weight (Theorem 2.1), which is the existence of an irreducible finite-dimensional representation with given highest weight. In fact, if λ is dominant and algebraically integral, then $L(\lambda + \delta)$ is an irreducible representation with highest weight λ, and all that has to be proved is the finite-dimensionality.

The topic of the third main theorem on representation theory in this section is characters, which we treat for now as formal exponential sums. We continue with $\mathfrak{g}$ as a semisimple Lie algebra, $\mathfrak{h}$ as a Cartan subalgebra, Δ as the set of roots, and $W(\Delta)$ as the Weyl group. Introduce a lexicographic ordering, and let $\alpha_1, \ldots, \alpha_l$ be the simple roots.

We regard the set $\mathbb{Z}^{\mathfrak{h}^*}$ of functions from $\mathfrak{h}^*$ to $\mathbb{Z}$ as an abelian group under pointwise addition. We write elements f of $\mathbb{Z}^{\mathfrak{h}^*}$ as $f = \sum_{\lambda\in\mathfrak{h}^*} f(\lambda)e^\lambda$. The **support** of such an f is defined to be the set of $\lambda \in \mathfrak{h}^*$ for which $f(\lambda) \neq 0$. Within $\mathbb{Z}^{\mathfrak{h}^*}$, let $\mathbb{Z}[\mathfrak{h}^*]$ be the subgroup of all f of finite support. The subgroup $\mathbb{Z}[\mathfrak{h}^*]$ has a natural commutative ring structure, which is determined by $e^\lambda e^\mu = e^{\lambda+\mu}$.

We introduce a larger ring, $\mathbb{Z}\langle\mathfrak{h}^*\rangle$. Let

$$Q^+ = \Big\{\sum_{i=1}^{l} n_i\alpha_i \mid \text{all } n_i \geq 0,\ n_i \in \mathbb{Z}\Big\}.$$

Then $\mathbb{Z}\langle\mathfrak{h}^*\rangle$ consists of all $f \in \mathbb{Z}^{\mathfrak{h}^*}$ whose support is contained in the union of finitely many sets $\nu_i - Q^+$ with each $\nu_i \in \mathfrak{h}^*$. Then we have inclusions

$$\mathbb{Z}[\mathfrak{h}^*] \subseteq \mathbb{Z}\langle\mathfrak{h}^*\rangle \subseteq \mathbb{Z}^{\mathfrak{h}^*}.$$

Multiplication in $\mathbb{Z}\langle\mathfrak{h}^*\rangle$ is given by

$$\Big(\sum_{\lambda\in\mathfrak{h}^*} c_\lambda e^\lambda\Big)\Big(\sum_{\mu\in\mathfrak{h}^*} \tilde{c}_\mu e^\mu\Big) = \sum_{\nu\in\mathfrak{h}^*}\Big(\sum_{\lambda+\mu=\nu} c_\lambda \tilde{c}_\mu\Big)e^\nu.$$

If V is a representation of $\mathfrak{g}$ (not necessarily finite-dimensional), we say that V **has a character** (for present purposes) if V is the direct sum of its weight spaces under $\mathfrak{h}$, i.e., $V = \bigoplus_{\mu\in\mathfrak{h}^*} V_\mu$, and if $\dim V_\mu < \infty$ for $\mu \in \mathfrak{h}^*$. In this case the **character** is

$$\operatorname{char}(V) = \sum_{\mu\in\mathfrak{h}^*} (\dim V_\mu)e^\mu$$

as a member of $\mathbb{Z}^{\mathfrak{h}^*}$. This definition is meaningful if V is finite-dimensional or if V is a Verma module.

The **Weyl denominator** is the member $d = e^\delta \prod_{\alpha\in\Delta^+}(1 - e^{-\alpha})$ of $\mathbb{Z}[\mathfrak{h}^*]$. In this expression, δ is again half the sum of the positive roots.

The **Kostant partition function** $\mathcal{P}$ is the function from Q^+ to the nonnegative integers that tells the number of ways, apart from order, that a member of Q^+ can be written as the sum of positive roots. By convention, $\mathcal{P}(0) = 1$. Define $K = \sum_{\gamma\in Q^+} \mathcal{P}(\gamma)e^{-\gamma} \in \mathbb{Z}\langle\mathfrak{h}^*\rangle$.

Lemma. *In the ring $\mathbb{Z}\langle\mathfrak{h}^*\rangle$, $Ke^{-\delta}d = 1$. Hence d^{-1} exists in $\mathbb{Z}\langle\mathfrak{h}^*\rangle$.*

REFERENCE. [K3, Lemma 5.72].

Now we come to the third main theorem.

Theorem 2.7 (Weyl Character Formula). *Let V be an irreducible finite-dimensional representation of the complex semisimple Lie algebra $\mathfrak{g}$ with highest weight λ. Then*

$$\operatorname{char}(V) = d^{-1} \sum_{w\in W(\Delta)} (\det w)e^{w(\lambda+\delta)}.$$

REFERENCE. [K3, Theorem 5.75].

3. Compact Lie Groups and Real Forms of Complex Lie Algebras

This section deals with the structure theory of compact Lie groups and with the existence of compact real forms of complex semisimple Lie algebras. Some references for this material are [He], [K1], [K3], and [V].

Throughout this section, $\mathfrak{g}$ will denote a finite-dimensional complex Lie algebra, and $\mathfrak{g}_0$ will denote a finite-dimensional real Lie algebra. Let $Z_{\mathfrak{g}_0}$ be the center of $\mathfrak{g}_0$.

Let $\operatorname{Aut}\mathfrak{g}_0$ be the automorphism group of $\mathfrak{g}_0$ as a Lie algebra. This is a closed subgroup of $GL(\mathfrak{g}_0)$, hence a Lie subgroup. Its Lie algebra is $\operatorname{Der}\mathfrak{g}_0$. Let $\operatorname{Int}\mathfrak{g}_0$ be the analytic subgroup of $\operatorname{Aut}\mathfrak{g}_0$ with Lie algebra $\operatorname{ad}\mathfrak{g}_0$. If G is a connected Lie group with Lie algebra $\mathfrak{g}_0$, then $\operatorname{Ad}(G)$ is an analytic subgroup of $GL(\mathfrak{g}_0)$ with Lie algebra $\operatorname{ad}\mathfrak{g}_0$, hence equals $\operatorname{Int}\mathfrak{g}_0$. Thus $\operatorname{Int}\mathfrak{g}_0$ provides a way of forming $\operatorname{Ad}(G)$ without using a particular G. It is the group of inner automorphisms of G or $\mathfrak{g}_0$.

We begin with a discussion of real forms. If we regard $\mathfrak{g}$ as a real Lie algebra, then a real Lie subalgebra $\mathfrak{g}_0$ such that $\mathfrak{g} = \mathfrak{g}_0 \oplus i\mathfrak{g}_0$ as vector spaces is called a **real form** of $\mathfrak{g}$. To a real form $\mathfrak{g}_0$ of $\mathfrak{g}$ is associated a **conjugation** of $\mathfrak{g}$, which is the $\mathbb{R}$ linear map that is 1 on $\mathfrak{g}_0$ and -1 on $i\mathfrak{g}_0$. This is an automorphism of $\mathfrak{g}$ as a real Lie algebra. If $\mathfrak{g}_0$ is given, then $\mathfrak{g}_0$ is a real form of its complexification $\mathfrak{g} = \mathfrak{g}_0 \otimes_{\mathbb{R}} \mathbb{C} = \mathfrak{g}_0 \oplus i\mathfrak{g}_0$. If $\mathfrak{g}_0$ is a real form of $\mathfrak{g}$, then $\mathfrak{g}_0$ is semisimple if and only if $\mathfrak{g}$ is semisimple, as a consequence of Cartan's criterion for semisimplicity (Theorem 1.1).

Examples.

1) $\mathfrak{sl}(n,\mathbb{R})$, $\mathfrak{su}(n)$, and $\mathfrak{su}(p,q)$ are real forms of $\mathfrak{sl}(n,\mathbb{C})$. Here $\mathfrak{su}(n)$ is the Lie algebra of n-by-n skew-Hermitian matrices of trace 0, and $\mathfrak{su}(p,q)$ consists of matrices $\begin{pmatrix} A & B \\ B^* & C \end{pmatrix}$ of trace 0 in which A and C are skew-Hermitian.

2) $\mathfrak{so}(n)$ is a real form of $\mathfrak{so}(n,\mathbb{C})$. Here $\mathfrak{so}(n)$ is the Lie algebra of n-by-n real skew-symmetric matrices.

3) $\mathfrak{so}(p,q)$ is isomorphic to a real form of $\mathfrak{so}(p+q,\mathbb{C})$ under conjugation by the block diagonal matrix $\begin{pmatrix} 1 & 0 \\ 0 & i \end{pmatrix}$. Here $\mathfrak{so}(p,q)$ consists of real matrices $\begin{pmatrix} A & B \\ B^t & C \end{pmatrix}$ in which A and C are skew-symmetric. When we complexify and then conjugate by $\begin{pmatrix} 1 & 0 \\ 0 & i \end{pmatrix}$, we obtain $\mathfrak{so}(p+q,\mathbb{C})$.

4) $\mathfrak{sp}(n,\mathbb{R})$ and $\mathfrak{sp}(n,\mathbb{C}) \cap \mathfrak{u}(2n)$ are real forms of $\mathfrak{sp}(n,\mathbb{C})$.

The Lie algebra $\mathfrak{g}_0$ is said to be **reductive** if to each ideal $\mathfrak{a}_0$ in $\mathfrak{g}_0$ corresponds an ideal $\mathfrak{b}_0$ in $\mathfrak{g}_0$ with $\mathfrak{g}_0 = \mathfrak{a}_0 \oplus \mathfrak{b}_0$.

Theorem 3.1. *The Lie algebra $\mathfrak{g}_0$ is reductive if and only if $\mathfrak{g}_0 = [\mathfrak{g}_0,\mathfrak{g}_0] \oplus Z_{\mathfrak{g}_0}$ with $[\mathfrak{g}_0,\mathfrak{g}_0]$ semisimple and $Z_{\mathfrak{g}_0}$ abelian.*

REFERENCE. [K3, Corollary 1.53].

Now we consider the Lie algebra of a compact Lie group.

Theorem 3.2. *If G is a compact Lie group and $\mathfrak{g}_0$ is its Lie algebra, then*

(a) $\operatorname{Int}\mathfrak{g}_0$ *is compact.*
(b) $\mathfrak{g}_0$ *is reductive.*
(c) *the Killing form of $\mathfrak{g}_0$ is negative semidefinite.*

Furthermore let Z_G be the center of G, and let G_{ss} be the analytic subgroup of G with Lie algebra $[\mathfrak{g}_0,\mathfrak{g}_0]$. Then

(d) *G_{ss} has finite center.*
(e) *$(Z_G)_0$ and G_{ss} are closed subgroups.*
(f) *G is the commuting product $G = (Z_G)_0 G_{ss}$.*

REFERENCE. [K3, §IV.4].

REMARKS. Conclusions (b) and (c) use the existence of a G invariant inner product on $\mathfrak{g}_0$, which is constructed using Haar measure on G. Conclusion (d) uses that G may be regarded as a Lie group of matrices; this fact is a consequence of the Peter-Weyl Theorem, which we do not review. See [K3, §IV.3].

Lemma. *If* $\mathfrak{g}_0$ *is semisimple, then* $\operatorname{Der}\mathfrak{g}_0 = \operatorname{ad}\mathfrak{g}_0$. *Hence* $\operatorname{Int}\mathfrak{g}_0 = (\operatorname{Aut}\mathfrak{g}_0)_0$, *and* $\operatorname{Int}\mathfrak{g}_0$ *is a closed subgroup of* $GL(\mathfrak{g}_0)$.

REFERENCE. [K3, Proposition 1.98].

REMARK. Since $\operatorname{Int}\mathfrak{g}_0$ is the group of inner automorphisms of $\mathfrak{g}_0$ and since $\operatorname{Int}\mathfrak{g}_0$ has Lie algebra $\operatorname{ad}\mathfrak{g}_0$, it is helpful to think of this lemma as saying that every derivation is inner.

Theorem 3.3. *If the Killing form of* $\mathfrak{g}_0$ *is negative definite, then* $\operatorname{Int}\mathfrak{g}_0$ *is compact.*

REFERENCE. [K3, Proposition 4.27].

Next we discuss compact real forms.

Theorem 3.4. *If* $\mathfrak{g}_0$ *is semisimple, then the following conditions are equivalent:*

(a) $\mathfrak{g}_0$ *is the Lie algebra of some compact Lie group.*
(b) $\operatorname{Int}\mathfrak{g}_0$ *is compact.*
(c) *the Killing form of* $\mathfrak{g}_0$ *is negative definite.*

PROOF. If G is compact connected with Lie algebra $\mathfrak{g}_0$, then $\operatorname{Ad}(G)$ is compact; hence (a) implies (b). Conversely if (b) holds, then $\operatorname{Int}\mathfrak{g}_0$ is a compact Lie group with Lie algebra $\operatorname{ad}\mathfrak{g}_0$. Since $\mathfrak{g}_0$ is semisimple, $\operatorname{ad}\mathfrak{g}_0$ is isomorphic to $\mathfrak{g}_0$; thus (b) implies (a). If (b) holds, then the Killing form is negative semidefinite by Theorem 3.2, and it must be negative definite by Cartan's criterion for semisimplicity (Theorem 1.1). Thus (b) implies (c). Conversely (c) implies (b) by Theorem 3.3.

Let $\mathfrak{g}$ be semisimple. A real form $\mathfrak{g}_0$ of $\mathfrak{g}$ is said to be **compact** if the equivalent conditions of Theorem 3.4 hold. Here are some examples.

Examples. $\mathfrak{su}(n)$ is a compact real form of $\mathfrak{sl}(n,\mathbb{C})$, $\mathfrak{so}(n)$ is a compact real form of $\mathfrak{so}(n,\mathbb{C})$, and $\mathfrak{sp}(n,\mathbb{C})\cap\mathfrak{u}(2n)$ is a compact real form of $\mathfrak{sp}(n,\mathbb{C})$.

Theorem 3.5. *Each complex semisimple Lie algebra has a compact real form.*

REFERENCE. [K3, Theorem 6.11].

This result is fundamental. The first step in the proof is to extend the vector space isomorphism $\varphi = -1$ of $\mathfrak{h}$ to an automorphism $\tilde{\varphi}$ of $\mathfrak{g}$, using the Isomorphism Theorem (Theorem 1.6). Then $\tilde{\varphi}$ is used to adjust the structural constants to produce a real form for which the Killing form is negative definite. Application of Theorem 3.4 completes the argument.

The next topic is maximal tori. The setting is that G is a compact connected Lie group, $\mathfrak{g}_0$ is its Lie algebra, $\mathfrak{g}$ is the complexification of $\mathfrak{g}_0$, and B is the negative of any $\operatorname{Ad}(G)$ invariant inner product on $\mathfrak{g}_0$. The **maximal tori** in G are defined to be the subgroups maximal with respect to the property of being compact connected abelian. The theorem below lists the first facts about maximal tori.

Theorem 3.6. *If G is a compact connected Lie group, then*

(a) *the maximal tori in G are exactly the analytic subgroups corresponding to the maximal abelian subalgebras of $\mathfrak{g}_0$.*

(b) *any two maximal abelian subalgebras of $\mathfrak{g}_0$ are conjugate via $\mathrm{Ad}(G)$ and hence any two maximal tori in G are conjugate via G.*

REFERENCE. [K3, Proposition 4.30 and Theorem 4.34].

Here are some standard examples of maximal tori.

Examples.

1) Let $G = SU(n)$, the special unitary group. The complexified Lie algebra is $\mathfrak{g} = \mathfrak{sl}(n, \mathbb{C})$. A maximal torus, its Lie algebra, and its complexified Lie algebra are

$$\begin{aligned} T &= \mathrm{diag}(e^{i\theta_1}, \ldots, e^{i\theta_n}) \\ \mathfrak{t}_0 &= \mathrm{diag}(i\theta_1, \ldots, i\theta_n) \\ \mathfrak{t} &= \text{standard Cartan subalgebra of } \mathfrak{sl}(n, \mathbb{C}). \end{aligned}$$

2) Let $G = SO(2n+1)$, the rotation group. The complexified Lie algebra is $\mathfrak{g} = \mathfrak{so}(2n+1, \mathbb{C})$. A maximal torus and its complexified Lie algebra are

$$\begin{aligned} T &\text{ from 2-by-2 blocks } \begin{pmatrix} \cos\theta_j & \sin\theta_j \\ -\sin\theta_j & \cos\theta_j \end{pmatrix} \text{ and a single 1-by-1 block } (1) \\ \mathfrak{t} &= \text{standard Cartan subalgebra of } \mathfrak{so}(2n+1, \mathbb{C}). \end{aligned}$$

3) Let $G = Sp(n, \mathbb{C}) \cap U(2n)$. Here $Sp(n, \mathbb{C}) = \{x \in GL(2n, \mathbb{C}) \mid x^t J x = J\}$, where $J = \begin{pmatrix} 0 & I \\ -I & 0 \end{pmatrix}$ as earlier. The complexified Lie algebra of G is $\mathfrak{g} = \mathfrak{sp}(n, \mathbb{C})$. A maximal torus and its complexified Lie algebra are

$$\begin{aligned} T &= \mathrm{diag}(e^{i\theta_1}, \ldots, e^{i\theta_n}, e^{-i\theta_1}, \ldots, e^{-i\theta_n}) \\ \mathfrak{t} &= \text{standard Cartan subalgebra of } \mathfrak{sp}(n, \mathbb{C}). \end{aligned}$$

4) Let $G = SO(2n)$, the rotation group. The complexified Lie algebra is $\mathfrak{g} = \mathfrak{so}(2n, \mathbb{C})$.

$$\begin{aligned} T &\text{ from 2-by-2 blocks } \begin{pmatrix} \cos\theta_j & \sin\theta_j \\ -\sin\theta_j & \cos\theta_j \end{pmatrix} \\ \mathfrak{t} &= \text{standard Cartan subalgebra of } \mathfrak{so}(2n, \mathbb{C}). \end{aligned}$$

The theory of Cartan subalgebras for the complex semisimple case extends to a complex reductive Lie algebras $\mathfrak{g}$ by just saying that the center of $\mathfrak{g}$ is to be adjoined to a Cartan subalgebra of the semisimple part of $\mathfrak{g}$.

Now let us extend the theory of Cartan subalgebras from the complex reductive case to the real reductive case. If $\mathfrak{g}_0$ is a real reductive Lie algebra, we call a Lie subalgebra of $\mathfrak{g}_0$ a **Cartan subalgebra** if its complexification is a Cartan subalgebra of $\mathfrak{g} = (\mathfrak{g}_0)^{\mathbb{C}}$. Using condition (c) in the definition of Cartan subalgebra for the complex semisimple Lie algebra, we readily see that if $\mathfrak{g}_0$ is the Lie algebra of a compact connected Lie group G and if $\mathfrak{t}_0$ is a maximal abelian subspace of $\mathfrak{g}_0$, then $\mathfrak{t}_0$ is a Cartan subalgebra. In this setting, we can form a root-space decomposition

$$\mathfrak{g} = \mathfrak{t} \oplus \bigoplus_{\alpha \in \Delta} \mathfrak{g}_\alpha.$$

Here $\mathfrak{g} = Z_\mathfrak{g} \oplus [\mathfrak{g}, \mathfrak{g}]$, $\mathfrak{t} = Z_\mathfrak{g} \oplus (\mathfrak{t} \cap [\mathfrak{g}, \mathfrak{g}])$, and the root spaces $\mathfrak{g}_\alpha$ lie in $[\mathfrak{g}, \mathfrak{g}]$. Moreover, each root is the complexified differential of a multiplicative character ξ_α of the maximal torus T that corresponds to $\mathfrak{t}_0$, with

$$\operatorname{Ad}(t)X = \xi_\alpha(t)X \qquad \text{for } X \in \mathfrak{g}_\alpha.$$

The next results concern centralizers of tori. These results give the main control over connectedness of subgroups of semisimple and reductive groups.

Theorem 3.7. *If G is a compact connected Lie group and T is a maximal torus, then each element of G is conjugate to a member of T.*

REFERENCE. [K3, Theorem 4.36].

This is a deep theorem. For $SU(n)$, it just amounts to the Spectral Theorem, but it becomes progressively more complicated for more complicated G. We list three immediate consequences.

Corollary.

(a) *Every element of a compact connected Lie group G lies in some maximal torus.*
(b) *The center Z_G of a compact connected Lie group lies in every maximal torus.*
(c) *For any compact connected Lie group G, the exponential map is onto G.*

With a supplementary argument and Theorem 3.7, we obtain

Theorem 3.8. *Let G be a compact connected Lie group, and let S be a torus of G. If g in G centralizes S, then there is a torus S' in G containing both S and g.*

REFERENCE. [K3, Theorem 4.50].

This theorem is normally applied in either of the two forms in the following corollary.

Corollary.

(a) *In a compact connected Lie group, the centralizer of a torus is connected.*
(b) *A maximal torus in a compact connected Lie group is equal to its own centralizer.*

Let us introduce Weyl groups in this context. The notation is unchanged: G is compact connected, $\mathfrak{g}_0$ is the Lie algebra of G, $\mathfrak{g}$ is the complexification, T is a maximal torus, $\mathfrak{t}_0$ is the Lie algebra of T, $\mathfrak{t}$ is the complexification, $\Delta(\mathfrak{g}, \mathfrak{t})$ is the set of roots, and B is the negative of a G invariant inner product on $\mathfrak{g}_0$. Define $\mathfrak{t}_\mathbb{R} = i\mathfrak{t}_0$. Roots are real on $\mathfrak{t}_\mathbb{R}$, hence are in $\mathfrak{t}_\mathbb{R}^*$. The form B, when extended to be complex bilinear, is positive definite on $\mathfrak{t}_\mathbb{R}$, yielding an inner product $\langle \cdot\,, \cdot \rangle$ on $\mathfrak{t}_\mathbb{R}^*$.

Let the root reflection s_α be defined on $\mathfrak{t}_\mathbb{R}^*$ by $s_\alpha(\lambda) = \lambda - \dfrac{2\langle \lambda, \alpha \rangle}{|\alpha|^2}\alpha$. The Weyl group $W(\Delta(\mathfrak{g}, \mathfrak{t}))$ is the group generated by all s_α for $\alpha \in \Delta(\mathfrak{g}, \mathfrak{t})$. This is a finite group.

We define $W(G,T)$ as the quotient of normalizer by centralizer

$$W(G,T) = N_G(T)/Z_G(T) = N_G(T)/T.$$

This also is a finite group. It follows from Theorems 3.7 and 3.6b that the conjugacy classes in G are parametrized by $T/W(G,T)$. (See [K3, Proposition 4.53].)

Theorem 3.9. *The group $W(G,T)$, when considered as acting on $\mathfrak{t}_{\mathbb{R}}^*$, coincides with $W(\Delta(\mathfrak{g},\mathfrak{t}))$.*

REFERENCE. [K3, Theorem 4.54].

Continuing with notation as above, we work with two notions of integrality. It is easy to see that the following two conditions on a member λ of $\mathfrak{t}^*$ are equivalent:

(1) Whenever $H \in \mathfrak{t}_0$ satisfies $\exp H = 1$, then $\lambda(H)$ is in $2\pi i\mathbb{Z}$.
(2) There is a multiplicative character ξ_λ of T with $\xi_\lambda(\exp H) = e^{\lambda(H)}$ for all $H \in \mathfrak{t}_0$.

When (1) and (2) hold, λ is said to be **analytically integral**. As before, we say that λ is **algebraically integral** if $\dfrac{2\langle\lambda,\alpha\rangle}{|\alpha|^2}$ is in $\mathbb{Z}$ for all $\alpha \in \Delta(\mathfrak{g},\mathfrak{t})$.

Theorem 3.10. *Analytic and algebraic integrality have the following eight properties:*

(a) *Weights of finite-dimensional representations of G are analytically integral. In particular, every root is analytically integral.*
(b) *Analytically integral implies algebraically integral.*
(c) *Fix a simple system of roots $\{\alpha_1,\dots,\alpha_l\}$. Then $\lambda \in \mathfrak{t}^*$ is algebraically integral if and only if $2\langle\lambda,\alpha_i\rangle/|\alpha_i|^2$ is in $\mathbb{Z}$ for each simple root α_i.*
(d) *If $\tilde{G}$ is a finite covering group of G, then the index of the group of analytically integral forms for G in the group of analytically integral forms for $\tilde{G}$ equals the order of the kernel of the covering homomorphism $\tilde{G} \to G$.*
(e) *The subgroup of $\mathbb{Z}$ combinations of roots in $\mathfrak{t}_{\mathbb{R}}^*$ is contained in the lattice of analytically integral forms, which in turn is contained in the subgroup of algebraically integral forms. If G is semisimple, all three subgroups are lattices.*
(f) *If G is semisimple, then the index of the lattice of $\mathbb{Z}$ combinations of roots in the lattice of algebraically integral forms is exactly the determinant of the Cartan matrix.*
(g) *If G is semisimple and Z_G is trivial, then every analytically integral form is a $\mathbb{Z}$ combination of roots.*
(h) *If G is simply connected and semisimple, then algebraically integral implies analytically integral.*

REFERENCE. [K3, §§IV.7 and V.8].

REMARKS. In the semisimple case, conclusion (e) identifies containments among three lattices in $\mathfrak{t}_{\mathbb{R}}^*$, and (f) says that the index of the smallest in the largest is the determinant of the Cartan matrix. Conclusions (g) and (h) give circumstances under which the middle lattice is equal to the smallest or the largest. The proof of (h) uses the existence result in the Theorem of the Highest Weight.

Theorem 3.11 (Weyl's Theorem). *If G is a compact semisimple Lie group, then the fundamental group of G is finite. Consequently the universal covering group of G is compact.*

REFERENCE. [K3, Theorem 4.69].

Combining Weyl's Theorem with Theorem 3.10, we obtain the following consequence.

Corollary. *In a compact semisimple Lie group G,*

(a) *the order of the fundamental group of G equals the index of the group of analytically integral forms for G in the group of algebraically integral forms.*
(b) *if G is simply connected, then the order of the center Z_G of G equals the determinant of the Cartan matrix.*

Let us now rephrase the results about representations of complex semisimple Lie algebras as results about compact connected Lie groups. (See [K3, §V.8].)

Theorem 3.12 (Theorem of the Highest Weight). *Let G be a compact connected Lie group with complexified Lie algebra $\mathfrak{g}$, let T be a maximal torus with complexified Lie algebra $\mathfrak{t}$, and let $\Delta^+(\mathfrak{g},\mathfrak{t})$ be a positive system for the roots. Apart from equivalence the irreducible finite-dimensional representations Φ of G stand in one-one correspondence with the dominant analytically integral linear functionals λ on $\mathfrak{t}$, the correspondence being that λ is the highest weight of Φ.*

In the context of representations of the compact connected group G, we can regard characters $\operatorname{char}(V)=\sum(\dim V_\lambda)e^\lambda$ as functions on $\mathfrak{t}_0$. The algebraic theory gives

$$d\operatorname{char}(V)=\sum_{w\in\Delta(\mathfrak{g},\mathfrak{t})}(\det w)e^{w(\lambda+\delta)}$$

in $\mathbb{Z}[\mathfrak{t}^*]$ for the semisimple case.

We can pass from the algebraic result in $\mathbb{Z}[\mathfrak{t}^*]$ to the group case for G semisimple by using the evaluation homormorphism at each point of $\mathfrak{t}_0$ and addressing analytic integrality. Then we can extend the group result to general compact connected G. One shows that the element $\delta\in\mathfrak{t}^*$ (half the sum of the positive roots) has $2\langle\delta,\alpha_i\rangle/|\alpha_i|^2=1$ for simple α_i, hence is algebraically integral. Nevertheless, δ is not always analytically integral; it is not analytically integral in $SO(3)$, for example. A sufficient compensation for this failure is that $\delta-w\delta$ is always analytically integral for all w. Consequently we are able to obtain the following group version of the Weyl Character Formula.

Theorem 3.13 (Weyl Character Formula). *Let G be a compact connected Lie group, let T be a maximal torus, let $\Delta^+=\Delta^+(\mathfrak{g},\mathfrak{t})$ be a positive system for the roots, and let $\lambda\in\mathfrak{t}^*$ be analytically integral and dominant. Then the character χ_λ of the irreducible finite-dimensional representation of G with highest weight λ is given by*

$$\chi_\lambda=\frac{\sum_{w\in W}(\det w)\xi_{w(\lambda+\delta)-\delta}(t)}{\prod_{\alpha\in\Delta^+}(1-\xi_{-\alpha}(t))}$$

at every $t\in T$ where no ξ_α takes the value 1 on t. If G is simply connected, then this formula can be rewritten as

$$\chi_\lambda=\frac{\sum_{w\in W}(\det w)\xi_{w(\lambda+\delta)}(t)}{\xi_\delta(t)\prod_{\alpha\in\Delta^+}(1-\xi_{-\alpha}(t))}.$$

Before concluding the treatment of compact groups, let us mention that much of the theory for compact connected Lie groups can be obtained directly, without first addressing complex semisimple Lie algebras. Weyl carried out such a program, using integration as the tool. Here is the formula that Weyl used.

Theorem 3.14 (Weyl Integration Formula). *Let T be a maximal torus of the compact connected Lie group G, and let invariant measures on G, T, and G/T be normalized so that*

$$\int_G f(x)\,dx = \int_{G/T}\Big[\int_T f(xt)\,dt\Big]\,d(xT)$$

for all continuous f on G. Then every Borel function $F \geq 0$ on G has

$$\int_G F(x)\,dx = \frac{1}{|W(G,T)|}\int_T\Big[\int_{G/T} F(gtg^{-1})\,d(gT)\Big]\,|D(t)|^2\,dt,$$

where

$$|D(t)|^2 = \prod_{\alpha\in\Delta^+} |1-\xi_\alpha(t^{-1})|^2.$$

REFERENCE. [K3, Theorem 8.60].

4. Structure Theory of Noncompact Semisimple Groups

This section deals with the structure theory of noncompact semisimple Lie groups and with the definition and first properties of reductive Lie groups. Some references for this material are [He], [K1], [K3], and [W].

The theory begins with the development of Cartan involutions. Let $\mathfrak{g}_0$ be a real semisimple Lie algebra, and let B be the Killing form. (Later we shall allow other forms in place of the Killing form.) A source of many examples of real semisimple Lie algebras is as follows.

Theorem 4.1. *If $\mathfrak{g}_0$ is a real Lie algebra of real or complex or quaternion matrices closed under conjugate transpose, then $\mathfrak{g}_0$ is reductive. If also $Z_{\mathfrak{g}_0} = 0$, then $\mathfrak{g}_0$ is semisimple.*

REFERENCE. [K3, Proposition 1.56].

Examples. The following examples are classical Lie algebras that satisfy the hypotheses of Theorem 4.1 for all n, p, and q. For appropriate values of n, p, and q, these examples are semisimple.

1) Compact Lie algebras: $\mathfrak{su}(n)$, $\mathfrak{so}(n)$, and $\mathfrak{sp}(n,\mathbb{C})\cap\mathfrak{u}(2n)\cong\mathfrak{sp}(n)$.

2) Complex Lie algebras: $\mathfrak{sl}(n,\mathbb{C})$, $\mathfrak{so}(n,\mathbb{C})$, and $\mathfrak{sp}(n,\mathbb{C})$.

3) Other Lie algebras: $\mathfrak{sl}(n,\mathbb{R})$, $\mathfrak{sl}(n,\mathbb{H})$, $\mathfrak{sp}(n,\mathbb{R})$, $\mathfrak{so}(p,q)$, $\mathfrak{su}(p,q)$, $\mathfrak{sp}(p,q)$, and $\mathfrak{so}^*(2n)$. Here $\mathfrak{sl}(n,\mathbb{H})$ refers to quaternion matrices for which the real part of the trace is 0, and $\mathfrak{sp}(p,q)$ refers to quaternion matrices preserving a Hermitian form of signature (p,q).

An involution θ of $\mathfrak{g}_0$ (understood to respect brackets) such that the symmetric bilinear form

$$B_\theta(X,Y) = -B(X,\theta Y)$$

is positive definite is called a **Cartan involution** of $\mathfrak{g}_0$. Correspondingly there is a **Cartan decomposition** of $\mathfrak{g}_0$ given by

$$\mathfrak{g}_0 = \mathfrak{k}_0 \oplus \mathfrak{p}_0.$$

The subspaces $\mathfrak{k}_0$ and $\mathfrak{p}_0$ are understood to be the $+1$ and -1 eigenspaces of θ; they satisfy the bracket relations

$$[\mathfrak{k}_0, \mathfrak{k}_0] \subseteq \mathfrak{k}_0, \quad [\mathfrak{k}_0, \mathfrak{p}_0] \subseteq \mathfrak{p}_0, \quad [\mathfrak{p}_0, \mathfrak{p}_0] \subseteq \mathfrak{k}_0.$$

Moreover, B is negative on $\mathfrak{k}_0$, B is positive on $\mathfrak{p}_0$, and $B(\mathfrak{k}_0, \mathfrak{p}_0) = 0$.

Examples.

1) If $\mathfrak{g}_0$ is as in the list of examples above, then θ can be taken to be negative conjugate transpose.

2) Let $\mathfrak{g}$ be a complex semisimple Lie algebra, let $\mathfrak{u}_0$ be a compact real form of $\mathfrak{g}$, and let τ be the corresponding conjugation of $\mathfrak{g}$. If $\mathfrak{g}$ is regarded as a real Lie algebra, then τ is a Cartan involution of $\mathfrak{g}$.

The main tool for handling Cartan involutions is Theorem 4.2 below. This is a result of Berger that improves on the original result of Cartan.

Theorem 4.2. *Let θ be a Cartan involution of $\mathfrak{g}_0$, and let σ be any involution. Then there exists φ in* $\operatorname{Int}\mathfrak{g}_0$ *such that $\varphi\theta\varphi^{-1}$ commutes with σ.*

REFERENCE. [K3, Theorem 6.16].

Corollary.

(a) *$\mathfrak{g}_0$ has a Cartan involution.*
(b) *Any two Cartan involutions of $\mathfrak{g}_0$ are conjugate via* $\operatorname{Int}\mathfrak{g}_0$.
(c) *If $\mathfrak{g}$ is a complex semisimple Lie algebra, then any two compact real forms of $\mathfrak{g}$ are conjugate via* $\operatorname{Int}\mathfrak{g}$.
(d) *If $\mathfrak{g}$ is a complex semisimple Lie algebra and is considered as a real Lie algebra, then the only Cartan involutions of $\mathfrak{g}$ are the conjugations with respect to the compact real forms of $\mathfrak{g}$.*

REFERENCE. [K3, §VI.2].

SKETCH OF PROOF. For (a), Theorem 4.2 is applied to $\mathfrak{g}$ made real, using θ from a compact real form and σ from conjugation of $\mathfrak{g}$ with respect to $\mathfrak{g}_0$. Conclusion (b) is immediate, and (c) is a special case of (b). Conclusion (d) follows from (b) and the fact that such a conjugation exists (Theorem 3.5).

If $\mathfrak{g}_0 = \mathfrak{k}_0 \oplus \mathfrak{p}_0$ is a Cartan decomposition of $\mathfrak{g}_0$, then $\mathfrak{k}_0 \oplus i\mathfrak{p}_0$ is a compact real form of $\mathfrak{g} = (\mathfrak{g}_0)^{\mathbb{C}}$. Conversely Theorem 3.3 shows that if $\mathfrak{h}_0$ and $\mathfrak{q}_0$ are the $+1$ and -1 eigenspaces of an involution σ, then σ is a Cartan involution if the real form $\mathfrak{h}_0 \oplus i\mathfrak{q}_0$ of $\mathfrak{g} = (\mathfrak{g}_0)^{\mathbb{C}}$ is compact.

These considerations allow B to be generalized a little. Fix an involution θ of $\mathfrak{g}_0$, and let $\mathfrak{g}_0 = \mathfrak{k}_0 \oplus \mathfrak{p}_0$ be the eigenspace decomposition relative to θ. We suppose that B is any nondegenerate symmetric invariant bilinear form on $\mathfrak{g}_0$ with $B(\theta X, \theta Y) = B(X, Y)$ such that $B_\theta(X, Y) = -B(X, \theta Y)$ is positive definite. Then B is negative definite on $\mathfrak{k}_0 \oplus i\mathfrak{p}_0$, and it follows that $\mathfrak{k}_0 \oplus i\mathfrak{p}_0$ is compact. Consequently θ is a Cartan involution. In this setting we allow B to be used in place of the Killing form.

Notice in this case that B is negative definite on a maximal abelian subspace of $\mathfrak{k}_0 \oplus i\mathfrak{p}_0$, hence positive definite on the real subspace of a Cartan subalgebra of $(\mathfrak{g}_0)^{\mathbb{C}}$ where roots are real-valued. Therefore B has the correct "sign" on $(\mathfrak{g}_0)^{\mathbb{C}}$ for the theory of complex semisimple Lie algebras to be applicable.

By a **semisimple Lie group**, we mean a connected Lie group whose Lie algebra is semisimple. The next theorem gives the **global Cartan decomposition** of a semisimple Lie group.

Theorem 4.3. *Let G be a semisimple Lie group, let θ be a Cartan involution of its Lie algebra $\mathfrak{g}_0$, let $\mathfrak{g}_0 = \mathfrak{k}_0 \oplus \mathfrak{p}_0$ be the corresponding Cartan decomposition, and let K be the analytic subgroup of G with Lie algebra $\mathfrak{k}_0$. Then*

(a) *there exists a Lie group automorphism Θ of G with differential θ, and Θ has $\Theta^2 = 1$.*
(b) *the subgroup of G fixed by Θ is K.*
(c) *the mapping $K \times \mathfrak{p}_0 \to G$ given by $(k, X) \mapsto k \exp X$ is a diffeomorphism onto.*
(d) *K is closed.*
(e) *K contains the center Z of G.*
(f) *K is compact if and only if Z is finite.*
(g) *when Z is finite, K is a maximal compact subgroup of G.*

REFERENCE. [K3, Theorem 6.31].

Example. When G is an analytic group of matrices and θ is negative conjugate transpose, Θ is conjugate transpose inverse. The content of (c) is that G is stable under the polar decomposition of matrices. Thus (c) of the theorem may be regarded as a generalization of the polar decomposition to all semisimple Lie groups.

This completes the discussion of Cartan involutions. For most of the remainder of this section, we shall use the following notation. Let G be a semisimple Lie group, let $\mathfrak{g}_0$ be its Lie algebra, let $\mathfrak{g}$ be the complexification of $\mathfrak{g}_0$, let θ be a Cartan involution of $\mathfrak{g}_0$, and let $\mathfrak{g}_0 = \mathfrak{k}_0 \oplus \mathfrak{p}_0$ be the corresponding Cartan decomposition. Let B as above be a θ invariant nondegenerate symmetric bilinear form on $\mathfrak{g}_0$ such that B_θ is positive definite.

The next topic will be restricted roots and the Iwasawa decomposition. Let $\mathfrak{a}_0$ be a maximal abelian subspace of $\mathfrak{p}_0$. **Restricted roots** are the nonzero $\lambda \in \mathfrak{a}_0^*$ such that the space $(\mathfrak{g}_0)_\lambda$ defined as

$$\{X \in \mathfrak{g}_0 \mid (\operatorname{ad} H)X = \lambda(H)X \text{ for all } H \in \mathfrak{a}_0\}$$

is nonzero. Let Σ be the set of restricted roots. Define $\mathfrak{m}_0 = Z_{\mathfrak{k}_0}(\mathfrak{a}_0)$. Restricted roots and the corresponding restricted-root spaces have the following elementary properties:

(a) $\mathfrak{g}_0 = \mathfrak{a}_0 \oplus \mathfrak{m}_0 \oplus \bigoplus_{\lambda \in \Sigma} (\mathfrak{g}_0)_\lambda$,
(b) $[(\mathfrak{g}_0)_\lambda, (\mathfrak{g}_0)_\mu] \subseteq (\mathfrak{g}_0)_{\lambda+\mu}$,
(c) $\theta(\mathfrak{g}_0)_\lambda = (\mathfrak{g}_0)_{-\lambda}$,
(d) Σ is a root system in $\mathfrak{a}_0^*$.

Introduce a lexicographic ordering in $\mathfrak{a}_0^*$, and define

$$\Sigma^+ = \{\text{positive restricted roots}\}$$

$$\mathfrak{n}_0 = \bigoplus_{\lambda \in \Sigma^+} (\mathfrak{g}_0)_\lambda.$$

The subspace $\mathfrak{n}_0$ of $\mathfrak{g}_0$ is a nilpotent Lie subalgebra.

Theorem 4.4 (Iwasawa decomposition of Lie algebra). *The semisimple Lie algebra $\mathfrak{g}_0$ is a vector-space direct sum $\mathfrak{g}_0 = \mathfrak{k}_0 \oplus \mathfrak{a}_0 \oplus \mathfrak{n}_0$. Here $\mathfrak{a}_0$ is abelian, $\mathfrak{n}_0$ is nilpotent, $\mathfrak{a}_0 \oplus \mathfrak{n}_0$ is a solvable Lie subalgebra of $\mathfrak{g}_0$, and $\mathfrak{a}_0 \oplus \mathfrak{n}_0$ has $[\mathfrak{a}_0 \oplus \mathfrak{n}_0, \mathfrak{a}_0 \oplus \mathfrak{n}_0] = \mathfrak{n}_0$.*

REFERENCE. [K3, Proposition 6.43].

Theorem 4.5 (Iwasawa decomposition of Lie group). *Let G be a semisimple group, let $\mathfrak{g}_0 = \mathfrak{k}_0 \oplus \mathfrak{a}_0 \oplus \mathfrak{n}_0$ be an Iwasawa decomposition of the Lie algebra $\mathfrak{g}_0$ of G, and let A and N be the analytic subgroups of G with Lie algebras $\mathfrak{a}$ and $\mathfrak{n}$. Then the multiplication map $K \times A \times N \to G$ given by $(k, a, n) \mapsto kan$ is a diffeomorphism onto. The groups A and N are simply connected.*

REFERENCE. [K3, Theorem 6.46].

Roots and restricted roots are related to each other. If $\mathfrak{t}_0$ is a maximal abelian subspace of $\mathfrak{g}_0$, then $\mathfrak{h}_0 = \mathfrak{a}_0 \oplus \mathfrak{t}_0$ is a Cartan subalgebra of $\mathfrak{g}_0$ (see [K3, Proposition 6.47]). Roots are real-valued on $\mathfrak{a}_0$ and imaginary-valued on $\mathfrak{t}_0$. The nonzero restrictions to $\mathfrak{a}_0$ of the roots turn out to be the restricted roots (see [K3, §VI.4]). Roots and restricted roots can be ordered compatibly by taking $\mathfrak{a}_0$ before $i\mathfrak{t}_0$.

The next theorem describes the effect of altering the choices that have been made in obtaining the Iwasawa decomposition.

Theorem 4.6.

(a) *If $\mathfrak{a}_0$ and $\mathfrak{a}_0'$ are two maximal abelian subspaces of $\mathfrak{p}_0$, then there is a member k of K with $\operatorname{Ad}(k)\mathfrak{a}_0' = \mathfrak{a}_0$. Consequently the space $\mathfrak{p}_0$ satisfies $\mathfrak{p}_0 = \bigcup_{k\in K} \operatorname{Ad}(k)\mathfrak{a}_0$.*
(b) *Any two choices of $\mathfrak{n}_0$ are conjugate by* Ad *of a member of $N_K(\mathfrak{a}_0)$.*
(c) *Define $W(G, A) = N_K(\mathfrak{a}_0)/Z_K(\mathfrak{a}_0)$. The Lie algebra of the normalizer $N_K(\mathfrak{a}_0)$ is $\mathfrak{m}_0$, and therefore $W(G, A)$ is a finite group.*
(d) *$W(G, A)$ coincides with $W(\Sigma)$.*

REFERENCE. [K3, §VI.5].

REMARKS. Already we know from the Corollary to Theorem 4.2 that any two Cartan decompositions of $\mathfrak{g}_0$ are conjugate via Int $\mathfrak{g}_0$. Therefore any two choices of K are conjugate in G. Conclusion (a) of the theorem says that with K fixed, any two choices of $\mathfrak{a}_0$ are conjugate, and conclusion (b) says that with K and $\mathfrak{a}_0$ fixed, any two choices of $\mathfrak{n}_0$ are conjugate. Therefore any two Iwasawa decompositions are conjugate.

Now let us study Cartan subalgebras and subgroups. We know that $\mathfrak{g}_0$ always has a Cartan subalgebra. Namely if $\mathfrak{t}_0$ is any maximal abelian subspace of $\mathfrak{m}_0$, then $\mathfrak{h}_0 = \mathfrak{a}_0 \oplus \mathfrak{t}_0$ is a Cartan subalgebra of $\mathfrak{g}_0$. However, Cartan subalgebras are not necessarily unique up to conjugacy, as the following example shows.

Example. The Lie algebra $\mathfrak{g}_0 = \mathfrak{sl}(2,\mathbb{R})$ has two Cartan subalgebras nonconjugate via $\operatorname{Int}\mathfrak{g}_0$, namely all $\begin{pmatrix} x & 0 \\ 0 & -x \end{pmatrix}$ and all $\begin{pmatrix} 0 & y \\ -y & 0 \end{pmatrix}$. Every Cartan subalgebra of $\mathfrak{g}_0$ is conjugate via $\operatorname{Int}\mathfrak{g}_0$ to one of these.

In a complex Lie algebra $\mathfrak{g}$, any two Cartan subalgebras are conjugate via $\operatorname{Int}\mathfrak{g}$. Therefore, despite the nonconjugacy, any two Cartan subalgebras of $\mathfrak{g}_0$ have the same dimension. This dimension is called the **rank** of $\mathfrak{g}_0$.

Let us mention some properties of Cartan subalgebras of $\mathfrak{g}_0$ (see [K3, §VI.6]). Any Cartan subalgebra is conjugate via $\operatorname{Int}\mathfrak{g}_0$ to a θ stable Cartan subalgebra.

If $\mathfrak{h}_0$ is a θ stable Cartan subalgebra, we can decompose $\mathfrak{h}_0$ according to $\mathfrak{g}_0 = \mathfrak{k}_0 \oplus \mathfrak{p}_0$ as $\mathfrak{h}_0 = \mathfrak{t}_0 \oplus \mathfrak{a}_0$ with $\mathfrak{t}_0 \subseteq \mathfrak{k}_0$ and $\mathfrak{a}_0 \subseteq \mathfrak{p}_0$. It is appropriate to think of $\mathfrak{t}_0$ as the compact part of $\mathfrak{h}_0$ and $\mathfrak{a}_0$ as the noncompact part. Define $\mathfrak{h}_0$ to be **maximally compact** if its compact part has maximal dimension among all θ stable Cartan subalgebras, or to be **maximally noncompact** if its noncompact part has maximal dimension. The Cartan subalgebra $\mathfrak{h}_0$ constructed after the Iwasawa decomposition is maximally noncompact. If $\mathfrak{t}_0$ is a maximal abelian subspace of $\mathfrak{k}_0$, then $\mathfrak{h}_0 = Z_{\mathfrak{g}_0}(\mathfrak{t}_0)$ is maximally compact.

Among θ stable Cartan subalgebras $\mathfrak{h}_0$ of $\mathfrak{g}_0$, the maximally noncompact ones are all conjugate via K, and the maximally compact ones are all conjugate via K. Hence the constructions in the previous paragraph yield all maximally compact and maximally noncompact θ stable Cartan subalgebras.

Up to conjugacy by $\operatorname{Int}\mathfrak{g}_0$, there are only finitely many Cartan subalgebras of $\mathfrak{g}_0$. In fact, any θ stable Cartan subalgebra, up to conjugacy, can be transformed into any other θ stable Cartan subalgebra by a sequence of **Cayley transforms**, which change a Cartan subalgebra of $\mathfrak{g}_0$ only within a subalgebra $\mathfrak{sl}(2,\mathbb{R})$. Within the $\mathfrak{sl}(2,\mathbb{R})$, the change is essentially the change between the two types in the example above. The relevant $\mathfrak{sl}(2,\mathbb{R})$'s for the Cayley transforms are the ones corresponding to particular kinds of roots.

By definition a **Cartan subgroup** of G is the centralizer in G of a Cartan subalgebra of $\mathfrak{g}_0$. In order to analyze noncompact semisimple groups, one wants an analog of the result Theorem 3.7 in the compact case that every element is conjugate to a member of a maximal torus.

For this purpose we introduce the regular elements of G. Let l be the common dimension of all Cartan subalgebras of $\mathfrak{g}_0$, and write

$$\det((\lambda+1)1_n - \operatorname{Ad}(x)) = \lambda^n + \sum_{j=0}^{n-1} D_j(x)\lambda^j.$$

We call $x \in G$ **regular** if $D_l(x) \neq 0$. Let G' be the set of all regular elements in G.

Theorem 4.7. *Let $(\mathfrak{h}_1)_0, \dots, (\mathfrak{h}_r)_0$ be a maximal set of nonconjugate θ stable Cartan subalgebras of $\mathfrak{g}_0$, and let $H_1, \dots, H_r$ be the corresponding Cartan subgroups of G. Then*

(a) $G' \subseteq \bigcup_{i=1}^r \bigcup_{x\in G} xH_ix^{-1}$.
(b) *each member of G' lies in just one Cartan subgroup of G.*

REFERENCE. [K3, Theorem 7.108].

REMARKS. By the theorem the regular elements are conjugate to members of Cartan subgroups. This fact turns out to be good enough to give an analog of the Weyl Integration Formula for noncompact semisimple groups. We omit the details.

This completes our discussion of Cartan subalgebras and Cartan subgroups. We turn now to the topic of parabolic subalgebras and parabolic subgroups. The notation remains unchanged.

First we introduce two subgroups M and N^-. The group N^- is often called $\overline{N}$ in the literature. The subgroup M of G is defined by $M = Z_K(\mathfrak{a}_0)$. Its Lie algebra is $\mathfrak{m}_0 = Z_{\mathfrak{k}_0}(\alpha_0)$, and M normalizes each restricted-root space $(\mathfrak{g}_0)_\lambda$.

It follows from the Iwasawa decomposition (Theorem 4.5) that MAN is a closed subgroup of G. It and its conjugates in G are called **minimal parabolic** subgroups. Its Lie algebra is $\mathfrak{m}_0 \oplus \mathfrak{a}_0 \oplus \mathfrak{n}_0$, a **minimal parabolic** subalgebra of $\mathfrak{g}_0$.

Let $\mathfrak{n}_0^- = \bigoplus_{\lambda\in\Sigma^+}(\mathfrak{g}_0)_{-\lambda} = \theta\mathfrak{n}_0$, and let $N^- = \Theta N$ be the corresponding analytic subgroup of G. Here is a handy integral formula used in analysis on G; for $g = SL(2,\mathbb{R})$, it amounts to an arctangent substitution for passing from the circle to the line.

Theorem 4.8. *Write elements of $G = KAN$ as $g = \kappa e^{H(g)}n$. Let 2ρ be the sum of the members of Σ^+ with multiplicities counted. Then there exists a normalization of Haar measures such that*

$$\int_K f(k)\,dk = \int_{N^-} f(\kappa(\bar{n}))e^{-2\rho H(\bar{n})}\,d\bar{n}$$

for all continuous f on K that are right invariant under M.

REFERENCE. [K3, Proposition 8.46].

The next theorem gives the double-coset decomposition of G relative to the subgroup MAN.

Theorem 4.9 (Bruhat decomposition). *Let $\{\tilde{w}\}$ be a set of representatives in K for the members w of $W(G,A)$, and let $[\tilde{w}]$ be the image of $\tilde{w}$ in $W(G,A)$. Then*

$$G = \bigcup_{[\tilde{w}]\in W(G,A)} MAN\tilde{w}MAN$$

disjointly.

REFERENCE. [K3, Theorem 7.40].

The existence half of the following decomposition is an immediate consequence of the global Cartan decomposition (Theorem 4.3) and the conjugacy of the various choices for $\mathfrak{a}_0$ (Theorem 4.6).

Theorem 4.10 (KAK decomposition). *Every element in G has a decomposition as k_1ak_2 with $k_1, k_2 \in K$ and $a \in A$. In this decomposition, a is uniquely determined up to conjugation by a member of $W(G,A)$. If a is fixed as $\exp H$ with $H \in \mathfrak{a}_0$ and if $\lambda(H) \neq 0$ for all $\lambda \in \Sigma$, then k_1 is unique up to right multiplication by a member of M.*

Before considering general parabolic subalgebras and subgroups, we mention special features of the "complex case." Suppose that the real semisimple Lie algbra

Lie algebra $\mathfrak{g}_0$ is actually complex, i.e., that there exists a linear map $J : \mathfrak{g}_0 \to \mathfrak{g}_0$ such that $J[X,Y] = [JX,Y] = [X,JY]$ and $J^2 = -1$. The corresponding group G then has an invariant complex structure and is called a **complex semisimple group**. Any choice of $\mathfrak{k}_0$ is a compact real form of $\mathfrak{g}_0$, and $\mathfrak{p}_0 = J\mathfrak{k}_0$. The Lie algebra $\mathfrak{m}_0$ is $J\mathfrak{a}_0$, and $\mathfrak{a}_0 \oplus J\mathfrak{a}_0$ is a complex Cartan subalgebra of the complex Lie algebra $\mathfrak{g}_0$. Each restricted root space has real dimension 2 and is a root space for $\mathfrak{a}_0 \oplus J\mathfrak{a}_0$. The group M is connected, all Cartan subalgebras are complex and are conjugate, and all Cartan subgroups are connected.

Returning to an arbitrary real semisimple Lie algebra $\mathfrak{g}_0$, let us now give the definitions of general parabolic subalgebras and subgroups. A **Borel subalgebra** of our complex semisimple Lie algebra $\mathfrak{g}$ is defined to be a subalgebra of the form $\mathfrak{h} \oplus \bigoplus_{\alpha\in\Delta^+} \mathfrak{g}_\alpha$, where $\mathfrak{h}$ is a Cartan subalgebra and Δ^+ is a positive system of roots. A **parabolic subalgebra** of $\mathfrak{g}$ is a subalgebra containing a Borel subalgebra.

Theorem 4.11. *The parabolic subalgebras containing a given Borel subalgebra may be parametrized as follows. Let Π be the set of simple roots defining the set Δ^+ of positive roots that determine the Borel subalgebra. If Π' is any subset of Π, then there is a parabolic subalgebra corresponding to Π', namely*

$$\begin{aligned}\mathfrak{p}_{\Pi'} &= \Big(\mathfrak{h} \oplus \bigoplus_{\alpha\in\operatorname{span}(\Pi')} \mathfrak{g}_\alpha\Big) \oplus \Big(\bigoplus_{\substack{\text{other}\\ \alpha\in\Delta^+}} \mathfrak{g}_\alpha\Big)\\ &= \textbf{Levi subalgebra} \oplus \textbf{nilpotent radical}\ .\end{aligned}$$

All parabolic subalgebras containing the given Borel subalgebra are of this form.

REFERENCE. [K3, Proposition 5.90].

Now let us consider $\mathfrak{g}_0$. Suppose above that $\mathfrak{h} = (\mathfrak{h}_0)^{\mathbb{C}}$ with $\mathfrak{h}_0$ constructed from the Iwasawa decomposition and with Δ^+ consistent with Σ^+. Then one can show that the parabolic subalgebras of $\mathfrak{g}$ that are complexifications are the complexifications of all subalgebras of $\mathfrak{g}_0$ containing a minimal parabolic $\mathfrak{q}_0 = \mathfrak{m}_0 \oplus \mathfrak{a}_0 \oplus \mathfrak{n}_0$.

We can parametrize these by subsets of simple restricted roots as follows. The formulas look similar to those in Theorem 4.11. Let Φ be a subset of simple restricted roots. Define

$$\begin{aligned}(\mathfrak{q}_\Phi)_0 &= \Big(\mathfrak{m}_0 \oplus \mathfrak{a}_0 \oplus \bigoplus_{\lambda\in\operatorname{span}(\Phi)} (\mathfrak{g}_0)_\lambda\Big) \oplus \Big(\bigoplus_{\substack{\text{other}\\ \lambda\in\Sigma^+}} (\mathfrak{g}_0)_\lambda\Big)\\ &= ((\mathfrak{m}_\Phi)_0 \oplus (\mathfrak{a}_\Phi)_0) \oplus (\mathfrak{n}_\Phi)_0,\end{aligned}$$

where $(\mathfrak{a}_\Phi)_0 = \bigcap_{\lambda\in\Phi} \ker\lambda$ and $(\mathfrak{m}_\Phi)_0$ is the orthocomplement of $(\mathfrak{a}_\Phi)_0$ in $(\mathfrak{m}_\Phi)_0 \oplus (\mathfrak{a}_\Phi)_0$. See [K3, §VII.7]. The decomposition $(\mathfrak{q}_\Phi)_0 = ((\mathfrak{m}_\Phi)_0 \oplus (\mathfrak{a}_\Phi)_0) \oplus (\mathfrak{n}_\Phi)_0$ is called the **Langlands decomposition** of $(\mathfrak{q}_\Phi)_0$.

The corresponding parabolic subgroup is the normalizer $Q_\Phi = N_G((\mathfrak{q}_\Phi)_0)$. This is a closed subgroup of G, being a normalizer. It has a **Langlands decomposition** $Q_\Phi = M_\Phi A_\Phi N_\Phi$, with the factors defined as follows: $(M_\Phi)_0$, A_Φ, N_Φ are to be connected, and $M_\Phi = M(M_\Phi)_0$. See [K3, §VII.7].

Finally we mention reductive Lie groups. Any representation theory done for the semisimple group G needs to be done also for all M_Φ, but M_Φ is not necessarily connected and $(M_\Phi)_0$ is not necessarily semisimple. One wants a class of groups containing interesting semisimple groups and closed under passage to the M_Φ's. Such groups are usually called **reductive Lie groups**.

There are various definitions, depending on the author. Here is the definition of G in the **Harish-Chandra class**:

(a) $\mathfrak{g}_0$ is reductive,
(b) G has finitely many components,
(c) the analytic subgroup of G corresponding to $[\mathfrak{g}_0, \mathfrak{g}_0]$ has finite center, and
(d) the action of every $\mathrm{Ad}(g)$ on $(\mathfrak{g}_0)^{\mathbb{C}}$ is in $\mathrm{Int}\,\mathfrak{g}$.

These groups have a number of important properties that we state in a qualitative form. First, $\mathfrak{g}_0$ has a Cartan involution θ. Second, G has a corresponding global Cartan decomposition. Third, the centralizer in G of any abelian θ stable subalgebra of $\mathfrak{g}_0$ is again in the class. Fourth, M meets every component of G. Fifth, the basic decompositions extend from the semisimple finite-center case to the reductive case. See [K3, §VII.2].

References

[He] S. Helgason, *Differential Geometry, Lie Groups, and Symmetric Spaces*, Academic Press, New York, 1978.

[Hu] J. E. Humphreys, *Introduction to Lie Algebras and Repressentation Theory*, Springer-Verlag, New York, 1972.

[J] N. Jacobson, *Lie Algebras*, Interscience Publishers, New York, 1962; second edition, Dover Publications, New York, 1979.

[K1] A. W. Knapp, *Representation Theory of Semisimple Groups: An Overview Based on Examples*, Princeton University Press, Princeton, N.J., 1986.

[K2] A. W. Knapp, *Lie Groups, Lie Algebras, and Cohomology*, Princeton University Press, Princeton, N.J., 1988.

[K3] A. W. Knapp, *Lie Groups Beyond an Introduction*, Birkhäuser, Boston, 1996.

[V] V. S. Varadarajan, *Lie Groups, Lie Algebras, and Their Representations*, Prentice-Hall, Englewood Cliffs, N.J., 1974; second edition, Springer-Verlag, New York, 1984.

[W] G. Warner, *Harmonic Analysis on Semi-Simple Lie Groups I*, Springer-Verlag, New York, 1972.

Department of Mathematics, State University of New York, Stony Brook, New York 11794, U.S.A.

E-mail address: aknapp@ccmail.sunysb.edu

Proceedings of Symposia in Pure Mathematics
Volume **61** (1997), pp. 29–49

Characters of Representations and Paths in $\mathfrak{H}_{\mathbb{R}}^*$

Peter Littelmann

Introduction

The aim of this note is to give an introduction to a new combinatorial tool in representation theory, the path model. The model is an extension of the usual weight theory of representations of a connected complex semisimple Lie group G, it can also be viewed as a generalization of the classical Young tableaux theory for the group $SL_n(\mathbb{C})$ to arbitrary connected complex semisimple Lie groups.

To construct objects like the tableaux in such a general setting, we consider piecewise linear paths $\pi : [0,1] \to \Lambda_{\mathbb{R}}$ in the real span of the weight lattice Λ of G. The idea is to associate to an irreducible representation V a set of paths $\mathbb{B}$ starting in the origin and ending in an integral weight, such that the character $\operatorname{Char} V$ of V reads as the sum $\sum e^{\eta(1)}$ over all paths in $\mathbb{B}$. The advantage of this approach, in comparison with the usual weight theory, is that we can speak of the "individual" contribution of a path to the character. This makes it possible to avoid the alternating sums in classical formulas like Steinberg's tensor product decomposition formula. In fact, as a consequence of this theory, we get a very simple decomposition formula for tensor products of representations, which can be seen as a generalization of the classical Littlewood-Richardson formula.

The main motivation for the construction of the path model came from the observation [13] of a connection between the work of Lakshmibai and Seshadri on standard monomial theory (see for example [9] for an overview of this work related to the geometry of Schubert varieties), and the work of Kashiwara on crystal bases of representations of quantum groups (see for example [4,5,6,7], or a book on quantum groups, for example [1,3,16]).

The model itself is a purely elementary construction; only some basic knowledge in weight theory and in the combinatoric of Weyl groups is required. In this note we consider only complex semisimple Lie algebras, though (with the appropriate reformulation) the statements hold more generally for arbitrary symmetrizable Kac-Moody algebras (see [12] for an overview).

The restriction to the case of complex semisimple Lie algebras enables us to give complete proofs of the most important statements in this note. The proofs given here are different from those in [10,11] and, I hope, much simpler.

1991 *Mathematics Subject Classification.* Primary 17B10, 05E10.

1. The Paths

The paths considered in this note live in the real span of the weight lattice of a semisimple complex Lie algebra. We assume for simplicity always that the paths are piecewise linear, but it is easy to see (by approximation by piecewise linear paths) that the theorems stated in the following hold also for piecewise smooth or, more generally, rectifiable paths.

Let $\mathfrak{g}$ be a complex semisimple Lie algebra, fix a compact form $\mathfrak{g}_o \subset \mathfrak{g}$ and a Cartan subalgebra $\mathfrak{H} \subset \mathfrak{g}$ such that $\mathfrak{H} = \mathfrak{H}_0 \oplus i\mathfrak{H}_0$, where $\mathfrak{H}_0 := \mathfrak{H} \cap \mathfrak{g}_o$. Recall that the restriction of the Killing form $(\cdot\,,\cdot)$ to $\mathfrak{H}_{\mathbb{R}} := i\mathfrak{H}_0$ is a positive definite form. Denote by $\Delta \subset \mathfrak{H}_{\mathbb{R}}^*$ the set of roots and by $\Lambda \subset \mathfrak{H}_{\mathbb{R}}^*$ the lattice of integral weights. Corresponding to the choice of a set of positive roots Δ^+, let Λ^+ be the set of dominant weights.

Definition 1.1. A **piecewise linear path** in $\mathfrak{H}_{\mathbb{R}}^*$ is a piecewise linear, continuous map $\pi : [0,1] \to \mathfrak{H}_{\mathbb{R}}^*$. We consider two paths as identical if there exists a piecewise linear, nondecreasing, continuous, surjective map $\phi : [0,1] \to [0,1]$ such that $\pi = \eta \circ \phi$. Denote by Π the set of all piecewise linear paths such that $\pi(0) = 0$ and $\pi(1) \in \Lambda$.

Example 1.2. For $\lambda \in \mathfrak{H}_{\mathbb{R}}^*$ set $\pi_\lambda(t) := t\lambda$. We often write just λ for the path π_λ. Then $\pi_\lambda \in \Pi$ if and only if $\lambda \in \Lambda$.

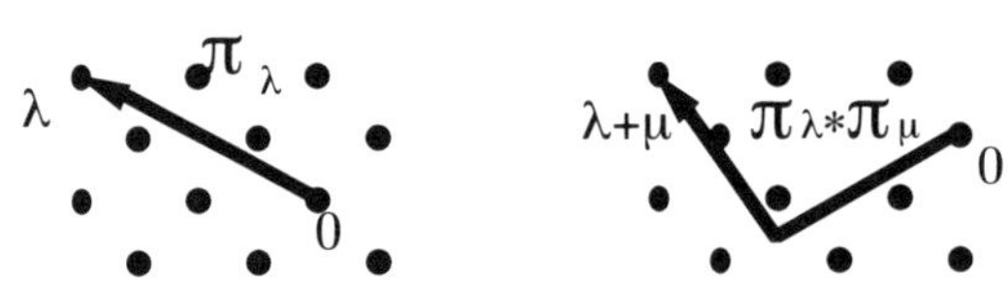

FIGURE 1. The straight line and the concatenation of paths

Example 1.3. Let π_1, π_2 be two piecewise linear paths starting in 0. By the **concatenation** $\pi := \pi_1 * \pi_2$ of the paths π_1 and π_2, we mean the path defined by

$$\pi(t) := \begin{cases} \pi_1(2t) & \text{if } 0 \le t \le 1/2, \\ \pi_1(1) + \pi_2(2t-1) & \text{if } 1/2 \le t \le 1. \end{cases}$$

Example 1.4. The piecewise linear paths can be identified with certain finite sequences of elements in $\mathfrak{H}_{\mathbb{R}}^*$: Let $\underline{\lambda} = (\lambda_1, \dots, \lambda_s)$ be such a finite sequence and set $\pi_{\underline{\lambda}} := \lambda_1 * \dots * \lambda_s$. This is the path that joins successively the weights $0, \lambda_1, \lambda_1 + \lambda_2$, etc. Of course, $\pi_{\underline{\lambda}} \in \Pi$ if and only if $\lambda_1 + \dots + \lambda_s \in \Lambda$. Note that, up to reparametrization, all paths in Π are of this form.

Example 1.5. For $\mathfrak{g} = \mathfrak{sl}_n$ let $\mathfrak{H}$ be the subalgebra of diagonal matrices of trace zero. A classical combinatorial tool in the representation theory of $\mathfrak{sl}_n$ is the semi-standard Young tableaux. These tableaux can be identified with certain paths as follows.

Fix a partition $p = (a_1, \dots, a_n)$; i.e., $a_1 \ge \dots \ge a_n \ge 0$ is a nonincreasing sequence of nonnegative integers. Recall that the **Young diagram** of shape p is a left justified sequence of rows of boxes with a_1 boxes in the first row, a_2 in the second, etc. A **semi-standard Young tableau T** of shape p is a filling of the boxes with numbers $1, \ldots, n$ such that the entries are not decreasing in the rows and strictly increasing in the columns.

Let $\epsilon_i : \mathfrak{H} \to \mathbb{C}$ be the projection of a diagonal matrix onto its i-th entry. For a given tableau $\mathbf{T}$ let $(i_1, \ldots, i_N)$ be the entries of the boxes, where we read the entries columnwise (from the top to the bottom of each column), starting with the right most column. We associate to $\mathbf{T}$ the path $\pi_{\mathbf{T}} := \epsilon_{i_1} * \ldots * \epsilon_{i_N}$.

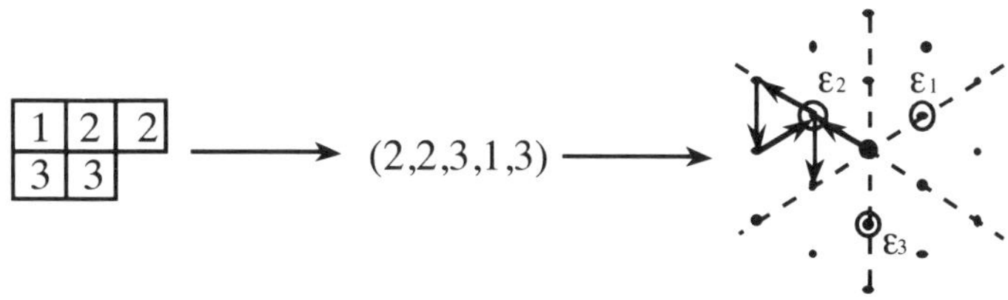

FIGURE 2. A tableau and its path

Example 1.6. We present two different procedures to create new paths from given ones: Fix $\eta \in \Pi$. By the **dual** path $\eta^* \in \Pi$ we mean the path defined by $\eta^*(t) := \eta(1-t) - \eta(1)$. By the **stretching** of paths we mean the multiplication of paths: For $n \in \mathbb{N}$ let $n\eta \in \Pi$ be the path defined by $(n\eta)(t) := n\eta(t)$.

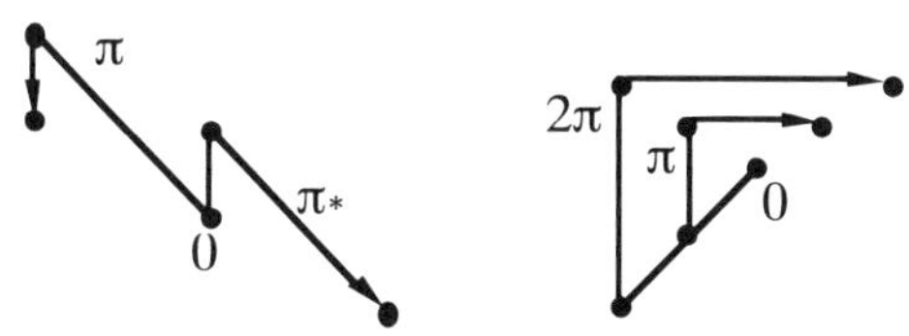

FIGURE 3. The dual path and the stretching of paths

Example 1.7. Here is an approach to the paths using the language of loop groups: Let G be a simply connected semisimple algebraic group with Lie algebra $\mathfrak{g}$. Fix a maximal compact subgroup K such that $\operatorname{Lie} K = \mathfrak{g}_o$, and let $H \subset K$ be a maximal torus such that $\operatorname{Lie} H = \mathfrak{H}_0$. The compact torus $\check{H} := \mathfrak{H}_{\mathbb{R}}/\Lambda$ is a maximal torus of the so-called dual group $\check{K}$ of K. Fix $\eta \in \Pi$, the map $t \mapsto \overline{\eta(t)} \in \check{H}$ induces a loop $\exp \eta : S^1 \to \check{H}$. This correspondence provides a bijection between the set of piecewise linear paths starting in 0 and ending in an integral weight, and the group of "piecewise linear" loops in $\check{H}$ at the identity.

2. The Root Operators

To obtain combinatorial character formulas and multiplicity formulas, we define **lowering** and **raising** operators f_α, e_α for each simple root. Let $(\cdot, \cdot)$ be the Killing-form. The definition of the operators is elementary; it is a cutting and gluing procedure. For convenience we introduce a special element θ, which is not a path but which has the abstract properties $\theta * \pi = \pi * \theta := \theta$ for all $\pi \in \Pi$. Fix $\pi \in \Pi$, let $\alpha^\vee := 2\alpha/(\alpha, \alpha)$ be the co-root of α, and denote by h_α the function:

$$h_\alpha : [0,1] \to \mathbb{R}, \qquad t \mapsto (\pi(t), \alpha^\vee).$$

Let m_α be the minimal value attained by this function. We define nondecreasing functions $l, r : [0,1] \to [0,1]$:

$l(t) := \min\{1, h_\alpha(s) - m_\alpha \mid t \le s \le 1\}$, $r(t) := 1 - \min\{1, h_\alpha(s) - m_\alpha \mid 0 \le s \le t\}$.

Note that $l(t) = 0$ for $0 \le t \le s$, where s is maximal such that $h(s) = m_\alpha$, and $r(t) = 1$ for $s' \le t \le 1$, where s' is minimal such that $h(s) = m_\alpha$.

Definition 2.1.

$$e_\alpha \pi := \begin{cases} t \mapsto \pi(t) + r(t)\alpha & \text{if } r(0) = 0; \\ \theta & \text{otherwise.} \end{cases}$$

Definition 2.2.

$$f_\alpha \pi := \begin{cases} t \mapsto \pi(t) - l(t)\alpha & \text{if } l(1) = 1; \\ \theta & \text{otherwise.} \end{cases}$$

We set $e_\alpha \theta = f_\alpha \theta := \theta$. For a path π let $s_\alpha(\pi)$ be defined by $s_\alpha(\pi)(t) := s_\alpha(\pi(t))$. If we think of a path as a concatenation of "smaller" paths $\pi = \pi_1 * \ldots * \pi_r$, then we can view e_α and f_α as operators that replace some of the π_j by $s_\alpha(\pi_j)$:

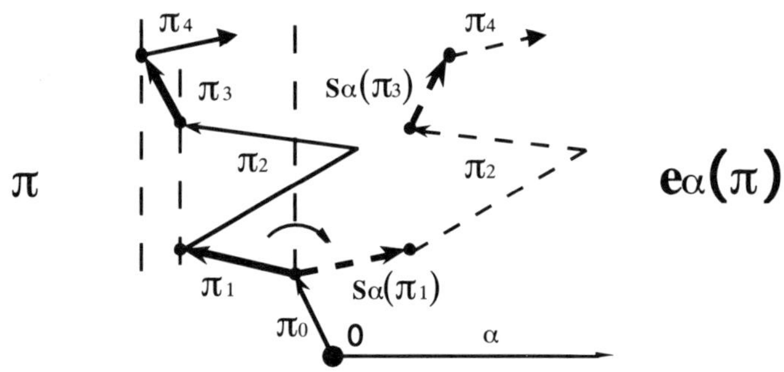

FIGURE 4. The part of $e_\alpha(\pi)$ different from π is drawn as a dashed line

Example 2.3. Suppose $\mathfrak{g} = \mathfrak{sl}_3$ and β is the highest root. The paths obtained from $\pi_\beta : t \mapsto t\beta$ by applying the operators f_α, e_α are the paths $\pi_\gamma(t) := t\gamma$, where γ is an arbitrary root, and for the two simple roots the paths $\eta_\alpha := \pi_{-\alpha/2} * \pi_{\alpha/2}$. The arrow $\stackrel{i}{\longrightarrow}$ indicates in the following picture that the operator f_{α_i} transforms the given path into the next one.

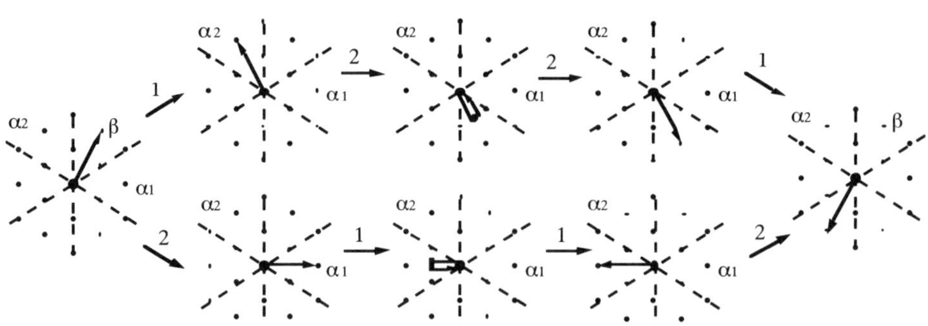

FIGURE 5. The paths generated by $\eta : t \to t\beta$

3. Some Simple Properties.

We list below some simple properties of the operators e_α, f_α that are easy to verify. These properties have also been the "guideline" for the definition of the operators; i.e., the operators are completely determined by these properties. Recall that we identify the weight lattice Λ with the paths of the form $t \mapsto t\lambda$.

i) MOVING, STRETCHING AND DUALIZING. The operators preserve the length of a path and move the endpoint by $\pm\alpha$. Whenever $e_\alpha\eta \neq \theta$ for some $\eta \in \Pi$, then $f_\alpha(e_\alpha\eta) = \eta$ and $(e_\alpha\eta)^* = f_\alpha(\eta^*)$. Similarly, if $f_\alpha\eta \neq \theta$ for some $\eta \in \Pi$, then $e_\alpha(f_\alpha\eta) = \eta$ and $(f_\alpha\eta)^* = e_\alpha(\eta^*)$. Further, the operators are compatible with stretching; i.e., for all $\eta \in \Pi$, we have

$$k(f_\alpha\eta) = f_\alpha^k(k\eta) \quad \text{and} \quad k(e_\alpha\eta) = e_\alpha^k(k\eta).$$

ii) α-STRINGS. For $\eta \in \Pi$ let n be maximal such that $f_\alpha^n\eta \neq \theta$, let m be maximal such that $e_\alpha^m\eta \neq \theta$, and let m_α be the minimum of the function h_α. Then

$$n - m = (\eta(1), \alpha^\vee), \quad m \leq |m_\alpha| < m + 1, \quad n \leq |(\eta(1), \alpha^\vee) - m_\alpha| < n + 1.$$

iii) REFLECTIONS. We define an action of the simple reflection s_α on Π. For $\eta \in \Pi$ set $k := (\eta(1), \alpha^\vee)$. Then:

$$s_\alpha(\eta) := f_\alpha^k(\eta) \text{ if } k > 0 \quad \text{and} \quad s_\alpha(\eta) := e_\alpha^{-k}(\eta) \text{ otherwise.}$$

Note that $s_\alpha^2 = id$, and the restriction of the action to $\Lambda \subset \Pi$ yields the usual action of the simple reflection on the weight lattice.

Remark 3.1. Let $\mathbb{B} \subset \Pi$ be a finite subset, such that $\mathbb{B} \cup \{\theta\}$ is stable under the root operators. Then (iii) implies that its character $\operatorname{Char}\mathbb{B} := \sum_{\eta\in\mathbb{B}} e^{\eta(1)}$ is W-stable.

iv) CONCATENATION. Let $\pi := \lambda_1 * \ldots * \lambda_r$ be such that $\lambda_1, \ldots, \lambda_r$ are integral weights. Set $a_i := (\lambda_1 + \ldots + \lambda_{i-1}, \alpha^\vee)$ and $a_0 := 0$. For the minimum m_α of the a_i fix p minimal with $m_\alpha = a_p$ and q maximal with $m_\alpha = a_q$. If $p = 0$, then $e_\alpha\pi = \theta$, and if $q = r$, then $f_\alpha\pi = \theta$. Otherwise we get for $x \leq \min\{a_i - m_\alpha \mid 0 \leq i \leq p-1\}$ and $y \leq \min\{a_i - m_\alpha \mid q+1 \leq i \leq r\}$:

$$e_\alpha^x\pi = \lambda_1 * \ldots * (e_\alpha^x\lambda_p) * \ldots * \lambda_r, \qquad f_\alpha^y\pi = \lambda_1 * \ldots * (f_\alpha^y\lambda_{q+1}) * \ldots * \lambda_r.$$

Let $\widetilde{\Pi}$ be the set of all piecewise linear paths η as before, but with a fixed parametrization. We define a distance on $\widetilde{\Pi}$:

$$d(\eta, \pi) := \max_{t\in[0,1]} \{\|\eta(t) - \pi(t)\|\}.$$

It is easy to see that if two paths are "close" with respect to $d(\cdot, \cdot)$, then the functions h_α and the functions l and r are close. More precisely:

v) CONTINUITY. The operators are "continuous": $\exists c \in \mathbb{R}$ (depending only on $\mathfrak{g}$) such that if $d(\eta, \pi) < y$ and $e_\alpha(\eta), e_\alpha(\pi) \neq \theta$ then $d(e_\alpha(\eta), e_\alpha(\pi)) < cy$, and if $f_\alpha(\eta), f_\alpha(\pi) \neq \theta$, then $d(f_\alpha(\eta), f_\alpha(\pi)) < cy$.

Proposition 3.2. *If $\{f'_\alpha, e'_\alpha \mid \alpha$ a simple root$\}$ is a set of maps $\Pi \to \Pi \cup \{\theta\}$ satisfying the properties* (i) *to* (v), *then $f_\alpha = f'_\alpha$ and $e_\alpha = e'_\alpha$ for all simple roots.*

PROOF. By a rational path $\eta \in \Pi$ we mean a path such that all turning points are rational weights, in other words: $n\eta = \lambda_1 * \ldots * \lambda_r$ for some $\lambda_1, \ldots, \lambda_r \in \Lambda$ and some $n \in \mathbb{N}$. The properties (ii) and (v) imply that the action on Π is a "continuous" extension of the action on the rational paths: Let m_α be as in (ii). For any given $\epsilon > 0$ we can approximate $\eta \in \Pi$ by a rational path π such that $d(\eta, \pi) < \epsilon$. Further, if η is such that $m_\alpha \geq k$ or $m_\alpha < k$ for some $k \in \mathbb{Z}$, then we can choose a rational approximation π with the same property. It is now easy to see that the definition of the operators is the "continuous" extension of the action on the rational paths.

So it is sufficient to consider rational paths and to prove $f_\alpha = f'_\alpha$ (property (i)). Fix $\lambda \in \Lambda$ such that $a := (\lambda, \alpha^\vee) > 0$. Now (iii) implies $f'^a_\alpha(\lambda) = s_\alpha(\lambda)$, and if $k \leq a$, then (iv) implies $f'^{ka}_\alpha(a\lambda) = (k f'^a_\alpha(\lambda)) * ((a-k)\lambda)$. It follows that

$$f'^k_\alpha(\lambda) = \big(\frac{k}{a} s_\alpha(\lambda)\big) * \big(\frac{a-k}{a}\lambda\big), \text{ and hence: } f'^k_\alpha(\lambda) = f^k_\alpha(\lambda).$$

If $\eta = \lambda_1 * \ldots * \lambda_r$ for some $\lambda_1, \ldots, \lambda_r \in \Lambda$, then the turning points are integral weights and the local minima of the function h_α are integers. Now (iv) and the definition of f_α imply that the local minima of the function $t \mapsto (f_\alpha\eta(t), \alpha^\vee)$ are integers. Since $f'^k_\alpha(\lambda) = f^k_\alpha(\lambda)$, it follows by (iv) that $f'^k_\alpha(\eta) = f^k_\alpha(\eta)$ and the local minima of $t \mapsto (f'_\alpha\eta(t), \alpha^\vee)$ are integers. Using stretching and (iv), it is now easy to see by induction that

$$f'^k_\alpha\eta = (f'^{k_1}_\alpha\lambda_1) * \ldots * (f'^{k_r}_\alpha\lambda_r) = (f^{k_1}_\alpha\lambda_1) * \ldots * (f^{k_r}_\alpha\lambda_r) = f^k_\alpha\eta.$$

Let $\eta \in \Pi$ be a rational path and fix n such that $n\eta = \lambda_1 * \ldots * \lambda_r$ for some $\lambda_1, \ldots, \lambda_r \in \Lambda$. Since $f_\alpha\eta = \frac{1}{n} f^n_\alpha(\lambda_1 * \ldots * \lambda_r) = \frac{1}{n} f'^n_\alpha(\lambda_1 * \ldots * \lambda_r) = f'_\alpha(\eta)$, the operators f_α, e_α and f'_α, e'_α coincide on rational paths, and hence on all paths.

4. A First Character Formula

Denote by $\Pi^+ \subset \Pi$ the set of paths η such that $\operatorname{Im}\eta$ is contained in the dominant Weyl chamber C, and let Π^+_0 be the set of paths such that $\operatorname{Im}\eta$ is in the interior of C (for $t > 0$). Let Λ^+ be the set of dominant weights and denote by $\rho \in \Lambda^+$ half the sum of the positive roots. If $\mathbb{B} \subset \Pi$ is a finite subset such that $\mathbb{B} \cup \theta$ is stable under the root operators e_α, f_α, then we have already seen that its character $\operatorname{Char}\mathbb{B} := \sum_{\eta\in\mathbb{B}} e^{\eta(1)}$ is stable under the action of W. In fact, $\operatorname{Char}\mathbb{B}$ can be computed by the following path version of Weyl's character formula:

Proposition 4.1.

$$\big(\sum_{w\in W} \operatorname{sgn}(w)e^{w(\rho)}\big) \operatorname{Char}\mathbb{B} = \sum_{\substack{\eta\in\mathbb{B} \\ \rho*\eta\in\Pi^+_0}} \big(\sum_{w\in W} \operatorname{sgn}(w)e^{w(\rho+\eta(1))}\big)$$

Corollary 1. *For $\mu \in \Lambda^+$ let V_μ be the corresponding irreducible representation of $\mathfrak{g}$. Then*

$$\operatorname{Char}\mathbb{B} = \sum_{\substack{\eta\in\mathbb{B} \\ \rho*\eta\in\Pi^+_0}} \operatorname{Char} V_{\eta(1)}$$

PROOF OF THE PROPOSITION. Both sides are stable under the Weyl group; so it is sufficient to compare the coefficients of the terms corresponding to dominant weights; i.e., we have to prove for $\Omega := \{(w,\pi) \mid w \in W, \pi \in \mathbb{B}, w(\rho)+\pi(1) \in \Lambda^+\}$:

$$\sum_{(w,\pi)\in\Omega} \operatorname{sgn}(w) e^{w(\rho)+\pi(1)} = \sum_{\eta\in\mathbb{B},\, \rho*\eta\in\Pi_0^+} e^{\rho+\eta(1)}.$$

Let Ω_0 be the set of pairs $(w,\pi) \in \Omega$ such that w is the identity and $\rho * \pi \in \Pi_0^+$. Set $\Omega' := \Omega - \Omega_0$. To prove the proposition we have to show:

$$\sum_{(w,\pi)\in\Omega'} \operatorname{sgn}(w) e^{w(\rho)+\pi(1)} = 0. \qquad (*)$$

We will define an involution $\varphi : \Omega' \to \Omega'$ such that $\varphi(w,\pi) = (w',\pi')$ has the property: $\operatorname{sgn}(w) = -\operatorname{sgn}(w')$ and $w(\rho)+\pi(1) = w'(\rho)+\pi'(1)$. This implies obviously $(*)$ and hence the proposition.

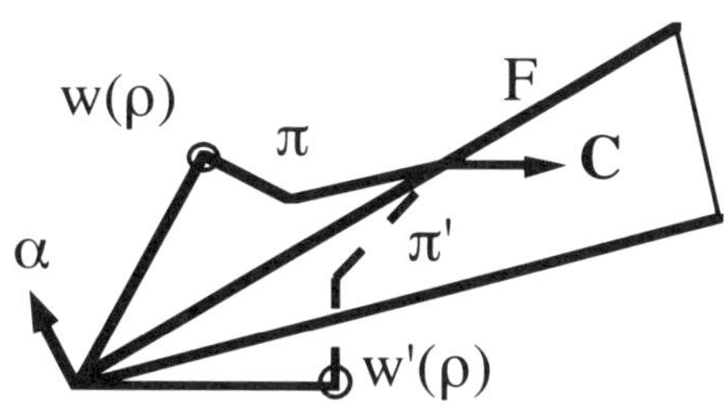

FIGURE 6. The involution φ

The construction of the involution: Suppose first $(w,\pi) \in \Omega'$ is such that w is not the identity. Since $w(\rho)+\pi(1) \in \Lambda^+$, the path $w(\rho) * \pi$ has to meet at least once a proper face of the dominant Weyl chamber C. If w is the identity, then $\rho * \pi$ also has to meet a proper face F of C. (The pair would otherwise be an element of Ω_0.)

For a proper face F of C denote by $\Omega'(F)$ the set of pairs $(w,\pi) \in \Omega'$ that meet F as the last face. More precisely: $w(\rho) * \pi$ meets F, and if $t_0 \in [0,1]$ is maximal with the property such that $w(\rho)+\pi(t_0) \in F$, then $w(\rho)+\pi(t_0)$ is in the interior of F, and $w(\rho)+\pi(t)$ is in the interior of C for all $t > t_0$.

The set Ω' is obviously the disjoint union of the $\Omega'(F)$; so it is sufficient to define the involution for such an $\Omega'(F)$. Let α be a simple root orthogonal to F. For $(w,\pi) \in \Omega'(F)$ set $n := (w(\rho), \alpha^\vee)$. Note that $n \neq 0$.

If $n > 0$, then the minimum m_α of the function $t \mapsto (\pi(t), \alpha^\vee)$ is at least $-n$ (since $w(\rho) * \pi$ meets F for some value of $t > 0$). It follows that $e_\alpha^n(\eta) \neq \theta$ and $w(\rho)+\pi(1) = s_\alpha w(\rho) + e_\alpha^n \pi(1)$. Further, if $t_0 \in [0,1]$ is maximal with the property that $w(\rho) * \pi(t_0) \in F$, then $w(\rho)+\pi(t) = s_\alpha w(\rho) + e_\alpha^n \pi(t)$ for all $t \geq t_0$, and hence

$$\varphi(w,\rho) := (s_\alpha w, e_\alpha^n(\pi)) \in \Omega'(F).$$

Similarly, if $n < 0$, then $f_\alpha^{|n|}(\eta) \neq \theta$, $w(\rho)+\pi(1) = s_\alpha w(\rho) + f_\alpha^{|n|}\pi(1)$, and

$$\varphi(w,\rho) := (s_\alpha w, f_\alpha^{|n|}(\pi)) \in \Omega'(F).$$

Property (i) in section 3 implies that φ is an involution, which finishes the proof.

5. Locally Integral Concatenations

The next aim is to describe the possible sets of paths $\mathbb{B}$ such that $\mathbb{B} \cup \{\theta\}$ is stable under the root operators. Of course, one is particulary interested in those sets such that Corollary 1 in section 4 provides a character formula for an irreducible representation V_λ. A good candidate for such a set is the following: Start with the path λ (recall, we identify the weight λ with the path $\pi_\lambda : t \mapsto t\lambda$), and let $\mathbb{B}_\lambda$ be the set of paths obtained from this line by applying the root operators. Since $\lambda \in \Pi^+$, it is evident that $\rho * \lambda \in \Pi_0^+$. So the character of V_λ will show up on the right side in the character formula for $\mathbb{B}_\lambda$. But, a priori, it is not at all evident that λ is the only path in $\mathbb{B}_\lambda$ with this property, and, even more important, so far it is not even clear that $\mathbb{B}_\lambda$ is a finite set.

To prove that $\mathbb{B}_\lambda$ has in fact these two properties, it turns out that it is much more natural to consider from the very beginning the following special class of paths: $\underline{\lambda} := \lambda_1 * \ldots * \lambda_r$, where the λ_i are rational dominant weights with the following properties: $\lambda := \lambda_1 + \ldots + \lambda_r \in \Lambda^+$, and for $1 \leq i \leq r-1$ there exist $p_i, q_i > 0$ with

$$\lambda_1 + \ldots + \lambda_{i-1} + p_i\lambda_i \in \Lambda^+, \quad q_i\lambda_{i+1} + \lambda_{i+2} + \ldots + \lambda_r \in \Lambda^+.$$

Examples of such paths are those of the form $\underline{\lambda} := \lambda_1 * \ldots * \lambda_r$, where $\lambda_1, \ldots, \lambda_r$ are dominant weights, or paths of the form $\underline{\lambda} := \lambda_1 * \lambda_2$, where $\lambda_1, \lambda_2 \in \Lambda_{\mathbb{Q}}^+$ are such that $\lambda := \lambda_1 + \lambda_2 \in \Lambda^+$. Let $\mathbb{P} \subset \Pi$ be the smallest set that contains all these paths and that has the property that $\mathbb{P} \cup \{\theta\}$ is stable under the operators e_α, f_α.

To give a more intrinsic description of $\mathbb{P}$, one associates to every turning point of a path a root system that "measures" the change of direction. We will show that the set $\mathbb{P}$ consists essentially of those paths for which all turning points are "integral points" for the associated root system.

Let "$\geq$" be the Bruhat order on W. If $\nu_1, \ldots, \nu_r \in \Lambda_{\mathbb{Q}}$ are rational weights, then let $\lambda_1, \ldots, \lambda_r \in \Lambda_{\mathbb{Q}}^+$ be the rational dominant weights such that $\nu_i \in W.\lambda_i$. We write:

$$\nu_1 \succeq \ldots \succeq \nu_r \iff \exists\, w_i \in W \text{ such that } \nu_i = w_i(\lambda_i) \text{ and } w_1 \geq \ldots \geq w_r.$$

Remark 5.1. Note that $\nu_1 \succeq \nu_2$ and $\nu_2 \succeq \nu_3$ does not necessarily imply $\nu_1 \succeq \nu_3$. Let ω_1, ω_2 be fundamental weights, and let α_1, α_2 be the corresponding simple roots. Set $\nu_1 := \omega_1$, $\nu_2 := \omega_2$, and $\nu_3 := s_{\alpha_1}(\omega_1)$. Then $\nu_1 \succeq \nu_2$ (choose $w_1 = w_2 = id$) and $\nu_2 \succeq \nu_3$ (choose $w_1 = w_2 = s_{\alpha_1}$), but, of course, $\nu_1 \not\succeq \nu_3$.

We use the length function $l(\,\cdot\,)$ on W also for the Weyl group orbits and cosets in W/W_μ: If $\nu \in W\mu$ for some $\mu \in \Lambda_{\mathbb{Q}}^+$, then let $\tau \in W/W_\mu$ be the unique element such that $\tau(\mu) = \nu$ and let $\tilde{\tau} \in W$ be the unique element of minimal length such that $\tilde{\tau}(\mu) = \nu$. We write then $l(\nu)$ and $l(\tau)$ for $l(\tilde{\tau})$. If β is positive root, then let $\beta^\vee = 2\beta/(\beta,\beta) \in \Phi^\vee$ be the dual root.

Suppose $\nu \succeq \mu$ are rational weights. Let $w \in W$ be the unique element of minimal length such that $w(\mu^+) = \mu$ for some $\mu^+ \in \Lambda_{\mathbb{Q}}^+$, and let $v \in W$ be the unique element of minimal length such that $v \geq w$ and $v(\nu^+) = \nu$ for some $\nu^+ \in \Lambda_{\mathbb{Q}}^+$ [2,9]. There exist positive roots $\beta_1, \ldots, \beta_r$ such that $w := s_{\beta_r} \ldots s_{\beta_1} v$ and

$$v_0 := v > v_1 := s_{\beta_1} v > \ldots > v_r := w, \quad l(v_0) = l(v_1) + 1 = \ldots = l(v_r) + r.$$

Definition 5.2. The root system $\Phi^\vee_{\nu,\mu} \subset \Phi^\vee$ spanned by the roots $\beta_1^\vee, \ldots, \beta_r^\vee$ is called the **root system of the pair** (ν, μ).

Note that this root system is independent of the choice of the β_i: This is evident if $l(v) - l(w) \leq 1$. Otherwise one proceeds by induction on $l(v) - l(w)$ and $l(v)$: Let α be a simple root such that $s_\alpha v < v$. If $s_\alpha w < w$, then it is easy to see that from any given sequence of roots for the pair (v, w) one can construct a sequence for the pair $(s_\alpha v, s_\alpha v)$ such that:

$$\Phi^\vee_{s_\alpha(\nu), s_\alpha(\mu)} = s_\alpha(\Phi^\vee_{\nu,\mu}).$$

Since the first is independent of the choice of the positive roots (by induction), so is the latter. Suppose now $s_\alpha w > w$, so that $v > s_\alpha v \geq w$. It is easy to see that, from any given sequence of roots for the pair (v, w), one can construct a sequence for the pair $(s_\alpha v, w)$ such that $\Phi^\vee_{\nu,\mu}$ is spanned by $\alpha^\vee$ and $\Phi^\vee_{s_\alpha v, w}$. Again, by induction, the latter is independent of the choice of the positive roots, and so is the first.

Definition 5.3. Suppose $\nu_1, \ldots, \nu_r$ are rational weights such that $\nu_1 + \ldots + \nu_r$ is in Λ and $\nu_1 \succeq \ldots \succeq \nu_r$. The path $\underline{\nu} := \nu_1 * \ldots * \nu_r \in \Pi$ is called a **locally integral concatenation** if it satisfies the following conditions for all $i = 2, \ldots, r$:

a) $(\nu_1 + \ldots + \nu_{i-1}, \beta^\vee) \in \mathbb{Z}$ for all $\beta^\vee \in \Phi^\vee_{\nu_{i-1}, \nu_i}$,
b) If there exists no $t_i > 0$ such that $w(t_i \nu_i) = \nu_{i+1}$ for some $w \in W$, then there exist $p_i, q_i > 0$ such that $\nu_1 + \ldots + \nu_{i-1} + p_i \nu_i$ and $q_i \nu_{i+1} + \nu_{i+2} + \ldots + \nu_r$ are in Λ.

Example 5.4. If $\lambda \in \Lambda^+$ or $\lambda_1, \lambda_2 \in \Lambda^+_{\mathbb{Q}}$ are such that $\lambda_1 + \lambda_2 \in \Lambda^+$, then the paths λ and $\lambda_1 * \lambda_2$ are locally integral concatenations.

Remark 5.5. It is easy to see that the property of being a locally integral concatenation is independent of the chosen parametrization.

Lemma 5.6. *Let $\underline{\nu} = \nu_1 * \ldots * \nu_r$ be a locally integral concatenation, and let α be a simple root. The local minima of the function $h_\alpha : t \mapsto (\underline{\nu}(t), \alpha^\vee)$ are integers.*

PROOF. Let $s \in [0,1]$ be such that h_α attains in s a local minimum. We may assume that $\underline{\nu}(s) = \nu_1 + \ldots + \nu_i$ for some $0 \leq i \leq r$. If $i = r$ or $i = 0$, then $\underline{\nu}(s)$ is an integral weight and hence $h_\alpha(s) \in \mathbb{Z}$. Suppose now $1 \leq i \leq r-1$. Since h_α attains a local minimum, we may further assume that $(\nu_i, \alpha^\vee) < 0$ and $(\nu_{i+1}, \alpha^\vee) \geq 0$. Let $v, w \in W$ for (ν_i, ν_{i+1}) be as in the definition of the associated root system. The first inequality implies $s_\alpha v < v$, and the latter implies $s_\alpha w > w$ by the minimality of w. So $v > s_\alpha v \geq w$ and hence $\alpha^\vee \in \Phi^\vee_{\nu_i, \nu_{i+1}}$, which implies $h_\alpha(s) \in \mathbb{Z}$.

Lemma 5.7. *If $\underline{\nu} = \nu_1 * \ldots * \nu_r$ is a locally integral concatenation such that $\rho * \underline{\nu}$ is in the interior of the dominant Weyl chamber (for $t > 0$), then $\nu_1, \ldots, \nu_r \in \Lambda^+_{\mathbb{Q}}$.*

PROOF. Let α be a simple root. If $\rho * \underline{\nu}$ is in the interior of the dominant Weyl chamber (for $t > 0$), then the minimum of the function $h_\alpha : t \mapsto (\underline{\nu}(t), \alpha^\vee)$ is > -1. But the minimum is an integer and hence is equal to 0; i.e., the image of $\underline{\nu}$ is in the dominant Weyl chamber. Suppose one of the $\nu_i \notin \Lambda^+_{\mathbb{Q}}$. We may assume that i is minimal with this property.

Let $w_1 \geq \ldots \geq w_r$ be such that $\nu_j = w_j(\nu_j^+)$, where ν_j^+ is a rational dominant weight. Choose a simple root α such that $(\nu_i, \alpha^\vee) < 0$. Note that this implies $s_\alpha w_i < w_i$ and hence $w_i \geq s_\alpha$. Since $w_1 \geq \ldots \geq w_i$, we know that $w_j \geq s_\alpha$ for

$1 \le j \le i$. But the ν_j are dominant. So $w_j \in W_{\nu_j}$ for $1 \le j < i$ and hence $(\nu_j, \alpha^\vee) = 0$ for $1 \le j < i$. But this would imply $(\nu_1 + \ldots + \nu_i, \alpha^\vee) = (\nu_i, \alpha^\vee) < 0$, in contradiction to the fact that the path is in the dominant Weyl chamber.

Proposition 5.8. *If $\underline{\nu} = \nu_1 * \ldots * \nu_r$ is a locally integral concatenation and $e_\alpha \underline{\nu} \neq 0$, then (after reparametrization) $\exists\, i, k$ such that $(\nu_j, \alpha^\vee) < 0$ for $i \le j \le k$ and*

$$e_\alpha \underline{\nu} = \nu_1 * \ldots * \nu_{i-1} * s_\alpha(\nu_i) * \ldots * s_\alpha(\nu_k) * \nu_{k+1} * \ldots * \nu_r.$$

If $f_\alpha \underline{\nu} \neq 0$, then $\exists\, i, k$ such that $(\nu_j, \alpha^\vee) > 0$ for $i \le j \le k$ and

$$f_\alpha \underline{\nu} = \nu_1 * \ldots * \nu_{i-1} * s_\alpha(\nu_i) * \ldots * s_\alpha(\nu_k) * \nu_{k+1} * \ldots * \nu_r.$$

PROOF. We consider only f_α. The proof for e_α is similar. Let s be maximal such that h_α attains in s its minimum m_α, and let i be such that $\underline{\nu}(s) = \nu_1 + \ldots + \nu_{i-1}$. Then $(\nu_{i-1}, \alpha^\vee) \le 0$ and $(\nu_i, \alpha^\vee) > 0$. Let $p > s$ be minimal such that $h_\alpha(t) \ge m_\alpha + 1$ for $t \ge p$. We may assume that $\underline{\nu}(p) = \nu_1 + \ldots + \nu_k$. Since the local minima of h_α are integers, the conditions on the choice of s and p imply that h_α is a nondecreasing function on the interval $[s, p]$. Thus $(\nu_j, \alpha^\vee) \ge 0$ for $i \le j \le k$.

Suppose now $(\nu_j, \alpha^\vee) = 0$. By the choice of i, k we know that $i < j < k$. So we may choose j such that $(\nu_{j+1}, \alpha^\vee) > 0$. If $\nu_j \neq w(x\nu_{j+1})$ for some $w \in W$ and $x > 0$, then there exists a $q > 0$ such that $\nu_1 + \ldots + \nu_{j-1} + q\nu_j$ is an integral weight. But this would imply that:

$$(\nu_1 + \ldots + \nu_{j-1} + \nu_j, \alpha^\vee) = (\nu_1 + \ldots + \nu_{j-1} + q\nu_j, \alpha^\vee) \in \mathbb{Z},$$

which is not possible. So $\nu_j = \tau(x\nu_{j+1})$ for some $\tau \in W$. Let $v, w \in W$ be chosen as in the definition of the associated root system. Now $\nu_j = \tau(x\nu_{j+1})$ implies that v, w are just the minimal elements in W such that $v(\nu_j^+) = \nu_j$ and $w(\nu_{j+1}^+) = \nu_{j+1}$ because $\nu_j^+ = x\nu_{j+1}^+$. But $(\nu_j, \alpha^\vee) = 0$ and $(\nu_{j+1}, \alpha^\vee) > 0$ implies $v \ge s_\alpha w > w$. Therefore $\alpha \in \Phi^\vee_{j,j+1}$ and $(\nu_1 + \ldots + \nu_j, \alpha^\vee) \in \mathbb{Z}$, which is not possible.

The proposition follows now by the definition of the operator f_α.

Proposition 5.9. *The set of locally integral concatenations is stable under the root operators.*

Corollary 1. *Let $\lambda_1, \ldots, \lambda_r \in \Lambda^+_{\mathbb{Q}}$ be such that $\underline{\lambda} := \lambda_1 * \ldots * \lambda_r \in \Pi^+$ is a locally integral concatenation. Then $\mathbb{B}_{\underline{\lambda}}$ is a finite set, and if $\underline{\nu} \in \mathbb{B}_{\underline{\lambda}}$ is such that $\rho * \underline{\nu}$ is (for $t > 0$) in the interior of the dominant Weyl chamber, then $\underline{\nu} = \underline{\lambda}$.*

As a consequence we get by Corollary 1, section 4, the following character formula. Note that this a special case of Theorem 6.1 in the next section.

Corollary 2. $\mathrm{Char}\, \mathbb{B}_{\underline{\lambda}} = \mathrm{Char}\, V_\lambda$.

Corollary 1 and the proposition above prove also the following characterization of the set of paths $\mathbb{P}$ introduced at the beginning of this section:

Corollary 3. *$\mathbb{P}$ coincides with the set of locally integral concatenations.*

PROOF OF COROLLARY 1. Suppose $\underline{\nu} = \nu_1 * \ldots * \nu_r \in \mathbb{B}_{\underline{\lambda}}$. We may assume (after a reparametrization of $\underline{\lambda}$) that the r is the same as for $\underline{\lambda}$ and the ν_i are Weyl group conjugates of the λ_i. If $\rho * \underline{\nu}$ is in the interior of the dominant Weyl chamber (for $t > 0$), then Lemma 5.7 implies that $\nu_1, \ldots, \nu_r$ are rational dominant weights.

So $\lambda_i = \nu_i$ and hence $\underline{\nu} = \underline{\lambda}$. As a consequence we know that all paths in $\mathbb{B}_{\underline{\lambda}}$ are of the form: $\pi = f_\alpha f_{\alpha'} \dots (\underline{\lambda})$, and therefore the possible endpoints are all of the form $\lambda - \sum a_\alpha \alpha$, where the a_α are nonnegative integers. Property (iii), section 3, then implies that all possible endpoints are in the convex hull of the Weyl group orbit of λ. So the number of possible endpoints is finite. Now for a given weight μ there are only a finite number of monomials in the f_α such that the endpoint of $\pi = f_\alpha f_{\alpha'} \dots (\underline{\lambda})$ is μ. So $\mathbb{B}_{\underline{\lambda}}$ is a finite set.

Proof of Proposition 5.9. Let $\underline{\nu} := \nu_1 * \dots * \nu_r$ be a locally integral concatenation. We will show that $f_\alpha \underline{\nu}$ is again a locally integral concatenation. The proof for the operator e_α is similar. By Proposition 5.8 we may assume that the parametrization of $\underline{\nu}$ is such that $(\nu_{i-1}, \alpha^\vee) \leq 0$ and $(\nu_j, \alpha^\vee) > 0$ for some $i \leq j \leq k$, and

$$\underline{\mu} := f_\alpha \underline{\nu} = \nu_1 * \dots * \nu_{i-1} * s_\alpha(\nu_i) * \dots * s_\alpha(\nu_k) * \nu_{k+1} * \dots * \nu_r = \mu_1 * \dots * \mu_r.$$

To check that $\mu_1 \succeq \dots \succeq \mu_r$ is a simple exercise in Weyl group combinatorics and is left to the reader. Denote by $P_l = \nu_1 + \dots + \nu_l$ (resp. $Q_l = \mu_1 + \dots + \mu_l$) the l-th turning point of $\underline{\nu}$ (resp. $\underline{\mu}$).

Suppose first $l < i - 1$ or $k < l < r$. Then $\nu_l = \mu_l$, $\nu_{l+1} = \mu_{l+1}$, $\Phi^\vee_{\mu_l, \mu_{l+1}} = \Phi^\vee_{\nu_l, \nu_{l+1}}$, and $P_l = Q_l$ or $P_l = Q_l + \alpha$, so that the conditions for a locally integral concatenation are obviously satisfied at these points.

If $l = k$, then $\nu_{l+1} = \mu_{l+1}$, $P_l = Q_l + \alpha$, and $\Phi^\vee_{\mu_l, \mu_{l+1}}$ is the root system spanned by $\Phi^\vee_{\nu_l, \nu_{l+1}}$ and $\alpha^\vee$. Since $(P_l, \alpha^\vee) \in \mathbb{Z}$ (Lemma 5.6), we know that $(Q_l, \alpha^\vee) \in \mathbb{Z}$. Thus part (a) of the condition for a locally integral concatenation is satisfied. If ν_k and ν_{k+1} are conjugate (up to multiplication by positive rational number), then so are μ_k and μ_{k+1} Suppose now $p > 0$ is such that $\nu := \nu_1 + \dots + \nu_{k-1} + p\nu_k \in \Lambda$. Then

$$\mu_1 + \dots + \mu_{k-1} + p\mu_k = \nu - (\nu - \mu_1 + \dots + \mu_{i-1}, \alpha^\vee)\alpha.$$

Since $(\nu - \mu_1 + \dots + \mu_{i-1}, \alpha^\vee)$ is an integer, it follows that $\mu_1 + \dots + \mu_{k-1} + p\mu_k \in \Lambda$.

If $l = i-1$, then $P_l = Q_l$, $\nu_l = \mu_l$, and either $\Phi^\vee_{\mu_l, \mu_{l+1}} = \Phi^\vee_{\nu_l, \nu_{l+1}}$ or $\Phi^\vee_{\nu_l, \nu_{l+1}}$ is the root system spanned by $\Phi^\vee_{\mu_l, \mu_{l+1}}$ and $\alpha^\vee$. Further, the same calculation as above shows that if $q\nu_{l+1} + \nu_{l+2} + \dots + \nu_k \in \Lambda$, then $q\mu_{l+1} + \mu_{l+2} + \dots + \mu_k \in \Lambda$. So the conditions for a locally integral concatenation are also satisfied in this point.

If $i \leq l < k$, then $\Phi_{\mu_l, \mu_{l+1}} = s_\alpha(\Phi_{\nu_l, \nu_{l+1}})$, and $Q_l = s_\alpha(P_l) \pm a\alpha$ for some $a \in \mathbb{N}$. So part (a) of the definition of a locally integral concatenation holds for $f_\alpha \underline{\nu}$. If ν_l, ν_{l+1} are conjugate under the Weyl group (up to multiplication by a positive rational number), then so are μ_l, μ_{l+1}. Suppose now x, y are positive rational numbers such that $\nu_1 + \dots + \nu_{l-1} + x\nu_l$ and $y\nu_{l+1} + \nu_{l+2} + \dots + \nu_r$ are integral weights. Since $(\nu_1 + \dots + \nu_{i-1}, \alpha^\vee)$ is an integer, it follows that $(\nu_i + \dots + \nu_{l-1} + x\nu_l, \alpha^\vee)$ is an integer, and hence:

$$\mu_1 + \dots + \mu_{l-1} + x\mu_l = \nu_1 + \dots + \nu_{i-1} + s_\alpha(\nu_i + \dots + \nu_{l-1} + x\nu_l) \in \Lambda.$$

Similarly, since $(\nu_{k+1} + \dots + \nu_r, \alpha^\vee)$ is an integer, it follows that $(y\nu_{l+1} + \dots + \nu_k, \alpha^\vee)$ is an integer, and hence:

$$y\mu_{l+1} + \mu_{l+2} + \dots + \mu_r = s_\alpha(y\nu_{l+1} + \nu_{l+2} + \dots + \nu_k) + \nu_{k+1} + \dots + \nu_r \in \Lambda,$$

which finishes the proof.

6. The General Case

For $\pi \in \Pi^+$ denote by $\mathbb{B}_\pi$ the set of all paths obtained from π by applying the root operators. In other words, $\mathbb{B}_\pi \cup \{\theta\}$ is the smallest set of paths that is stable under the root operators and contains π. Let $\lambda = \pi(1) \in \Lambda^+$ be the endpoint of π. In the following we present the most important properties of the set $\mathbb{B}_\pi$. Proofs will be given in the following sections.

Theorem 6.1. *$\mathbb{B}_\pi$ is a finite set, and if $\eta \in \mathbb{B}_\pi$ is such that $\rho * \eta \in \Pi_0^+$, then $\eta = \pi$.*

As an immediate consequence we get by Corollary 3.1:

Corollary 1. $\operatorname{Char} \mathbb{B}_\pi = \operatorname{Char} V_\lambda$

Example 6.2: Tableaux and paths. Let $p = (a_1, \ldots, a_n)$ be a partition, and denote by $\mathbf{T}_0$ the semi-standard Young tableau of shape p having only 1's as entry in the first row, 2's in the second row etc.

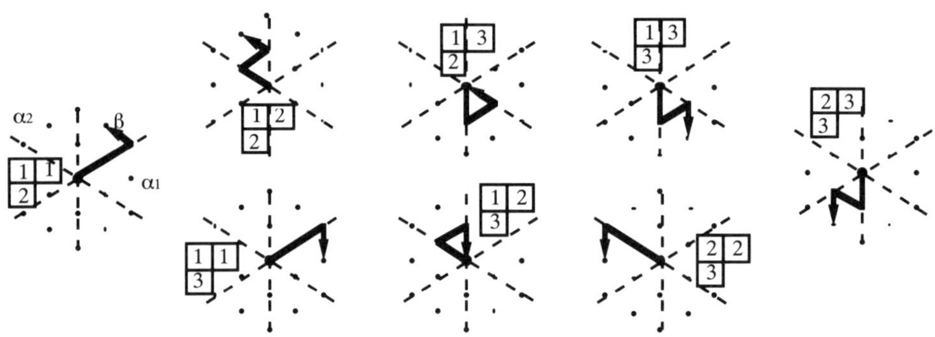

FIGURE 7. The tableaux and the associated paths for the adjoint representation of $\mathfrak{sl}_3$

If π_0 is the associated path (Example 1.5), then $\pi_0(1) = a_1\epsilon_1 + \ldots + a_n\epsilon_n$. The condition $a_1 \geq \ldots \geq a_n$ implies that the image of the path is contained in the dominant Weyl chamber. So $\pi_0 \in \Pi^+$. It is a nice exercise to check that

$$\mathbb{B}_{\pi_0} = \{\pi_{\mathbf{T}} \mid \mathbf{T} \text{ semi-standard Young tableau of shape } p\}.$$

The classical formula using semi-standard tableaux to calculate dimensions and characters of $\mathfrak{sl}_n$-modules can hence be considered as a special case of Corollary 1.

Theorem 6.1 characterizes the path $\pi \in \Pi^+$ as the unique path in $\mathbb{B}_\pi$ such that the image of $\rho * \pi$ is completely contained in the interior of the dominant Weyl chamber for $t > 0$. Since $\rho * \eta$ meets at least one of the walls for any other path $\eta \in \mathbb{B}_\pi$, this means that for any other path there exist at least one simple root α such that minimum of the function $t \mapsto (\eta(t), \alpha^\vee)$ is smaller or equal to -1. By property (iv), section 3, this implies that there exists at least one simple root α such that $e_\alpha(\eta) \neq \theta$. So we get a characterization of π and $\mathbb{B}_\pi$ resembling that of a highest weight vector and a highest weight module (without using the equality of characters above):

Corollary 2.

i) $\mathbb{B}_\pi = \{\eta \in \Pi \mid \exists i_1, \dots, i_s : \eta = f_{\alpha_{i_1}} \cdots f_{\alpha_{i_s}} \pi\}$, *and if* $\eta \in \mathbb{B}_\pi$ *is such that* $e_\alpha(\eta) = \theta$ *for all simple roots, then* $\eta = \pi$.

ii) *Set* $\lambda = \pi(1)$. *For every* $w \in W/W_\lambda$ *there exists a unique path* $\eta \in \mathbb{B}_\pi$ *such that* $\eta(1) = w(\lambda)$.

Since the character is independent of the choice of π, this means that for any choice of a path $\pi \in \Pi^+$ ending in λ we get a different combinatorial model for V_λ. So the next question is: *What do these models have in common?*

Definition 6.3. For $\pi \in \Pi^+$ let $\mathcal{G}_\pi$ be the colored, directed graph having as vertices the elements of $\mathbb{B}_\pi$. We put an arrow $\eta \xrightarrow{\alpha} \eta'$ with color a simple root α between $\eta, \eta' \in \mathbb{B}_\pi$ if and only if $f_\alpha(\eta) = \eta'$ (or, equivalently, $e_\alpha(\eta') = \eta$).

Remark 6.4. Corollary 2 implies that $\mathcal{G}_\pi$ is connected and has a special vertex: π is the unique vertex with no "incoming" arrow.

We call two such graphs $\mathcal{G}_\pi, \mathcal{G}_{\pi'}$ isomorphic if there exists a bijection $\phi : \mathcal{G}_\pi \to \mathcal{G}_{\pi'}$ of the vertices such that we have an arrow $\eta \xrightarrow{\alpha} \eta'$ with color α between $\eta, \eta' \in \mathbb{B}_\pi$ if and only if we have an arrow $\phi(\eta) \xrightarrow{\alpha} \phi(\eta')$ with color α between $\phi(\eta), \phi(\eta') \in \mathbb{B}_{\pi'}$.

Such an isomorphism maps necessarily the special vertices π and π' onto each other. Further, ϕ maps obviously the α-strings in the graphs onto each other. So property (ii), section 3, implies $(\pi(1), \alpha^\vee) = (\pi'(1), \alpha^\vee)$ for all simple roots, and hence $\pi(1) = \pi'(1)$. In fact, this condition is also sufficient:

Theorem 6.5. *The graphs* $\mathcal{G}_\pi$ *and* $\mathcal{G}_{\pi'}$ *are isomorphic if and only if* $\pi(1) = \pi'(1)$.

Remark 6.7. Since the graph depends only on the endpoint, it makes sense to write just $\mathcal{G}_\lambda$ for the graph $\mathcal{G}_\pi$, where $\lambda = \pi(1)$.

Example 6.8. Figure 5 in Example 2.3 is the graph associated to the adjoint representation of $\mathfrak{sl}_3$.

Example 6.9. We consider again the Lie algebra $\mathfrak{sl}_3$, and we take as highest weight $\lambda = 2\omega_1 + \omega_2$. Using the identification: Young tableaux $\leftrightarrow$ paths (see Example 1.5), we get the following graph $\mathcal{G}_{2\omega_1+\omega_2}$:

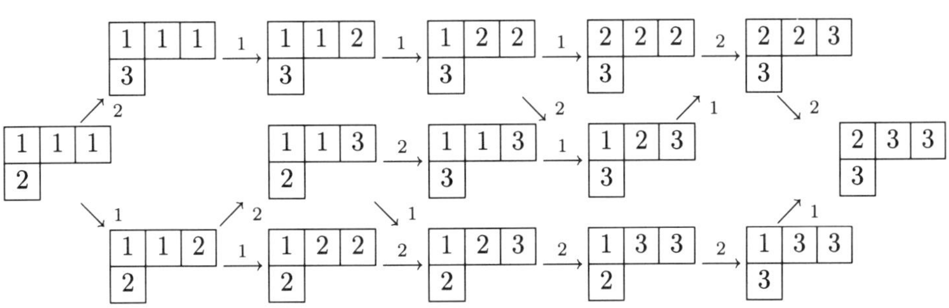

The following property is very important for the concept of the path model:

Definition 6.10. A path $\eta \in \Pi$ is called **integral** if the minimum of the function $h_\alpha : t \mapsto (\pi(t), \alpha^\vee)$ is an integer for all simple roots. We call a subset $\mathbb{B} \subset \Pi$ **integral** if all elements of $\mathbb{B}$ are integral.

Suppose $\mathbb{B} \subset \Pi$ is finite and integral, and fix $\eta \in \mathbb{B}$. Then $e_\alpha(\eta) = 0$ for all simple roots implies $h_\alpha(t) \geq 0$ and hence $\eta \in \Pi^+$. Theorem 6.1 and Corollary 2 hence imply:

Lemma 6.11. *If $\mathbb{B} \subset \Pi$ is integral and $\mathbb{B} \cup \{\theta\}$ is stable under the root operators, then $\mathbb{B}$ is the disjoint union $\mathbb{B} = \bigcup \mathbb{B}_\pi$, where the union is taken over all $\pi \in \mathbb{B} \cap \Pi^+$.*

Let $\mathbb{B}, \mathbb{B}' \subset \Pi$ be two integral subsets, and denote by $\mathbb{B} * \mathbb{B}'$ the set of all concatenations $\pi * \pi'$, where $\pi \in \mathbb{B}$, $\pi' \in \mathbb{B}'$. The set $\mathbb{B} * \mathbb{B}'$ is obviously again integral. The following lemma is a simple consequence of section 3, property (ii).

Lemma 6.12. *If $\mathbb{B}, \mathbb{B}' \subset \Pi$ are integral and stable under the root operators, then $\mathbb{B} * \mathbb{B}'$ is stable under the root operators too. More precisely: $e_\alpha(\pi * \eta) = \pi * (e_\alpha \eta)$ if $\exists\, n \geq 1$ such that $e_\alpha^n \eta \neq 0$ but $f_\alpha^n \pi = 0$, and $e_\alpha(\pi * \eta) = (e_\alpha \pi) * \eta$ otherwise, and*

$$f_\alpha(\pi * \eta) = \begin{cases} (f_\alpha \pi) * \eta, & \textit{if } \exists n \geq 1 \textit{ such that } f_\alpha^n \pi \neq 0 \textit{ but } e_\alpha^n \eta = 0; \\ \pi * (f_\alpha \eta), & \textit{otherwise.} \end{cases}$$

Theorem 6.13. *If $\pi \in \Pi^+$, then B_π is integral.*

Since $\operatorname{Char} \mathbb{B}_{\pi_1} * \mathbb{B}_{\pi_2} = \operatorname{Char} \mathbb{B}_{\pi_1} \operatorname{Char} \mathbb{B}_{\pi_2} = \operatorname{Char} V_{\pi_1(1)} \otimes V_{\pi_2(1)}$ for $\pi_1, \pi_2 \in \Pi^+$, we get as an immediate consequence:

Generalized Littlewood-Richardson rule. *For $\lambda, \mu \in \Lambda^+$ let $\pi_1, \pi_2 \in \Pi^+$ be such that $\pi_1(1) = \lambda$ and $\pi_2(1) = \mu$. Then the tensor product $V_\lambda \otimes V_\mu$ is isomorphic to the direct sum*

$$V_\lambda \otimes V_\mu \simeq \bigoplus V_{\lambda + \eta(1)},$$

where the sum runs over all paths $\eta \in \mathbb{B}_{\pi_2}$ such that $\pi_1 * \eta \in \Pi^+$.

Let $\mathfrak{l} \subset \mathfrak{g}$ be a Levi subalgebra associated to a subset of the set of simple roots. We denote by $C_\mathfrak{l} \supset C$ and $\Lambda_\mathfrak{l}^+$ the dominant Weyl chamber and the set of dominant weights for $\mathfrak{l}$. Let $\Pi_\mathfrak{l}^+$ be the set of all paths in Π such that the image is completely contained in the dominant Weyl chamber $C_\mathfrak{l}$ of $\mathfrak{l}$. For $\nu \in \Lambda_\mathfrak{l}^+$ let U_ν be the associated simple $\mathfrak{l}$-module. The same arguments as above prove:

Restriction formula. *For $\lambda \in \Lambda^+$ let $\pi \in \Pi^+$ be such that $\pi(1) = \lambda$. The simple $\mathfrak{g}$-module V_λ decomposes as $\mathfrak{l}$ module into the direct sum*

$$V_\lambda \simeq \bigoplus U_{\eta(1)},$$

where the sum runs over all paths $\eta \in \mathbb{B}_\pi$ such that $\eta \in \Pi_\mathfrak{l}^+$.

7. The Weyl Group Action

In section 3 we defined an action of the simple reflections on Π.

Proposition 7.1. *The action of the s_α extends to an action of W on Π such that $(w(\eta))(1) = w(\eta(1))$ for $\pi \in \Pi$, $w \in W$.*

PROOF. We have to prove that the braid relations are satisfied. Without loss of generality, we may hence assume that we are in the rank two case. Using the continuity property and approximation by rational paths, it is sufficient to prove that the relations hold for rational paths. By using the stretching property, we can even assume that the path we start with is of the form $\eta = \nu_1 * \nu_2 * \ldots * \nu_r$ for some $\nu_1, \ldots, \nu_r \in \Lambda$.

If $\lambda_1, \ldots, \lambda_r \in \Lambda^+$ are dominant weights such that ν_i is in the Weyl group orbit of λ_i, then $\eta \in \mathbb{B}_{\lambda_1} * \ldots * \mathbb{B}_{\lambda_r}$. But this implies that η and the paths obtained from η by applying the root operators are integral. So to prove the claim, we are reduced to prove the following: Let $\mathbb{B} \subset \Pi$ be integral and stable under the root operators. Then the action of the simple reflections on $\mathbb{B}$ extends to an action of the Weyl group.

Let α, γ be the simple roots. Now s_α, s_γ commute if and only if the roots are orthogonal to each other. The root operators commute in this case too; so there is nothing to be proved. We may hence assume that α is not orthogonal to γ.

For $\eta \in \mathbb{B}$ denote by η^n the path obtained by concatenating the path $\eta * \eta * \ldots * \eta$ with itself n-times. Suppose $(\eta(1), \alpha^\vee) > 0$ and $\pi \in \mathbb{B}$ is arbitrary. It is easy to see that for $n, k \in \mathbb{N}$ big enough we can find k_1, k_2 such that (Lemma 6.12):

$$f_\alpha^k(\pi * \eta^n) = f_\alpha^{k_1}(\pi) * s_\alpha(\eta) * \ldots * s_\alpha(\eta) * f_\alpha^{k_2}(\eta).$$

So if we choose n big enough for given π_1, π_2, then there exist $\pi_1', \pi_2' \in \Pi$ such that

$$s_\alpha(\pi_1 * \eta^n * \pi_2) = \pi_1' * s_\alpha(\eta)^k * \pi_2'$$

for some $k \in \mathbb{N}$, where the k depends linearly on n for $n \gg 0$.

To prove that the braid relations hold, it is sufficient to prove that $s_\alpha s_\gamma \cdots (\eta) = s_\gamma s_\alpha \cdots (\eta)$ for $\eta \in \mathbb{B}$ with $\eta(1) \in \Lambda^+$. Since $s_\alpha(\eta) = \eta$ (resp. $s_\gamma(\eta) = \eta$) if $(\eta(1), \alpha^\vee) = 0$ (resp. $(\eta(1), \gamma^\vee) = 0$), the relation holds trivially if $\eta(1)$ is a multiple of a fundamental weight. So we may assume that $\eta(1)$ is regular.

Let $\lambda \in \Lambda^+$ be a regular dominant weight. Then, for a given $n \in \mathbb{N}$, we can find a $k \in \mathbb{N}$ such that $k\lambda * \eta^n \in \Pi^+$. Now Corollary 2 in section 6 implies:

$$s_\alpha s_\gamma \cdots (k\lambda * \eta^n) = s_\gamma s_\alpha \cdots (k\lambda * \eta^n).$$

But the arguments above show that we can choose n, π_1, π_2 such that the left and right side are of the form:

$$\pi_1 * s_\alpha s_\gamma \cdots (\eta)^k * \pi_2 \quad \text{and} \quad \pi_1 * s_\gamma s_\alpha \cdots (\eta)^k * \pi_2$$

for some $k > 0$. It follows that $s_\alpha s_\gamma \cdots (\eta) = s_\gamma s_\alpha \cdots (\eta)$.

8. The Proofs

It remains to give the proofs for Theorem 6.1, 6.5, and 6.13. A first step is the proof of the following weaker version of Theorem 6.5:

Proposition 8.1. *If λ, μ are dominant weights, then the graphs $\mathcal{G}_{\lambda+\mu}$ and $\mathcal{G}_{\lambda*\mu}$ are isomorphic.*

PROOF. Consider the family of paths $\pi_s := ((1-s)\lambda) * (\mu + s\lambda)$. Note that $\pi_0 = \lambda * \mu$, $\pi_1 = \lambda + \mu$, and $\pi_s \in \Pi^+$ for all $s \in [0,1]$. The results in section 5 imply that for all rational $t \in [0,1]$:

$$\mathbb{B}_t := \mathbb{B}_{\pi_t} \text{ is integral and } \pi_t \text{ is the only path in } \mathbb{B}_t \text{ such that } \rho * \pi_t \in \Pi_0^+. \qquad (**)$$

We use now $(**)$ to prove that the graphs are isomorphic: If $s \in [0,1]$, then, for any $\epsilon > 0$, we can find a rational number $s' \in [0,1]$ such that (after choosing appropriate parametrizations) $d(\pi_s, \pi_{s'}) < \epsilon$. So by continuity, it follows that $\mathbb{B}_s$ is integral and π_s is the only path in $\mathbb{B}_s$ such that $\rho * \pi_s \in \Pi_0^+$.

Corollary 1, section 4, implies that $\operatorname{Char} \mathbb{B}_s = \operatorname{Char} V_{\lambda+\mu}$ for all $s \in [0,1]$. Fix $\eta_1 \in \mathbb{B}_1$, and let $\alpha_1, \dots, \alpha_r$ be such that $\eta_1 = f_{\alpha_1} \cdots f_{\alpha_r} \pi_1$. Since the $\mathbb{B}_s$ are integral, it follows by continuity (and property (ii), section 3) that $\eta_s = f_{\alpha_1} \cdots f_{\alpha_r} \pi_s \neq \theta$. Of course, the $\alpha_1, \dots, \alpha_r$ are not necessarily uniquely determined. Suppose $\gamma_1, \dots, \gamma_r$ are simple roots such that $\eta_1 = f_{\gamma_1} \cdots f_{\gamma_r} \pi_1$. Property (i), section 3, implies then $\pi_1 = e_{\gamma_r} \cdots e_{\gamma_1} \eta_1$. Again, since the $\mathbb{B}_s$ are integral, continuity and property (ii), section 3, implies that $e_{\gamma_r} \cdots e_{\gamma_1} \eta_t \neq 0$. The endpoint of this path is $\lambda + \mu$. So the character of $\mathbb{B}_s$ implies that this is the path π_s, and hence $\eta_s = f_{\gamma_1} \cdots f_{\gamma_r} \pi_s$.

This proves that the map $\mathbb{B}_1 \to \mathbb{B}_s$, $\eta_1 \mapsto \eta_s$, is well-defined for all $s \in [0,1]$. The same arguments prove $f_{\alpha_1} \cdots f_{\alpha_r} \pi_1 = f_{\gamma_1} \cdots f_{\gamma_r} \pi_1$ iff $f_{\alpha_1} \cdots f_{\alpha_r} \pi_s = f_{\gamma_1} \cdots f_{\gamma_r} \pi_s$. Therefore the map is bijective and induces an isomorphism of the graphs.

We prepare now the proofs of the theorems. Let λ be a dominant weight. The path $\lambda * (-\lambda)$ has the property that $f_\alpha(\lambda * (-\lambda)) = e_\alpha(\lambda * (-\lambda)) = \theta$. Further, if $\pi \in \Pi^+$, then the map $\mathbb{B}_\pi \to \mathbb{B}_{\pi * \lambda * (-\lambda)}$, $\eta \mapsto \eta * \lambda * (-\lambda)$ induces by Lemma 6.12 obviously an isomorphism of graphs $\mathcal{G}_\pi \to \mathcal{G}_{\pi * \lambda * (-\lambda)}$.

Suppose λ, μ, ν are dominant weights such that $\lambda + \mu = \nu$. By Proposition 8.1, we have an isomorphism $\mathcal{G}_{\lambda * \mu} \to \mathcal{G}_\nu$. Since $-\mu$ is an element of $\mathbb{B}_{-w_0(\mu)}$, the isomorphism above induces an inclusion and a bijection:

$$\mathbb{B}_{\nu * -\mu} \hookrightarrow \mathbb{B}_\nu * \mathbb{B}_{-w_0(\mu)} \xrightarrow{\sim} \mathbb{B}_{\lambda * \mu} * \mathbb{B}_{-w_0(\mu)},$$

such that the image of $\nu * -\mu$ is the path $\lambda * \mu * -\mu$. Since the locally integral concatenations are integral, Lemma 6.12 implies that this map induces an isomorphism of graphs $\mathcal{G}_{\nu * (-\mu)} \to \mathcal{G}_\lambda = \mathcal{G}_{\nu - \mu}$. An easy induction process shows:

Lemma 8.2. *If $\pi = \lambda_1 * \ldots * \lambda_r \in \Pi^+$ is such that the λ_i are in either Λ^+ or $-\Lambda^+$, then $\mathcal{G}_\pi$ is isomorphic to $\mathcal{G}_\lambda$, where $\lambda = \lambda_1 + \ldots + \lambda_r$.*

The proof of the general case will be reduced to paths of the form above. The next lemma is an important step in the reduction procedure:

Lemma 8.3. *If $\pi \in \Pi^+$ and $\lambda \in \Lambda^+$ are such that $\mathcal{G}_{n\pi}$ is isomorphic to $\mathcal{G}_{n\lambda}$ for $n \gg 0$, then $\mathbb{B}_\pi$ is integral and $\mathcal{G}_\pi$ is isomorphic to $\mathcal{G}_\lambda$.*

PROOF OF THE LEMMA. The stretching property ((i), section 3) implies that $\mathcal{G}_\lambda \subset \mathcal{G}_{n\lambda}$ and $\mathcal{G}_\pi \subset \mathcal{G}_{n\pi}$ can be recovered as the subgraphs associated to the operators f_α^n, e_α^n. So the isomorphism $\mathcal{G}_{n\pi} \to \mathcal{G}_{n\lambda}$ implies an isomorphism $\phi : \mathcal{G}_\pi \to \mathcal{G}_\lambda$.

Suppose $\eta \in \mathbb{B}_\pi$ and the minimum m_α of the function $h_\alpha : t \to (\eta(t), \alpha^\vee)$ is not an integer. By Lemma 5.6, we know that the minimum n_α of the function $t \mapsto (\phi(\eta)(t), \alpha^\vee)$ is an integer.

By property (ii), section 3, $|n_\alpha|$ is maximal such that $e_\alpha^{|n_\alpha|}\phi(\eta) \neq \theta$. By the isomorphism of graphs, this implies also that $|n_a|$ is maximal with the property that $e_\alpha^{|n_a|}\eta \neq \theta$ and hence $|n_\alpha| < |m_a|$. Choose $k \in \mathbb{N}$ such that $k|m_a| - k|n_\alpha| > 1$.

The minima of the functions $t \mapsto (k\phi(\eta)(t), \alpha^\vee)$ and $t \mapsto (k\eta(t), \alpha^\vee)$ are kn_α and km_α. This implies $e_\alpha^{|kn_\alpha|+1}\phi(\eta) = \theta$. Since $k|m_a| - k|n_\alpha| > 1$, we get $e_\alpha^{|kn_\alpha|+1}(\eta) \neq \theta$. But the graphs $\mathcal{G}_{k\pi}$ and $\mathcal{G}_{k\lambda}$ are isomorphic, and the stretching property implies that $k\eta$ is mapped onto $k\phi(\eta)$. So $e_\alpha^{|kn_\alpha|+1}\phi(\eta) = \theta$ implies $e_\alpha^{|kn_\alpha|+1}\eta = \theta$. It follows that $n_\alpha = m_\alpha$.

Proof of Theorem 6.5 and 6.13. Let $\pi \in \Pi^+$ be a rational path. By Lemma 8.3 it is sufficient to prove the theorems for $n\pi$, $n \gg 0$. Thus we may in fact assume that $\pi = \lambda_1 * \ldots * \lambda_r$ for some $\lambda_1, \ldots, \lambda_r \in \Lambda$. Set $\lambda = \pi(1)$.

If $\lambda_i \notin \Lambda^+$ and $-\lambda_i \notin \Lambda^+$, then fix $\mu_i, \nu_i \in \Lambda^+$ such that $\lambda_i = \mu_i - \nu_i$. Further, we may assume that for any simple root α we have either $(\nu_i, \alpha^\vee) = 0$ or $(\mu_i, \alpha^\vee) = 0$.

For $\eta \in \Pi$ denote by $\eta^k = \eta * \ldots * \eta$ the path obtained by concatenating η k-times. Note that for any given ϵ we can chose $k \gg 0$ such that (after choosing an appropriate parametrization of the paths) $d(\lambda_i, (\frac{\mu_i}{k} * \frac{-\nu_i}{k})^k) < \epsilon$.

So we can approximate π by paths of the form $\lambda_1 * \ldots * \lambda_r$ such that λ_i or $-\lambda_i$ is a rational dominant weight. By Lemma 8.2 and Lemma 8.3 we know that the corresponding paths are integral, and the graph is isomorphic to $\mathcal{G}_\lambda$. Then it follows by continuity that $\mathbb{B}_\pi$ is integral and $\mathcal{G}_\pi$ isomorphic to $\mathcal{G}_\lambda$.

An arbitrary path $\pi \in \Pi^+$ can be approximated by rational paths in Π^+. Since the structure of the graph is independent of the choice of the approximation, the continuity property implies that $\mathcal{G}_\pi$ is isomorphic to $\mathcal{G}_\lambda$ and $\mathbb{B}_\pi$ is integral.

Proof of Theorem 6.1. Suppose $\lambda = \pi(1)$. The isomorphism of graphs implies that the cardinalities of the sets $\mathbb{B}_\pi$ and $\mathbb{B}_\lambda$ are the same. Since the latter is equal to $\dim V_\lambda$ by Corollary 1, section 5, it follows that $\mathbb{B}_\pi$ is a finite set. Recall that λ is the only path in $\mathbb{B}_\lambda$ such that $\rho * \lambda$ is contained in the interior of the dominant Weyl chamber for $t > 0$ (Corollary 1, section 5). So for any $\eta' \in \mathbb{B}_\lambda$ there exists a simple root α such that $e_\alpha(\eta') \neq \theta$. The isomorphism of graphs hence implies that for any $\eta \in \mathbb{B}_\pi$, $\eta \neq \pi$, there exists a simple root α such that $e_\alpha(\eta) \neq 0$. It follows that there exists a simple root α such that the minimum of the function $t \mapsto (\eta(t), \alpha)$ is ≤ -1. Thus $\rho * \eta$ meets at least one of the walls of the dominant Weyl chamber.

9. A Demazure-Type Character Formula

For a simple root α denote by Λ_α the Demazure operator on the group ring $\mathbb{Z}[\Lambda]$:

$$\Lambda_\alpha(e^\mu) := \frac{e^{\mu+\rho} - e^{s_\alpha(\mu+\rho)}}{1 - e^{-\alpha}} e^{-\rho}.$$

In other words:

$$\Lambda_\alpha(e^\mu) = \begin{cases} e^\mu + e^{\mu-\alpha} + \ldots + e^{s_\alpha(\mu)} & \text{if } (\mu, \alpha^\vee) \geq 0; \\ 0 & \text{if } (\mu, \alpha^\vee) = -1; \\ -e^{\mu+\alpha} - \ldots - e^{s_\alpha(\mu)-\alpha} & \text{if } (\mu, \alpha^\vee) < -1. \end{cases}$$

Note that $\Lambda_\alpha \circ \Lambda_\alpha = \Lambda_\alpha$. So Λ_α applied to a root string $e^\mu + e^{\mu-\alpha} + \ldots + e^{s_\alpha(\mu)}$ just reproduces the string.

Fix $\lambda \in \Lambda^+$ and let $\underline{\nu} := \nu_1 * \ldots * \nu_r$ be a path in $\mathbb{B}_\lambda$. The "first direction" ν_1 is (up to multiplication by a positive rational number) of the form $\sigma(\lambda)$ for some $\sigma \in W/W_\lambda$. We define a map $i : \mathbb{B}_\lambda \longrightarrow W/W_\lambda$ by $i(\underline{\nu}) := \sigma$. For $w \in W/W_\lambda$ denote by $\mathbb{B}_\lambda(w)$ the subset

$$\mathbb{B}_\lambda(w) := \{\underline{\nu} \in \mathbb{B}_\lambda \mid i(\underline{\nu}) \leq w\}$$

Note that $\mathbb{B}_\lambda = \mathbb{B}_\lambda(w_0)$ for the longest word w_0 in the Weyl group.

Theorem 9.1. $\operatorname{Char} \mathbb{B}_\lambda(w) := \sum_{\underline{\nu} \in \mathbb{B}_\lambda(w)} e^{\underline{\nu}(1)} = \Lambda_{\alpha_1} \circ \cdots \circ \Lambda_{\alpha_r}(e^\lambda)$ *for any reduced decomposition* $w = s_{\alpha_1} \ldots s_{\alpha_r}$.

Comments about the proof. The details can be found in [10]. The main idea is to prove the following two properties.

First property: If $s_\alpha w > w$ in W/W_λ and $i(\underline{\nu}) = s_\alpha w$, then there exists an $k > 0$ such that $e_\alpha^{k+1}(\underline{\nu}) = \theta$, $e_\alpha^k(\underline{\nu}) \neq \theta$, $i(e_\alpha^k(\underline{\nu})) = w$, $i(e_\alpha^j(\underline{\nu})) = s_\alpha w$ for all $j < k$ and $i(f_\alpha^j(\underline{\nu})) = s_\alpha w$ for all j such that $f_\alpha^j(\underline{\nu}) \neq \theta$.

Second property: If $s_\alpha w > w$ in W/W_λ and $i(\underline{\nu}) = w$, then either $e_\alpha \underline{\nu} = \theta$ and $i(f_\alpha^j(\underline{\nu})) = s_\alpha w$ for all j such that $f_\alpha^j(\underline{\nu}) \neq \theta$, or $i(f_\alpha^j(\underline{\nu})) = i(e_\alpha^k(\underline{\nu})) = w$ for all j such that $f_\alpha^j(\underline{\nu}) \neq \theta$ and all k such that $e_\alpha^k(\underline{\nu}) \neq \theta$.

The theorem is obviously true for the class of the identity. We proceed now by induction on the length of w. Let α be a simple root such that $s_\alpha w > w$. The second property implies that we can decompose $\mathbb{B}_\lambda(w)$ into $\mathbb{B}_\lambda^+(w) \cup \mathbb{B}_\lambda^0(w)$, where $\mathbb{B}_\lambda^0(w)$ is the set of all paths in $\mathbb{B}_\lambda(w)$ such that $i(f_\alpha^j(\underline{\nu})) = i(e_\alpha^k(\underline{\nu})) = w$ for all j having $f_\alpha^j(\underline{\nu}) \neq \theta$ and all k having $e_\alpha^k(\underline{\nu}) \neq \theta$, and $\mathbb{B}_\lambda^+(w)$ is the set of all paths such that $e_\alpha \underline{\nu} = \theta$ and $i(f_\alpha^j(\underline{\nu})) = s_\alpha w$ for all j having $f_\alpha^j(\underline{\nu}) \neq \theta$.

Obviously $\operatorname{Char} \mathbb{B}_\lambda(w) = \operatorname{Char} \mathbb{B}_\lambda^+(w) + \operatorname{Char} \mathbb{B}_\lambda^0(w)$. Now $\operatorname{Char} \mathbb{B}_\lambda^0(w)$ is just a sum of α-strings. So $\Lambda_\alpha(\operatorname{Char} \mathbb{B}_\lambda^0(w)) = \operatorname{Char} \mathbb{B}_\lambda^0(w)$. The first property implies that

$$\mathbb{B}_\lambda(s_\alpha w) = \mathbb{B}_\lambda^0(w) \cup \{f_\alpha^k \underline{\nu} \mid \underline{\nu} \in \mathbb{B}_\lambda^+(w), 0 \leq k \leq (\underline{\nu}(1), \alpha^\vee)\}.$$

(Note: $f_\alpha^{(\underline{\nu}(1),\alpha^\vee)+1} \underline{\nu} = 0$ since $e_\alpha \underline{\nu} = 0$.) Since $\operatorname{Char}\{\underline{\nu}, f_\alpha \underline{\nu}, \ldots, \tilde{s}_\alpha(\underline{\nu})\} = \Lambda_\alpha(e^{\underline{\nu}(1)})$, we get: $\operatorname{Char} \mathbb{B}_\lambda(s_\alpha w) = \Lambda_\alpha(\operatorname{Char} \mathbb{B}_\lambda(w))$.

10. The P–R–V Conjecture

Consider the tensor product $V_\lambda \otimes V_\mu$ of two simple $\mathfrak{g}$-modules of highest weight λ and μ. The Parthasarathy–Ranga-Rao–Varadarajan conjecture (which has been proved independently in [8] and [17]) states:

Theorem 10.1. *If* $\tau_1, \tau_2 \in W$ *are such that* $\nu := \tau_1(\lambda) + \tau_2(\mu)$ *is a dominant weight, then the module* V_ν *occurs in* $V_\lambda \otimes V_\mu$.

Proof. Using the generalized Littlewood-Richardson rule, one can give a purely combinatorial proof. To say that $\nu := \tau_1(\lambda) + \tau_2(\mu)$ is a dominant weight is the same as to say that ν is the unique dominant weight in the Weyl group orbit of $\lambda + \sigma(\mu)$, where $\sigma \in W/W_\mu$ is such that $\sigma(\mu) = \tau_1^{-1}\tau_2(\mu)$.

We construct now a path $\eta \in \mathbb{B}_\mu$ such that $\lambda + \eta(1) = \nu$ and $\lambda * \eta \in \Pi^+$. The generalized Littlewood-Richardson rule implies then that V_ν occurs in $V_\lambda \otimes V_\mu$ with multiplicity at least one.

Let us start with the path $\eta := \sigma(\mu)$. If $\lambda * \eta \in \Pi^+$, then we are done. So suppose $\lambda * \eta \notin \Pi^+$. Then we can find a $t \in [0,1]$ such that $\lambda + \eta(s)$ is in the dominant Weyl chamber for $s \leq t$ and such that $\lambda + \eta(s)$ is outside the dominant Weyl chamber for $s > t$. We are going to fold the last part of the path back into the dominant Weyl chamber. Let Δ be the set of simple roots orthogonal to $\lambda + \eta(t)$.

Note that $(\lambda + t\sigma(\mu), \alpha^\vee) = 0$ implies $(1-t)(\sigma(\mu), \alpha^\vee) \in \mathbb{Z}$ for $\alpha \in \Delta$. So $(1-t)\sigma(\mu)$ is an integral weight for the sub-root system spanned by the simple roots in Δ. Further, since $(\lambda, \alpha^\vee) \geq 0$, it follows that $(1-t)(\sigma(\mu), \alpha^\vee) \leq 0$. Let w_0' be the longest element in the Weyl group generated by the reflections s_α, $\alpha \in \Delta$, and let $\sigma_1 \in W/W_\mu$ be such that $\sigma_1(\mu) = w_0'(\sigma(\mu))$.

Since $(1-t)\sigma(\mu)$ is an antidominant integral weight for the simple roots in Δ, it is easy to see that $\eta' := t\sigma(\mu) * (1-t)\sigma_1(\mu)$ is an element of $\mathbb{B}_\mu$.

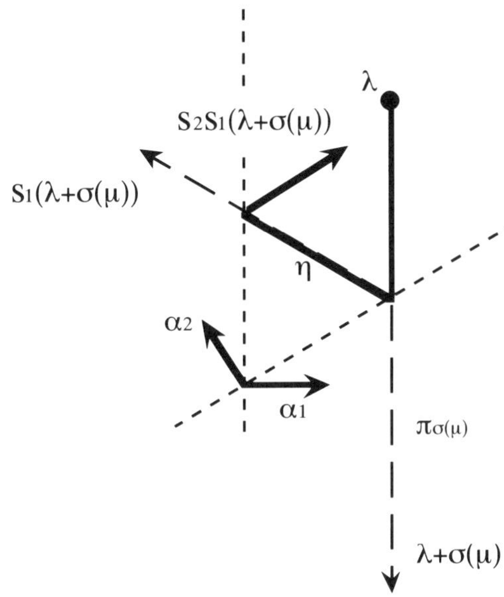

FIGURE 8. A proof of the P–R–V conjecture

Note that the new path is "better" then the old one: First of all, $\lambda + \eta'(1) = w_0'(\lambda + \eta(1))$, so that the endpoint of $\lambda * \eta'$ is in the Weyl group orbit of $\lambda + \sigma(\mu)$. Also the path $\lambda * \eta'$ stays longer in the dominant Weyl chamber than the path $\lambda * \eta$: For $s \leq t_1$ we have of course $\lambda * \eta'(s) = \lambda * \eta(s)$; so these points are in the dominant Weyl chamber. If $\alpha \in \Delta$ and $s > t$, then

$$(\lambda + \eta'(s), \alpha^\vee) = (\lambda + \eta(t), \alpha^\vee) + (s-t)(\sigma_1(\mu), \alpha^\vee) \geq 0.$$

And if $\alpha \notin \Delta$, then $(\lambda + \eta'(t), \alpha^\vee) > 0$; so we can choose $0 < r \ll 1$ such that $(\lambda + \eta'(s), \alpha^\vee) \geq 0$ for all $t \leq s \leq t + r$ and all $\alpha \notin \Delta$. It follows that $\lambda + \eta'(s)$ is in the dominant Weyl chamber for all $0 \leq s \leq t + r$.

If $\lambda * \eta' \in \Pi^+$, we are done; otherwise we proceed as in Figure 8; i.e., we fix $t' > t$ such that $\lambda + \eta'(s)$ is in the dominant Weyl chamber for all $s \leq t'$ and $\lambda + \eta'(s)$ is outside the dominant Weyl chamber for $s > t'$. Using the same procedure as above, we then fold another part of the path back into the dominant Weyl chamber.

Note that this is a finite procedure: The endpoint of the new path is always of the form the endpoint of the old path plus a sum of positive roots. Since the weights that can occur are all of the form $\lambda + \mu$ minus a sum of positive roots, this procedure has to end after a finite number of steps.

11. Paths, the Crystal Graph, and the Plactic Algebra

We would like to conclude this note with two remarks, one concerning the relation of the paths to the crystal graph and one remark on the so-called plactic algebra.

Let $U_q(\mathfrak{g})$ be the q-analogue of the enveloping algebra of $\mathfrak{g}$. For a detailed introduction we mention the books [1,3,16]. A finite-dimensional irreducible representation V_λ of $\mathfrak{g}$ admits a quantum deformation V_λ^q [15]. Kashiwara introduced in [7] the notion of a crystal graph of an $U_q(\mathfrak{g})$-representation (and, using the quantum deformation, one can of course associate such a graph to a $\mathfrak{g}$-representation). This graph can be considered as a refined version of the character of the representation.

The following connection between the paths and the crystal graph was found by Kashiwara [6] and Joseph [3]:

Theorem 11.1. *Suppose $\pi \in \Pi^+$, and set $\lambda := \pi(1)$. Then the crystal graph C_λ of the representation V_q^λ is isomorphic to the graph $\mathcal{G}_\lambda$.*

Using the tensor product of quantum representations, one can make the union of all crystal bases into an algebra. In terms of paths this would be the $\mathbb{C}$-vector space with basis $\bigcup_{\pi\in\Pi^+} \mathbb{B}_\pi$, with product the concatenation of paths, but where we factor out the relations obtained via the isomorphisms $\mathcal{G}_\pi \simeq \mathcal{G}_{\pi'}$ for $\pi(1) = \pi'(1)$.

This algebra contains a great deal of information about the tensor products of the representations. For the groups $GL_n(\mathbb{C})$ and $SL_n(\mathbb{C})$, such an algebra has been defined before by Lascoux and Schützenberger. Their idea was to define a product structure on the set of all semi-standard Young tableaux such that this product mimics the tensor product of GL_n-representations. It turns out that this "plactic algebra" (as they call it) is precisely the crystal or path algebra defined above.

A description of this algebra in terms of generators and relations (i.e., a description more in the style of [18]) can be found in [14].

References

1. V. Chari and A. Pressley, *A Guide to Quantum Groups*, Cambridge University Press, Cambridge, 1994; corrected reprint, 1995.
2. V. Deodhar, A splitting criterion for Bruhat orderings on Coxeter groups, *Commun. in Algebra* **15** (1987), 1889–1894.
3. A. Joseph, *Quantum Groups and Their Primitive Ideals*, Springer-Verlag, Berlin, 1995.
4. M. Kashiwara, Crystal base and Littelmann's refined Demazure character formula, *Duke Math. J.* **71** (1993), 839–858.
5. M. Kashiwara, Crystal bases of modified quantized enveloping algebras, *Duke Math. J.* **73** (1994), 383–413.
6. M. Kashiwara, Similarity of crystal bases, *Lie Algebras and Their Representations*, Contemporary Mathematics, vol. 194, 1996, pp. 177–186.
7. M. Kashiwara, Crystalizing the q-analogue of universal enveloping algebras, *Commun. in Math. Phys.* **133** (1990), 249–260.
8. S. Kumar, Proof of the Parthasarathy–Ranga Rao–Varadarajan Conjecture, *Invent. Math.* **93** (1988), 117–130.
9. V. Lakshmibai and C. S. Seshadri, Standard monomial theory, *Proceedings of the Hyderabad Conference on Algebraic Groups* (S. Ramanan, ed.), Manoj Prakashan, Madras, 1991, pp. 279–323.
10. P. Littelmann, A Littlewood-Richardson rule for symmetrizable Kac-Moody algebras, *Invent. Math.* **116** (1994), 329–346.

11. P. Littelmann, Paths and root operators in representation theory, *Annals of Math.* **142** (1995), 499–525.
12. P. Littelmann, The path model for representations of symmetrizable Kac-Moody algebras, *Proc. Intern. Congress Math., Zürich 1994* (S. I. Hariharan and T. H. Moulton, eds.), vol. 1, Birkhäuser, Basel, 1995, pp. 298–308.
13. P. Littelmann, Crystal graphs and Young tableaux, *J. Algebra* **175** (1995), 65–87.
14. P. Littelmann, A plactic algebra for semisimple Lie algebras, *Advances in Math.* **124** (1996), 312–331.
15. G. Lusztig, Quantum deformations of certain simple modules over enveloping algebras, *Advances in Math.* **70** (1988), 237–249.
16. G. Lusztig, *Introduction to Quantum Groups*, Progress in Mathematics, vol. 110, Birkhäuser, Boston, 1993.
17. O. Mathieu, Construction d'un groupe de Kac-Moody et applications, *Compositio Math.* **69** (1989), 37–60.
18. M. P. Schützenberger, La correspondance de Robinson, *Combinatoire et Représentation du Groupe Symétrique*, Lecture Notes in Mathematics, vol. 579, Springer-Verlag, New York, 1977, pp. 59–113.

Département de Mathématiques et IRMA, 7, rue René Descartes, Université Louis Pasteur, 67084 Strasbourg, France

E-mail address: littelma@math.u-strasbg.fr

Proceedings of Symposia in Pure Mathematics
Volume **61** (1997), pp. 51–59

Irreducible Representations of SL(2, R)

Robert W. Donley, Jr.*

1. Introduction

We review the theory of the irreducible representations of $SL(2, \mathbb{R})$. The finite-dimensional irreducible representations can be realized in spaces of homogeneous polynomials in two variables. We will consider natural alternatives to these realizations and show how they relate to the irreducible unitary representations. This will lead to a definition for a class of admissible representations that will not necessarily be irreducible. This failure for irreducibility will lead us to integral operators with interesting properties.

Many excellent sources exist for this material; we refer the reader to the list of references at the end of this paper.

2. Finite-Dimensional Representations

First we introduce some notation. Define

$$G = SL(2, \mathbb{R}) = \left\{ \begin{pmatrix} a & b \\ c & d \end{pmatrix} \in M_2(\mathbb{R}) \,\middle|\, ad - bc = 1 \right\}. \tag{2.1}$$

Let $\mathfrak{g}_0 = \mathfrak{sl}(2, \mathbb{R}) = \{\text{traceless matrices in } M_2(\mathbb{R})\}$. Denote the complexification of $\mathfrak{g}_0$ by $\mathfrak{g} = \mathfrak{sl}(2, \mathbb{C})$. The usual basis for $\mathfrak{g}_0$ over $\mathbb{R}$ (or $\mathfrak{g}$ over $\mathbb{C}$) is given by

$$e = \begin{pmatrix} 0 & 1 \\ 0 & 0 \end{pmatrix}, \qquad f = \begin{pmatrix} 0 & 0 \\ 1 & 0 \end{pmatrix}, \qquad h = \begin{pmatrix} 1 & 0 \\ 0 & -1 \end{pmatrix} \tag{2.2}$$

with relations

$$[h, e] = 2e, \qquad [h, f] = -2f, \qquad \text{and} \qquad [e, f] = h. \tag{2.3}$$

1991 *Mathematics Subject Classification.* Primary 22-02; Secondary 22E45.

* The author was supported by an Alfred P. Sloan Doctoral Dissertation Fellowship at the time of the conference. Currently he is supported by a National Science Foundation Mathematical Sciences Postdoctoral Research Fellowship. He also thanks the Institut Mittag-Leffler for their hospitality.

Note that $\mathbb{R}h$ is a Cartan subalgebra of $\mathfrak{g}_0$. Sometimes it will be convenient to use the Cartan subgroup and Cartan subalgebra

$$T = \begin{pmatrix} \cos\theta & \sin\theta \\ -\sin\theta & \cos\theta \end{pmatrix} \quad \text{and} \quad \mathfrak{t}_0 = \mathbb{R}t, \qquad \text{where } t = \begin{pmatrix} 0 & 1 \\ -1 & 0 \end{pmatrix}. \tag{2.4}$$

Weights are real-valued on the element it, and all weight computations are with respect to this element. We also often abbreviate elements of T by θ for $0 \le \theta < 2\pi$.

Fix a nonnegative integer N. Let V_2^N be the space of complex homogeneous polynomials in z_1, z_2 of degree N. G acts on V_2^N by

$$\left(\Phi_N\begin{pmatrix} a & b \\ c & d \end{pmatrix}P\right)\begin{pmatrix} z_1 \\ z_2 \end{pmatrix} = P\left(\begin{pmatrix} a & b \\ c & d \end{pmatrix}^{-1}\begin{pmatrix} z_1 \\ z_2 \end{pmatrix}\right) \tag{2.5}$$

This representation is irreducible of dimension $N+1$. An easy check shows that f sends $z_1^i z_2^{N-i}$ to a nonzero multiple of $z_1^{i+1}z_2^{N-i-1}$ for $0 \le i < N$, and the h-weight of $z_1^i z_2^{N-i}$ is $N-2i$. For comparison the t-weight vectors are of the form $(z_1+iz_2)^j(z_1-iz_2)^{N-j}$.

We note two alternate realizations of V_2^N. Suppose $P \in V_2^N$. Let

$$(\phi P)(z) = P\begin{pmatrix} 1 \\ z \end{pmatrix}; \tag{2.6}$$

ϕP is a polynomial of degree $\le N$. Note that

$$\begin{pmatrix} a & b \\ c & d \end{pmatrix}^{-1}\begin{pmatrix} z_1 \\ z_2 \end{pmatrix} = \begin{pmatrix} d & -b \\ -c & a \end{pmatrix}\begin{pmatrix} z_1 \\ z_2 \end{pmatrix} = \begin{pmatrix} dz_1 - bz_2 \\ -cz_1 + az_2 \end{pmatrix}. \tag{2.7}$$

Thus

$$P\left(\begin{pmatrix} a & b \\ c & d \end{pmatrix}^{-1}\begin{pmatrix} z_1 \\ z_2 \end{pmatrix}\right) = (dz_1 - bz_2)^N P\begin{pmatrix} 1 \\ \frac{az_2 - cz_1}{dz_1 - bz_2} \end{pmatrix}. \tag{2.8}$$

Set $z_1 = 1$ and $z_2 = z$. Then

$$(\phi(\Phi_N(g)P))(z) = (-bz+d)^N(\phi P)\Big(\frac{az-c}{-bz+d}\Big). \tag{2.9}$$

Thus we have an equivalent representation given in the space of polynomials of degree less than or equal to N with group action

$$[\Psi_N(g)Q](z) = (-bz+d)^N Q\Big(\frac{az-c}{-bz+d}\Big). \tag{2.10}$$

Note that the action on the Riemann sphere $\mathbb{C} \cup \infty$ given by

$$z \longrightarrow \frac{dz+c}{bz+a} \tag{2.11}$$

preserves the upper half-plane, the lower half-plane, and $\mathbb{R} \cup \infty$. Thus we can restrict Q to $\mathbb{R} \cup \infty$ to get another equivalent representation, with space given by complex-valued polynomials on $\mathbb{R}$ of degree less than or equal to N and group action

$$[\Psi_N^{\mathbb{R}}(g)Q](x) = (-bx+d)^N\, Q\Big(\frac{ax-c}{-bx+d}\Big)$$

$$(2.12)\qquad = \begin{cases} \operatorname{sgn}(-bx+d)\,|-bx+d|^N\, Q(\frac{ax-c}{-bx+d}), & N \text{ odd} \\ |-bx+d|^N\, Q(\frac{ax-c}{-bx+d}), & N \text{ even.} \end{cases}$$

3. Families of Unitary Representations for G

(I) **Principal series**: $\mathcal{P}^{+,\,iv}$ and $\mathcal{P}^{-,\,iv}$, v in $\mathbb{R}$

Fix $v \in \mathbb{R}$. The first natural place to look for a unitary representation is in complex-valued $L^2(\mathbb{R})$ with an action similar to (2.12). With respect to the usual norm on $L^2(\mathbb{R})$, a quick computation shows that a unitary group action is given by one of

(3.1)
$$\mathcal{P}^{\varepsilon,\,iv}\begin{pmatrix} a & b \\ c & d\end{pmatrix} f(x) = \begin{cases} |-bx+d|^{-1-iv}\, f(\frac{ax-c}{-bx+d}) & \text{if } \varepsilon = + \\ \operatorname{sgn}(-bx+d)\,|-bx+d|^{-1-iv}\, f(\frac{ax-c}{-bx+d}) & \text{if } \varepsilon = -. \end{cases}$$

The exponent is shifted because dx is not invariant under $x \leftrightarrow \frac{ax-c}{-bx+d}$.

These representations are irreducible except for $\mathcal{P}^{-,\,0}$. Unitary equivalences occur for

$$(3.2)\qquad \mathcal{P}^{+,\,iv} \cong \mathcal{P}^{+,\,-iv} \text{ and } \mathcal{P}^{-,\,iv} \cong \mathcal{P}^{-,\,-iv}.$$

(II) **Complementary Series**: $\mathcal{C}^u$, $0 < u < 1$.

Fix $0 < u < 1$. Another approach to using (2.12) is to find a space other than $L^2(\mathbb{R})$. Such a space (with norm) is given by

$$(3.3)\qquad \Big\{ f : \mathbb{R} \to \mathbb{C} \,\Big|\, f \in L^1_{loc}(\mathbb{R}),\ \|f\|_u^2 = \iint\limits_{\mathbb{R}\,\mathbb{R}} \frac{f(x)\overline{f(y)}}{|x-y|^{1-u}}\, dx\, dy < \infty \Big\}$$

with group action

$$(3.4)\qquad \mathcal{C}^u\begin{pmatrix} a & b \\ c & d\end{pmatrix} f(x) = |-bx+d|^{-1-u}\, f\Big(\frac{ax-c}{-bx+d}\Big).$$

These representations are irreducible unitary. It is not evident that $\|\cdot\|_u$ is an inner product; the usual norm on $L^2(\mathbb{R})$ can be written as

$$\int_{\mathbb{R}}\int_{\mathbb{R}} \delta(x-y) f(x)\overline{f(y)}\, dx\, dy.$$

The idea is to replace δ by a distribution that transforms correctly for (2.12); this is the technique in [Ba], where it is carried out for the circle rather than the line.

(III) **Discrete series** $\mathcal{D}_n^+$ and $\mathcal{D}_n^-$, $n \geq 2$ in $\mathbb{Z}$

Fix $n \geq 2$ in $\mathbb{Z}$. Here we use the action of (2.10) on the upper half plane

$$H = \{z \in \mathbb{C} \mid Im\, z > 0\}.$$

Let $z = x + iy$. The Hilbert space and norm for $\mathcal{D}_n^+$ are given by

$$(3.5)\quad L^2_{n,+}(H) = \Big\{f \text{ analytic for } Im\ z > 0 \,\Big|\, \|f\|^2 = \iint\limits_{Im\, z > 0} |f(z)|^2\, y^n\, \frac{dx\, dy}{y^2} < \infty\Big\}$$

with group action

$$\mathcal{D}_n^+\begin{pmatrix} a & b \\ c & d \end{pmatrix} f(z) = (-bz+d)^{-n}\, f\Big(\frac{az-c}{-bz+d}\Big). \tag{3.6}$$

The reader familiar with hyperbolic spaces will note that $\frac{dx\,dy}{y^2}$ is G-invariant. This space is nonzero since $(z+i)^{-n}$ is in it; for $m \geq 0$, a t-weight vector of weight $n+2m$ is given by

$$F_m(z) = \frac{(z-i)^m}{(z+i)^{n+m}}. \tag{3.7}$$

Later we shall see these functions account for all t-weight vectors in $L^2_{n,+}(H)$.

This representation is irreducible and unitary. It is also square-integrable; this means every matrix coefficient is in $L^2(G)$. Matrix coefficients are functions of G of the form

$$\psi_{h,f}(g) = \langle \mathcal{D}_n^+(g) f, h\rangle \tag{3.8}$$

where $f, h \in L^2_{n,+}(H)$.

The space $L^2_{n,-}(H)$ is the complex-conjugate space of $L^2_{n,+}(H)$ with new group action

$$\mathcal{D}_n^-\begin{pmatrix} a & b \\ c & d \end{pmatrix} f(z) = (\overline{-bz+d})^{-n}\, f\Big(\frac{az-c}{-bz+d}\Big). \tag{3.9}$$

Here the weight vectors are of the form $-n-2m$ where $m \geq 0$.

(IV) **Limits of Discrete Series** $\mathcal{D}_1^+$ and $\mathcal{D}_1^-$

The spaces $L^2_{1,+}(H)$ and $L^2_{1,-}(H)$ are analogs of the discrete series spaces but with a new norm

$$\|f\|^2 = \sup_{y>0} \int_{-\infty}^{\infty} |f(x+iy)|^2\, dx. \tag{3.10}$$

The group actions are given as in (3.6) and (3.9) but with $n = 1$. These representations are not square-integrable.

We note that (III) and (IV) have alternate realizations on the unit disc; we refer the reader to [La] for more details but note that F_m is carried to a multiple of $G_m(w) = w^m$ by the equivalence.

Families (I) through (IV), togther with the one-dimensional trivial representation, exhaust the irreducible unitary representations of G up to equivalence. This theorem was proved by Bargmann [Ba].

4. Other Irreducible Representations of G

If we widen our interest to nonunitary representations, the principal series have an obvious generalization by replacing iv with any $w \in \mathbb{C}$. These representations are the **nonunitary principal series** $\mathcal{P}^{+,w}$ and $\mathcal{P}^{-,w}$. Fix such a w. The space is given by complex-valued $L^2(\mathbb{R}, (1+x^2)^{Re\,w}\,dx)$ with G-action

(4.1)
$$\mathcal{P}^{\varepsilon,w}\begin{pmatrix} a & b \\ c & d \end{pmatrix} f(x) = \begin{cases} |-bx+d|^{-1-w}\, f\left(\frac{ax-c}{-bx+d}\right) & \text{if } \varepsilon = +, \\ \operatorname{sgn}(-bx+d)\,|-bx+d|^{-1-w}\, f\left(\frac{ax-c}{-bx+d}\right) & \text{if } \varepsilon = -. \end{cases}$$

The action is not unitary unless w is imaginary. When $0 < w < 1$, it becomes unitary by properly renorming the space; this is the case of the complementary series.

These representations are not always irreducible, but in fact we have already encountered all possibilities for reducibility. First (2.12) shows that

$$\Phi_n \subset \begin{cases} \mathcal{P}^{+,-(n+1)} & \text{if } n \text{ even} \\ \mathcal{P}^{-,-(n+1)} & \text{if } n \text{ odd.} \end{cases} \tag{4.2}$$

Similarly, restricting the functions in $L^2_{n,\varepsilon}(H)$ to $\mathbb{R}$ gives

$$\mathcal{D}_n^+ \oplus \mathcal{D}_n^- \subseteq \begin{cases} \mathcal{P}^{+,n-1} & \text{if } n \text{ even} \\ \mathcal{P}^{-,n-1} & \text{if } n \text{ odd.} \end{cases} \tag{4.3}$$

In particular,

$$\mathcal{P}^{-,0} \cong \mathcal{D}_1^+ \oplus \mathcal{D}_1^- \tag{4.4}$$

This accounts for all reducibility of $\mathcal{P}^{+,w}$ and $\mathcal{P}^{-,w}$. The quotient of a reducible $\mathcal{P}^{\varepsilon,-(n+1)}$ by Φ_n is essentially the sum of two discrete series representations, and the quotient of a reducible $\mathcal{P}^{\varepsilon,n-1}$ by the sum of two discrete series representations is finite-dimensional.

5. Alternate Realizations

The above nonunitary principal series are a special case of a construction for general semisimple groups. Facts about this construction are deduced by examining different decompositions of G. Several realizations of a given representation will arise. The representations in Section 4 occur in the realization known as the **noncompact picture**.

First we define four subgroups of G; let

$$\overline{N} = \begin{pmatrix} 1 & 0 \\ x & 1 \end{pmatrix}, \quad M = \{I, -I\}, \quad A = \begin{pmatrix} |a| & 0 \\ 0 & \frac{1}{|a|} \end{pmatrix}, \text{ and } N = \begin{pmatrix} 1 & y \\ 0 & 1 \end{pmatrix}. \tag{5.1}$$

If $a \neq 0$ and $\epsilon = \operatorname{sgn}(a)$,

$$\begin{pmatrix} a & b \\ c & d \end{pmatrix} = \begin{pmatrix} 1 & 0 \\ \frac{c}{a} & 1 \end{pmatrix}\begin{pmatrix} \epsilon & 0 \\ 0 & \epsilon \end{pmatrix}\begin{pmatrix} |a| & 0 \\ 0 & \frac{1}{|a|} \end{pmatrix}\begin{pmatrix} 1 & \frac{b}{a} \\ 0 & 1 \end{pmatrix}. \tag{5.2}$$

This decomposition is unique and the product $\overline{N}MAN$ is a dense open submanifold of G. Choose $w \in \mathbb{C}$ and γ an irreducible M-representation. Define

$$\operatorname{ind}_{MAN}^G(\gamma \otimes e^w \otimes 1) = \{f \in C^\infty(G, \mathbb{C}) \mid f(gman) = \gamma(m)^{-1} a_{11}^{-1-w} f(g)\}, \tag{5.3}$$

where $m \in M$, $a \in A$, $n \in N$, and a_{11} is the upper left entry of a. G acts by left translation: for $g, g' \in G$,

$$[U(\gamma, w)(g)f](g') = f(g^{-1}g'). \tag{5.4}$$

The space is completed with respect to the inner product

$$\|f\|^2 = \int_T |f(\theta)|^2 \, d\theta. \tag{5.5}$$

This realization is called the **induced picture**.

Let us relate this construction to (4.1). Note that, when defined, the decomposition relative to (5.2) of

$$\begin{pmatrix} d & -b \\ -c & a \end{pmatrix} \begin{pmatrix} 1 & 0 \\ x & 1 \end{pmatrix} = \begin{pmatrix} -bx+d & -b \\ ax-c & a \end{pmatrix} \tag{5.6}$$

has $\overline{N}$-variable $\frac{ax-c}{-bx+d}$, M-component $\operatorname{sgn}(-bx+d)I$, and a_{11}-variable $|-bx+d|$.

Now define the map

$$\mathcal{R}(\gamma, w) : \operatorname{ind}_{MAN}^{G}(\gamma \otimes e^{w} \otimes 1) \to \mathcal{P}^{\varepsilon, w} \tag{5.7}$$

by

$$(\mathcal{R}(\gamma, w)F)(x) = f\begin{pmatrix} 1 & 0 \\ x & 1 \end{pmatrix}. \tag{5.8}$$

The M-representation γ is trivial when $\varepsilon = +$, and nontrivial otherwise. This map intertwines the G-actions by (5.3) and (5.6). Up to a scalar, it preserves norms.

Another equivalent realization, the **compact picture**, arises from the Iwasawa decomposition. By reinterpreting the Gram-Schmidt orthogonalization procedure in matrix terms, we see that every element of G has a unique decomposition

$$g = \kappa(g)\alpha(g)n(g) \tag{5.9}$$

associated to

$$G = TAN \qquad \text{(Iwasawa decomposition)}.$$

To get a G-representation, one takes an element of (5.3) and restricts to T; this space is completed with respect to the same norm. To get an intertwining operator, the group action is defined by

$$[U_T(\gamma, w)(g)f](k) = \alpha(g)_{11}^{-1-w} f(\kappa(g^{-1}k)). \tag{5.10}$$

Note that the space is independent of w; it is a subspace of $L^2(T)$ and can be studied using Fourier analysis. The choice of M representation leads to

$$f(\theta + \pi) = f(\theta) \qquad \text{when } \gamma \text{ is trivial} \tag{5.11}$$

and

$$f(\theta + \pi) = -f(\theta) \qquad \text{when } \gamma \text{ is nontrivial}. \tag{5.12}$$

When γ is trivial, the space is given by the space of even Fourier series, and the set of t-weight vectors is given by $f_m(\theta) = e^{im\theta}$ for m even. When γ is nontrivial, the space is given by the space of odd Fourier series and the set of t-weight vectors is given by f_m with m odd. Note that each weight space has dimension one.

6. Integral Intertwining Operators

Consider the reducibility in Section 4 for a fixed positive integer n. For example suppose n is odd. At the level of T-representations, we see that the quotient of the nonunitary principal series $\mathcal{P}^{+,n}$ by its discrete series subrepresentations $\mathcal{D}^{\varepsilon}_{n+1}$ leads to a finite-dimensional representation of dimension n (which is in fact irreducible). We note that such a representation occurs as a subrepresentation in the principal series $\mathcal{P}^{+,-n}$.

This quotient operator can be exhibited by an integral intertwining operator. In the noncompact picture, define

$$A_{\varepsilon,w} : \mathcal{P}^{\varepsilon,w} \longrightarrow \mathcal{P}^{\varepsilon,-w}$$

by

$$(A_{+,w}f)(x) = \int_{-\infty}^{\infty} \frac{f(x-y)\,dy}{|y|^{1-w}} \tag{6.1}$$

and

$$(A_{-,w}f)(x) = \int_{-\infty}^{\infty} \frac{f(x-y)\,\mathrm{sgn(y)}\,dy}{|y|^{1-w}}. \tag{6.2}$$

These convolution operators converge when $\mathrm{Re}\, w > 0$. Taking the appropriate limit as $\mathrm{Re}\, w$ approaches 0, we obtain the operators that exhibit the equivalences in (3.2). Furthermore the positivity of the complementary series norm in (3.3) can be deduced from these operators. In fact for $f \in L^2(\mathbb{R}) \cap L^1(\mathbb{R})$

$$||f||_u^2 = \langle A_{+,u}f, f\rangle_{L^2} = \frac{1}{2\pi}\int_{-\infty}^{\infty} \widehat{|x|^{u-1}}\,|\widehat{f}\,|^2\,dx, \tag{6.3}$$

where $\widehat{\ }$ denotes Fourier transform. The last equality follows from the Plancherel formula. A homogeneity argument shows that $\widehat{|x|^{u-1}}$ is a multiple of $|x|^{-u}$, which defines a tempered distribution for $u < 1$. But

$$\begin{aligned}\widehat{|\cdot|^{u-1}}(1) &= \int_{-\infty}^{\infty} |x|^{u-1}e^{-ix}\,dx \\ &= 2\int_{0}^{\infty} x^{u-1}\cos(x)\,dx.\end{aligned}$$

This integral is convergent (and positive) when $0 < u < 1$.

The corresponding quotient operators in the induced picture have wider application. By altering the spaces appropriately, these operators emerge in a natural manner. Construct

$$\mathrm{ind}_{MA\overline{N}}^{G}(\gamma \otimes e^{w} \otimes 1)$$

as in (5.3) but with invariance condition

$$f(gma\overline{n}) = \gamma(m)^{-1}a_{11}^{1-w}f(g) \tag{6.4}$$

with $\overline{n} \in \overline{N}$. Let $\omega = \begin{pmatrix} 0 & 1 \\ -1 & 0 \end{pmatrix}$. Define

$$I(N:\overline{N}:\gamma:w):\mathrm{ind}_{MA\overline{N}}^{G}(\gamma\otimes e^{w}\otimes 1)\to \mathrm{ind}_{MAN}^{G}(\gamma\otimes e^{-w}\otimes 1)$$

by

$$(I(N:\overline{N}:\gamma:w)f)(g)=f(g\omega). \tag{6.5}$$

To see that this map is an equivalence of representations, one simply needs to analyze the effect of conjugation by ω on A and $\overline{N}$. We now seek an intertwining operator

$$A(\overline{N}:N:\gamma:w):\mathrm{ind}_{MAN}^{G}(\gamma\otimes e^{w}\otimes 1)\to \mathrm{ind}_{MA\overline{N}}^{G}(\gamma\otimes e^{w}\otimes 1).$$

Since we want to produce functions that are right $\overline{N}$-invariant, the first obvious guess, which is formally correct, is

$$[A(\overline{N}:N:\gamma:w)f](g)=\int_{\overline{N}} f(g\overline{n})\,d\overline{n}, \tag{6.6}$$

where $d\overline{n}$ is the left-invariant Haar measure on $\overline{N}$, normalized to coincide with Lebesgue measure on $\mathbb{R}$ in an obvious manner.

The composite operator $A(\gamma,w) = I(N:\overline{N}:\gamma:w)A(\overline{N}:N:\gamma:w)$ is a mapping

$$A(\gamma,w):\mathrm{ind}_{MAN}^{G}(\gamma\otimes e^{w}\otimes 1)\to \mathrm{ind}_{MAN}^{G}(\gamma\otimes e^{-w}\otimes 1)$$

that is given by

$$[A(\gamma,w)f](g)=\int_{\overline{N}} f(g\omega\overline{n})\,d\overline{n}, \tag{6.7}$$

and it is the desired operator in the induced picture. When $Re\ w>0$ and the T-span of f is finite-dimensional, this integral converges. When $Re\ w=0$, further analysis leads to operators that exhibit the equivalences in (3.2) and the quotient mappings for (4.4). We refer the reader to Ch. VII of [K2], which shows how these operators fit into the Langlands classification.

As a final exercise, we show that this map has the predicted behavior for $w=n$ as above; we will compute the integral on the weight vectors f_m (which are extended to G using (5.3) and (5.9)). First m must be of the form $n+2l-1$ for some $l\in\mathbb{Z}$. Also note that the element

(6.8)

$$\begin{pmatrix} 1 & 0 \\ x & 1 \end{pmatrix} = \begin{pmatrix} \frac{1}{(1+x^2)^{1/2}} & \frac{-x}{(1+x^2)^{1/2}} \\ \frac{x}{(1+x^2)^{1/2}} & \frac{1}{(1+x^2)^{1/2}} \end{pmatrix} \begin{pmatrix} (1+x^2)^{1/2} & 0 \\ 0 & (1+x^2)^{-1/2} \end{pmatrix} n \begin{pmatrix} 1 & 0 \\ x & 1 \end{pmatrix}.$$

Thus for $\theta \in T$ and $\kappa(x) = \kappa \begin{pmatrix} 1 & 0 \\ x & 1 \end{pmatrix}$,

$$
\begin{aligned}
(6.9) \qquad [A(\gamma, n) f_m](\theta) &= \int_{\mathbb{R}} f_m(\theta\omega \begin{pmatrix} 1 & 0 \\ x & 1 \end{pmatrix}) \, dx \\
&= \int_{\mathbb{R}} f_m(\theta\omega\kappa(x)) \, (1+x^2)^{-\frac{n+1}{2}} \, dx \\
&= \int_{\mathbb{R}} f_m(\theta + \tfrac{\pi}{2}) \left(\frac{(1-ix)}{(1+x^2)^{\frac{1}{2}}} \right)^m (1+x^2)^{-\frac{n+1}{2}} \, dx \\
&= f_m(\theta) \int_{\mathbb{R}} \frac{(x+i)^{l-1}}{(x-i)^{n+l}} \, dx.
\end{aligned}
$$

We compute the last integral by a contour integration in the upper half plane. When $l \le -n$, the integrand is analytic and the integral is zero. When $l \ge 1$, the residue at i is zero and again the integral vanishes. For $-n+1 \le l \le 0$, the right-hand side is $\frac{\pi i^m}{2^{n-1}} \binom{n-1}{|l|} f_m(\theta)$.

Similar computations for general w involve gamma functions; we refer the reader to [Wa] as a starting point.

References

[Ba] Bargmann, V., Irreducible unitary representations of the Lorentz group, *Annals of Math.* **48** (1947), 568–640.

[HT] Howe, R., and E.-C. Tan, *Non-Abelian Harmonic Analysis: Applications of* $SL(2, \mathbf{R})$, Springer-Verlag, New York, 1992.

[K1] Knapp, A. W., Representations of $\mathrm{GL}_2(R)$ and $\mathrm{GL}_2(C)$, *Automorphic Forms, Representations, and L-functions*, Proc. Symp. Pure Math., vol. 33, Part I, American Mathematical Society, Providence, 1979, pp. 87–91.

[K2] Knapp, A. W., *Representation Theory of Semisimple Lie Groups: An Overview Based on Examples*, Princeton University Press, Princeton, 1986.

[KuS] Kunze, R. A., and E. M. Stein, Uniformly bounded representations and harmonic analysis of the 2×2 real unimodular group, *Amer. J. Math.* **82** (1960), 1–62.

[La] Lang, S., $SL_2(\mathbf{R})$, Addison-Wesley, Reading, Mass., 1975; second edition, Springer-Verlag, New York, 1985.

[Sa] Sally, P. J., Analytic continuation of the irreducible unitary representations of the universal covering group of $\mathrm{SL}(2, R)$, *Memoirs Amer. Math. Soc.* **69** (1967).

[Su] Sugiura, M., *Unitary Representations and Harmonic Analysis: An Introduction*, Wiley, New York, 1975.

[Wa] Wallach, N. R., Representations of reductive Lie groups, *Automorphic Forms, Representations, and L-functions*, Proc. Symp. Pure Math., vol. 33, Part I, American Mathematical Society, Providence, 1979, pp. 71–86.

School of Mathematics, Institute for Advanced Study, Princeton, NJ 08540, U.S.A.

Current address: Department of Mathematics, Massachusetts Institute of Technology, Cambridge, MA 02139, U.S.A.

E-mail address: donley@math.ias.edu

Proceedings of Symposia in Pure Mathematics
Volume **61** (1997), pp. 61–72

General Representation Theory of Real Reductive Lie Groups

M. Welleda Baldoni

0. Introduction

Lie algebras play a fundamental role in the representation theory of compact Lie groups. Such a role is best expressed by the theorem of the highest weight, which characterizes the finite-dimensional irreducible representations of such a group in terms of representations of the corresponding Lie algebra. For noncompact groups, one can try to proceed in the same way, exploiting, as much as possible, the interplay between group representations and Lie algebra representations. This approach, by the infinitesimal method, is indeed the one used by Harish-Chandra when he studied infinite-dimensional representations of semisimple Lie groups by looking at the Lie algebra action.

The passage to the noncompact case requires some care, and the distinction between differentiable vectors and analytic vectors is important. These notes will discuss the details, the final outcome being the description of a class of representations, the $(\mathfrak{g}, K)$-modules, that are the "basic" algebraic models for understanding the core of the representation theory. The material exposed is by now "classical" and appears in many texts, often as the starting point for further development in the field. The references at the end include books and one paper ([4], [12], [13], [14], [15], [17]) covering this material, as well as one of the basic papers of Harish-Chandra on the subject ([3]). Other articles ([2], [7], [10], [11]) are also mentioned because of their relevance to the the theorems as stated.

1. Representations and Differentiable Vectors

To begin we assume that G is a Lie group and we fix a left invariant measure dg on G. Let $(H, \langle \cdot, \cdot \rangle)$ be a complex, separable Hilbert space and denote by $G\ell(H)$ the group of bounded linear operators on H with bounded inverse.

We briefly recall the notion of a representation and some related concepts.

Definition 1. A **representation** π of G on H is a homomorphism from G into $G\ell(H)$ such that the map $G \times H \to H$ given by $(g, v) \to \pi(g)v$ is continuous. We often refer to the representation as (π, H).

1991 *Mathematics Subject Classification.* Primary 22E46.

If (π, H) is a representation of G, a subspace V of H is said to be **invariant** for π, or simply G-invariant, if $\pi(g)V \subset V$ for all $g \in G$. We say that (π, H) is **irreducible** if the only closed invariant subspaces are H and $\{0\}$. Also (π, H) is **unitary** if $\pi(g)$ is a unitary operator on H for all $g \in G$.

Definition 2. Given representations (π_1, H_1) and (π_2, H_2) of G, define

$$Hom_G(H_1, H_2) = \left\{ L : H_1 \to H_2 \;\middle|\; \begin{array}{l} L \text{ is continuous and linear,} \\ \pi_2(g)L = L\pi_1(g) \text{ for all } g \in G \end{array} \right\}.$$

$Hom_G(H_1, H_2)$ is the space of **intertwining operators** between π_1 and π_2. Two representations π_1 and π_2 are said to be **equivalent** if there exists an invertible intertwining operator between them, i.e., an operator L such that L and L^{-1} are both in $Hom_G(H_1, H_2)$.

Let $\mathfrak{g}$ denote a Lie algebra over $\mathbb{C}$ or $\mathbb{R}$. If V is a complex vector space, denote by $End(V)$ the space of the linear maps from V into itself.

Definition 3. A **representation** π of $\mathfrak{g}$ on V is an homomorphism (of Lie algebras) from $\mathfrak{g}$ into $End(V)$. We will say that (π, V) is a representation of $\mathfrak{g}$, or we shall simply call V a **$\mathfrak{g}$-module**. Such a representation extends uniquely to an associative algebra homomorphism of $U(\mathfrak{g})$ into $End(V)$, $U(\mathfrak{g})$ being the universal enveloping algebra of $\mathfrak{g}$.

A subspace W of V is said to be **invariant** for π, or $\mathfrak{g}$-invariant, if $\pi(X)W \subset W$ for all $X \in \mathfrak{g}$. The representation (π, V) is said to be **irreducible** if the only invariant subspaces are V and $\{0\}$.

If V and W are $\mathfrak{g}$-modules then we denote by $Hom_{\mathfrak{g}}(V, W)$ the space of $\mathfrak{g}$-module homomorphisms from V into W (i.e., all the linear maps from V to W commuting with the $\mathfrak{g}$ action). We say that V and W are **equivalent** if there exists an invertible element in $Hom_{\mathfrak{g}}(V, W)$.

We now list some properties of a representation.

Lemma 1. *Let π be a homomorphism from G into $G\ell(H)$. Then (π, H) is a representation of G if and only if π satisfies the following two conditions:*

1. *If $\Omega \subset G$ is compact, then there exists $C_\Omega < \infty$ such that $\|\pi(g)\| \leq C_\Omega$ for all $g \in \Omega$.*
2. *The map given by $g \to \langle \pi(g)v, w \rangle$ is continuous on G for all $v, w \in H$.*

PROOF. If π ia a representation of G, then conclusion (1) follows from the principle of uniform boundedness and (2) is obvious since strong continuity implies weak continuity. For the converse cf. [17], Proposition 4.2.2.1. In the reference just mentioned it is shown that only condition (2) is needed.

Given the Lie group G, denote by $\mathfrak{g}$ the corresponding Lie algebra.

If (π, H) is a representation of G with H finite-dimensional, then the following diagram commutes

$$\begin{array}{ccc} G & \xrightarrow{\ \pi\ } & G\ell(H) \\ \uparrow{\scriptstyle \exp} & & \uparrow{\scriptstyle \exp} \\ \mathfrak{g} & \xrightarrow{\ d\pi\ } & \mathfrak{gl}(H) \end{array}$$

where $d\pi(X) = \left.\frac{d}{dt}\right|_{t=0} \pi(\exp tX)$.

In other words, in the finite-dimensional case, we can pass, by differentiating, from a representation of a group to one of the corresponding Lie algebra, so that the latter carries enough information about the representation of the identity component of the group we started from. Conversely, if we have a representation ϕ of the Lie algebra then we can lift ϕ locally to a representation of G by defining $\pi(\exp X) = \exp(\phi(X))$ for $X \in \mathfrak{g}$. If G is connected and simply connected, this correspondence is a bjiection, and it is clear from the formulas that the representation π of G is irreducible if and only if the representation $d\pi$ of $\mathfrak{g}$ is irreducible. So instead of working with representations of G, one wants to work with representations of the Lie algebra, thereby linearizing problems.

Harish-Chandra used this same infinitesimal method in his approach to the study of infinite-dimensional representations. Briefly, in the infinite-dimensional case one would like, given a representation of a group, to construct a representation of the Lie algebra encoding all the relevant original information. To have a good correspondence, it will be necessary to restrict the class of representations that one considers on the group. This necessity will lead in a natural way to the concept of admissible representation and to the theory of $(\mathfrak{g}, K)$-modules. We thus proceed following the above outline.

In the infinite-dimensional case, it is not reasonable to expect that the function $g \to \pi(g)v$ is differentiable for all $v \in H$. To generalize the notion of differential of a representation, it is natural to give the following definition. Let (π, H) be a representation of G.

Definition 4. A vector v in the Hilbert space H is called a C^∞ vector, or a **differentiable vector**, if the function from G to H defined by $g \to \pi(g)v$ is of class C^∞. It is known that this is equivalent to require for each $w \in H$ that the complex-valued function $g \to \langle \pi(g)v, w\rangle$ is of class C^∞. The proof that weak differentiability implies differentiability is due to Grothendieck; for discussion and references cf. [9].

Let H^∞ be the space of differentiable vectors. We can then define, as in the finite-dimensional case,

$$\pi(X)v = \left.\frac{d}{dt}\right|_{t=0} \pi(\exp tX)v, \qquad \text{for } X \in \mathfrak{g} \text{ and } v \in H^\infty.$$

We shall see that the space of differentiable vectors is dense and that it carries a representation of the Lie algebra.

Proposition 1. *Let (π, H) be a representation of G. Then*

1. *H^∞ is G-invariant*
2. *H^∞ is $\mathfrak{g}$- invariant*
3. *$\pi[X, Y] = [\pi(X), \pi(Y)]$ on H^∞*
4. *(π, H^∞) is a representation of $\mathfrak{g}$.*

PROOF. For (1) if g_0 is in G and v is a differentiable vector, then the map $g \to \pi(gg_0)v$ is C^∞ as a function of g, being the composition of C^∞ functions.

For (2) let $v \in H^\infty$, and for $X \in \mathfrak{g}$ denote by $\widetilde{X}$ the left invariant vector field corresponding to X. The function $f_v(g) = \pi(g)v$ is differentiable on G and satisfies $\pi(g)\pi(X)v = (\widetilde{X} f_v)(g)$. Because both $\widetilde{X}$ and f_v are differentiable, (2) follows.

For (3) one can make a direct computation using integral curves or use ideas from (2) and proceed as follows. From the proof of (2), we have $\pi(g)\pi(X)v = (\widetilde{X}f_v)(g)$. Putting $g = \exp tY$ with $Y \in \mathfrak{g}$, we obtain $\pi(\exp tY)\pi(X)v = (\widetilde{X}f_v)(\exp tY)$. In particular by (2)

$$\pi(Y)\pi(X)v = \frac{d}{dt}\bigg|_{t=0} \pi\,(\exp tY)\pi(X)v = \frac{d}{dt}\bigg|_{t=0} (\widetilde{X}f_v)(\exp tY) = \widetilde{Y}(\widetilde{X}f_v)(e).$$

Interchanging Y and X and subtracting, we obtain

$$\pi(X)\pi(Y)v - \pi(Y)\pi(X)v = (\widetilde{X}(\widetilde{Y}f_v) - (\widetilde{Y}(\widetilde{X}f_v))(e).$$

For $\lambda \in H^*$, we have $\lambda \cdot (\widetilde{[X,Y]}\, f_v) = \widetilde{[X,Y]}\,(\lambda \cdot f_v) = (\widetilde{X}\widetilde{Y} - \widetilde{Y}\widetilde{X})\,(\lambda \cdot f_v) = \lambda \cdot (\widetilde{X}\widetilde{Y} - \widetilde{Y}\widetilde{X})\,(f_v)$. Thus $\widetilde{[X,Y]}\, f_v = (\widetilde{X}\widetilde{Y} - \widetilde{Y}\widetilde{X})\, f_v$,. Evaluating both sides at the identity, we obtain (3). Then (4) is now obvious.

So far we have succeeded in passing from a representation of G to a representation of $\mathfrak{g}$, but we have changed the space. Although it turns out that the space of differentiable vectors has the nice property of being dense, the correspondence between group representations on Hilbert spaces and Lie algebra representations on spaces of differentiable vectors is not good, as the next example shows.

Example. Let $G = \mathbb{R}$, and let $(\pi, L^2(\mathbb{R}))$ be the regular representation of G. Set $V = \{f \in C_c^\infty \mid \text{supp } f \subset [0,1]\}$. Then V is $\mathfrak{g}$-invariant, but neither V nor its closure is G-invariant.

Definition 5. Let (π, H) be a representation of G. For v and w in H and f in $C_c(G)$, define $\pi(f)$ by

$$\langle \pi(f)v, w\rangle = \int_G f(g)\langle \pi(g)v, w\rangle\, dg.$$

Note that $\pi(f)$ is well defined since $|\langle \pi(f)v, w\rangle| \le C_\Omega \|v\| \|w\| \|f\|_1$ if Ω is any compact subset of G containing supp f. Here C_Ω is given by (1) of Lemma 1, and $\|f\|_1$ is the L^1 norm of f. In particular for $w = \pi(f)v$, we obtain $\|\pi(f)v\| \le C_\Omega \|v\| \|f\|_1$, which shows that $\pi(f)$ is in $End(H)$.

If U is an open subset of G with compact closure $\overline{U}$, then define

$$L^1(U) = \{f \in L^1(G) \mid \text{supp } f \subset U\}.$$

The above discussion shows that π defines a continuous linear map of $L^1(U)$ into $End(H)$ with the property that $\|\pi(f)\| \le C_{\overline{U}} \|f\|_1$.

Lemma 2 (Gårding).

1. *If $f \in C_c^\infty$, then $\pi(f)v$ is in H^∞ for any $v \in H$*
2. *The Gårding subspace, i.e., the linear span of $\{\pi(f)v \mid f \in C_c^\infty$ and $v \in H\}$, is dense in H*
3. *H^∞ is dense in H.*

PROOF. For (1) one easily computes that $\pi(g)\pi(f)v = \pi(L_g f)v$, L_g being the left regular representation of G. Since f is C^∞ and has compact support, we can differentiate under the integral sign in the formula

$$\langle \pi(L_g f)v, w\rangle = \int_G f(g^{-1}x)\langle \pi(x)v, w\rangle \, dx,$$

and we see that $\pi(g)\pi(f)v$ is of class C^∞ in g.

To prove (2) let $v \in H$ and $\epsilon > 0$. By strong continuity, we may find a compact neighborhood V of $g = e$ in G with $\|\pi(g)v - v\| < \epsilon$, for $g \in V$. Choose a C^∞ function $f \geq 0$ such that $\operatorname{supp} f \subset V$ and $\int_G f(g)\, dg = 1$. Then

$$\|\pi(f)v - v\| = \left\|\int_G f(g)(\pi(g)v - v)\, dg\right\| \leq \int_V f(g)\|\pi(g)v - v\|\, dg < \epsilon.$$

Hence v can be approximated by vectors in the Gårding subspace, as was to be proved.

Finally (3) is a consequence of (1) and (2).

NOTE. Dixmier and Malliavin proved in [1] that the Gårding subspace is exactly H^∞.

As we have seen, (π, H^∞) is still not good enough if we want the closure of a $\mathfrak{g}$-invariant subspace to be G-invariant. We are indeed missing one of the important features that connects representations of Lie groups and Lie algebras in the finite dimensional case. To restore this correspondence we introduce the $\mathfrak{g}$-invariant subspace of H^∞ given by the K-finite C^∞ vectors, where K is a compact subgroup of G. Even if this subspace will almost never be G-invariant, nevertheless it has many analytic vectors when π is admissible, and this is the main ingredient for the correspondence to which we alluded before. To proceed we need to recall some results for compact groups and to set up some further notation.

2. $(\mathfrak{g}, K)$-modules

We assume for the moment that G is compact. Let

$$\widehat{G} = \{\text{classes of finite-dimensional irreducible unitary representations of } G\}.$$

If $\gamma \in \widehat{G}$, denote by (τ_γ, V_γ) an element in the class of γ. Set $d_\gamma = \dim V_\gamma$, $\chi_\gamma(g) = tr\, \tau_\gamma(g)$, and $\alpha_\gamma(g) = d_\gamma \bar{\chi}_\gamma(g)$. If (π, H) is a representation of the compact group G, let $H(\gamma)$ be the closure of the sum of the irreducible subspaces of type γ; $H(\gamma)$ is called the γ **isotypic component** of H. Put $E_\gamma = \pi(\alpha_\gamma)$.

Theorem 1. *Let (π, H) be a unitary representation of a compact group G. Then*

1. $H(\gamma) = E_\gamma(H)$
2. $H = \overline{\bigoplus_{\gamma \in \widehat{G}}} H(\gamma)$ *(Hilbert space direct sum)*

PROOF. For γ and γ' in $\widehat{G}$, one easily checks the following:

(a) $E_\gamma E_{\gamma'} = E_{\gamma'} E_\gamma = 0$ if $\gamma \neq \gamma'$
(b) $E_\gamma^2 = E_\gamma$
(c) $\langle E_\gamma v, w\rangle = \langle v, E_\gamma w\rangle$
(d) $\pi(g)E_\gamma v = E_\gamma \pi(g)v$
(e) if $v \in V_\gamma$, then $E_\gamma v = v$.

In particular E_γ is an orthogonal projection.

For (1) we have $H(\gamma) \subset E_\gamma(H)$ by (e). Thus we need to prove that $E_\gamma(H) \subset H(\gamma)$. Let $v \in E_\gamma(H)$, and let $P_\gamma = L(\alpha_\gamma)$ be the projection according to the type γ in the left regular representation L of G on $L^2(G)$. For $w \in H$, consider the function defined by $f_{v,w}(g) = \langle \pi(g^{-1})v, w \rangle$. One computes that $(P_\gamma f_{v,w})(g) = f_{E_\gamma v,w}(g) = f_{v,w}(g)$. Thus $f_{v,w}$ is in $P_\gamma(L^2(G))$, i.e., is in the isotypic component of type γ. It is known that $P_\gamma(L^2(G)) = V_\gamma \otimes V_\gamma^*$, where V_γ^* is the contragradient representation. This function space has finite dimension d_γ^2. Let Z be the linear span of $\{\pi(g)v,\ g \in G\}$. For $z = \pi(g)v \in Z$, $f_{z,w} = L_g f_{v,w}$ is in $P_\gamma(L^2(G))$. Hence $\dim\{\text{span}\{f_{z,w} \mid z \in Z,\ w \in H\}\} \le d_\gamma^2$. Consider the map from H to $P_\gamma(L^2(G))$ defined by $w \to \lambda_w$, where $\lambda_w = f_{v,w}$. Since $\lambda_w = 0$ if and only if $w \in Z^\perp$, we obtain $\dim(H/Z^\perp) \le d_\gamma^2$. Thus $\dim \overline{Z} < \infty$. This implies that $\dim Z < \infty$. Decomposing Z into irreducibles, one sees that only γ types occur in Z. Thus the result follows.

To prove (2), use (1) and the fact that every unitary representation of a compact group is completely reducible (a consequence of the Peter-Weyl Theorem), together with (a), (b), and (c).

We now turn to the general setting, with G an arbitrary Lie group. Let K be a compact subgroup. If (π, H) is a representation of G, we let (π_K, H) be the representation of K defined by restriction: $\pi_K(k) = \pi(k)$ for $k \in K$. By integrating over K the function $\langle \pi(k)v, \pi(k)w \rangle$ one obtains a new inner product on H that gives the same topology and with respect to which the action of K is unitary. We may thus assume, without loss of generality, that (π_K, H) is unitary (cf. [15], 1.4.8). Then, because of Theorem 1, we may decompose H according to K as a Hilbert space direct sum:

$$H = \overline{\bigoplus_{\gamma \in \widehat{K}}} H(\gamma) \qquad \text{with } H(\gamma) = E_\gamma(H).$$

The representation of $\mathfrak{g}$ to which we alluded at the end of the previous section will be on the *algebraic* direct sum

$$H_F = \bigoplus_{\gamma \in \widehat{K}} H(\gamma) \cap H^\infty.$$

To prove that $\mathfrak{g}$ acts, we proceed as follows.

Proposition 2.

1. $E_\gamma H^\infty \subset H^\infty$
2. $E_\gamma H^\infty = H(\gamma) \cap H^\infty$
3. $\overline{E_\gamma H^\infty} = H(\gamma) = E_\gamma(H)$.

PROOF. (1) follows by standard results on differentiation under the integral sign, and (2) is a simple calculation. Finally (3) is a consequence of the density of H^∞ in H and the boundness of E_γ.

Definition 6. A vector $v \in H$ is said to be K-**finite** if the linear span of $\{\pi(k)v \mid k \in K\}$ is finite-dimensional.

Write H_F for the set of K-finite differentiable vectors.

Theorem 2.

1. $H_F = \bigoplus_{\gamma \in \widehat{K}} E_\gamma H^\infty$
2. *H_F is dense in H*
3. *(π, H_F) is a $\mathfrak{g}$-module under the action restricted from H^∞*
4. *In terminology to be defined below, (π, H_F) is a $(\mathfrak{g}, K)$-module.*

PROOF. Let $A = \bigoplus_{\gamma \in \widehat{K}} E(\gamma)H^\infty$. The inclusion $H_F \subset A$ is easy. Indeed let v be in H_F. Then by the K-finiteness of v one can assume that v is in some V_γ, i.e., $v = E_\gamma v$. Thus $H_F \subset A$. To prove the reverse inclusion, let v be in A. We may assume that v is in $E_\gamma H^\infty$. Because of (1) in Proposition 2, we need to prove that v is K-finite, i.e., that $W = \sum_{k \in K} \mathbb{C}\pi(k)v$ is finite-dimensional. For $w \in H$, consider the function $f_{v,w}(k) = \langle \pi(k^{-1})v, w \rangle$. Since v is in $E_\gamma H^\infty$, $f_{v,w}$ is in $C^\infty(K)$. Arguing as in the proof of (2) in Theorem 1, we find that $f_{v,w} \in P_\gamma(L^2(K))$, and hence W is finite-dimensional.

So v is K-finite, and this proves (1). Conclusion (2) is obvious by what we have already seen. For (3) we are to prove that $H_F \subset H^\infty$ is a $\mathfrak{g}$-invariant subspace, i.e., that $\pi(X)H_F \subset H_F$ for $X \in \mathfrak{g}$. For $v \in H_F$ and $V = \text{span}\{\pi(k)v \mid k \in K\}$, consider the linear map $B : \mathfrak{g} \otimes V \to H^\infty$ given by $X \otimes w \mapsto \pi(X)w$. Then

$$\pi(k)(B(X \otimes w)) = \pi(k)(\pi(X)w) = \frac{d}{dt}\bigg|_{t=0} \pi(k \exp tX)w$$

$$= \frac{d}{dt}\bigg|_{t=0} \pi(\exp tAd(k)X)\pi(k)w = \pi(Ad(k)X)\pi(k)w = B(Ad(k)X \otimes \pi(k)w),$$

where $Ad(k)X \in \mathfrak{g}$ and $\pi(k)w \in V$. So B commutes with the action of K, and we see that $\pi(k)(\pi(X)w)$ lies in the finite-dimensional space $B(\mathfrak{g} \otimes V)$.

Remark 1. The results of section 1 and Theorem 2 are still valid (cf. [16], Th. 3.1 and cf. [17], Th. 4.4.3.1) for Fréchet representations, i.e., representations π of G on a Fréchet space V. Here by **Fréchet space** we mean a complete locally convex linear topological space defined by a countable separating family of seminorms. An example of a Fréchet representation is (π, H^∞), with (π, H) a representation of G. The family of seminorms for H^∞ is defined as follows. If $X \in U(\mathfrak{g}_\mathbb{C})$, $\mathfrak{g}_\mathbb{C}$ being the complexification of $\mathfrak{g}$, define the seminorm $p_X(v) = ||\pi(X)v||$, for $v \in H$. If $X_1, \ldots.., X_n$ is a basis of $\mathfrak{g}_\mathbb{C}$, then $\{X_{i_1} \ldots.. X_{i_k} \mid 1 \le i_1 \le \ldots. \le i_k \le n\}$ form a basis of $\mathfrak{g}_\mathbb{C}$ and H^∞ is a Fréchet space with the topology induced by the seminorms $p_{X_{i_1} \ldots.. X_{i_k}}$. To conclude we would like to mention that Harish-Chandra wrote his paper [3] in the setting of Banach space representations.

Remark 2. On $V = H_F$ we have two actions: a $\mathfrak{g}$ action as above and a K action by restriction of the original action of G. These two actions are compatible, as in the following definition. We drop π in the notation for convenience.

Definition 7. Let G be a real Lie group with Lie algebra $\mathfrak{g}$, let K be a compact subgroup of G, and let $\mathfrak{k}$ be the Lie algebra of K. If V is a complex vector space with a Lie algebra representation of $\mathfrak{g}$ and a group representation of K, then V is called a **$(\mathfrak{g}, K)$-module** if these actions satisfy the following three compatibility conditions:

1. $k \cdot (X \cdot v) = (Ad(k)X) \cdot k \cdot v$ for $k \in K$, $X \in \mathfrak{g}$, $v \in H_F$

2. For all $v \in H_F$, $\{k \cdot v \mid k \in K\}$ spans a finite-dimensional space W_v and the action on K on W_v is continuous, hence C^∞.
3. For all $Y \in \mathfrak{k}$ and $v \in H_F$, $\left.\frac{d}{dt}\right|_{t=0} (\exp tY) \cdot v = Y \cdot v$.

In the case of $V = H_F$, property (1) was verified in the proof of Theorem 2.

For $(\mathfrak{g}, K)$-modules we can make the usual definitions in representation theory of invariant subspaces and so on, with everything defined completely in terms of algebra. For instance, **invariant subspace** means invariant for $\mathfrak{g}$ and for K, and **irreducible** means having no proper invariant subspaces for both $\mathfrak{g}$ and K. If V and W are $(\mathfrak{g}, K)$-modules, then we denote by $Hom_{\mathfrak{g},K}(V,W)$ the space of $\mathfrak{g}$ homomorphisms that are also K homomorphisms, and the notion of **equivalence** is with respect to $Hom_{\mathfrak{g},K}(V,W)$. We will also say that a $(\mathfrak{g}, K)$-module is **unitary** if there exists a positive definite Hermitian form on the space of the action on which $\mathfrak{g}$ has a skew Hermitian action and K a unitary action. Finally we say that (π_1, H_1) is **infinitesimally equivalent** to (π_2, H_2) if the corresponding $(\mathfrak{g}, K)$-modules are equivalent. If K is connected, the role of K is really limited in the above; for instance $(\mathfrak{g}, K)$-module maps are the same as $\mathfrak{g}$-module maps.

Remark 3.

1. Any $(\mathfrak{g}, K)$-module decomposes as a K-module as
$$V = \bigoplus_{\gamma \in \widehat{K}} E_\gamma V.$$
$V(\gamma) = E_\gamma V$ is called the **isotypic component** of type γ.
2. If (π, H) is a representation of G, then (π, H_F) is a $(\mathfrak{g}, K)$-module.

Definition 8. A representation (π, H) of G is called **admissible** if $\dim H(\gamma) < \infty$ for every $\gamma \in \widehat{K}$. Similarly a $(\mathfrak{g}, K)$-module V is **admissible** if $\dim V(\gamma) < \infty$ for every $\gamma \in \widehat{K}$.

We call (π, H_F) the **underlying** $(\mathfrak{g}, K)$**-module** of (π, H). Note that the group representation (π, H) is admissible if and only if (π, H_F) is admissible as a $(\mathfrak{g}, K)$-module.

The class of representations for which we have a good correspondence is the class of admissible representations. To have a rich supply of admissible representations, we shall impose further hypotheses on G and K.

3. General Theory for Real Reductive Lie Groups

For the remaining part we assume that G is real reductive Lie group in the Harish-Chandra class, and we let K be a maximal compact subgroup. For the definition we refer to [5] in this volume. An example of such a group is a connected semisimple Lie group with finite center.

In this situation the class of admissible representations contains many interesting representations as the next theorem shows.

Theorem 3. *The irreducible unitary representations and the representations induced from parabolic subgroups by admissible representations of Levi subgroups are admissible.*

For the first part of Theorem 3 cf. [15], Th. 3.4.10. The second statement is an immediate consequence of the Frobenius reciprocity theorem (cf. the discussion following 11.42 in [6]). We will be using the admissibility of irreducible unitary representations in the proof of Theorem 5.

To proceed with our analysis, we need to introduce the concept of analytic vectors.

Definition 9. If (π, H) is a representation of G, then we say that $v \in H$ is an **analytic vector** if the function $g \to \langle \pi(g)v, w\rangle$ is real analytic for all $w \in H$.

This notion, as for differentiable vectors, agrees with other standard terminology since weak analyticity implies strong analyticity (cf. [9]).

Let H^ω be the set of analytic vectors.

Remark 4. Harish-Chandra proved that the analytic vectors are dense in H. Unlike the case with the Gårding lemma (Lemma 2), the proof is quite difficult this time, because the representation is not assumed to be admissible. By contrast the proof is relatively easy when admissibility is assumed (see Theorem 4 below). The fact that the set of analytic vectors is dense was later generalized by Nelson [8] to the class of all connected Lie groups.

One can also prove a completely analogous theorem to Theorem 2 with H_F replaced by the set of K-finite analytic vectors and with H^∞ replaced by H^ω. Moreover H^ω is G-invariant.

We now proceed to prove

Theorem 4. *Let (π, H) be an admissible representation of G. Then*

1. *Every K-finite vector is differentiable*
2. *Every K-finite vector is analytic.*

Proof. For (1) recall that $\overline{E_\gamma H^\infty} = E_\gamma H = H(\gamma)$. Thus the C^∞ vectors in each isotypic component are dense in the isotypic component. But this space is finite-dimensional, and hence $E_\gamma H^\infty = H(\gamma)$. Now if v is K-finite, then $\{\pi(k)v \mid v \in K\}$ spans a finite-dimensional space. By decomposing it into irreducibles, we may assume that v is in $H(\gamma)$, for some $\gamma \in \widehat{K}$. Then the result follows.

For (2) we need to show that the function $f(g) = \langle \pi(g)v, w\rangle$ is real analytic. The proof relies on the fact that this function is annihilated by an elliptic differential operator with real analytic coefficients. If X is in $\mathfrak{g}$ and we regard X as acting as a left invariant vector field, then

$$X\langle \pi(g)v, w\rangle = \left.\frac{d}{dt}\right|_{t=0} \langle \pi(g \exp tX)v, w\rangle.$$

So

$$(*) \qquad D\langle \pi(g)v, w\rangle = \langle \pi(g)\pi(D)v, w\rangle \qquad \text{for all } D \in U(\mathfrak{g}).$$

In particular all the derivatives of f at $g = e$ can be computed from the formula

$$D(f)(e) = \langle \pi(D)v, w\rangle,$$

i.e., in terms of the action of $U(\mathfrak{g})$. Let now v be in H_F. We may assume, without loss of generality, that v is in some isotypic component, say $H(\gamma)$ for some $\gamma \in \widehat{K}$. If Ω is the Casimir operator for $\mathfrak{g}$, then $\pi(\Omega)$ preserves the isotypic components of each type, since it commutes with $\pi(K)$. Since $H(\gamma)$ is finite-dimensional, we conclude that there exists a monic polynomial p such that $p(\pi(\Omega))$ is zero on $H(\gamma)$. By Schur's lemma, applied to the K irreducible representation V_γ, we conclude that the Casimir operator $\pi(\Omega_K)$ of K acts on $H(\gamma)$ as a scalar.

Set $D = \Omega - 2\Omega_K$. Then D is an elliptic differential operator with real analytic coefficients. By the above considerations we may conclude that there exists a monic polynomial q such that $q(\pi(D))$ is zero on $H(\gamma)$. By $(*)$ one immediately obtains $q(\pi(D))f = 0$. The analytic elliptic regularity theorem then implies the result.

Theorem 4 implies that if π is admissible, then H_F consists of all the K-finite vectors.

Theorem 5. *Let (π, H) be an admissible representation of G, and let (π, H_F) be the underlying $(\mathfrak{g}, K)$-module. Then there is a one-one correspondence*

$$\{\textit{closed } G\textit{-invariant subspaces of } H\} \longleftrightarrow \{\mathfrak{g} \textit{ and } K\textit{-invariant subspaces of } H_F\}$$

given by

$$U \longrightarrow U \cap H_F \qquad \textit{and} \qquad \overline{W} \longleftarrow W.$$

PROOF. We will prove that if $W \subset H_F$ is a $(\mathfrak{g}, K)$-invariant subspace then its closure is G-invariant, all the rest being fairly straightforward. To prove that $\pi(g)\overline{W} \subset \overline{W}$, it is enough to show that $\pi(g)W \subset \overline{W}$ for all $g \in G$. If we denote by G^0 the connected component of the identity of G, then $G = KG^0$ and the inclusion will follow if we show that $\pi(g)W \subset \overline{W}$ for all $g \in G^0$. So we may assume that G is connected. Let u be in W, and let v be in $\overline{W}^\perp$. The function $f(g) = \langle \pi(g)u, v\rangle$ is real analytic, since u is in H_F. Expanding it as the sum of its Taylor series about the identity, we obtain:

$$\langle \pi(\exp X)u, v\rangle = \sum_{k=0}^{\infty} \frac{1}{k!}\,(X^k\langle \pi(g)u, v\rangle)_{g=e}$$

for X sufficiently small in $\mathfrak{g}$. Thus, by $(*)$,

$$\langle \pi(\exp X)u, v\rangle = \sum \frac{1}{k!}\,\langle \pi(X^k)u, v\rangle.$$

The right side is zero since W is $\mathfrak{g}$-invariant and v is in $\overline{W}^\perp$. Thus $\langle \pi(g)u, v\rangle$ vanishes in a neighborhood of the identity in G. Being real analytic on a connected set, it vanishes everywhere. So $\pi(g)u$ is in $(\overline{W}^\perp)^\perp = \overline{W}$.

Remark 5. The same argument applies to show that in the space of analytic vectors any $\mathfrak{g}$-invariant subspace has G-invariant closure if G is connected.

From Theorem 5 we immediately obtain:

Corollary 1. *Let (π, H) be an admissible representation of G. Then (π, H) is irreducible if and only if its underlying $(\mathfrak{g}, K)$-module (π, H_F) is irreducible.*

Theorem 6. *Let (π_1, H_1) and (π_2, H_2) be irreducible unitary representations of G. Then π_1 and π_2 are infinitesimally equivalent if and only if they are unitarily equivalent.*

PROOF. Necessity is trivial. Consider the sufficiency: Let T be an invertible element in $Hom_{(\mathfrak{g},K)}((H_1)_F, (H_2)_F)$.

Since T is a K intertwining map and a bijection, we readily check that $T(H_1(\gamma)) = H_2(\gamma)$, i.e., T sends each isotypic component onto the corresponding isotypic component. Because we are dealing with admissible representations, each isotypic component is finite-dimensional. We can thus define $T^* : (H_2)_F \to (H_1)_F$ by requiring that on each isotypic component T^* be the adjoint of T, i.e.

$$\langle Tv, w\rangle = \langle v, T^*w\rangle \qquad \text{for } K\text{-finite } v \text{ and } w.$$

Then T^* is also a $(\mathfrak{g}, K)$ map. Arguing as before, we see that T^*T stabilizes each isotypic component and therefore it must have a nonzero eigenvalue. Therefore, the usual argument of Schur's Lemma shows that, up to a multiplicative constant for T, we may assume that T^*T is the identity on $(H_1)_F$. It follows that T is the restriction to $(H_1)_F$ of a unitary isomorphism of H_1 with H_2 that intertwines the action of K. Finally we observe that the functions $\langle \pi_1(g)u, w\rangle$ and $\langle T^{-1}\pi_2(g)Tu, w\rangle$ for u and w in $(H_1)_F$ are real analytic, coincide at $g = e$, and have the same derivatives at $g = e$ (use $(*)$ to compute the derivatives). Therefore they are identical on G^0. Since $G = KG^0$, T is the required unitary equivalence.

We conclude with the following theorem that completes this brief exposition and should complete the idea that on one side the irreducible admissible representations and on the other side the irreducible $(\mathfrak{g}, K)$-modules are basic objects to study in representation theory. In fact, the theorem characterizes the irreducible $(\mathfrak{g}, K)$-modules as the underlying modules of the irreducible admissible representations, and, within them, the irreducible unitary $(\mathfrak{g}, K)$-modules as the underlying modules of the irreducible unitary representations.

Theorem 7.

1. *Every irreducible $(\mathfrak{g}, K)$-module is the underlying $(\mathfrak{g}, K)$-module of an irreducible admissible representation.*
2. *Let V be a unitary irreducible $(\mathfrak{g}, K)$-module. Then there exists an irreducible unitary representation π of G on a Hilbert space H such that H_F is equivalent to V. The representation π is unique up to unitary equivalence.*

Uniqueness in the situation of (2) was given in Theorem 5.

References

1. J. Dixmier and P. Malliavin, Factorisations de fonctions et de vecteurs indéfiniment différentiables, *Bull. des Sci. Math.* **102** (1978), 305–330.
2. L. Gårding, Note on continuous representations of Lie groups, *Proc. Nat. Acad. Sci. USA* **33** (1947), 331–332.
3. Harish-Chandra, Representations of a semisimple Lie group on a Banach space I, *Trans. Amer. Math. Soc.* **75** (1953), 185–243.
4. A. W. Knapp, *Representation Theory of Semisimple Groups: An Overview Based on Examples*, Princeton University Press, Princeton, 1986.
5. A. W. Knapp, Structure theory of semisimple Lie groups, *these* PROCEEDINGS, pp. 1–27.
6. A. W. Knapp and D. A. Vogan, *Cohomological Induction and Unitary Representations*, Princeton University Press, Princeton, 1995.
7. J. Lepowsky, Algebraic results on representations of semisimple Lie groups, *Trans. Amer. Math. Soc.* **176** (1973), 1-44.
8. E. Nelson, Analytic vectors, *Annals of Math.* **70** (1959), 572–615.

9. J. B. Neto, Spaces of vector valued real analytic functions, *Trans. Amer. Math. Soc.* **112** (1964), 381–391.
10. I. E. Segal, A class of operator algebras which are determined by groups, *Duke Math. J.* **18** (1951), 221–265.
11. I. E. Segal, Hypermaximality of certain operators on Lie groups, *Proc. Amer. Math. Soc.* **3** (1952), 13–15.
12. V. S. Varadarajan, *Harmonic Analysis on Real Reductive Groups*, Lecture Notes in Math., Vol. 576, Springer-Verlag, Berlin, 1977.
13. V. S. Varadarajan., *An Introduction to Harmonic Analysis on Semisimple Lie Groups*, Cambridge University Press, Cambridge, 1989.
14. N. R. Wallach, Representations of semisimple Lie groups and Lie algebras, *Lie Theories and Their Applications, Proceedings of the 1977 Annual Seminar of the Canadian Mathematical Congress*, Queen's Papers in Pure and Applied Mathematics, No. 48, Queen's University, Kingston, Ontario, 1978, pp. 154–245.
15. N. R. Wallach, *Real Reductive Groups I*, Pure and applied Mathematics, Vol. 132, Academic Press Inc., Boston, 1988.
16. N. R. Wallach, C^∞-vectors, *Representations of Lie Groups and Quantum Groups*, Pitman Research Notes in Mathematics, vol. 311, Longman Scientific & Technical, Harlow, UK, 1994, pp. 205–270.
17. G. Warner, *Harmonic Analysis on Semi-Simple Lie Groups I*, Springer-Verlag, New York, 1972.

Dipartimento di Matematica, Università di Roma "Tor Vergata," Via della Ricerca Scientifica, 00133 Roma, Italy
E-mail address: baldoni@mat.utovrm.it

Proceedings of Symposia in Pure Mathematics
Volume **61** (1997), pp. 73–81

Infinitesimal Character and Distribution Character of Representations of Reductive Lie Groups

Patrick Delorme

1. Infinitesimal Character: Definition

Let G be a Lie group with Lie algebra $\mathfrak{g}$, (π, V) a continuous representation of G on a complete, locally convex, complex linear topological space. Let V^∞ be the space of smooth vectors of this representation; i.e., V^∞ is the space of elements v of V such that the map $g \mapsto \pi(g)v$ from G to V is smooth. This linear space has a natural topology, which we will use in the sequel.

The lectures [B] by M. W. Baldoni in this volume discuss smooth vectors in detail. The Lie algebra $\mathfrak{g}$ and the universal enveloping algebra $U(\mathfrak{g}_\mathbb{C})$ of $\mathfrak{g}_\mathbb{C}$ act on V^∞ via differentiation of the action of G. The corresponding representation is still denoted by π.

Definition 1. The representation π is said to have an **infinitesimal character** if, for any element Z of the center $Z(\mathfrak{g}_\mathbb{C})$ of $U(\mathfrak{g}_\mathbb{C})$, $\pi(Z)$ acts by a scalar $\chi(Z)$ on V^∞. Then χ defines a character of $Z(\mathfrak{g}_\mathbb{C})$, i.e., a morphism from the algebra $Z(\mathfrak{g}_\mathbb{C})$ with unit into $\mathbb{C}$. This is the **infinitesimal character** of π.

The algebra $Z(\mathfrak{g}_\mathbb{C})$ is an interesting algebra that is often larger than its subalgebra $U(\mathfrak{z}_\mathbb{C})$, where $\mathfrak{z}$ denotes the center of $\mathfrak{g}$. For $\mathfrak{g}$ semisimple, $\mathfrak{z}$ is reduced to $\{0\}$ but $Z(\mathfrak{g}_\mathbb{C})$ always contains the Casimir element.

Definition 1 extends similarly to representations of the Lie algebra that do not come from group representations.

Theorem 1.

(i) *If V is an irreducible $\mathfrak{g}$-module over $\mathbb{C}$, it has an infinitesimal character.*

(ii) *If (π, V) is a unitary irreducible representation of G, with G connected, then π has an infinitesimal character.*

References. For (i), cf. [Di], Proposition 2.6.8. For (ii), cf. [War], Corollary 4.4.1.6.

1991 *Mathematics Subject Classification.* Primary 22E45.

2. Center of the Enveloping Algebra of a Complex Semisimple Lie Algebra

Let $\mathfrak{g}$ be a complex semisimple Lie algebra, $\mathfrak{h}$ a Cartan subalgebra of $\mathfrak{g}$, $\Delta(\mathfrak{g},\mathfrak{h})$ the root system of $\mathfrak{h}$ in $\mathfrak{g}$, and W the Weyl group of $\Delta(\mathfrak{g},\mathfrak{h})$. More precisely $\Delta(\mathfrak{g},\mathfrak{h})$ is the set of nonzero elements α of the dual $\mathfrak{h}^*$ of $\mathfrak{h}$ for which there exists a nonzero element X_α of $\mathfrak{g}$ such that, for all H in $\mathfrak{h}$, one has

$$[H, X_\alpha] = \alpha(H)X_\alpha.$$

The union of a basis $H_1, \dots, H_l$ of $\mathfrak{h}$ with the set of X_α, $\alpha \in \Delta(\mathfrak{g},\mathfrak{h})$, gives a basis of $\mathfrak{g}$. Choose a set of positive roots of $\Delta(\mathfrak{g},\mathfrak{h})$, $\Delta^+(\mathfrak{g},\mathfrak{h})$ (or Δ^+), and let the positive roots be $\alpha_1, \dots, \alpha_n$. The theorem of Poincaré-Birkhoff-Witt (cf. [Di], Theorem 2.1.11) asserts that the monomials

$$X_{-\alpha_1}^{q_1} \cdots X_{-\alpha_n}^{q_n} H_1^{m_1} \cdots H_l^{m_l} X_{\alpha_1}^{p_1} \cdots X_{\alpha_n}^{p_n} \tag{1}$$

form a basis of $U(\mathfrak{g})$ when the indices q_i, m_j, p_k vary through integers ≥ 0.

Thus $U(\mathfrak{h})$ may be viewed as a subalgebra of $U(\mathfrak{g})$. If $U(\mathfrak{g})\mathfrak{n}$ is the left ideal of $U(\mathfrak{g})$ generated by the subalgebra $\mathfrak{n} := \sum_{\alpha\in\Delta^+(\mathfrak{g},\mathfrak{h})} \mathbb{C}X_\alpha$ of $\mathfrak{g}$, one gets easily

$$U(\mathfrak{g})\mathfrak{n} = \sum_{\alpha\in\Delta^+(\mathfrak{g},\mathfrak{h})} U(\mathfrak{g})X_\alpha. \tag{2}$$

Recall that a representation (π, V) of $\mathfrak{g}$ is said to be a representation of **highest weight** $\lambda \in \mathfrak{h}^*$ with respect to Δ^+ if there is a nonzero element v of V such that:

$$\pi(U(\mathfrak{g}))v = V \tag{3}$$

$$\pi(H)v = \lambda(H)v, \quad H \in \mathfrak{h} \tag{4}$$

$$\pi(X)v = 0, \quad X \in \mathfrak{n}. \tag{5}$$

Lemma 1.

(i) *The space $U(\mathfrak{h}) \cap U(\mathfrak{g})\mathfrak{n}$ is reduced to $\{0\}$.*
(ii) *$Z(\mathfrak{g})$ is contained in $U(\mathfrak{h}) \oplus U(\mathfrak{g})\mathfrak{n}$.*

PROOF. Let (π, V) be a finite-dimensional representation of $\mathfrak{g}$ with highest weight λ. Notice from the theory of finite-dimensional representations of $\mathfrak{g}$ that such a representation exists if and only if λ is an integral dominant weight (cf. [Di], Chapter 7, §2). Let v be a nonzero element of V satisfying (4) and (5). Then $\pi(U(\mathfrak{g})\mathfrak{n})v = 0$.

On the other hand, as $\mathfrak{h}$ is commutative, $U(\mathfrak{h})$ is the symmetric algebra of $\mathfrak{h}$, and we may identify $U(\mathfrak{h})$ with the algebra of polynomial functions on $\mathfrak{h}^*$. With this identification, one sees easily that every element D of $U(\mathfrak{h})$ satisfies $\pi(D)v_0 = D(\lambda)v_0$ (start with D in $\mathfrak{h}$). Hence, if $D \in U(\mathfrak{h}) \cap U(\mathfrak{g})\mathfrak{n}$, one has $D(\lambda) = 0$ for all λ in the set of integral dominant weights. But this set is Zariski dense in $\mathfrak{h}^*$ (in rank one, this reduces to the fact that the nonnegative integers are Zariski dense in $\mathbb{C}$). Thus $D = 0$, which proves (i).

Now let Z be an element of $Z(\mathfrak{g})$, and expand it in our basis (1) of $U(\mathfrak{g})$. Writing that $[H, Z] = 0$ for any $H \in \mathfrak{h}$, one sees that, for each nonzero term in this expansion, one has $\sum_i p_i\alpha_i - \sum_i q_i\alpha_i = 0$. This comes from the computation of the bracket of H with the elements of the basis. Each basis vector is an eigenvector under this bracket, and only those corresponding to zero eigenvalues can actually contribute to our expansion of Z. For the corresponding terms, there is a nonzero p_i if and only if there is a nonzero q_j, and in this case the terms are elements of $U(\mathfrak{g})\mathfrak{n}$. The other terms are in $U(\mathfrak{h})$. This finishes the proof of the lemma.

Using the preceeding lemma, one defines a linear map $p_0^{\Delta^+} : Z(\mathfrak{g}) \to U(\mathfrak{h})$ by projecting on $U(\mathfrak{h})$ along $U(\mathfrak{g})\mathfrak{n}$. One defines also an automorphism τ_{Δ^+} of the algebra $S(\mathfrak{h})$ by setting:

$$\tau_{\Delta^+}(D)(\lambda) := D(\lambda - \rho) \qquad \text{for } \lambda \in \mathfrak{h}^*, \tag{6}$$

where ρ is the half sum of the elements in Δ^+.

Theorem 2.

(i) *The linear map $\gamma := \tau_{\Delta^+} \circ p_0^{\Delta^+}$ from $Z(\mathfrak{g})$ to $S(\mathfrak{h})$ is an algebra isomorphism onto the subalgebra $S(\mathfrak{h})^W$ of invariant elements of $S(\mathfrak{h})$ under W. Moreover γ is independent of the choice of Δ^+. (It is called the* **Harish-Chandra isomorphism** *relative to $\mathfrak{h}$ and is sometimes denoted by $\gamma_\mathfrak{h}$ or by $\gamma_{\mathfrak{g},\mathfrak{h}}$.)*

(ii) *Let (π, V) be representation of $\mathfrak{g}$ with highest weight $\lambda - \rho$, where λ is an element of $\mathfrak{h}^*$. Then (π, V) has an infinitesimal character χ_λ (denoted also $\chi_\lambda^\mathfrak{h}$), where χ_λ is defined by:*

$$\chi_\lambda(Z) = \gamma(Z)(\lambda) \qquad \textit{for } Z \in Z(\mathfrak{g}).$$

(iii) *All characters of $Z(\mathfrak{g})$ are of the form χ_λ for some $\lambda \in \mathfrak{h}^*$.*

(iv) *The equality $\chi_\lambda = \chi_\mu$ holds if and only if λ is an element of the W-orbit of μ in $\mathfrak{h}^*$.*

REFERENCES. Cf. [Di], Chapter 7, §4, or [Kn], Chapter 8, §5.

Remark 1.

(i) The algebra $S(\mathfrak{h})^W$ is isomorphic with the algebra of polynomials over $\mathbb{C}$ with $l = \dim \mathfrak{h}$ variables. This a direct consequence of a theorem of Chevalley (cf. [War], Theorem 2.1.3.1)

(ii) The theorem extends to complex reductive Lie algebras.

3. Infinitesimal Character of Generalized Principal Series of a Reductive Lie Group

In the following G will denote, unless otherwise mentioned, a linear connected reductive Lie group, i.e., a closed connected group of real or complex matrices that is stable under conjugate transpose.

But all the results that we will present are true for more general classes of groups like the Harish-Chandra class (see [H-C] for a precise definition) or the class of real reductive Lie groups of [Wal], §2.1.1.

Let K be a maximal compact subgroup of G.

Definition 2. An **admissible representation** π of G in a Hilbert space H_π is a continuous representation in H_π such that

(i) the restriction of π to K, denoted $\pi_{|K}$, is unitary and
(ii) the multiplicity of any irreducible representation of K in $\pi_{|K}$ is finite.

Admissible representations are introduced in [B] in this volume.

For an admissible representation (π, H_π) of G, the space of K-finite vectors, $(H_\pi)_{(K)}$, is the space of elements v of H_π such that the family $\pi(k)v, k \in K$, spans a finite-dimensional subspace of H_π. This space is a $\mathfrak{g}$ and K invariant subspace of the space of C^∞-vectors of H_π. With the action of both $\mathfrak{g}$ and K, $(H_\pi)_{(K)}$ is called the **underlying** $(\mathfrak{g}, K)$**-module** of π.

Theorem 3.

(i) *Irreducible continuous unitary representation of G are admissible.*

(ii) *Admissible irreducible representations (i.e., without nontrivial closed invariant subspaces) have an infinitesimal character.*

REFERENCE. For (i), cf. [Kn], Theorem 8.1. For (ii), cf. [Kn], Corollary 8.14.

Let MAN be the Langlands decomposition of a parabolic subgroup P of G. Let $\rho_P \in \mathfrak{a}^*$ be defined by $\rho_P(X) := \frac{1}{2}Tr(ad\, X_{|\mathfrak{n}})$, $X \in \mathfrak{a}$. If (δ, V_δ) is a unitary irreducible representation of M and $\lambda \in \mathfrak{a}^*_{\mathbb{C}}$, one defines the smooth **generalized principal series** as follows. This is a representation $(\pi_{\delta,\lambda}, I_{\delta,\lambda})$ of G with

$$I_{\delta,\lambda} := \left\{ \varphi : G \to V_\delta^\infty \;\middle|\; \begin{array}{l} \varphi \text{ is } C^\infty \text{ and } \varphi(gman) = a^{-\lambda-\rho_P}\delta(m^{-1})\varphi(g) \\ \text{for } g \in G,\, m \in M,\, a \in A,\, n \in N \end{array} \right\} \tag{7}$$

$$(\pi_{\delta,\lambda}(g)\varphi)(x) := \varphi(g^{-1}x) \qquad \text{for } \varphi \in I_{\delta,\lambda} \text{ and } g, x \in G. \tag{8}$$

Equip $I_{\delta,\lambda}$ with the prehilbertian norm

$$\|\varphi\|^2 := \int_K \|\varphi(k)\|^2_{V_\delta}\, dk. \tag{9}$$

Then G acts continuously on the completion $H_{\delta,\lambda}$ of $I_{\delta,\lambda}$ with respect to this norm, by a representation denoted by $\overline{\pi}_{\delta,\lambda}$. One has $I_{\delta,\lambda} = H^\infty_{\delta,\lambda}$.

Theorem 4.

(i) *The representation $\overline{\pi}_{\delta,\lambda}$ of G is admissible.*

(ii) *Let $\mathfrak{h}_M$ be a Cartan subalgebra of the complexified Lie algebra $\mathfrak{m}_{\mathbb{C}}$ of the Lie algebra $\mathfrak{m}$ of M. Suppose that δ has an infinitesimal character $\chi_\Lambda^{\mathfrak{h}_M}$ for some $\Lambda \in \mathfrak{h}_M^*$. Then $\mathfrak{h} := \mathfrak{h}_M \oplus \mathfrak{a}_{\mathbb{C}}$ is a Cartan subalgebra of $\mathfrak{g}_{\mathbb{C}}$, and $\overline{\pi}_{\delta,\lambda}$ and $\pi_{\delta,\lambda}$ have $\chi^{\mathfrak{h}}_{\Lambda+\lambda}$ as infinitesimal character.*

REFERENCES. For (i) (resp. (ii)) cf. [Kn], Proposition 8.4 (resp. Proposition 8.22).

SKETCH OF PROOF OF (ii) FOR G SPLIT AND P MINIMAL. Suppose that G is split and P is a minimal parabolic subgroup of G. We recall the description of $U(\mathfrak{g}_{\mathbb{C}})$ as the convolution algebra of (complex-valued) distributions on G supported at the neutral element e of G. If X is an element of $\mathfrak{g}$, the corresponding distribution $\widetilde{X}$ is defined by $f \mapsto (d/dt)(f(\exp tX))_{|t=0}$. From this point of view, the action of X on a smooth function F as a right-invariant differential operator is given by the convolution $\widetilde{X} * F$. More precisely one has $\widetilde{X} * F(g) = (d/dt)F((\exp -tX)g)_{|t=0}$. A similar formula holds when F is replaced by a distribution. The action of elements of $U(\mathfrak{g}_{\mathbb{C}})$ as right-invariant differential operators on G acting on distributions can also be described by convolution.

Now, our hypothesis that G is split and P is a minimal parabolic subgroup of G is equivalent to $\mathfrak{m} = \{0\}$. Thus $\mathfrak{a}_{\mathbb{C}}$ is a Cartan subalgebra of $\mathfrak{g}_{\mathbb{C}}$ and Λ is equal to zero. Moreover, as G is connected and linear, M is abelian and δ is one dimensional. Let $\Delta^+(\mathfrak{g}_{\mathbb{C}}, \mathfrak{a}_{\mathbb{C}})$ (or simply Δ^+) be the set of the opposites of weights of $\mathfrak{a}_{\mathbb{C}}$ in $\mathfrak{n}_{\mathbb{C}}$, and denote by $\mathfrak{n}_{\mathbb{C}}^-$ the sum of the corresponding weight spaces in $\mathfrak{g}_{\mathbb{C}}$.

Let Z be an element of $U(\mathfrak{g}_{\mathbb{C}})$ and φ an element of $I_{\delta,\lambda}$. Then one has $\pi_{\delta,\lambda}(Z))\varphi = \widetilde{Z} * \varphi$. Now assume that $Z \in Z(\mathfrak{g}_{\mathbb{C}})$. This implies that convolution by $\widetilde{Z}$ commutes with left translations. Thus, one sees that $\pi_{\delta,\lambda}(Z)\varphi = \varphi * \widetilde{Z}$.

The definition of $p_0^{\Delta^+}$, after Lemma 1, and the proof of this lemma shows that $Z = p_0^{\Delta^+}(Z) + Z_1$ with $p_0^{\Delta^+}(Z) \in S(\mathfrak{a}_{\mathbb{C}})$ and $Z_1 \in U(\mathfrak{g}_{\mathbb{C}})\mathfrak{n}_{\mathbb{C}}^- \cap \mathfrak{n}_{\mathbb{C}}U(\mathfrak{g}_{\mathbb{C}})$. Hence $\pi_{\delta,\lambda}(Z)\varphi$ is equal to the sum of $\varphi * (p_0^{\Delta^+}(Z))^\sim$ with $\varphi * (Z_1)^\sim$. But φ is right invariant by $\mathfrak{n}$. Using the associativity of the convolution product, we see that the second term is zero. Now $\varphi * (p_0^{\Delta^+}(Z))^\sim$ is easily computed from the covariance property of φ under the right action of A. Together with the definition of γ and χ_λ, this finishes the proof of (ii).

4. Distribution Character

A **trace class operator** T on a Hilbert space H is a bounded linear operator such that for some orthonormal basis (v_i) and for every bounded bijective linear operator B in H, $\sum_i |(B^{-1}TBv_i, v_i)| < \infty$. In this case the sum is finite for every orthonormal basis. Then $Tr\,T := \sum_i (Tv_i, v_i)$ does not depend on the orthonormal basis and is called the **trace** of T.

The trace class operators form a two-sided ideal of the algebra of bounded operators on H.

The following criterion will be used later:

> If an operator on H satisfies $\sum_{i,j} |(Tv_i, v_j)| < \infty$ for one orthonormal basis (v_i), it is of trace class.

A detailed discussion of trace class operators may be found in [L].

Definition 3. A (not necessarily unitary) continuous representation π of a Lie group G on a Hilbert space H is said to have a **distribution** (or **global**) **character** if and only if

(i) for every element f of $C_c^\infty(G)$, the operator $\pi(f)$ is of trace class and
(ii) the map $f \mapsto Tr(\pi(f))$ is a distribution on G (denoted by Θ_π and called the **distribution character** of π).

Then Θ_π is invariant under conjugacy, i.e., $\Theta_\pi(f) = \Theta_\pi(f^g)$ for $g \in G$, where $f^g(x) = f(gxg^{-1})$ for $x \in G$. In fact $\pi(f^g) = \pi(g)\pi(f)\pi(g^{-1})$, and the trace of both sides of this equality are equal.

For the remainder of the paper, we return to our assumption that G is a connected linear reductive group.

Theorem 5. *Every irreducible admissible representation of G has a distribution character. More generally, if the underlying $(\mathfrak{g}, K)$-module of an admissible representation π of G has finite length, the representation π has a distribution character.*

REFERENCES. Cf [Kn], Theorem 10.2, or [Wal], Chapter 8, §1.

SKETCH OF PROOF. For each element Z of $Z(\mathfrak{k}_\mathbb{C})$ and for each unitary irreducible representation μ of K, Z acts by multiplication by a scalar that we denote $\mu(Z)$. Fix two elements Z and Z' of $Z(\mathfrak{k}_\mathbb{C})$. Let v (resp. v') be a unit vector in the isotypic component of type μ (resp. μ'), and let f be an element of $C_c^\infty(G)$. Then, using the fact that the orthogonal projection on every isotypic component commutes with $\pi(Z)$ and $\pi(Z')$, one has:

$$(\pi(\widetilde{Z'} * f * \widetilde{Z})v, v') = \mu(Z)\mu'(Z')(\pi(f)v, v').$$

The left hand side is bounded by the norm of the operator $\pi(\widetilde{Z'} * f * \widetilde{Z})$. Choose Z such that $\mu(Z)$ grows rapidly with the norm of a highest weight of μ (use Theorem 2 to exhibit Z). Choose Z' in the same way. The preceding equation gives a bound on $|(\pi(f)v, v')|$ independent of v and v' in the given isotypic components. The bound depends only on the norm of the highest weights, the support of f, and the supremum of finitely many derivatives of f. Then, the fact that $\pi(f)$ is of trace class follows from a bound on the multiplicities of the irreducible representations of K in π.

Theorem 6. *If π and π' are admissible irreducible representations of G, their distribution characters are equal if and only if the underlying $(\mathfrak{g}, K)$-modules are isomorphic.*

REFERENCES. Cf. [Kn], Proposition 10.5 and Theorem 10.6, or [Wal], Chapter 8.

Remark 2. This theorem points out that what is canonical is the underlying $(\mathfrak{g}, K)$-module. There are also canonical G-modules attached functorially to a $(\mathfrak{g}, K)$-module of finite length (cf. [C], [S], and [Wal], Chapter 11).

We have seen that distribution characters are invariant, and they have another important and elementary property. If an admissible representation π has an infinitesimal character χ and a distribution character Θ, the equality $\pi(Zf) = \chi(Z)\pi(f)$ for $Z \in Z(\mathfrak{g}_\mathbb{C})$ and $f \in C_c^\infty(G)$ implies that $\Theta(Zf) = \chi(Z)\Theta(f)$. Thus Θ is also an eigendistribution under $Z(\mathfrak{g}_\mathbb{C})$.

5. Invariant Eigendistributions

Let G^{reg} be the set of regular elements in G. This is the set of elements x of G such that $Ad\, x$ has the eigenvalue 1 with the minimal multiplicity (which is equal to the dimension of any Cartan subalgebra of $\mathfrak{g}$). Recall that a **Cartan subalgebra** of $\mathfrak{g}$ is a subalgebra whose complexification is a Cartan subalgebra of $\mathfrak{g}_\mathbb{C}$. A Cartan subgroup of G is the centralizer in G of a Cartan subalgebra of $\mathfrak{g}$. If H is a Cartan subgroup of G, the Weyl group $W(G, H)$ of H in G is the quotient of the normalizer of H in G by its centralizer. Denoting by $Car\, G$ a set of representatives of conjugacy classes of Cartan subgroups of G, and, for $H \in Car\, G$, defining $H^{reg} := H \cap G^{reg}$, one has:

$$G^{reg} = \bigcup_{H \in Car G,\, g \in G} g H^{reg} g^{-1}. \tag{10}$$

Let H be an element of $Car\, G$, let $(H^{reg})^G$ be the union of the conjugates of H^{reg}, and consider the map $p : G/H \times H^{reg} \to (H^{reg})^G$ defined by $(gH, h) \mapsto ghg^{-1}$.

This map is a local diffeomorphism onto the open set $(H^{reg})^G$. For each element y of $(H^{reg})^G$, $p^{-1}(y)$ is an orbit of $W(G,H)$.

If Θ is an invariant distribution on G, Θ lifts to a distribution $\widetilde{\Theta}$ on $G/H\times(H^{reg})$ by the formula $\widetilde{\Theta}(F)=\Theta(\varphi)$ for $F\in C_c^\infty(G/H\times(H^{reg}))$, where φ is the element of $C_c^\infty((H^{reg})^G)$ defined by

$$\varphi(y)=Card(W(G,H))^{-1}\sum_{x\in p^{-1}(y)}F(x).$$

But p is G-equivariant when G acts on $G/H\times H^{reg}$ by left translation on the first factor and on $(H^{reg})^G$ by conjugacy. The distribution Θ being invariant, the distribution $\widetilde{\Theta}$ is invariant by this action of G on $G/H\times H^{reg}$. Then it is easy to see that

$$\widetilde{\Theta}=1\otimes\Theta_H,$$

where 1 is the constant function on G/H (identified with a distribution on G/H with the help of a measure on G/H left invariant by G) and Θ_H is a distribution on H^{reg} invariant under the action of $W(G,H)$. We call Θ_H the **restriction** of Θ to H^{reg}. When Θ is given by an invariant smooth function, Θ_H is given by the restriction of this function to H^{reg}.

Assume moreover that Θ is an eigendistribution under $Z(\mathfrak{g}_\mathbb{C})$. Then Θ satisfies partial differential equations, and this implies that Θ_H satisfies corresponding "radial" partial differential equations analogous to an equation from the radial part of the Laplacian. If the dimension of H is one, there is essentially one equation coming from the Casimir element of $\mathfrak{g}$, and explicit computations show that Θ_H is annihilated by a differential operator of order two with real analytic coefficients, the coefficient in front of the second order derivative never vanishing (ellipticity). More generally, one gets a system with a similar property of ellipticity. This implies

Theorem 7. *The restriction to the set of regular elements of every Cartan subgroup of G of every invariant eigendistribution on G is a real analytic function.*

REFERENCES. Cf. [Kn], Chapter 10, §§4, 5, 6, or [Wal], Chapter 8, §4. See also [V].

Let Θ be an invariant eigendistribution on G. From (10) and (11) one sees that the restriction of Θ to the dense open set G^{reg} is determined by Θ_H for $H\in Car\,G$.

The radial differential equations can be described in terms of the Harish-Chandra isomorphism, and this implies a more precise form of Θ_H. Fix a set of positive roots of the system of roots of $\mathfrak{h}_\mathbb{C}$ in $\mathfrak{g}_\mathbb{C}$. Let ρ be the half sum of the positive roots and define:

$$D_H(h)=h^\rho\prod_{\alpha>0}(1-h^{-\alpha})\qquad\text{for } h\in H.$$

Here $h^{-\alpha}$ is the scalar by which h is acting on the space of weight $-\alpha$. To define h^ρ correctly, it is necesary to make further assumptions. For example, one can assume that G is semisimple and is contained in a connected and simply connected Lie group $G_\mathbb{C}$ with Lie algebra $\mathfrak{g}_\mathbb{C}$. From now on, we will make this assumption.

Theorem 8. *Let H be in $Car\,G$. Let Θ be an invariant eigendistribution on G for the eigencharacter χ_λ of $Z(\mathfrak{g}_\mathbb{C})$, where λ is in $\mathfrak{h}_\mathbb{C}^*$. Let h be an element of H and let C be a connected component of the set $\{X\in\mathfrak{h}\mid h\exp X\in G^{reg}\}$. For every* *element w of the Weyl group $W(\mathfrak{g}_\mathbb{C}, \mathfrak{h}_\mathbb{C})$ of the system of roots of $\mathfrak{h}_\mathbb{C}$ in $\mathfrak{g}_\mathbb{C}$, there exists a polynomial $p_{w,C}$ on $\mathfrak{h}$ such that*

$$\Theta_H(h \exp X)) = (D_H(h \exp X))^{-1} \sum_{w \in W(\mathfrak{g}_\mathbb{C}, \mathfrak{h}_\mathbb{C})} p_{w,C}(X) e^{w\lambda(X)} \quad \textit{for } X \in C.$$

Moreover the degree of every $p_{w,C}$ is strictly less than the cardinal of the stabilizer of λ in $W(\mathfrak{g}_\mathbb{C}, \mathfrak{h}_\mathbb{C})$.

REFERENCES. Cf. [Kn], Chapter 10, §7, or [Wal]. See also [V].

COMMENT. The radial differential equations imply that the real analytic function on C given by $X \mapsto u(X) := (D_H \Theta_H)(h \exp X)$ satisfies

$$\gamma_{\mathfrak{h}_\mathbb{C}}(Z) u = \chi_\lambda(Z) u \qquad \text{for } Z \in Z(\mathfrak{g}_\mathbb{C}),$$

where the elements of $S(\mathfrak{h}_\mathbb{C})$ are viewed as differential operators on $\mathfrak{h}$ with constant coefficients. The theorem follows easily from the theory of differential equations with constant coefficients.

The following important and difficult theorem shows that the restriction of Θ to G^{reg} determines Θ completely.

Theorem 9. *Every invariant eigendistribution on G is a locally integrable function on G.*

REFERENCES. This is a deep theorem of Harish-Chandra. Cf. [Wal], Chapter 8, §§3, 4, or [V], Part II, Chapter 4.

6. Distribution Character of Generalized Principal Series

We retain the notation of §3. The distribution character of $\overline{\pi}_{\delta,\lambda}$ exists, essentially in view of Theorem 4(i) and Theorem 5. There is a way to compute this character in terms of the distribution character of δ. The starting point is to realize the principal series in a space of functions on K with values in V_δ by restricting the elements of $H_{\delta,\lambda}$ to K. In this (compact) realization, $\overline{\pi}_{\delta,\lambda}(f)$, for $f \in C_c^\infty(G)$, appears as an integral operator given by a smooth kernel $\mathcal{K}$ on K (i.e., a smooth function on $K \times K$) with values in trace class operators on V_δ. More precisely the operator associates to every element φ of our compact realization, the function on K with values in V_δ given by $k \mapsto \int_K \mathcal{K}(k,x)\varphi(x)\,dx$. Here dx is a Haar measure on K. Then the trace of $\overline{\pi}_{\delta,\lambda}(f)$ is given by the integral $\int_K Tr(\mathcal{K}(x,x))\,dx$. Use of various integral formulae, including a generalization of the Weyl integral formula for compact Lie groups, leads to an explicit formula for the trace.

Here is an example. We assume that P is a minimal parabolic subgroup of G. Let B be a Cartan subgroup of M. Then $J := BA$ is a Cartan subgroup of G and its Lie algebra $\mathfrak{j}$ satisfies $\mathfrak{j} = \mathfrak{b} \oplus \mathfrak{a}$. Choose a set Δ^+ of positive roots of $\mathfrak{j}_\mathbb{C}$ in $\mathfrak{g}_\mathbb{C}$ such that the restriction to $\mathfrak{a}$ of every positive root is either zero or is a weight of $\mathfrak{a}$ in $\mathfrak{n}$. We use this set of positive roots to define D_J. Let Δ_R^+ be the set of positive roots that are real on $\mathfrak{j}$ (i.e., zero on $\mathfrak{b}$, as M is compact). Define

$$D'_{J,R}(j) = \prod_{\alpha \in \Delta_R^+} (1 - j^{-\alpha}), \qquad j \in J,$$

$$\varepsilon_{J,R}(j) = \operatorname{sgn} D'_{J,R}(j), \qquad j \in J^{reg}.$$

Using the set of positive roots Δ_M^+ of $\mathfrak{j}$ in $\mathfrak{m} \oplus \mathfrak{a}$, given by the set of roots in Δ^+ that are zero on $\mathfrak{a}$, one defines similarly functions on J, D_J^{MA}, and $\varepsilon_{J,R}^{MA}$. If $j = ba$ with $b \in B$ and $a \in A$, define

$$(\varepsilon_R D)_J^{G/MA}(j) := a^\rho j^{-\rho} j^{\rho_M} (\varepsilon_{J,R} D_J / \varepsilon_{J,R}^{MA} D_J^{MA})(j).$$

Here ρ_M is the half sum of the elements of Δ_M^+ and j^{ρ_M} has been defined like $j^{-\rho}$. If δ is an element of the set $\widehat{M}$ of equivalence classes of unitary irreducible representations of M (called the **unitary dual** of M), we denote its distribution character by Θ_δ. As M is compact, δ is finite dimensional and Θ_δ is a continuous function on M. The normalizer of A in K, $N_K(A)$, normalizes M, hence acts on the unitary dual of M. As the centralizer of A in K, $Z_K(A)$, is equal to M, this action goes through the quotient to an action of $W(G,A) = N_K(A)/Z_K(A)$. Also $W(G,A)$ acts on $\mathfrak{a}_{\mathbb{C}}^*$.

Theorem 10. *Let λ be an element of $\mathfrak{a}_{\mathbb{C}}^*$, let δ be a unitary irreducible representation of M, and let $[\delta]$ be the equivalence class of δ in $\widehat{M}$. Then the distribution character of the principal series $\overline{\pi}_{\delta,\lambda}$ is given by the locally integrable function $\Theta_{\overline{\pi}_{\delta,\lambda}}$ on G characterized by its conjugacy invariance and by*

$$\Theta_{\overline{\pi}_{\delta,\lambda}}(g) = 0 \text{ if } g \text{ is not conjugate to an element of } J,$$

$$\Theta_{\overline{\pi}_{\delta,\lambda}}(j) = \frac{\sum_{w \in W(G,A)} a^{w\lambda} \Theta_{w[\delta]}(b)}{(\varepsilon_R D)_J^{G/MA}(j)} \quad \text{for } a \in A,\ b \in B, \text{ and } j = ba \in J^{reg}.$$

REFERENCE. Cf. [Kn], Proposition 10.18.

The reader will find many interesting historical notes in the books given as references.

References

[B] M. W. Baldoni, General representation theory of real reductive Lie groups, *these* PROCEEDINGS, pp. 61–72.

[C] W. Casselman, Canonical extensions of Harish-Chandra modules to representations of G, *Canad. J. Math.* **41** (1989), 385–438.

[Di] J. Dixmier, *Enveloping Algebras*, North-Holland, Amsterdam, 1977.

[H-C] Harish-Chandra, Harmonic analysis on real reductive groups, I, *J. Func. Anal.* **19** (1975), 104–204.

[Kn] A. W. Knapp, *Representation Theory of Semisimple Lie Groups: An Overview Based on Examples*, Princeton University Press, Princeton, 1986.

[L] S. Lang, $SL_2(R)$, Addison-Wesley, Reading, Mass., 1975; second edition, Springer-Verlag, New York, 1985.

[S] W. Schmid, Boundary value problems for group invariant differential equations, *Élie Cartan et les mathématiques d'aujourd'hui (Lyon, juin 1984)*, Astérisque, Numéro hors série, 1985, pp. 311–321.

[V] V. S. Varadarajan, *Harmonic Analysis on Real Reductive Groups*, Lecture Notes in Mathematics, vol. 576, Springer-Verlag, Berlin, 1977.

[Wal] N. Wallach, *Real Reductive Groups*, vol. I, II, Academic Press, Boston, 1988, 1992.

[War] G. Warner, *Harmonic Analysis on Semi-Simple Lie Groups*, vol. I, II, Springer-Verlag, Berlin, 1972.

INSTITUT DE MATHÉMATIQUES DE LUMINY, U.P.R. 9016 DU C.N.R.S., FACULTÉ DES SCIENCES DE LUMINY, 163 AVENUE DE LUMINY, CASE 930, 13288 MARSEILLE CEDEX 09, FRANCE

E-mail address: delorme@iml.univ-mrs.fr

Proceedings of Symposia in Pure Mathematics
Volume **61** (1997), pp. 83–113

Discrete Series

Wilfried Schmid

Notes by Vernon Bolton

This article contains notes of five lectures on a common theme: the geometric realization of representations. The topics we cover include the Borel-Weil Theorem, the Borel-Weil-Bott Theorem, discrete series, realizations in L^2 cohomology, and realizations in sheaf cohomology.

The first named author would like to thank the note-taker for his diligent transcription of the lecture notes and his compilation of the biblography.

1. Borel-Weil Theorem

The Borel-Weil Theorem gives a geometric realization of each irreducible representation of a compact connected semisimple Lie group, or equivalently, of each irreducible holomorphic representation of a complex connected semisimple Lie group. The realization is in the space of holomorphic sections of a holomorphic line bundle over the flag variety of the group. This first lecture explains the underlying geometric notions, frames the Borel-Weil Theorem as a special case of the Borel-Weil-Bott Theorem, and proves the special case. For an illustration of the Borel-Weil Theorem, see Donley's lecture [Do2].

Throughout let $\mathfrak{g}$ be a complex semisimple Lie algebra, and let G be a connected complex Lie group with Lie algebra $\mathfrak{g}$.

A **Borel subalgebra** is a maximal solvable subalgebra of $\mathfrak{g}$. All Borel subalgebras $\mathfrak{b}$ are of the form $\mathfrak{b} = \mathfrak{h} \oplus \mathfrak{n}$, where $\mathfrak{h}$ is a Cartan subalgebra of $\mathfrak{g}$, $\mathfrak{n}$ is $\bigoplus_{\alpha\in\Phi^+} \mathfrak{g}^{-\alpha}$, Φ^+ is a system of positive roots of $\mathfrak{h}$ in $\mathfrak{g}$, and $\mathfrak{g}^{-\alpha}$ is the root space for the root $-\alpha$. Any two Borel subalgebras are conjugate via $\mathrm{Ad}(G)$. In fact, from the point of view of Lie groups, it is known that any two Cartan subalgebras of $\mathfrak{g}$ are conjugate via $\mathrm{Ad}(G)$ and that each element of the Weyl group of the root system has a representative in G; thus the conjugacy follows from the formula $\mathfrak{b} = \mathfrak{h} \oplus \mathfrak{n}$. Alternatively, from the point of view of algebraic groups, the conjugacy of Borel subgroups – and therefore also the conjugacy of Borel subalgebras – is a basic fact

---

1991 *Mathematics Subject Classification.* Primary 22E45, 22E46.
Supported in part by an NSF grant.

([Bor2, Theorem 11.1]), from which the conjugacy of Cartan subgroups and Cartan subalgebras can be deduced.

To define the notion of Borel subgroups, let us consider a particular Borel subalgebra $\mathfrak{b} \subset \mathfrak{g}$. Its normalizer in G,

$$B = N_G(\mathfrak{b}) = \{x \in G \mid \mathrm{Ad}(x)\mathfrak{b} \subseteq \mathfrak{b}\},$$

is connected and has Lie algebra $\mathfrak{b}$. Groups of this type are called **Borel subgroups** of G. It should be remarked that the connectedness of Borel subgroups depends crucially on the assumption that the ambient group G is complex.

As a set, the **flag variety** X of $\mathfrak{g}$ is the collection of all Borel subalgebras of $\mathfrak{g}$. The solvable subalgebras of a given dimension constitute a closed subvariety in a Grassmannian for $\mathfrak{g}$, hence X has a natural structure of complex projective variety. Since any two Borel subalgebras are conjugate via $\mathrm{Ad}(G)$, G acts transitively on X, with isotropy group $N_G(\mathfrak{b}) = B$ at the point at $\mathfrak{b}$. Consequently we may make the identification $X \simeq G/B$. Every complex algebraic variety is smooth (i.e., nonsingular) outside a proper subvariety. But G acts transitively on X, so the flag variety cannot have any singularities: it is a smooth complex projective variety.

Example. Let $\mathfrak{g} = \mathfrak{sl}(n, \mathbb{C})$. Then X is (naturally isomorphic to) the variety of all complete flags in $\mathbb{C}^n$, i.e., nested sequences of linear subspaces of $\mathbb{C}^n$, one in each dimension:

$$X \simeq \{F = (F_j) \mid 0 \subset F_1 \subset F_2 \subset \cdots \subset F_n = \mathbb{C}^n \text{ and } \dim F_j = j\}.$$

To see this, we assign to the complete flag $F = (F_j)$ its stabilizer in $\mathfrak{sl}(n, \mathbb{C})$, which turns out to be a Borel subalgebra $\mathfrak{b}$; this can be checked by looking at any particular flag F, since any two are conjugate under the action of $G = GL(n, \mathbb{C})$. Using the transitivity of the G-action on the set of complete flags once more, we get the identification between this set and $G/N_G(\mathfrak{b}) \simeq G/B \simeq X$.

In the discussion of the Borel-Weil theorem below, we shall regard the smooth projective variety X as a compact complex manifold and use terminology and methods from complex analysis (see [We], for example, for background). One could equally well work in the algebraic category, but the analytic setting will help to bring out the analogy between the compact and noncompact cases.

Fix a particular Borel subalgebra of the form $\mathfrak{b} = \mathfrak{h} \oplus \mathfrak{n}$ with $\mathfrak{n} = \bigoplus_{\alpha \in \Phi^+} \mathfrak{g}^{-\alpha}$, and again let $B = N_G(\mathfrak{b})$. The centralizer $H = Z_G(\mathfrak{h})$ is a Cartan subgroup of G. It is connected since G is complex. Define

$$\widehat{H} = \mathrm{Hom}_{holo}(H, \mathbb{C}^\times),$$

the group of holomorphic homomorphisms from H to the multiplicative group $\mathbb{C}^\times$. It is an abelian group, which we identify with the **weight lattice** $\Lambda \subset \mathfrak{h}^*$, i.e., the lattice of linear functionals on $\mathfrak{h}$ whose values on the **unit lattice**

$$L = \{Z \in \mathfrak{h} \mid \exp(Z) = e\}$$

are integral multiples of $2\pi i$. Explicitly, the identification $\Lambda \simeq \widehat{H}$ is given by

$$\lambda \leftrightarrow e^\lambda, \qquad \text{with } e^\lambda(\exp Z) = e^{\langle \lambda, Z \rangle} \text{ for } Z \in \mathfrak{h};$$

here $\langle \lambda, Z \rangle$ refers to the canonical pairing between $\mathfrak{h}^*$ and $\mathfrak{h}$.

Each member e^λ of $\widehat{H}$ lifts to a holomorphic character $e^\lambda : B \to \mathbb{C}^\times$ via the isomorphism $H \simeq B/[B,B]$. Consider the fiber product $\mathcal{L}_\lambda = G \times_B \mathbb{C}_\lambda$, where $\mathbb{C}_\lambda$ denotes $\mathbb{C}$, equipped with the B-action via the character e^λ; by definition, the fiber product is the quotient $\mathcal{L}_\lambda = G \times \mathbb{C}_\lambda / \sim$ under the equivalence relation $(gb, z) \sim (g, e^\lambda(b)z)$. The natural projection $G \times \mathbb{C}_\lambda \to G$ induces a well defined G-equivariant holomorphic map $\mathcal{L}_\lambda \to G/B \simeq X$, which exhibits $\mathcal{L}_\lambda$ as a G-equivariant holomorphic line bundle over X – i.e., a holomorphic line bundle with a holomorphic G-action (by bundle maps) that lies over the action of G on the base space X.

Let $\mathcal{O}(\mathcal{L}_\lambda)$ be the sheaf of germs of holomorphic sections of this line bundle. The cohomology $H^*(X, \mathcal{O}(\mathcal{L}_\lambda))$ is a graded, finite-dimensional complex vector space (finite-dimensional since X is compact) with a linear G action. The resulting representation is holomorphic, since it is induced by holomorphic actions of G on X and on the line bundle $\mathcal{L}_\lambda$.

Let ρ be half the sum of the members of Φ^+. Then 2ρ lies in the weight lattice Λ, and ρ itself lies in Λ if G is simply connected. The line bundle $\mathcal{L}_{-2\rho}$ is the **canonical bundle** on X, whose local holomorphic sections are the top-degree holomorphic differential forms. For λ in Λ, define

$$p(\lambda + \rho) = \#\{\alpha \in \Phi^+ \mid (\lambda + \rho, \alpha) < 0\},$$

where $(\,\cdot\,,\,\cdot\,)$ is the bilinear form on $\mathfrak{h}^*$ induced by the Killing form.

Theorem (Borel-Weil-Bott).

1) *$H^p(X, \mathcal{O}(\mathcal{L}_\lambda)) = 0$ if either $\lambda + \rho$ is singular or $p \neq p(\lambda + \rho)$.*
2) *If $\lambda + \rho$ is regular and $p = p(\lambda + \rho)$, then the holomorphic representation of G on $H^p(X, \mathcal{O}(\mathcal{L}_\lambda))$ is nonzero and irreducible of highest weight $w(\lambda + \rho) - \rho$, where w is the unique member of the Weyl group that makes $w(\lambda + \rho)$ dominant.*

REMARKS.

1) The first assertion implies in particular that $H^0(X, \mathcal{O}(\mathcal{L}_\lambda)) \neq 0$ if and only if $\lambda + \rho$ is dominant regular, and this happens if and only if λ is dominant. In this case the element w in the second assertion is $w = 1$. Thus, for λ dominant, the theorem says that

$$H^0(X, \mathcal{O}(\mathcal{L}_\lambda)) \text{ is nonzero, irreducible, of highest weight } \lambda.$$

This is the statement of the **Borel-Weil Theorem**, which appeared originally in [Se], [Ti], and [HC2]. Because of the Theorem of the Highest Weight, it produces a concrete, geometric realization for every irreducible holomorphic representation.

2) According to our convention, $\mathfrak{b}$ is built from the root spaces for the negative roots. This has the effect of making the line bundle $\mathcal{L}_\lambda$ "positive" in the sense of complex analysis (see [We, p. 223], for example) precisely when the parameter λ is dominant. The opposite convention, which uses the root spaces for positive roots, lets positive line bundles correspond to antidominant weights and makes $H^0(X, \mathcal{O}(\mathcal{L}_\lambda))$, for antidominant λ, the G-module with *lowest* weight λ.

3) One explanation of the ρ-shift in the statement is that it is the shift in the parameter which makes the result compatible with Serre duality [We, p. 170], as it has to be.

4) Since the Borel-Weil Theorem already gives a geometric realization of all irreducible holomorphic representations of G, why should one look also at the higher

cohomology groups? Before Bott's theorem – i.e., the description of the higher cohomology groups of the $\mathcal{L}_\lambda$ – the dimensions of these cohomology groups were not known, but they were of obvious interest in algebraic geometry; Bott's theorem directly implies a description of the cohomology groups of every holomorphic line bundle on every complex projective manifold on which G acts transitively (see [Do2] for a discussion of the relationship between the Borel-Weil Theorem and line bundles over complex projective space). On a more fundamental level, perhaps, one's understanding of a left exact functor, such as H^0, is not complete without the knowledge of the right derived functors, so the Borel-Weil theorem really cries out for a characterization of the higher cohomology. What may not have been clear in the beginning is the extent to which the Borel-Weil-Bott theorem would become indispensable in the study of infinite-dimensional representations.

SKETCH OF PROOF OF BOREL-WEIL THEOREM. Let $G_\mathbb{R} \subset G$ be a **compact real form**, i.e., a compact Lie subgroup with Lie algebra $\mathfrak{g}_\mathbb{R}$ such that $\mathfrak{g} = \mathfrak{g}_\mathbb{R} \oplus i\mathfrak{g}_\mathbb{R}$. Recall from the lectures [Kn3], for example, that

(a) such subgroups $G_\mathbb{R}$ exist,
(b) they are maximal among compact subgroups of G,
(c) they are connected, and
(d) any two such are conjugate in G.

We can choose the Cartan subalgebra $\mathfrak{h}$ of $\mathfrak{g}$ so that it is the complexification of a subalgebra $\mathfrak{h}_\mathbb{R}$ of $\mathfrak{g}_\mathbb{R}$; all we have to do is take $\mathfrak{h}_\mathbb{R}$ to be any maximal abelian subspace of $\mathfrak{g}_\mathbb{R}$. Then $H_\mathbb{R} = G_\mathbb{R} \cap H$ is a Cartan subgroup of $G_\mathbb{R}$, i.e., a maximal torus; its Lie algebra is $\mathfrak{h}_\mathbb{R}$.

The $G_\mathbb{R}$-orbit of the point $\mathfrak{b}$ of X is a closed submanifold because $G_\mathbb{R}$ is compact, and it is open in X by a dimension count. Therefore $G_\mathbb{R}$ acts transitively on X. To compute the isotropy subgroup at $\mathfrak{b}$, we observe that

$$G_\mathbb{R} \cap B = G_\mathbb{R} \cap B \cap \overline{B} = G_\mathbb{R} \cap H = H_\mathbb{R},$$

hence

$$X \simeq G/B \simeq G_\mathbb{R}/(G_\mathbb{R} \cap B) = G_\mathbb{R}/H_\mathbb{R}.$$

If we identify $X \simeq G_\mathbb{R}/H_\mathbb{R}$, we see that $\mathcal{L}_\lambda$, as $G_\mathbb{R}$-equivariant C^∞ complex line bundle, is given by

$$\mathcal{L}_\lambda \simeq G_\mathbb{R} \times_{H_\mathbb{R}} \mathbb{C}_\lambda\,;$$

here $\mathbb{C}_\lambda$ is the one dimensional $H_\mathbb{R}$-module on which $H_\mathbb{R}$ acts via the character e^λ. This leads to the following description of the space of C^∞ sections of $\mathcal{L}_\lambda$:

$$C^\infty(X, \mathcal{L}_\lambda) \simeq$$
$$\simeq \{f \in C^\infty(G_\mathbb{R}) \mid f(gh) = e^{-\lambda}(h) f(g) \text{ for all } h \in H_\mathbb{R}\} \simeq (C^\infty(G_\mathbb{R}) \otimes \mathbb{C}_\lambda)^{H_\mathbb{R}};$$

here $(C^\infty(G_\mathbb{R}) \otimes \mathbb{C}_\lambda)^{H_\mathbb{R}}$ denotes the space of $H_\mathbb{R}$-invariants in $C^\infty(G_\mathbb{R}) \otimes \mathbb{C}_\lambda$, relative to the action by right translation on $C^\infty(G_\mathbb{R})$ and by e^λ on $\mathbb{C}_\lambda$ (for a discussion of these isomorphisms, see [Do2]). How can one characterize the *holomorphic* sections among the C^∞ sections – in other words, what are the Cauchy-Riemann equations? Suppose that $U \subseteq X \simeq G_\mathbb{R}/H_\mathbb{R}$ is open and that $\tilde{U} \subseteq G_\mathbb{R}$ is its inverse image. Then

$$(*) \qquad C^\infty(U, \mathcal{L}_\lambda) \simeq \{f \in C^\infty(\tilde{U}) \mid f(gh) = e^{-\lambda}(h) f(g) \text{ for } h \in H_\mathbb{R}\}$$

by specialization of the previous isomorphism to U, and our question is answered by:

Lemma. *Under the isomorphism* $(*)$, *a function* f *on* $\tilde{U}$ *corresponds to a holomorphic section of* $\mathcal{L}_\lambda$ *over* U *if and only if* $r(Z)f = 0$ *for all* $Z \in \mathfrak{n} = \bigoplus_{\alpha\in\Phi^+} \mathfrak{g}^{-\alpha}$, *where* $r(Z)$ *denotes infinitesimal right translation on* $G_\mathbb{R}$ *by* $Z \in \mathfrak{g} = \mathfrak{g}_\mathbb{R} \oplus i\mathfrak{g}_\mathbb{R}$.

The lemma is readily proved by starting from the Cauchy-Riemann equations on G – see [Gr-Sch. pp. 258–259], for example. Using it, we can identify the space of global holomorphic sections as

$$H^0(X, \mathcal{O}(\mathcal{L}_\lambda)) \simeq \{f \in C^\infty(G_\mathbb{R}) \mid r(\mathfrak{n})f = 0 \text{ and } f(gh) = e^{-\lambda}(h)f(g) \text{ for } h \in H_\mathbb{R}\},$$

and this isomorphism is an isomorphism of representations of $G_\mathbb{R}$. The space $C^\infty(G_\mathbb{R})$ is contained in $L^2(G_\mathbb{R})$, which we can identify by the Peter-Weyl Theorem as a Hilbert space direct sum $\sum_{i\in\widehat{G_\mathbb{R}}} V_i \otimes V_i^*$. Here $G_\mathbb{R}$ acts on V_i by left translation, and on V_i^* by right translation (see [Sch8, §1], for example). The subspace of $C^\infty(G_\mathbb{R})$ corresponding to $H^0(X, \mathcal{O}(\mathcal{L}_\lambda))$ is finite-dimensional and $G_\mathbb{R}$-invariant, hence contained in the *algebraic* direct sum $\bigoplus_{i\in\widehat{G_\mathbb{R}}} V_i \otimes V_i^*$. We conclude that

$$\begin{aligned} H^0(X, \mathcal{O}(\mathcal{L}_\lambda)) &\simeq \{f \in \bigoplus_i V_i \otimes V_i^* \mid r(\mathfrak{n})f = 0 \text{ and } f(gh) = e^{-\lambda}(h)f(g)\} \\ &\simeq \bigoplus_i V_i \otimes \{v \in (V_i^* \otimes \mathbb{C}_\lambda)^H \mid \mathfrak{n}v = 0\}. \end{aligned}$$

The condition $\mathfrak{n}v = 0$ picks out the lowest weight space since $\mathfrak{b}$ is built from the root spaces for the negative roots. Therefore the right side is

$$\bigoplus_{\substack{V_i^* \text{ has lowest} \\ \text{weight } -\lambda}} V_i \otimes (\text{lowest weight space in } V_i^*).$$

At this point, the Borel-Weil Theorem follows from the Theorem of the Highest Weight.

2. Borel-Weil-Bott Theorem

In §1 we stated the Borel-Weil-Bott Theorem and proved the special case known as the Borel-Weil Theorem. In this section we shall derive the Borel-Weil-Bott Theorem from the Borel-Weil Theorem. The argument is similar to that in Bott's original paper [Bott].

We continue with the notation of §1. The Lie algebra $\mathfrak{g}$ is assumed complex semisimple, and G is a complex connected Lie group with Lie algebra $\mathfrak{g}$. Let X be the flag variety of $\mathfrak{g}$. To identify X explicitly as a homogeneous space of G, we fix a Borel subalgebra $\mathfrak{b} = \mathfrak{h} \oplus \mathfrak{n}$, where $\mathfrak{n} = \bigoplus_{\alpha\in\Phi^+} \mathfrak{g}^{-\alpha}$, and we let $B = N_G(\mathfrak{b})$. Then $X \simeq G/B$.

We shall introduce a fibration of X. Fix a simple root $\alpha \in \Phi^+$, and define $\mathfrak{p}_\alpha = \mathfrak{b} \oplus \mathfrak{g}^\alpha$; this is a parabolic subalgebra of $\mathfrak{g}$. The corresponding parabolic subgroup $P_\alpha = N_G(\mathfrak{p}_\alpha)$ of G is connected since G is complex, P_α is closed, and has Lie algebra $\mathfrak{p}_\alpha$. Define $X_\alpha = G/P_\alpha$. This is a particular kind of generalized flag variety. For $\mathfrak{g} = \mathfrak{sl}(n, \mathbb{C})$, it is the set of all incomplete flags $F = (F_j)$ for which F_j is specified for all but one value of j, which depends on the choice of α.

The inclusion $B \subset P_\alpha$ induces a G equivariant holomorphic fiber bundle

$$X \simeq G/B \xrightarrow{\pi} G/P_\alpha = X_\alpha$$

with fiber
$$P_\alpha/B \simeq SL(2,\mathbb{C})\Big/\left\{\begin{pmatrix} * & 0 \\ * & * \end{pmatrix}\right\}$$
$$\simeq \mathbb{P}^1$$
$$= \text{flag variety of } \mathfrak{sl}(2,\mathbb{C}).$$

To prove the Borel-Weil-Bott Theorem, we shall argue by induction. The induction step uses the theorem for $SL(2,\mathbb{C})$. In that special case, one can verify the assertion directly – for example, by combining the Borel-Weil theorem with Serre duality. The connection between The Borel-Weil-Bott theorem and $\mathfrak{n}$-cohomology, which will be commented on below, can also be used to give a simple argument in the case of the group $SL(2,\mathbb{C})$.

Sketch of proof of Borel-Weil-Bott Theorem. The holomorphic fibration $X \to X_\alpha$ yields a Leray spectral sequence for computing $H^*(X, \mathcal{O}(\mathcal{L}_\lambda))$, with E_2 term
$$E_2^{p,q} = H^q(X_\alpha, \mathcal{O}_{X_\alpha}(\mathcal{V}^p(\lambda))),$$
where $\mathcal{V}^p(\lambda)$ is the G equivariant holomorphic vector bundle whose fiber at $y \in X_\alpha$ is
$$\mathcal{V}^p(\lambda)_y = H^p(\pi^{-1}(y), \mathcal{O}_{\pi^{-1}(y)}(\mathcal{L}_\lambda|_{\pi^{-1}(y)})).$$
Here the range of p is $0 \le p \le \dim_{\mathbb{C}} \pi^{-1}(y) = 1$.

The Borel-Weil-Bott Theorem for $SL(2,\mathbb{C})$ implies that $\mathcal{V}^p(\lambda) \neq 0$ for at most one p. Thus $E_2^{p,*} \neq 0$ for at most one p, the spectral sequence collapses, and we obtain
$$H^k(X, \mathcal{O}(\mathcal{L}_\lambda)) = \bigoplus_{p+q=k} E_2^{p,q} = \bigoplus_{p+q=k} H^q(X_\alpha, \mathcal{O}_{X_\alpha}(\mathcal{V}^p(\lambda))).$$

Let s_α denote reflection in α. Application of the same argument to $\mathcal{L}_{s_\alpha(\lambda+\rho)-\rho}$ gives
$$H^k(X, \mathcal{O}(\mathcal{L}_{s_\alpha(\lambda+\rho)-\rho})) = \bigoplus_{p+q=k} H^q(X_\alpha, \mathcal{O}_{X_\alpha}(\mathcal{V}^p(s_\alpha(\lambda+\rho)-\rho))).$$

Set $m = 2(\lambda,\alpha)/|\alpha|^2$. The Borel-Weil-Bott Theorem for $SL(2,\mathbb{C})$ shows that if $m \ge -1$,
$$\mathcal{V}^0(\lambda) \simeq \mathcal{V}^1(s_\alpha(\lambda+\rho)-\rho)\,, \qquad \mathcal{V}^1(\lambda) = \mathcal{V}^0(s_\alpha(\lambda+\rho)-\rho) = 0\,,$$
and that all four of these vector bundles vanish if $m = -1$. Therefore
$$(**) \qquad H^p(X, \mathcal{O}(\mathcal{L}_\lambda)) \simeq H^{p+1}(X, \mathcal{O}(\mathcal{L}_{s_\alpha(\lambda+\rho)-\rho}))$$
as representations of G if $(\lambda+\rho,\alpha) \ge 0$. When $(\lambda+\rho,\alpha) = 0$, both sides are 0.

The first conclusion we obtain from $(**)$ is the vanishing of the higher cohomology groups $H^p(X, \mathcal{O}(\mathcal{L}_\lambda))$, $p > 0$, if λ is dominant. Let w_0 denote the "longest" element of the Weyl group. It has a minimal expansion as a product of $\dim \mathfrak{n} = \dim_{\mathbb{C}} X$ simple reflections. Thus, when λ is dominant, we can iterate $(**)$ to obtain
$$H^{p+\dim X}(X, \mathcal{O}(\mathcal{L}_{w_0(\lambda+\rho)-\rho})) \simeq H^p(X, \mathcal{O}(\mathcal{L}_\lambda))\,.$$
This forces $H^p(X, \mathcal{O}(\mathcal{L}_\lambda)) = 0$ if $p > 0$, since $H^k(X, \mathcal{O}(\mathcal{L}_{w_0(\lambda+\rho)-\rho}))$ must vanish when k exceeds the complex dimension of X.

The second conclusion from $(**)$ is the full Borel-Weil-Bott Theorem. Suppose $\lambda+\rho$ is regular and $w(\lambda+\rho)$ is dominant. If w is written minimally as the product of simple reflections, then the number of factors will be $p(\lambda+\rho)$, and $(**)$ and the first conclusion above give

$$\begin{aligned} &H^{p(\lambda+\rho)+q}(X,\mathcal{O}(\mathcal{L}_\lambda)) \\ &\qquad\simeq H^q(X,\mathcal{O}(\mathcal{L}_{w(\lambda+\rho)-\rho})) \\ &\qquad\simeq \begin{cases} \text{representation of highest weight } w(\lambda+\rho)-\rho & \text{if } q=0 \\ 0 & \text{if } q\neq 0. \end{cases} \end{aligned}$$

Next suppose $\lambda+\rho$ is singular and $w(\lambda+\rho)$ is dominant. Then there exists a simple root β such that $(w(\lambda+\rho),\beta)=0$. From what we have already seen, $H^p(X,\mathcal{O}(\mathcal{L}_{w(\lambda+\rho)-\rho}))=0$ for all p. Writing w as a product of simple reflections and iterating $(**)$, we obtain $H^p(X,\mathcal{O}(\mathcal{L}_\lambda))=0$ for all p.

This completes the sketch of the proof of the Borel-Weil-Bott Theorem. What we have given is essentially Bott's argument in [Bott], except that Bott made use of the Kodaira Vanishing Theorem. We return to this point at the end of this section.

We shall now relate the Borel-Weil-Bott Theorem to a computation of $\mathfrak{n}$-cohomology that is the subject of work of Kostant [Kos2]. This relationship was already pointed out in Bott's paper [Bott].

Let $\mathcal{A}^p(\mathcal{L}_\lambda)$ be the sheaf of (germs of) smooth, $\mathcal{L}_\lambda$-valued $(0,p)$ forms on X, and $\bar\partial:\mathcal{A}^p(\mathcal{L}_\lambda)\to\mathcal{A}^{p+1}(\mathcal{L}_\lambda)$ the Dolbeault operator. According to the Dolbeault lemma, the complex of sheaves $(\mathcal{A}^\cdot(\mathcal{L}_\lambda),\bar\partial)$ resolves the sheaf $\mathcal{O}(\mathcal{L}_\lambda)$. These are sheaves of $\mathcal{C}^\infty_X$-modules, hence cohomologically trivial. Thus, by basic homological algebra, the **Dolbeault complex**, i.e., the complex of global sections $A^\cdot(\mathcal{L}_\lambda)=\Gamma\mathcal{A}^\cdot(\mathcal{L}_\lambda)$, computes the cohomology of $\mathcal{O}(\mathcal{L}_\lambda)$:

$$H^p(X,\mathcal{O}(\mathcal{L}_\lambda))\simeq H^p(A^\cdot(\mathcal{L}_\lambda),\bar\partial).$$

This isomorphism is the complex analytic analogue of the de Rham isomorphism. It is natural, hence G-equivariant. For details on the Dolbeault isomorphism, see [We, pp. 33–35], for example.

Earlier we observed that $X\simeq G_\mathbb{R}/H_\mathbb{R}$. The description of the Cauchy-Riemann equations on X implies that the antiholomorphic cotangent bundle on X is the $G_\mathbb{R}$-equivariant C^∞ vector bundle modeled on $\mathfrak{n}^*$. Thus the Dolbeault complex is given by

$$A^\cdot(\mathcal{L}_\lambda)\simeq(C^\infty(G_\mathbb{R})\otimes\mathbb{C}_\lambda\otimes\textstyle\bigwedge^\cdot\mathfrak{n}^*)^{H_\mathbb{R}}.$$

Here $H_\mathbb{R}$ acts on $C^\infty(G_\mathbb{R})$ by right translation, on $\mathfrak{n}^*$ via Ad^*, and on $\mathbb{C}_\lambda$ via e^λ. Under this identification, $\bar\partial$ corresponds to the coboundary operator d of the standard complex

$$(\mathrm{Hom}(\textstyle\bigwedge^\cdot\mathfrak{n},C^\infty(G_\mathbb{R})\otimes\mathbb{C}_\lambda))^{H_\mathbb{R}}$$

with $\mathfrak{n}$ acting trivially on $\mathbb{C}_\lambda$ by e^λ, and on $C^\infty(G_\mathbb{R})$ via right translation, as before.

Arguing as in the proof of the Borel-Weil Theorem, we have the embedding

$$C^\infty(G_\mathbb{R})\subset L^2(G_\mathbb{R})\simeq\sum_{i\in\widehat{G_\mathbb{R}}}V_i\otimes V_i^* \qquad \text{(Hilbert space direct sum)},$$

from which we obtain

$$(***)\qquad H^*(X,\mathcal{O}(\mathcal{L}_\lambda))\simeq H^*(A^\cdot(\mathcal{L}_\lambda),\bar\partial)\simeq\bigoplus V_i\otimes(H^*(\mathfrak{n},V_i^*)\otimes\mathbb{C}_\lambda)^{H_\mathbb{R}}.$$

Thus the Borel-Weil-Bott Theorem is directly equivalent to the description of the $\mathfrak{n}$-cohomology groups of all irreducible representations of $G_{\mathbb{R}}$, including their structure as $H_{\mathbb{R}}$-modules.

Since $G_{\mathbb{R}}$ is compact, $\mathcal{L}_\lambda$ and the cotangent bundle of X carry $G_{\mathbb{R}}$-invariant hermitian metrics. Thus we can form $\bar{\partial}^*$ as the formal adjoint of $\bar{\partial}$, and $\bar{\partial}^*$ will then also be $G_{\mathbb{R}}$-invariant. The corresponding **Laplace-Beltrami operator** is

$$\square = \bar{\partial}^* \bar{\partial} + \bar{\partial} \bar{\partial}^*.$$

The Hodge Theorem implies that

$$H^p(X, \mathcal{O}(\mathcal{L}_\lambda)) \simeq \text{kernel of } \square \text{ on } A^p(\mathcal{L}_\lambda).$$

In turn, the ellipticity of $\square$ implies that this kernel is

$$\begin{aligned} &= \text{kernel of self-adjoint extension of } \square \text{ on the space} \\ &\quad \text{of } \mathcal{L}_\lambda\text{-valued } (0,p) \text{ forms with } L^2 \text{ coefficients.} \end{aligned}$$

We shall come back to this description of $H^p(X, \mathcal{O}(\mathcal{L}_\lambda))$ when we discuss the discrete series.

Kostant [Kos2] calculates the $\mathfrak{n}$-cohomology of any irreducible $G_{\mathbb{R}}$-module V by identifying the kernel of the analogue of $\square$ acting on $\mathrm{Hom}(\bigwedge^{\cdot}\mathfrak{n}, V)$, the standard complex of Lie algebra cohomology. In a nutshell, Kostant's argument shows that the $\mathfrak{n}$-cohomology of V is computed by a certain subcomplex of $\mathrm{Hom}(\bigwedge^{\cdot}\mathfrak{n}, V)$ which has trivial differential, so the subcomplex is isomorphic to $H^*(\mathfrak{n}, V)$. Bott's original argument can also be adapted to the calculation of $H^*(\mathfrak{n}, V)$, with the Hochschild-Serre spectral sequence taking the place of the Leray spectral sequence. Finally, we should mention the paper [Cas-O] of Casselman-Osborne, which formulates and proves a property of the $\mathfrak{h}$-module structure of the $\mathfrak{n}$-cohomology of an arbitrary, possibly infinite dimensional, $\mathfrak{g}$-module. They point out that their result, in the case of a finite dimensional irreducible $\mathfrak{g}$-module V, leads to the description of $H^*(\mathfrak{n}, V)$ – in effect, they use homological algebra instead of the computation of $\square$ to identify $H^*(\mathfrak{n}, V)$ with the same trivial subcomplex of the standard complex which comes up in Kostant's argument.

The vanishing of cohomology that we have seen above is an instance of a more general phenomenon: Let X be a compact complex manifold, and let $\mathcal{L} \to X$ be a holomorphic line bundle with a hermitian metric. Suppose that the curvature form of $\mathcal{L}$ has everywhere signature (p, q) with $p + q = \dim_{\mathbb{C}} X$. Then $\mathcal{L}$ tends to have cohomology only in degree q. The relationship between curvature and vanishing theorems was explored in the early 1950s by Bochner and Yano [Boc-Y] and by Kodaira [Kod]. See [We, p. 226] for an exposition of the Kodaira Vanishing Theorem. Bott [Bott] proved the vanishing in the Borel-Weil-Bott Theorem as a special case of Kodaira's result. The papers [An-Ve], [Gr-Sch] pursue vanishing theorems of this kind in the noncompact case and will play a role in §4 below.

3. Discrete Series

In this section we begin a discussion of the discrete series of a noncompact semisimple Lie group. We continue with $\mathfrak{g}$ as a complex semisimple Lie algebra and with G as a complex connected Lie group with Lie algebra $\mathfrak{g}$. Let $\mathfrak{g}_{\mathbb{R}} \subset \mathfrak{g}$ be a real form of $\mathfrak{g}$; what we have in mind is a noncompact real form, though we do not formally exclude the case a compact real form. We let $G_{\mathbb{R}}$ denote the connected subgroup of G with Lie algebra $\mathfrak{g}_{\mathbb{R}}$.

Lemma. *If (π, V_π) is an irreducible unitary representation of $G_\mathbb{R}$, then the following conditions on (π, V_π) are equivalent:*

a) *for some pair of nonzero u and v in V_π, $g \mapsto (\pi(g)u, v)$ is in $L^2(G_\mathbb{R})$;*
b) *for any pair of nonzero u and v in V_π, $g \mapsto (\pi(g)u, v)$ is in $L^2(G_\mathbb{R})$;*
c) *V_π embeds in $G_\mathbb{R}$-equivariant unitary fashion into the left regular representation of $G_\mathbb{R}$ on $L^2(G_\mathbb{R})$;*
d) *the Plancherel measure for the decomposition of $L^2(G_\mathbb{R})$ assigns strictly positive mass to the one-point set $\{\pi\}$ in the unitary dual $\widehat{G}_\mathbb{R}$.*

If π satisfies the equivalent conditions in the lemma, π is said to be **square-integrable**, and we say that π is a **discrete series representation**. By definition, the **discrete series** of $G_\mathbb{R}$ is the set of isomorphism classes of irreducible, square-integrable, unitary representations.

REFERENCES. The equivalence of (a) through (c) in the lemma is due to Godement [Go]. Godement's proof uses functional analysis and is valid for all locally compact unimodular groups. For an exposition in the semisimple case, see [Kn1, §IX.3]. The proof of the equivalence of (a) and (b) shows that $\|(\pi(g)u, v)\|^2 = d_\pi^{-1}\|u\|^2\|v\|^2$ for a constant d_π independent of u and v. This constant is called the **formal degree** of π. The equivalence of (d) with (a) through (c) is due to Harish-Chandra, and the statement of this equivalence in [HC1] is given in the strong form that the Plancherel measure assigns mass d_π to the one-point set $\{\pi\}$.

We recall some facts which were covered in the lectures of Delorme [De]. A linear operator on a Hilbert space V is **Hilbert-Schmidt** if the sum of the squares of the absolute values of its matrix elements relative to an orthonormal basis is finite. One shows easily that this sum does not depend on the particular choice of orthonormal basis. It is denoted by $\|\cdot\|_{\mathrm{HS}}^2$, and is the norm squared for a hermitian inner product $(\,\cdot\,,\,\cdot\,)_{\mathrm{HS}}$ which turns the space of Hilbert-Schmidt operators into a Hilbert space. A linear operator T is said to be of **trace class** if the sum of its diagonal matrix entries converges absolutely for every orthonormal basis; in that case, the sum is not affected by the choice of orthonormal basis. One calls the sum the trace of the operator, denoted $\operatorname{Tr} T$. Equivalently, the trace class operators can be characterized by the property that they can be written as the composition of two Hilbert-Schmidt operators; if S and T are Hilbert-Schmidt operators, then $\operatorname{Tr}(ST) = (S, T^*)_{\mathrm{HS}}$. Let (π, V_π) be an irreducible admissible representation on the Hilbert space V_π, e.g., an irreducible unitary representation. Then

1) for each $f \in C_0^\infty(G_\mathbb{R})$, $\pi(f) = \int_{G_\mathbb{R}} f(g)\pi(g)\,dg$ is a well defined bounded linear operator (we write $\pi(f) \in \operatorname{End}(V_\pi)$);
2) $\pi(f)$ is a trace class operator, and $f \mapsto \Theta_\pi(f) = \operatorname{Tr}\pi(f)$ is a distribution on $G_\mathbb{R}$ (the **global character** of π).

The global character of a discrete series representation has a better boundedness property than an arbitrary distribution. To get at this property, we define Sobolev spaces on $G_\mathbb{R}$. Typically, Sobolev spaces are used in analysis to quantify the regularity of functions locally. In contrast, here we use them to measure the decay of discrete series characters at infinity. Let $U_n(\mathfrak{g})$ be the subspace of elements of order $\leq n$ in $U(\mathfrak{g})$. We introduce on $C_0^\infty(G_\mathbb{R})$ the system of seminorms $\|r(Z)f\|_2$, $Z \in U_n(\mathfrak{g})$; here $r(Z)$ refers to infinitesimal right translation by Z. Since $U_n(\mathfrak{g})$ is finite-dimensional, finitely many of these seminorms describe this space. The completion is the n-th left global **Sobolev space** on $G_{\mathbb{R}}$, and we may think of it as

$$S_n(G_{\mathbb{R}}) = \{f \in L^2(G_{\mathbb{R}}) \mid r(Z)f \in L^2(G_{\mathbb{R}}) \text{ for all } Z \in U_n(\mathfrak{g})\}.$$

The space $S_n(G_{\mathbb{R}})$ is a Banach space on which $G_{\mathbb{R}}$ acts by left translation, and $C_0^\infty(G_{\mathbb{R}})$ is a dense subspace.

Proposition 1. *If π is a discrete series representation, then its global character Θ_π extends to a bounded linear functional on $S_n(G_{\mathbb{R}})$ if n is sufficiently large.*

Characters are conjugation invariant, so boundedness on the n-th left Sobolev space implies, and is implied by, boundedness on the corresponding right Sobolev space. The proof of the proposition will make use of the following Important Fact.

Important Fact. *If π is a discrete series representation, then the map $f \mapsto \pi(f)$ extends to a bounded linear map from $L^2(G_{\mathbb{R}})$ into the space End$(V_\pi)_{\mathrm{HS}}$ of Hilbert-Schmidt operators on V_π.*

EXPLANATION. When $G_{\mathbb{R}}$ is compact, the Peter-Weyl Theorem allows us to write

$$L^2(G_{\mathbb{R}}) \simeq \sum_{i \in \widehat{G_{\mathbb{R}}}} V_i \otimes V_i^* \simeq \sum_{i \in \widehat{G_{\mathbb{R}}}} \mathrm{End}(V_i),$$

the sums being Hilbert space direct sums. For $G_{\mathbb{R}}$ not necessarily compact, the analogue is a direct integral decomposition

$$L^2(G_{\mathbb{R}}) \simeq \int_{i \in \widehat{G_{\mathbb{R}}}} \mathrm{End}(V_i)_{\mathrm{HS}}\, d\mu(i),$$

where μ is the Plancherel measure. In particular, this formula implies that when any L^2 function f is expanded according to the right side, the integrand is finite almost everywhere. According to the lemma, μ assigns positive mass to the class of any discrete series π_i. Then it follows that

$$\|\pi_i(f)\|_{\mathrm{HS}}^2 \le \mu(i)^{-1} \|f\|_2^2 ,$$

as required.

PROOF OF PROPOSITION 1. For any irreducible admissible representation (π, V_π) of $G_{\mathbb{R}}$ on a Hilbert space V_π, Harish-Chandra's proof of the existence of the global character produces a $Z \in U(\mathfrak{g})$ and $T \in \mathrm{End}(V_\pi)_{\mathrm{HS}}$, such that

$$\pi(f) = \pi(r(Z)f)T$$

for all $f \in C_0^\infty(G_{\mathbb{R}})$; see [At2] or [De] for details. Consequently

$$\mathrm{Tr}\, \pi(f) = (\pi(r(Z)f), T^*)_{\mathrm{HS}}$$

and

$$|\mathrm{Tr}\, \pi(f)| \le \|\pi(r(Z)f)\|_{\mathrm{HS}}\, \|T\|_{\mathrm{HS}}.$$

If π is in the discrete series, we can combine this inequality with the Important Fact to obtain

$$|\mathrm{Tr}\, \pi(f)| \le \mathrm{Const}(\pi) \|r(Z)f\|_2,$$

The proposition follows, with n equal to the order of Z.

REFERENCE. For a direct proof of Proposition 1 that does not use direct integrals and the Plancherel decomposition, see [HC6, p. 88].

We shall now work toward a statement of Harish-Chandra's celebrated classification theorem of discrete series representations in terms of their global characters. References for this work are [HC4], [HC5], and [HC6]. Our exposition is based on [At-Sch]. We shall need the following facts about characters, all due to Harish-Chandra, which were discussed in Delorme's lectures [De]; see [At2] for another exposition.

1) The character of an irreducible admissible representation is an **invariant eigendistribution** on $G_{\mathbb{R}}$. That is, it is invariant under group conjugation, and each member of the center of the universal enveloping algebra acts on it by a scalar.
2) Any invariant eigendistribution on $G_{\mathbb{R}}$ is (integration against) a locally integrable function on $G_{\mathbb{R}}$, and the locally integrable function may be taken to be real analytic on the subset $G'_{\mathbb{R}}$ of regular semisimple elements.
3) Because of (2), it is meaningful to restrict an invariant eigendistribution to a function on the set of regular elements of each Cartan subgroup of $G_{\mathbb{R}}$.
4) Each invariant eigendistribution is completely determined by restriction as in (3) to the Cartan subgroups, and it is enough to choose one Cartan subgroup from each of the finitely many conjugacy classes.

Among these, only the regularity statement (2) is difficult. It was originally proved in [HC3]; a quite different argument is sketched in [At2].

Let $H_{\mathbb{R}} \subset G_{\mathbb{R}}$ be a Cartan subgroup. Relative to a positive system Φ^+ of roots, the Weyl denominator is formally the expression

$$\Delta = \prod_{\alpha \in \Phi^+} (e^{\alpha/2} - e^{-\alpha/2}).$$

To give meaning to this expression, we can rewrite it as $\Delta = e^{\rho} \prod_{\alpha \in \Phi^+} (1 - e^{-\alpha})$. Since 2ρ is a weight, $|\Delta|$ is a well defined function on $H_{\mathbb{R}}$, independent of the choice of positive system. Following Harish-Chandra, we say that $G_{\mathbb{R}}$ is **acceptable** if ρ lies in the weight lattice. This is always the case if the complex group G is simply connected. For most purposes, and for the present discussion in particular, there is no loss of generality in assuming that $G_{\mathbb{R}}$ is acceptable: if it fails to be so, its inverse image in an appropriate 2-fold covering of the complex group G will be. Let us assume the acceptability of $G_{\mathbb{R}}$, to make Δ a well defined function on $H_{\mathbb{R}}$. Then every invariant eigendistribution Θ has the following additional properties:

5) The function $\Theta|_{H'_{\mathbb{R}}}\Delta$ on each component of the regular set $H'_{\mathbb{R}}$ is a linear combination of exponentials with polynomial coefficients (exponential, respectively polynomial, when pulled back to the Lie algebra $\mathfrak{h}_{\mathbb{R}}$ via exp).
6) If $H_{\mathbb{R}}$ is maximally compact, then $\Theta|_{H'_{\mathbb{R}}}\Delta$ extends to a C^{∞} function on all of $H_{\mathbb{R}}$.
7) The restriction of $\Theta\Delta$ to two Cartan subgroups that are related by a simple Cayley transform satisfy certain **matching conditions** due to Hirai [Hirai] and modeled on corresponding conditions in the Lie algebra case discovered by Harish-Chandra [HC3] (see [Kn1, §XI.7] for an exposition of the matching conditions).

8) If Θ is an irreducible character, then the polynomial coefficients in (5) are always constants.

Definition. An invariant eigendistribution Θ will be said to be **bounded at** ∞ if, for each Cartan subgroup $H_{\mathbb{R}} \subset G_{\mathbb{R}}$, the function $\Theta|_{H_{\mathbb{R}}}|\Delta|$ is bounded on $H_{\mathbb{R}}$. The invariant eigendistribution Θ will be said to **decay at** ∞ if, for each $H_{\mathbb{R}}$, the function $\Theta|_{H_{\mathbb{R}}}|\Delta|$ tends to 0 outside of compact subsets of $H_{\mathbb{R}}$.

"Bounded at ∞" implies "tempered" in the sense of Harish-Chandra (i.e., extending continuously from C_0^∞ to Harish-Chandra's Schwartz space). The converse implication, for characters only, follows from the fact (8) above. Decay at ∞ is distinctly stronger than the condition of being bounded at ∞.

Proposition 2. *If an invariant eigendistribution Θ extends continuously to some global Sobolev space $S_n(G_{\mathbb{R}})$, then Θ decays at ∞.*

IDEA OF PROOF. There is no loss of generality in assuming that $G_{\mathbb{R}}$ is acceptable. Using the Weyl Integration Formula and the hypothesis of the proposition, one shows that the distribution $\Theta|_{H'_{\mathbb{R}}}\Delta$ extends to be continuous on a classical Sobolev space $S_{n'}(H'_{\mathbb{R}})$. Taking into account that $\Theta|_{H'_{\mathbb{R}}}\Delta$ on each component of $H'_{\mathbb{R}}$ is a linear combination of exponentials with polynomial coefficients, we see that all the exponentials must decay at infinity.

Proposition 3. *If an invariant eigendistribution $\Theta \neq 0$ decays at ∞, then there exists a compact Cartan subgroup $T_{\mathbb{R}} \subset G_{\mathbb{R}}$ such that $\Theta|_{T'_{\mathbb{R}}} \neq 0$.*

IDEA OF PROOF. Again there is no loss of generality in assuming that $G_{\mathbb{R}}$ is acceptable. The idea is to use the matching conditions (7) for invariant eigendistributions. Because of the assumed decay at ∞, the matching conditions imply that Θ is completely determined by its restriction to a maximally compact Cartan subgroup $H_{\mathbb{R}}$. By (6), $\Theta|_{H'_{\mathbb{R}}}\Delta$ extends to a smooth function on $H_{\mathbb{R}}$. Because of the decay condition and the reasoning for Proposition 2, each exponential term that occurs must vanish everywhere at ∞. If $H_{\mathbb{R}}$ is noncompact, this forces its coefficient to be zero.

Corollary. *If $G_{\mathbb{R}}$ has discrete series representations, then $G_{\mathbb{R}}$ has a compact Cartan subgroup.*

PROOF. This follows immediately by combining Propositions 1, 2, and 3.

For the remainder of the discussion, we may therefore assume that $G_{\mathbb{R}}$ has a compact Cartan subgroup $T_{\mathbb{R}}$. As a compact Cartan subgroup, $T_{\mathbb{R}}$ is unique up to conjugacy. Also it is connected, hence is a torus. Fix a maximal compact subgroup $K_{\mathbb{R}} \subset G_{\mathbb{R}}$. Any two choices of $K_{\mathbb{R}}$ are conjugate, but $K_{\mathbb{R}}$ becomes uniquely determined when we require that $K_{\mathbb{R}} \supset T_{\mathbb{R}}$, as we shall. Let $\mathfrak{t}_{\mathbb{R}}$ and $\mathfrak{k}_{\mathbb{R}}$ be the Lie algebras of $T_{\mathbb{R}}$ and $K_{\mathbb{R}}$, let $\mathfrak{t}$ and $\mathfrak{k}$ be the complexifications inside $\mathfrak{g}$, and let T and K be the connected subgroups of G with Lie algebras $\mathfrak{t}$ and $\mathfrak{k}$.

Let Φ be the root system of $(\mathfrak{g}, \mathfrak{t})$. A root $\alpha \in \Phi$ is a **compact root** if $\mathfrak{g}^\alpha \subset \mathfrak{k}$ and is a **noncompact root** if $\mathfrak{g}^\alpha$ is orthogonal to $\mathfrak{k}$ with respect to the Killing form. Let Φ_c and Φ_n be the sets of compact and noncompact roots, respectively. Then $\Phi = \Phi_c \cup \Phi_n$. Also Φ_c is the root system of $(\mathfrak{k}, \mathfrak{t})$, hence is a root subsystem of Φ.

We work with two Weyl groups for $\mathfrak{t}$ in $\mathfrak{g}$. The group $W = N_G(\mathfrak{t})/T$ is the **complex Weyl group** and may be identified with the Weyl group of the root system $(\mathfrak{g}, \mathfrak{t})$. The group $W_{\mathbb{R}} = N_{G_{\mathbb{R}}}(\mathfrak{t})/T_{\mathbb{R}}$ is the **real Weyl group**. Every member of $W_{\mathbb{R}}$ has a representative in $K_{\mathbb{R}}$, and $W_{\mathbb{R}}$ may be identified with the Weyl group of the root system $(\mathfrak{k}, \mathfrak{t})$. Thus $W_{\mathbb{R}} \subset W$.

Let $\widehat{T_{\mathbb{R}}}$ be the character group of $T_{\mathbb{R}}$. This is the same as the group $\widehat{T}$ of algebraic characters of T and is isomorphic to the weight lattice $\Lambda \subset i\mathfrak{t}_{\mathbb{R}}^*$, the isomorphism being $e^{\lambda} \leftrightarrow \lambda$, just as in the case of a maximal torus in a compact real form.

Let Θ be an invariant eigendistribution, and suppose that Θ is not 0 on $T_{\mathbb{R}}$. First suppose G is acceptable. Then (6) and the compactness of $T_{\mathbb{R}}$ imply that $\Theta|_{T_{\mathbb{R}}}\Delta$ is a $\mathbb{C}$-linear combination of expressions e^{μ} with μ in the weight lattice. If e^{μ} occurs with a nonzero coefficient, then Θ has infinitesimal character μ, i.e., the center of $U(\mathfrak{g})$ acts on Θ by $z\Theta = \chi_{\mu}(z)\Theta$. Let Φ^+ be a system of positive roots, and let ρ be half the sum of the positive roots. Then we can write

$$e^{\rho}\Big(\prod_{\alpha\in\Phi^+}(1-e^{-\alpha})\Big)\Theta|_{T_{\mathbb{R}}} = \sum_{w\in W} a_w e^{w\mu}$$

for some constants a_w. If G is not necessarily acceptable, we can make sense of this expression by multiplying through by $e^{-\rho}$, and we see that $w\mu - \rho$ must be a weight. Putting $\mu = \lambda + \rho$, we see for any $G_{\mathbb{R}}$, acceptable or not, that there exists $\lambda \in \Lambda$ such that $\Theta|_{T_{\mathbb{R}}}$ is given by the well defined expression

$$\Theta|_{T_{\mathbb{R}}} = \frac{\sum_{w\in W} a_w e^{w(\lambda+\rho)}}{\prod_{\alpha\in\Phi^+}(e^{\alpha/2} - e^{-\alpha/2})}.$$

The restriction $\Theta|_{T_{\mathbb{R}}}$ must be $W_{\mathbb{R}}$-invariant, and the Weyl denominator is $W_{\mathbb{R}}$-skew. Thus we can rewrite the above expression as follows: Choose $\lambda_1, \dots, \lambda_k$ so that $\{w(\lambda+\rho) \mid w \in W\}$ is the disjoint union of the $\{w(\lambda_i+\rho) \mid w \in W_{\mathbb{R}}\}$. Then there exist $a_1, \dots, a_k$ such that

$$\Theta|_{T_{\mathbb{R}}} = \sum_i a_i \frac{\sum_{w\in W_{\mathbb{R}}} \varepsilon(w) e^{w(\lambda_i+\rho)}}{\prod_{\alpha\in\Phi^+}(e^{\alpha/2} - e^{-\alpha/2})}.$$

Moreover, if $\lambda_i + \rho$ happens to be Φ_c-singular, $\sum_{w\in W_{\mathbb{R}}} \varepsilon(w)e^{w(\lambda_i+\rho)}$ vanishes, so a_i can be chosen to be zero in that case.

As a result, we have a bound on the number of discrete series representations. Any such representation has infinitesimal character $\chi_{\lambda+\rho}$ for some $\lambda \in \Lambda$. For fixed infinitesimal character, the global characters must be linearly independent on $T_{\mathbb{R}}$, and the dimension of their span must be $\le |W/W_{\mathbb{R}}|$. Hence there are at most $|W/W_{\mathbb{R}}|$ discrete series representations with infinitesimal character $\chi_{\lambda+\rho}$. More specifically, Harish-Chandra's theorem below asserts that there are exactly $|W/W_{\mathbb{R}}|$ discrete series representations with infinitesimal character $\chi_{\lambda+\rho}$ if $\lambda+\rho$ is regular, and no such representations if $\lambda+\rho$ is singular.

Fix a system of positive roots, and use it to define ρ.

Theorem (Harish-Chandra [HC6]). *Let G have a compact Cartan subgroup $T_{\mathbb{R}}$. Suppose that λ is in Λ with $\lambda+\rho$ regular. Then there exists a unique invariant eigendistribution $\Theta_{\lambda+\rho}$ such that*

(a) *$\Theta_{\lambda+\rho}$ decays at ∞ and*

(b) $\Theta_{\lambda+\rho}|_{T_{\mathbb{R}}} = (-1)^q \dfrac{\sum_{w\in W_{\mathbb{R}}} \varepsilon(w)e^{w(\lambda+\rho)}}{\prod_{\alpha\in\Phi \text{ with } (\lambda+\rho,\alpha)>0} (e^{\alpha/2} - e^{-\alpha/2})}$,
where $q = \frac{1}{2}\dim G_{\mathbb{R}}/K_{\mathbb{R}}$.

Every discrete series character is one of the $\Theta_{\lambda+\rho}$ *with* $\lambda+\rho$ *regular, and conversely every* $\Theta_{\lambda+\rho}$ *is a discrete series character.*

Because discrete series characters are determined by their restrictions to $T_{\mathbb{R}}$, it is implicit in the statement of the theorem that $\Theta_{\lambda_1+\rho} = \Theta_{\lambda_2+\rho}$ if and only if $w(\lambda_1+\rho) = \lambda_2+\rho$ for some $w \in W_{\mathbb{R}}$.

We shall sketch a proof of this theorem in §§4–5, using geometric realizations. For the remainder of this section, we shall prepare the setting.

Let X be the flag variety of $\mathfrak{g}$. It is known (see [Wo]) that $G_{\mathbb{R}}$ acts on X with only finitely many orbits. Consequently there must be open orbits. Let us see what the open orbits are under the assumption that $G_{\mathbb{R}}$ has a compact Cartan subgroup $T_{\mathbb{R}}$.

Let Φ be the root system of $(\mathfrak{g},\mathfrak{t})$, and let Φ^+ be a positive system for Φ. Let $\mathfrak{b}$ be the Borel subalgebra $\mathfrak{b} = \mathfrak{t}\oplus\mathfrak{n}$, where $\mathfrak{n} = \bigoplus_{\alpha\in\Phi^+}\mathfrak{g}^{-\alpha}$, and define $B = N_G(\mathfrak{b})$.

Fact. *Let* $D = D(\Phi^+)$ *be the* $G_{\mathbb{R}}$*-orbit of* $\mathfrak{b}$ *in* X. *Then* D *is open, and every open orbit arises in this way by choosing a suitable positive system. If* $\widetilde{\Phi^+}$ *is a second positive system, then* $D(\Phi^+) = D(\widetilde{\Phi^+})$ *if and only if* Φ^+ *and* $\widetilde{\Phi^+}$ *are conjugate under the action of* $W_{\mathbb{R}}$.

As in the compact case (§1), the isotropy subgroup of $G_{\mathbb{R}}$ at $\mathfrak{b} = \mathfrak{t}\oplus\mathfrak{n}$ is

$$G_{\mathbb{R}}\cap B = G_{\mathbb{R}}\cap B\cap\overline{B} = G_{\mathbb{R}}\cap T = T_{\mathbb{R}}.$$

Therefore $D \simeq G_{\mathbb{R}}/T_{\mathbb{R}}$ has the same dimension as X, and thus must be open. If $\mathfrak{b}$ and $\tilde{\mathfrak{b}}$ are two Borel subalgebras of $\mathfrak{g}$, both normalized by $T_{\mathbb{R}}$ and conjugate under $G_{\mathbb{R}}$, they must be conjugate even under the normalizer of $T_{\mathbb{R}}$ in $G_{\mathbb{R}}$ since $\mathfrak{g}_{\mathbb{R}}\cap\mathfrak{b} = \mathfrak{t} = \mathfrak{g}_{\mathbb{R}}\cap\tilde{\mathfrak{b}}$; this implies that $D(\Phi^+) = D(\widetilde{\Phi^+})$ if and only if $\widetilde{\Phi^+}$ is $W_{\mathbb{R}}$-conjugate to Φ^+. To see that orbits of this type are the only open $G_{\mathbb{R}}$-orbits in X, one can argue as follows. Any Borel subalgebra $\mathfrak{b}$ intersects its complex conjugate $\bar{\mathfrak{b}}$ in a subalgebra of $\mathfrak{g}$ which is defined over $\mathbb{R}$ and has the same rank as $\mathfrak{g}$. Thus $\mathfrak{b}\cap\bar{\mathfrak{b}}$ contains a Cartan subalgebra $\mathfrak{h}_{\mathbb{R}}$ of $\mathfrak{g}_{\mathbb{R}}$. One can show that the root system $\Phi(\mathfrak{g},\mathfrak{h})$ contains at least one real root, provided the Cartan subgroup $H_{\mathbb{R}}$ with Lie algebra $\mathfrak{h}_{\mathbb{R}}$ is not fundamental, i.e., not maximally compact – in our situation, if $H_{\mathbb{R}}$ is not $G_{\mathbb{R}}$-conjugate to $T_{\mathbb{R}}$. When $\Phi(\mathfrak{g},\mathfrak{h})$ contains real roots, the real dimension of $\mathfrak{g}_{\mathbb{R}}\cap\mathfrak{b}$ exceeds the rank of $\mathfrak{g}$, so the $G_{\mathbb{R}}$-orbit of $\mathfrak{b}$ cannot be open.

Let us consider a particular open orbit $D = D(\Phi^+) \simeq G_{\mathbb{R}}/T_{\mathbb{R}}$. To any λ in Λ, we associate the $G_{\mathbb{R}}$-equivariant holomorphic line bundle $\mathcal{L}_\lambda = G_{\mathbb{R}}\times_{T_{\mathbb{R}}}\mathbb{C}_\lambda$ over D, where $T_{\mathbb{R}}$ acts on $\mathbb{C}_\lambda$ by e^λ. This is a $G_{\mathbb{R}}$-equivariant holomorphic line bundle, being the restriction of the G-equivariant holomorphic line bundle over X constructed in §1. In discussing geometric realizations of discrete series and the proof of Harish-Chandra's theorem, we shall look at two types of cohomology groups associated to the line bundle $\mathcal{L}_\lambda$.

The first type is **sheaf cohomology** $H^*(D,\mathcal{O}(\mathcal{L}_\lambda))$. We use the Dolbeault isomorphism to topologize this vector space. Dolbeault cohomology is the quotient of the space of cocycles by the subspace of coboundaries. The C^∞ topology on differential forms turns the space of cocycles into a Fréchet space. There is no a priori reason why the $\bar\partial$ operator must have closed range, so the quotient might not be Hausdorff. But in any event, $G_{\mathbb{R}}$ acts continuously with respect to the quotient topology on $H^*(D, \mathcal{O}(\mathcal{L}_\lambda))$.

The second type is the **separated L^2 cohomology** of $\mathcal{L}_\lambda$, which we denote by $H^*_{(2)}(D, \mathcal{O}(\mathcal{L}_\lambda))$. To define it, we put $G_{\mathbb{R}}$-invariant hermitian metrics on D and on the line bundle $\mathcal{L}_\lambda$. These choices give meaning to $\bar\partial^*$, the formal adjoint of the $\bar\partial$ operator, and to the Laplace-Beltrami operator $\square = \bar\partial^*\bar\partial + \bar\partial\bar\partial^*$. We regard $\bar\partial$, $\bar\partial^*$, and $\square$ as unbounded operators on the Hilbert space

$$L^*(\mathcal{L}_\lambda) \;=\; L^2 \text{ closure of } \; A^*_c(\mathcal{L}_\lambda)\,,$$

by taking the closures of these operators acting on $A^*_c(\mathcal{L}_\lambda)$, the space of compactly supported, smooth Dolbeault forms. By definition,

$$\begin{aligned} H^*_{(2)}(D, \mathcal{O}(\mathcal{L}_\lambda)) \;&=\; \text{Ker } \square \text{ acting on } L^*(\mathcal{L}_\lambda) \\ &=\; (\,\text{Ker } \bar\partial \text{ acting on } L^*(\mathcal{L}_\lambda)\,) \cap (\,\text{Ker } \bar\partial^* \text{ acting on } L^*(\mathcal{L}_\lambda)\,)\,. \end{aligned}$$

This is a Hilbert space, on which $G_{\mathbb{R}}$ operates continuously and unitarily. In the present situation, the definition of L^2 cohomology coincides with the most naive one, namely as the kernel of $\square$ acting on $L^*(\mathcal{L}_\lambda) \cap A^*(\mathcal{L}_\lambda)$: the metric on D is invariant, hence complete, and this implies that the largest and smallest closed extensions of $\square$ coincide [An-Ve].

Since $\square$ is elliptic, every $\phi \in H^*_{(2)}(D, \mathcal{O}(\mathcal{L}_\lambda))$ is a $\bar\partial$-closed, smooth Dolbeault form, and as such determines a Dolbeault cohomology class. In other words, there exists a natural map

$$H^*_{(2)}(D, \mathcal{O}(\mathcal{L}_\lambda)) \;\longrightarrow\; H^*(D, \mathcal{O}(\mathcal{L}_\lambda))\,.$$

In the case of a compact manifold, the Hodge theorem asserts that this natural map from L^2 cohomology to sheaf cohomology is an isomorphism. In general, on a noncompact manifold, this map may be neither injective nor surjective; broadly speaking, the L^2 cohomology reflects more the curvature of the manifold and of the bundle than their cohomological properties.

4. Realizations in L^2 Cohomology

We continue to assume that $T_{\mathbb{R}} \subset K_{\mathbb{R}} \subset G_{\mathbb{R}} \subset G$. We fix a positive system Φ^+ of roots of $(\mathfrak{g}, \mathfrak{t})$ and let $\mathfrak{b} = \mathfrak{b}(\Phi^+) = \mathfrak{t} \oplus \mathfrak{n}$ with $\mathfrak{n}$ built from the root spaces for the negative roots. Let $X \simeq G/B$ be the flag variety, and let $D = D(\Phi^+)$ be the (open) $G_{\mathbb{R}}$-orbit of $\mathfrak{b}$ in X. We have $D \simeq G_{\mathbb{R}}/T_{\mathbb{R}}$. For $\lambda \in \Lambda$, the G-equivariant holomorphic line bundle $\mathcal{L}_\lambda$ on X restricts to a $G_{\mathbb{R}}$-equivariant holomorphic line bundle on the complex manifold D. In this section we study $H^*_{(2)}(D, \mathcal{O}(\mathcal{L}_\lambda))$.

Define

$$p(\lambda+\rho) = \#\{\alpha \in \Phi^+_c \mid (\lambda+\rho, \alpha) < 0\} + \#\{\beta \in \Phi^+_n \mid (\lambda+\rho, \beta) > 0\}.$$

Note that this definition is consistent with the definition in §1 in the case that $G_{\mathbb{R}}$ is compact. Whether or not $G_{\mathbb{R}}$ is compact, this integer has a geometric interpretation as the number of strictly negative eigenvalues of the curvature form of $\mathcal{L}_{\lambda+\rho}$ relative to a $G_{\mathbb{R}}$-invariant hermitian metric [Gr-Sch].

Theorem (Langlands Conjecture). *Let λ be in Λ.*

(a) $H^p_{(2)}(D, \mathcal{O}(\mathcal{L}_\lambda)) = 0$ *if* $p \neq p(\lambda+\rho)$ *or if* $\lambda+\rho$ *is singular.*

(b) *If* $\lambda+\rho$ *is regular and* $p = p(\lambda+\rho)$, *then* $H^p_{(2)}(D, \mathcal{O}(\mathcal{L}_\lambda))$ *is nonzero, is irreducible, is square integrable, and has character* $\Theta_{\lambda+\rho}$.

REMARKS.

1) Recall that L^2 cohomology and sheaf cohomology are isomorphic when $D \simeq G_{\mathbb{R}}/T_{\mathbb{R}}$ is compact, which is the case precisely when $G_{\mathbb{R}}$ is a compact real form. Also for $G_{\mathbb{R}}$ compact, all irreducible representations belong to the discrete series. Hence, in the compact case, the Langlands Conjecture reduces to the Borel-Weil-Bott Theorem.

2) Conclusion (b) of the theorem gives all discrete series representations, even under the additional hypothesis that $(\lambda+\rho, \alpha) < 0$ for all $\alpha \in \Phi^+$. To obtain all of the discrete series in this way, one needs to use all positive root systems Φ^+ modulo the action of $W_{\mathbb{R}}$, or equivalently, all open orbits.

Historically the Langlands Conjecture came about as follows. While visiting Berkeley in 1965/66, Langlands talked with Griffiths, who had just calculated the curvature of homogeneous holomorphic vector bundles over classifying spaces for Hodge structures. These calculations also applied to open $G_{\mathbb{R}}$-orbits in flag manifolds. In that case, they led to the conclusion that $H^p_{(2)}(D, \mathcal{O}(\mathcal{L}_\lambda))$ can be nonzero only in degree $p = p(\lambda+\rho)$, at least if $\lambda+\rho$ is "very regular", i.e., if $|(\alpha, \lambda+\rho)| \gg 0$ for every $\alpha \in \Phi$. Langlands had observed that a purely formal application of the Atiyah-Bott fixed point formula to the action of $T_{\mathbb{R}}$ on $H^*_{(2)}(D, \mathcal{O}(\mathcal{L}_\lambda))$ produced Harish-Chandra's formula for the discrete series character $\Theta_{\lambda+\rho}$. Combining this observation with the vanishing theorem for L^2 cohomology in the very regular case, Langlands arrived at his conjecture.

The remainder of this section will be devoted to a sketch of the proof of the Langlands Conjecture and of Harish-Chandra's theorem in §3.

For $\lambda \in \Lambda$, let $\mathbb{C}_\lambda$ denote the one-dimensional $T_{\mathbb{R}}$-module with action e^λ. For any $T_{\mathbb{R}}$-module U, let $U_{-\lambda} = (U \otimes \mathbb{C}_\lambda)^{T_{\mathbb{R}}}$ be the $-\lambda$ weight space of U under the action of $T_{\mathbb{R}}$. Recall the abstract Plancherel decomposition for $G_{\mathbb{R}}$:

$$L^2(G_{\mathbb{R}}) \simeq \int_{i \in \widehat{G_{\mathbb{R}}}} V_i \mathbin{\widehat{\otimes}} V_i^* \, d\mu(i) ,$$

where $V_i \mathbin{\widehat{\otimes}} V_i^*$ is the completed tensor product of the Hilbert spaces V_i, V_i^*. Note that the natural inclusion $V_i \otimes V_i^* \hookrightarrow \mathrm{End}(V_i)$ of the algebraic tensor product into $\mathrm{End}(V_i)$ extends isometrically to a canonical isomorphism $V_i \mathbin{\widehat{\otimes}} V_i^* \simeq \mathrm{End}(V_i)_{\mathrm{HS}}$.

The arguments will depend on three crucial lemmas. For $i \in \widehat{G_{\mathbb{R}}}$, V_i^∞ will denote the space of C^∞ vectors in V_i.

Lemma 1. *For all* $i \in \widehat{G_{\mathbb{R}}}$ *and* $\lambda \in \Lambda$, *the inclusion* $(V_i)_{K_{\mathbb{R}}\text{-finite}} \hookrightarrow V_i^\infty$ *induces an isomorphism*

$$H^*(\mathfrak{n}, (V_i)_{K_{\mathbb{R}}\text{-finite}})_{-\lambda} \simeq H^*(\mathfrak{n}, V_i^\infty)_{-\lambda} .$$

These weight spaces are finite-dimensional.

Lemma 2. *The Plancherel decomposition induces a* $G_{\mathbb{R}}$*-equivariant isomorphism*

$$H^p_{(2)}(D, \mathcal{O}(\mathcal{L}_\lambda)) \simeq \int_{i \in \widehat{G_{\mathbb{R}}}} V_i \otimes H^p(\mathfrak{n}, V_i^{*\,\infty})_{-\lambda} \, d\mu(i) ,$$

which becomes an isometry with respect to an appropriately chosen inner product on $H^p(\mathfrak{n}, V_i^{*\,\infty})_{-\lambda}$.

Lemma 3. *For any* $i \in \widehat{G_{\mathbb{R}}}$,

$$\Big(\prod_{\alpha\in\Phi^+}(1-e^{\alpha})\Big)\Theta(V_i)|_{T_{\mathbb{R}}} = \mathrm{char}_{T_{\mathbb{R}}}\Big(\sum_p (-1)^p H^p(\mathfrak{n}, (V_i)_{K_{\mathbb{R}}\text{-finite}})\Big),$$

where Θ_i *is the global character of* V_i.

REMARKS.

1) It is instructive to see what these lemmas say when $G_{\mathbb{R}}$ is compact. Lemma 1 becomes a tautology, since then $K_{\mathbb{R}} = G_{\mathbb{R}}$, so every vector is both $K_{\mathbb{R}}$-finite and differentiable. In the compact case the left side in the formula of Lemma 2 is isomorphic with sheaf cohomology, and the conclusion of Lemma 2 is a formal one that already appeared in §2 as (∗∗∗). For Lemma 3, one uses the Euler-Poincaré principle: the Euler characteristic of a finite dimensional complex agrees with that of its cohomology; applied to the weight spaces in the standard complex, this implies Lemma 3 if $G_{\mathbb{R}}$ is compact.

2) The proofs of Lemma 1 and 2 are related. Let $d_{i^*} : \mathrm{Hom}(\bigwedge\mathfrak{n}, (V_i^*)_{K_{\mathbb{R}}\text{-finite}}) \to \mathrm{Hom}(\bigwedge\mathfrak{n}, (V_i^*)_{K_{\mathbb{R}}\text{-finite}})$ denote the coboundary operator in the standard complex for $\mathfrak{n}$-cohomology of the Harish-Chandra module $(V_i^*)_{K_{\mathbb{R}}\text{-finite}}$. Let $(d_{i^*})^*$ be its formal adjoint, relative to the inner product on $(V_i^*)_{K_{\mathbb{R}}\text{-finite}}$ coming from V_i^* and the inner product on $\mathfrak{n}$ which the hermitian metric of D induces on

$$\mathfrak{n} \simeq \text{antiholomorphic tangent space of } D \simeq G_{\mathbb{R}}/T_{\mathbb{R}} \text{ at the identity coset.}$$

The closed extension of $\Box_{i^*} = d_{i^*}(d_{i^*})^* + (d_{i^*})^* d_{i^*}$ from $\mathrm{Hom}(\bigwedge\mathfrak{n}, (V_i^*)_{K_{\mathbb{R}}\text{-finite}})$ to the Hilbert space $\mathrm{Hom}(\bigwedge\mathfrak{n}, V_i^*)$ is a self-adjoint operator. Let $\mathcal{H}^p(\mathfrak{n}, V_i^*)$ denote the kernel of $\Box_i^*$. Then, arguing as in the compact case, one finds

$$H^p_{(2)}(D, \mathcal{O}(\mathcal{L}_\lambda)) \simeq \int_{i\in\widehat{G}_{\mathbb{R}}} V_i \mathbin{\widehat{\otimes}} \mathcal{H}^p(\mathfrak{n}, V_i^*)_{-\lambda}\, d\mu(i)\,.$$

In effect, the Laplace-Beltrami operator $\Box$ is the direct integral of the operators $1_{V_i} \mathbin{\widehat{\otimes}} \Box_{i^*}$, restricted to the $(-\lambda)$-weight space, and the kernels decompose accordingly.

3) One can study the harmonic space $\mathcal{H}^p(\mathfrak{n}, V_i^*)_{-\lambda}$ using the following heuristic dictionary:

$$V_i^* \leftrightarrow L^2 \text{ functions on a compact complex manifold}$$

$$(V_i^*)^\infty \leftrightarrow C^\infty \text{ function on a compact complex manifold}$$

$$\mathrm{Hom}(\textstyle\bigwedge\mathfrak{n}, \cdot\,) \leftrightarrow \text{Dolbeault complex} \begin{cases} \text{with } L^2 \text{ coefficients for } V_i^* \\ \text{with } C^\infty \text{ coefficients for } (V_i^*)^\infty. \end{cases}$$

Following the pattern of the proof of the Hodge theorem, this gives a natural isomorphism $\mathcal{H}^*(\mathfrak{n}, V_i^*)_{-\lambda} \simeq H^*(\mathfrak{n}, (V_i^*)^\infty)_{-\lambda}$.

4) The mechanism of the Hochschild-Serre spectral sequence for the pair $(\mathfrak{n}\cap\mathfrak{k}, \mathfrak{n})$ produces two spectral sequences which compute, respectively, $H^*(\mathfrak{n}, (V_i)_{K_{\mathbb{R}}\text{-finite}})_{-\lambda}$ and $H^*(\mathfrak{n}, (V_i)^\infty)_{-\lambda}$. The inclusion $(V_i)_{K_{\mathbb{R}}\text{-finite}} \hookrightarrow V_i^\infty$ induces a morphism of spectral sequences, which becomes an isomorphism at the E_2 level. What one uses here is an argument similar to the one in Remark 3, but applied to $(\mathfrak{n}\cap\mathfrak{k})$-cohomology. The isomorphism at E_2 implies $H^*(\mathfrak{n}, (V_i)_{K_{\mathbb{R}}\text{-finite}})_{-\lambda} \simeq H^*(\mathfrak{n}, (V_i)^\infty)_{-\lambda}$. The finite

dimensionality of these spaces becomes visible already at the E_2 level. Lemma 1 follows. This last step, combined with Remarks 2 and 3, also implies Lemma 2.

5) The proof of Lemma 3 also uses the Hochschild-Serre spectral sequence for the pair $(\mathfrak{n} \cap \mathfrak{k}, \mathfrak{n})$. Finite dimensionality at E_2 implies, for purely formal reasons, the assertion of the Lemma when the global character $\Theta(V_i)$ is replaced by the $K_{\mathbb{R}}$-character of V_i, i.e., by the formal sum of the $K_{\mathbb{R}}$-characters of the $K_{\mathbb{R}}$-irreducible constituents of V_i. This sum converges in the sense of distributions to a limit whose restriction to $K_{\mathbb{R}} \cap G'_{\mathbb{R}}$ ($G'_{\mathbb{R}} =$ set of regular semisimple elements in $G_{\mathbb{R}}$) is a real analytic function – however, unlike the global character, which is a locally L^1 function on $G_{\mathbb{R}}$, the $K_{\mathbb{R}}$-character is not locally L^1 as function on $K_{\mathbb{R}}$; in particular, the $K_{\mathbb{R}}$-character is not determined by its restriction to $K_{\mathbb{R}} \cap G'_{\mathbb{R}}$. According to a result of Harish-Chandra [HC2], which has analytic content but is not very difficult, the global character and the $K_{\mathbb{R}}$-character agree as functions on $K_{\mathbb{R}} \cap G'_{\mathbb{R}}$. Lemma 3 follows.

A lemma of Casselman-Osborne [Cas-O] asserts that if $H^p(\mathfrak{n}, V_i^{\infty})_\lambda \neq 0$, then V_i has infinitesimal character $\chi_{\lambda-\rho}$. Thus, by duality, if $H^p(\mathfrak{n}, V_i^{*\,\infty})_{-\lambda}$ is not 0, then V_i has infinitesimal character $\chi_{\lambda+\rho}$. There are only finitely many irreducible representations V_i with this property, and hence the integrand on the right side in Lemma 2 is nonzero on only a finite set. The only contribution to the integral can therefore be from representations with $\mu(\{i\}) > 0$, and these are discrete series representations by the Lemma in §3. We state the result as a corollary.

Corollary. *$H^p_{(2)}(D, \mathcal{O}(\mathcal{L}_\lambda))$ is a finite direct sum of discrete series representations, all having infinitesimal character $\chi_{\lambda+\rho}$.*

It will be convenient now to fix a normalization of Haar measure for $G_{\mathbb{R}}$. Take a compact real form $U_{\mathbb{R}}$ of G whose Lie algebra is invariant under the Cartan involution for $\mathfrak{g}_{\mathbb{R}}$, and normalize Haar measure on $U_{\mathbb{R}}$ to have total mass 1. Express the Haar measure du as the restriction to $U_{\mathbb{R}}$ of a holomorphic differential form of top degree on G, and then restrict this differential form from G to $G_{\mathbb{R}}$. The resulting differential form gives a specific Haar measure on $G_{\mathbb{R}}$. In turn, the normalization of Haar measure determines a specific normalization of the Plancherel measure μ.

We recall the discussion of formal degree in connection with the Lemma in §3. If i is in the discrete series, then its formal degree is given by $d_i = \mu(\{i\})$.

We shall work with the expression $\sum_p (-1)^p \dim H^p(\mathfrak{n}, V_i^{*\,\infty})_{-\lambda}$. Application of Lemma 2 yields the following observation.

Observation. *$\sum_p (-1)^p \dim H^p(\mathfrak{n}, V_i^{*\,\infty})_{-\lambda}$ is the multiplicity of V_i in the virtual $G_{\mathbb{R}}$-module $\sum_p (-1)^p H^p_{(2)}(D, \mathcal{O}(\mathcal{L}_\lambda))$.*

We shall use Atiyah's L^2 Index Theorem, which we state later in this section. This theorem implies that

$$\sum_p \sum_{\substack{i \in \text{ discrete series,} \\ \text{inf. char. } \chi_{\lambda+\rho}}} d_i (-1)^p \dim H^p(\mathfrak{n}, V_i^{*\,\infty})_{-\lambda}$$

$$= (-1)^q \sum_p (-1)^p \dim H^p(X, \mathcal{O}(\mathcal{L}_\lambda)),$$

where $q = \frac{1}{2}\dim G_{\mathbb{R}}/K_{\mathbb{R}}$. The Borel-Weil-Bott Theorem and the Weyl Dimension Formula for holomorphic representations of G show that the right side of this identity is $(-1)^q \prod_{\alpha\in\Phi^+}\big((\lambda+\rho,\alpha)/(\rho,\alpha)\big)$. Let us multiply both sides of the identity by $e^{-\lambda}$, replace λ by $w(\lambda+\rho)-\rho$, sum on $w\in W$, and multiply by $e^{-\rho}$. Lemma 3 shows that the resulting left side is

$$e^{-\rho}\Big(\prod_{\alpha\in\Phi^+}(1-e^{\alpha})\Big)\sum_i d_i\Theta(V_i^*)|_{T_{\mathbb{R}}}.$$

The resulting right side is

$$(-1)^q\Big(\prod_{\alpha\in\Phi^+}\frac{(\lambda+\rho,\alpha)}{(\rho,\alpha)}\Big)\sum_{w\in W}\varepsilon(w)e^{-w(\lambda+\rho)}.$$

Taking the complex conjugate of our equality therefore gives

$$(\dagger)\qquad \sum_{\substack{i\in\text{ discrete series,}\\ \text{inf. char. }\chi_{\lambda+\rho}}} d_i\Theta(V_i)|_{T_{\mathbb{R}}} = (-1)^q\Big(\prod_{\alpha\in\Phi^+}\frac{(\lambda+\rho,\alpha)}{(\rho,\alpha)}\Big)\frac{\sum_{w\in W}\varepsilon(w)e^{w(\lambda+\rho)}}{\prod_{\alpha\in\Phi^+}(e^{\alpha/2}-e^{-\alpha/2})}.$$

The right side of (†) vanishes if $\lambda+\rho$ is singular. Since restriction to $T_{\mathbb{R}}$ is injective on the linear span of the discrete series characters, it follows that no discrete series representation has singular infinitesimal character. In view of our earlier corollary, this in turn implies

$$H^*_{(2)}(D,\mathcal{O}(\mathcal{L}_\lambda))=0 \qquad \text{if } \lambda+\rho \text{ is singular.}$$

We have established both Harish-Chandra's theorem and Langlands' conjecture in the special case when $\lambda+\rho$ is singular. From now on, then, let us suppose that $\lambda+\rho$ is regular.

Lemma 4. *Let i be a class in the discrete series, λ a weight with $\lambda+\rho$ regular, and p an integer such that $H^p(\mathfrak{n},V_i^\infty)_{-\lambda}\neq 0$. Then*

(a) $p=p(\lambda+\rho)$;
(b) $\dim H^p(\mathfrak{n},V_i^\infty)_{-\lambda}=1$;
(c) $H^*(\mathfrak{n},V_i^\infty)_{-\lambda'}\neq 0$, *for* $\lambda'\in\Lambda \implies \lambda'+\rho$ *is* $W_{\mathbb{R}}$*-conjugate to* $\lambda+\rho$.

Sketch of Proof. First, let us suppose that $\lambda+\rho$ is sufficiently regular, in the sense that $|(\lambda+\rho,\alpha)|>C$ for every root α, with $C>0$ to be specified in the course of the argument. If C is large enough, a curvature argument in the style of Bochner, Yano, and Kodaira shows that $H^p_{(2)}(D,\mathcal{O}(\mathcal{L}_\lambda))=0$ if $p\neq p(\lambda+\rho)$ [Gr-Sch]. The same argument, applied not to the L^2 Dolbeault complex for $\mathcal{L}_\lambda$, but to the formal complex $\operatorname{Hom}(\bigwedge\mathfrak{n},V_i)$ mentioned in Remarks 2 and 3 above, shows

$$(1)\qquad \mathcal{H}^p(\mathfrak{n},V_i)_{-\lambda}\neq 0 \quad\implies\quad p=p(\lambda+\rho).$$

What matters here is the unitarity of V_i, not the fact that i belongs to the discrete series. At this point, Remark 3 implies assertion (a), provided $\lambda+\rho$ is sufficiently regular.

The curvature estimate that leads to (1) proves more. Let

$$\mu=-\lambda-\textstyle\sum\{\alpha\in\Phi_c\cap\Phi^+ \mid (\lambda,\alpha)<0\}-\sum\{\beta\in\Phi_n\cap\Phi^+\mid(\lambda,\beta)>0\},$$

choose $w \in W_{\mathbb{R}}$ so that $w\mu$ is dominant with respect to $\Phi_c \cap \Phi^+$, and let U^ν denote the irreducible $K_{\mathbb{R}}$-module of highest weight ν if $\nu \in \Lambda$ is $(\Phi_c \cap \Phi^+)$-dominant. Then, again under the hypothesis of sufficient regularity, for $p = p(\lambda + \rho)$,

$$\dim \mathcal{H}^p(\mathfrak{n}, V_i)_{-\lambda} = \dim \operatorname{Hom}_{K_{\mathbb{R}}}(U^{w\mu}, V_i), \quad \text{and}$$

$$\text{(2)} \qquad \mathcal{H}^p(\mathfrak{n}, V_i)_{-\lambda} \neq 0 \implies \begin{cases} \operatorname{Hom}_{K_{\mathbb{R}}}(U^{w(\mu-B)}, V_i) = 0 \quad \text{for every nonzero} \\ \text{sum } B \text{ of distinct roots in } \{\beta \in \Phi_n \mid (\lambda, \beta) < 0\}. \end{cases}$$

We shall see in the next section that the occurrence of $U^{w\mu}$ and simultaneous non-occurrence of $U^{w(\mu-B)}$ in V_i, for all B as in (2), determines V_i up to isomorphism, and further implies

$$\dim \operatorname{Hom}_{K_{\mathbb{R}}}(U^{w\mu}, V_i) = 1, \quad \text{and}$$

$$\text{(3)} \qquad \operatorname{Hom}_{K_{\mathbb{R}}}(U^{\nu}, V_i) \neq 0 \text{ only if } \begin{cases} \nu \text{ is of the form } w(\mu + C), \text{ where } C \text{ is} \\ \text{a sum of (not necessarily distinct) roots} \\ \text{in } \{\beta \in \Phi_n \mid (\lambda, \beta) < 0\}. \end{cases}$$

The assertions (3) are not so difficult to establish in the sufficiently regular case; a self-contained argument can be found in [At-Sch, Appendix]. Appealing once more to Remark 3, we get assertion (b), still in the sufficiently regular case. Assertion (c) also follows: the cone in which the $K_{\mathbb{R}}$-spectrum of V_i must lie according to (3) determines $\lambda + \rho$ up to $K_{\mathbb{R}}$-conjugacy.

We now use (the most naive version of) the Jantzen-Zuckerman translation principle [Zu] to remove the hypothesis of sufficient regularity. According to this principle, if n a is positive integer, there exists a natural bijection $i \leftrightarrow \tilde{i} = \tilde{i}(i, n)$ between discrete series characters Θ_i with infinitesimal character $\chi_{-\lambda-\rho}$ and discrete series characters $\Theta_{\tilde{i}}$ with infinitesimal character $\chi_{n(-\lambda-\rho)}$. Loosely speaking, the correspondence is characterized by the formulas for the "Weyl numerators" $\Delta\Theta_i$ and $\Delta\Theta_{\tilde{i}}$, on every connected component of the regular set $H'_{\mathbb{R}}$ in every Cartan subgroup $H_{\mathbb{R}} \subset G_{\mathbb{R}}$: the formulas for $\Delta\Theta_{\tilde{i}}$ are obtained from those for $\Delta\Theta_i$ by consistently substituting $-n(\lambda + \rho)$ for $-(\lambda + \rho)$. Under the process of translation from $-(\lambda+\rho)$ to $-n(\lambda+\rho)$, $\mathfrak{n}$-cohomology of the underlying Harish-Chandra module behaves exactly like the global character; this follows from the Casselman-Osborne lemma [Cas-O]. In particular,

$$H^*(\mathfrak{n}, (V_{\tilde{i}})_{K_{\mathbb{R}}\text{-finite}})_{-n(\lambda+\rho)+\rho} \simeq H^*(\mathfrak{n}, (V_i)_{K_{\mathbb{R}}\text{-finite}})_{-\lambda}.$$

One can make $n(\lambda+\rho)$ sufficiently regular by choosing n large enough, so Lemma 4 in the sufficiently regular case, together with Lemma 1, implies Lemma 4 in general.

It is now a simple matter to complete the proof of Harish-Chandra's theorem and of Langlands' conjecture. Let λ be a weight such that $\lambda + \rho$ is regular, and let i be a class in the discrete series with infinitesimal character $\chi_{\lambda+\rho}$. Note that the integer $p(\lambda + \rho) - p(w(\lambda + \rho))$, for $w \in W$, has even parity if $\varepsilon(w) = 1$, odd parity if $\varepsilon(w) = -1$. The four lemmas in this section apply uniformly for all positive root systems $\Phi^+ \subset \Phi(\mathfrak{g}, \mathfrak{t})$ and all weights λ, with ρ and the orbit D determined by Φ^+. Thus Lemma 4, applied to the dual class, and Lemmas 1 and 3 show

$$(\dagger\dagger) \qquad H^p(\mathfrak{n}, (V_i^*)^\infty)_{-\lambda} \neq 0 \iff \Delta\Theta_i|_{T_{\mathbb{R}}} = (-1)^p \sum_{w \in W_{\mathbb{R}}} \varepsilon(w) e^{w(\lambda+\rho)}.$$

Recall that restriction to $T_{\mathbb{R}}$ is injective on the linear span of the discrete series characters. Thus, again by Lemmas 1 and 3,

$$(\dagger\dagger\dagger) \qquad H^*(\mathfrak{n}, (V_i^*)^\infty)_{-w\lambda} \neq 0 \quad \text{for some } w \in W\,.$$

If $\lambda+\rho$ is dominant with respect to Φ^+, the integer $q = \frac{1}{2}\dim_{\mathbb{R}} G_{\mathbb{R}}/K_{\mathbb{R}} = |\Phi_n \cap \Phi^+|$ equals $p(\lambda+\rho)$. Comparing $(\dagger\dagger)$ and $(\dagger\dagger\dagger)$ with the earlier displayed statement $(\dagger)$, we get Harish-Chandra's formula for the restriction of the discrete series characters to $T_{\mathbb{R}}$, as well as his formula for the formal degree,

$$d_i = \prod_{(\lambda+\rho,\alpha)>0} \frac{(\lambda+\rho,\alpha)}{(\rho,\alpha)}\,.$$

Langlands' conjecture now follows from Lemma 2, Lemma 4, and $(\dagger\dagger)$.

We turn to Atiyah's L^2 Index Theorem. Let M be a compact manifold, and let $\tilde{M} \to M$ be a regular covering with Γ as group of deck transformations, so that $M = \Gamma\backslash\tilde{M}$. Suppose that D is an elliptic operator on M, mapping sections of one smooth vector bundle over M to sections of another such bundle, and let $\tilde{D}$ be its lifting to $\tilde{M}$. Impose Γ-invariant metrics on $\tilde{M}$ and the lifted vector bundles. Since M is compact, the notion of square integrability on $\tilde{M}$ is not affected by the particular choice of metrics on M and the bundles. Define $\tilde{D}^*$ to be the formal adjoint of $\tilde{D}$.

Let $\mathcal{H}^+$ and $\mathcal{H}^-$ be the L^2-kernels of $\tilde{D}$ and $\tilde{D}^*$, respectively. The orthogonal projections from the spaces of L^2 sections to $\mathcal{H}^+$ and $\mathcal{H}^-$ are integral operators with smooth kernel functions $k^\pm(x,y)$ that are Γ-invariant; the functions $k^\pm(x,y)$ take values in End of the respective bundles. Define

$$\dim_\Gamma \mathcal{H}^\pm = \int_{\Gamma\backslash\tilde{M}} \operatorname{Tr} k(x,x)\,dx$$

and

$$\operatorname{index}_\Gamma \tilde{D} = \dim_\Gamma \mathcal{H}^+ - \dim_\Gamma \mathcal{H}^-\,.$$

Theorem. *Under the above hypotheses,* $\operatorname{index}_\Gamma \tilde{D}$ *equals the usual index of* D *on* M*.*

Now suppose that $\tilde{M}$ is a homogeneous space of $G_{\mathbb{R}}$ with compact isotropy subgroup, and that the operator $\tilde{D}$ is $G_{\mathbb{R}}$-invariant. Suppose that Γ is a discrete cocompact subgroup of $G_{\mathbb{R}}$ with no elements of finite order, and set $M = \Gamma\backslash\tilde{M}$. Then it is possible to define a $G_{\mathbb{R}}$-index, which is related to the Γ-index by the formula

$$\operatorname{index}_\Gamma \tilde{D} = \operatorname{vol}(\Gamma\backslash G_{\mathbb{R}})\,\operatorname{index}_{G_{\mathbb{R}}} \tilde{D}.$$

This $G_{\mathbb{R}}$-index can be computed in terms of the Plancherel decompositions of $\ker\tilde{D}$ and $\ker\tilde{D}$: if

$$\mathcal{H}^\pm \simeq \int_{i\in\widehat{G}_{\mathbb{R}}} V_i \otimes U_i^\pm\,d\mu(i)\,,$$

then the spaces $U_i^\pm$, which quantify the multiplicities with which V_i occurs in the direct integrals, are finite dimensional almost everywhere, and

$$\operatorname{index}_{G_{\mathbb{R}}} \tilde{D} = \int_{i\in\widehat{G}_{\mathbb{R}}} (\dim U_i^+ - \dim U_i^-)\,d\mu(i)\,.$$

The relationship between the Γ-index and the $G_{\mathbb{R}}$-index and the formula expressing the $G_{\mathbb{R}}$-index in terms of the Plancherel decomposition are established by connecting all of these quantities to traces in the appropriate von Neumann algebras.

We apply these considerations with $\tilde{M} = G_{\mathbb{R}}/T_{\mathbb{R}}$, Γ discrete and cocompact in $G_{\mathbb{R}}$ with no elements of finite order, $M = \Gamma\backslash G_{\mathbb{R}}/T_{\mathbb{R}}$, and $\tilde{D} = \bar{\partial}+\bar{\partial}^*$, going from $\mathcal{L}_\lambda$-valued Dolbeault forms of even degree to forms of odd degree. A subgroup Γ with the required properties exists by a theorem of Borel [Bor1]. We need to compute the index of $\bar{\partial}+\bar{\partial}^*$ on $\Gamma\backslash G_{\mathbb{R}}/T_{\mathbb{R}}$, and we do so using the Hirzebruch proportionality principle [Hirz]. The computation compares $\Gamma\backslash G_{\mathbb{R}}/T_{\mathbb{R}}$ with $U_{\mathbb{R}}/T_{\mathbb{R}} = X$, where $U_{\mathbb{R}}$ is the compact form of G compatible with the Cartan involution of $G_{\mathbb{R}}$. The Atiyah-Singer index theorem expresses the index of $\bar{\partial}+\bar{\partial}^*$ on $\Gamma\backslash G_{\mathbb{R}}/T_{\mathbb{R}}$ as the value of a certain cohomology class on the fundamental cycle. This Atiyah-Singer class can be represented by a $G_{\mathbb{R}}$-invariant differential form on $G_{\mathbb{R}}/T_{\mathbb{R}}$, computable in terms of the curvature of $G_{\mathbb{R}}$-invariant line bundles; from $G_{\mathbb{R}}/T_{\mathbb{R}}$, the form descends to the quotient $\Gamma\backslash G_{\mathbb{R}}/T_{\mathbb{R}}$, where it can then be integrated to give the value of the Atiyah-Singer class on the fundamental cycle. Entirely analogously, the index of $\bar{\partial}+\bar{\partial}^*$ on $U_{\mathbb{R}}/T_{\mathbb{R}} = X$ can be computed by integrating a $U_{\mathbb{R}}$-invariant Atiyah-Singer form over X. Because of the invariance, the (top-degree components of the) two types of Atiyah-Singer forms are multiples of the natural volume forms on the two spaces. A comparison of the curvature forms of invariant line bundles on $G_{\mathbb{R}}/T_{\mathbb{R}}$ and $U_{\mathbb{R}}/T_{\mathbb{R}} = X$ shows

$$\begin{aligned}\text{index of } \bar{\partial}+\bar{\partial}^* \text{ on } \Gamma\backslash G_{\mathbb{R}}/T_{\mathbb{R}} \;&= \\ &(-1)^q \operatorname{vol}(\Gamma\backslash G_{\mathbb{R}}) \;\times\; \text{index of } \bar{\partial}+\bar{\partial}^* \text{ on } \Gamma\backslash G_{\mathbb{R}}/T_{\mathbb{R}}\,.\end{aligned}$$

When the above identities are combined, the volume of $\Gamma\backslash G_{\mathbb{R}}$ cancels out, and one arrives at the conclusion stated earlier in this section.

References. Harish-Chandra's main work on discrete series appears in [HC4], [HC5], and [HC6]. The Langlands Conjecture was stated in [La] and proved in [Sch7]. Atiyah's L^2 Index Theorem appears in [At1]. The use of a cocompact discrete subgroup $\Gamma \subset G_{\mathbb{R}}$ can be avoided: Connes and Moscovici [Co-M] calculate the $G_{\mathbb{R}}$-index of any $G_{\mathbb{R}}$-invariant elliptic operator on $G_{\mathbb{R}}/H_{\mathbb{R}}$, for $H_{\mathbb{R}}$ compact in $G_{\mathbb{R}}$, in terms of a $G_{\mathbb{R}}$-invariant Atiyah-Singer form on $G_{\mathbb{R}}/H_{\mathbb{R}}$. For other material in this section, see [Sch4], [Sch6], and [At-Sch].

5. Realizations in Sheaf Cohomology

We continue with the notation of §§3-4. In this section we discuss the $K_{\mathbb{R}}$-structure of discrete series representations. Our tool will be the sheaf cohomology $H^*(D, \mathcal{O}(\mathcal{L}_\lambda))$, which is also of independent interest.

Fix a positive system Φ^+, let $\lambda \in \Lambda$ be given, and suppose that $\lambda+\rho$ is dominant regular. We shall study the decomposition under $K_{\mathbb{R}}$ of the discrete series representation with global character $\Theta_{\lambda+\rho}$. According to Harish-Chandra's theorem, all discrete series arise in this way for some choice of Φ^+. If we let ρ_n be half the sum of the noncompact positive roots, we can expand $\Theta_{\lambda+\rho}$ formally as

$$\begin{aligned}\Theta_{\lambda+\rho}|_{T_{\mathbb{R}}} &= (-1)^q \frac{\sum_{w\in W_{\mathbb{R}}} \varepsilon(w) e^{w(\lambda+\rho)}}{\prod_{\alpha\in\Phi^+}(e^{\alpha/2}-e^{-\alpha/2})} \\ &= (-1)^q \sum_{w\in W_{\mathbb{R}}} w\, \frac{e^{\lambda+\rho+\rho_n}}{\prod_{\alpha\in\Phi_c^+}(e^{\alpha/2}-e^{-\alpha/2}) \prod_{\beta\in\Phi_n^+}(e^{\beta}-1)}\end{aligned}$$

$$
\begin{aligned}
&= \sum_{w \in W_{\mathbb{R}}} w \frac{e^{\lambda+\rho+\rho_n}}{\prod_{\alpha\in\Phi_c^+}(e^{\alpha/2}-e^{-\alpha/2})\prod_{\beta\in\Phi_n^+}(1-e^{-\beta})} \\
&= \sum_{w \in W_{\mathbb{R}}} \sum_{\text{all } n_\beta \geq 0} w \frac{e^{\lambda+\rho+\rho_n+\sum_{\beta\in\Phi_n^+} n_\beta\beta}}{\prod_{\alpha\in\Phi_c^+}(e^{\alpha/2}-e^{-\alpha/2})} \\
&= \sum_{\text{all } n_\beta \geq 0} \sum_{w \in W_{\mathbb{R}}} \frac{\varepsilon(w) e^{w\left(\lambda+\rho+\rho_n+\sum_{\beta\in\Phi_n^+} n_\beta\beta\right)}}{\prod_{\alpha\in\Phi_c^+}(e^{\alpha/2}-e^{-\alpha/2})}.
\end{aligned}
$$

According to Weyl's character formula for $K_{\mathbb{R}}$, each term $\sum_{w\in W_{\mathbb{R}}} \cdots$ is either 0 or $\pm$ an irreducible character of $K_{\mathbb{R}}$. As the sum of coefficients $\sum_{\beta\in\Phi_n^+} n_\beta$ tends to ∞, so does the length of the weight $\lambda+\rho+\rho_n+\sum_{\beta\in\Phi_n^+} n_\beta\beta$, because the positive roots lie on one side of a hyperplane in $it_{\mathbb{R}}^*$. It follows that every irreducible character χ_j, $j \in \widehat{K_{\mathbb{R}}}$, occurs only finitely often in the formal expansion. Let $n_j(\lambda+\rho)$ denote this multiplicity, so that

$$
\Theta_{\lambda+\rho}|_{K_{\mathbb{R}}} = \sum_{j\in\widehat{K_{\mathbb{R}}}} n_j(\lambda+\rho)\,\chi_j\,,
$$

again purely formally. It is not at all obvious, for example, whether the integers $n_j(\lambda+\rho)$ are all nonnegative. Recall the definition of the $K_{\mathbb{R}}$-character of an admissible representation π of $G_{\mathbb{R}}$, as the formal sum of the $K_{\mathbb{R}}$-irreducible constituents of $\pi|_{K_{\mathbb{R}}}$, each taken with the appropriate multiplicity.

Theorem 1. *The discrete series representation with global character $\Theta_{\lambda+\rho}$ has $K_{\mathbb{R}}$-character $\sum_{j\in\widehat{K_{\mathbb{R}}}} n_j(\lambda+\rho)\,\chi_j$.*

Formally this statement is analogous to Kostant's formula [Kos1] for the multiplicity of a weight in a finite-dimensional representation. Shortly after Harish-Chandra had constructed the discrete series characters, Blattner observed that the characters could be expanded formally as described above, and he communicated his observation to Harish-Chandra. The integers $n_j(\lambda+\rho)$ first appeared as $K_{\mathbb{R}}$-multiplicities of actual $G_{\mathbb{R}}$-representations in [Sch1], namely as the $K_{\mathbb{R}}$-multiplicities in the Dolbeault cohomology groups $H^*(D,\mathcal{O}(\mathcal{L}_\lambda))$. At that point, the connection between the Dolbeault cohomology groups and the discrete series had been established only in special cases, but seemed highly likely in general. The conjectured form of Theorem 1 was explicitly stated in [Sch2], as **Blattner's conjecture**, and was first proved in [He-Sch]. We shall comment on other proofs of Blattner's conjecture at the end of this section.

Corollary. *In the discrete series representation with global character $\Theta_{\lambda+\rho}$, with Φ^+ chosen so as to make $\lambda+\rho$ dominant,*

(i) *the $K_{\mathbb{R}}$-type of highest weight $\lambda+2\rho_n$ occurs exactly once;*
(ii) *no $K_{\mathbb{R}}$-type has highest weight $\lambda+2\rho_n-B$ with B a nonempty sum of distinct positive noncompact roots;*
(iii) *all $K_{\mathbb{R}}$-types have highest weights $\lambda+2\rho_n+C$, where C is a sum of (not necessarily distinct) positive noncompact roots.*

Conclusions (i) and (ii) are immediate combinatorial consequences of the multiplicity formula in Theorem 1. Conclusion (iii) is a similar such consequence if C is asserted merely to be a sum of (not necessarily distinct) positive roots. That the roots can be taken to be noncompact requires using tools that go into the proof of the theorem.

By a **Harish-Chandra module** we shall mean an admissible $(\mathfrak{g}, K)$-module of finite length.

Theorem 2. *Up to infinitesimal equivalence, there exists only one irreducible Harish-Chandra module satisfying the following two conditions:*

(i) *it contains the $K_{\mathbb{R}}$-type with highest weight $\lambda + 2\rho_n$ at least once;*
(ii) *it contains no $K_{\mathbb{R}}$-type with highest weight $\lambda + 2\rho_n - B$, where B is a nonempty sum of distinct positive noncompact roots.*

Because of the Corollary to Theorem 1, the Harish-Chandra module satisfying the conditions (i), (ii) underlies the discrete series representation with character $\Theta_{\lambda+\rho}$. Theorem 2 is a formal analogue of the Theorem of the Highest Weight for finite dimensional representations. Like Theorem 1, the original proof came from an understanding of sheaf cohomology. A number of other proofs have appeared since. See [Sch6] and [Vo1], for example.

We turn our attention now to representations in sheaf cohomology. Let $D = D(\Phi^+)$ be an open orbit in X, and let $\lambda \in \Lambda$ be arbitrary. Recall the natural map

$$H^*_{(2)}(D, \mathcal{O}(\mathcal{L}_\lambda)) \longrightarrow H^*(D, \mathcal{O}(\mathcal{L}_\lambda))$$

from L^2 cohomology to sheaf cohomology. It can be seen from its definition that the map is $G_{\mathbb{R}}$-invariant and continuous with respect to the natural topologies of these two spaces.

Theorem 3. *The topology on $H^*(D, \mathcal{O}(\mathcal{L}_\lambda))$ is Hausdorff and therefore Fréchet. The resulting representation of $G_{\mathbb{R}}$ is admissible, of finite length, and has infinitesimal character $\lambda + \rho$. If $\lambda + \rho$ is antidominant and regular, then the natural map $H^*_{(2)}(D, \mathcal{O}(\mathcal{L}_\lambda)) \to H^*(D, \mathcal{O}(\mathcal{L}_\lambda))$ is injective, has dense image, and induces an isomorphism of the underlying Harish-Chandra modules.*

Under the hypotheses of the theorem, the representation in $H^p_{(2)}(D, \mathcal{O}(\mathcal{L}_\lambda))$ is nonzero in degree $p = p(\lambda + \rho) = \dim(\mathfrak{n} \cap \mathfrak{k})$ and vanishes in all other degrees. As a formal consequence of the theorem, $H^p(D, \mathcal{O}(\mathcal{L}_\lambda))$ is then also concentrated in degree $p = \dim(\mathfrak{n} \cap \mathfrak{k})$. However, unlike the L^2 cohomology, which jumps into a different degree when the integer $p(\lambda + \rho)$ changes, the Dolbeault cohomology remains concentrated in degree $p = \dim(\mathfrak{n} \cap \mathfrak{k})$ as long as $\lambda + \rho$ is antidominant with respect to $\Phi^+ \cap \Phi_c$ only. The proof of Theorem 3 involves the following ingredients.

a) A vanishing theorem of Andreotti and Grauert [An-Grt] that generalizes Cartan's Theorem B. This theorem applies to the cohomology of any coherent sheaf of $\mathcal{O}$-modules over certain "partially pseudoconvex" complex manifolds.
b) The demonstration that the Andreotti-Grauert vanishing theorem, applied to the open $G_{\mathbb{R}}$-orbits $D \subset X$, forces the cohomology of coherent sheaves to vanish above degree $\dim(\mathfrak{n} \cap \mathfrak{k})$. This involves the construction of a "partially pseudoconvex" exhaustion function for any open orbit.
c) Expansion of cohomology of $\mathcal{O}(\mathcal{L}_\lambda)$ around $K_{\mathbb{R}}/T_{\mathbb{R}} \subset G_{\mathbb{R}}/T_{\mathbb{R}} \simeq D$, which turns out to be a subvariety of D, isomorphic to the flag variety of $\mathfrak{k}$.

d) The Borel-Weil-Bott Theorem for $K_{\mathbb{R}}$, which in the presence of the vanishing of cohomology asserted by (a) and (b) implies that there is no obstruction to extending cohomology from the compact subvariety to all of D.
e) A version of the Leray spectral sequence for the (nonholomorphic, in general) fibration $G_{\mathbb{R}}/T_{\mathbb{R}} \to G_{\mathbb{R}}/K_{\mathbb{R}}$.
f) The translation principle that was mentioned in §4.

REFERENCES. The parts of Theorem 3 about sheaf cohomology in the "very regular, antidominant" case, except for the identification of the representation as the discrete series representation with character $\Theta_{\lambda+\rho}$, are proved in [Sch1]. Still in the very regular, antidominant situation, [Sch1] also proves that the natural map from L^2 cohomology to sheaf cohomology is injective. The actual identification of the representation in the very regular, antidominant case is in [Sch2]. Aguilar-Rodriguez [Ag] extended the theorem to the form stated here.

Besides the antidominant case, there are certain other pairs $(D, \mathcal{L}_\lambda)$ for which the natural map from L^2 cohomology to sheaf cohomology induces an infinitesimal equivalence, i.e., an isomorphism of underlying Harish-Chandra modules. One extreme case is of particular interest.

Observation. *For given Φ^+, the following are equivalent:*
1) *there exists a regular $\lambda + \rho$ with $\lambda \in \Lambda$ such that $p(\lambda + \rho) = 0$;*
2) *$G_{\mathbb{R}}/K_{\mathbb{R}}$ is hermitian symmetric, i.e., has an invariant complex structure, and the natural fibration $D(\Phi^+) \simeq G_{\mathbb{R}}/T_{\mathbb{R}} \to G_{\mathbb{R}}/K_{\mathbb{R}}$ is holomorphic.*

When condition (1) is satisfied, $\widetilde{\Phi^+} = \{\alpha \in \Phi \mid (\alpha, \lambda + \rho) > 0\}$ is a second positive root system, which gives the same notion of positivity as Φ^+ for compact roots, and the opposite notion for noncompact roots. It follows that the subspaces $\mathfrak{p}^+ = \oplus_{\alpha \in \Phi_n \cap \Phi^+} \mathfrak{g}^\alpha$, $\mathfrak{p}^- = \oplus_{\alpha \in \Phi_n \cap \Phi^+} \mathfrak{g}^{-\alpha}$ satisfy

$$\mathfrak{g} = \mathfrak{k} \oplus \mathfrak{p}^+ \oplus \mathfrak{p}^-\,, \qquad \mathfrak{p}^- = \text{ complex conjugate of } \mathfrak{p}^+\,,$$
$$[\mathfrak{k}, \mathfrak{p}^+] \subset \mathfrak{p}^+\,, \qquad [\mathfrak{k}, \mathfrak{p}^-] \subset \mathfrak{p}^-\,.$$

This in turn implies

$$[\mathfrak{p}^+, \mathfrak{p}^+] = 0 = [\mathfrak{p}^-, \mathfrak{p}^-]\,.$$

Conversely, if ad($\mathfrak{k}$)-invariant subspaces $\mathfrak{p}^\pm \subset \mathfrak{g}$ with these properties exist, one can show that $K_{\mathbb{R}}$ contains a torus $T_{\mathbb{R}}$ which is a Cartan subgroup of $G_{\mathbb{R}}$, and one can then produce a positive root system $\Phi^+ \subset \Phi(\mathfrak{g}, \mathfrak{t})$ satisfying the condition (1). By a criterion of Nijenhuis, a splitting of $\mathfrak{g}$ as above determines a $G_{\mathbb{R}}$-invariant complex structure on $G_{\mathbb{R}}/K_{\mathbb{R}}$ so that $\mathfrak{p}^-$ corresponds to the antiholomorphic tangent space at the identity coset. This makes the $G_{\mathbb{R}}$-invariant fibration $G_{\mathbb{R}}/T_{\mathbb{R}} \to G_{\mathbb{R}}/K_{\mathbb{R}}$ holomorphic. Harish-Chandra [HC2] constructs an equivariant holomorphic embedding of the resulting complex manifold $G_{\mathbb{R}}/K_{\mathbb{R}}$ into a generalized flag variety for $\mathfrak{g}$, so that the image lies as a bounded open set in an open Schubert cell isomorphic to $\mathfrak{p}^+$; in this way, he exhibits $G_{\mathbb{R}}/K_{\mathbb{R}}$ as a bounded symmetric domain, $K_{\mathbb{R}}$-equivariantly embedded in $\mathfrak{p}^+$. An exposition of these matters can be found in [Kn2, §VII.9].

Long before his general construction of the discrete series, Harish-Chandra associated discrete series representations to orbits D which fiber holomorphically over a hermitian symmetric quotient $G_{\mathbb{R}}/K_{\mathbb{R}}$ [HC1,HC2]. The resulting representations constitute the so-called **holomorphic discrete series**. They arise as square integrable, holomorphic sections of line bundles $\mathcal{L}_\lambda$ over open orbits $D = D(\Phi^+)$ of this very special type, with $\lambda+\rho$ regular and $p(\lambda+\rho) = 0$. In the present situation, the Fréchet property of $H^0(D, \mathcal{O}(\mathcal{L}_\lambda))$ is obvious, as is the injectivity of the map

$$H^0_{(2)}(D, \mathcal{O}(\mathcal{L}_\lambda)) \longrightarrow H^0(D, \mathcal{O}(\mathcal{L}_\lambda))\,;$$

the density of the image is not difficult to establish, either. It follows that the natural map induces an infinitesimal equivalence. Both types of cohomology vanish above degree zero when $p(\lambda + \rho) = 0$. A simple spectral sequence argument gives an alternative realization of $H^0_{(2)}(D, \mathcal{O}(\mathcal{L}_\lambda))$ as a space of square integrable, holomorphic sections of a $G_\mathbb{R}$-equivariant holomorphic vector bundle over the hermitian symmetric space $G_\mathbb{R}/K_\mathbb{R}$. The same type of spectral sequence then leads back to the situation covered by Theorem 3.

In many ways, holomorphic discrete series representations have a much simpler structure than the others. For example, there exists a simple explicit formula for the holomorphic discrete series characters on every Cartan subgroup, compact or not [Ma,He], and results in [Sch3] lead to a far more concrete description of the $K_\mathbb{R}$-structure of holomorphic discrete series representations than is provided by Blattner's conjecture in the general case. For more on the holomorphic discrete series, see [Kn1, Ch. VI], for example. Among simple groups containing a compact Cartan subgroup, $SL(2,\mathbb{R})$ (or the isomorphic group $SU(1,1)$) is special in that all of its discrete series representations belong to the holomorphic discrete series; see Donley's lecture [Do1] for a detailed discussion of this group.

Both Theorem 3 and the situation just described give embeddings of unitary representations into infinitesimally equivalent Fréchet representations. These are particular instances of a much more general phenomenon. Let $\mathcal{R}(G_\mathbb{R})$ be the (additive) category whose objects are the continuous $G_\mathbb{R}$-representations, satisfying the conditions of admissiblility and finite length, on complete, locally convex Hausdorff topological vector spaces; its morphisms are the continuous, linear, $G_\mathbb{R}$-invariant maps. Let $\mathcal{H}(\mathfrak{g}, K)$ be the category of all Harish-Chandra modules and $(\mathfrak{g}, K)$-invariant linear maps between them. Passage to the underlying Harish-Chandra module defines a natural faithful, covariant functor

$$\mathrm{HC}\ :\ \mathcal{R}(G_\mathbb{R}) \longrightarrow \mathcal{H}(\mathfrak{g}, K)\,.$$

A theorem of Casselman [Cas1] asserts that this functor is surjective: every Harish-Chandra module has a **globalization**, i.e., it underlies a continuous, admissible, finite length representation of $G_\mathbb{R}$ on a complete, locally convex Hausdorff topological vector space. The globalization is far from unique, however. Principal series representations of the group $SU(1,1)$, for example, can be realized on a variety of function spaces on the circle, such as $L^p(S^1)$ with $1 \leq p < \infty$, $C^\infty(S^1)$, $C^{-\infty}(S^1)$ (=space of distributions), $C^\omega(S^1)$ (=space of real analytic functions), and $C^{-\omega}(S^1)$ (=space of hyperfunctions). All of these topological realizations have the same space of $K_\mathbb{R}$-finite vectors, so they all globalize the same Harish-Chandra module.

It is natural to ask whether functorial globalizations exist – in other words, if the functor HC has a functorial right inverse. Functorial globalizations were first constructed by Casselman-Wallach [Cas2]; these are the C^∞ and $C^{-\infty}$ globalizations which bound, in a very precise sense, the possible Banach globalizations from below and above – see the comment following the statement of Theorem 5 below. Banach globalizations, we should remark, cannot be functorial, as can be seen already in simple examples. What we have done in this section is clarified by looking at the minimal and maximal globalization functors

$$\mathrm{mg} \;:\; \mathcal{H}(\mathfrak{g}, K) \longrightarrow \mathcal{R}(G_{\mathbb{R}}), \qquad \mathrm{MG} \;:\; \mathcal{H}(\mathfrak{g}, K) \longrightarrow \mathcal{R}(G_{\mathbb{R}}),$$

which can be characterized as the left, respectively right adjoint of the functor HC [Sch9, Ka-Sch]. The maximal globalization $\mathrm{MG}(V)$ of any Harish-Chandra module V is a nuclear Fréchet space, and the minimal globalization $\mathrm{mg}(V)$ a DNF ("dual of nuclear Fréchet") space. The adjointness properties of the two functors can be rephrased as follows: for any (π, V_π) in $\mathcal{R}(G_{\mathbb{R}})$, the identity map on $\mathrm{HC}(V_\pi)$ induces continuous $G_{\mathbb{R}}$-invariant linear maps

$$\mathrm{mg}(\mathrm{HC}(V_\pi)) \hookrightarrow V_\pi \hookrightarrow \mathrm{MG}(\mathrm{HC}(V_\pi)).$$

This is the reason for the names "minimal globalization" and "maximal globalization". Much deeper than existence are the next two statements.

Theorem 4. *The functors* MG *and* mg *are topologically exact.*

Theorem 5. *Suppose that (π, V_π) is a Banach representation in $\mathcal{R}(G_{\mathbb{R}})$, and let $V_\pi^\omega \in \mathcal{R}(G_{\mathbb{R}})$ be the space of analytic vectors. Then the natural inclusion* $\mathrm{mg}(\mathrm{HC}(V_\pi)) \hookrightarrow V_\pi^\omega$ *is an isomorphism. If V_π is reflexive and $(\,\cdot\,)'$ denotes strong dual, then* $((V_\pi')^\omega)' \hookrightarrow \mathrm{MG}(\mathrm{HC}(V_\pi))$ *is an isomorphism.*

The C^∞ and $C^{-\infty}$ globalizations of Casselman-Wallach are also exact, and they satisfy the analogue of Theorem 5, with the space of C^∞ vectors V_π^∞ in place of the space of analytic vectors V_π^ω.

The following result can be deduced easily from the proof of Theorem 3.

Theorem 6. *For every open $G_{\mathbb{R}}$-orbit $D \subset X$ and every $\lambda \in \Lambda$, $H^*(D, \mathcal{O}(\mathcal{L}_\lambda))$ is the maximal globalization of its underlying Harish-Chandra module.*

Corollary. *If $\lambda + \rho$ is regular antidominant, $H^p(D, \mathcal{O}(\mathcal{L}_\lambda))$ with $p = \dim(\mathfrak{n} \cap \mathfrak{k})$ is the maximal globalization of the discrete series module with global character $\Theta_{\lambda+\rho}$.*

The inclusion of L^2 cohomology into Dolbeault cohomology, we now see, is simply the canonical map from the unitary realization of a discrete series representation into the maximal globalization of its Harish-Chandra module.

The minimal and maximal globalizations have good cohomological properties. To mention the most important example, let us consider a discrete subgroup $\Gamma \subset G_{\mathbb{R}}$. It was pointed out by Bunke and Olbrich that results in [Ka-Sch] imply:

Theorem 7. *Let V' denote the Harish-Chandra module dual to $V \in \mathcal{H}(\mathfrak{g}, K)$. Then*

$$\mathrm{Ext}^*_{(\mathfrak{g},K)}(V, C^\infty(\Gamma\backslash G_{\mathbb{R}})_{K_{\mathbb{R}}-\mathrm{fini}}) \;\simeq\; H^*(\Gamma, \mathrm{MG}(V'))\,.$$

Specialized to degree zero, this is a version of Frobenius reciprocity: the space of embeddings of a Harish-Chandra module V into $C^\infty(\Gamma\backslash G_{\mathbb{R}})_{K_{\mathbb{R}}\text{-finite}}$ is naturally isomorphic to the space of Γ-invariants in the maximal globalization of the dual Harish-Chandra module V'. In the case of the discrete series, realized on Dolbeault cohomology, embeddings of discrete series representations into $C^\infty(\Gamma\backslash G_{\mathbb{R}})$ correspond to Γ-invariant Dolbeault cohomology classes, i.e., to "automorphic cohomology". All of this is understood for cocompact discrete subgroups [Sch4], but questions remain in the – much more difficult – situation of a general discrete subgroup of finite covolume.

Let us conclude these lectures with a very quick overview of other approaches to the discrete series. There are two geometric constructions of discrete series representations similar in spirit to the realization on the L^2 cohomology of line bundles over $D \simeq G_{\mathbb{R}}/T_{\mathbb{R}}$. First, on the L^2 cohomology of $G_{\mathbb{R}}$-equivariant holomorphic vector bundles over $G_{\mathbb{R}}/K_{\mathbb{R}}$ when this quotient has invariant complex structures. Narasimhan and Okamoto [Na-Ok] produce all discrete series representations with "sufficiently regular" infinitesimal character for groups $G_{\mathbb{R}}$ of this type. Parthasarathy's construction [Pa] uses L^2 harmonic spinors on $G_{\mathbb{R}}/K_{\mathbb{R}}$, which can be defined whether or not the quotient carries an invariant complex structure; when an invariant complex structure does exist, L^2 harmonic spinors can be identified with L^2 cohomology classes of holomorphic vector bundles. Like [Sch4, Sch7], the papers [N-O] and [Pa] depend on Harish-Chandra's construction of the discrete series. Our approach in §§3,4 follows [At-Sch], which works with the realization in terms of L^2 spinors to give an independent proof of existence and exhaustion of the discrete series; in these lectures, of course, the arguments of [At-Sch] have been translated into the setting of line bundles over $G_{\mathbb{R}}/T_{\mathbb{R}}$. Now, after the fact, it is not difficult to go back and forth between these various L^2 realizations of the discrete series.

An entirely different approach is due to Flensted-Jensen [Fl], who produces $K_{\mathbb{R}}$-finite eigenfunctions for $\mathcal{Z}(\mathfrak{g})$ (= center of the universal enveloping algebra) on quotients $G_{\mathbb{R}}/H_{\mathbb{R}}$ of $G_{\mathbb{R}}$ by $H_{\mathbb{R}}$, the (typically noncompact) fixed point group of an involutive automorphism of $G_{\mathbb{R}}$. By an ingenious argument, he is able to estimate the growth of these eigenfunctions. When applied to the quotient $G_{\mathbb{R}} \times G_{\mathbb{R}}/\text{diagonal} \simeq G_{\mathbb{R}}$, his method gives discrete series representations with "sufficiently regular" infinitesimal character. One can then apply the Jantzen-Zuckerman translation principle to get all of the discrete series, and exhaustion can be proved using Harish-Chandra's original method; see [Kn1] for details.

It is possible to construct the Harish-Chandra modules underlying discrete series representations by algebraic methods. Both Enright-Varadarajan [En-Va] and Zuckerman produce these Harish-Chandra modules; to identify them as discrete series modules requires a tool like Theorem 2, or else an analytic argument which establishes the square integrability of matrix coefficients directly. Zuckerman's construction – carried out in detail in [Vo2] – was a conscious and successful effort to mimic algebraically the mechanism of Dolbeault cohomology, thereby circumventing a number of technical difficulties. The Zuckerman modules visibly have the $K_{\mathbb{R}}$-structure predicted by Blattner's conjecture. Thus, when one identifies them with discrete series modules, one obtains another proof of Blattner's conjecture.

Beilinson and Bernstein [Be-Be1], [Be-Be2] have discovered a very powerful method for studying those modules over the universal enveloping algebra $\mathcal{U}(\mathfrak{g})$ of a semisimple Lie algebra $\mathfrak{g}$ that have an infinitesimal character. Roughly speaking, Beilinson and Bernstein set up an equivalence of categories between $\mathcal{U}(\mathfrak{g})$-modules with infinitesimal character χ_λ on the one hand, and on the other, (quasi-coherent) sheaves of modules over $\mathcal{D}_{X,\lambda}$, the sheaf of linear differential operators, with algebraic coefficient, on the flag variety X, now viewed as algebraic variety rather than as complex manifold. The subscript λ signifies twisting by an equivariant line bundle when λ is integral, or by a "fractional line bundle" in general. This brings to bear the arsenal of algebraic geometry on the study of $\mathcal{U}(\mathfrak{g})$-modules. Various properties of $\mathcal{U}(\mathfrak{g})$-modules – such as being a Harish-Chandra module – translate immediately into geometric properties of the corresponding $\mathcal{D}_{X,\lambda}$-module

– such as equivariance under the action of K, the complexification of $K_{\mathbb{R}}$. In the Beilinson-Bernstein picture, discrete series modules are attached to closed K-orbits in X, though it takes an (at least residually) analytic argument to make the connection to square-integrability of the resulting representations. The $K_{\mathbb{R}}$-structure of discrete series modules is almost obvious from this point of view. Irreducible Harish-Chandra modules ouside the discrete series correspond to other K-orbits in X; temperedness, reducibility of standard modules, vanishing theorems for the Zuckerman functor, the classification of irreducible Harish-Chandra modules all can be read off from the geometry [HMSW].

References

[Ag] Aguilar-Rodriguez, R., *Connections between representations of Lie groups and sheaf cohomology*, Ph.D. Thesis, Harvard University, 1987.

[An-Grt] Andreotti, A., and H. Grauert, Théorèmes de finitude pour la cohomologie des espaces complexes, *Bull. Soc. Math. France* **90** (1962), 193–259.

[An-Ve] Andreotti, A., and E. Vesentini, Carleman-estimates for the Laplace-Beltrami operator on complex manifolds, *I.H.E.S. Publications Mathématiques* **25** (1965), 81–130; Erratum, **27** (1965), 153–156.

[At1] Atiyah, M., Elliptic operators, discrete groups and von Neumann algebras, *Astérisque* **32/33** (1976), 43–72.

[At2] Atiyah, M., Characters of semi-simple Lie groups, *Michael Atiyah, Collected Works*, vol. 4, Clarendon Press, Oxford, 1988, pp. 489–557.

[At-Sch] Atiyah, M. F., and W. Schmid, A geometric construction of the discrete series for semi-simple Lie groups, *Invent. Math.* **42** (1977), 1–62; Erratum, **54** (1979), 189–192.

[Be-Be1] Beilinson, A., and J. N. Bernstein, Localization de $\mathfrak{g}$-modules, *C. R. Acad. Sci. Paris* **292** (1981), 15–18.

[Be-Be2] Beilinson, A., and J. N. Bernstein, A generalization of Casselman's submodule theorem, *Representation Theory of Reductive Groups*, Progress in Mathematics, vol. 40, Birkhäuser, Boston, 1983, pp. 35–52.

[Boc-Y] Bochner, S., and K. Yano, *Curvature and Betti Numbers*, Princeton University Press, Princeton, 1953.

[Bor1] Borel, A., Compact Clifford-Klein forms on symmetric spaces, *Topology* **2** (1963), 112–122.

[Bor2] Borel, A., *Linear Algebraic Groups*, W. A. Benjamin, New York, 1969.

[Bott] Bott, R., Homogeneous vector bundles, *Annals of Math.* **66** (1957), 203–248.

[Car] Cartier, P., Remarks on "Lie algebra cohomology and the generalized Borel-Weil theorem," by B. Kostant, *Annals of Math.* **74** (1961), 388–390.

[Cas1] Casselman, W., Jacquet modules for real reductive groups, *Proceedings of the International Congress of Mathematicians*, Helsinki, 1978, pp. 557–563.

[Cas2] Casselman, W., Canonical extensions of Harish-Chandra modules to representations of G, *Canadian Jour. of Math.* **41** (1989), 385–438.

[Cas-O] Casselman, W., and M. S. Osborne, The $\mathfrak{n}$-cohomology of representations with an infinitesimal character, *Compositio Math.* **31** (1975), 219–227.

[Co-M] Connes, A., and H. Moscovici, The L^2-index theorem for homogeneous spaces of Lie groups, *Annals of Math.* **115** (1982), 291–330.

[De] Delorme, P., Infinitesimal character and distribution character of representations of reductive Lie groups, *these* PROCEEDINGS, pp. 73–81.

[Do1] Donley, R. W., Irreducible representations of SL(2,R), *these* PROCEEDINGS, pp. 51–59.

[Do2] Donley, R. W., The Borel-Weil theorem for $U(n)$, *these* PROCEEDINGS, pp. 115–121.

[En-Va] Enright, T. J., and V. S. Varadarajan, On an infinitesimal characterization of the discrete series, *Annals of Math.* **102** (1975), 1–15.

[Fl] Flensted-Jensen, M., Discrete series for semisimple symmetric spaces, *Annals of Math.* **111** (1980), 253–311.

[Go] Godement, R., Sur les relations d'orthogonalité de V. Bargmann, I and II, *C. R. Acad. Sci. Paris* **225** (1947), 521–523 and 657–659.

[Gr-Sch] Griffiths, P., and W. Schmid, Locally homogeneous complex manifolds, *Acta Math.* **123** (1969), 253–302.

[HC1] Harish-Chandra, Integrable and square-integrable representations of a semisimple Lie group, *Proc. Nat. Acad. Sci. U.S.A.* **41** (1955), 314–317.

[HC2] Harish-Chandra, Representations of semisimple Lie groups IV, *Amer. J. Math.* **77** (1955), 743–777; V, **78** (1956), 1–41; VI, **78** (1956), 564–628.

[HC3] Harish-Chandra, Invariant eigendistributions on a semisimple Lie algebra, *I.H.E.S. Publications Mathématiques* **27** (1965), 5–54.

[HC4] Harish-Chandra, Discrete series for semisimple Lie groups I, *Acta Math.* **113** (1965), 241–318.

[HC5] Harish-Chandra, Two theorems on semi-simple Lie groups, *Annals of Math.* **83** (1966), 74–128.

[HC6] Harish-Chandra, Discrete series for semisimple Lie groups II, *Acta Math.* **116** (1966), 1–111.

[He] Hecht, H., Characters of some representations of Harish-Chandra, *Math. Annalen* **219** (1976), 213–226.

[He-Sch] Hecht, H., and W. Schmid, A proof of Blattner's conjecture, *Invent. Math.* **31** (1975), 129–154.

[Hirai] Hirai, T., Invariant eigendistributions of Laplace operators on real simple Lie groups II, *Japanese J. Math.* **2** (1976), 27–89.

[Hirz] Hirzebruch, F., Automorphe Formen und der Satz von Riemann-Roch, *Symposium Internacional de Topología Algebraica (México 1956)*, Universidad Nacional de México, 1958, pp. 129–144.

[HMSW] Hecht, H., D. Miličić, W. Schmid and J. A. Wolf, Localization and standard modules for real semisimple Lie groups I: The duality theorem, *Invent. Math.* **90** (1987), 297–332; II: Irreducibility, vanishing theorems and classification, preprint.

[Ka-Sch] Kashiwara, M., and W. Schmid, Quasi-equivariant $\mathcal{D}$-modules, equivariant derived category, and representations of reductive Lie groups, *Lie Theory and Geometry, in Honor of Bertram Kostant*, Progress in Mathematics, vol. 123, Birkhäuser, Boston, 1994, pp. 457–488.

[Kn1] Knapp, A. W., *Representation Theory of Semisimple Groups: An Overview Based on Examples*, Princeton University Press, Princeton, N.J., 1986.

[Kn2] Knapp, A. W., *Lie Groups Beyond an Introduction*, Progress in Mathematics, vol. 140, Birkhäuser, Boston, 1996.

[Kn3] Knapp, A. W., Structure theory of semisimple Lie groups, *these* PROCEEDINGS, pp. 1–27.

[Kn-Vo] Knapp, A. W., and D. A. Vogan, *Cohomological Induction and Unitary Representations*, Princeton University Press, Princeton, N.J., 1995.

[Kod] Kodaira, K., On a differential-geometric method in the theory of analytic stacks, *Proc. Nat. Acad. Sci. U.S.A.* **39** (1953), 1268–1273.

[Kos1] Kostant, B., A formula for the multiplicity of a weight, *Trans. Amer. Math. Soc.* **93** (1959), 53–73.

[Kos2] Kostant, B., Lie algebra cohomology and the generalized Borel-Weil theorem, *Annals of Math.* **74** (1961), 329–387.

[La] Langlands, R. P., Dimension of spaces of automorphic forms, *Algebraic Groups and Discontinuous Subgroups*, Proc. Symp. Pure Math., vol. 9, American Mathematical Society, Providence, 1966, pp. 253–257.

[Ma] Martens, S., The characters of the holomorphic discrete series, *Proc. Nat. Sci. U.S.A.* **72** (1975), 3275–3276.

[Na-Ok] Narasimhan, M. S., and K. Okamoto, An analogue of the Borel-Weil-Bott theorem for Hermitian symmetric pairs of noncompact type, *Annals of Math.* **91** (1970), 486–511.

[Pa] Parthasarathy, R., Dirac operators and the discrete series, *Invent. Math.* **96** (1972), 1–30.

[Sch1] Schmid, W., *Homogeneous complex manifolds and representations of semisimple Lie groups*, Ph.D. Thesis, University of California, Berkeley, 1967; in P. J. Sally and D. A. Vogan (eds.), *Representation Theory and Harmonic Analysis on Semisimple Lie Groups*, Math. Surveys and Monographs, vol. 31, American Mathematical Society, Providence, 1989, pp. 223–286.

[Sch2] Schmid, W., On the realization of the discrete series of a semisimple Lie group, *Complex Analysis, 1969* (L. Resnikoff and R. O. Wells, eds.), Rice University Studies, vol. 56, No. 2, 1970, pp. 99–108.

[Sch3] Schmid, W., Die Randwerte holomorpher Funktionen auf hermitesch symmetrischen Räumen, *Invent. Math.* **9** (1969), 61–80.

[Sch4] Schmid, W., On a conjecture of Langlands, *Annals of Math.* **93** (1971), 1–42.

[Sch5] Schmid, W., On the characters of the discrete series: the Hermitian symmetric case, *Invent. Math.* **30** (1975), 47–144.

[Sch6] Schmid, W., Some properties of square-integrable representations of semisimple Lie groups, *Annals of Math.* **102** (1975), 535–564.

[Sch7] Schmid, W., L^2-cohomology and the discrete series, *Annals of Math.* **103** (1976), 375–394.

[Sch8] Schmid, W. (notes by B. F. Steer), Representations of semi-simple Lie groups, *Representation Theory of Lie Groups: Proceedings of the SRC/LMS Research Symposium on Representations of Lie Groups, Oxford, 28 June–15 July 1977*, London Mathematical Society Lecture Notes Series, vol. 34, Cambridge University Press, Cambridge, 1979, pp. 185–235.

[Sch9] Schmid, W., Boundary value problems for group invariant differential equations, *Élie Cartan et les mathématiques d'aujourd'hui (Lyon, juin 1984)*, Astérisque, Numéro hors série, 1985, pp. 311–321.

[Se] Serre, J.-P., Représentations linéaires et espaces homogènes Kählériennes des groupes de Lie compacts, *Séminaire Bourbaki, 6e année, 1953/54*, Exposé 100, Inst. Henri Poincaré, Paris, 1954; reprinted with corrections, 1965.

[Ti] Tits, J., Sur certaines classes d'espaces homogènes de groupes de Lie, *Acad. Roy. Belg. Cl. Sci. Mém. Coll.* **29** (1955), No. 3.

[Vo1] Vogan, D. A., The algebraic structure of the representation of semisimple Lie groups I, *Annals of Math.* **109** (1979), 1–60.

[Vo2] Vogan, D. A., *Representations of Real Reductive Groups*, Progress in Mathematics, vol. 15, Birkhäuser, Boston, 1981.

[Wa] Wallach, N., *Real Reductive Groups*, vol. I, Academic Press, Boston, 1988.

[We] Wells, R. O., *Differential Analysis on Complex Manifolds*, Prentice-Hall, Englewood Cliffs, N.J., 1973; second edition, Springer, New York, 1980.

[Wo] Wolf, J. A., The action of a real semisimple Lie group on a complex flag manifold I, *Bull. Amer. Math. Soc.* **75** (1969), 1121–1237.

[Zu] Zuckerman, G., Tensor products of finite and infinite dimensional representations of semisimple Lie groups, *Annals of Math.* **106** (1977), 295–308.

Department of Mathematics, Harvard University, Cambridge, MA 02138, U.S.A.

Proceedings of Symposia in Pure Mathematics
Volume **61** (1997), pp. 115–121

The Borel-Weil Theorem for $U(n)$

Robert W. Donley, Jr.*

1. Introduction

Schmid's lectures [Sc] include a treatment of the Borel-Weil Theorem for a general compact semisimple Lie group, and the present lecture shows how that theorem takes on a classical form for certain representations of the unitary group $U(n)$. First we set up standard notation for $U(n)$:

$$\begin{aligned} U(n) &= \{g \in GL(n,\mathbb{C}) \mid g\bar{g}^t = I_n\} \\ \mathfrak{u}(n) &= \{X \in M_n(\mathbb{C}) \mid X = -\overline{X}^t\} \\ \mathfrak{gl}(n,\mathbb{C}) &= \mathfrak{u}(n) \oplus i\mathfrak{u}(n) \\ \mathfrak{h} &= \{\text{diagonal matrices in } \mathfrak{gl}(n,\mathbb{C})\} \\ E_{ij} &= \text{matrix with } (i,j)^{\text{th}} \text{ entry} = 1 \text{ and with } 0 \text{ elsewhere} \\ H_i &= E_{ii} \\ \mathfrak{h}_0 &= \{\text{diagonal matrices in } \mathfrak{u}(n)\} = \bigoplus_{j=1}^{n} iH_j. \end{aligned}$$

Define $e_i \in \mathfrak{h}^*$ by $e_i(H_j) = \delta_{ij}$. For $i \neq j$, the vector E_{ij} is a root vector for the root $e_i - e_j$. The set of all roots is

$$\Delta = \Delta(\mathfrak{gl}(n,\mathbb{C}),\mathfrak{h}) = \{e_i - e_j \mid 1 \leq i,j \leq n,\, i \neq j\}.$$

We order $(i\mathfrak{h}_0)^*$ via the **lexicographical ordering**:

$$f > 0 \quad \text{iff} \quad \begin{aligned} &f(H_1) > 0 \\ \text{or } &f(H_1) = 0 \text{ and } f(H_2) > 0, \text{ etc.} \end{aligned}$$

This ordering yields the set of positive roots $\Delta^+ = \{e_i - e_j \mid i < j\}$.

We consider a special case of the Borel-Weil Theorem that reflects the geometry of $(n-1)$-dimensional complex projective space $\mathbb{C}P^{n-1}$. The statement of the

---

1991 *Mathematics Subject Classification.* Primary 22-02, Secondary 22E45.

* The author was supported by an Alfred P. Sloan Doctoral Dissertation Fellowship at the time of the conference. Currently he is supported by a National Science Foundation Mathematical Sciences Postdoctoral Research Fellowship. He also thanks the Institut Mittag-Leffler for their hospitality.

Borel-Weil Theorem can be found in [Sc]. Actually we are specializing the Borel-Weil Theorem in the case of a general parabolic subgroup; this theorem (included in the corresponding Borel-Weil-Bott Theorem) may be found in section 6 of [GS].

Special Case of Borel-Weil Theorem. *For each weight $\lambda = -Ne_n$ with $N \in \mathbb{Z}$, let $\mathcal{L}_\lambda$ be the homogeneous holomorphic line bundle constructed as in the examples of section 4 below, and identify the space of holomorphic sections $\Gamma(\mathcal{O}(\mathcal{L}_\lambda))$ as a representation of $U(n)$. When λ is dominant, this space is equivalent to the irreducible representation with highest weight λ. Otherwise it is zero.*

2. Geometry of $\mathbb{C}P^{n-1}$

Let us build $\mathbb{C}P^{n-1}$ from basics in order to see exactly how the group $U(n)$ enters. We define

$$\mathbb{C}P^{n-1} = (\mathbb{C}^n - \{0\})/\sim \tag{2.1}$$

where $\sim$ is the equivalence relation given by

$$v \sim w \text{ if and only if } v = cw \text{ for some } c \in \mathbb{C}^*. \tag{2.2}$$

The equivalence class of the vector $(z_1, \ldots, z_n) \in \mathbb{C}^n - \{0\}$ is denoted $[z_1 : \cdots : z_n]$.

The space $\mathbb{C}P^{n-1}$ is a compact complex manifold of complex dimension $n-1$, and its holomorphic structure is given by the n inverse coordinate maps

$$\phi_i : \mathbb{C}^{n-1} \to U_i \subset \mathbb{C}P^{n-1},$$

where

$$\phi_i(z_1, \ldots, z_{n-1}) = [z_1 : \cdots : z_{i-1} : 1 : z_i : \cdots : z_{n-1}]. \tag{2.3}$$

We define U_i as the image of ϕ_i, and we readily check that the sets $\{U_i\}$ cover $\mathbb{C}P^{n-1}$. With $k = j$ if $j < i$ and with $k = j-1$ otherwise, the transition function

$$\psi_{j,i}(= \phi_j^{-1} \circ \phi_i) : \mathbb{C}^{n-1} - \{z_k = 0\} \to \mathbb{C}^{n-1} \tag{2.4}$$

is given by

$$\psi_{j,i}(z_1, \ldots, z_{n-1}) = \begin{cases} \frac{1}{z_j}(z_1, \ldots, \widehat{z_j}, \ldots, z_{i-1}, 1, z_i, \ldots, z_n) & \text{if } i > j, \\ \frac{1}{z_{j-1}}(z_1, \ldots, z_{i-1}, 1, z_i, \ldots, \widehat{z}_{j-1}, \ldots, z_n) & \text{if } i < j, \end{cases}$$

where $\widehat{\ }$ denotes deletion. The coordinates of $\psi_{j,i}$ are clearly holomorphic on the domain. Other details are left to the reader.

3. Group Theoretic Interpretation

To introduce groups into the picture, note that $GL(n, \mathbb{C})$ acts transitively on $\mathbb{C}P^{n-1}$ by

$$g \cdot [v] = [g \cdot v] \tag{3.1}$$

where $g \in GL(n, \mathbb{C})$ and $v \in \mathbb{C}^n - \{0\}$. (Note that although we write vectors as rows, matrices will always act upon them as if they were columns.)

Let $v_n = (0, \ldots, 0, 1)$. Let Q' be subset of $GL(n, \mathbb{C})$ that fixes v_n, and let Q be the subset that fixes $[v_n]$. Then

$$Q' = \begin{pmatrix} * & 0 \\ * & 1 \end{pmatrix} \text{ and } Q = \begin{pmatrix} * & 0 \\ * & c \end{pmatrix}.$$

Here the leftmost $(n \times (n-1))$ submatrices are arbitrary, the 0 represents a column of $n-1$ zeros, and $c \in \mathbb{C}^*$. (Representing matrices in this manner will be standard for us; we will usually be interested only in the rightmost column.)

We note that since $GL(n,\mathbb{C})$ is an open subset of $\mathbb{C}^{n^2}$, it can be given the structure of a complex manifold using its coordinate functions. The subgroups Q and Q' both inherit the structure of a complex manifold since their coordinate functions come from those on $GL(n,\mathbb{C})$.

The interesting point is how the transition functions behave in terms of matrices. The holomorphic diffeomorphism

$$I : GL(n,\mathbb{C})/Q \to \mathbb{C}P^{n-1}$$

defined by

$$I(gQ) = [g \cdot v_n] \tag{3.2}$$

can be factored in an obvious way through $GL(n,\mathbb{C})/Q'$. Consider the map

$$q_1 : GL(n,\mathbb{C})/Q' \to \mathbb{C}^n - \{0\}$$

defined by

$$q_1(gQ') = g \cdot v_n; \tag{3.3}$$

in matrix terms,

$$q_1 \begin{pmatrix} * & z_1 \\ * & \vdots \\ * & z_n \end{pmatrix} = (z_1, \dots, z_n). \tag{3.4}$$

This clearly exhibits q_1 as a holomorphic diffeomorphism. Define

$$q_2 : \mathbb{C}^n - \{0\} \to \mathbb{C}P^{n-1}$$

by

$$q_2(v) = [v].$$

The quotient map $gQ' \to gQ$ corresponds precisely to the map q_2.

Note that we can now factor each coordinate map ϕ_i as $\phi_i = q_2 \circ q_1 \circ \phi_i'$, where

$$\phi_i' : \mathbb{C}^{n-1} \to GL(n,\mathbb{C})/Q'$$

is defined by

$$\phi_i'(z_1, \dots, z_{n-1}) = (* \quad z) \tag{3.5}$$

and z is the column $(z_1, \dots, z_{i-1}, 1, z_i, \dots, z_{n-1})$. Here $*$ is any $(n \times (n-1))$ matrix whose i^{th} row is 0 and whose remaining part is invertible.

When $j < i$ and $z_j \neq 0$, the transition function $\psi_{j,i}$ corresponds to the factorization of matrices in $GL(n,\mathbb{C})$ given by

$$(* \quad z) = (* \quad z') \begin{pmatrix} I_{n-1} & 0 \\ 0 & z_j \end{pmatrix}, \tag{3.6}$$

where z is as in the previous paragraph, z' is the column

$$\left(\frac{z_1}{z_j}, \dots, \frac{z_j}{z_j}, \dots, \frac{z_{i-1}}{z_j}, \frac{1}{z_j}, \frac{z_i}{z_j}, \dots, \frac{z_n}{z_j} \right),$$

and I_{n-1} is the $(n-1)$-sized identity matrix.

Since $U(n)$ also acts transitively on $\mathbb{C}P^{n-1}$, $\mathbb{C}P^{n-1}$ is isomorphic to $U(n)/L$, where

$$L = Q \cap U(n) = \left\{ \begin{pmatrix} k & 0 \\ 0 & \omega \end{pmatrix} \,\middle|\, k \in U(n-1),\, \omega \in U(1) \right\} = U(n-1) \times U(1).$$

4. Homogeneous Holomorphic Line Bundles on $\mathbb{C}P^{n-1}$

In this section we construct the homogeneous holomorphic vector bundle $\mathcal{L}_\lambda$, describe its sections abstractly, and produce some explicit holomorphic sections.

Let χ be a one-dimensional representation of L. Then χ is trivial on $U(n-1)$ and has differential $\lambda = -Ne_n$ for some $N \in \mathbb{Z}$. The linear functional λ is dominant with respect to Δ^+ when $N \geq 0$. Extend χ holomorphically to $L^{\mathbb{C}}$ and trivially to the rest of Q. Explicitly, if $q = \begin{pmatrix} * & 0 \\ * & c \end{pmatrix}$ is in Q, then $\chi(q) = c^{-N}$. The space for this one-dimensional representation will be denoted by $\mathbb{C}_\lambda$.

Define

$$\mathcal{L}_\lambda = GL(n,\mathbb{C}) \times_Q \mathbb{C}_\lambda = (GL(n,\mathbb{C}) \times \mathbb{C}_\lambda)/\sim,$$

where

$$(gq, z) \sim (q, \chi(q)z) \text{ for } g \in GL(n,\mathbb{C}),\ q \in Q, \text{ and } z \in \mathbb{C}_\lambda. \tag{4.1}$$

The group $GL(n,\mathbb{C})$ acts on $\mathcal{L}_\lambda$ by $g' \cdot [(g,z)] = [(g'g, z)]$.

Next we describe smooth and holomorphic sections of $\mathcal{L}_\lambda$. Consider the section $\varphi : G/Q \to \mathcal{L}_\lambda$ defined by $\varphi(gQ) = [(g, \gamma(g))]$. $GL(n,\mathbb{C})$ acts on such a section by

$$[\pi'_\lambda(g)\varphi](g') = g \cdot \varphi(g^{-1}g'). \tag{4.2}$$

For φ to be well defined, we must have

$$[(gq, \gamma(gq))] = [(g, \chi(q)\gamma(gq))].$$

Thus sections φ of $\mathcal{L}_\lambda$ correspond to functions γ on $GL(n,\mathbb{C})$ such that

$$\gamma(gq) = \chi(q)^{-1}\gamma(g). \tag{4.3}$$

Smoothness or holomorphicity of φ corresponds to smoothness or holomorphicity of γ, and the latter can be easily checked. $GL(n,\mathbb{C})$ acts compatibly by left translation on γ; that is, for $g, g' \in GL(n,\mathbb{C})$,

$$[\pi_\lambda(g)\gamma](g') = \gamma(g^{-1}g') \tag{4.4}$$

For fixed $N \geq 0$, consider the space V_n^N of homogeneous holomorphic polynomials of degree N in n variables $z_1, \ldots, z_n$. For $P \in V_n^N$, set

$$\Phi_P(g) = P(g \cdot v_n).$$

Note that for $q = \begin{pmatrix} * & 0 \\ * & c \end{pmatrix} \in Q$,

$$\Phi_P(gq) = P(gq \cdot v_n) = P(g \cdot cv_n) = c^N P(g \cdot v_n) = \chi(q)^{-1}\Phi_P(g). \tag{4.5}$$

We also have that $\Phi_P(g) = P(q_1(gQ'))$. Thus we see that the homogeneous holomorphic polynomials of degree N in n variables yield holomorphic sections of $\mathcal{L}_\lambda$ (in a one-one fashion).

5. Homogeneous Polynomials of Degree N

We investigate properties of the space V_n^N of homogeneous holomorphic polynomials of degree N in n variables as a representation of $U(n)$. Specifically, we identify it as the space of an irreducible representation with highest weight $\lambda = -Ne_n$.

For $P \in V_n^N$, we let $U(n)$ act by left translation; that is, for $z \in \mathbb{C}^n$ and $k \in U(n)$,

$$(\pi_N(k)P)(z) = P(k^{-1} \cdot z). \tag{5.1}$$

It is clear that the operation that produces sections of $\mathcal{L}_\lambda$ intertwines π_N and π_λ.

A basis for V_n^N is given by the monomials:

$$\Big\{P_{i_1,\ldots,i_n}(z) = z_1^{i_1} \cdots z_n^{i_n} \,\Big|\, \sum_{j=1}^{n} i_j = N\Big\}. \tag{5.2}$$

Thus dim $V_n^N = \binom{N+n-1}{n-1}$; counting the set of monomials is equivalent to the problem of counting the ways to place N identical balls into n boxes.

To see that V_n^N is irreducible as a $\mathfrak{gl}(n,\mathbb{C})$-representation, we compute for $j > k$ and $i_j > 0$,

$$\begin{aligned} E_{jk} \cdot P_{i_1,\ldots,i_n}(z) &= \frac{d}{dt}\big(\exp(tE_{jk}) \cdot z_1^{i_1} \cdots z_n^{i_n}\big)\big|_{t=0} \\ &= \frac{d}{dt}\big(z_1^{i_1} \cdots (z_j - tz_k)^{i_j} \cdots z_n^{i_n}\big)\big|_{t=0} \\ &= -i_j\, P_{i_1,\ldots,i_k+1,\ldots,i_j-1,\ldots,i_n}(z). \end{aligned} \tag{5.3}$$

Thus we see that every monomial can be obtained by successive applications of certain E_{jk} to $P_{0,\ldots,0,N} = z_n^N$. Hence V_n^N is cyclic, generated by $P_{0,\ldots,0,N}$.

A similar computation using each H_i shows that the monomials are $\mathfrak{h}$-weight vectors, where the $\mathfrak{h}$-weight of $P_{i_1,\ldots,i_n}(z) = z_1^{i_1} \cdots z_n^{i_n}$ is $-\sum_{i=1}^{n} i_k e_k$. Thus the highest weight is given by $-Ne_n$. The vector $P_{0,\ldots,0,N}$ has this weight. Since $P_{0,\ldots,0,N}$ is cyclic, the representation is irreducibile with highest weight $\lambda = -Ne_n$.

6. Exhaustion of Holomorphic Sections

In this section all holomorphic sections are shown to arise from holomorphic homogeneous polynomials as in Section 4. Our strategy rests on the following fact:

(6.1) Every holomorphic function of several variables on $\mathbb{C}^n$ can be represented (uniquely) by a convergent power series.

Let φ be a holomorphic section of $\mathcal{L}_\lambda$ with associated function γ. When $z_{j-1} \neq 0$, γ satisfies (by (4.3))

$$\gamma\begin{pmatrix} * & 1 \\ * & z_1 \\ \vdots & \vdots \\ * & z_n \end{pmatrix} = \gamma\left(\begin{pmatrix} * & \frac{1}{z_{j-1}} \\ * & \frac{z_1}{z_{j-1}} \\ \vdots & \vdots \\ * & \frac{z_n}{z_{j-1}} \end{pmatrix}\begin{pmatrix} I_{n-1} & 0 \\ 0 & z_{j-1} \end{pmatrix}\right) = z_{j-1}^N \gamma\begin{pmatrix} * & \frac{1}{z_{j-1}} \\ * & \frac{z_1}{z_{j-1}} \\ \vdots & \vdots \\ * & \frac{z_n}{z_{j-1}} \end{pmatrix}. \tag{6.2}$$

Noting that $\frac{z_{j-1}}{z_{j-1}} = 1$ occurs in the $(j,n)^{\text{th}}$ entry of the last matrix and recalling (3.5), we can translate (6.2) into

$$(6.3)\quad (\gamma\circ\phi_1')(z_1,\ldots,z_{n-1}) = z_{j-1}^N(\gamma\circ\phi_j')\Big(\frac{1}{z_{j-1}},\frac{z_1}{z_{j-1}},\ldots,\frac{z_{j-2}}{z_{j-1}},\frac{z_j}{z_{j-1}},\ldots,\frac{z_{n-1}}{z_{j-1}}\Big).$$

We note that $\gamma\circ\phi_i'$ is well defined since γ is right invariant under Q'. Now each $P_i = \gamma\circ\phi_i'$ is a holomorphic function on $\mathbb{C}^{n-1}$ with associated power series expansion

$$(6.4)\qquad P_i(z_1,\ldots,z_{n-1}) = \sum_{\text{all } i_k\geq 0} a^{(i)}_{i_1,\ldots,i_{n-1}} z_1^{i_1}\cdots z_{n-1}^{i_{n-1}}.$$

When these expansions are placed in (6.3), it follows immediately that P_j must be a polynomial in $n-1$ variables of total degree less than or equal to N. If not, then z_{j-1} must occur with negative exponent in some monomial on the left side of (6.3), which is absurd. Note also that when $N < 0$, there can be no holomorphic sections.

The dimension of the space of polynomials in $n-1$ variables with degree less than or equal to N is $\binom{N+n-1}{n-1}$. This can be seen by induction and the fact that this space is a direct sum of the spaces of polynomials of degree less than or equal to $N-1$ and the homogeneous polynomials of degree N.

Alternatively, if $N > 0$, the homogeneous polynomial can be recovered by "homogenizing":

$$(6.5)\qquad \mathcal{P}_\varphi(z_1,\ldots,z_n) = z_i^N P_i\Big(\frac{z_1}{z_i},\ldots,\frac{z_{i-1}}{z_i},\frac{z_{i+1}}{z_i},\ldots,\frac{z_n}{z_i}\Big).$$

(When $N = 0$, φ is a constant.) We leave it to the reader to investigate the independence of the choices involved.

Thus we see that there are at most $\binom{N+n-1}{n-1}$ holomorphic sections and the Borel-Weil Theorem holds.

7. Weyl Dimension Formula

Central to the above discussion was the dimension of an irreducible representation of a given highest weight. We calculated this dimension directly in section 5 and then used the result in section 6 to complete the proof of irreducibility of the Borel-Weil realization.

The Weyl dimension formula calculates this dimension using only the highest weight and the root data. In our application it bypasses the need for calculating with root vectors in section 5. The formula follows from the Weyl character formula; see [K1], Ch. 4, for more details. The formula is

$$(7.1)\qquad d_\lambda = \prod_{\alpha\in\Delta^+} \frac{\langle\lambda+\rho,\alpha\rangle}{\langle\rho,\alpha\rangle},$$

where ρ is half the sum of the positive roots. In our case, $\rho = \frac{1}{2}\sum_{i=1}^{n}(n-2i+1)e_i$ and $\lambda = -Ne_n$.

Note that roots orthogonal to λ cancel by division. Thus we need only consider only roots of form $e_i - e_n$. The contribution from $e_i - e_n$ is $\frac{N+n-i}{n-i}$. Collecting these factors gives

$$d_\lambda = \prod_{i=1}^{n-1} \frac{N+n-i}{n-i} = \binom{N+n-1}{n-1}. \tag{7.2}$$

8. The Borel-Weil-Bott Theorem for $SU(2)$

Schmid's lectures [Sc] address also Bott's generalization of the Borel-Weil Theorem, which replaces holomorphic sections with suitable $\bar{\partial}$ cohomology sections. This theorem was first discovered in the general case as Theorem IV′ in [Bo]. Precise statements of the theorem can also be found in [BE], [GS], and [K2].

The above discussion for $U(2)$ also handles degree 0 cohomology (holomorphic sections) for $SU(2)$, and the only remaining case for $SU(2)$ is degree 1. The full result for $SU(2)$ is used in conjunction with a spectral sequence to compute the theorem for a general compact connected Lie group.

The degree 1 result for $SU(2)$ can be obtained as follows. Referring to [We], we see that Hodge theory reduces matters to identifying strongly harmonic forms (those in the kernels of both $\bar{\partial}$ and $\bar{\partial}^*$). Explicit formulas for $\bar{\partial}$ and $\bar{\partial}^*$ may be found in [GS]. The condition "strongly harmonic" in this context reduces to conjugate analytic. Using the techniques of previous sections, one can show that a conjugate analytic section arises from the complex conjugate of a holomorphic polynomial, homogeneous of degree $-N-2$.

References

[BE] Baston, R. J., and M. G. Eastwood, *The Penrose Transform: Its Interaction with Representation Theory*, Oxford University Press, Oxford, 1989.

[Bo] Bott, R., Homogeneous vector bundles, *Annals of Math.* **66** (1957), 203–248.

[GS] Griffiths, P., and W. Schmid, Locally homogeneous complex manifolds, *Acta Math.* **123** (1970), 253–302.

[K1] Knapp, A. W., *Representation Theory of Semisimple Lie Groups: An Overview Based on Examples*, Princeton University Press, Princeton, 1986.

[K2] Knapp, A. W., *Lie Groups, Lie Algebras, and Cohomology*, Princeton University Press, Princeton, 1988.

[Sc] Schmid, W., Discrete series, *these* PROCEEDINGS, pp. 83–113.

[We] Wells, R. O., *Differential Analysis on Complex Manifolds*, Prentice-Hall, Englewood Cliffs, N.J., 1973; second edition, Springer, New York, 1980.

SCHOOL OF MATHEMATICS, INSTITUTE FOR ADVANCED STUDY, PRINCETON, NJ 08540, U.S.A.

Current address: Department of Mathematics, Massachusetts Institute of Technology, Cambridge, MA 02139, U.S.A.

E-mail address: donley@math.ias.edu

Proceedings of Symposia in Pure Mathematics
Volume **61** (1997), pp. 123–155

Induced Representations and the Langlands Classification

E. P. van den Ban

ABSTRACT. In these notes we discuss the concept of induction and some of its applications to the representation theory of a real semisimple Lie group. In particular, we give an introduction to parabolic induction, Bruhat theory, the asymptotic behavior of matrix coefficients, the subrepresentation theorem, characterization of discrete series and tempered representations, and, finally, the Langlands classification of irreducible admissible representations.

1. Induced Representations

1.1. Homogeneous vector bundles

The process of induction allows us to create representations of a Lie group, starting from representations of a subgroup.

We start by recalling the notion of an associated vector bundle. Let G be a Lie group, H a closed subgroup, and (ξ, V) a finite-dimensional continuous representation of H (here and in the following, representation spaces are always assumed to be complex linear). The group H acts freely and properly on the product $G \times V$ by

$$h \cdot (g, v) = (gh^{-1}, \xi(h)v).$$

The associated quotient space $G \times_H V := (G \times V)/H$ therefore has a unique structure of smooth manifold turning $G \times V \to G \times_H V$ into a principal fiber bundle with structure group H. Projection onto the first coordinate induces a smooth map

$$p : G \times_H V \to G/H. \tag{1}$$

Now p is a fiber bundle; for each $g \in G$ the map $v \mapsto (g, v)$ induces a bijection φ_g from V onto the fiber $p^{-1}(gH)$ over gH. The requirement that every φ_g be linear determines a unique structure of a (complex) vector bundle on the fiber bundle (1). This vector bundle is said to be **associated** with the representation ξ; we shall also denote it by $\mathcal{V} := G \times_H V$.

The natural action of G on $G \times V$ by left multiplication on the first coordinate induces a smooth action of G on $\mathcal{V}$. In this way $p : \mathcal{V} \to G/H$ becomes a homogeneous vector bundle. Here we recall that a **homogeneous vector bundle** over

1991 *Mathematics Subject Classification.* Primary 22E45; Secondary 43A65.

G/H is a vector bundle $q : \mathcal{W} \to G/H$ together with a smooth action of G on $\mathcal{W}$ such that for each $g \in G$ the following two conditions are fulfilled:

(a) the following diagram commutes:

$$\begin{array}{ccc} \mathcal{W} & \xrightarrow{g\cdot} & \mathcal{W} \\ {\scriptstyle q}\downarrow & & \downarrow{\scriptstyle q} \\ G/H & \xrightarrow{l_g} & G/H \end{array}$$

(in particular $g\cdot$ maps each fiber $\mathcal{W}_x := q^{-1}(x)$ onto the fiber $\mathcal{W}_{gx}$);

(b) for every $x \in G/H$, the map $g\cdot : \mathcal{W}_x \to \mathcal{W}_{gx}$ is linear.

Any homogeneous vector bundle $q : \mathcal{W} \to G/H$ is associated with a continuous representation of H. Indeed the fiber $V := q^{-1}(eH)$ is invariant under the action of H; we thus obtain a continuous representation ξ of H in V. The smooth G-map $G \times V \to \mathcal{W}$ with $(g, v) \mapsto g \cdot v$ factors to an isomorphism of the associated homogeneous vector bundle $\mathcal{V} = G \times_H V$ onto $\mathcal{W}$.

It follows from the above that the category of continuous finite-dimensional representations of H is equivalent to the category of G-homogeneous vector bundles on G/H. The equivalence is established by the above construction of the associated vector bundle, an inverse by restriction to the fiber above the origin eH of G/H.

If ξ and $\mathcal{V}$ are as above, then by $C(\mathcal{V})$ and $C^\infty(\mathcal{V})$, we denote the spaces of continuous and smooth sections of $\mathcal{V}$, respectively. The group G has a natural representation π in $C(\mathcal{V})$, given by the rule

$$[\pi(g)s](x) = g \cdot [s(g^{-1}x)],$$

for $s \in C(\mathcal{V})$, $x \in G/H$, and $g \in G$. The representation π is called the representation of G **induced** from the representation ξ of H; it is denoted by

$$\pi = \mathrm{ind}_H^G(\xi).$$

Note that the space $C(\mathcal{V})$, equipped with the topology of uniform convergence on compact sets, is a Fréchet space. The induced representation is continuous for this topology. Depending on the context it is sometimes convenient to work with a different representation space. For instance the action of G on the space of smooth sections $C^\infty(\mathcal{V})$ is a continuous representation $\pi_\circ$ of G in a Fréchet space as well. Moreover, $\pi_\circ$ is the restriction of π to the G-invariant subspace $C^\infty(\mathcal{V})$ of $C(\mathcal{V})$. By density of this subspace, the representation π is completely determined by $\pi_\circ$. In this sense we are justified also to call $\pi_\circ$ the induced representation.

1.2. The induced picture

For the purpose of representation theory it is often convenient to realize the induced representation $\mathrm{ind}_H^G(\xi)$ on a space of vector-valued functions rather than sections of a bundle.

We identify V with the fiber of $\mathcal{V}$ above eH via the linear isomorphism induced by the map $v \mapsto (e, v)$. By $C(G, V)$ we denote the space of continuous functions $G \to V$. Given a section $s \in C(\mathcal{V})$, we define the function $\varphi = \varphi_s \in C(G, V)$ by $\varphi(g) = g^{-1} \cdot s(gH)$. Then φ transforms according to the rule

$$\varphi(gh) = \xi(h)^{-1}\varphi(g) \qquad (g \in G,\ h \in H).$$

The space of functions $\varphi \in C(G,V)$ transforming according to the above rule is denoted by $C(G,\xi) = C(G,H,\xi)$. Let R denote the representation of H on $C(G)$ by right translation. Then via the natural identification $C(G,V) \simeq C(G) \otimes V$ we have an isomorphism

$$C(G,\xi) \simeq [C(G) \otimes V]^H,$$

where the superscript H indicates that the subspace of invariants for the representation $R \otimes \xi$ of H has been taken.

It is easily seen that the map $s \mapsto \varphi_s$ is a topological linear isomorphism from $C(\mathcal{V})$ onto $C(G,\xi)$ (equipped with the natural structure of a Fréchet space). Indeed, if $\varphi \in C(G,\xi)$, then the associated section $s = s_\varphi$ is given by

$$s_\varphi(gH) = [(g, \varphi(g))].$$

Note that this definition is unambiguous by the transformation property of φ.

By transference under the isomorphism $C(\mathcal{V}) \simeq C(V,\xi)$ we may realize the representation $\pi = \mathrm{ind}_H^G(\xi)$ of G in $C(G,\xi)$. It is then given by

$$[\pi(g)\varphi](x) = \varphi(g^{-1}x) \qquad (g, x \in G).$$

In future references we shall call this realization of $\mathrm{ind}_H^G(\pi)$ the "induced picture" (to distinguish it from the "geometric picture"). One of the advantages of the induced picture is that it allows a straightforward generalization to infinite-dimensional representations ξ.

Remark 1.1. In the above induced picture the representation space is characterized by means of transformation properties from the right. Equivalently one may of course use transformation properties from the left. Indeed, let ${}^{`}C(G,\xi)$ denote the space of continuous functions $\varphi : G \to V$ transforming according to the rule

$$\varphi(hx) = \xi(h)\varphi(x) \qquad (x \in G,\ h \in H).$$

Moreover, let ${}^{`}\pi$ be the representation of $G$ in ${}^{`}C(G,\xi)$ coming from the right regular action. If $\varphi \in C(G,\xi)$, then the function ${}^{`}\varphi : x \mapsto \varphi(x^{-1})$ belongs to ${}^{`}C(G,\xi)$, and the map $\varphi \mapsto {}^{`}\varphi$ defines an equivalence of the representations $\pi$ and ${}^{`}\pi$.

1.3. Frobenius reciprocity

Let H be a closed subgroup of G, and let ξ be a continuous (not necessarily unitary) representation of H in a Hilbert space $V = V_\xi$. We drop the assumption that V is finite-dimensional and define $C(G,\xi)$ and $\mathrm{ind}_H^G(\xi)$ by the formulas of the induced picture. The following result is known as **Frobenius reciprocity**.

Lemma 1.2 (Frobenius reciprocity). *Let (δ, V_δ) be a finite-dimensional continuous representation of G. Then the map $\varphi : T \mapsto \mathrm{ev}_e \circ T$ defines a natural isomorphism*

$$\mathrm{Hom}_G(V_\delta, \mathrm{ind}_H^G(\xi)) \simeq \mathrm{Hom}_H(V_\delta, V_\xi). \tag{2}$$

PROOF. If T belongs to the space on the left, put $\varphi(T) = \mathrm{ev}_e \circ T$. Then $\varphi(T) : V_\delta \to V$ is a linear map, which is readily seen to be H-equivariant. By G-equivariance, if $v \in V_\delta$ and $g \in G$, then $T(v)(g) = T(\delta(g)^{-1}v)(e) = \varphi(T)(\delta(g)^{-1})$, from which the injectivity of φ follows. If S belongs to the space on the right then the map $T : V_\delta \to C(G,V)$ defined by $T(v)(g) = S(\delta(g^{-1})v)$ is readily checked to belong to the space on the left-hand side of (2), and $S = \varphi(T)$. Hence φ is surjective as well. □

1.4. The bundle of densities

If V is an n-dimensional real-linear space, then a **density** on V is a map $\omega : V^n \to \mathbb{C}$ transforming according to the rule

$$T^*\omega := \omega \circ T^n = |\det T|\,\omega \qquad (T \in \mathrm{End}(V)).$$

In these notes the (complex-linear) space of densities on V is denoted by $\mathcal{D}V$.

If φ is a linear isomorphism from V onto a real-linear space W, then the map $\varphi^* : \omega \mapsto \omega \circ \varphi^n$ is a linear isomorphism $\mathcal{D}W \to \mathcal{D}V$ of the associated spaces of densities. The space $\mathcal{D}V$ is one-dimensional; in fact, if $v_1, \dots, v_n$ is a basis of V, then the map $T \mapsto T(v_1, \dots, v_n)$ is a linear isomorphism from $\mathcal{D}V$ onto $\mathbb{C}$.

If X is a smooth manifold, then by T_xX we denote the tangent space of X at a point x. By a well known procedure we may define the bundle $\mathcal{D}TX$ of densities on X; it is a complex line bundle with fiber $(\mathcal{D}TX)_x \simeq \mathcal{D}(T_xX)$. If φ is a diffeomorphism of X onto a manifold Y, then we define the map $\varphi^* : C(\mathcal{D}TY) \to C(\mathcal{D}TX)$ by $(\varphi^*\omega)(x) = D\varphi(x)^*\omega(\varphi(x))$.

Let $e_1, \dots, e_n$ be the standard basis of $\mathbb{R}^n$. The density $\lambda \in \mathcal{D}\mathbb{R}^n$ given by $\lambda(e_1, \dots, e_n) = 1$ is called the **standard density** on $\mathbb{R}^n$. Let $U \subset \mathbb{R}^n$ be an open subset. Then by triviality of the tangent bundle $TU \simeq U \times \mathbb{R}^n$, the map $f \mapsto f\lambda$ defines a linear isomorphism from $C^\infty(U)$ onto $C^\infty(\mathcal{D}TU)$. If $f \in C_c(U)$ we define the integral

$$\int_U f\lambda := \int_{\mathbb{R}^n} f(x)\,dx,$$

where dx denotes Lebesgue measure. If φ is a diffeomorphism from U onto a second open subset $V \subset \mathbb{R}^n$, then we have $\varphi^*(g\lambda)(x) = g(\varphi(x))|\det D\varphi(x)|\lambda(\varphi(x))$ for $g \in C_c(V)$. Thus, by the substitution of variables theorem,

$$\int_U \varphi^*\omega = \int_V \omega \qquad (\omega \in C_c\mathcal{D}TV). \tag{3}$$

This observation allows us to extend the notion of integral to any compactly supported continuous density on any smooth manifold. The extension involves reduction to charts by using partitions of unity, exactly as in the definition of integration of differential forms of top dimension. Note that integration of forms depends on an orientation, whereas the present integration of densities does not.

The following result is a consequence of these definitions.

Proposition 1.3. *Let $\varphi : U \to V$ be a diffeomorphism of C^∞-manifolds. Then* (3) *holds.*

Half densities. If V is an n-dimensional real-linear space and $\alpha \in \mathbb{C}$ a complex number, then an α-density on V is a function $\nu : V^n \to \mathbb{C}$ transforming according to the rule $\nu \circ T^n = |\det T|^\alpha \nu$, for every $T \in \mathrm{End}(V)$. The space of α-densities on V is denoted by $\mathcal{D}^\alpha V$. Thus $\mathcal{D}V = \mathcal{D}^1 V$. The elements of $\mathcal{D}^{1/2}V$ are called **half densities** on V.

The product of two densities is a density; multiplication induces a linear isomorphism from $\mathcal{D}^\alpha V \otimes \mathcal{D}^\beta V$ onto $\mathcal{D}^{\alpha+\beta} V$ $(\alpha, \beta \in \mathbb{C})$. Note that 0-densities are constant functions; hence $\mathcal{D}^0 V \simeq \mathbb{C}$. The natural isomorphism $\mathcal{D}^{-\alpha}V \otimes \mathcal{D}^\alpha V \simeq \mathbb{C}$ induces a natural identification $(\mathcal{D}^\alpha V)^* \simeq \mathcal{D}^{-\alpha} V$.

If X is a manifold, then the bundle of α-densities on X is denoted by $\mathcal{D}^\alpha TX$. Fiberwise multiplication induces, for $\alpha, \beta \in \mathbb{C}$, an isomorphism $\mathcal{D}^\alpha TX \otimes \mathcal{D}^\beta TX \simeq \mathcal{D}^{\alpha+\beta}TX$ of vector bundles. The dual vector bundle $(\mathcal{D}^\alpha TX)^*$ is naturally isomorphic to $\mathcal{D}^{-\alpha}TX$.

Densities and generalized sections. If X is a smooth manifold, and $p : \mathcal{V} \to X$ a vector bundle on X, then by $C^\infty(\mathcal{V})$ we denote the space of smooth sections of $\mathcal{V}$. This space is equipped with a Fréchet topology in the usual way. The space of compactly supported smooth sections of $\mathcal{V}$ is denoted by $C_c^\infty(\mathcal{V})$. It is equipped with the structure of a complete locally convex Hausdorff space by means of the inductive limit topology.

The space $C^{-\infty}(\mathcal{V})$ of generalized sections of $\mathcal{V}$ is defined by

$$C^{-\infty}(\mathcal{V}) := C_c^\infty(\mathcal{V}^* \otimes \mathcal{D}TX)'.$$

This definition has the effect of allowing a natural embedding $\iota : C^\infty(\mathcal{V}) \hookrightarrow C^{-\infty}(\mathcal{V})$. Indeed, let $\langle \cdot\,, \cdot \rangle$ denote the (pointwise defined) natural bilinear map $C^\infty(\mathcal{V}) \times C_c^\infty(\mathcal{V}^* \otimes \mathcal{D}TX) \to C_c^\infty \mathcal{D}TX$. Then the embedding ι is given by

$$\iota(s) : \sigma \mapsto \int_X \langle s, \sigma \rangle.$$

Here we adopt the convention of denoting the full linear dual of a linear space and the dual of a vector bundle by a star. The topological linear dual of a topological linear space is denoted by a prime.

A special case of the above is the definition of the space $C^{-\infty}(X)$ of $\mathbb{C}$-valued generalized functions on X by

$$C^{-\infty}(X) = C_c^\infty(\mathcal{D}TX)'.$$

Since $(\mathcal{D}^{1/2}TX)^* \otimes \mathcal{D}TX \simeq \mathcal{D}^{1/2}TX$ naturally, it follows from the above that

$$C^{-\infty}(\mathcal{V} \otimes \mathcal{D}^{1/2}TX) \simeq C_c^\infty(\mathcal{V}^* \otimes \mathcal{D}^{1/2}TX)'. \tag{4}$$

1.5. Densities on G/H

Let H be a closed subgroup of a Lie group G. The tangent bundle $T(G/H)$ is a G-homogeneous vector bundle on G/H; the action of G on the space $CT(G/H)$ of continuous vector fields on G/H is given by

$$g \cdot v(x) = Dl_g(x)v(g^{-1}x).$$

Let $\mathfrak{g}$ and $\mathfrak{h}$ denote the Lie algebras of G and H respectively. Here we adopt the convention that Lie groups are denoted by italic capitals, their Lie algebras by the corresponding Gothic lower case letters. The projection $G \to G/H$ induces a natural isomorphism $\mathfrak{g}/\mathfrak{h} \simeq T_{eH}(G/H)$. Accordingly, $T(G/H)$ is the homogeneous bundle associated with the representation $\xi = \mathrm{Ad}_{\mathfrak{g}/\mathfrak{h}}$ of H on $\mathfrak{g}/\mathfrak{h}$ defined by $\xi(h)(X + \mathfrak{h}) = \mathrm{Ad}(h)X + \mathfrak{h}$.

In a similar way we see that the bundle of densities $\mathcal{D}T(G/H)$ is G-homogeneous; the action of G on the associated space of continuous densities is given by $g \cdot \omega = l_g^{-1*}\omega$, for $\omega \in C\mathcal{D}T(G/H)$. This is the homogeneous bundle associated with the character δ of H given by $\delta(h) = |\det \mathrm{Ad}_{\mathfrak{g}/\mathfrak{h}}(h)^{*-1}|$. Hence

$$\delta(h) = |\det \mathrm{Ad}_{\mathfrak{g}/\mathfrak{h}}(h)|^{-1} \qquad (h \in H). \tag{5}$$

For $\alpha \in \mathbb{C}$, the bundle of α-densities on G/H is homogeneous as well; it is associated with the character δ^α.

Generalized sections. If (σ, V) is a finite-dimensional continuous representation of H, let $\mathcal{V}$ denote the associated vector bundle. Then $\mathcal{V} \otimes \mathcal{D}^{1/2}T(G/H)$ is naturally isomorphic to the homogeneous bundle associated with the tensor product representation of H in $V \otimes \mathcal{D}^{1/2}(\mathfrak{g}/\mathfrak{h})$. The isomorphism (4) then naturally corresponds to an isomorphism

$$C^{-\infty}(G, \sigma \otimes \delta^{1/2}) \simeq C_c^\infty(G, \sigma^\vee \otimes \delta^{1/2})', \tag{6}$$

where $\sigma^\vee$ denotes the representation contragredient to σ.

Densities and measures. Let H be a closed subgroup of the Lie group G. If Ω is a continuous density on G/H then the map $\mu_\Omega : C_c(G/H) \to \mathbb{C}$ defined by

$$\mu_\Omega(f) = \int_{G/H} f\,\Omega \qquad (f \in C_c(G/H))$$

is continuous linear, hence defines a Radon measure on G/H. It follows from Proposition 1.3 that

$$\mu_{l_g^{*-1}\Omega} = \mu_\Omega \circ l_g^*.$$

One now readily sees that $\Omega \mapsto \mu_\Omega$ defines a linear isomorphism from the space $C(\mathcal{D}T(G/H))^G$ of G-invariant densities on G/H onto the space of G-invariant Radon measures on G/H.

On the other hand, the map $\Omega \mapsto \Omega(eH)$ is an isomorphism from $C(\mathcal{D}T(G/H))^G$ onto $[\mathcal{D}(\mathfrak{g}/\mathfrak{h})]^H$. We thus see that there exists a natural isomorphism from $[\mathcal{D}(\mathfrak{g}/\mathfrak{h})]^H$ onto the space of G-invariant Radon measures on G/H. In particular, the latter space is nontrivial if and only if the character (5) is identically 1.

1.6. Normalized induction

Let H be a closed subgroup of the Lie group G, and assume that the quotient space G/H is compact. Let ξ be a (not necessarily unitary) representation of H in a (possibly infinite-dimensional) Hilbert space V. We denote the inner product by $\langle\,\cdot\,,\,\cdot\,\rangle_\xi$; inner products on complex Hilbert spaces are always assumed to be conjugate-linear in the second variable.

The induced representation $\pi = \mathrm{ind}_H^G(\xi)$ is said to be **unitarizable** if the representation space $C(G, \xi)$ allows a pre-Hilbert structure such that π extends to a unitary representation in the associated Hilbert completion.

Unitarity of the representation ξ does not necessarily imply unitarizability of $\mathrm{ind}_H^G(\xi)$. However, as we will see, by twisting with half densities we may normalize the induction so that unitarity is preserved.

Let $\xi \otimes \delta^{1/2}$ denote the tensor product representation of H in $V \otimes \mathcal{D}^{1/2}(\mathfrak{g}/\mathfrak{h})$. We now observe that $(\lambda_1, \lambda_2) \mapsto \lambda_1\bar{\lambda}_2$ defines a sesquilinear map from $\mathcal{D}^{1/2}(\mathfrak{g}/\mathfrak{h}) \times \mathcal{D}^{1/2}(\mathfrak{g}/\mathfrak{h})$ onto $\mathcal{D}(\mathfrak{g}/\mathfrak{h})$. Given $v_1, v_2 \in V$ and $\lambda_1, \lambda_2 \in \mathcal{D}^{1/2}(\mathfrak{g}/\mathfrak{h})$, we define $(v_1 \otimes \lambda_1, v_2 \otimes \lambda_2) = \langle v_1, v_2\rangle_\xi \lambda_1\bar{\lambda}_2$ and extend this to a sesquilinear pairing $V \otimes \mathcal{D}^{1/2}(\mathfrak{g}/\mathfrak{h}) \times V \otimes \mathcal{D}^{1/2}(\mathfrak{g}/\mathfrak{h}) \to \mathcal{D}(\mathfrak{g}/\mathfrak{h})$.

Now assume that ξ is unitary. If $\varphi, \psi \in C(G, \xi \otimes \delta^{1/2})$, then the function $(\varphi, \psi) : g \mapsto (\varphi(g), \psi(g))$ belongs to $C(G, \delta)$. It may therefore be canonically identified with a density on G/H, which in turn may be integrated. We put

$$\langle \varphi, \psi\rangle := \int_{G/H} (\varphi, \psi) \qquad (\varphi, \psi \in C(G, \xi \otimes \delta^{1/2})). \tag{7}$$

Lemma 1.4. *Let ξ be a unitary representation of H. Then the sesquilinear pairing (7) defines a pre-Hilbert structure on $C(G, \xi \otimes \delta^{1/2})$, which is G-equivariant. The induced representation $\mathrm{ind}_H^G(\xi \otimes \delta^{1/2})$ extends to a unitary representation in the associated Hilbert completion. In particular it is unitarizable.*

PROOF. We denote the induced representation by π. One readily verifies that the given pairing defines a pre-Hilbert structure. If $\varphi, \psi \in C(G, \xi \otimes \delta^{1/2})$, then $(\pi(g)\varphi, \pi(g)\psi) = (l_g^{-1*}\varphi, l_g^{-1*}\psi) = l_g^{-1*}(\varphi, \psi)$. The integral of the latter density over G/H equals the integral of (φ, ψ) over G/H, by Proposition 1.3, whence the equivariance of the pre-Hilbert structure.

It follows that for every $g \in G$ the map $\pi(g)$ extends uniquely to a unitary endomorphism of the Hilbert completion $\mathcal{H}$ of $C(G, \xi \otimes \delta^{1/2})$. One readily sees that the pre-Hilbert structure is a continuous form on $C(G, \xi \otimes \delta^{1/2})$; hence the latter space embeds continuously into $\mathcal{H}$. From this and the unitarity of each $\pi(g)$ for $g \in G$ it easily follows that the extension of π is a continuous representation of G in $\mathcal{H}$. □

In view of the above result the representation $\mathrm{ind}_H^G(\xi \otimes \delta^{1/2})$ is said to be obtained from ξ by **normalized induction**. In these notes this induced representation will also be denoted by $\mathrm{Ind}_H^G(\xi)$. In fact, the notation $\mathrm{Ind}_H^G(\xi)$ will be used for $\mathrm{ind}_H^G(\xi \otimes \delta^{1/2})$ even if ξ is not unitary.

2. Parabolically Induced Representations

2.1. Basic notions

From now on we assume that G is a real reductive group of Harish-Chandra's class. For purposes of induction this class is more convenient than the slightly smaller class of connected semisimple groups with finite center (see [18], II.1, p. 192).

Let K be a maximal compact subgroup of G, and θ the associated Cartan involution. The infinitesimal involution of the Lie algebra $\mathfrak{g}$ of G associated with θ is denoted by the same symbol; the associated Cartan decomposition is denoted by $\mathfrak{g} = \mathfrak{k} \oplus \mathfrak{p}$. Thus $\mathfrak{k}$ and $\mathfrak{p}$ are the eigenspaces for θ for the eigenvalues 1 and -1, respectively. We fix a nondegenerate bilinear form B on $\mathfrak{g}$ that is $\mathrm{Ad}(G)$- and θ-invariant. Then $\mathfrak{k} \perp \mathfrak{p}$ relative to B.

Let $\mathfrak{a}$ be a maximal abelian subspace of $\mathfrak{p}$ and let $\Sigma \subset \mathfrak{a}^*$ be the (possibly nonreduced) root system of $\mathfrak{a}$ in $\mathfrak{g}$. Its Weyl group W is naturally isomorphic with $N_K(\mathfrak{a})/M$, where $N_K(\mathfrak{a})$ and M denote the normalizer and the centralizer, respectively, of $\mathfrak{a}$ in K.

We fix a positive system Σ^+ for Σ and denote the associated system of simple roots by Δ. Let $\mathfrak{n}$ be the sum of the root spaces $\mathfrak{g}_\alpha$ for $\alpha \in \Sigma^+$, and let $N = \exp \mathfrak{n}$ and $A = \exp \mathfrak{a}$. We recall that G decomposes according to the Iwasawa decomposition

$$G = KAN;$$

here the product map $K \times A \times N \to G$ is a diffeomorphism.

Every parabolic subgroup of G is K-conjugate to a standard parabolic subgroup (relative to Σ^+). We recall ([18], Theorem II.6.9, p. 285) that the standard parabolic subgroups are in one-to-one correspondence with the collection of subsets of Δ. For $F \subset \Delta$, the associated standard parabolic subgroup Q_F can be described as follows. Let $\mathfrak{a}_F$ be the intersection of the root spaces $\ker \alpha$ for $\alpha \in F$, and let M_{1F} be the centralizer of $\mathfrak{a}_F$ in G. Furthermore, let $\mathfrak{n}_F$ be the sum of the root spaces $\mathfrak{g}_\alpha$ for $\alpha \in \Sigma^+ \setminus \mathrm{span}(F)$, and let $N_F = \exp \mathfrak{n}_F$. Then $Q_F = M_{1F}N_F$.

The group M_{1F} is stable under θ, hence decomposes as $M_{1F} = K_F \exp(\mathfrak{m}_{1F} \cap \mathfrak{p})$, where $K_F := K \cap M_{1F}$. Let $\mathfrak{m}_F$ denote the B-orthocomplement of $\mathfrak{a}_F$ in $\mathfrak{m}_{1F}$; it is the Lie algebra of the group $M_F = K_F \exp(\mathfrak{m}_F \cap \mathfrak{p})$. The latter group is again of Harish-Chandra's class and has compact center (use [18], Theorem II.6.13, p. 286): this allows induction on the dimension of G as a method of proof. Note that $M_{1F} = M_F A_F$. Hence

$$Q_F = M_F A_F N_F;$$

this is the Langlands decomposition of Q_F. Note that $A_\emptyset = A$ and $N_\emptyset = N$; moreover, $M_\emptyset$ equals the centralizer M of $\mathfrak{a}$ in K. Hence the minimal standard parabolic subgroup of G is given by $Q_\emptyset = Q = MAN$.

We recall that exp maps $\mathfrak{a}$, the Lie algebra of A, diffeomorphically onto A; the inverse of $\exp : \mathfrak{a} \to A$ is denoted by log. Moreover, given $\lambda \in \mathfrak{a}^*_{\mathbb{C}} := \mathrm{Hom}(\mathfrak{a}, \mathbb{C})$ and $a \in A$, we write

$$a^\lambda := e^{\lambda \log a}.$$

The (complexified) bilinear form B naturally defines a linear isomorphism of the complexification $\mathfrak{g}_{\mathbb{C}}$ with its dual $\mathfrak{g}^*_{\mathbb{C}}$. Accordingly we may identify $\mathfrak{a}^*_{F\mathbb{C}}$ with a linear subspace of $\mathfrak{a}^*_{\mathbb{C}}$.

If σ is a continuous representation of M_F in a Hilbert space $\mathcal{H}_\sigma$, and $\lambda \in \mathfrak{a}^*_{F\mathbb{C}}$, then a representation $\sigma \otimes \lambda \otimes 1$ of Q_F in $\mathcal{H}_\sigma$ is defined by

$$(\sigma \otimes \lambda \otimes 1)(man) = a^\lambda \sigma(m) \qquad (m \in M_F,\ a \in A_F,\ n \in N_F).$$

This is indeed a representation since M_F centralizes A_F and $M_F A_F$ normalizes N_F.

Normalized induction from Q_F to G involves the function $\delta_F : Q_F \to \mathbb{C}$ defined by

$$\delta_F(man) = |\det \mathrm{Ad}_{\mathfrak{g}/\mathfrak{q}_F}(man)|^{-1} \qquad (m \in M_F,\ a \in A_F,\ n \in N_F).$$

Now $\mathrm{Ad}(m)$ and $\mathrm{Ad}(n)$ act by determinant 1 on $\mathfrak{g}$ and $\mathfrak{q}_F$, and $\mathrm{Ad}(a)$ preserves the spaces $\mathfrak{q}_F$ and $\bar{\mathfrak{n}}_F = \theta\mathfrak{n}_F$. Note that $\bar{\mathfrak{n}}_F$ is the sum of the root spaces $\mathfrak{g}_{-\alpha}$ for $\alpha \in \Sigma^+ \setminus \mathrm{span}(F)$. Thus $\mathfrak{g} = \bar{\mathfrak{n}}_F \oplus \mathfrak{q}_F$ as a linear space, and it follows that $\mathrm{Ad}(a)$ acts by determinant $\det[\mathrm{Ad}(a)|\bar{\mathfrak{n}}_F]$ on the quotient $\mathfrak{g}/\mathfrak{q}_F$. Hence

$$\delta_F(man) = a^{2\rho_F},$$

where $\rho_F \in \mathfrak{a}^*_F$ is defined by $\rho_F(X) := \frac{1}{2}\mathrm{tr}(\mathrm{ad}(X)|\mathfrak{n}_F)$.

We define the representation $\pi_{\sigma,\lambda}$ of G by

$$\pi_{\sigma,\lambda} := \mathrm{ind}_{Q_F}^G(\sigma \otimes (\lambda + \rho_F) \otimes 1).$$

The underlying representation space

$$C(G, \sigma, \lambda) := C(G, \sigma \otimes (\lambda + \rho_F) \otimes 1)$$

is defined as in Section 1.2. Thus it consists of the continuous functions $\varphi : G \to \mathcal{H}_\sigma$ transforming according to the rule

$$\varphi(xman) = a^{-\lambda - \rho_F}\, \sigma(m)^{-1}\, \varphi(x)$$

for $x \in G$ and $(m, a, n) \in M_F \times A_F \times N_F$. The action is by left translation.

If σ is unitary and $\lambda \in i\mathfrak{a}^*_F$, then $\xi_{\sigma,\lambda} := \sigma \otimes \lambda \otimes 1$ is a unitary representation of Q_F and $\xi_{\sigma,\lambda} \otimes \delta_F^{1/2} \simeq \sigma \otimes (\lambda + \rho_F) \otimes 1$. Hence $\pi_{\sigma,\lambda}$ is the representation obtained by normalized induction from $\xi_{\sigma,\lambda}$. It follows that $\pi_{\sigma,\lambda}$ is unitarizable when σ is unitary and λ is imaginary. Accordingly we write, also when λ is not imaginary,

$$\pi_{\sigma,\lambda} = \mathrm{Ind}_{Q_F}^{G}(\sigma \otimes \lambda \otimes 1).$$

Now assume that σ is unitary, but $\lambda \in \mathfrak{a}^*_{F\mathbb{C}}$ general. Let dk be the normalized Haar measure on K. If $\varphi \in C(G,\sigma,\lambda)$ and $\psi \in C(G,\sigma,-\bar{\lambda})$, we define

$$\langle \varphi, \psi \rangle := \int_K (\varphi(k), \psi(k))_\sigma \, dk; \tag{8}$$

here $(\,\cdot\,,\,\cdot\,)_\sigma$ denotes the inner product of $\mathcal{H}_\sigma$.

Lemma 2.1. *Let σ be a unitary representation of M_F and let $\lambda \in \mathfrak{a}^*_{F\mathbb{C}}$. Then (8) defines a G-equivariant nondegenerate sesquilinear pairing*

$$C(G,\sigma,\lambda) \times C(G,\sigma,-\bar{\lambda}) \to \mathbb{C}.$$

In particular, if λ is imaginary then the pairing defines a G-equivariant pre-Hilbert structure on $C(G,\sigma,\lambda)$. The representation $\pi_{\sigma,\lambda}$ extends to a unitary representation in the associated Hilbert completion.

PROOF. From the Iwasawa decomposition $G = KAN$ it follows that $G/Q_F \simeq K/K \cap Q_F = K/K_F$. Hence restriction to K induces a continuous linear isomorphism

$$C(G,\sigma,\lambda) \xrightarrow{\simeq} C(K,K_F,\sigma_F),$$

where σ_F denotes the restriction of σ to K_F. The latter space is the representation space of the induced representation $\mathrm{ind}_{K_F}^{K}(\sigma_F)$. It now follows straightforwardly that the pairing is nondegenerate, sesquilinear, and K-equivariant; the G-equivariance remains to be established.

Let $\omega \in \mathcal{D}(\mathfrak{k}/\mathfrak{k}_F) \simeq \mathcal{D}(\mathfrak{g}/\mathfrak{q}_F)$ be the density corresponding to the normalized K-invariant measure $d(kK_F)$ on K/K_F (see Section 1.5). If $f : G \to \mathbb{C}$ is a continuous function transforming according to the character δ_F of Q_F on the right, i.e., $f \in C(G,\delta_F)$, then $f \otimes \omega \in C(G, \mathcal{D}(\mathfrak{g}/\mathfrak{q}_F))$ may be identified canonically with a density on G/Q_F. Its integral equals

$$\int_{G/Q_F} f \otimes \omega = \int_{K/K_F} f(k)\, d(kK_F) = \int_K f(k)\, dk.$$

If $\varphi \in C(G,\sigma,\lambda)$, $\psi \in C(G,\sigma,-\bar{\lambda})$, and $g \in G$, then the function $f = (\varphi,\psi)_\sigma$, defined by $f(x) = (\varphi(x),\psi(x))_\sigma$, belongs to $C(G,\delta_F)$, and so does the function $L_g f : x \mapsto f(g^{-1}x)$. Hence by the above observation we obtain

$$\langle \pi_{\sigma,\lambda}(g)\varphi, \pi_{\sigma,-\bar{\lambda}}(g)\psi \rangle = \int_{G/Q_F} L_g f \otimes \omega = \int_{G/Q_F} l_g^{*-1}(f \otimes \omega) = \int_{G/Q_F} f \otimes \omega = \langle \varphi, \psi \rangle.$$

We conclude that the pairing (8) is G-equivariant. If λ is imaginary, the pairing defines a (continuous) pre-Hilbert structure on $C(G,\sigma,\lambda)$ that is G-equivariant. It follows that $\pi_{\sigma,\lambda}$ extends to a continuous unitary representation of G in the Hilbert completion of $C(G,\sigma,\lambda)$. (See also the argument at the end of the proof of Lemma 1.4.) □

2.2. The three pictures for the induced representation

The induced picture. We assume that σ is a unitary representation. If λ is in $\mathfrak{a}_{F\mathbb{C}}^*$, then we equip $C(G, \sigma, \lambda)$ (the space for $\pi_{\sigma,\lambda}$ in the induced picture) with the pre-Hilbert structure defined by

$$(\varphi_1, \varphi_2) = \int_K (\varphi_1(k), \varphi_2(k))_\sigma \, dk$$

for $\varphi_j \in C(G, \sigma, \lambda)$. We denote by $\mathcal{H}_{\sigma,\lambda}$ the completion of this pre-Hilbert space. The representation $\pi_{\sigma,\lambda}$ has a unique extension to a continuous representation of G in $\mathcal{H}_{\sigma,\lambda}$, which we denote by the same symbol. Indeed, for imaginary λ this follows from Lemma 2.1 (and then the extension is unitary); for general λ it is best seen in the compact picture discussed below.

Alternatively $\mathcal{H}_{\sigma,\lambda}$ may be described as the space of measurable (almost everywhere defined) functions $\varphi : G \to \mathcal{H}_\sigma$ such that

(a) $\varphi(xman) = a^{-\lambda-\rho_F}\sigma(m)^{-1}\varphi(x)$ for $x \in G$, $m \in M_F$, $a \in A_F$, $n \in N_F$;
(b) $\varphi|K \in L^2(K, \mathcal{H}_\xi)$.

In this picture the induced representation is given by the formula

$$[\pi_{\sigma,\lambda}(g)\varphi](x) = \varphi(g^{-1}x) \qquad (x, g \in G).$$

The compact picture. We denote by $L^2(K, \mathcal{H}_\sigma)$ the space of $\mathcal{H}_\sigma$-valued L^2-functions relative to the Haar measure dk. Then restriction to K induces a surjective isometry

$$\mathcal{H}_{\sigma,\lambda} \xrightarrow{\simeq} L^2(K, \sigma_F), \tag{9}$$

where $L^2(K, \sigma_F)$ denotes the Hilbert space of functions $\varphi \in L^2(K, \mathcal{H}_\sigma)$ transforming according to the rule $\varphi(km) = \sigma(m)^{-1}\varphi(k)$ for $k \in K$, $m \in K_F$.

By transference under the isometry (9), the induced representation $\pi_{\sigma,\lambda}$ may be realized in the Hilbert space $L^2(K, \sigma_F)$, which has the advantage that it is independent of λ. We call this realization of the induced representation the "compact picture." It may described as follows.

The multiplication map $K \times \exp(\mathfrak{m}_F \cap \mathfrak{p}) \times A_F \times N_F \to G$ is a diffeomorphism. Accordingly we may define analytic maps κ_F, μ_F, H_F, ν_F from G to the spaces K, $\exp(\mathfrak{m}_F \cap \mathfrak{p})$, $\mathfrak{a}_F$, N_F, respectively, such that, for all $x \in G$,

$$x = \kappa_F(x)\mu_F(x) \exp H_F(x)\nu_F(x).$$

If $\varphi \in \mathcal{H}_{\sigma,\lambda}$ then for $x \in G$, $k \in K$ we have

$$\varphi(x^{-1}k) = \varphi(\kappa_F(x^{-1}k)\mu_F(x^{-1}k) \exp H_F(x^{-1}k)),$$

and hence in the compact picture the representation $\pi_{\sigma,\lambda}$ is described by

$$[\pi_{\sigma,\lambda}(x)\varphi](k) = e^{(-\lambda-\rho_F)H_F(x^{-1}k)}\,\sigma(\mu_F(x^{-1}k))^{-1}\,\varphi(\kappa_F(x^{-1}k)).$$

The noncompact picture. It is known that the inclusion $\bar{N}_F \to G$ induces a diffeomorphism j from $\bar{N}_F$ onto an open dense subset of G/Q_F. Let Ω be the K-invariant density on $G/Q_F \simeq K/K_F$ corresponding to the normalized K-invariant Radon measure on K/K_F (see 1.5). Let λ be the $\bar{N}_F$-invariant density on $\bar{N}_F$ determined by $\lambda(e) = j^*(\Omega)(e)$. Then

$$\lambda(\bar{n}) = Dl_{\bar{n}}(e)^{*-1}j^*(\Omega)(e).$$

Let $d\bar{n}$ be the Haar measure on $\bar{N}_F$ corresponding to the density λ (use 1.5 with $\bar{N}_F$, $\{e\}$ in place of G, H). Then

Lemma 2.2. $j^*(d\bar{k}) = e^{-2\rho_F H_F(\bar{n})}\, d\bar{n}$.

PROOF. Let $\bar{n} \in \bar{N}_F$ and put $t(\bar{n}) = \mu_F(\bar{n}) \exp H_F(\bar{n}) \nu_F(\bar{n})$. Then by K-invariance of Ω it follows that $l_{\bar{n}}^*\Omega = l_{t(\bar{n})}^*\Omega$. Now $l_{t(\bar{n})}$ preserves the origin of G/Q_F; its tangent map at the origin is the isomorphism of $\mathfrak{g}/\mathfrak{q}_F$ induced by $\mathrm{Ad}(t(\bar{n}))$, hence has determinant $\exp(-2\rho_F H_F(\bar{n}))$. It follows that

$$l_{\bar{n}}^*\Omega(e) = e^{-2\rho_F H_F(\bar{n})}\Omega(e).$$

Hence

$$j^*(\Omega)(\bar{n}) = Dl_{\bar{n}}(e)^{*-1}[l_{\bar{n}}^* j^*\Omega(e)] = Dl_{\bar{n}}(e)^{*-1} j^*(l_{\bar{n}}^*\Omega)(e) = e^{-2\rho_F H_F(\bar{n})}\,\lambda(\bar{n}).$$

The equality between the densities in the first and last member in the above display corresponds to the equality of the measures in the assertion of the lemma. □

It follows from the lemma that, for $\varphi_1, \varphi_2 \in C(G, \sigma, \lambda)$,

$$\begin{aligned}(\varphi_1, \varphi_2) &= \int_{K/K_F} (\varphi_1(k), \varphi_2(k))_\sigma\, d(kk_F)\\ &= \int_{\bar{N}_F} (\varphi_1(\kappa_F(\bar{n}), \varphi_2(\kappa_F(\bar{n}))_\sigma\, e^{-2\rho_F H(\bar{n})}\, d\bar{n}\\ &= \int_{\bar{N}_F} (\varphi_1(\bar{n}), \varphi_2(\bar{n}))_\sigma\, e^{-2\mathrm{Re}\,\lambda H_F(\bar{n})}\, d\bar{n}. \end{aligned} \tag{10}$$

Hence restriction to $\bar{N}_F$ induces an isometry

$$\mathcal{H}_{\sigma,\lambda} \simeq L^2(\bar{N}_F, \mathcal{H}_\sigma, e^{-2\mathrm{Re}\,\lambda H(\bar{n})}\, d\bar{n}).$$

In the latter Hilbert space (the "noncompact picture") the representation $\pi_{\sigma,\lambda}$ can be realized as follows. Let the analytic maps $\bar{n}_F, m_F, a_F, n_F$ from $\bar{N}_F Q_F$ to $\bar{N}_F, M_F, A_F, N_F$, respectively, be defined by

$$y = \bar{n}_F(y) m_F(y) a_F(y) n_F(y).$$

Then the representation $\pi_{\sigma,\lambda}$ is given by

$$\pi_{\sigma,\lambda}(x)\varphi(\bar{n}) = a_F(x^{-1}\bar{n})^{-\lambda-\rho_F}\sigma(m_F(x^{-1}\bar{n}))^{-1}\varphi(\bar{n}_F(x^{-1}\bar{n}))$$

for $\varphi \in L^2(\bar{N}_F, \mathcal{H}_\sigma, e^{-2\mathrm{Re}\,\lambda H(\bar{n})}\, d\bar{n})$, $x \in G$, and $\bar{n} \in \bar{N}_F$.

Remark 2.3. For later purposes we also describe the sesquilinear pairing of Lemma 2.1 in the noncompact picture. If $\varphi_1 \in C(G, \sigma, \lambda)$ and $\varphi_2 \in C(G, \sigma, -\bar{\lambda})$, then by the same substitution of variables $k = \kappa_F(\bar{n})$ as in (10) we obtain

$$\langle \varphi_1, \varphi_2 \rangle = \int_{\bar{N}_F} (\varphi_1(\bar{n}), \varphi_2(\bar{n}))_\sigma\, d\bar{n}.$$

2.3. *K*-finite vectors

If π is a continuous representation of G in a complete locally convex space V, we denote by V^∞ the space of C^∞-vectors, and by V_K the space of K-finite vectors of the representation. We recall that $V_K \cap V^\infty$ is a dense subspace of V that is invariant for K and the infinitesimal action of $\mathfrak{g}$ (cf., e.g., [18], Lemma II.7.10, p. 312).

Let $\widehat{K}$ denote the set of (equivalence classes of) irreducible finite-dimensional representations of K. If $\delta \in \widehat{K}$, we denote by $V(\delta)$ the space of vectors in V_K that are K-isotypical of type δ. Let V_δ be a representation space for δ; then we recall that the map $(T, v) \mapsto T(v)$ induces a natural isomorphism

$$\mathrm{Hom}_K(V_\delta, V) \otimes V_\delta \xrightarrow{\simeq} V(\delta) \tag{11}$$

of K-modules; here K acts on the tensor product by $I \otimes \delta$. The representation π is called **admissible** if $V(\delta) < \infty$ for all $\delta \in \widehat{K}$. If π is admissible, it can be shown that $V_K \subset V^\infty$ ([18], Theorem II.7.14); thus, by what was said earlier, V_K is a module for both K and $\mathfrak{g}$.

Let $Q_F = M_F A_F N_F$ be the standard parabolic subgroup determined by a set $F \subset \Delta$. The group M_F is of Harish-Chandra's class, and $K_F = K \cap M_F$ is a maximal compact subgroup of M_F. Let $(\sigma, \mathcal{H}_\sigma)$ be a unitary representation of M_F.

Lemma 2.4. *If σ is admissible (for M_F, K_F), then for every $\lambda \in \mathfrak{a}_{F\mathbb{C}}^*$ the induced representation $\mathrm{Ind}_{Q_F}^G(\sigma \otimes \lambda \otimes 1)$ is admissible for G, K.*

PROOF. Let (δ, V_δ) be a finite-dimensional irreducible representation of K. Then we must show that $\dim \mathcal{H}_{\sigma,\lambda}(\delta) < \infty$. In view of (11) this is equivalent to

$$\dim \mathrm{Hom}_K(V_\delta, \mathcal{H}_{\sigma,\lambda}) < \infty.$$

By the compact picture the K-module $\mathcal{H}_{\sigma,\lambda}$ is isomorphic to $L^2(K, \sigma_F)$, the representation space for $\mathrm{Ind}_{K_F}^{M_F}(\sigma_F)$. By Frobenius reciprocity (Lemma 1.2) we have

$$\mathrm{Hom}_K(V_\delta, L^2(K, \sigma_F)) \simeq \mathrm{Hom}_{K_F}(V_\delta, \mathcal{H}_\sigma),$$

and the latter space is finite-dimensional since σ is admissible. □

2.4. The infinitesimal character

Let $\mathcal{Z}(\mathfrak{g})$ be the center of the universal enveloping algebra $U(\mathfrak{g})$ of the complexication $\mathfrak{g}_\mathbb{C}$ of $\mathfrak{g}$. In this section we investigate the action of $\mathcal{Z}(\mathfrak{g})$ on parabolically induced representations.

If $\mathfrak{h} \subset \mathfrak{g}$ is a Cartan subalgebra, then by $I(\mathfrak{h})$ we denote the space of Weyl group invariants in $S(\mathfrak{h})$, the symmetric algebra of $\mathfrak{h}_\mathbb{C}$. Moreover, by $\gamma = \gamma_\mathfrak{h}^\mathfrak{g}$ we denote the canonical (Harish-Chandra) isomorphism from $\mathcal{Z}(\mathfrak{g})$ onto $I(\mathfrak{h})$.

A continuous representation (π, V) of G is said to have **infinitesimal character** $\Lambda \in \mathfrak{h}_\mathbb{C}^*$ if, for all $v \in V^\infty$,

$$\pi(Z)v = \gamma_\mathfrak{h}^\mathfrak{g}(Z, \Lambda) \qquad (Z \in \mathcal{Z}(\mathfrak{g})).$$

In the following we let $F \subset \Delta$ and assume that $\mathfrak{h} \subset \mathfrak{g}$ is a Cartan subalgebra of $\mathfrak{g}$ containing $\mathfrak{a}_F$. Since $\mathfrak{m}_{1F} = \mathfrak{m}_F + \mathfrak{a}_F$ is the centralizer of $\mathfrak{a}_F$ in $\mathfrak{g}$, it follows that

$$\mathfrak{h} = \mathfrak{h}_{M_F} \oplus \mathfrak{a}_F, \tag{12}$$

where $\mathfrak{h}_{M_F} := \mathfrak{h} \cap \mathfrak{m}_F$ is a Cartan subalgebra of $\mathfrak{m}_F$. As mentioned before we use the bilinear form B to identify the dual spaces $\mathfrak{h}_{M_F}^*$ and $\mathfrak{a}_F^*$ (as well as their complexifications) with subspaces of $\mathfrak{h}_\mathbb{C}^*$. Since the decomposition (12) is B-perpendicular, $\mathfrak{h}_{M_F}^*$ corresponds to the subspace of functionals in $\mathfrak{h}^*$ that vanish on $\mathfrak{a}_F$; similarly, $\mathfrak{a}_F^*$ corresponds to the space of functionals in $\mathfrak{h}^*$ that vanish on $\mathfrak{h}_{M_F}$.

Lemma 2.5. *Let $(\sigma, \mathcal{H}_\sigma)$ be a unitary representation of M_F with infinitesimal character $\Lambda_\sigma \in \mathfrak{h}^*_{M_F\mathbb{C}}$. Then for every $\lambda \in \mathfrak{a}^*_{F\mathbb{C}}$ the induced representation $\mathrm{Ind}_{Q_F}^G(\sigma \otimes \lambda \otimes 1)$ has infinitesimal character $\Lambda_\sigma + \lambda$.*

PROOF. Let $\varphi \in C^\infty(G, \sigma, \lambda)$. Then $\varphi(e)$ is a smooth vector in $\mathcal{H}_\sigma$. Hence for $Z_M \in \mathcal{Z}(\mathfrak{m})$ we have $\sigma(Z_M)\varphi(e) = \gamma^{\mathfrak{m}}_{\mathfrak{h}_{M_F}}(Z_M, \Lambda_\sigma)\varphi(e)$. Let $Z \in \mathcal{Z}(\mathfrak{g})$. Then

$$[\pi_{\sigma,\lambda}(Z)\varphi](e) = [L(Z)\varphi](e) = [R(Z^\vee)\varphi](e), \tag{13}$$

where $Z \mapsto Z^\vee$ denotes the anti-automorphism of $U(\mathfrak{g})$ induced by the anti-automorphism $X \mapsto -X$ of $\mathfrak{g}$. There exists a unique $Z_0 \in U(\mathfrak{m}_F + \mathfrak{a}_F)$ such that $Z \simeq Z_0 \mod \mathfrak{n}_F U(\mathfrak{g})$. The element Z_0 belongs to $\mathcal{Z}(\mathfrak{m}_{1F})$. It follows from the characterization of the Harish-Chandra isomorphisms of $(\mathfrak{g}, \mathfrak{h})$ and $(\mathfrak{m}_{1F}, \mathfrak{h})$ in terms of the root structure that for every $\mu \in \mathfrak{h}^*_{\mathbb{C}}$

$$\gamma^{\mathfrak{m}_{1F}}_{\mathfrak{h}}(Z_0, \mu + \rho_F) = \gamma^{\mathfrak{g}}_{\mathfrak{h}}(Z, \mu).$$

From the Q_F-behavior of φ on the right it now follows that the right-hand side of (13) equals

$$[\sigma \otimes (\lambda + \rho_F)](Z_0)\,\varphi(e) = \gamma^{\mathfrak{m}_F + \mathfrak{a}_F}_{\mathfrak{h}}(Z_0, \Lambda_\sigma + \lambda + \rho_F)\varphi(e) = \gamma^{\mathfrak{g}}_{\mathfrak{h}}(Z, \Lambda_\sigma + \lambda)\varphi(e).$$

Hence $\pi_{\sigma,\lambda}(Z)\varphi = \gamma^{\mathfrak{g}}_{\mathfrak{h}}(Z, \Lambda_\sigma + \lambda)\varphi$ at the identity element; since G centralizes $\mathcal{Z}(\mathfrak{g})$, the identity holds at every point of G. □

2.5. Irreducibility

The following result on the irreducibility of parabolically induced representations is due to F. Bruhat [1] for a minimal parabolic subgroup, and to Harish-Chandra [8] in general. Proofs of the general result following the original ideas of Bruhat can be found in the recent papers [2], Appendix B, and [14].

Theorem 2.6 (Bruhat, Harish-Chandra). *Let σ be an irreducible unitary representation of M_F, and let $\lambda \in i\mathfrak{a}^*_F$. Assume that*

(a) *σ has real infinitesimal character (in other words, $\langle \Lambda_\sigma, \alpha \rangle \in \mathbb{R}$ for every $(\mathfrak{m}_{1F\mathbb{C}}, \mathfrak{h}_{\mathbb{C}})$-root α);*
(b) *$\langle \lambda, \beta \rangle \neq 0$ for every $\mathfrak{a}_F$-weight β in $\mathfrak{n}_F$.*

Then the induced representation $\mathrm{Ind}_{Q_F}^G(\sigma \otimes \lambda \otimes 1)$ is irreducible.

IDEAS OF PROOF. We sketch the ideas of Bruhat's proof in the case that $F = \emptyset$; then $Q_F = Q = MAN$ is minimal. Since M is compact, σ is finite-dimensional; this simplifies the functional analysis involved in the argument.

Let σ, λ fulfill the hypotheses of the theorem. Since $\pi_{\sigma,\lambda}$ is unitary it suffices to show that the space $\mathrm{End}_G(\mathcal{H}_{\sigma,\lambda})$ of continuous self-intertwining operators of $\pi_{\sigma,\lambda}$ consists of the scalar multiples of the identity. Let $T \in \mathrm{End}_G(\mathcal{H}_{\sigma,\lambda})$. Then by equivariance T maps the space $\mathcal{H}^\infty_{\sigma,\lambda}$ of smooth vectors continuously and equivariantly into itself. Now $\mathcal{H}^\infty_{\sigma,\lambda} = C^\infty(G, \sigma, \lambda)$; hence the evaluation map $\mathrm{ev}_e : \varphi \mapsto \varphi(e)$ is a continuous linear map from $\mathcal{H}^\infty_{\sigma,\lambda}$ to $\mathcal{H}_\sigma$. Put $u_T = \mathrm{ev}_e \circ T | \mathcal{H}^\infty_{\sigma,\lambda}$. Then

$$u_T \in [(\mathcal{H}^\infty_{\sigma,\lambda})' \otimes \mathcal{H}_\sigma]^Q, \tag{14}$$

where the prime indicates that the topological linear dual has been taken, and where the superscript Q indicates that the space of invariants for the tensor product of $\pi'_{\sigma,\lambda}|Q$ and $\sigma \otimes (\lambda + \rho_F) \otimes 1$ has been taken.

The G-module $\mathcal{H}^\infty_{\sigma,\lambda}$ is isomorphic to the space of smooth sections in the G-homogeneous vector bundle associated with the representation $\sigma\otimes(\lambda+\rho_F)\otimes 1$ of Q in $\mathcal{H}_\sigma$; accordingly its linear topological dual may be identified with the G-module of generalized sections in the bundle associated to the representation $\sigma^\vee\otimes(-\lambda+\rho_F)\otimes 1$ (see (6)). We denote the latter space by $C^{-\infty}(G,\sigma^\vee,-\lambda)$. The projection $p: G \to G/Q$ induces a natural embedding of $C^{-\infty}(G,\sigma^\vee,-\lambda)$ into $C^{-\infty}(G)\otimes\mathcal{H}_\sigma$, with image the space of Q-invariants for the tensor product $R\otimes[\sigma^\vee\otimes(-\lambda+\rho_F)\otimes 1]$. Thus we see that the space in (14) is naturally isomorphic to

$$[C^{-\infty}(G)\otimes(\mathcal{H}_\sigma^*\otimes\mathcal{H}_\sigma)]^{Q\times Q}, \tag{15}$$

where superscript $Q\times Q$ indicates the subspace of invariants for the following action of $Q\times Q$. The action of $Q\times Q$ on $C^{-\infty}(G)$ is $L\hat{\otimes}R$, the exterior tensor product of the left and right regular actions. The action of $Q\times Q$ on $\mathcal{H}_\sigma^*\otimes\mathcal{H}_\sigma$ is by the exterior tensor product $[\sigma^\vee\otimes(-\lambda+\rho_F)]\hat{\otimes}[\sigma\otimes(\lambda+\rho_F)]$. Finally, in (15) the tensor product of these two actions of $Q\times Q$ has been taken.

It follows from the above that $\operatorname{supp} u_T$ is a union of double cosets for the $Q\times Q$-action on G. Thus the **Bruhat decomposition** comes into play. Let $W = N_K(\mathfrak{a})/M$, where $N_K(\mathfrak{a})$ denotes the normalizer of $\mathfrak{a}$ in K. Then W is naturally isomorphic to the Weyl group of the root system Σ. We recall that

$$G = \bigcup_{s\in W} QsQ \qquad \text{(disjoint union)}.$$

There is a unique open double coset, which is dense in G; it corresponds to the longest element in W (relative to Σ^+). For details, see, e.g., [18], p. 300.

Suppose that $s\in W$ is such that QsQ is maximal among the $Q\times Q$-orbits in $\operatorname{supp} u_T$. First we assume that QsQ is open, that is, s is the longest Weyl group element. The generalized function u_T restricts to a smooth $Q\times Q$-invariant $\mathcal{H}_\sigma^*\otimes\mathcal{H}_\sigma$-valued function on this open orbit; its value $u_T(s)$ at s must be fixed under the stabilizer $\operatorname{Stab}(s)$ of s in $Q\times Q$. The latter group equals

$$\operatorname{Stab}(s) = \{(q_1,q_2)\in Q\times Q \mid q_1 s q_2^{-1} = s\} = \{ma, s^{-1}mas) \mid m\in M,\, a\in A\}.$$

Hence

$$(\mathcal{H}_\sigma^*\otimes\mathcal{H}_\sigma)^{\operatorname{Stab}(s)} \simeq \operatorname{Hom}_{MA}(\sigma\otimes(\lambda-\rho_F), s\sigma\otimes s(\lambda+\rho_F)).$$

The latter space is trivial because $\lambda\neq s\lambda$ by the regularity assumption on λ. It follows that $u_T(s)=0$. Hence u_T is supported by the lower-dimensional $Q\times Q$-orbits.

Now assume that QsQ is not open. Then by an analysis in the same spirit as above, but with the additional complication that transversal derivatives to the orbit QsQ have to be taken into account, it follows again that $\sigma\sim s\sigma$ and $\lambda = s\lambda$. Because of the condition on λ this can only happen when $s=1$. It follows from this that u_T has to be supported at e and hence is a derivative of a Dirac function with coefficients in $\operatorname{End}(\mathcal{H}_\sigma)$. By a further analysis one can show that u_T must have order 0 and is equal to $\delta_e\otimes A_T$, with δ_e a Dirac function in e and A_T an element of the space $\operatorname{Hom}_M(\mathcal{H}_\sigma)$. The latter space is one-dimensional by the irreducibility of σ.

Finally, it follows from the above that the map $\operatorname{End}_G(\mathcal{H}_{\sigma,\lambda})\to\operatorname{Hom}_M(\mathcal{H}_\sigma)$ given by $T\mapsto A_T$ is injective. Hence $\operatorname{End}_G(\mathcal{H}_{\sigma,\lambda})$ is one-dimensional. □

3. Asymptotic Behavior of Matrix Coefficients

3.1. Matrix coefficients

Let π be a continuous representation of G in a complete locally convex Hausdorff space V. By a **matrix coefficient** of π we mean a function $G \to \mathbb{C}$ of the form

$$m_{v,v'} : x \mapsto \langle \pi(x)v, v' \rangle,$$

where $v \in V$ and where v' belongs to V', the topological linear dual of V. The following lemma is easily verified.

Lemma 3.1. *Let $v' \in V'$.*

(a) *The map $v \mapsto m_{v,v'}$ intertwines π with the right regular action R of G.*
(b) *If v is in the space V^∞ of C^∞ vectors, then $m_{v,v'} \in C^\infty(G)$.*
(c) *The map $v \mapsto m_{v,v'}$ of $V^\infty \to C^\infty(G)$ intertwines the $U(\mathfrak{g})$-actions induced by π and R.*

If the representation π has an infinitesimal character $\Lambda \in \mathfrak{h}_{\mathbb{C}}^*$, then it follows from the above that every matrix coefficient $m = m_{v,v'}$, with v a smooth vector, is a function in $C^\infty(G)$ satisfying the following system of differential equations:

$$R(Z)m = \gamma(Z, \Lambda)m \qquad (Z \in \mathcal{Z}(\mathfrak{g})); \tag{16}$$

here R denotes the right regular representation of $U(\mathfrak{g})$ in $C^\infty(G)$.

Example 3.2. With $Q_F = M_F A_F N_F$ as in Section 2.1, let σ be an irreducible unitary representation of M, and $\lambda \in \mathfrak{a}_{F\mathbb{C}}^*$. Then the sesquilinear pairing $\mathcal{H}_{\sigma,\lambda} \times \mathcal{H}_{\sigma,-\bar{\lambda}} \to \mathbb{C}$ defined by (8) is nondegenerate, hence induces a conjugate-linear embedding $\mathcal{H}_{\sigma,-\bar{\lambda}} \subset \mathcal{H}_{\sigma,\lambda}'$. For elements $\varphi \in \mathcal{H}_{\sigma,\lambda}$ and $\psi \in \mathcal{H}_{\sigma,-\bar{\lambda}}$ the corresponding matrix coefficient of $\pi_{\sigma,\lambda}$ is given by

$$m_{\varphi,\psi}(x) = \int_K \langle \varphi(x^{-1}k), \psi(k) \rangle \, dk \qquad (x \in G).$$

If φ (or ψ) is a smooth vector, then the matrix coefficient $m_{\varphi,\psi}$ is a smooth function on G. Moreover, by (16) and Lemma 2.5 it satisfies the following system of differential equations

$$R(Z)m_{\varphi,\psi} = \gamma(Z, \Lambda_\sigma + \lambda)m_{\varphi,\psi} \qquad (Z \in \mathcal{Z}(\mathfrak{g}));$$

here Λ_σ denotes the infinitesimal character of σ.

Example 3.3. Let notation be as in the above example, but now assume that $F = \emptyset$, i.e., Q_F is the minimal standard parabolic subgroup $Q = MAN$. Then M is compact, and hence σ is finite-dimensional. The representations $\pi_{\sigma,\lambda}$ are said to belong to the **minimal principal series**. From the compact picture one sees that $\pi_{\sigma,\lambda}$ has a K-fixed vector if and only if $\sigma = 1$. The representations $\pi_\lambda = \pi_{1,\lambda}$ constitute the **spherical principal series**. The space of K-fixed (or spherical) vectors in $\mathcal{H}_\lambda = \mathcal{H}_{1,\lambda}$ is equal to $\mathbb{C}\mathbf{1}_\lambda$, where $\mathbf{1}_\lambda$ is determined by $\mathbf{1}_\lambda|K = 1$. In the induced picture the vector $\mathbf{1}_\lambda$ is the function $G \to \mathbb{C}$ described by

$$\mathbf{1}_\lambda(x) = \mathbf{1}_\lambda(\kappa(x) \exp H(x) \nu(x)) = e^{(-\lambda-\rho)H(x)} \qquad (x \in G),$$

where we have suppressed the index $F = \emptyset$ in the notation.

By equivariance of the pairing (8), the matrix coeffient $\varphi_\lambda := m_{\mathbf{1}_\lambda, \mathbf{1}_{-\bar\lambda}}$ is given by the formula

$$\varphi_\lambda(x) = \langle \mathbf{1}_\lambda, \pi_{-\bar\lambda}(x^{-1})\mathbf{1}_{-\bar\lambda}\rangle = \int_K e^{(\lambda-\rho)H(xk)}\, dk \qquad (x \in G).$$

This is Harish-Chandra's formula for the elementary (or zonal) **spherical function** associated with the Riemannian symmetric space G/K. The function φ_λ is a bi-K-invariant smooth function on G satisfying a system of differential equations of the form (16), coming from the action of the center of $U(\mathfrak{g})$.

Note also that $U(\mathfrak{g})^K$, the algebra of K-invariants in $U(\mathfrak{g})$, preserves the space $(\mathcal{H}_\lambda)^K = \mathbb{C}\mathbf{1}_\lambda$. Hence there exists an algebra homomorphism $\chi_\lambda : U(\mathfrak{g})^K \to \mathbb{C}$ such that

$$X\mathbf{1}_\lambda = \chi_\lambda(X)\mathbf{1}_\lambda \qquad (X \in U(\mathfrak{g})^K).$$

By the equivariance stated in Lemma 3.1(c) it now follows that φ_λ satisfies the system of differential equations

$$R(X)\varphi_\lambda = \chi_\lambda(X)\varphi_\lambda \qquad (X \in U(\mathfrak{g})^K).$$

The space $C^\infty(G/K)$ of C^∞-functions on G/K is canonically isomorphic with the space of right K-invariant C^∞-functions on G. Accordingly, if $X \in U(\mathfrak{g})^K$, then $R(X)$ acts on $C^\infty(G/K)$ as an element from $\mathbb{D}(G/K)$, the algebra of left G-invariant differential operators on G/K. It is known that $X \mapsto R(X)$ is a surjective algebra homomorphism from $U(\mathfrak{g})^K$ onto $\mathbb{D}(G/K)$, with kernel $U(\mathfrak{g})^K \cap U(\mathfrak{g})\mathfrak{k}$. Hence it induces an algebra isomorphism from $U(\mathfrak{g})^K/U(\mathfrak{g})^K \cap U(\mathfrak{g})\mathfrak{k}$ onto $\mathbb{D}(G/H)$. We thus see that the zonal spherical function φ_λ is a simultaneous eigenfunction for the algebra $\mathbb{D}(G/K)$.

The function $\Xi := \varphi_0$, given by the formula

$$\Xi(x) = \int_K e^{-\rho H(xk)}\, dk \qquad (x \in G)$$

plays a fundamental role in harmonic analysis on G.

3.2. A cofinite ideal

Let $(\pi, \widetilde{V})$ be an admissible representation of G. We recall from Section 2.3 that $V := \widetilde{V}_K$ is a module for K and the infinitesimal action of $\mathfrak{g}$. One readily sees that the $\mathfrak{g}$- and K-module V is a $(\mathfrak{g}, K)$-module. This means that (suppressing π in the notation)

(a) for every $v \in V$, the span V_0 of the vectors kv for $k \in K$ is finite-dimensional, and the action of K on V_0 is continuous;
(b) $kXv = [\mathrm{Ad}(k)X]kv$ for $k \in K$, $X \in \mathfrak{g}$, and $v \in V$;
(c) $Xv = d/dt((\exp tX)\,v)|_{t=0}$ for $X \in \mathfrak{k}$ and $v \in V$.

See [19], I, 3.3, for more details. A $(\mathfrak{g}, K)$-module is called **admissible** if $\dim V(\delta) < \infty$ for all $\delta \in \widehat{K}$; an admissible $(\mathfrak{g}, K)$-module that is finitely generated is called a **Harish-Chandra module**.

The $(\mathfrak{g}, K)$-module $V = \widetilde{V}_K$ is admissible, since $V(\delta) = \widetilde{V}(\delta)$ for all $\delta \in \widehat{K}$. We assume that V is finitely generated (this is automatic if π is irreducible) and call it the Harish-Chandra module associated with π. The space $V' := (V^*)_K$ is readily seen to be an admissible $(\mathfrak{g}, K)$-module, naturally isomorphic to $[\widetilde{V}']_K$. This dual

$(\mathfrak{g}, K)$-module is also finitely generated, but this is not so obvious; see, e.g., [19], Lemma 4.3.2.

The goal of this section is to describe the asymptotic behavior, for $v \in V$ and $v' \in V'$, of the matrix coefficient $m_{v,v'}(x)$, as x tends to infinity in G.

Since v and v' are K-finite, the matrix coefficient $m_{v,v'}$ is (real) analytic and behaves finitely under the actions of K from the left and from the right. In view of the Cartan decomposition

$$G = K\mathrm{cl}(A^+)K \tag{17}$$

it is therefore sufficient to study the behavior of $m_{v,v'}(a)$ as $a \to \infty$ in $\mathrm{cl}(A^+)$. Here $\mathrm{cl}(A^+)$ denotes the closure of $A^+ = \exp(\mathfrak{a}^+)$, where $\mathfrak{a}^+$ is the open positive Weyl chamber in $\mathfrak{a}$. We recall that if $x \in G$, then $x \in KaK$ with $a \in \mathrm{cl}(A^+)$ uniquely determined.

The matrix coefficient's behavior along $\mathrm{cl}(A^+)$ will turn out to be severely restricted by a system of differential equations it satisfies. Define the ideal I of $\mathcal{Z}(\mathfrak{g})$ by

$$I = \{Z \in \mathcal{Z}(\mathfrak{g}) \mid \pi(Z) = 0 \text{ on } V\}.$$

Then it follows from the equivariance formulated in Lemma 3.1(c) that

$$R(Z)m_{v,v'} = 0 \qquad (Z \in I). \tag{18}$$

An ideal $\mathcal{I}$ of an algebra $\mathcal{A}$ over $\mathbb{C}$ is said to be **cofinite** if $\mathcal{A}/\mathcal{I}$ is a finite-dimensional complex vector space. The following result expresses that the system (18) is large.

Lemma 3.4. *The ideal I is cofinite in the algebra $\mathcal{Z}(\mathfrak{g})$.*

PROOF. For $\vartheta \subset \widehat{K}$ a finite subset, the finite-dimensional space

$$V(\vartheta) := \bigoplus_{\delta \in \vartheta} V(\delta)$$

is invariant for the action of $\mathcal{Z}(\mathfrak{g})$; let $\nu = \nu_\vartheta : \mathcal{Z}(\mathfrak{g}) \to \mathrm{End}(V(\vartheta))$ be the induced homomorphism of algebras. Then $\ker \nu$ is an ideal of $\mathcal{Z}(\mathfrak{g})$, containing I, and of finite codimension at most $\dim \mathrm{End}(V(\vartheta))$.

Since V is finitely generated as a $(\mathfrak{g}, K)$-module, we may fix a finite set $\vartheta \subset \widehat{K}$ such that $U(\mathfrak{g})V(\vartheta) = V$. Since $\mathcal{Z}(\mathfrak{g})$ is central it follows that $\ker \nu = I$; hence $I = \ker \nu$ and we see that I is cofinite. $\square$

Remark 3.5. If V has an infinitesimal character $\Lambda \in \mathfrak{h}_{\mathbb{C}}^*$, then the associated ideal I is the kernel of the character $\gamma(\,\cdot\,, \Lambda)$ of $\mathcal{Z}(\mathfrak{g})$, hence of codimension 1. In this case, (18) is a system of eigenequations.

3.3. Spherical functions

In view of the K-finiteness of $m_{v,v'}$, the restriction of the function $m_{v,v'}$ satisfies a system of differential equations on the group A (which is diffeomorphic to the vector space $\mathfrak{a}$). We shall arrive at this system essentially by applying the method of separation of variables. For this it is convenient to introduce the notion of τ-spherical functions.

Let τ be a (continuous) representation of $K \times K$ on a finite-dimensional complex-linear space E. We agree to write $\tau_1(k_1)v\tau_2(k_2) = \tau(k_1, k_2^{-1})v$ for $v \in V$ and

$k_1, k_2 \in K$. A continuous function $\psi : G \to E$ is said to be τ-**spherical** if it transforms according to the rule

$$\psi(k_1 x k_2) = \tau(k_1)\psi(x)\tau(k_2) \qquad \text{for } x \in G \text{ and } k_1, k_2 \in K.$$

The space of all such functions is denoted by $C(G,\tau)$, the space of all analytic τ-spherical functions by $\mathcal{A}(G,\tau)$; the spaces $C^\infty(G,\tau), C_c^\infty(G,\tau)$ are defined similarly. Note that

$$C(G,\tau) \simeq (C(G)\otimes E)^{K\times K};$$

where $K \times K$ acts on $C(G)$ by the left times right action.

The K-finite matrix coeffient $m_{v,v'}$ is expressible in terms of a spherical function as follows. Let $\vartheta \subset \widehat{K}$ be a finite subset containing the K-types occurring in v and v'. Define the representation τ of $K \times K$ on $E := \mathrm{End}(V(\vartheta))$ by $\tau(k_1,k_2)A = \pi(k_1)\circ A \circ \pi(k_2)^{-1}$. Let $\iota_\vartheta : V(\vartheta) \to V$ be the inclusion map and $P_\vartheta : V \to V(\vartheta)$ the K-equivariant projection map. Then the function $\varphi : G \to E$ defined by

$$\varphi(x) = P_\vartheta \circ \pi(x) \circ \iota_\vartheta$$

is τ-spherical. Moreover, let $\eta = \eta_{v,v'}$ be the linear functional on E defined by $\eta(A) = \langle Av, v'\rangle$. Then

$$m_{v,v'} = \eta \circ \varphi.$$

Note that the function φ belongs to the space $\mathcal{A}(G,\tau)$ and satisfies the system (18) as well. We denote the space of all such functions by $\mathcal{A}(G,\tau,I)$ and proceed by studying the asymptotic behavior at infinity of the elements of $\mathcal{A}(G,\tau,I)$.

3.4. The radial differential equations

The restriction $\mathrm{Res}\,\varphi$ to A^+ of a function $\varphi \in C^\infty(G,\tau)$ has values in the space

$$E^M := \{v \in E \mid v = \tau(m)v\tau(m)^{-1} \text{ for all } m \in M\};$$

where M denotes the centralizer of A in K. Indeed, this follows from the observation that $\varphi(a) = \varphi(mam^{-1}) = \tau_1(m)\varphi(a)\tau_2(m)^{-1}$ for $a \in A$ and $m \in M$.

The map $(k_1,k_2,a) \mapsto k_1 a k_2$ induces a diffeomorphism of $K \times_M K \times A^+$ onto an open subset of G. Hence if $f \in C_c^\infty(A^+, E^M)$, then there exists a unique function in $C_c^\infty(G,\tau)$, denoted $\mathrm{Lift}\, f$, whose restriction to A^+ is f.

If $Z \in \mathcal{Z}(\mathfrak{g})$, then the operator

$$\Pi_\tau(Z) := \mathrm{Res} \circ R(Z) \circ \mathrm{Lift} \tag{19}$$

from $C_c^\infty(A^+, E^M)$ to $C^\infty(A^+, E^M)$ is readily seen to be continous linear and support preserving. Hence it is a differential operator on A^+ with smooth $\mathrm{End}(E^M)$-valued coefficients. We denote the algebra of all such operators by $\mathcal{D}^\infty$. One readily verifies that $Z \mapsto \Pi_\tau(Z)$ is a homomorphism of algebras from $\mathcal{Z}(\mathfrak{g})$ to $\mathcal{D}^\infty$. The differential operator (19) is called the τ-**radial component** of Z.

Let $S(\mathfrak{a})$ denote the symmetric algebra of $\mathfrak{a}_\mathbb{C}$. The right regular representation of A in $C^\infty(A)$ induces an isomorphism of $S(\mathfrak{a})$ onto the algebra of invariant differential operators on A. Accordingly we identify elements of $S(\mathfrak{a})$ with differential operators; in particular if $H \in \mathfrak{a}$ and $f \in C^\infty(A)$, then $Hf(a) = d/dt(f(a\exp tH))_{t=0}$. The above identification induces a linear isomorphism

$$C^\infty(A^+) \otimes \mathrm{End}(E^M) \otimes S(\mathfrak{a}) \simeq \mathcal{D}^\infty, \tag{20}$$

by which we shall identify these spaces from now on. An element $c \otimes L \otimes v$ of (20) thus acts on $C^\infty(A^+, E^M)$ according to the formula

$$[(c \otimes L \otimes v)f](a) := c(a)L(vf(a)).$$

In particular, the tensor product on the left-hand side of (20) is equipped with the structure of an algebra. Its multiplication law is readily determined by using the Leibniz rule for differentiation.

Let $\mathcal{R}$ be the ring of functions $A^+ \to \mathbb{C}$ generated by 1 and the functions $a \mapsto a^{-\alpha}$ and $a \mapsto (1 - a^{-2\beta})^{-1}$ for $\alpha, \beta \in \Sigma^+$. This ring is stable under differentiation by elements from $S(\mathfrak{a})$. Hence

$$\mathcal{D} := \mathcal{R} \otimes \operatorname{End}(E^M) \otimes S(\mathfrak{a})$$

defines a subalgebra of $\mathcal{D}^\infty$. The following result can be proved by a computation in $U(\mathfrak{g})$; see [4], Proposition 3.2.

Lemma 3.6. *The map Π_τ is an algebra homomorphism from $\mathcal{Z}(\mathfrak{g})$ into $\mathcal{D}$.*

It follows from the above discussion that if $\varphi \in \mathcal{A}(G, \tau, I)$, then its restriction $f = \operatorname{Res}\varphi$ to A^+ is a smooth function $A^+ \to E^M$ satisfying the following system of radial differential equations:

$$\Pi_\tau(Z)f = 0 \qquad (Z \in I). \tag{21}$$

Example 3.7. We consider the group $G = \mathrm{SL}(2, \mathbb{R})$. As a basis of its Lie algebra we take the following standard $\mathfrak{sl}(2)$-triple:

$$H = \begin{pmatrix} 1 & 0 \\ 0 & -1 \end{pmatrix}, \quad X = \begin{pmatrix} 0 & 1 \\ 0 & 0 \end{pmatrix}, \quad Y = \begin{pmatrix} 0 & 0 \\ 1 & 0 \end{pmatrix}.$$

Let $U := X - Y$, and put $k_\varphi = \exp(\varphi U)$. Then

$$k_\varphi = \begin{pmatrix} \cos\varphi & \sin\varphi \\ -\sin\varphi & \cos\varphi \end{pmatrix},$$

and we see that $K = \exp(\mathbb{R}U) = \mathrm{SO}(2)$ is a maximal compact subgroup of G. Put $\mathfrak{a} = \mathbb{R}H$. Then $\Sigma = \{\alpha, -\alpha\}$, where α is determined by $\alpha(H) = 2$. Fix $\Sigma^+ = \{\alpha\}$, and put $H_\alpha = \frac{1}{2}H$. Note that $\mathfrak{n} = \mathbb{R}X$ and $\bar{\mathfrak{n}} = \mathbb{R}Y$. The center $\mathcal{Z}(\mathfrak{g})$ of $U(\mathfrak{g})$ is the polynomial algebra generated by the following element C (which is a multiple of the Casimir):

$$C = H_\alpha^2 + H_\alpha + YX.$$

Using the identity $Y = \varphi_1(a)U^a + \varphi_2(a)U$, where $U^a = \operatorname{Ad}(a^{-1})U$ and

$$\varphi_1(a) = \frac{-a^{-\alpha}}{(1 - a^{-2\alpha})^2} \quad \text{and} \quad \varphi_2(a) = \frac{a^{-2\alpha}}{(1 - a^{-2\alpha})^2},$$

we obtain, for every $a \in A^+$,

$$C = H_\alpha^2 + \frac{1 + a^{-2\alpha}}{1 - a^{-2\alpha}} H_\alpha + \frac{a^{-2\alpha}}{(1 - a^{-2\alpha})^2}\left([U^a]^2 + U^2\right) - \frac{a^{-\alpha}(1 + a^{-2\alpha})}{(1 - a^{-2\alpha})^2} U^a U. \tag{22}$$

Let the representation $\tau = (\tau_1, \tau_2)$ of $K \times K$ in $E = \mathbb{C}$ be defined by $\tau_1(k_\varphi) = e^{in\varphi}$ and $\tau_2(k_\varphi) = e^{im\varphi}$. Then for the associated representations of $\mathfrak{k}$ we have $\tau_1(U) = n$

and $\tau_2(U) = m$. Hence from (22) we see that the radial component of C is given by

$$\Pi_\tau(C) = H_\alpha^2 + \frac{1+a^{-2\alpha}}{1-a^{-2\alpha}} H_\alpha + (n^2+m^2)\frac{a^{-2\alpha}}{(1-a^{-2\alpha})^2} + nm\frac{a^{-\alpha}(1+a^{-2\alpha})}{(1-a^{-2\alpha})^2}.$$

Here H_α is identified with a first order differential operator on A via the right regular representation, in the usual way.

The system of radial differential equations is cofinite in the following sense:

Proposition 3.8. *Let J be the left ideal of $\mathcal{D}$ generated by $\Pi_\tau(I)$. Then $\mathcal{D}/J$ is finitely generated as a left module over $\mathcal{R} \otimes \mathrm{End}(E^M)$.*

The proof of this proposition, which we shall not give here, relies on the cofiniteness of the ideal I and on the following lemma, which will be useful at a later stage as well; see [19], I, 3.7, p. 95 for its proof.

Lemma 3.9. *There exists a finite-dimensional subspace $\mathcal{E} \subset U(\mathfrak{a})$ such that*

$$U(\mathfrak{g}) = U(\bar{\mathfrak{n}})\,\mathcal{E}\,\mathcal{Z}(\mathfrak{g})\,U(\mathfrak{k}).$$

In the following we write Δ for the collection of simple roots in Σ^+, and we assume that Δ is a basis of $\mathfrak{a}^*$. This assumption, which is equivalent to the assumption that G has a compact center, is only made to simplify the notation. The basis of $\mathfrak{a}$ dual to Δ is denoted by $(H_\alpha \mid \alpha \in \Delta)$. Its elements may be viewed as differential operators on A, in the fashion described above.

As an immediate consequence of Proposition 3.8 we obtain

Corollary 3.10. *There exist finitely many operators $D_1 = 1, D_2, \dots, D_n \in \mathcal{D}$ and functions $g_\alpha \in \mathcal{R} \otimes \mathrm{End}((E^M)^n)$ for $\alpha \in \Delta$ such that the function*

$$F := \begin{pmatrix} D_1 f \\ \vdots \\ D_n f \end{pmatrix} \qquad \textit{satisfies } H_\alpha F = g_\alpha F \textit{ for } \alpha \in \Delta. \tag{23}$$

Example 3.11. We return to the situation discussed in Example 3.7, and assume that I is an ideal of codimension one in $\mathcal{Z}(\mathfrak{g})$; then it is generated by $C - s$, for some $s \in \mathbb{C}$. Under this assumption the system of radial differential equations consists of one eigenequation: $\Pi_\tau(C)f = sf$. Now $E^M = \mathbb{C}$, and the assertion of the above corollary is valid with $n = 2$, $D_1 = 1$, and $D_2 = H_\alpha$. This corresponds to the usual reduction of a second order differential equation to a system of two first order differential equations.

Remark 3.12. The results of the present section are essentially due to Harish-Chandra [6], but his results remained unpublished for a long time. In [4] it was observed that the system (23) is of the regular singular type at infinity; this allowed a simplification of Harish-Chandra's original theory. Our presentation of the theory follows [4] rather closely.

That the system (23) is of the regular singular type at infinity is seen by using the coordinates $z_\alpha = a^{-\alpha}$ $(\alpha \in \Delta)$ on A^+. More precisely, define the map $\underline{z}$ from A^+ onto the Δ-fold Cartesian product of intervals $]0,1[^\Delta$ by

$$\underline{z}(a)_\alpha = a^{-\alpha}.$$

Then $\underline{z}$ is an analytic bijection. We denote its inverse by $\underline{a}$. In the following a function f on A^+ will be identified with the corresponding function $\tilde{f} := f \circ \underline{a}$ on $]0,1[^\Delta$. Note that every function $g \in \mathcal{R}$ corresponds to a real analytic function on $]0,1[^\Delta$ that has a unique extension to a holomorphic function on D^Δ; here D denotes the complex unit disc $\{z \in \mathbb{C} \mid |z| < 1\}$. Thus we may view $\mathcal{R}$ as a subring of $\mathcal{O}(D^\Delta)$, the ring of holomorphic functions $D^\Delta \to \mathbb{C}$. In the coordinates z_α the system (23) becomes

$$z_\alpha \frac{\partial}{\partial z_\alpha} F(z) = g_\alpha(z) F(z) \qquad (\alpha \in \Delta). \tag{24}$$

Here the functions g_α belong to $\mathcal{O}(D^\Delta) \otimes \mathrm{End}((E^M)^n)$; the system has regular singularities of the simple type along the coordinate hyperplanes $z_\alpha = 0$.

Example 3.13. We return to the situation of Example 3.11. The radial component takes the following form in the variable $z = a^{-\alpha}$:

$$\Pi_\tau(C) = \left(z \frac{d}{dz}\right)^2 + \frac{1+z^2}{1-z^2}\, z \frac{d}{dz} + \frac{(n^2+m^2)z^2 + nmz(1+z^2)}{(1-z^2)^2};$$

this operator is of the regular singular type at $z = 0$. The reduction of the eigenequation $\Pi_\tau(C) f = sf$ to a first order system now takes the form $F = (f, z\, df/dz)$, as in the classical theory.

By a variation on the monodromy arguments of the classical one variable theory (see [5], p. 109) the following lemma can be proved (see [4], Appendix). If $\lambda \in \mathfrak{a}^*_{\mathbb{C}}$, we write $\lambda_\alpha := \lambda(H_\alpha)$; thus $\lambda = \sum_{\alpha \in \Delta} \lambda_\alpha \alpha$. Moreover, if $m \in \mathbb{N}\Delta$ and $\xi \in \mathfrak{a}^*_{\mathbb{C}}$, we define multivalued holomorphic functions on $(D^*)^\Delta := (D \setminus \{0\})^\Delta$ by the formulas

$$(\log z)^m := \prod_{\alpha \in \Delta} (\log z_\alpha)^{m_\alpha}, \qquad z^\xi := \prod_{\alpha \in \Delta} (z_\alpha)^{\xi_\alpha}.$$

Lemma 3.14. *Every solution of* (24) *is a multivalued holomorphic function on* $(D^*)^\Delta$ *of the form*

$$\sum_{\xi, m} (\log z)^m z^\xi F_{\xi,m}(z), \tag{25}$$

where (ξ, m) *ranges over a finite subset of* $\mathfrak{a}^*_{\mathbb{C}} \times \mathbb{N}\Delta$, *and where the* $F_{\xi,m}$ *are* $(E^M)^n$-*valued holomorphic functions on* D^Δ.

If $m \in \mathbb{N}\Delta$, we write $|m| = \sum_{\alpha \in \Delta} m_\alpha$ and

$$(\log a)^m = \prod_{\alpha \in \Delta} \alpha(\log a)^{m_\alpha};$$

thus $z \mapsto \log \underline{a}(z)^m$ is a branch over $]0,1[^\Delta$ of the multivalued holomorphic function $z \mapsto (\log z)^m$. Similarly, $z \mapsto \underline{a}(z)^{-\xi}$ is a branch of $z \mapsto z^\xi$, for $\xi \in \mathfrak{a}^*_{\mathbb{C}}$.

Proposition 3.15. *There exists a finite set* $X \subset \mathfrak{a}^*_{\mathbb{C}}$ *and a* $d \in \mathbb{N}$ *such that every* $\varphi \in \mathcal{A}(G, \tau, I)$ *admits an absolutely convergent expansion of the form*

$$\varphi(a) = \sum_{\substack{\xi \in X - \mathbb{N}\Delta \\ |m| \le d}} (\log a)^m a^\xi c_{\xi,m} \qquad (a \in A^+) \tag{26}$$

with uniquely determined coefficients $c_{\xi,m} \in E^M$. *Here* $X - \mathbb{N}\Delta$ *denotes the collection of elements* $\xi - \mu$ *for* $\xi \in X$ *and* $\mu \in \mathbb{N}\Delta$.

PROOF. As before we denote the restriction of φ to A^+ by f. In the z variables the function $F = (D_1 f, \dots, D_n f)$ has an expression of the form (25). In particular its first component $D_1 f = f$ does. Expanding the holomorphic functions $F_{\xi,m}$ into power series around 0 and rewriting the resulting series in terms of functions of the form $(\log a)^m a^\xi$ on A^+ one obtains existence of the above expansion for f.

We also give a sketch of the argument that establises uniqueness of the expansion; it is in the spirit of [4]. In the following we put $c_{\xi,m} = 0$ when $\xi \notin X - \mathbb{N}\Delta$ or $|m| > d$. There exists a set $S \subset \mathfrak{a}_{\mathbb{C}}^*$ such that $X - \mathbb{N}\Delta \subset S - \mathbb{N}\Delta$ and for $s, s' \in S$ we have $s - s' \in \mathbb{Z}\Delta \Rightarrow s = s'$. For $s \in S$ put $f_{s,m}(a) = \sum_{\mu \in \mathbb{N}\Delta} c_{s-\mu,m} a^{-\mu}$. Then the series for $f_{s,m}$ converges absolutely on A^+, and hence the corresponding power series $\sum_\mu z^\mu c_{s-\mu,m}$ converges absolutely on D^Δ. It follows that the $f_{s,m}$ may be viewed as holomorphic functions on D^Δ. Moreover, in the coordinates z the function f is given by the finite sum

$$f(z) = \sum_{\substack{s \in S \\ |m| \le d}} (\log z)^m z^{-s} f_{s,m}(z) \tag{27}$$

on $]0,1[^\Delta$, with real-valued branches for the occurring multivalued functions. The function $F = (D_1 f, \dots, D_n f)$ satisfies the system (24). It follows that F, hence also $f = F_1$, admits a multivalued analytic extension to $(D^*)^\Delta$; by analytic continuation the expression (27) holds on $(D^*)^\Delta$ as well. From the monodromy behavior around the coordinate axes $z_\alpha = 0$ it now follows that an expression like (27) is uniquely determined (once S is prescribed). It follows that the coefficients $c_{\xi,m}$ are unique.□

Remark 3.16. In the above proof the series (26) is rewritten as (27). The occurring functions $f_{s,m}$ are holomorphic on $\mathcal{O}(D^\Delta)$, hence admit power series expansions on D^Δ. It follows from this that the series in (26) converges in a much stronger sense than stated in the lemma. In particular the convergence allows term by term application of differential operators from $\mathcal{O}(D^\Delta) \otimes S(\mathfrak{a})$.

Let $\varphi \in \mathcal{A}(G, \tau, I)$; an element $\xi \in X - \mathbb{N}\Delta$ for which there exists an $m \in \mathbb{N}\Delta$ such that $c_{\xi,m} \neq 0$ is called an **exponent** for φ (along A^+). The set of all exponents of φ is denoted by $\mathcal{E}(\varphi)$. Let the partial ordering $\preceq$ on $\mathfrak{a}_{\mathbb{C}}^*$ be defined by

$$\xi_1 \preceq \xi_2 \iff \xi_2 - \xi_1 \in \mathbb{N}\Delta.$$

The $\preceq$-maximal elements in $\mathcal{E}(\varphi)$ are called the **leading exponents** of φ; the set of these is denoted by $\mathcal{E}_L(\varphi)$.

Theorem 3.17. *There exists a finite set $\mathcal{E}_I \subset \mathfrak{a}_{\mathbb{C}}^*$, depending only on the cofinite ideal I, such that $\mathcal{E}_L(\varphi) \subset \mathcal{E}_I$ for every τ and all $\varphi \in \mathcal{A}(G, \tau, I)$.*

IDEA OF PROOF. There exists a system of polynomial equations, depending only on I, that is the appropriate analogue of the classical indicial equation. The (finite) set of solutions for this system determines $\mathcal{E}_I$. □

3.5. The subrepresentation theorem

Let $(\pi, \widetilde{V})$, V, and V' be as in the beginning of Section 3.2. The asymptotic theory of the previous sections implies the following.

Corollary 3.18. *There exist unique bilinear maps* $c_{\xi,m} : V \times V' \to \mathbb{C}$ *defined for* $\xi \in \mathcal{E}_I - \mathbb{N}\Delta$ *and* $m \in \mathbb{N}^\Delta$ *such that, for* $v \in V$ *and* $v' \in V'$,

$$m_{v,v'}(a) = \sum_{\xi,m} c_{\xi,m}(v,v')\, a^\xi \,(\log a)^m \qquad (a \in A^+)$$

with absolutely converging series.

To $v \in V$ we may now associate the set of exponents

$$\mathcal{E}(v) := \{\xi \in \mathcal{E}_I - \mathbb{N}\Delta \mid \exists m \in \mathbb{N}\Delta \text{ and } \exists v' \in V' \text{ with } c_{\xi,m}(v,v') \neq 0\}.$$

The union $\mathcal{E}(V) := \bigcup_{v\in V} \mathcal{E}(v)$ of these sets is called the set of exponents of the Harish-Chandra module V; its $\preceq$-maximal elements are called the leading exponents of V. The set of these is denoted by $\mathcal{E}_L(V)$.

Lemma 3.19. *Let* $v' \in V'$. *Then for* $\xi \in \mathcal{E}_L(V)$ *the map*

$$\gamma : v \mapsto \sum_m c_{\xi,m}(v,v')a^\xi(\log a)^m$$

factors to a nontrivial $\mathfrak{a}$*-module homomorphism* $V/\bar{\mathfrak{n}}V \to C^\infty(A)$.

PROOF. If $v \in V$ and $X \in \mathfrak{g}$, then

$$m_{Xv,v'}(a) = \frac{d}{dt}\bigg|_{t=0} m_{v,v'}(a \exp tX) = m_{v,-\mathrm{Ad}(a)Xv'}(a). \tag{28}$$

If $\alpha \in \Sigma^+$ and $X \in \mathfrak{g}_{-\alpha}$, then $\mathrm{Ad}(a)X_\alpha = a^{-\alpha}X_\alpha$. So $m_{Xv,v'}(a) = -a^{-\alpha}m_{v,Xv'}$. Hence if $X \in \bar{\mathfrak{n}}$, no exponent $\xi \in \mathcal{E}_L(V)$ occurs in $m_{Xv,v'}$. Therefore each map $v \mapsto c_{\xi,m}(v,v')$ is zero on $\bar{\mathfrak{n}}V$. From (28) and the fact that term by term differentiations are allowed (Remark 3.16) it follows that the map γ is an $\mathfrak{a}$-module homomorphism. □

The group M normalizes $\bar{\mathfrak{n}}$; hence $V/\bar{\mathfrak{n}}V$ is a $(\mathfrak{m}+\mathfrak{a}, M)$-module.

Lemma 3.20. *The* $(\mathfrak{m}+\mathfrak{a}, M)$*-module* $V/\bar{\mathfrak{n}}V$ *is nontrivial and is finite-dimensional.*

PROOF. The nontriviality follows from Lemma 3.19, the finite-dimensionality from Lemma 3.9. □

The following result is due to Casselman. Let $\bar{Q} = MA\bar{N}$ be the minimal parabolic subgroup of G opposite to $Q = MAN$.

Theorem 3.21 (Subrepresentation theorem). *Let* V *be irreducible. Then there exist* $\sigma \in \widehat{M}$ *and* $\lambda \in \mathfrak{a}^*_{\mathbb{C}}$ *such that* V *occurs as a* $(\mathfrak{g},K)$*-submodule of* $\mathrm{Ind}_{\bar{Q}}^G(\sigma\otimes\lambda\otimes 1)_K$.

PROOF. Fix an irreducible quotient H of the finite-dimensional nontrivial $(\mathfrak{m}+\mathfrak{a}, M)$-module $V/\bar{\mathfrak{n}}V$. Let $\sigma \otimes (\lambda - \rho_Q)$ be the associated representation of MA. Then

$$\mathrm{Hom}_{\mathfrak{m}+\mathfrak{a},M}(V/\bar{\mathfrak{n}}V, \sigma\otimes(\lambda-\rho_Q)) \neq 0.$$

Hence by the Frobenius theorem formulated in Lemma 3.22 below, there exists a nontrivial $(\mathfrak{g},K)$-module homomorphism T from V into $\mathrm{Ind}_{\bar{Q}}^G(\sigma\otimes\lambda\otimes 1)_K$. Since V is irreducible, $\ker T = 0$. □

Lemma 3.22 (Frobenius reciprocity). *Let σ be an irreducible representation of M, and $\lambda \in \mathfrak{a}_{\mathbb{C}}^*$. Let V be a $(\mathfrak{g}, K)$-module. Then*

$$\operatorname{Hom}_{\mathfrak{g},K}(V, \operatorname{Ind}_Q^G(\sigma \otimes \lambda \otimes 1)_K) \simeq \operatorname{Hom}_{\mathfrak{m}+\mathfrak{a},M}(V/\bar{\mathfrak{n}}V, \sigma \otimes (\lambda - \rho_Q)).$$

PROOF. One readily checks that $T \mapsto \mathrm{ev}_e \circ T$ provides the isomorphism. (See also the proof of Lemma 1.2.) □

Remark 3.23. Let V be an irreducible admissible $(\mathfrak{g}, K)$-module as above. It follows from Frobenius reciprocity that the collection of parameters $\lambda \in \mathfrak{a}_{\mathbb{C}}^*$ such that $V \hookrightarrow \operatorname{Ind}_Q^G(\sigma \otimes (\lambda - \rho) \otimes 1)$ for some σ is equal to the collection $\mathcal{E}(V, \bar{\mathfrak{n}})$ of $\mathfrak{a}$-weights in $V/\bar{\mathfrak{n}}V$. Moreover, it follows from Lemma 3.19 that

$$\mathcal{E}_L(V) \subset \mathcal{E}(V, \bar{\mathfrak{n}}).$$

This inclusion can be proper. However, it can be shown that $\mathcal{E}_L(V)$ equals the set of $\preceq$-maximal elements in $\mathcal{E}(V, \bar{\mathfrak{n}})$ (see [16], Theorem II.2.1, p. 74). Thus we may regard $\mathcal{E}(V, \bar{\mathfrak{n}})$ as a set of algebraic asymptotic exponents associated with the module V.

Remark 3.24. The observation that the $\mathfrak{a}$-weights of $V/\bar{\mathfrak{n}}V$ play a role in the asymptotics of the matrix coefficients of the module V is the starting point of another approach to asymptotics, via the theory of Jacquet modules. We refer the reader to [3], [19], I, Ch. 4, [10] for more details.

Remark 3.25. It can be shown that every finitely generated admissible $(\mathfrak{g}, K)$-module V is the module of K-finite vectors for some G-module $\widetilde{V}$ (see, e.g., [19], I, 4.2.4). For more information on such "globalizations" of V we refer the reader to [19], II, Ch. 11, [17].

It is important to note that the K-finite matrix coefficients are independent of the globalization under consideration. See [4], Theorem 8.7, for details.

3.6. Asymptotic behavior along the walls

The asymptotic theory along A^+ described so far is strong enough to obtain uniform estimates for matrix coefficients on subsets of $\mathrm{cl}(A^+)$ of the form

$$A^+(R) = \{a \in A \mid a^\alpha \geq R \text{ for } \alpha \in \Delta\},$$

with $R > 1$. In this section we shall briefly indicate how such estimates may be extended to $A^+(1) = \mathrm{cl}(A^+)$, by using converging expansions "along the walls."

Let $\varphi \in \mathcal{A}(G, \tau, I)$ (see Section 3.3). We will say that a functional $\omega \in \mathfrak{a}^*$ **dominates** $\mathcal{E}_L(\varphi)$ on $\mathfrak{a}^+$ if, for each $\xi \in \mathcal{E}_L(\varphi)$,

$$\operatorname{Re}\xi \leq \omega \qquad \text{on } \mathfrak{a}^+.$$

One readily sees that the above estimate is equivalent to the collection of estimates

$$(\operatorname{Re}\xi)_\alpha \leq \omega_\alpha \qquad (\alpha \in \Delta).$$

Lemma 3.26. *Let $\varphi \in \mathcal{A}(G, \tau, I)$ and assume that ξ dominates $\mathcal{E}_L(\varphi)$. Then there exist constants $d \in \mathbb{N}$ and $C > 0$ such that*

$$\varphi(a) \leq C(1 + |\log a|)^d a^\xi \qquad (a \in \mathrm{cl}(A^+)). \tag{29}$$

SOME IDEAS IN THE PROOF. This result can be established by using asymptotic expansions along *walls* of A^+, using the method of [4], of which we shall give only a sketch.

To $F \subset \Delta$ we associate the wall

$$A_F^+ := \{a \in A \mid a^\alpha = 1 \text{ for } \alpha \in F \text{ and } a^\beta > 1 \text{ for } \beta \in \Delta \setminus F\}.$$

Note that $\mathrm{cl}(A^+)$ is the disjoint union of the walls A_F^+ for $F \subset \Delta$.

By a "grouping of terms" (see [4]) one may rewrite the expansion (26) of the function φ as an "expansion along the wall A_F^+." For every $\epsilon > 0$ this expansion converges uniformly absolutely on the set

$$A^+(F, \epsilon) := \{a \in A \mid 1 \leq a^\alpha \leq 1 + \epsilon \text{ for } \alpha \in F \text{ and } a^\beta > 1 + \epsilon \text{ for } \beta \in \Delta \setminus F\}.$$

The leading exponents of φ along A_F^+ are elements of $\mathfrak{a}_{F\mathbb{C}}^*$. From the grouping of terms procedure one reads off that on $\mathfrak{a}_F^+$ the leading exponents are dominated by the element $\xi|\mathfrak{a}_F$. From this one obtains, for every $\epsilon > 0$, that the estimate (29) holds on $A^+(F, \epsilon)$ (with a constant C depending on ϵ.) The proof is completed by the observation that for every fixed $\epsilon > 0$ the sets $A^+(F, \epsilon)$ $(F \subset \Delta)$ cover $\mathrm{cl}(A^+)$.□

4. Tempered Representations

In this section we give several characterizations of tempered representations in terms of their asymptotic exponents.

4.1. The discrete series

Our first goal is to give different characterizations of square integrable representations, i.e., representations whose matrix coefficients belong to the space $L^2(G)$ of functions that are square integrable with respect to a bi-invariant Haar measure on G (which we assume to be fixed from now on).

The left and right regular representations (denoted L and R respectively) of G in $L^2(G)$ are unitary, by invariance of the Haar measure. For a proof of the following lemma, see [18], II.15, p. 435.

Lemma 4.1. *If π is an irreducible unitary representation of G, then the following conditions are equivalent:*

(a) *π is unitarily equivalent to an irreducible closed subrepresentation of $(R, L^2(G))$.*
(b) *π has a nonzero matrix coefficient that belongs to $L^2(G)$.*
(c) *Every matrix coefficient of π belongs to $L^2(G)$.*

If an irreducible unitary representation π of G satisfies any of the above conditions, it is said to belong to the **discrete series** of G.

Our next goal is to characterize discrete series representations in terms of their leading exponents along A^+. For this we need the following lemma. Let dx be a choice of Haar measure on G, and let dk be normalized Haar measure on K. We put $m_\alpha = \dim(\mathfrak{g}_\alpha)$, for $\alpha \in \Sigma$, and define the function $J : A \to [0, \infty[$ by

$$J(a) = \prod_{\alpha \in \Sigma^+} |a^\alpha - a^{-\alpha}|^{m_\alpha}. \tag{30}$$

Lemma 4.2. *There exists a (unique) choice of Haar measure da on A such that for all* $f \in C_c(G)$

$$\int_G f(x)dx = \int_K \int_{A^+} \int_K f(k_1 a k_2)\, J(a)\, dk_1\, da\, dk_2,$$

This lemma can be proved by substitution of variables; the function J occurs as a Jacobian.

For $R > 1$ we put

$$A^+(R) = \{a \in A \mid a^\alpha > R \text{ for } \alpha \in \Delta\}.$$

From (30) we readily see that for every $R > 1$ there exists a constant $C_R > 0$ such that

$$C_R\, a^{2\rho} \leq J(a) \leq a^{2\rho} \qquad (a \in A^+(R)). \tag{31}$$

In the following we assume that G has compact center. Then Δ, the collection of simple roots in Σ^+, is a basis of $\mathfrak{a}^*$. Let $\omega_\alpha \in \mathfrak{a}^*$, for $\alpha \in \Delta$, be the associated fundamental weights, i.e., $2\langle\omega_\alpha, \beta\rangle / \langle\beta, \beta\rangle = \delta_{\alpha\beta}$ for $\alpha, \beta \in \Delta$.

Proposition 4.3. *Let V be the Harish-Chandra module of an irreducible unitary representation π. Then the following conditions are equivalent:*

(a) *π belongs to the discrete series of G;*

(b) *each $\xi \in \mathcal{E}_L(V)$ satisfies the estimates*

$$\langle \mathrm{Re}\xi + \rho, \omega_\alpha\rangle < 0 \qquad \textit{for every } \alpha \in \Delta; \tag{32}$$

(c) *each $\xi \in \mathcal{E}(V, \bar{\mathfrak{n}})$ satisfies the estimates* (32).

Sketch of proof. It follows from Remark 3.23 that (b) $\iff$ (c). We shall sketch the proof of (a) $\iff$ (b).

Assume (a) and let $\xi \in \mathcal{E}_L(V)$. Then $\xi \in \mathcal{E}_L(\varphi)$ for a spherical function φ associated with π as in the end of Section 3.3. Now $\varphi \in L^2(G)$, since π belongs to the discrete series. Fix $R > 1$. Then from Lemma 4.2 and the estimate (31) we see that the function $\psi : A^+(R) \to \mathbb{R}$ defined by $\psi(a) = \|\varphi(a)\|^2 a^{2\rho}$ has a bounded L^1-norm $\|\psi\|_1$ with respect to da (note that $\|\varphi\|$ is bi-K-invariant).

Fix $H \in \mathrm{cl}(\mathfrak{a}^+)$ with $H \neq 0$. Then for every $a \in A^+(R)$ the ray $l_{a,H} = a\exp(\mathbb{R}_+ H)$ is contained in $A^+(R)$. The expansion (26) describes the asymptotic behavior of $\psi(a\exp(tH))$ as $t \to \infty$, locally uniformly in $a \in A^+(R)$. It follows that $\psi(a\exp(tH)) \sim C_a e^{tr}$, where $r = 2\max_{\eta\in\mathcal{E}(\varphi)} \mathrm{Re}\,(\eta + \rho)(H)$, and where $C_a > 0$ is a constant that may be chosen locally independent of a. If $\mathcal{K}$ is a compact subset of A^+ then the L^1 norm of ψ's restriction to the union of the rays $l_{a,H}$ for $a \in \mathcal{K}$ is bounded by $\|\psi\|_1$. This implies that $r < 0$. It follows from the above that $\mathrm{Re}\,\xi + \rho < 0$ on $\mathrm{cl}(\mathfrak{a}^+) \setminus \{0\}$. Then (32) follows.

We have now established the implication (a) $\Rightarrow$ (b) by using the uniform absolute convergence of the series (29) on sets of the form $A^+(R)$, with $R > 1$. To prove the converse implication we need to invoke Lemma 3.26, obtained from "asymptotics along the walls." Assume (b). Then we may fix $r > 0$ sufficiently small so that, for all $\xi \in \mathcal{E}_L(V)$,

$$\mathrm{Re}\,(\xi + \rho)(H) \leq -r\rho(H) \qquad (H \in \mathrm{cl}(\mathfrak{a}^+)).$$

In other words, $-(1 + r)\rho$ dominates $\mathcal{E}_L(V)$. Let φ be any τ-spherical function associated with π as in the end of Section 3.3. Then $\mathcal{E}(\varphi) \subset \mathcal{E}(V)$; hence by

Lemma 3.26 it follows that

$$J(a)\|\varphi(a)\|^2 \leq a^{2\rho}\|\varphi(a)\|^2 \leq C(1+|\log a|)^{2d}a^{-2r\rho} \qquad (a \in A^+),$$

with $C > 0$ a suitable constant. In view of Lemma 4.2 this implies that $\varphi \in L^2(G)$. By Lemma 4.1 we conclude that π belongs to the discrete series. □

4.2. Tempered representations

An admissible representation $(\pi, \widetilde{V})$ is said to be **tempered** if its K-finite matrix coefficients belong to $L^{2+\epsilon}(G)$, for every $\epsilon > 0$.

Let V be the Harish-Chandra module associated with π. Then by an asymptotic analysis as in the previous section we conclude that π is tempered if and only if every $\xi \in \mathcal{E}(V)$ satisfies the estimates $\langle(2+\epsilon)\mathrm{Re}\,\xi + 2\rho, \omega_\alpha\rangle < 0$ for $\alpha \in \Delta$. From this we readily obtain:

Lemma 4.4. *Let π be an admissible representation with associated Harish-Chandra module V. Then the following conditions are equivalent.*

(a) *π is tempered;*
(b) *every $\xi \in \mathcal{E}(V)$ satisfies the estimates*

$$\langle \mathrm{Re}\,\xi + \rho, \omega_\alpha\rangle \leq 0 \qquad (\alpha \in \Delta); \tag{33}$$

(c) *every $\xi \in \mathcal{E}(V, \bar{\mathfrak{n}})$ satisfies the estimates* (33).

4.3. Embedding of tempered representations

The following result, due to Langlands ([15], Lemma 4.10), is the main result of this section. For $F \subset \Delta$ let $\bar{Q}_F = \theta Q_F$ be the parabolic subgroup opposite to the standard parabolic subgroup Q_F.

Theorem 4.5. *Let π be an irreducible tempered representation, with associated Harish-Chandra module V. Then there exists a standard parabolic subgroup Q_F, a discrete series representation σ of M_F and a $\lambda \in i\mathfrak{a}_F^*$, such that the $(\mathfrak{g}, K)$-module V allows an embedding*

$$V \hookrightarrow \mathrm{Ind}_{\bar{Q}_F}^G(\sigma \otimes \lambda \otimes 1).$$

Note that it follows from this theorem that π is infinitesimally (i.e., on the level of the Harish-Chandra modules) equivalent to a unitary representation. Therefore, in the sequel we may, and shall, restrict our attention to unitary tempered representations.

We shall explain the main ideas that enter the proof of Theorem 4.5. Let us first recall some facts about standard parabolic subgroups, meanwhile fixing notation. If $F \subset \Delta$, let ${}^*\mathfrak{a}_F$ be the B-orthocomplement of $\mathfrak{a}_F$ in $\mathfrak{a}$; this is a maximal abelian subspace of $\mathfrak{m}_F \cap \mathfrak{p}$. Then $\mathfrak{a} = {}^*\mathfrak{a}_F \oplus \mathfrak{a}_F$. Via B we identify the dual spaces ${}^*\mathfrak{a}_F^*$ and $\mathfrak{a}_F^*$ with subspaces of $\mathfrak{a}^*$; thus ${}^*\mathfrak{a}_F^*$ is the space of linear functionals in $\mathfrak{a}^*$ that vanish on $\mathfrak{a}_F$, and vice versa. These spaces and embeddings are naturally complexified.

The set $\Sigma_F := \mathbb{Z}F \cap \Sigma$ is naturally identified with the system $\Sigma(\mathfrak{m}_F, {}^*\mathfrak{a}_F)$ of restricted roots of ${}^*\mathfrak{a}_F$ in $\mathfrak{m}_F$. We note that

$${}^*\mathfrak{a}_F^* = \mathrm{span}(F), \qquad \mathfrak{a}_F^* = \mathrm{span}\{\omega_\beta \mid \beta \in \Delta \setminus F\}.$$

Let ${}^*\mathfrak{n}_F := \mathfrak{n} \cap \mathfrak{m}_F$. Then $\mathfrak{m} + {}^*\mathfrak{a}_F + {}^*\mathfrak{n}_F$ is a minimal parabolic subalgebra of $\mathfrak{m}_F$; the associated ρ is denoted by ${}^*\rho_F$. Note that $\mathfrak{n} = {}^*\mathfrak{n}_F \oplus \mathfrak{n}_F$ as a direct sum of vector spaces. Since $\mathfrak{a}_F$ centralizes ${}^*\mathfrak{n}_F$, whereas $\mathfrak{m}_F \supset {}^*\mathfrak{a}_F$ acts with 0 trace on $\mathfrak{n}_F$, we have

$$\rho = {}^*\rho_F + \rho_F.$$

Lemma 3.20 has the following generalization; see [19], I, 4.3.1, p. 114, for a proof.

Lemma 4.6. *Let V be a finitely generated admissible $(\mathfrak{g}, K)$-module. Then $V/\bar{\mathfrak{n}}_F V$ is a nontrivial, finitely generated, admissible $(\mathfrak{m}_{1F}, K_F)$-module.*

As in the proof of the subrepresentation theorem a key role is played by the following generalization of the Frobenius reciprocity result of Lemma 3.22, for σ an admissible representation of M_F and $\lambda \in \mathfrak{a}^*_{F\mathbb{C}}$. Let $V_{\sigma\otimes(\lambda-\rho_F)}$ denote the representation space of σ, equipped with the action of A_F by the character $\lambda - \rho_F$.

Proposition 4.7 (Frobenius reciprocity). *Let V be a $(\mathfrak{g}, K)$-module. Then*

$$\mathrm{Hom}_{\mathfrak{g},K}(V, \mathrm{Ind}_{\bar{Q}_F}^G(\sigma\otimes\lambda\otimes 1)_K) \simeq \mathrm{Hom}_{\mathfrak{m}_{1F},K_F}(V/\bar{\mathfrak{n}}_F V, (V_{\sigma\otimes(\lambda-\rho_F)})_{K_F}).$$

PROOF. As before the map $T \mapsto \mathrm{ev}_e \circ T$ provides the isomorphism. □

Thus, in order to prove the main result, we must find a subset $F \subset \Delta$ such that $V/\bar{\mathfrak{n}}_F V$ has a quotient that is square integrable. For this we need the following lemmas.

Since $V/\bar{\mathfrak{n}}_F V$ is a finitely generated admissible $(\mathfrak{m}_{1F}, K_F)$-module, the central subalgebra $\mathfrak{a}_F$ of $\mathfrak{m}_{1F}$ acts globally finitely. Hence if $\mu \in \mathfrak{a}^*_{F\mathbb{C}}$, then the associated generalized weight space $(V/\bar{\mathfrak{n}}_F V)_\mu$ is a finitely generated admissible sub $(\mathfrak{m}_{1F}, K_F)$-module of $V/\bar{\mathfrak{n}}_F V$.

Lemma 4.8. *The algebra $\mathfrak{a}_F$ acts (globally) finitely on $V/\bar{\mathfrak{n}}_F V$, with a set of generalized weights equal to $\mathcal{E}(V, \bar{\mathfrak{n}})|\mathfrak{a}_F$. If $\mu \in \mathcal{E}(V, \bar{\mathfrak{n}})|\mathfrak{a}_F$, then*

$$\mathcal{E}((V/\bar{\mathfrak{n}}_F V)_\mu, {}^*\bar{\mathfrak{n}}_F) + \mu \subset \mathcal{E}(V, \bar{\mathfrak{n}}).$$

PROOF. From the direct sum decomposition $\bar{\mathfrak{n}} = {}^*\bar{\mathfrak{n}}_F + \bar{\mathfrak{n}}_F$ it follows that

$$0 \to {}^*\bar{\mathfrak{n}}_F(V/\bar{\mathfrak{n}}_F V) \to V/\bar{\mathfrak{n}}_F V \to V/\bar{\mathfrak{n}} V \to 0.$$

is a short exact sequence of $\mathfrak{a}$-modules. The assertion about the weights follows from inspection of this sequence. □

Remark 4.9. The set $\mathcal{E}(V, \bar{\mathfrak{n}})|\mathfrak{a}_F$ governs the asymptotic exponents of V along the wall A_F^+. This is analogous to what was said in Remark 3.23.

PROOF OF THEOREM 4.5. By Frobenius reciprocity it suffices to find $F \subset \Delta$, $\lambda \in i\mathfrak{a}^*_{F\mathbb{C}}$, and an irreducible quotient U of the $(\mathfrak{m}_F, K_F)$-module $(V/\bar{\mathfrak{n}}_F V)_{\lambda-\rho_F}$ that is square integrable. (See also the proof of Theorem 3.21.) By Lemma 4.1 applied to the group M_F, the requirement that U is square integrable is equivalent to the requirement that every ${}^*\eta \in \mathcal{E}(U, {}^*\bar{\mathfrak{n}}_F)$ satisfies the estimates

$$\langle \mathrm{Re}\, {}^*\eta + {}^*\rho_F, \omega_\alpha \rangle < 0 \qquad (\alpha \in F). \tag{34}$$

For each $\xi \in \mathcal{E}(V, \bar{\mathfrak{n}})$, we define

$$\Delta_\xi = \{\alpha \in \Delta \mid \langle \mathrm{Re}\, \xi + \rho, \omega_\alpha \rangle < 0\}.$$

Fix $\xi \in \mathcal{E}(V, \bar{\mathfrak{n}})$ such that Δ_ξ has a minimal number of elements, and put $F = \Delta_\xi$. Then

$$\langle \mathrm{Re}\,\xi + \rho, \omega_\beta \rangle = 0 \qquad (\beta \in \Delta \setminus F),$$

by temperedness of π and minimality of Δ_ξ. The weights ω_β for $\beta \in \Delta \setminus F$ span $\mathfrak{a}_F^*$; hence if we put $\mu = \xi|\mathfrak{a}_F$, then $\mathrm{Re}\,\mu + \rho_F = (\mathrm{Re}\,\xi + \rho_F)|\mathfrak{a}_F = 0$, and we see that $\lambda := \mu + \rho_F$ belongs to $i\mathfrak{a}_F^*$.

It follows from Lemma 4.8 that $(V/\bar{\mathfrak{n}}_F V)_\mu \neq 0$. We claim that every ${}^*\eta \in \mathcal{E}((V/\bar{\mathfrak{n}}_F V)_\mu, {}^*\bar{\mathfrak{n}}_F)$ satisfies the estimates (34). To see this, fix such a weight ${}^*\eta \in {}^*\mathfrak{a}_{F\mathbb{C}}^*$. Then by Lemma 4.8 we have ${}^*\eta + \mu \in \mathcal{E}(V, \bar{\mathfrak{n}})$. Now obviously

$$\mathrm{Re}\,({}^*\eta + \mu + \rho) = \mathrm{Re}\,({}^*\eta + \lambda + {}^*\rho_F) = \mathrm{Re}\,({}^*\eta + {}^*\rho_F). \tag{35}$$

Hence $\langle \mathrm{Re}\,({}^*\eta + \mu + \rho), \omega_\beta \rangle = 0$ for $\beta \in \Delta \setminus F$, and we see that $\Delta_{{}^*\eta+\mu} \subset F$. By minimality of $|F|$ the latter inclusion is actually an equality. In view of (35) this implies the estimates (34), and the claim follows.

We now select any irreducible quotient U of the finitely generated admissible $(\mathfrak{m}_F, K_F)$-module $(V/\bar{\mathfrak{n}}_F V)_\mu$. This quotient satifies (34); hence U is square integrable. □

The following result asserts that all induced representations occurring in Theorem 4.5 are tempered. See [19], I, Lemma 5.2.8, p. 143, for a proof.

Proposition 4.10. *If $F \subset \Delta$, σ a discrete series representation of M_F, and $\lambda \in i\mathfrak{a}_F^*$, then*

$$\mathrm{Ind}_{Q_F}^G(\sigma \otimes \lambda \otimes 1) \tag{36}$$

is a tempered representation.

It follows from Theorem 4.5 and the above proposition that the irreducible summands of the induced representations (36) *exhaust* the (unitary) tempered representations. For G connected linear semisimple, the *classification* of the irreducible (unitary) tempered representation has been achieved in [13].

5. The Langlands Classification

In this section we describe the classification of the irreducible admissible $(\mathfrak{g}, K)$-modules, which is due to R.P. Langlands ([15]).

By **Langlands data** we shall mean a triple (Q_F, σ, λ) with $F \subset \Delta$, Q_F the associated standard parabolic subgroup, σ an irreducible tempered representation of M_F, and λ an element of $\mathfrak{a}_{F\mathbb{C}}^*$ satisfying

$$\langle \mathrm{Re}\,\lambda, \alpha \rangle > 0 \qquad (\alpha \in \Delta \setminus F).$$

Here we recall that $\mathfrak{a}_{F\mathbb{C}}^*$ is embedded in $\mathfrak{a}_{\mathbb{C}}^*$, in the fashion described after Theorem 4.5.

Theorem 5.1 (Langlands).

(a) *If (Q_F, σ, λ) are Langlands data, then the $(\mathfrak{g}, K)$-module $\mathrm{Ind}_{Q_F}^G(\sigma \otimes \lambda \otimes 1)_K$ has a unique irreducible quotient $J(Q_F, \sigma, \lambda)$ (the* **Langlands quotient***).*

(b) *Assume $(Q_{F_j}, \sigma_j, \lambda_j)$ are Langlands data for $j = 1, 2$. If the associated Langlands quotients $J(Q_{F_j}, \sigma_j, \lambda_j)$ for $j = 1, 2$ are equivalent, then $F_1 = F_2$, the representations σ_1 and σ_2 are equivalent, and $\lambda_1 = \lambda_2$.*

(c) *Every irreducible admissible $(\mathfrak{g}, K)$-module is equivalent to a Langlands quotient.*

Remark 5.2. In [15] in the text preceding Lemma 3.13, Langlands defines $J = J(Q_F, \sigma, \lambda)$ as the quotient of $I = \mathrm{Ind}_{Q_F}^G(\sigma \otimes \lambda \otimes 1)_K$ by the kernel of the intertwining operator introduced in the lemmas below (see also Corollary 5.8). In [16] it is observed that J is actually the unique irreducible quotient of I.

The rest of this section is devoted to a sketch of some of the main ideas that enter the proof of Theorem 5.1. The complete proof may be found in [19], I, Ch. 5. We start with a crucial lemma. From the description of the noncompact picture in Section 2.2 we recall that $\bar{N}_F$ is equipped with a bi-invariant Haar measure $d\bar{n}$.

Lemma 5.3. *Let (Q_F, σ, λ) be Langlands data. Then for every function $f \in (\mathcal{H}_{Q_F,\sigma,\lambda})_K$ and every $x \in G$ the integral*

$$A(\bar{Q}_F, Q_F, \sigma, \lambda) f(x) := \int_{\bar{N}_F} f(x\bar{n})\, d\bar{n} \tag{37}$$

converges absolutely.

This result is proved by careful estimation of the integrand.

Lemma 5.4. *Let (Q_F, σ, λ) be Langlands data. Then the map*

$$A = A(\bar{Q}_F, Q_F, \sigma, \lambda)$$

defined by (37) *is a nonzero* $(\mathfrak{g}, K)$*-map*

$$A : (\mathcal{H}_{Q_F,\sigma,\lambda})_K \to (\mathcal{H}_{\bar{Q}_F,\sigma,\lambda})_K. \tag{38}$$

Sketch of proof. In the compact picture (see Section 2.2) this operator is readily seen to be an integral operator with a nontrivial integral kernel; hence the operator is nontrivial.

The following reasoning can be justified by showing that the occurring integrals all converge absolutely. Let $f \in (\mathcal{H}_{Q_F,\sigma,\lambda})_K$ and $x \in G$. Then for $(m, a, \bar{n}_0)$ in $M_F \times A_F \times \bar{N}_F$ we have

$$\begin{aligned}
Af(xma\bar{n}_0) &= \int_{\bar{N}_F} f(xma\bar{n}_0\bar{n})\, d\bar{n} \\
&= \int_{\bar{N}_F} f(xma\bar{n}(ma)^{-1}ma)\, d\bar{n} \\
&= a^{2\rho_F} \int_{\bar{N}_F} f(x\bar{n}'ma)\, d\bar{n}' \\
&= a^{-\lambda+\rho_F}\sigma(m)^{-1} \int_{\bar{N}_F} f(x\bar{n})\, d\bar{n} \\
&= a^{-\lambda+\rho_F}\sigma(m)^{-1} Af(x).
\end{aligned}$$

In the above sequence of equations the second equality follows from the left invariance of $d\bar{n}$. The endomorphism $\mathrm{Ad}(ma)$ normalizes $\bar{\mathfrak{n}}_F$, and has determinant $a^{-2\rho_F}$. Hence the third equality follows by the substitution of variables $\bar{n}' = ma\bar{n}(ma)^{-1}$. Finally, the fourth equality follows from the transformation properties of f under the action by Q_F (on the right).

It follows that A maps $(\mathcal{H}_{Q_F,\sigma,\lambda})_K$ into $(\mathcal{H}_{\bar{Q}_F,\sigma,\lambda})_K$. The $(\mathfrak{g}, K)$ actions are on the left; hence formally the $(\mathfrak{g}, K)$-equivariance of A is obvious. □

Remark 5.5. The operator $A(\bar{Q}_F, Q_F, \sigma, \lambda)$ is called the **standard intertwining operator** from $\mathrm{Ind}_{Q_F}^G(\sigma \otimes \lambda \otimes 1)$ to $\mathrm{Ind}_{\bar{Q}_F}^G(\sigma \otimes \lambda \otimes 1)$. In a well defined sense this operator has a meromorphic continuation in the parameter $\lambda \in \mathfrak{a}_{F\mathbb{C}}^*$. For more information on the standard intertwining operator and its role in harmonic analysis we refer the reader to [12], [9], [19], Vol II, Ch. 10.

The following result is crucial in the proof of the Langlands classification. It relates the intertwining operator to the asymptotic behavior of matrix coefficients. We assume that (Q_F, σ, λ) are Langlands data.

Proposition 5.6. *Let $f \in (\mathcal{H}_{Q_F,\sigma,\lambda})_K$ and $g \in (\mathcal{H}_{Q_F,\sigma,-\bar{\lambda}})_K$. Then, for $X \in \mathfrak{a}_F^+$,*

$$\lim_{t\to\infty} e^{(-\lambda+\rho_F)(tX)} \langle \pi(\exp tX)f, g\rangle = \langle [A(\bar{Q}_F, Q_F, \sigma, \lambda)f](e), g(e)\rangle_\sigma. \tag{39}$$

Sketch of proof. Put $a_t = \exp tX$. Then by equivariance of the pairing $\mathcal{H}_{\sigma,\lambda} \times \mathcal{H}_{\sigma,-\bar{\lambda}} \to \mathbb{C}$ and by Remark 2.3 the expression under the limit in (39) may be rewritten as

$$a_t^{-\lambda+\rho_F} \langle f, \pi(a_t^{-1})g\rangle = \int_{\bar{N}_F} \langle f(\bar{n}), g(a_t \bar{n} a_t^{-1})\rangle_\sigma \, d\bar{n}.$$

If $t \to \infty$, then the integrand on the right-hand side tends to $\langle f(\bar{n}), g(e)\rangle_\sigma$ pointwise. The result now follows by an application of the dominated convergence theorem.□

Corollary 5.7. *Suppose that $U \subset (\mathcal{H}_{Q_F,\sigma,\lambda})_K$ is a proper $(\mathfrak{g}, K)$-submodule. Then $U \subset \ker A(\bar{Q}_F, Q_F, \sigma, \lambda)$.*

Proof. The orthocomplement $U^\perp$ of U in $(\mathcal{H}_{Q_F,\sigma,-\bar{\lambda}})_K$ (with respect to the nondegenerate pairing (8)) is a nontrivial $(\mathfrak{g}, K)$-submodule. Let $f \in U$ and $g \in U^\perp$. Then for $X \in \mathfrak{a}_F^+$ and $t \in \mathbb{R}$ we have $\pi(\exp tX)f \in \bar{U}$ (the closure in $\mathcal{H}_{Q_F,\sigma,\lambda}$); hence $\pi(\exp tX)f \perp g$. By taking the limit for $t \to \infty$ it follows that

$$\langle Af(e), g(e)\rangle_\sigma = 0. \tag{40}$$

The map $\mathrm{ev}_e : g \mapsto g(e)$ from $(\mathcal{H}_{Q_F,\sigma,-\bar{\lambda}})_K$ to $(\mathcal{H}_\sigma)_{K_F}$ is a homomorphism of $(\mathfrak{m}_F, K_F)$-modules. Its image $\mathcal{H}_0$ is either 0 or $(\mathcal{H}_\sigma)_{K_F}$, by irreducibility of σ. If $g \in U^\perp \setminus \{0\}$, then $g(k) \neq 0$ for some $k \in K$. Now $\pi(k^{-1})g \in U^\perp$ and $\mathrm{ev}_e(\pi(k^{-1})g) = g(k) \neq 0$. Hence $\mathcal{H}_0$ is nontrivial; it must therefore be equal to $\mathcal{H}_\sigma$. From (40) it now follows that $Af(e) = 0$ for all $f \in U$. By K_F-equivariance of A this implies that $Af = 0$ on K_F, hence on G. Hence $Af = 0$ for all $f \in U$. □

As an immediate consequence of the last corollary we obtain:

Corollary 5.8. *The kernel of the operator* (38) *is the unique maximal proper submodule of the $(\mathfrak{g}, K)$-module $(\mathcal{H}_{Q_F,\sigma,\lambda})_K$.*

Assertion (a) of Theorem 5.1 is an immediate consequence of this corollary. Assertion (b) follows from a careful analysis of the asymptotic behavior of the matrix coefficients of induced representations.

Corollary 5.9. *The $(\mathfrak{g}, K)$-module $(\mathcal{H}_{\bar{Q}_F,\sigma,\lambda})_K$ has a unique irreducible submodule. The standard intertwining operator $A = A(\bar{Q}_F, Q_F, \sigma, \lambda)$ factors to an isomorphism from $J(Q_F, \sigma, \lambda)$ onto this unique irreducible submodule.*

PROOF. We first note that $(\bar{Q}_F, \sigma, -\bar{\lambda})$ are Langlands data with respect to the positive system $-\Sigma^+$. By assertion (a) of Theorem 5.1 for these Langlands data, it follows that $(\mathcal{H}_{\bar{Q}_F,\sigma,-\bar{\lambda}})_K$ has a unique irreducible quotient. By nondegeneracy and equivariance of the sesquilinear pairing $\mathcal{H}_{\bar{Q}_F,\sigma,\lambda} \times \mathcal{H}_{\bar{Q}_F,\sigma,-\bar{\lambda}} \to \mathbb{C}$ it follows that $(\mathcal{H}_{\bar{Q}_F,\sigma,\lambda})_K$ has a unique irreducible submodule.

By Corollary 5.8 the intertwining operator A factors to an isomorphism from $J(Q_F, \sigma, \lambda)$ onto a submodule of $(\mathcal{H}_{\bar{Q}_F,\sigma,\lambda})_K$. The latter submodule must be irreducible, since $J(Q_F, \sigma, \lambda)$ is. The final assertion now follows. □

Let V be an irreducible admissible $(\mathfrak{g}, K)$-module. Then by an asymptotic analysis in the spirit of the previous section one can establish the existence of Langlands data (Q_F, σ, λ) such that the $(\mathfrak{m}_F, K_F)$-module $V/\bar{\mathfrak{n}}_F V$ has $\sigma \otimes \lambda \otimes 1$ as an irreducible quotient. By Frobenius reciprocity (Proposition 4.7) this implies that

$$\mathrm{Hom}_{\mathfrak{g},K}(V, \mathrm{Ind}_{\bar{Q}_F}^G(\sigma \otimes \lambda \otimes 1)) \neq 0,$$

and by Corollary 5.9 it now follows that

$$V \simeq J(Q_F, \sigma, \lambda).$$

References

1. F. Bruhat, Sur les représentations induites des groupes de Lie, *Bull. Soc. Math. France* **84** (1956), 97–205.
2. J. Carmona and P. Delorme, Base méromorphe de vecteurs distributions H-invariants pour les séries principales généralisées d'espaces symétriques réductifs: Equation fonctionelle, *J. Func. Anal.* **122** (1994), 152–221.
3. W. Casselman, Jacquet modules for real reductive groups, *Proc. Intern. Congress Math., Helsinki 1978*, vol. 2, Academia Scientiarum Fennica, Hensinki, 1980, pp. 567–573.
4. W. Casselman and D. Miličić, Asymptotic behavior of matrix coefficients of admissible representations, *Duke Math. J.* **49** (1982), 869–930.
5. E. A. Coddington and N. Levinson, *Theory of Ordinary Differential Equations*, McGraw-Hill, New York, 1955.
6. Harish-Chandra, Some results on differential equations, unpublished manuscript (1960); see *Collected Papers*, vol. III (1984), Springer-Verlag, New York, 7–48.
7. Harish-Chandra, Differential equations and semisimple Lie groups, unpublished manuscript (1960); see *Collected Papers*, vol. III (1984), Springer-Verlag, New York, 57–120.
8. Harish-Chandra, Letter to G. van Dijk, October 1, 1983.
9. Harish-Chandra, Harmonic analysis on real reductive groups III. The Maass-Selberg relations and the Plancherel formula, *Annals of Math.* **104** (1976), 117–201.
10. H. Hecht and W. Schmid, On the asymptotics of Harish-Chandra modules, *J. Reine Angew. Math.* **343** (1983), 169–183.
11. A. W. Knapp, *Representation Theory of Semisimple Groups: An Overview Based on Examples*, Princeton University Press, Princeton, 1986.
12. A. W. Knapp and E. M. Stein, Intertwining operators for semisimple groups II, *Invent. Math.* **60** (1980), 9–84.
13. A. W. Knapp and G. J. Zuckerman, Classification of irreducible tempered representations of semisimple groups, *Annals of Math.* **116** (1982), 389–501, and **119** (1984), 639.
14. J. A. C. Kolk and V. S. Varadarajan, On the transverse symbol of vectorial distributions and some applications to harmonic analysis, *Indag. Math.* **7** (1996), 67–96.
15. R. P. Langlands, On the classification of irreducible representations of real algebraic groups, mimeographed notes, Institute for Advanced Study, Princeton, NJ, 1973, *Representation Theory and Harmonic Analysis on Semisimple Lie Groups* (P. J. Sally and D. A. Vogan, eds.), Mathematical Surveys and Monographs, vol. 31, American Mathematical Society, Providence, 1989, pp. 101–170.
16. D. Miličić, Asymptotic behaviour of matrix coefficients of the discrete series, Duke Math. J. **44** (1977), 59–88.

17. W. Schmid, Boundary value problems for group invariant differential equations, *Élie Cartan et les mathématiques d'aujourd'hui (Lyon, juin 1984)*, Astérisque, Numéro hors série, 1985, pp. 311–321.
18. V. S. Varadarajan, *Harmonic Analysis on Real Reductive Groups*, Lecture Notes in Mathematics, vol. 576, Springer-Verlag, Berlin, 1977.
19. N. R. Wallach, *Real Reductive Groups*, vol. I and II, Academic Press, Boston, 1988 and 1992.

MATHEMATISCH INSTITUUT, UNIVERSITEIT UTRECHT, P. O. BOX 80 010, 3508 TA UTRECHT, NETHERLANDS

E-mail address: `ban@math.ruu.nl`

Proceedings of Symposia in Pure Mathematics
Volume **61** (1997), pp. 157–166

Representations of GL(n) over the Real Field

C. Mœglin

The representation theory of $\mathrm{GL}(n,\mathbb{R})$ is simpler than for general real reductive groups because a number of special techniques apply and a number of special results are valid. One associates (see van den Ban's lectures [B]) to each parabolic subgroup $Q = MAN$ of a reductive Lie group G a series of induced representations:

$$ind_Q^G(\sigma \otimes \lambda \otimes 1),$$

where σ is a discrete series or limit of discrete series representation of M and λ is a complex-valued linear functional on the Lie algebra of A. The general theory of representations analyzes the reducibility of these representations for arbitrary λ, and it completely characterizes the irreducible constituents and nature of reducibility when λ is imaginary. The Langlands classification describes all irreducible admissible representations of G in terms of "Langlands quotients" of these induced representations.

When $G = \mathrm{GL}(n,\mathbb{R})$, the above induced representations corresponding to imaginary λ are all irreducible. In the case that Q is minimal parabolic, M is a finite abelian group and σ is simply a character. For this situation Gelfand-Graev [G-G] gave an elementary proof of the irreducibility using abelian Fourier analysis, and their proof works even for $\mathrm{GL}(n)$ over an arbitrary local field. For a general Q, limits of discrete series on M give redundant realizations of representations and may be ignored. When λ is imaginary and the representation of M is in the discrete series, the induced representation is still irreducible, as Jacquet [J] showed with a short argument that is valid over arbitrary local fields.

For $G = \mathrm{GL}(n,\mathbb{R})$, Speh [Sp1] gave a necessary and sufficient condition for reducibility of an induced representation with Q and λ general. The Speh argument reduces matters to a maximal parabolic subgroup of a subgroup $G' = \mathrm{GL}(n',\mathbb{R})$ for which the M of the parabolic has discrete series, and one shows readily that n' is 2, 3, or 4. The difficult step is then to handle this subgroup G'. If Q is a minimal parabolic subgroup of $\mathrm{GL}(n,\mathbb{R})$, then the reduction is ultimately to $G' = SL(2,\mathbb{R})$, and a short direct computation is possible. Speh-Vogan [S-V] later extended Speh's theorem to general reductive groups G, at the same time shortening the proof.

In this paper we shall use the above theorem of Speh's as a vehicle for discussing some of the representation theory of $\mathrm{GL}(n,\mathbb{R})$. We shall not necessarily indicate the

---

1991 *Mathematics Subject Classification.* Primary 22E45, 22E46.

most economical possible proofs but instead shall indicate what general theory and what computations are needed to prove Speh's theorem. We state this theorem in section 1 for the case that Q is minimal parabolic, and sections 2 to 8 are devoted to its proof. But we have in mind the results of [V2], and we hope that this lecture is also an introduction to [V2]. This is the reason for section 9, where we recite, in a very rough form, the classification due to Vogan [V2] of the irreducible unitary representations of $\mathrm{GL}(n, \mathbb{R})$.

I would like to thank A. Knapp for helpful comments.

GENERAL NOTATION. Denote by B a Borel subgroup of $\mathrm{GL}(n) = \mathrm{GL}(n, \mathbb{R})$; to fix the ideas, let B consist of the upper triangular matrices with nonzero determinant. Denote by U the unipotent radical of B and by T a split Cartan subgroup (for example the diagonal torus). Let W be the Weyl group, identified with the group of permutations on n letters. Denote by $O(n)$ the orthogonal group; it is a maximal compact subgroup of $\mathrm{GL}(n)$.

1. Principal Series

Let $\underline{\chi}$ be a character of T, not necessarily unitary. In other words $\underline{\chi}$ is a collection of n characters of $\mathbb{R}^*$:

$$\underline{\chi} = (\chi_1, \cdots, \chi_n).$$

To fix the notation, for all $i \in [1, n]$, denote by ϵ_i the character of $\mathbb{R}^*$ of order two (ϵ_i is the sign or the trivial character) and by s_i the element of $\mathbb{C}$ such that:

$$\chi_i(x) = \epsilon_i(x)|x|^{s_i} \qquad \text{for } \forall x \in \mathbb{R}^*. \tag{1}$$

As usual, one extends $\underline{\chi}$ to a character of B trivial on U. Denote by $I(\underline{\chi})$ the induced representation, more precisely the associated $(\mathfrak{g}, O(n))$-module, where $\mathfrak{g}$ is the complexified Lie algebra $End_{\mathbb{C}}(\mathbb{C}^n)$ of $\mathrm{GL}(n)$; see [B] for a precise definition of this induced representation. The general theory asserts that such an induced representation is admissible and of finite length. Sections 2–8 of this paper will be built around the proof of the following theorem:

Theorem 1 (Speh [Sp1]). *The Harish-Chandra module $I(\underline{\chi})$ is irreducible if and only if, for all $i, j \in [1, n]$:*

$$\text{if } \epsilon_i = \epsilon_j, \text{ then } s_i - s_j \notin 1 + 2\mathbb{Z}$$
$$\text{if } \epsilon_i \neq \epsilon_j, \text{ then } s_i - s_j \notin 2\mathbb{Z} - \{0\}.$$

2. Induced Representations and Characters

Fix a permutation $w \in W$. Then $I(w\underline{\chi})$ is well defined.

Proposition 2. *$I(\underline{\chi})$ and $I(w\underline{\chi})$ have the same character. In other words, the two Harish-Chandra modules define the same element in the Grothendieck group associated to the category of Harish-Chandra modules of finite length of* $\mathrm{GL}(n)$.

See [K1, Theorem 10.2] for the definition of character, [K1, Proposition 10.18] for the computation of the characters in our case, and [K1, Theorem 10.6] for the end of the proposition. Characters are discussed in Delorme's lectures [D].

A consequence of this proposition is that to prove Theorem 1, we can and do assume the following condition that $\underline{\chi}$ is **positive**:

$$\forall\, i, j \in [1, n],\ \text{if } i < j \text{ then } Re(s_i - s_j) \geq 0. \tag{2}$$

We say that $\underline{\chi}$ is **strictly positive** if

$$\forall\, i, j \in [1, n],\ \text{if } i < j \text{ then } Re(s_i - s_j) > 0. \tag{3}$$

3. Tempered Principal Series

A special case of Theorem 1 is needed in the proof of the general case:

Proposition 3. *Assume that* $\underline{\chi}$ *satisfies:*

$$Re\, s_i = Re\, s_j \qquad \text{for } \forall\, i, j \in [1, n].$$

Then $I(\underline{\chi})$ *is irreducible.*

This theorem is due to Gelfand-Graev [G-G]. Their method of proof, which is similar to an argument of Gelfand-Naimark, is sketched in [K, Proposition 7.1]. A proof by more general methods will be explained in section 6.

4. Langlands Theorem

Using this proposition, we can reformulate part of the Langlands classification theorem in the following way. Denote by w_0 the longest element of W. Then:

$$w_0\underline{\chi} = (\chi_n, \cdots, \chi_1).$$

Theorem 4 (Special case of Langlands theorem for GL(n)).

(i) *If* $\underline{\chi}$ *is positive in the sense of* (2), *then* $I(\underline{\chi})$ *has a unique irreducible quotient, denoted* $J(\underline{\chi})$. *Moreover* $I(w_0\underline{\chi})$ *has a unique irreducible submodule and this submodule is isomorphic to* $J(\underline{\chi})$.

(ii) *For any* $I(\underline{\chi})$, *if* $w \in W$ *is chosen so that* $w\underline{\chi}$ *is positive in the sense of* (2), *then* $J(w\underline{\chi})$ *appears with multiplicity one as a subquotient in* $I(\underline{\chi})$.

See [K1], Theorems 8.54 and 7.24 as well as Proposition 8.61. See also Miličić [M]. The proof uses Proposition 3; the proposition is unnecessary if the hypothesis "positive" is changed to "strictly positive." The full Langlands theorem for GL(n) will be stated in section 10.

5. Intertwining Operators

In the notation of (1), let us fix ϵ_i and let the s_i vary, for $i \in [1, n]$. Changing notation, we denote by $\chi_{\underline{s}}$, for $\underline{s} = (s_1, \cdots, s_n) \in \mathbb{C}^n$, the collection of the n characters $\epsilon_i | \cdot |^{s_i}$, for $i \in [1, n]$.

We denote by $\mathfrak{X}_{\underline{\epsilon}}$ the space of right $O(n)$-finite functions, ϕ, on $O(n)$ such that:

$$\phi(tk) = \prod_{i \in [1,n]} \epsilon_i(t_i)\phi(k) \qquad \text{for } \forall k \in O(n) \text{ and } \forall t = (t_1, \cdots, t_n) \in T \cap O(n).$$

We define $\mathfrak{X}_{w_0\underline{\epsilon}}$ in a similar way.

Fix $w \in W$. For any $\underline{s} \in \mathbb{C}^n$, restriction of functions from $\mathrm{GL}(n)$ to $O(n)$ gives isomorphisms:

$$\begin{array}{lll} i_{\chi_{\underline{s}}} & : I(\chi_{\underline{s}}) & \to \mathfrak{X}_{\underline{\epsilon}} \\ i_{w\chi_{\underline{s}}} & : I(w\chi_{\underline{s}}) & \to \mathfrak{X}_{w\underline{\epsilon}}. \end{array} \tag{4}$$

Now define, first of all formally:

$$A(w, \chi_{\underline{s}})\phi(k) := \int_{U \cap w^{-1}Uw \backslash U} (i_{\chi_{\underline{s}}}^{-1}\phi)(wuk)\, du \qquad \text{for } \forall \phi \in \mathfrak{X}_{\underline{\epsilon}} \text{ and } \forall k \in O(n),$$

where du is a Haar measure on the domain of integration.

Theorem 5 (Kunze-Stein [Ku-S2], Schiffmann [Sc]). *The above integral converges absolutely if $\underline{s}$ is strictly positive (see* (3) *in section* 2 *above). So $A(w, \chi_{\underline{s}})$ is defined when $\underline{s}$ is strictly positive as an operator from $\mathfrak{X}_{\underline{\epsilon}}$ to $\mathfrak{X}_{w\underline{\epsilon}}$. As a function of $\underline{s}$, the operator $A(w, \chi_{\underline{s}})$ has a meromorphic continuation to $\underline{s} \in \mathbb{C}^n$. If $w = w_1 w_2$ with $w_1, w_2 \in W$ and $\ell(w_1) + \ell(w_2) = \ell(w)$ (ℓ is the length in W), then we have an equality of meromorphic functions:*

$$A(w, \chi_{\underline{s}}) = A(w_1, w_2\chi_{\underline{s}})A(w_2, \chi_{\underline{s}}).$$

See [K], Corollary 7.13.

6. Intertwining Algebra and R-group

As stated in the introduction, an important fact for $\mathrm{GL}(n)$ is that induction preserves irreducibility when we start from discrete series on M with an imaginary parameter on A. For a general reductive group G, to study the reducibility of such an induced representation, one uses the theory of the R-group. More precisely, let ρ be a discrete series of a Levi subgroup of G (with *unitary* central character) and denote by π the $(\mathfrak{g}, K)$-module induced by ρ. (Here $\mathfrak{g}$ is the complexified Lie algebra of G, and K is a maximal compact subgroup of G.) Then Knapp and Stein, [K-S1] and [K-S2], have constructed a particular subgroup $R = R(\rho)$ of the Weyl group associated to the parabolic that gives linearly independent operators in the commuting algebra; this is done using the intertwining operators. Using a result of Harish-Chandra [H-C] (the completeness theorem for C-functions), Wallach [W] (see also [K-S3] and [A]) proves:

$$\dim \mathbb{C}[R] = \dim End_G(\pi).$$

In fact, we have more. Knapp, [K2] and [K3, section 7], constructs an isomorphism

$$\mathbb{C}[R] \simeq End_G(\pi).$$

Vogan has given an algebraic way to compute R using minimal K-types; see [V1, 6.6.7].

In general, one knows that R is a product of copies of $\mathbb{Z}/2\mathbb{Z}$ and that a nontrivial R group appears already in $SL(2, \mathbb{R})$; look at the induced representation from the sign character of the diagonal torus (extended as usual to a Borel subgroup).

So Proposition 3 is equivalent to:

Proposition 6. *For* $\mathrm{GL}(n)$, *let* $\underline{s}_0 \in i\mathbb{R}^n \subset \mathbb{C}^n$. *Then the R-group for* $I(\chi_{\underline{s}_0})$ *is trivial.*

If we use Vogan's approach, it is enough to prove that $I(\chi_{\underline{s}})$ has a unique minimal $O(n)$-type for any $\underline{s}$; this can be done by hand. This uniqueness of a minimal $O(n)$-type remains true for any $(\mathfrak{g}, O(n))$-module induced from a parabolic subgroup with a discrete series representation on the Levi subgroup.

If we use the approach of Harish-Chandra and Knapp-Stein, it is enough to prove that for all imaginary $\underline{s}_0$ and all nontrivial $w \in W$ that fix $\chi_{\underline{s}_0}$, the intertwining operator, suitably normalized, is a nonzero scalar at $\underline{s} = \underline{s}_0$. A special feature of $\mathrm{GL}(n)$ is that the elements $w \in W$ that fix $\chi_{\underline{s}_0}$ are generated by the root reflections fixing $\chi_{\underline{s}_0}$ (for a good choice of positive $\chi_{\underline{s}_0}$ in its conjugacy class). Thus it is enough to prove that if w is a root reflection fixing $\chi_{\underline{s}_0}$, the normalized $A(w, \chi_{\underline{s}})$ is scalar at $\underline{s} = \underline{s}_0$. In other words, to prove irreducibility, it is enough to prove that for all root reflections w fixing $\chi_{\underline{s}_0}$, the unnormalized $A(w, \chi_{\underline{s}})$ has a pole at $\underline{s} = \underline{s}_0$. This can be proved using the formulas of section 7 below, and thus Proposition 3 is proved.

7. Intertwining Operator for GL(2)

The intertwining operator for GL(2) has been known from a long time; see [Ku-S1], [Sa1], and [Sa2].

Here $n = 2$. For $x \in \mathbb{R}$, denote:

$$\tilde{x} := \begin{pmatrix} 1 & x \\ 0 & 1 \end{pmatrix},$$

and denote by

$$\sigma := \begin{pmatrix} 0 & 1 \\ -1 & 0 \end{pmatrix}$$

a representative of the unique nontrivial element in the Weyl group. Then:

$$\begin{aligned}\sigma\tilde{x} &= \begin{pmatrix} 0 & 1 \\ -1 & 0 \end{pmatrix}\begin{pmatrix} 1 & x \\ 0 & 1 \end{pmatrix} \\ &= \begin{pmatrix} 1 & -x(1+x^2)^{-1} \\ 0 & 1 \end{pmatrix}\begin{pmatrix} (1+x^2)^{-1/2} & 0 \\ 0 & (1+x^2)^{1/2} \end{pmatrix}\begin{pmatrix} -x(1+x^2)^{-1/2} & (1+x^2)^{-1/2} \\ -(1+x^2)^{-1/2} & -x(1+x^2)^{-1/2} \end{pmatrix}.\end{aligned}$$

Denote by k_x the last element written; k_x is in $SO(2)$. Put:

$$s := s_1 - s_2.$$

The positivity condition (2) in section 2 is that $Re\, s \geq 0$. Let $\phi \in \mathfrak{X}_{\epsilon_1,\epsilon_2}$. Then:

$$(A(\sigma, \chi_{\underline{s}})\phi)(k) = \int_{\mathbb{R}} (1+x^2)^{-(s+1)/2}\phi(k_x k)\, dx \qquad \text{for } \forall k \in O(2).$$

Put:

$$\eta := \begin{pmatrix} 1 & 0 \\ 0 & -1 \end{pmatrix}.$$

The subgroup of $O(2)$ generated by η and $SO(2)$ is $O(2)$ itself. Denote by $\mathfrak{X}_{\epsilon_1\epsilon_2}$ the space of right $SO(2)$-finite functions, ϕ, on $SO(2)$ satisfying:

$$\phi\left(\begin{pmatrix} -1 & 0 \\ 0 & -1 \end{pmatrix}\right) = \epsilon_1(-1)\epsilon_2(-1)\phi(k) \qquad \text{for } \forall k \in SO(2).$$

Now for $\phi \in \mathfrak{X}_{\epsilon_1,\epsilon_2}$, we have:

$$\phi(\eta k) = \epsilon_2(-1)\phi(k) \qquad \text{for } \forall k \in O(2).$$

This proves that restriction of functions from $O(2)$ to $SO(2)$ gives a bijection between $\mathfrak{X}_{\epsilon_1,\epsilon_2}$ and $\mathfrak{X}_{\epsilon_1\epsilon_2}$.

For $\ell \in \mathbb{Z}$, let ϕ_ℓ be the character of $SO(2)$:

$$\phi_\ell\left(\begin{pmatrix} \cos\theta & \sin\theta \\ -\sin\theta & \cos\theta \end{pmatrix}\right) = e^{\ell i\theta} \qquad \text{for } \forall\theta \in \mathbb{R}.$$

It is easy to see:

$$\phi_\ell \in \mathfrak{X}_{\epsilon_1\epsilon_2} \Longleftrightarrow \begin{cases} \ell \in 2\mathbb{Z} & \text{if } \epsilon_1\epsilon_2 = 1 \\ \ell \in 1 + 2\mathbb{Z} & \text{if } \epsilon_1\epsilon_2 \neq 1. \end{cases}$$

We can compute:

$$A(\sigma, \chi_{\underline{s}})\phi_\ell(k) = \lambda_\ell(\underline{s})\phi_\ell(k) \qquad \text{for } \forall \ell \in \mathbb{Z} \text{ and } k \in SO(2).$$

Here $\lambda_\ell(\underline{s})$ is given by:

$$\begin{aligned} \lambda_\ell(\underline{s}) = & \int_{\mathbb{R}} (1+x^2)^{-(s+1)/2}\left((-x+i)(1+x^2)^{-1/2}\right)^\ell dx \\ &= i^\ell \int_{\mathbb{R}} (1+x^2)^{-(s+1+\ell)/2}(1+ix)^\ell \, dx \\ &= i^\ell \int_{\mathbb{R}} (1-ix)^{-(s+1+\ell)/2}(1+ix)^{-(s+1-\ell)/2} \, dx \\ &= \pi i^\ell 2^{-s+1}\,\Gamma\Big(\frac{s+1}{2} - \frac{\ell}{2}\Big)^{-1}\Gamma\Big(\frac{s+1}{2} + \frac{\ell}{2}\Big)^{-1}\Gamma(s) \\ &= 2^{-s+1} i^\ell \sin\pi\Big(\frac{s+1}{2} - \frac{\ell}{2}\Big)\Gamma(s)\,\Gamma\Big(1 - \frac{s+1}{2} + \frac{\ell}{2}\Big)\Gamma\Big(\frac{s+1}{2} + \frac{\ell}{2}\Big)^{-1}. \end{aligned}$$

We have used that $\Gamma(s)\Gamma(1-s) = \pi/\sin\pi s$; see [S-Z, p. 315]. It is known (see [S-Z, sec. 9.1]) that the Γ function has no zero and has precisely simple poles at the negative integers (including 0). We obtain:

Proposition 7.

(i) *$A(\sigma, \chi_{\underline{s}})$ is holomorphic for $\underline{s}$ strictly positive.*

(ii) *$\Gamma(s_1 - s_2)^{-1}A(\sigma, \chi_{\underline{s}})$ is holomorphic for all $\underline{s}$, and this operator is bijective if and only if:*

$$\begin{aligned} s_1 - s_2 &\notin 1 + 2\mathbb{Z} \quad & \text{if } \epsilon_1 = \epsilon_2 \\ s_1 - s_2 &\notin 2\mathbb{Z} \quad & \text{if } \epsilon_1 \neq \epsilon_2. \end{aligned}$$

If $s_1 = s_2$, the operator is 0 if $\epsilon_1 \neq \epsilon_2$ and is the scalar 2 if $\epsilon_1 = \epsilon_2$.

(iii) *If $s_1 = s_2$ and $\epsilon_1 \neq \epsilon_2$, then $A(\sigma, \chi_{\underline{s}})$ is holomorphic and bijective.*

(iv) *If $\epsilon_1 = \epsilon_2$, then $A(\sigma, \chi_{\underline{s}})$ has a pole at $s_1 = s_2$.*

A common normalization of the intertwining operator is to make this operator be the identity on a minimal K-type; this gives (see [S-Z, 1.12 on p. 315]):

$$N(\sigma, \chi_{\underline{s}}) := A(\sigma, \chi_{\underline{s}}) \cdot \pi^{-1/2} i^{-\epsilon}\,\Gamma\Big(\frac{s+\epsilon}{2}\Big)^{-1}\Gamma\Big(\frac{s+1+\epsilon}{2}\Big),$$

where $\epsilon = 0$ if $\epsilon_1 = \epsilon_2$ and $\epsilon = 1$ if $\epsilon_1 \neq \epsilon_2$. I use $N(\cdot)$ for normalized; there seems to be no standard notation. With such a normalization, the operator $N(\sigma, \chi_{\underline{s}})$ is holomorphic for $\underline{s}$ positive and is nowhere the 0 operator on this set. Also we have the product formula:

$$N(\sigma, \sigma\chi_{\underline{s}})N(\sigma, \chi_{\underline{s}}) = 1.$$

Normalization of intertwining operators for reductive groups is due to [K-S4] and [K-S5] and is helpful in addressing reducibility questions and existence of complementary series. Shahidi [Sh] showed how to give a normalization that is consistent with the Langlands formalism.

8. Proof of Theorem 1

Fix $\underline{s}_0 \in \mathbb{C}^n$ and look at the irreducibility of $I(\chi_{\underline{s}_0})$. By Proposition 2, we may assume $\underline{s}_0$ is positive. For all $w \in W$, define:

$$A'(w, \chi_{\underline{s}}) := A(w, \chi_{\underline{s}})\Big(\prod_{\substack{i \text{ and } j \text{ with} \\ i<j,\, w(i)>w(j),\, \epsilon_i=\epsilon_j}} \Gamma(s_i - s_j)\Big)^{-1}.$$

Because of the isomorphisms (4), this defines for all $\underline{s}$ an intertwining operator from $I(\chi_{\underline{s}})$ to $I(w\chi_{\underline{s}})$. It follows from Proposition 7 that this operator is holomorphic for all positive $\underline{s}$ and is never the 0 operator. Moreover it is bijective at $\underline{s} = \underline{s}_0$ for all w, if and only if the conditions of Theorem 1 are fullfilled. This proves that these conditions are necessary for irreducibility. They are also sufficient because they imply that $A'(w, \chi_{\underline{s}_0})$ gives an isomorphism between $I(\chi_{\underline{s}_0})$ and $I(w_0\chi_{\underline{s}_0})$ and the Langlands theorem then gives an isomorphism with $J(\chi_{\underline{s}_0})$, which is irreducible by definition.

9. Unitarity

Theorem 9 (Vogan).

(i) *Let $\underline{\chi}$ be as in section* 1, *and suppose that $Re(s_i - s_j) \in \{0, 1, 2, \dots\}$ for all $i, j \in [1, n]$. Then $J(\underline{\chi})$ is unitary if and only if there exist a partition $n = n_1 + \cdots + n_r$ ($r \in \{1, 2, 3, \dots\}$) and unitary characters η_i of $\mathbb{R}^*$ for $i \in [1, r]$ such that $J(\underline{\chi})$ is isomorphic to the induced representation from the unitary character $\bigotimes_{i\in[1,r]} \eta_i(\det_{\mathrm{GL}(n_i)})$ of the Levi subgroup $\prod_{i\in[1,r]} \mathrm{GL}(n_i)$ of* $\mathrm{GL}(n)$.

(ii) *Any unitary representation of* $\mathrm{GL}(n)$ *is obtained from the representations in* (i) *by complementary series (see for example* [Sp-V] *for a definition) and cohomological induction (see* [V] *and* [K-V]*).*

This is a qualitative version of the main theorem of Vogan [V2]. It is easy to see when the condition on $\underline{\chi}$ in (i) is satisfied: If we take $n = \sum_{i\in[1,r]} n_i$ and η_i as in the statement of (i), then $J(\underline{\chi})$ is isomorphic to the induced representation specified in (i) if and only if $\underline{\chi}$ is conjugate to the collection of n characters:

$$\{\eta_i|\cdot|^{(n_i-2k+1)/2};\ i \in [1, r],\, k \in [1, n_i]\}.$$

10. Langlands Classification for GL(n)

A basic result of Harish-Chandra's is that a reductive group admits discrete series if and only if it has a compact Cartan subgroup (modulo the center). For $\mathrm{GL}(n)$, the conjugacy classes of Cartan subgroups are parametrized by the set $\{r \in \{0,1,2,\dots\};\ 2r \leq n\}$. Fix such an r and consider $C_r := (\mathbb{C}^*)^r \times (\mathbb{R}^*)^{n-2r}$. To any isomorphism:

$$\mathbb{C}^{2r} \times \mathbb{R}^{n-2r} \simeq \mathbb{R}^n,$$

corresponds an imbedding of C_r in $\mathrm{GL}(n)$; in this realization C_r is a Cartan subgroup. For r fixed, all these embeddings are conjugate, and they describe the conjugacy class associated to r.

In particular, $\mathrm{GL}(n)$ has discrete series if and only if $n = 1$ or 2.

Let $t \in \{1,2,3,\dots\}$. For each $j \in [1,t]$, fix $s_j \in \mathbb{C}$ and fix $n_j = 1$ or 2 such that

$$\sum_{j \in [1,t]} n_j = n.$$

If $n_j = 1$, fix also a character σ_j of order 2, and, if $n_j = 2$, an integer $p_j \in \{1,2,3,\dots\}$. In this last case, write σ_j for the unique discrete series of $\mathrm{GL}(2)$ with infinitesimal character $(p_j/2, -p_j/2)$. In other words, σ_j is the unique irreducible submodule of $I(\epsilon|\cdot|^{p_j/2} \times |\cdot|^{-p_j/2})$, where ϵ is the sign character of $\mathbb{R}^*$ if p_j is odd and is the trivial character otherwise. Write $\underline{\chi}$ for the set of triples:

$$\underline{\chi} := \{(n_j, s_j, \sigma_j)_{j \in [1,t]}\}.$$

Denote by $I(\underline{\chi})$ the $(\mathfrak{g}, O(n))$-module induced from the representation

$$\bigotimes_{j \in [1,t]} (\sigma_j \otimes |\det{}_{\mathrm{GL}(n_j)}|^{s_j})$$

of the Levi subgroup $\prod_{j \in [1,t]} \mathrm{GL}(n_j)$ of $\mathrm{GL}(n)$. (Of course, we have to choose a parabolic subgroup with this Levi subgroup, and we choose the one containing B.) Such a $\underline{\chi}$ is said to be **positive** if for all $i, j \in [1,t]$ with $i < j$, $\mathit{Re}\, s_i \geq \mathit{Re}\, s_j$. This definition is consistent with (2) in section 2.

The general Langlands theorem (see references given in section 4) for $\mathrm{GL}(n)$ asserts:

Theorem 10a.

(i) *Let $\underline{\chi}$ be as above and assume that $\underline{\chi}$ is positive. Then $I(\underline{\chi})$ has a unique irreducible quotient, $J(\underline{\chi})$.*

(ii) *If π is an irreducible $(\mathfrak{g}, O(n))$-module for $\mathrm{GL}(n)$, then there exists $\underline{\chi}$ positive such that $\pi \simeq J(\underline{\chi})$. Moreover $\underline{\chi}$ is unique except that permutation of $i, j \in [1,t]$ with $\mathit{Re}\, s_i = \mathit{Re}\, s_j$ is allowed.*

Theorem 10b ([Sp1]). *If $\underline{\chi}$ is as above, then $I(\underline{\chi})$ is irreducible if and only if for all $i, j \in [1,t]$ with $n_i \geq n_j$, either $s_i - s_j \notin \mathbb{R}$ or the appropriate one of the following conditions holds:*

(i) *if $n_i = n_j = 1$, $|s_i - s_j|$ is not an even (resp. odd) nonzero integer if $\sigma_i \neq \sigma_j$ (resp. $\sigma_i = \sigma_j$),*

(ii) *if $n_i = 2$ and $n_j = 1$ (so that p_i is defined), $-p_i/2 + |s_i - s_j| \notin \{1,2,3,\dots\}$,*

(iii) *if $n_i = n_j = 2$ (so that p_i and p_j are defined), $-|p_i - p_j|/2 + |s_i - s_j| \notin \{1,2,3,\dots\}$.*

To prove this last theorem, using the method explained here, one reduces to GL(a) with $a = 2$, 3, or 4. These cases are done in [Sp1] (see also [Sp2] and, for more general results, [Sp-V]).

References

[A] Arthur, J. G., *Intertwining integrals for cuspidal parabolic subgroups*, duplicated notes, Yale University, New Haven, CT, 1974.

[B] Ban, E. P. van den, Induced representations and the Langlands classification, *these* PROCEEDINGS, pp. 123–155.

[D] Delorme, P., Infinitesimal character and distribution character of representations of reductive Lie groups, *these* PROCEEDINGS, pp. 73–81.

[G-G] Gelfand, I. M., and M. I. Graev, Unitary representations of the real unimodular group (Russian), *Izv. Akad. Nauk USSR Ser. Math.* **17** (1953), 189-248; *Translations Amer. Math. Soc.* (2) **2** (1956), 147–205.

[H-C] Harish-Chandra, Harmonic analysis on real reductive groups III. The Maass-Selberg relations and the Plancherel formula, *Annals of Math.* **104** (1976), 117–201.

[J] Jacquet, H., Generic representations, *Non-Commutative Harmonic Analysis (Actes Colloq., Marseille-Luminy, 1976)*, Lecture Notes in Mathematics, vol. 587, Springer-Verlag, Berlin, 1977, pp. 91–101.

[K1] Knapp, A. W., *Representation Theory of Semisimple Groups: An Overview Based on Examples*, Princeton University Press, 1986, 773 pp.

[K2] Knapp, A. W., Commutativity of intertwining operators, *Bull. Amer. Math. Soc.* **79** (1973), 1016–1018.

[K3] Knapp, A. W., Commutativity of intertwining operators for semisimple groups, *Compositio Math.* **46** (1982), 33–84.

[K-S1] Knapp, A. W., and E. M. Stein, Irreducibility theorems for the principal series, *Conference on Harmonic Analysis*, Lecture Notes in Mathematics, vol. 266, Springer-Verlag, Berlin, 1972, pp. 197–214.

[K-S2] Knapp, A. W., and E. M. Stein, Singular integrals and the principal series IV, *Proc. Nat. Acad. Sci.* **72** (1975), 2459–2461.

[K-S3] Knapp, A. W., and E. M. Stein, Singular integrals and the principal series III, *Proc. Nat. Acad. Sci.* **71** (1974), 4622–4624.

[K-S4] Knapp, A. W., and E. M. Stein, Intertwining operators for semisimple groups, *Annals of Math.* **93** (1971), 489–578.

[K-S5] Knapp, A. W., and E. M. Stein, Intertwining operators for semisimple groups II, *Invent. Math.* **60** (1980), 9–84.

[K-V] Knapp, A. W., and D. A. Vogan, *Cohomological Induction and Unitary Representations*, Princeton University Press, Princeton, 1995, 948 pp.

[Ku-S1] Kunze, R. A., and E. M. Stein, Uniformly bounded representations and harmonic analysis of the 2×2 real unimodular group, *Amer. J. Math.* **82** (1960), 1–62.

[Ku-S2] Kunze, R. A., and E. M. Stein, Uniformly bounded representations II. Analytic continuation of the principal series of representations of the $n \times n$ complex unimodular group, *Amer. J. Math* **83** (1961), 723–786.

[M] Miličić, D., Asymptotic behaviour of matrix coefficients of the discrete series, *Duke Math. J.* **44** (1977), 59–88.

[S-Z] Saks, S., and A. Zygmund, *Fonctions Analytiques*, Masson, Paris, 1970, 389 pp.

[Sa1] Sally, P. J., Analytic continuation of the irreducible unitary representations of the universal covering group of SL($2, R$), *Memoirs Amer. Math. Soc.* **69** (1967).

[Sa2] Sally, P. J., Intertwining operators and the representations of SL($2, R$), *J. Func. Anal.* **6** (1970), 441–453.

[Sc] Schiffmann, G., Intégrales d'entrelacement et fonctions de Whittaker, *Bull. Soc. Math. France* **99** (1971), 3–72.

[Sh] Shahidi, F., Local coefficients as Artin factors for real groups **52** (1985), 973–1007.

[Sp1] Speh, B., *Some results on principal series for* GL(n, R), Ph. D. Dissertation, M.I.T., Cambridge, MA, June 1977.

[Sp2] Speh, B., The unitary dual for GL($3, R$) and GL($4, R$), *Math. Annalen* **258** (1981), 113–133.

[Sp-V] Speh, B., and D. A. Vogan, Reducibility of generalized principal series representations, *Acta Math.* **145** (1980), 227–299.

[V1] Vogan, D. A., *Representations of Real Reductive Lie Groups*, Progress in Math., vol. 15, Birkhäuser, Boston, 1981, 754 pp.

[V2] Vogan, D. A., The unitary dual of GL(n) over an archimedean field, *Invent. Math.* **83** (1986), 449–505.

[W] Wallach, N. R., On Harish-Chandra's generalized C-functions, *Amer. J. Math.* **97** (1975), 386–403.

Département de Mathématiques, Université de Paris VII, F-75 251 Paris cedex 05, France

E-mail address: moeglin@math.jussieu.fr

Proceedings of Symposia in Pure Mathematics
Volume **61** (1997), pp. 167–189

Orbital Integrals, Symmetric Fourier Analysis, and Eigenspace Representations

Sigurdur Helgason

ABSTRACT. These lectures give an informal exposition of the three topics in the title. Although the topics are closely related and share notational conventions the three sections should be readable independently.

The first section describes Harish-Chandra's Plancherel formula for semisimple Lie groups G, which is based on the study of the integrals of functions over conjugacy classes in G. The second section deals with the Fourier transform on the symmetric space $X = G/K$ associated with G and selected applications of this transform to differential equations. In the last section we discuss irreducibility properties of representations of G on eigenspaces of invariant differential operators on various homogeneous spaces of G.

§1. Orbital Integrals and Plancherel Formula

1.1. The Plancherel formula

In this section I shall attempt to describe in an informal way the approach to the Plancherel formula on a semisimple Lie group via orbital integral theory.

The following notation will be used: $\mathbf{C}$, $\mathbf{R}$, and $\mathbf{Z}$ denote the complex numbers, real numbers, and integers; $\mathbf{R}^+ = \{t \in \mathbf{R} : t \geq 0\}$, and $\mathbf{Z}^+ = \mathbf{R}^+ \cap \mathbf{Z}$. For a topological space X, $C(X)$ and $C_c(X)$ denote the spaces of continuous functions, the subscript c denoting compact support. If X has a metric d and $x \in X$, $B_r(x)$ denotes the ball $\{y \in X : d(x,y) < r\}$ and $S_r(x)$ the sphere $\{y \in X : d(x,y) = r\}$. If X is a manifold we use the notation $\mathcal{E}(X)$ for $C^\infty(X)$ and $\mathcal{D}(X)$ for $C_c^\infty(X)$.

Lie groups will be denoted by capital letters, $A, B, \ldots$ and the corresponding Lie algebras by corresponding German letters $\mathfrak{a}, \mathfrak{b}, \ldots$. The adjoint representations of A and $\mathfrak{a}$ are denoted by Ad_A (or Ad) and $ad_\mathfrak{a}$ (or ad).

If A is an abelian group with character group $\widehat{A}$ (and Haar measure dx) the Fourier transform on A is given by (see [W])

$$\tilde{f}(\chi) = \int_A f(x)\chi(x^{-1})\,dx, \qquad \chi \in \widehat{A}. \tag{1}$$

1991 *Mathematics Subject Classification.* Primary 22E45, 22E46, 43A85; Secondary 35L05.

With the Haar measure $d\chi$ suitably normalized one has the inversion formula

$$f(x) = \int_{\widehat{A}} \tilde{f}(\chi)\chi(x)\,d\chi \tag{2}$$

for f in a suitable dense subspace of $L^1(A)$. We can write (2) in the form

$$\delta = \int_{\widehat{A}} \chi\,d\chi, \tag{3}$$

where δ is the delta function of A at e and χ denotes the measure $f \to \tilde{f}(\chi)$ on A.

Next let G be a semisimple Lie group with finite center, $\widehat{G}$ the set of equivalence classes of irreducible unitary representations of G. With $\pi \in \widehat{G}$ operating in the Hilbert space $\mathcal{H}_\pi$ the **Fourier transform** of a function f on G is defined by

$$\tilde{f}(\pi) = \int_G f(x)\pi(x)\,dx, \tag{4}$$

dx being the Haar measure on G. Thus the Fourier transform assigns to f a family of operators on different Hilbert spaces $\mathcal{H}_\pi$. For π in the principal series of the complex classical groups, Gelfand and Naimark showed [GN] that π always has a character defined almost everywhere on the group. This was completed by Harish-Chandra [HC1], who showed for any G and any $f \in C_c^\infty(G)$ and $\pi \in \widehat{G}$ that the operator $\tilde{f}(\pi)$ has a trace $\chi_\pi(f)$ and that the functional $f \to \chi_\pi(f)$ is a distribution on G, the **character** of π. This distribution is a real analytic function on the set of regular elements in G; this extends the result of [GN]. The principal step in his proof is showing that if K is a maximal compact subgroup of G and $\delta \in \widehat{K}$ then the restriction $\pi|K$ contains δ at most $N \dim \delta$ times (N being a constant). The objective is then, in analogy with (2), to find a measure $d\pi$ on $\widehat{G}$ such that

$$f(e) = \int_{\widehat{G}} \chi_\pi(f)\,d\pi, \qquad f \in C_c^\infty(G). \tag{5}$$

If we use this on the function $f * f^*$ where $f^*(x) = \overline{f(x^{-1})}$ we would get

$$\int_G |f(x)|^2\,dx = \int_{\widehat{G}} \chi_\pi(f * f^*)\,d\pi.$$

However $\chi_\pi(f * f^*) = \operatorname{Tr}\left(\tilde{f}(\pi)\tilde{f}(\pi)^*\right) = \|\tilde{f}(\pi)\|^2$ ($\|\cdot\|$ = Hilbert-Schmidt norm) so we have the Plancherel formula

$$\int_G |f(x)|^2\,dx = \int_{\widehat{G}} \|\tilde{f}(\pi)\|^2\,d\pi. \tag{6}$$

The existence of a measure $d\pi$ satisfying (6) (for G locally compact unimodular and $\|\cdot\|$ a more abstract operator algebra norm) had been proved by Segal [Se]. However (5) is a more precise decomposition of $\delta = \delta_e$ into characters:

$$\delta = \int_{\widehat{G}} \chi_\pi\,d\pi. \tag{7}$$

1.2. Compact groups

Following Harish-Chandra let us restate the Peter-Weyl theorem in this framework for G compact and simply connected. The theorem can be written

$$f = \sum_{\pi \in \widehat{G}} d_\pi (\chi_\pi * f), \qquad f \in C^\infty(G), \tag{8}$$

where d_π is the degree and $*$ denotes convolution. Let $T \subset G$ denote a maximal torus in G and $\mathfrak{t} \subset \mathfrak{g}$ their Lie algebras, $\mathfrak{t}^c \subset \mathfrak{g}^c$ their complexifications. Consider the weight lattice

$$\Lambda = \left\{ \lambda \in \mathfrak{t}_c^* \ : \ 2\frac{\langle \lambda, \alpha \rangle}{\langle \alpha, \alpha \rangle} \in \mathbf{Z} \ \text{ for } \alpha \in \Delta \right\},$$

$\langle \cdot , \cdot \rangle$ denoting the Killing form and Δ the set of roots of $(\mathfrak{g}^c, \mathfrak{t}^c)$. Under the bijection $\mu \to e^\mu$ $\big(e^\mu(\exp H) = e^{\mu(H)},\ H \in \mathfrak{t}\big)$ the lattice Λ is identified with the character group $\widehat{T}$. By the highest weight theory the set $\Lambda(+)$ of dominant weights is identified with $\widehat{G}$. Thus if Δ^+ denotes the set of positive roots,

$$\widehat{G} = \Lambda(+) = \{\lambda \in \Lambda \ : \ \langle \lambda, \alpha \rangle \geq 0 \ \text{ for } \alpha \in \Delta^+\} = \Lambda/W = \widehat{T}/W,$$

where W denotes the Weyl group.

If π has highest weight λ we have Weyl's formulas for the degree $d_\lambda = d_\pi$ and the character $\chi_\lambda = \chi_\pi$:

$$d_\pi = \prod_{\alpha \in \Delta^+} \frac{\langle \lambda + \rho, \alpha \rangle}{\langle \rho, \alpha \rangle}, \qquad \chi_\pi(\exp H) = \frac{\Sigma_s \det s\, e^{s(\lambda+\rho)(H)}}{\Sigma_s \det s\, e^{s\rho(H)}}, \tag{9}$$

where s runs through W. Denoting the denominator in the χ_π formula by $D(t)$ $(t = \exp H)$ we have the integral formula

$$|W| \int_G f(g)\, dg = \int_T |D(t)|^2 \int_G f(gtg^{-1})\, dg\, dt. \tag{10}$$

We consider now the orbital integral

$$F_f(t) = D(t) \int_G f(gtg^{-1}) dg, \qquad t \in T.$$

If Ω denotes the differential operator $\Pi_{\alpha \in \Delta^+} H_\alpha$ ($H_\alpha \in \mathfrak{t}^c$ corresponding to α) we have for the Fourier transform $F \to \widetilde{F}$ on T

$$(\Omega F)^\sim(\mu) = \omega(\mu) \widetilde{F}(\mu), \qquad \mu \in \widehat{T} = \Lambda,$$

where ω is the polynomial $\prod_{\alpha \in \Delta^+} \alpha$. The value $F(e)$ being the sum of the Fourier coefficients of a smooth F, we have

$$(\Omega F_f)(e) = \sum_{\mu \in \Lambda} \omega(\mu) \widetilde{F}_f(\mu).$$

Here we can restrict μ to the set Λ' of regular elements in Λ. Since $\omega \widetilde{F}_f$ is W-invariant the sum becomes

$$|W| \sum_{\mu \in \Lambda' \cap \Lambda(+)} \omega(\mu) \widetilde{F}_f(\mu).$$

Since $\Lambda' \cap \Lambda(+) = \rho + \Lambda(+)$ and $\omega(\lambda+\rho) = d_\lambda \prod_{\alpha\in\Delta^+} \langle \rho, \alpha\rangle$, this implies (with a constant c)

$$c(\Omega F_f)(e) = \sum_{\lambda\in\Lambda(+)} d_\lambda \widetilde{F}_f(\lambda+\rho).$$

Integrating $f(g)\chi_\lambda(g^{-1})$ over G we obtain, from (9) and (10),

$$(\chi_\lambda * f)(e) = \widetilde{F}_f(\lambda+\rho), \tag{11}$$

a formula that relates the Fourier transforms on G and on T. Thus (8) implies the orbital integral formula

$$f(e) = c(\Omega F_f)(e), \qquad c^{-1} = |W| \prod_{\alpha\in\Delta^+} \langle \rho, \alpha\rangle. \tag{12}$$

The point is now, that since $D(t)$ vanishes at $t = e$ to the same order as the degree of Ω, formula (12) is quite immediate. Going backwards, one can then derive the Peter-Weyl decomposition (8).

1.3. Orbital integrals

The orbital integral problem amounts to determining the value of a function on G at e in terms of its integrals over (generic) conjugacy classes. For $G = \mathbf{SL}(n, \mathbf{C})$ this problem was solved by Gelfand and Naimark [GN] and used to prove the Plancherel formula (6). Once the characters of the principal series are determined by a formula analogous to (9), the method above for G compact illustrates the underlying idea of [GN] for proving (6) for $\mathbf{SL}(n, \mathbf{C})$. Their method was shortly afterwards generalized by Harish-Chandra to all complex semisimple G [HC2].

Recognizing the importance of the orbital integral problem, Gelfand and Graev [GG1] found in 1953 a new elegant solution for all complex classical G. We sketch this method.

Consider a quadratic form $\omega(x_1, \dots, x_m)$ and the generalized Riesz integral

$$R(\lambda) = \int_{\omega\geq 0} f(x_1, \dots, x_m)\omega(x_1, \dots, x_m)^{\lambda/2}\, dx, \qquad \lambda \in \mathbf{C},\ f \in C_c^\infty(\mathbf{R}^n).$$

Assume ω has signature (s,t) with s odd, t even. Then $R(\lambda)$ is meromorphic in $\mathbf{C}$ *with simple poles at*

$$\lambda = -m - 2k \qquad (k = 0, 1, 2, \dots)$$

and corresponding residues

$$\operatorname{Res}_{\lambda=-m-2k} R(\lambda) = c(L^k f)(0), \tag{13}$$

where $c \neq 0$ is a constant and L the Laplacian corresponding to ω.

If $H \subset G$ is a Cartan subgroup with Lie algebra $\mathfrak{h}$, let $\Delta \geq 0$ be determined by

$$\int_G f(g)\, dg = \int_H \Delta(h)^2 \left(\int_{G/H} f(ghg^{-1})\, dg_H \right) dh,$$

so that

$$\int_G f(g)\, dg = \int_H F_f(h)\Delta(h)\, dh, \qquad f \in C_c^\infty(G), \tag{14}$$

where F_f is the **orbital integral**

$$F_f(h) = \Delta(h) \int_{G/H} f(ghg^{-1})\, dg_H. \tag{15}$$

While the integral in (15) is at first defined only for $h \in H$ regular, the factor $\Delta(h)$ makes F_f extendable to a smooth function on all of H. Let f have support in a neighborhood of e that is the diffeomorphic image of a neighborhood of zero in $\mathfrak{g}$ under exp. Let $J(X) = dg/dX$ and ω the Killing form. Then we have by (14)

$$\int_{\mathfrak{g}} f(\exp X)\omega(X,X)^{\lambda/2} J(X)\, dX = \int_{\mathfrak{h}} F_f(\exp H)\Delta(\exp H)\omega(H,H)^{\lambda/2}\, dH. \tag{16}$$

In order to use (13) we actually need two such formulas, namely for $\omega \geq 0$ and for $\omega < 0$. Ignoring this complication, let us use (16) for $G = \mathbf{SL}(n, \mathbf{C})$, calculating residues for $\lambda = -\dim G$ $(= -2n^2 + 2)$. On the left hand side we get $cf(e)$ with $c \neq 0$. Since $\dim \mathfrak{h} = 2n - 2 = \dim \mathfrak{g} - 2(n(n-1))$ we get on the right hand side

$$\left(L^{n(n-1)}(F_f\Delta)\right)(e).$$

After some manipulation this gives the orbital integral formula

$$f(e) = (\Omega F_f)(e), \tag{17}$$

where Ω is an explicit invariant differential operator on H. A minor complication is that the necessary parity condition on (s, t) in the residue theorem is not satisfied here. This is remedied by going over to the group $G_1 = \{g \in \mathbf{GL}(n, \mathbf{C}) : \det g \in \mathbf{R}\}$ of dimension $2n^2 - 1$.

In [HC4] Harish-Chandra extended (17) to all semisimple G, real or complex. (See also [HC3] for $\mathbf{SL}(2, \mathbf{R})$ and Gelfand-Graev [GG2] for $\mathbf{SL}(n, \mathbf{R})$.) For this theorem H has to be a fundamental Cartan subgroup, that is, one for which the compact part has maximal dimension. This method was based on the Fourier transform on $\mathfrak{g}$ relative to the indefinite form ω. In [HC7], (1975) he gave yet another method based on fundamental solutions for powers of the Casimir polynomial ω on $\mathfrak{g}$. I shall indicate this remarkable method for the case (including G complex) when all the Cartan subgroups are conjugate.

Let $m = [\dim \mathfrak{g}/2]$ and Ξ a specific fundamental solution of $\partial(\omega^m)$, locally integrable on $\mathfrak{g}$ and smooth where $\omega \neq 0$. Let $\delta_{\mathfrak{g}}$ and $\delta_{\mathfrak{h}}$ denote the delta functions on $\mathfrak{g}$ and $\mathfrak{h}$, respectively, and $\Xi_{\mathfrak{h}}$ and $\omega_{\mathfrak{h}}$ the restrictions of Ξ and ω to $\mathfrak{h}$. The choice of Ξ is such that not only is

$$\partial(\omega^m)\Xi = \delta_{\mathfrak{g}} \tag{18}$$

but $\Xi_{\mathfrak{h}}$ is locally integrable on $\mathfrak{h}$ and if $k = [\dim \mathfrak{h}/2]$ we have also

$$\partial(\omega_{\mathfrak{h}}^k)\Xi_{\mathfrak{h}} = c\delta_{\mathfrak{h}}, \qquad c \neq 0 \text{ constant.} \tag{19}$$

Given a function f on $\mathfrak{g}$ we consider the Lie algebra orbital integral

$$\varphi_f(Z) = \pi(Z) \int_{G/H} f(Ad(g)(Z))\, dg_H, \qquad Z \in \mathfrak{h}, \tag{20}$$

where π is the product of the positive roots. From (18) we have

$$f(0) = \int_{\mathfrak{g}} \left(\Xi\partial(\omega^m)f\right)(X)\, dX. \tag{21}$$

Since $dX = |\pi(Z)|^2 dg_H\, dZ$ and $\overline{\pi(Z)} = \mathrm{const}\, \pi(Z)$, we deduce from (20)

$$f(0) = \mathrm{const} \int_{\mathfrak{h}} \Xi_{\mathfrak{h}}(Z)\pi(Z)\phi_{f_m}(Z)\, dZ, \tag{22}$$

where $f_m = \partial(\omega^m)f$. However,

$$\phi_{f_m} = \partial(\omega_{\mathfrak{h}}^m)\phi_f, \tag{23}$$

and a direct calculation shows that

$$\pi \circ \partial(\omega_{\mathfrak{h}}^m) = \partial(\omega_{\mathfrak{h}}^k) \circ \eta, \tag{24}$$

where η is a differential operator on $\mathfrak{h}$ with polynomial coefficients whose local expression η_0 at zero is a constant multiple of $\partial(\pi)$. Substituting (23) into (22) we get

$$f(0) = \mathrm{const} \int_{\mathfrak{h}} \Xi_{\mathfrak{h}}(Z)\partial(\omega_{\mathfrak{h}}^k)(\eta\phi_f)(Z)\, dZ,$$

so by (24) and the mentioned formula $\eta_0 = \mathrm{const}\, \partial(\pi)$, we get

$$f(0) = C\left(\partial(\pi)\phi_f\right)(0), \qquad C = \mathrm{const}. \tag{25}$$

Separate arguments are then needed to lift this formula to the group.

For G complex the principal series $\pi_{\widehat{h}}$ is parameterized by characters $\widehat{h}$ of the Cartan subgroup $H \subset G$. Because of the formula indicated for the character $T_{\widehat{h}}$ of $\pi_{\widehat{h}}$ one has in analogy with (11),

$$T_{\widehat{h}}(f) = (\widetilde{F}_f)(\widehat{h}), \tag{26}$$

and since $(\Omega F_f)^{\sim} = \widetilde{\Omega}\widetilde{F}_f$ for $\widetilde{\Omega}$ a certain polynomial on $\widehat{H}$, formula (17) implies the Plancherel formula

$$\delta = \int_{\widehat{H}} T_{\widehat{h}}\widehat{\Omega}(\widehat{h})\, d\widehat{h}. \tag{27}$$

The argument of course relies on the smoothness of F_f, and this remains valid if G has just one conjugacy class of Cartan subgroups. The argument for (27) can still be carried out (with $\widetilde{\Omega}$ a polynomial).

Dropping the assumption that all Cartan subgroups are conjugate presents major obstacles that were eventually overcome by Harish-Chandra [HC6] through the results sketched below.

While (17) remains valid for general G provided H is fundamental, its use is complicated by the fact that F_f is no longer smooth, in fact its jumps are related to orbital integrals for other Cartan subgroups. (See (32) below for $\mathbf{SL}(2, \mathbf{R})$.)

Let $H_1, \dots, H_r$ be a maximal family of nonconjugate Cartan subgroups, all invariant under a fixed Cartan involution of G. Then (14) is replaced by

$$\int_G f(g)\, dg = \sum_i \frac{1}{|W_i|} \int_{H_i} F_f^i(h)\Delta_i(h)\, dh, \tag{28}$$

where F_f^i is the orbital integral relative to the group H_i and W_i is the corresponding Weyl group. If θ is a conjugacy invariant distribution that equals F on the regular set of G then

$$\theta(f) = \sum_i \frac{1}{|W_i|} \int_{H_i} F_f^i(h)\Delta_i(h)F(h)\, dh. \tag{29}$$

Let H be one of the H_i, A its vector part (with Lie algebra $\mathfrak{a}$) and $P = MAN$ a parabolic subgroup for which MA is the centralizer of A in G and $M \cap A = \{e\}$. With an arbitrary character $a \to e^{i\lambda(\log a)}$ of A ($\lambda \in \mathfrak{a}^*$) and an arbitrary discrete series representation σ of M, we obtain a representation $\pi_{\sigma,\lambda}$ of G unitarily induced by the representation $man \to \sigma(m)e^{i\lambda(\log a)}$ of P. For $\lambda \in \mathfrak{a}^*$ regular, $\pi_{\sigma,\lambda}$ is irreducible. Let $\theta_{\sigma,\lambda}$ denote its character. Denoting by $\mathcal{S}(G)$ the L^2 Schwartz space of G, let $\mathcal{S}_H(G)$ denote the subspace of functions orthogonal to the matrix coefficients of $\pi_{\sigma,\lambda}$ coming from the other Cartan subgroups $H_i \neq H$. Then ([HC6])

$$\mathcal{S}(G) = \bigoplus_i \mathcal{S}_{H_i}(G), \tag{30}$$

and for each individual $\mathcal{S}_H(G)$ one has

$$f(e) = \int_{\widehat{M}_d \times \mathfrak{a}^*} \theta_{\sigma,\lambda}(f)\, d\mu_H(\sigma,\lambda), \qquad f \in \mathcal{S}_H(G), \tag{31}$$

where $\widehat{M}_d$ is the discrete series of M and μ_H a certain explicitly determined positive measure. Combining (30) and (31) we get Harish-Chandra's formula (5) in an explicit form.

Example. $G = \mathbf{SL}(2,\mathbf{R})$. This case was settled already in 1952 by Harish-Chandra, [HC3]. The Lie algebra $\mathfrak{sl}(2,\mathbf{R})$ is given by

$$\{X = x_1X_1 + x_2X_2 + x_3X_3 \ : \ x_1, x_2, x_3 \in \mathbf{R}\},$$

where

$$X_1 = \begin{pmatrix} 0 & 1 \\ -1 & 0 \end{pmatrix}, \qquad X_2 = \begin{pmatrix} 1 & 0 \\ 0 & -1 \end{pmatrix}, \qquad X_3 = \begin{pmatrix} 0 & 1 \\ 1 & 0 \end{pmatrix}.$$

It has the two nonconjugate Cartan subalgebras

$$\mathfrak{b} = \mathbf{R}X_1, \qquad \mathfrak{a} = \mathbf{R}X_2$$

invariant under the Cartan involution $X \to -{}^tX$. We write ψ_f^B and ψ_f^A for the corresponding φ_f in (20). The Cartan subgroups H_1, H_2 are now

$$B = \left\{k_\theta = \begin{pmatrix} \cos\theta & \sin\theta \\ -\sin\theta & \cos\theta \end{pmatrix}\right\}, \qquad A = \left\{\epsilon a_t : a_t = \begin{pmatrix} e^t & 0 \\ 0 & e^{-t} \end{pmatrix}, \quad \epsilon = \pm 1\right\}.$$

The adjoint action of G on $\mathfrak{sl}(2,\mathbf{R})$ is $X \to gXg^{-1}$. Under the mapping $X \to (x_1, x_2, x_3)$ the orbits

$$G \cdot \begin{pmatrix} 0 & \theta \\ -\theta & 0 \end{pmatrix} \ (\theta > 0), \qquad G \cdot \begin{pmatrix} 0 & \theta \\ -\theta & 0 \end{pmatrix} \ (\theta < 0), \qquad G \cdot \begin{pmatrix} t & 0 \\ 0 & -t \end{pmatrix}$$

are respectively

$$\begin{aligned} H_+ &: \text{hyperboloid } x_1^2 - x_2^2 - x_3^2 = \theta^2 && (x_1 > 0), \\ H_- &: \text{hyperboloid } x_1^2 - x_2^2 - x_3^2 = \theta^2 && (x_1 < 0), \\ H &: \text{hyperboloid } x_1^2 - x_2^2 - x_3^2 = -t^2, \end{aligned}$$

which are indicated in the figure. Let $C_\pm$ denote the upper and lower light cone, respectively, and $C = C_+ \cup C_-$.

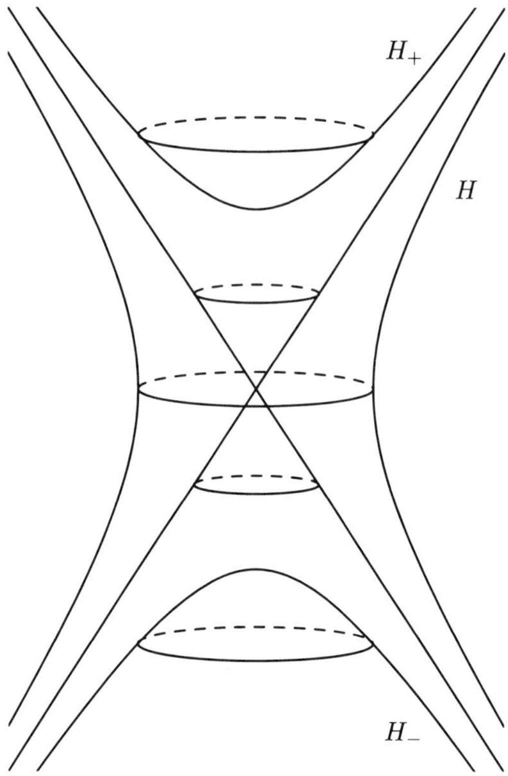

FIGURE 1. Hyperboloids and light cone

Then, as the geometry suggests,

$$\lim_{\theta\to 0+} \psi_f^B(\theta) = \int_{C_+} f$$

$$\lim_{\theta\to 0-} \psi_f^B(\theta) = -\int_{C_-} f,$$

so for the jump of ψ_f^B we have

$$\psi_f^B(0^+) - \psi_f^B(0^-) = \int_C f = \psi_f^A(0). \tag{32}$$

On the other hand, $d\psi_f^B/d\theta$ is continuous at $\theta = 0$, and since the Cartan subgroup B is fundamental,

$$\frac{d\psi_f^B}{d\theta}(0) = \text{const}\, f(0). \tag{33}$$

Formulas similar to (32) and (33) hold for the group orbital integrals (where $x^g = gxg^{-1}$):

$$F_f^B(k_\theta) = \left(e^{i\theta} - e^{-i\theta}\right) \int_G f(k_\theta^g)\, dg, \qquad \theta \notin \pi\mathbf{Z}.$$

$$F_f^A(\epsilon a_t) = \left|e^t - e^{-t}\right| \int_{G/A} f(\epsilon a_t^g)\, dg_A, \qquad t \neq 0.$$

Next we write down the characters of the representations $\pi_{\sigma,\lambda}$ above.

For $H_1 = B$ there is no vector part so we just have the discrete series of G. It is parameterized by $k \in \mathbf{Z} - \{0\}$, and the characters are given by

$$\theta_k(a_t) = (-1)^{k-1}\theta_k(-a_t) = \frac{e^{-|kt|}}{|e^t - e^{-t}|}, \qquad \theta_k(k_\theta) = -(\operatorname{sign} k)\frac{e^{ik\theta}}{e^{i\theta} - e^{-i\theta}}.$$

On the other hand, the characters $\theta_{\pm,s}$ of the representations of G induced by the representations

$$\epsilon a_t \to e^{st} \qquad \text{and} \qquad \epsilon a_t \to (\operatorname{sign}\epsilon)e^{st}$$

of AN are given by

$$\theta_{+,s}(a_t) = \theta_{+,s}(-a_t) = \frac{e^{st} + e^{-st}}{|e^t - e^{-t}|}, \qquad \theta_{+,s}(k_\theta) = 0,$$

$$\theta_{-,s}(a_t) = -\theta_{-,s}(-a_t) = \frac{e^{st} + e^{-st}}{|e^t - e^{-t}|}, \qquad \theta_{-,s}(k_\theta) = 0.$$

The Weyl groups W_B and W_A have orders 1 and 2, respectively, and the sign of $\Delta_i(h)$ is determined by the positivity of the map $f \to F_f^i\Delta_i$. Thus we conclude from (28), A being two copies of $\mathbf{R}$,

$$\int_G f(g)\, dg = \frac{1}{2}\int_{\mathbf{R}} \left|e^t - e^{-t}\right| \left(F_f^A(a_t) + F_f^A(-a_t)\right) dt$$
$$+ \frac{1}{2\pi}\int_0^{2\pi} \left(e^{-i\theta} - e^{i\theta}\right) F_f^B(k_\theta)\, d\theta.$$

Applying (29) to $\theta = \theta_k$ we obtain

$$\theta_k(f) = \frac{1}{2}\int_{\mathbf{R}} e^{-|kt|}\left(F_f^A(a_t) + (-1)^{k-1}F_f^A(-a_t)\right) dt \tag{34}$$
$$+ (\operatorname{sign} k)\,\frac{1}{2\pi}\int_0^{2\pi} e^{ik\theta} F_f^B(k_\theta)\, d\theta.$$

For $\theta = \theta_{\pm,s}$ we have, since $F_f^A(\pm a_t) = F_f^A(\pm a_{-t})$,

$$\theta_{\pm,s}(f) = \int_{\mathbf{R}} \left[F_f^A(a_t) \pm F_f^A(-a_t)\right] e^{st}\, dt. \tag{35}$$

Formulas (34) and (35) relate the characters to the orbital integrals, the latter formula being analogous to (26). Finally, one uses (33) combined with (34) to get the decomposition of δ into characters. Here the jumps of F_f^B for $\theta = 0$ and π come into play. The full details of the calculation can be found in [HC3] and in [K], §XI.3 and in [L].

The final result is as follows, [HC8].

Theorem 1.1. *Let $G = \mathbf{SL}(2, \mathbf{R})$. Then we have, for $f \in C_c^\infty(G)$,*

$$8\pi f(e) = \sum_{m \neq 0} m\theta_m(f) + \int_0^\infty \frac{\lambda}{2} \tanh\left(\frac{\pi\lambda}{2}\right) \theta_{+,i\lambda}(f)\, d\lambda$$
$$+ \int_0^\infty \frac{\lambda}{2} \coth\left(\frac{\pi\lambda}{2}\right) \theta_{-,i\lambda}(f)\, d\lambda.$$

All the characters represent irreducible representations except $\theta_{-,0}$ which is a sum of two irreducible characters.

The argument can be used also to show that each conjugacy invariant eigendistribution of the Casimir operator on G is a linear combination of irreducible characters ([S], [H5]).

For another proof of the theorem in which Rossmann's formula [Ro1] for the orbital integral plays a prominent role, see a beautiful account by Vergne [Ve3]. See also [Ve1] and [Ve2]. Yet another approach is in [SW], [He2], and [HW]. See also Varadarajan's article [V] for a lucid description of Harish-Chandra's original proof.

For further work on the orbital integral problem for semisimple G see e.g. Herb [He1], Bouaziz [Bo], and Shelstad [Sl]. When one views G as $G \times G$/diagonal, the orbital problem seems of considerable interest for a semisimple symmetric space G/H, although it remains to be seen whether it will play a role in harmonic analysis as in the group case. For G/H Lorentzian of constant curvature the inversion problem is solved in [H1], and for G/H of rank one by Orloff [Or].

In [Ha] Harinck investigates these orbital integrals for $G^{\mathbf{C}}/G$ and obtains the Plancherel formula for the corresponding spherical transform.

§2. Analysis on Riemannian Symmetric Spaces

2.1. The Fourier transform

The Fourier transform on a semisimple Lie group G

(1) $$\tilde{f}(\pi) = \int_G f(x)\pi(x)dx, \qquad \pi \in \widehat{G},$$

has a nice Plancherel formula

$$\int_G |f(x)|^2\, dx = \int_{\widehat{G}} \|\tilde{f}(\pi)\|^2\, d\pi,$$

$\|\cdot\|$ being the Hilbert-Schmidt norm. An explicit description of the measure $d\pi$ in terms of the structure of G was given by Harish-Chandra. Range questions for the Fourier transform $f \to \tilde{f}$ have been investigated by many people, primarily Arthur (see [A1], [A2], and references there.)

Classical Fourier analysis in $\mathbf{R}^n$, which originated in the study of the heat equation, certainly has one of its principal applications in the theory of partial differential equations. For semisimple G the Fourier transform is actually a rather unwieldy gadget in that it associates to a function f on G an object $\tilde{f}$ given by a family of operators on different Hilbert spaces. Thus it is not immediately suited for applications to differential equations.

Nevertheless, there has been certain amount of activity in studying invariant differential equations on G. It was proved in [H7] that if D is a bi-invariant differential operator on G then D is locally solvable, i.e., there exists an open neighborhood V of the identity e in G such that for each $f \in \mathcal{D}(V)$ there is a $u \in \mathcal{E}(V)$ satisfying

$$Du = f.$$

Such a result had been proved by Raïs [R] for nilpotent groups, was extended to solvable groups by Rouvière [Ro] and Duflo and Raïs [DR], and was obtained by Duflo [D] for arbitrary Lie groups. The Fourier transform, however, was never involved in the proofs.

On the symmetric space $X = G/K$ (G semisimple with finite center, $K \subset G$ maximal compact) one can however define a Fourier transform [H4] that is scalar-valued and concrete enough to be directly applicable to differential equations.

Consider the Iwasawa decompositions of our semisimple G, namely

$$G = NAK \qquad \text{and} \qquad \mathfrak{g} = \mathfrak{n} + \mathfrak{a} + \mathfrak{k} \tag{2}$$

with N, A, and K nilpotent, abelian, and compact, respectively, and with $\mathfrak{n}$, $\mathfrak{a}$, and $\mathfrak{k}$ their respective Lie algebras. Let θ be the Cartan involution of G with fixed group K. Let $\sum$ denote the set of roots for $(\mathfrak{g}, \mathfrak{a})$; for $\alpha \in \sum$ let $\mathfrak{g}_\alpha$ denote the corresponding root space and m_α its dimension. Then $\mathfrak{n} = \bigoplus_{\alpha \in \sum^+} \mathfrak{g}_\alpha$, where $\sum^+$ is the set of positive roots. Let $\mathfrak{a}^+$ denote the Weyl chamber where the $\alpha \in \sum^+$ are positive. Let $\mathfrak{a}^*$ (resp. $\mathfrak{a}_c^*$) denote the space of $\mathbf{R}$-linear maps $\mathfrak{a} \to \mathbf{R}$ (resp. $\mathfrak{a}_c^* \to \mathbf{C}$), and let $\rho \in \mathfrak{a}^*$ be given by $2\rho = \mathrm{Tr}((ad\, H)|\mathfrak{n})$, $H \in \mathfrak{a}$. Let M denote the centralizer of A in K and put $B = K/M$. We define $A(g) \in \mathfrak{a}$ in terms of (2) by $g = n \exp A(g) k$, and we define the vector-valued "inner product"

$$A : X \times B \to \mathfrak{a}$$

by

$$A(gK, kM) = A(k^{-1}g).$$

The **Fourier transform** $f \to \tilde{f}$ on X is then defined [H4] by

$$\tilde{f}(\lambda, b) = \int_X f(x) e^{(-i\lambda+\rho)(A(x,b))}\, dx \tag{3}$$

for all $\lambda \in \mathfrak{a}_c^*$, $b \in B$ for which the integral converges. Note that in contrast to (1) $\tilde{f}(\lambda, b)$ is scalar-valued.

The basic results for this Fourier transform are ([H4], [H6], 1970):

a) Inversion. *If $f \in \mathcal{D}(X)$ then*

$$f(x) = \frac{1}{w} \int_{\mathfrak{a}^* \times B} \tilde{f}(\lambda, b) e^{(i\lambda+\rho)(A(x,b))}\, d\mu(\lambda, b),$$

where w is the order of the Weyl group $W = W(\mathfrak{g}, \mathfrak{a})$ and $d\mu = |c(\lambda)|^{-2}\, d\lambda\, db$.

Here $d\lambda$ and db are suitably normalized invariant measures on $\mathfrak{a}^*$ and B, respectively, and $c(\lambda)$ the Harish-Chandra c-function given by the following integral over $\overline{N} = \theta N$:

$$c(\lambda) = \int_{\overline{N}} e^{(i\lambda+\rho)(A(\bar{n}))}\, d\bar{n}.$$

This can be evaluated in terms of Γ functions ([GK]).

b) Plancherel Theorem. *The map* $f \to \tilde{f}$ *extends to an isometry of* $L^2(X)$ *onto* $L^2(\mathfrak{a}_+^* \times B, d\mu)$, $\mathfrak{a}_+^*$ *being the positive Weyl chamber in* $\mathfrak{a}^*$.

If f is K-invariant, i.e., if $f(k \cdot x) \equiv f(x)$, these results reduce to the principal results in Harish-Chandra's theory of spherical functions [HC5]. However, for applications to analysis on X, the K-invariance condition is of course too restrictive.

2.2. Applications

As is very familiar to analysts the characterization of the Fourier transform space $\mathcal{D}(\mathbf{R}^n)$ as the space of entire functions of exponential type (the Paley-Wiener theorem) is an important tool in partial differential equation theory. The analog for X is the following [H7].

c) Paley-Wiener type theorem. *The range* $\mathcal{D}(X)^\sim$ *is identical to the space*

$$\left\{ \varphi \in \mathcal{E}(\mathfrak{a}^* \times B) : \begin{array}{ll} \text{(i)} & \lambda \to \varphi(\lambda, b) \text{ is holomorphic in } \mathfrak{a}_c^* \\ & \text{of exponential type uniformly in } b. \\ \text{(ii)} & \int_B \varphi(\lambda, b) e^{(i\lambda+\rho)(A(x,b))}\, db \text{ is } W\text{-invariant in } \lambda. \end{array} \right\}$$

The first application is the following existence theorem for members D of the algebra $\mathbf{D}(G/K)$ of differential operators on $X = G/K$ that are invariant under the action of G.

Theorem 2.1. *Each nonzero* $D \in \mathbf{D}(G/K)$ *is surjective on* $\mathcal{E}(X)$, *i.e.*,

$$D\mathcal{E}(X) = \mathcal{E}(X). \tag{4}$$

The first step is to get a fundamental solution for D, i.e., a distribution J such that $DJ = \delta$. This can be done by means of (c) in the K-invariant case, which then can be extended to distributions.

Once the existence of J has been established one can by fairly general functional analysis methods reduce the problem to the proof of the following implication, V being the closure of a ball $B_R(o)$ in X (R arbitrary):

$$f \in \mathcal{D}(X), \qquad \operatorname{supp}(Df) \subset V \Longrightarrow \operatorname{supp}(f) \subset V. \tag{5}$$

This is an easy consequence of (c). In fact the functions $x \to e^{\nu(A(x,b))}$ are eigenfunctions of each $D \in \mathbf{D}(G/K)$, and

$$(Df)^\sim(\lambda, b) = p(\lambda)\tilde{f}(\lambda, b),$$

where $p(\lambda)$ is a polynomial. The conditions $f \in \mathcal{D}(X)$ and $\operatorname{supp}(Df) \subset V$ imply that $p(\lambda)\tilde{f}(\lambda, b)$ is entire and of exponential type $\leq R$. From complex variable theory this implies that $\tilde{f}(\lambda, b)$ has exponential type $\leq R$. Hence $\operatorname{supp}(f) \subset V$, verifying (5).

The Paley-Wiener theorem in (c) asserts that $f \to \tilde{f}$ is a bijection between two function spaces. However, although $\mathcal{D}(X)$ has a natural topology we have not introduced any natural topology on the other side. Using the induced topology of $\mathcal{E}(X \times \mathfrak{a}^*)$ would not make $f \to \tilde{f}$ a homeomorphism.

This can be remedied by specializing the theorem to the subspace of $\mathcal{D}(X)$ consisting of functions of a given K-type. For $\delta \in \widehat{K}$ acting on V_δ with V_δ^M the fixed space under $\delta(M)$, consider the subspace

$$\mathcal{D}_{\check{\delta}}(X) \subset \mathcal{D}(X) \qquad (\check{\delta} = \text{contragredient to } \delta) \tag{6}$$

consisting of $f \in \mathcal{D}(X)$ of K-type $\check{\delta}$. Then

$$\tilde{f}(\lambda, kM) = \mathrm{Tr}\left(\delta(k)\tilde{f}(\lambda)\right),$$

where $\tilde{f}(\lambda)$ is the vector-valued Fourier transform

$$\tilde{f}(\lambda) = d(\delta) \int_X f(x)\Phi_{\bar{\lambda},\delta}(x)^* \, dx \tag{7}$$

with $\Phi_{\lambda,\delta}$ the **generalized spherical function** (or **Eisenstein integral**)

$$\Phi_{\lambda,\delta}(x) = \int_K e^{(i\lambda+\rho)(A(x,kM))}\delta(k)\, dk. \tag{8}$$

In (7) $*$ refers to adjoint and $\bar{\lambda}$ is the complex conjugate of the $\mathbf{R}$-linear function λ. For $a \in A$, $\Phi_{\lambda,\delta}(a)$ is a linear transformation of V_δ^M into itself. If we expand $\Phi_{\bar{\lambda},\delta}(a)^*$ near $a = e$ using the simple roots $\alpha_1, \dots, \alpha_l$, we have

$$\Phi_{\bar{\lambda},\delta}(a)^* = \sum_n P_{(n)}(\lambda)\alpha_1(\log a)^{n_1} \cdots \alpha_\ell(\log a)^{n_\ell},$$

and each $P_{(n)}(\lambda)$ is divisible by a certain polynomial matrix $Q^{\check{\delta}}(\lambda)$. Then we have the following result ([H6], II).

Theorem 2.2. *The Eisenstein integral satisfies the following functional equations:*

$$Q^{\check{\delta}}(\lambda)^{-1}\Phi_{\bar{\lambda},\delta}(a)^* \qquad \text{is } W\text{-invariant in } \lambda. \tag{9}$$

From this one can deduce a stronger topological version of (c). Let $H(\mathfrak{a}_c^*, \mathrm{Hom}(V_\delta, V_\delta^M))$ denote the space of holomorphic functions on $\mathfrak{a}_c^*$ of exponential type with values in $\mathrm{Hom}(V_\delta, V_\delta^M)$, let $\mathcal{I}^\delta(\mathfrak{a}^*)$ be the subspace of W-invariants, and let $\mathcal{H}^\delta(\mathfrak{a}^*)$ be the subspace $Q^{\check{\delta}}\mathcal{I}^\delta(\mathfrak{a}^*)$ with the induced topology.

Theorem 2.3. *The Fourier transform $f \to \tilde{f}$ given by* (7) *is a homeomorphism of $\mathcal{D}_{\check{\delta}}(X)$ onto $Q^{\check{\delta}}\mathcal{I}^\delta(\mathfrak{a}^*)$.*

Corollary 2.4. *The K-finite joint eigenfunctions of $\mathbf{D}(G/K)$ are precisely the integrals*

$$f(x) = \int_B e^{(i\lambda+\rho)(A(x,b))}F(b)\, db, \tag{10}$$

where F is a K-finite continuous function on B.

Sketch of proof of the corollary ([H6], II). Fix $\delta \in \widehat{K}$ and let f be a joint eigenfunction of $\mathbf{D}(G/K)$. Consider the Harish-Chandra isomorphism $\Gamma : \mathbf{D}(G/K) \to I(\mathfrak{a})$, where $I(\mathfrak{a})$ denotes the set of W-invariants in $S(\mathfrak{a})$. Then

$$Df = \Gamma(D)(i\lambda)f \tag{11}$$

for some $\lambda \in \mathfrak{a}_c^*$. Let $\mathcal{E}_\delta(B)$ be the space of K-finite continuous functions on B of type δ, $\mathcal{E}_\lambda(X)$ the space of all $f \in \mathcal{E}(X)$ satisfying (11), and $\mathcal{E}_{\lambda,\delta}(X)$ the space of K-finite elements in $\mathcal{E}_\lambda(X)$ of type δ. The **Poisson transform**

$$\mathcal{P}_\lambda : F(b) \to f(x) = \int_B e^{(i\lambda+\rho)(A(x,b))}F(b)\, db \tag{12}$$

maps $\mathcal{E}(B)$ into $\mathcal{E}_\lambda(X)$ and $\mathcal{E}_\delta(B)$ into $\mathcal{E}_{\lambda,\delta}(X)$. For at least one $s \in W$, $\mathcal{P}_{s\lambda}$ is injective (see Theorem 3.3 below). Thus since $\mathcal{E}_{s\lambda,\delta}(X) = \mathcal{E}_{\lambda,\delta}(X)$ for $s \in W$ we may assume $\mathcal{P}_\lambda$ injective. Then

$$\dim \mathcal{E}_{\lambda,\delta}(X) \geq \dim \mathcal{E}_\delta(B) = \dim V_\delta \dim V_\delta^M. \tag{13}$$

The corollary will be concluded by proving the converse inequality. For this let $h \in \mathcal{E}_{\lambda,\delta}(X)$. We view h as a distribution on X. Define its Fourier transform $\tilde{h}$ as a linear form on $\mathcal{H}^\delta(\mathfrak{a}^*)$ by

$$\tilde{h}(\tilde{f}) = h(f) = \int_X h(x)f(x)\,dx, \qquad f \in \mathcal{D}_{\check{\delta}}(X).$$

Then the map $\psi \in \mathcal{I}^\delta(\mathfrak{a}^*) \to \tilde{h}(Q^{\check{\delta}}\psi))$ is a continuous linear functional. Thus by Theorem 2.3 (for $\delta = I$) this map is given by the Fourier transform of a K-invariant distribution j on X, i.e.,

$$\tilde{j}(\widetilde{\varphi}) = \tilde{h}(Q^{\check{\delta}}\widetilde{\varphi}) = j(\varphi) \tag{14}$$

for all K-invariant φ in $\mathcal{D}(X, \operatorname{Hom}(V_\delta, V_\delta^M))$. Put $p_D(\mu) = \Gamma(D)(-i\mu)$; then formula (11) implies easily

$$p_D\tilde{h} = p_D(-\lambda)\tilde{h}.$$

Combining this with (14), one proves $p_D\tilde{j} = p_D(-\lambda)\tilde{j}$, which in turn implies $Dj = \Gamma(D)(i\lambda)j$. But j is also K-invariant so $j = \varphi_\lambda A$ where φ_λ is the zonal spherical function ((8) for $\delta = I$) and $A \in \operatorname{Hom}(V_\delta, V_\delta^M)$. This proves the converse of (13) and the corollary.

An alternative proof was later found by Ban and Schlichtkrull [BS].

2.3. Multitemporal wave equations

We now explain another application of Theorem 2.3, namely to the system of differential equations

$$Du = \partial(\Gamma(D))u, \qquad D \in \mathbf{D}(G/K). \tag{15}$$

Here u is in $\mathcal{E}(X \times \mathfrak{a})$, D operates on the first argument, and $\partial(p)$ is the constant coefficient operator on $\mathfrak{a}$ corresponding to $p \in S(\mathfrak{a})$.

The following result is easily established.

Proposition 2.5. *Let $f \in C^2(A)$ and put*

$$u(x,H) = f(\exp(A(x,b) + H))e^{-\rho(H)}, \qquad b \in B.$$

Then u is a solution to (15).

Next we impose initial conditions on the system (15). We choose a real homogeneous basis $p_1 = 1, p_2, \ldots, p_w$ of the W-harmonic polynomials on $\mathfrak{a}$. Given $f_1, \ldots, f_w \in \mathcal{D}(X)$ we impose the initial conditions

$$(\partial(p_i)u)(x,0) = f_i(x), \qquad 1 \leq i \leq w, \tag{16}$$

on the solution to (15). The system (15)–(16) was first considered by [STS] and then by [Sh] and [PS].

If G/K has rank 1, the Laplacian L_X on X generates $\mathbf{D}(G/K)$, and $\Gamma(L_X) = L_A - |\rho|^2$ where L_A is the Laplacian on the 1-dimensional space $\mathfrak{a}$. Also $p_1 = 1$ and $p_2 = H$ so that the system (15)–(16) reduces to the shifted wave equation

$$(17) \qquad (L_X + |\rho|^2)u = \frac{\partial^2 u}{\partial t^2}, \qquad u(x,0) = f_1(x), \qquad u_t(x,0) = f_2(x).$$

On the quotient field $C(S(\mathfrak{a}))$ we consider the bilinear form

$$(a,b) = \sum_{\sigma \in W} a^\sigma b^\sigma,$$

which has values in $C(I(\mathfrak{a}))$. We determine $q^j \in C(S(\mathfrak{a}))$ by $(q^j, p_i) = \delta_{ij}$. If π denotes the product of the positive indivisible roots, it is known from [HC5] I, §3 that $\pi q^j \in S(\mathfrak{a})$. Again let θ denote the Cartan involution. The matrix $A = (A_{ij})$, where

$$(18) \qquad A_{ij} = \left(\pi q^j, \theta(\pi q^i)\right), \qquad 1 \leq i,j \leq w,$$

has entries in $I(\mathfrak{a})$ so that we can consider the matrix $\mathcal{A} = (\mathcal{A}_{ij})$ with entries in $\mathbf{D}(G/K)$ such that $\Gamma(\mathcal{A}_{ij}) = A_{ij}$. Given $u, v \in \mathcal{E}(X \times \mathfrak{a})$ define the column vectors μ and ν by

$$(19) \qquad \mu_i(x,H) = (\partial(p_i)u)(x,H), \qquad 1 \leq i \leq w,$$

$$(20) \qquad \nu_i(x,H) = (\partial(p_i)v)(x,H), \qquad 1 \leq i \leq w.$$

With ${}^t\mu$ denoting the transpose of μ, the **energy** is defined by

$$(21) \qquad E(u,v;H) = \int_X ({}^t\mu \mathcal{A} \bar{\nu})(x,H)\,dx$$

whenever the integral converges. For the special case (17) this reduces to the usual energy

$$E(u,u;0) = c\int_X \left(-f_1(L_X + |\rho|^2)\bar{f}_1 + |f_2|^2\right)\,dx.$$

As proved by Shahshahani [Sh], if u is a solution to (15)–(16) then $E(u,u;H)$ is independent of H. This disagrees with Proposition 3, §1.1 in [STS], where the same statement is made with a different definition of the energy, namely with θ missing in (18).

Let F denote the row vector

$$F(x) = (f_1(x), \ldots, f_w(x)),$$

and for each $\sigma \in W$, $\lambda \in \mathfrak{a}_c^*$, $b \in B$ let μ^σ be the column vector with components

$$\mu_j^\sigma(x,H;\lambda,b) = \partial(p_j)_H(e^{i\lambda(H)+(i\sigma\lambda+\rho)(A(x,b))}).$$

We consider then the linear map (modifying (5) in [STS])

$$\mathcal{E}^\sigma : F(x) \to \int_X (F\mathcal{A}\bar{\mu}^\sigma)(x,0;\lambda,b)\,dx,$$

which maps $\mathcal{D}(X) \times \ldots \times \mathcal{D}(X)$ (w times) into a function space on $\mathfrak{a}_c^* \times B$.

Theorem 2.6. *For each $\sigma \in W$ the map $\mathcal{E}^\sigma$ is an injective norm-preserving map of $\mathcal{D}(X) \times \ldots \times \mathcal{D}(X)$ onto a dense subspace of $L^2(\mathfrak{a}^* \times B, d\lambda db / |\pi(\lambda)c(\lambda)|^2)$. Thus*

$$\int_{\mathfrak{a}^* \times B} \mathcal{E}^\sigma(F)\overline{\mathcal{E}^\sigma(F)}(\lambda, b) \frac{d\lambda\, db}{|\pi(\lambda)c(\lambda)|^2} = \int_X (F\mathcal{A}\, {}^t\overline{F})(x)\, dx.$$

For $\sigma = e$ a rather complicated proof of this was given in [Sh]. Our more general result is based on Theorem 2.3 and the following new identity, which relates $\mathcal{E}^\sigma$ to the Fourier transform $f \to \tilde{f}$.

Theorem 2.7. *For each $\sigma \in W$,*

$$\mathcal{E}^\sigma(F)(\lambda, b) = \pi(\lambda)^2 \sum_{k=1}^{w} q^k(i\lambda)\tilde{f}_k(\sigma\lambda, b).$$

For $H \in \mathfrak{a}$ let U_H denote the operator

$$F(x) \to {}^t\mu(x, H),$$

with ${}^t\mu$ as in (21). Then the translation invariance of (15) implies that U_{H_o} maps ${}^t\mu(x, H)$ to ${}^t\mu(x, H_0 + H)$. The mapping $\mathcal{E}^\sigma$ in Theorem 2.7 is then easily shown to have the following property.

Theorem 2.8. *For $H \in \mathfrak{a}$ let $e(H)$ denote the endomorphism*

$$e(H) : \varphi(\lambda, b) \to e^{i\lambda(H)}\varphi(\lambda, b) \quad \text{of} \quad L^2\big(\mathfrak{a}^* \times B, d\lambda\, db / |\pi(\lambda)c(\lambda)|^2\big).$$

Then

$$\mathcal{E}^\sigma \circ U_{H_o} = e(H_o)\mathcal{E}^\sigma, \qquad H_o \in \mathfrak{a}.$$

This means that the wave motion $u(x, H) \to u(x, H + H_o)$ corresponds under $\mathcal{E}^\sigma$ to the simple map $e(H_o)$.

§3. Eigenspace Representations

3.1. Generalities

Spherical harmonics are by definition the eigenfunctions on the unit sphere $\mathbf{S}^{n-1}$ of the Laplacian $L = L_{\mathbf{S}^{n-1}}$. The name comes from the fact that these eigenfunctions are precisely the restrictions to $\mathbf{S}^{n-1}$ of homogeneous harmonic polynomials on $\mathbf{R}^n$. If $c \in \mathbf{C}$ the eigenspace

$$E_c = \left\{f \in \mathcal{E}\left(\mathbf{S}^{n-1}\right) : L_{\mathbf{S}^{n-1}} f = cf\right\} \tag{1}$$

is invariant under each orthogonal transformation of $\mathbf{S}^{n-1}$. This gives a representation of $\mathbf{O}(n)$ on E_c. The space E_c is $\neq 0$ if and only if

$$c = -k(k + n - 2) \qquad \text{for some } k \in \mathbf{Z}^+$$

and $\mathbf{O}(n)$ acts irreducibly on E_c. Also, since $\mathbf{S}^{n-1}$ is two-point homogeneous under $\mathbf{O}(n)$, the only differential operators on $\mathbf{S}^{n-1}$ that are invariant under $\mathbf{O}(n)$ are the polynomials in $L_{\mathbf{S}^{n-1}}$ ([H1]).

This example motivates the definition of a fairly general class of representations that I called **eigenspace representations** in [H6]. Given a Lie group L and a closed subgroup H, let $\mathbf{D}(L/H)$ denote the algebra of differential operators on L/H

that are invariant under L. Given a homomorphism $\chi : \mathbf{D}(L/H) \to \mathbf{C}$, consider the joint eigenspace

$$E_\chi = \{f \in \mathcal{E}(L/H) \ : \ Df = \chi(D)f \ \text{ for } D \in \mathbf{D}(L/H)\} \tag{2}$$

with the topology induced by that of $\mathcal{E}(L/H)$. Let T_χ denote the natural representation of L on E_χ, i.e., $(T_\chi(\ell)f)(xH) = f(\ell^{-1}xH)$.

If $\mathbf{D}(L/H)$ is not commutative it would be natural to replace it by a commutative subalgebra in order to have a rich supply of joint eigenfunctions. It might also be natural to pass from eigenfunctions in (2) to eigendistributions. Further natural generalization is obtained by replacing functions in (2) by sections of vector bundles.

Coming back to T_χ above, we are led naturally to the following problem:

For which χ is T_χ irreducible and what are the representations of L so obtained?

Note that in our setup there is no Hilbert space in sight; in particular there is no particular emphasis on unitary representations.

Example. Consider $\mathbf{R}^n$ as the quotient space $\mathbf{M}(n)/\mathbf{O}(n)$ where $\mathbf{M}(n)$ is the group of isometries of $\mathbf{R}^n$. Here $\mathbf{D}(\mathbf{M}(n)/\mathbf{O}(n))$ consists of the polynomials in the Laplacian $L = L_{\mathbf{R}^n}$. In this case our problem has a simple solution ([H8]). Given $\lambda \in \mathbf{C}$ consider the eigenspace

$$\mathcal{E}_\lambda(\mathbf{R}^n) = \left\{f \in \mathcal{E}(\mathbf{R}^n) : Lf = -\lambda^2 f\right\}, \tag{3}$$

and let T_λ denote the corresponding eigenspace representation.

Theorem 3.1. *T_λ is irreducible if and only if $\lambda \neq 0$.*

PROOF (SKETCH). If $\lambda = 0$ the space of harmonic polynomials of degree $\leq k$ is a closed invariant subspace for each k. For the converse let $\lambda \neq 0$ and consider the Poisson transform

$$\mathcal{P}_\lambda : F \in \mathcal{E}\left(\mathbf{S}^{n-1}\right) \to f \in \mathcal{E}_\lambda\left(\mathbf{R}^n\right) \tag{4}$$

given by

$$f(x) = \int_{\mathbf{S}^{n-1}} e^{i\lambda(x,\omega)} F(\omega)\, d\omega.$$

This mapping commutes with the $\mathbf{O}(n)$ action and is injective for $\lambda \neq 0$. Using PDE techniques one can prove ([H6] '70) that there exists a sphere $S = S_r(0)$ (r depending on λ) such that the restriction map

$$f \in \mathcal{E}_\lambda\left(\mathbf{R}^n\right) \to f|S \tag{5}$$

is injective. Let the subscript δ refer to the spaces of $\mathbf{O}(n)$-finite functions of type δ. Then from the injectivity of the maps (4) and (5) we deduce

$$\dim\left(\mathcal{E}(S^{n-1})_\delta\right) \leq \dim\left(\mathcal{E}_\lambda(\mathbf{R}^n)_\delta\right) \leq \dim\left(\mathcal{E}(S)_\delta\right). \tag{6}$$

The extremes having the same dimension, equality holds in (6) so

$$\mathcal{E}_\lambda\left(\mathbf{R}^n\right)_\delta = \mathcal{P}_\lambda\left(\mathcal{E}(\mathbf{S}^{n-1})_\delta\right). \tag{7}$$

To conclude the proof, consider the Hilbert space

$$\mathcal{H}_\lambda = \left\{f(x) = \int_{\mathbf{S}^{n-1}} e^{i\lambda(x,\omega)} F(\omega)\, d\omega \ : \ F \in L^2(S^{n-1})\right\},$$

the norm of f taken as the L^2 norm of F. Expanding $f \in \mathcal{E}_\lambda(\mathbf{R}^n)$ according to its δ-components, $f = \Sigma_\delta f_\delta$ with $f_\delta \in \mathcal{E}_\lambda(\mathbf{R}^n)_\delta$, we see that $\mathcal{H}_\lambda$ is dense in $\mathcal{E}_\lambda(\mathbf{R}^n)$. The action of $\mathbf{M}(n)$ on $\mathcal{H}_\lambda$ is easy to analyze and irreducibility follows quickly. Using the density of $\mathcal{H}_\lambda$ in $\mathcal{E}_\lambda(\mathbf{R}^n)$, we easily deduce the irreducibility of T_λ.

3.2. The symmetric space case

Consider now the case of a symmetric space G/K of the noncompact type. We adopt the notation from §2 and let Σ_0 denote the set of indivisible roots. With Γ as in §2 consider for each $\lambda \in \mathfrak{a}_c^*$ the joint eigenspace

$$\mathcal{E}_\lambda(X) = \{f \in \mathcal{E}(X) \;:\; Df = \Gamma(D)(i\lambda)f \quad \text{for } D \in \mathbf{D}(G/K)\}.$$

Each joint eigenspace is of this form for some $\lambda \in \mathfrak{a}_c^*$, and $\mathcal{E}_{s\lambda}(X) = \mathcal{E}_\lambda(X)$ for each $s \in W$. Let T_λ denote the eigenspace representation of G on $\mathcal{E}_\lambda(X)$. When is it irreducible?

Consider the product

$$\Gamma_X(\lambda) = \prod_{\alpha \in \Sigma_0} \Gamma\left(\tfrac{1}{2}\left(\tfrac{1}{2}m_\alpha + 1 + \langle i\lambda, \alpha_0\rangle\right)\right)\Gamma\left(\tfrac{1}{2}\left(\tfrac{1}{2}m_\alpha + m_{2\alpha} + \langle i\lambda, \alpha_0\rangle\right)\right)$$

"the Gamma function of X" where α_0 stands for $\alpha/\langle\alpha,\alpha\rangle$. This $\Gamma_X(\lambda)$ is the denominator in the Gindikin-Karpelevič formula for $c(\lambda)c(-\lambda)$. The counterpart to Theorem 3.1 for the symmetric space G/K is the following result ([H6]).

Theorem 3.2. *Let $\lambda \in \mathfrak{a}_c^*$. Then T_λ is irreducible if and only if*

$$\frac{1}{\Gamma_X(\lambda)} \neq 0.$$

The proof involves a study of the Poisson transform

$$(\mathcal{P}_\lambda F)(x) = \int_B e^{(i\lambda+\rho)(A(x,b))} F(b)\, db.$$

Here the principal property is the following result ([H6], I and II).

Theorem 3.3. *Let $\lambda \in \mathfrak{a}_c^*$. Then*

$$\mathcal{P}_\lambda \text{ is injective} \Longleftrightarrow \frac{1}{\Gamma_X^+(\lambda)} \neq 0.$$

Here $\Gamma_X^+(\lambda)$ equals $\Gamma_X(\lambda)$ except that the product ranges only over $\Sigma_0^+ = \Sigma_0 \cap \Sigma^+$. We call λ **simple** if $\mathcal{P}_\lambda$ is injective. Thus Theorem 3.2 states that T_λ is irreducible if both λ and $-\lambda$ are simple. Here is a sketch of the proof, using Corollary 2.4.

Suppose first both λ and $-\lambda$ are simple. Let $V \subset \mathcal{E}_\lambda(X)$ be a closed invariant subspace. Then clearly $\varphi_\lambda \in V$. Since $-\lambda$ is simple the functions

$$b \to \sum_j a_j e^{(-i\lambda+\rho)(A(g_j\cdot 0,b))}, \qquad a_j \in \mathbf{C},\; g_j \in G,$$

form a dense subspace of $L^2(B)$. On the other hand φ_λ has the following symmetry property ([H6], I, p. 116):

$$\varphi_\lambda(g^{-1}\cdot x) = \int_B e^{(i\lambda+\rho)(A(x,b))} e^{(-i\lambda+\rho)(A(g\cdot 0,b))}\, db.$$

Thus we can conclude that V contains the space $\mathcal{H}_\lambda$ of functions

$$f(x) = \int_B e^{(i\lambda+\rho)(A(x,b))} F(b)\,db, \qquad F \in L^2(B).$$

On the other hand the proof of Corollary 2.4 showed that

$$\mathcal{E}_{\lambda,\delta}(X) = \mathcal{P}_\lambda(\mathcal{E}(B)_\delta)$$

so we conclude that V contains each $\mathcal{E}_{\lambda,\delta}(X)$. However, each $f \in \mathcal{E}(X)$ can be expanded in a Fourier series according to K,

$$f = \sum_{\delta \in \widehat{K}} f_\delta,$$

and then $f_\delta \in \mathcal{E}_{\lambda,\delta}(X)$. Consequently V is dense in $\mathcal{E}_\lambda(X)$ whence $V = \mathcal{E}_\lambda(X)$, proving the irreducibility. The converse, T_λ irreducible $\Rightarrow$ λ and $-\lambda$ simple, involves similar ideas.

For semisimple symmetric spaces G/H (where H is the fixed point group of any involution) the eigenspace representations have been relatively little studied. However for G/H isotropic the irreducibility question was completely answered by Schlichtkrull (by classification) and the composition series determined for each case ([Sc]). For the real hyperbolic spaces related results had been obtained by Rossmann ([Ro2]).

3.3. The principal series

Next I shall discuss the principal series from the point of view of eigenspace representations.

Given $\delta \in \widehat{M}$ operating on the vector space V_δ and $\lambda \in \mathfrak{a}_c^*$, consider the space $\Gamma_{\delta,\lambda}$ of smooth functions $f : G \to V_\delta$ satisfying

$$f(gman) \equiv \delta(m^{-1}) e^{(i\lambda-\rho)(\log a)} f(g). \tag{8}$$

Let $\tau_{\delta,\lambda}$ denote the representation of G on $\Gamma_{\delta,\lambda}$ given by

$$(\tau_{\delta,\lambda}(g_1)f)(g_2) = f(g_1^{-1}g_2). \tag{9}$$

This family of representations is called the **principal series**.

Let us first consider the case when G is complex. In this case MA is a Cartan subgroup H of G and $\dim V_\delta = 1$. Let $\mathbf{D}(H)$ (respectively, $\mathbf{D}(G/N)$) denote the algebras of left invariant differential operators on H (respectively G-invariant differential operators on G/N). Given $U \in \mathbf{D}(H)$, we can define the differential operator D_U on G/N by

$$(D_U f)(gN) = \{U_h(f(ghN))\}_{h=e}\,. \tag{10}$$

The operator D_U is well defined since $hNh^{-1} \subset N$, and it is clearly a G-invariant differential operator on G/N. We now have the following result relating these algebras ([H3], [H6], I).

Theorem 3.4. *The mapping* $U \to D_U$ *is an isomorphism of* $\mathbf{D}(H)$ *onto* $\mathbf{D}(G/N)$. *In particular,* $\mathbf{D}(G/N)$ *is commutative.*

As a consequence the joint eigenspaces are precisely the spaces

$$E_\omega = \{f \in \mathcal{E}(G/N) : f(ghN) \equiv \omega(h)f(gN)\} \tag{11}$$

as ω runs through the C^∞ characters of H. Comparing with (8) we therefore conclude

Corollary 3.5 (G complex). *The principal series representations of* G *are precisely the eigenspace representations for* G/N.

In order to treat real G in the same spirit it is convenient to use the familiar connection between induced representations and vector bundles. Consider the representation $\delta \otimes 1$ of MN, which defines a vector bundle

$$G \times_{MN} V_\delta .$$

The sections of this bundle are the maps $F : G \to V_\delta$ satisfying

$$F(gmn) \equiv \delta(m^{-1})F(g). \tag{12}$$

Let $\mathbf{D}(A)$ (respectively, $\mathbf{D}(G/MN)$) denote the algebras of left invariant differential operators on A (respectively G-invariant differential operators on G/MN). Given $U \in \mathbf{D}(A)$ we define the G-invariant differential operator D_U on G/MN by

$$(D_U f)(gMN) = \{U_a(f(gaMN))\}_{a=e} .$$

In analogy with Theorem 3.4 we have ([H3]):

Proposition 3.6. *The mapping* $U \to D_U$ *is an isomorphism of* $\mathbf{D}(A)$ *onto* $\mathbf{D}(G/MN)$.

The operator D_U operates also on the sections (12) of the bundle $G \times_{MN} V_\delta$ by

$$(D_U F)(gmn) = \{U_a\left(F(gamn)\right)\}_{a=e} .$$

Then in fact

$$(D_U F)(gmn) = \delta(m^{-1})(D_U F)(g)$$

and the functions (8) are the joint eigenfunctions of $\mathbf{D}(G/MN)$. This proves

Theorem 3.7. *Let* $\delta \in \widehat{M}$. *Then the principal series representations* $\tau_{\lambda,\delta}$ *for* $\lambda \in \mathfrak{a}_c^*$ *are the eigenspace representations for the algebra* $\mathbf{D}(G/MN)$ *acting on the sections of the bundle* $G \times_{MN} V_\delta$.

3.4. The discrete series

In the case when rank $G =$ rank K, G has a discrete series ([HC6]). This has been displayed in several models, but Hotta's realization of the discrete series (or most of it) fits best in the above framework ([Ho]). Schmid's earlier model ([Sm]) is set up in similar spirit.

Consider the irreducible representation δ_λ of K on V_λ with lowest weight $\lambda + 2\rho_k$ ($2\rho_k =$ sum of the positive compact roots) relative to a compact Cartan subalgebra $\mathfrak{h} \subset \mathfrak{k}$. Consider the corresponding vector bundle $G \times_K V_\lambda$. Then the Casimir operator Ω is an invariant differential operator on this bundle. Let Γ_{δ_λ} denote the space of corresponding smooth square integrable sections and put

$$\mathcal{E}_\lambda = \{f \in \Gamma_{\delta_\lambda} : \Omega f = (\langle \lambda + \rho, \lambda + \rho \rangle - \langle \rho, \rho \rangle) f\}.$$

Assume $\langle \lambda + \rho, \alpha \rangle < 0$ for all roots $\alpha > 0$. Then there exists a constant $a > 0$ such that if $|\langle \lambda + \rho, \beta \rangle| > a$ for all noncompact roots β then $\mathcal{E}_\lambda$ realizes the discrete series representation whose character on a maximal torus T of K is given by

$$\varepsilon \frac{\sum_{s \in W_G} (\det s) e^{s(\lambda+\rho)}}{\prod_{\alpha \in P} \left(e^{\frac{\alpha}{2}} - e^{-\frac{\alpha}{2}}\right)}.$$

Here $W_G = \widetilde{T}/T$, where $\widetilde{T}$ is the normalizer of T in G, P is the set of positive roots for (G, T), and ε is a power of (-1).

Thus the discrete series arises as eigenspace representations for the algebra generated by Ω.

3.5. G/H with G nilpotent

Here the eigenspace representations have been analyzed rather completely by Stetkaer and Jacobsen. This goes beyond the Kirillov theory in that the results are not restricted to unitary representations. They have also extended this to some solvable groups.

References to this work and other results on eigenspace representations can be found in [H9], especially Chapter VI.

Acknowledgement. I am indebted to Erik van den Ban for very useful comments and to Toby Bailey and particularly Tony Knapp for help in the preparation of the manuscript.

References

[A1] J. Arthur, A Paley-Wiener theorem for real reductive groups, *Acta Math.* **150** (1983), 1–89.

[A2] J. Arthur, *Harmonic analysis of the Schwartz space on a reductive Lie group I, II*, Mathematical Surveys and Monographs, American Mathematical Society (to appear).

[BS] E. P. van den Ban and H. Schlichtkrull, Asymptotic expansions and boundary values of eigenfunctions on a Riemannian symmetric space, *J. Reine Angew. Math.* **380** (1987), 108–165.

[Bo] A. Bouaziz, Intégrales orbitales surles groupes de Lie réductifs, *Annales Scient. Ecole Norm. Sup.* **27** (1994), 573–609.

[D] M. Duflo, Opérateurs differentiels bi-invariant sur un groupe de Lie, *Annales Scient. Ecole Norm. Sup.* **10** (1977), 265–288.

[DR] M. Duflo and M. Raïs, Sur l'analyse harmonique sur les groupes de Lie résolubles, *Annales Scient. Ecole Norm. Sup.* **9** (1976), 107–144.

[DV] M. Duflo and M. Vergne, La formule de Plancherel des groupes de Lie semisimples réels, *Advanced Studies in Pure Math.* **14** (1988), 289–336.

[GG1] I. M. Gelfand and M. I. Graev, Analog of the Plancherel formula for the classical groups, *Trudy Moscow Mat. Obšč.* **4** (1955), 375–404.

[GG2] I. M. Gelfand and M. I. Graev, The analog of Plancherel's theorem for real unimodular groups, *Dokl. Akad. Nauk USSR* **92** (1953), 461–464.

[GN] I. M. Gelfand and M. A. Naimark, An analog of Plancherel's formula for the complex unimodular group, *Dokl. Akad. Nauk USSR* **63** (1948), 609–612.

[GK] S. G. Gindikin and F. I. Karpelevič, Plancherel measure of Riemannian symmetric spaces of non-positive curvature, *Dokl. Akad. Nauk. USSR* **145** (1962), 252–255.

[Ha] P. Harinck, Inversion des intégrales orbitales et formule de Plancherel pour $G_{\mathbf{C}}/G_{\mathbf{R}}$, *C. R. Acad. Sci. Paris* **320** (1995), 1295–1298.

[HC1] Harish-Chandra, Representations of semisimple Lie groups III, *Proc. Nat. Acad. Sci. USA* **37** (1951), 366–369.

[HC2] Harish-Chandra, Plancherel formula for complex semisimple Lie groups, *Proc. Nat. Acad. Sci. USA* **37** (1951), 813–818.

[HC3] Harish-Chandra, Plancherel formula for the 2×2 real unimodular group, *Proc. Nat. Acad. Sci. USA* **38** (1952), 337–342.

[HC4] Harish-Chandra, A formula for semisimple Lie groups, *Proc. Nat. Acad. Sci. USA* **42** (1956), 538–540.

[HC5] Harish-Chandra, Spherical functions on a semisimple Lie groups I, *Amer. J. Math.* **80** (1958), 241–310; II, *ibid.*, 553–613.

[HC6] Harish-Chandra, Harmonic analysis on semisimple Lie groups, *Bull. Amer. Math. Soc.* **76** (1970), 529–551.

[HC7] Harish-Chandra, Harmonic analysis on real reductive groups I, II, III, *J. Func. Anal.* **19** (1975), 104–204; *Invent. Math.* **36** (1976), 1–55; *Annals of Math.* **104** (1976), 117–201.

[HC8] Harish-Chandra, Harmonic analysis on semisimple Lie groups (1966), *Collected Papers*, vol. III, pp. 649–654.

[H1] S. Helgason, Differential operators on homogeneous spaces, *Acta Math.* **102** (1959), 239–299.

[H2] S. Helgason, Fundamental solutions of invariant differential operators on symmetric spaces, *Amer. J. Math.* **86** (1964), 565–601.

[H3] S. Helgason, Duality and Radon transforms for symmetric spaces, *Amer. J. Math.* **85** (1963), 667–692.

[H4] S. Helgason, Radon-Fourier transforms on symmetric spaces and related group representations, *Bull. Amer. Math. Soc.* **71** (1965), 757–763.

[H5] S. Helgason, *Analysis on Lie Groups and Homogeneous Spaces*, Conference Board of Mathematical Sciences Regional Conference Series in Mathematics, vol. 14, American Mathematical Society, Providence, 1972.

[H6] S. Helgason, A duality for symmetric spaces with applications to group representations, I, *Advances in Math.* **5** (1970), 1–154; II, *Advances in Math.* **22** (1976), 187–219.

[H7] S. Helgason, The surjectivity of invariant differential operators on symmetric spaces, *Annals of Math.* **98** (1973), 451–480.

[H8] S. Helgason, Eigenspace of the Laplacian; integral representation and irreducibility, *J. Func. Anal.* **17** (1974), 328–353.

[H9] S. Helgason, *Geometric Analysis on Symmetric Spaces*, Mathematical Surveys and Monographs, vol. 39, American Mathematical Society, Providence, 1994.

[He1] R. Herb, Fourier inversion of invariant integrals on semisimple real Lie groups, *Trans. Amer. Math. Soc.* **249** (1979), 281–302.

[He2] R. Herb, Fourier inversion and the Plancherel theorem for semisimple real Lie groups, *Amer. J. Math.* **104** (1982), 9–58.

[HW] R. A. Herb and J. A. Wolf, The Plancherel theorem for general semisimple Lie groups, *Compositio Math.* **57** (1986), 271–355.

[Ho] R. Hotta, On a realization of the discrete series for semisimple Lie groups, *J. Math. Soc. Japan* **23** (1971), 384–407.

[K] A. W. Knapp, *Representation Theory of Semisimple Groups: An Overview Based on Examples*, Princeton University Press, Princeton, 1986.

[L] S. Lang, $\mathbf{SL}_2(\mathbf{R})$, Addison-Wesley, Reading, Mass., 1975; second edition, Springer-Verlag, New York, 1985.

[Or] J. Orloff, Orbital integrals on symmetric spaces, *Noncommutative Harmonic Analysis and Lie Groups*, Lecture Notes in Mathematics, vol. 1243, Springer-Verlag, New York, 1987, pp. 198–239.

[PS] R. S. Phillips and M. Shahshahani, Scattering theory for symmetric spaces of the noncompact type, *Duke Math. J.* **72** (1993), 1–29.

[R] M. Raïs, Solutions élémentaires des opérateurs differentiels bi-invariants sur un groupe de Lie nilpotent, *C. R. Acad. Sci. Paris* **273** (1971), 495–498.

[Ro1] W. Rossmann, Kirillov's character formula for reductive Lie groups, *Invent. Math.* **48** (1978), 207–220.

[Ro2] W. Rossmann, Analysis on real hyperbolic spaces, *J. Func. Anal.* **30** (1978), 448–477.

[Ro] F. Rouvière, Sur la résolubilité locale des opérateurs bi-invariants, *Annali della Scuola Norm. Sup. Pisa* **3** (1976), 231–244.

[SW] P. J. Sally and G. Warner, The Fourier transform on semisimple Lie groups of real rank one, *Acta Math.* **131** (1973), 1–26.

[S] G. Schiffmann, Distributions centrales sur **SL**(2, **R**), *Annales Inst. Henri Poincaré* **13** (1970), 229–240.

[Sc] H. Schlichtkrull, Eigenspaces of the Laplacian on hyperbolic spaces; composition series and integral transforms, *J. Func. Anal.* **70** (1987), 194–219.

[Sm] W. Schmid, On a conjecture of Langlands, *Annals of Math.* **93** (1971), 1–42.

[Se] I. E. Segal, An extension of Plancherel's formula to separable unimodular groups, *Annals of Math.* **52** (1950), 272–292.

[STS] M. A. Semenov-Tjan-Shanski, Harmonic analysis on Riemannian symmetric spaces of negative curvature and scattering theory, *Math. USSR Izvestija* **10** (1976), 535–563.

[Sh] M. Shahshahani, Invariant hyperbolic systems on symmetric spaces, *Differential Geometry* (R. Brooks et. al., eds.), Birkhäuser, Boston, 1983, pp. 203–233.

[Sl] D. Shelstad, Characters and inner forms of a quasi-split group over **R**, *Compositio Math.* **39** (1979), 11–45.

[V] V. S. Varadarajan, Introduction to Harish-Chandra's *Collected Papers*, Springer Verlag, New York, 1984.

[Ve1] M. Vergne, On Rossmann's character formula for discrete series, *Invent. Math.* **54** (1979), 11–14.

[Ve2] M. Vergne, A Poisson-Plancherel formula for semisimple Lie groups, *Annals of Math.* **115** (1982), 639–666.

[Ve3] M. Vergne, Representations of Lie groups and the orbit method, *Emmy Noether in Bryn Mawr*, Springer, New York, 1983, pp. 59–101.

[W] A. Weil, *L'Intégration dans les Groupes Topologiques et ses Applications*, Hermann, Paris, 1940.

Department of Mathematics, Massachusetts Institute of Technology, Cambridge, MA 02139, U.S.A.

E-mail address: helgason@mit.edu

Proceedings of Symposia in Pure Mathematics
Volume **61** (1997), pp. 191–217

Harmonic Analysis on Semisimple Symmetric Spaces: A Survey of Some General Results

E. P. van den Ban, M. Flensted-Jensen, and H. Schlichtkrull

Abstract. We give a survey of the status of some of the fundamental problems in harmonic analysis on semisimple symmetric spaces, including the description of the discrete series, the definition of the Fourier transform, the inversion formula, the Plancherel formula, and the Paley-Wiener theorem.

1. Introduction

The rich and beautiful theory of harmonic analysis on $\mathbb{R}$ and $\mathbb{T} = \mathbb{R}/\mathbb{Z}$ has become a powerful tool, widely used in other branches of mathematics, in physics, engineering, etc. From our point of view all the basic questions are completely and explicitly solved: The Fourier transform is defined, there exist a Plancherel formula and an inversion formula for it, and (for $\mathbb{R}$) there is a Paley-Wiener theorem, describing the image of the space of smooth compactly supported functions.

There exist many generalizations of this theory. Let us mention a few of these, based on various ways of viewing the exponential function $x \mapsto e^{\lambda x}$ on $\mathbb{R}$ ($\lambda \in i\mathbb{R}$) and on $\mathbb{T}$ ($\lambda \in 2\pi i\mathbb{Z}$):

- On $\mathbb{R}$, the exponential functions are eigenfunctions for d/dx. This point of view leads to: Spectral theory for differential operators. Sturm-Liouville theory. Expansion in orthogonal polynomials.
- The exponential functions are characters for the topological groups $\mathbb{R}$ and $\mathbb{T}$. This point of view leads to: Fourier analysis on locally compact Abelian groups. The Peter-Weyl theory for Fourier analysis on compact groups.
- The exponential functions generate one-dimensional representations of the Lie groups $\mathbb{R}$ and $\mathbb{T}$. This point of view leads to: The representation theory for compact Lie groups (the Cartan-Weyl classification, Weyl's character formula, etc.). Representation theory for general Lie groups (semisimple, reductive, nilpotent, solvable, etc.).

1991 *Mathematics Subject Classification.* Primary 43A85; Secondary 22E45.

Key words and phrases. Harmonic analysis, Symmetric space, Fourier transform, Discrete series, Plancherel formula, Paley-Wiener theorem.

We are grateful to Patrick Delorme and Sigurdur Helgason for fruitful discussions.

- The manifolds $\mathbb{R}$ and $\mathbb{T}$ are homogeneous spaces for the Lie groups $\mathbb{R}$ and $\mathbb{T}$, respectively (the action being translation), and the exponential functions are simultaneous eigenfunctions for the algebras of invariant differential operators on these manifolds. This point of view leads to: Harmonic analysis on homogeneous spaces of Lie groups.

As an aspect of the last viewpoint we could mention the theory of spherical harmonic expansion on the n-sphere S^n, which is a homogeneous space for the rotation group $O(n+1)$. The spherical harmonics are eigenfunctions for the Laplace operator, which is rotation invariant.

Here we take this last mentioned viewpoint. We claim that inside the class of smooth manifolds the class of (not necessarily Riemannian) reductive symmetric spaces constitutes an appropriate framework for generalization of harmonic analysis: On the one hand this class of manifolds is wide enough to contain very many important spaces of relevance in other branches of mathematics and in physics. On the other hand it is restrictive enough to make feasible a theory of harmonic analysis, with explicit parametrizations and descriptions of representations, explicit Plancherel formulae, etc. The irreducible members of the class of reductive symmetric spaces are either one-dimensional flat, i.e. $\mathbb{R}$ or $\mathbb{T}$, or semisimple. In this paper we discuss the semisimple symmetric spaces. The exposition in the present paper consists of a rewriting and updating of parts of [8], extended with a description of recent developments.

2. Semisimple Symmetric Spaces

2.1. Definition and structure

We define a semisimple symmetric space as follows:

Definition. Let G be a connected Lie group and H a closed subgroup. The pair (G, H) is called a **semisimple symmetric pair** if G is semisimple and H is an open subgroup of the group of fixed points for an involution σ of G. If (G, H) is a semisimple symmetric pair we say that the homogeneous space $M = G/H$ is a **semisimple symmetric space**.

Notice that the phrase "Let $M = G/H$ be a semisimple symmetric space" thus essentially means "Let (G, H) be a semisimple symmetric pair, and let $M = G/H$." This way of defining the notion of a semisimple symmetric space is slightly unsatisfactory because in general the "same space" will correspond to many different pairs (G, H). For example, a set consisting of a single point can be regarded as a homogeneous space for the trivial action of any group, and thus it will be the semisimple symmetric space G/G for any semisimple connected Lie group G. The question that we have not addressed here is in which category of spaces we require equality in order that two semisimple symmetric spaces $M = G/H$ and $M' = G'/H'$ coincide. However, for this exposition we do not need to address this question, and we refer to [34] for a more satisfactory definition.

We are going to introduce only the most necessary aspects and technicalities of the general theory of semisimple symmetric spaces. For a more complete treatment and some of the details we refer to [34], [63], [42, Part II] and the references cited there.

An important case is when M is a semisimple Lie group G_1, i.e., when G is the product $G_1 \times G_1$ and its action on G_1 is the left times right action $(x, y)z = xzy^{-1}$.

The involution of G is given by $\sigma(x, y) = (y, x)$, and H is the diagonal $d(G_1)$. We shall call this the **group case**.

Our goal in this paper is to indicate the state of the art for harmonic analysis on semisimple symmetric spaces. From now on we assume that $M = G/H$ is such a space.

For simplicity of exposition we assume (as we may, up to coverings of M) that G is a closed subgroup of $\mathrm{GL}(n, \mathbb{R})$ for some n and that G is stable under transposition. Let $K = G \cap \mathrm{SO}(n)$, or equivalently $K = G^\theta$, where $\theta(x) = {}^t x^{-1}$. Then K is a maximal compact subgroup of G. We may choose the base point such that $\theta(H) = H$, or equivalently such that $\sigma \circ \theta = \theta \circ \sigma$.

Let $M = G/H$ be a semisimple symmetric space. The kernel of the action of G on M is a closed normal subgroup of G. Hence the quotient of G by this subgroup is a semisimple Lie group, which we shall denote $\mathrm{A}(G)$. (The group $\mathrm{A}(G)$ can be regarded as a subgroup of the group of diffeomorphisms of the manifold M.) We define $\mathrm{A}(H)$ similarly. Then $(\mathrm{A}(G), \mathrm{A}(H))$ is a semisimple symmetric pair (and M would naturally coincide with $\mathrm{A}(G)/\mathrm{A}(H)$ if we had defined such a notion of coincidence). We call $M = G/H$ **irreducible** if either $\mathrm{A}(G)$ is simple or if G/H is a group case for which $G = G_1 \times G_1$ with G_1 simple.

We shall distinguish three types of irreducible semisimple symmetric spaces $M = G/H$. In general the type of G/H will be the same as that of $\mathrm{A}(G)/\mathrm{A}(H)$; we may thus assume that $G = \mathrm{A}(G)$.

- M is of the **compact type** if $G = K$, or equivalently if all geodesic curves have compact closures.
- M is of the **noncompact type** if $H = K$, or equivalently if all geodesic curves have noncompact closures.
- M is of the **non-Riemannian type** if $G \neq K$ and $K \neq H$, or equivalently if there exist geodesic curves of both types.

If M is of one of the first two types we say that it is of the **Riemannian type**, because it then has a natural structure as a Riemannian manifold. In the third case the natural structure is only pseudo-Riemannian. Notice that a simple group G_1, considered as a symmetric space, is either of the compact type or of the non-Riemannian type. In general we say that M belongs to a given one of these types if all its irreducible constituents belong to that type.

2.2. Examples

The irreducible symmetric spaces have been classified by M. Berger [17]. Compared with the list of Riemannian symmetric spaces (see [46, Ch. X]), Berger's list is considerably longer.

There is (up to coverings) one two-dimensional space of each of the three types:

- The compact type: The **2-sphere** $\mathrm{S}^2 = \mathrm{SO}(3)/\mathrm{SO}(2)$.
- The noncompact type: The **hyperbolic 2-space** $M = \mathrm{H}^2$. This has several isomorphic realizations: As $\mathrm{SL}(2, \mathbb{R})/\mathrm{SO}(2)$, as $\mathrm{SU}(1,1)/\mathrm{S}(\mathrm{U}(1) \times \mathrm{U}(1))$, or as $\mathrm{SO}_e(2,1)/\mathrm{SO}(2)$, corresponding to, respectively, the upper half plane in $\mathbb{C}$, the unit disc in $\mathbb{C}$, or a sheet of the two-sheeted hyperboloid in $\mathbb{R}^3$.
- The non-Riemannian type: The **one-sheeted hyperboloid** in $\mathbb{R}^3$, $H^{1,1} = \mathrm{SO}_e(2,1)/\mathrm{SO}_e(1,1)$, which can also be realized as $\mathrm{SL}(2, \mathbb{R})/\mathrm{SO}(1,1)$. It has the two-fold cover $\mathrm{SL}(2, \mathbb{R})/\mathrm{SO}_e(1,1)$.

In higher dimensions there exist several "families" of symmetric spaces, many of which have one of the spaces above as their lowest-dimensional member. For example we could mention:

- The **n-sphere**: $\mathrm{S}^n = \mathrm{SO}(n+1)/\mathrm{SO}(n)$.
- The space of **positive definite quadratic forms** in $\mathbb{R}^n$, which is given by $M = \mathrm{SL}(n,\mathbb{R})/\mathrm{SO}(n)$.
- The space of **quadratic forms of signature** (p,q) in $\mathbb{R}^n$, where $n = p+q$: $M = \mathrm{SL}(n,\mathbb{R})/\mathrm{SO}(p,q)$.
- The **hyperboloids** in $\mathbb{R}^{n+1}$:
$$M = \mathrm{H}^{p,q} = \{x \in \mathbb{R}^{n+1} \mid x_1^2 + \cdots + x_p^2 - x_{p+1}^2 - \cdots - x_{p+q+1}^2 = -1\},$$
where $p+q=n$. (If $q=0$, one must take a connected component.) Here $M = \mathrm{SO}_e(p,q+1)/\mathrm{SO}_e(p,q)$.

Similarly, one can take the corresponding spaces over the complex numbers or over the quaternions.

2.3. Some basic notation

Let G, H, K, σ, and θ be as above. Let $\mathfrak{g}$ be the (real) Lie algebra of G, and let $\mathfrak{h}$ and $\mathfrak{k}$ be the subalgebras corresponding to H and K, and $\mathfrak{q}$ and $\mathfrak{p}$ their respective orthocomplements with respect to the Killing form. Then

$$\mathfrak{g} = \mathfrak{h} \oplus \mathfrak{q} = \mathfrak{k} \oplus \mathfrak{p}$$

is the decomposition of $\mathfrak{g}$ into the ± 1 eigenspaces for σ and θ, respectively. Since θ and σ commute we also have the joint decomposition

$$\mathfrak{g} = \mathfrak{h} \cap \mathfrak{k} \oplus \mathfrak{h} \cap \mathfrak{p} \oplus \mathfrak{q} \cap \mathfrak{k} \oplus \mathfrak{q} \cap \mathfrak{p}. \tag{1}$$

Notice that there is a natural identification of $\mathfrak{q}$ with the tangent space $T_{x_o}(M)$ at the base point $x_o = eH$. We denote by $\mathfrak{g}_{\mathbb{C}}, \mathfrak{h}_{\mathbb{C}}$, etc. the complexifications of $\mathfrak{g}$, $\mathfrak{h}$, etc.

A **Cartan subspace** $\mathfrak{b}$ for G/H is a maximal Abelian subspace of $\mathfrak{q}$ that consists of semisimple elements. (If we assume, as we may in the following, that $\mathfrak{b}$ is θ-invariant, then all its elements are automatically semisimple, once $\mathfrak{b}$ is maximal Abelian). All Cartan subspaces have the same dimension, which we call the **rank** of M. The number of H-conjugacy classes of Cartan subspaces is finite. Geometrically, a Cartan subspace is the tangent space of a maximally flat regular subsymmetric space.

We say that a Cartan subspace $\mathfrak{b}$ is **fundamental** if the intersection $\mathfrak{b} \cap \mathfrak{k}$ is maximal Abelian in $\mathfrak{q} \cap \mathfrak{k}$, and that it is **maximal split** if the intersection $\mathfrak{b} \cap \mathfrak{p}$ is maximal Abelian in $\mathfrak{q} \cap \mathfrak{p}$. There exist, up to conjugation by $K \cap H$, a unique fundamental and a unique maximal split Cartan subspace. If the fundamental Cartan subspace is contained in $\mathfrak{k}$, it is called a **compact Cartan subspace**. The dimension of the $\mathfrak{p}$-part of a maximal split Cartan subspace is called the **split rank** of M.

Let $\mathbb{D}(G/H)$ denote the algebra of G-invariant differential operators on G/H. There is a natural isomorphism (the Harish-Chandra isomorphism) χ of this algebra with the algebra $S(\mathfrak{b})^W$ of W-invariant elements in the symmetric algebra of any Cartan subspace $\mathfrak{b}_{\mathbb{C}}$. Here W is the reflection group of the root system of $\mathfrak{b}_{\mathbb{C}}$ in $\mathfrak{g}_{\mathbb{C}}$. In particular, $\mathbb{D}(G/H)$ is commutative, and its characters are parametrized up to W-conjugation by $D \mapsto \chi_\lambda(D) = \chi(D)(\lambda) \in \mathbb{C}$. It is known (see [2]) that the symmetric elements of $\mathbb{D}(G/H)$ have self-adjoint closures as operators on $L^2(G/H)$.

3. Basic Harmonic Analysis

3.1. Harmonic analysis on $\mathbb{R}^n$

We want to generalize some basic notions and results from harmonic analysis on $\mathbb{R}^n$. These are:

- FOURIER TRANSFORM: $f \mapsto f^\wedge(\lambda) = (2\pi)^{-n/2} \int_{\mathbb{R}^n} f(t)e^{-i\lambda\cdot t}\,dt$, $f \in C_c^\infty(\mathbb{R}^n)$.
- INVERSION FORMULA: If $f \in C_c^\infty(\mathbb{R}^n)$ then

$$f(x) = (2\pi)^{-n/2} \int_{\mathbb{R}^n} f^\wedge(\lambda)e^{i\lambda\cdot x}\,d\lambda.$$

- PLANCHEREL THEOREM: $f \mapsto f^\wedge$ extends to an isometry of $L^2(\mathbb{R}^n)$ onto $L^2(\mathbb{R}^n)$.
- PALEY-WIENER THEOREM: $f \mapsto f^\wedge$ is a bijection of $C_c^\infty(\mathbb{R}^n)$ onto $\mathrm{PW}(\mathbb{R}^n)$, where $\mathrm{PW}(\mathbb{R}^n)$ is the space of rapidly decreasing entire functions of exponential type. More precisely, a complex function ψ on $\mathbb{R}^n$ belongs to $\mathrm{PW}(\mathbb{R}^n)$ if and only if it extends to an entire function on $\mathbb{C}^n$ for which there exists $R > 0$ such that the following holds for all $N \in \mathbb{N}$:

$$\sup_{\lambda\in\mathbb{C}^n} (1 + |\lambda|)^N e^{-R|\mathrm{Im}\,\lambda|}|\psi(\lambda)| < +\infty. \tag{2}$$

The aim of the basic harmonic analysis on G/H is to obtain analogues of these notions and results.

3.2. Abstract harmonic analysis on a semisimple symmetric space

If G and H are as above, then $M = G/H$ has an invariant measure, and the action of G by translations gives a unitary representation $\mathcal{L}$ in the associated Hilbert space $L^2(G/H)$. From general representation theory it is known (since G is "type 1") that this representation can be decomposed as a direct integral of irreducible unitary representations:

$$\mathcal{L} \simeq \int_{G^\wedge}^{\oplus} m_\pi\, \pi\, d\mu(\pi), \tag{3}$$

where $G^\wedge$ is the unitary dual of G. The measure $d\mu$ (whose class is uniquely determined) is called the **Plancherel measure**, and m_π (which is unique almost everywhere) is the **multiplicity** of π. Moreover, only the so-called "H-spherical representations" can occur in this decomposition. By definition, an irreducible unitary representation $(\pi, \mathcal{H}_\pi)$ of G is H-**spherical** if the space $(\mathcal{H}_\pi^{-\infty})^H$ of its H-fixed distribution vectors is nontrivial. Here we denote by $\mathcal{H}_\pi^\infty$ and $\mathcal{H}_\pi^{-\infty}$, respectively, the C^∞ vectors and the distribution vectors for $\mathcal{H}_\pi$ such that $\mathcal{H}_\pi^\infty \subset \mathcal{H}_\pi \subset \mathcal{H}_\pi^{-\infty}$. We write

$$\mathcal{V}_\pi = (\mathcal{H}_\pi^{-\infty})^H.$$

It is known (see [2]) that $m_\pi \le \dim \mathcal{V}_\pi < +\infty$; in particular, all multiplicities are finite. Denote by $G_H^\wedge$ the set of (equivalence classes) of H-spherical representations; then it follows that the Plancherel measure $d\mu$ is carried by $G_H^\wedge$.

The **abstract Fourier transform** $f \mapsto f^\wedge(\pi)$ for G/H is now defined by

$$f^\wedge(\pi)(\eta) = \pi(f)\eta = \int_{G/H} f(x)\pi(x)\eta\,dx \in \mathcal{H}_\pi^\infty$$

for $\pi \in G_H^\wedge, \eta \in \mathcal{V}_\pi$, and $f \in C_c^\infty(G/H)$. Thus

$$f^\wedge(\pi) \in \mathrm{Hom}_{\mathbb{C}}(\mathcal{V}_\pi, \mathcal{H}_\pi^\infty) \simeq \mathcal{H}_\pi^\infty \otimes \mathcal{V}_\pi^*.$$

Notice that the integral over G/H makes sense only because η is H-invariant. One can prove (using [60] and [31]) that there exists for almost all $\pi \in G_H^\wedge$ a subspace $\mathcal{V}_\pi^o$ (of dimension m_π) of $\mathcal{V}_\pi$, equipped with the structure of a Hilbert space, such that if $f^\wedge(\pi)$ is restricted to $\mathcal{V}_\pi^o$ for almost all π, then $f \mapsto f^\wedge$ extends to an isometry of $L^2(G/H)$ onto $\int_{G_H^\wedge}^\oplus \mathrm{Hom}_{\mathbb{C}}(\mathcal{V}_\pi^o, \mathcal{H}_\pi)\, d\mu(\pi)$. Here the norm on $\mathrm{Hom}_{\mathbb{C}}(\mathcal{V}_\pi^o, \mathcal{H}_\pi)$ is given by

$$\|\varphi\|_\pi^2 = \mathrm{Tr}(\varphi^* \circ \varphi) = \sum_i \|\varphi(v_i)\|^2, \qquad \varphi \in \mathrm{Hom}_{\mathbb{C}}(\mathcal{V}_\pi^o, \mathcal{H}_\pi),$$

where φ^* is the adjoint of φ and $\{v_i\}_{i=1,\dots,m_\pi}$ is an orthonormal basis in $\mathcal{V}_\pi^o$.

We thus have the **Plancherel formula**

$$\|f\|_2^2 = \int_{G_H^\wedge} \|f^\wedge(\pi)\|_\pi^2\, d\mu(\pi), \qquad f \in L^2(G/H).$$

Similarly, there is the **inversion formula** (for suitably nice functions f)

$$f(e) = \int_{G_H^\wedge} \sum_{i=1}^{m_\pi} \langle f^\wedge(\pi) v_i \mid v_i \rangle\, d\mu(\pi). \tag{4}$$

Here $\langle \cdot \mid \cdot \rangle$ denotes the inner product on $\mathcal{H}_\pi$, as well as the naturally associated pairing $\mathcal{H}_\pi^\infty \times \mathcal{H}_\pi^{-\infty} \to \mathbb{C}$. Consequently we also have, for suitable f,

$$f(x) = \int_{G_H^\wedge} \sum_{i=1}^{m_\pi} \langle f^\wedge(\pi) v_i \mid \pi(x) v_i \rangle\, d\mu(\pi).$$

The basic problems in making abstract harmonic analysis concrete are these:

(a) Describe (parametrize) $G_H^\wedge$, or at least μ-almost all of it.
(b) For μ-almost all $\pi \in G_H^\wedge$ describe (parametrize) $\mathcal{V}_\pi^o$ and its Hilbert space structure.
(c) Determine $d\mu$ explicitly.

A **Paley-Wiener theorem** would amount to an intrinsic description of the Fourier image of $C_c^\infty(G/H)$ in terms of $G_H^\wedge$. We add this as a fourth basic problem:

(d) Describe $C_c^\infty(G/H)^\wedge$ in terms of the parametrizations and possible holomorphic extensions.

For each $\pi \in G_H^\wedge$ we have that $\mathcal{V}_\pi$ is a $\mathbb{D}(G/H)$-module in a natural way. Using that the symmetric elements of $\mathbb{D}(G/H)$ are essentially selfadjoint operators on $L^2(G/H)$, one can show (with the arguments in [31]) that $\mathcal{V}_\pi^o$ can be chosen to be invariant and diagonalizable for this action. Thus $\mathcal{V}_\pi^o$ is spanned by its joint eigenvectors for $\mathbb{D}(G/H)$. Let $\mathfrak{b} \subset \mathfrak{q}$ be a Cartan subspace. Then such an eigenvector satisfies

$$\pi(D)v = \chi_\lambda(D)v, \qquad D \in \mathbb{D}(G/H),$$

for some $\lambda \in \mathfrak{b}_{\mathbb{C}}^*$. We say that v is a **spherical vector of type** λ, and that the orthonormal basis $\{v_i\}_{i=1,\dots,m_\pi}$ in $\mathcal{V}_\pi^o$ is **spherical**, if its members are spherical.

The maps $\xi_{\pi,i}\colon f \mapsto \langle f^\wedge(\pi)v_i \mid v_i\rangle$ in (4) are H-invariant distributions on G/H. As distributions on G they are positive definite and extreme (see [31]). With a spherical basis $\{v_i\}$ each $\xi_{\pi,i}$ is also a **spherical distribution**, that is, an H-invariant eigendistribution for $\mathbb{D}(G/H)$. The solution of Problem (b) is then closely related to the study of the spherical distributions.

3.3. Results for specific classes of symmetric spaces

Here we give some brief remarks concerning the above problems for some specific classes of semisimple symmetric spaces.

3.3.1. The compact type. For a homogeneous space G/H with a compact group G the abstract formulation above follows easily from the Peter-Weyl theorem and the Schur orthogonality relations. In particular, $\mathcal{V}_\pi^o = \mathcal{V}_\pi = \mathcal{H}_\pi^H$, and if we give $\mathcal{V}_\pi^o$ the subspace norm from $\mathcal{H}_\pi$, we have $d\mu(\pi) = \dim(\pi)$. For the symmetric spaces of compact type we then have the following explicit solutions to the above concrete problems (see [26], [47, § V.4]):

(a) $G_H^\wedge$ is parametrized by a subset of the set of dominant weights.
(b) $\dim \mathcal{V}_\pi^o = 1$ for $\pi \in G_H^\wedge$.
(c) $d\mu$ is given by Weyl's dimension formula.
(d) The smooth functions are determined by a certain growth condition on the Fourier transforms (see [66]).

3.3.2. The noncompact type. We write M as G/K. The four questions were settled beautifully by the work of Harish-Chandra, Helgason, and others. See [47, § IV.7] and [48, Ch. III]. Let $\mathfrak{a}$ be a maximal Abelian subspace of $\mathfrak{p}$.

(a) A sufficient subset of $G_K^\wedge$ is parametrized (up to conjugacy by the Weyl group) by means of the spherical functions φ_λ for $\lambda \in i\mathfrak{a}^*$ and the corresponding spherical principal series representations $(\pi_\lambda, \mathcal{H}_\lambda)$.
(b) For $\pi = \pi_\lambda \in G_K^\wedge$ we have $\mathcal{V}_\pi^o = \mathcal{H}_\lambda^K$ and $\dim(\mathcal{V}_\pi^o) = 1$. We can then use the subspace norm from $\mathcal{H}_\lambda$.
(c) The Plancherel measure is given by $d\mu(\pi_\lambda) = |\mathbf{c}(\lambda)|^{-2}\, d\lambda$ on $i\mathfrak{a}^*/W$. Here $\mathbf{c}(\lambda)$ is Harish-Chandra's c-function, which is explicitly given in terms of the structure of G/K by the formula of Gindikin-Karpelevič.
(d) We have $C_c^\infty(K\backslash G/K)^\wedge = \mathrm{PW}(\mathfrak{a})^W$. Here $\mathrm{PW}(\mathfrak{a})^W$ is the space of W-invariant functions in the image space $\mathrm{PW}(\mathfrak{a})$ for the Fourier transform

$$f \mapsto f^\wedge(\lambda) = \int_{\mathfrak{a}} f(X)e^{-\lambda(X)}dX, \qquad \lambda \in \mathfrak{a}_{\mathbb{C}}^*,\ f \in C_c^\infty(\mathfrak{a}), \tag{5}$$

that is, the space of rapidly decreasing entire functions of exponential type on $\mathfrak{a}_{\mathbb{C}}^*$. (Compare with Section 3.1, but note that since the imaginary unit i is not present in the exponent in (5), one has to replace $\operatorname{Im}\lambda$ by $\operatorname{Re}\lambda$ in (2).) Helgason has extended the Paley-Wiener theorem to the space $C_c^\infty(K;G/K)$ of K-finite functions in $C_c^\infty(G/K)$, and also to the full space $C_c^\infty(G/K)$. See [48, Ch. III, Theorems 5.1 and 5.11].

3.3.3. The group case, $M = G_1$. This case is settled by the work of Harish-Chandra [40] and others (for expositions, see e.g. [49], [68]).

(a) The map $\pi_1 \mapsto \pi_1 \otimes \pi_1^*$ is a bijective correspondence from the unitary dual $G_1^\wedge$ onto $G_H^\wedge$. A sufficient subset of $G_1^\wedge$ is described by the discrete series and different families of (cuspidal) principal series.
(b) For $\pi_1 \in G_1^\wedge$ and $\pi = \pi_1 \otimes \pi_1^*$ we have $\mathcal{V}_\pi = (\mathcal{H}_\pi^{-\infty})^H = \mathbb{C}\, 1_{\pi_1}$, where 1_{π_1} is the identity operator on $\mathcal{H}_{\pi_1}$. Notice however that in this case $\mathcal{V}_\pi \not\subset \mathcal{H}_\pi$, since the latter space can be identified with the space of Hilbert-Schmidt operators on $\mathcal{H}_{\pi_1}$. We take $\mathcal{V}_\pi^o = \mathcal{V}_\pi$ and use on it the Hilbert space structure obtained from the identification with $\mathbb{C}$ in which $1_{\pi_1} = 1$.
(c) With the above choice one can give $d\mu$ explicitly in terms of the formal degrees of discrete series and certain c-functions.
(d) A Paley-Wiener theorem for the K-finite functions on G_1 has been established in [22] for split rank one and [1] in general. In particular, the Paley-Wiener space is determined by the minimal principal series only. The extension of the Paley-Wiener theorem to the full space $C_c^\infty(G_1)$ is still an open problem.

3.3.4. The non-Riemannian type, rank one. There is an extensive literature dealing with the questions (a)-(c) on specific classes of rank one symmetric spaces of the non-Riemannian type. See for example [32], [31], [55]. Common for all these spaces is that the decomposition of $L^2(G/H)$ contains a discrete series as well as a continuous part.

3.3.5. Type $G_{\mathbb{C}}/G_{\mathbb{R}}$. When G is complex and H is a real form of it, precise solutions to questions (a)-(c) have been given by P. Harinck. See [20], [35], [36], [37].

3.4. Results for general semisimple symmetric spaces

The listed basic problems have been solved in a general setting for semisimple symmetric spaces. In the following sections we outline the solution, with precise references to the literature.

By analogy with the group case one expects in general that the left regular representation $\mathcal{L}$ on $L^2(G/H)$ can be decomposed into several "series" of representations, one series for each H-conjugacy class of Cartan subspaces for $\mathfrak{q}$. The most extreme of these would then be the "most continuous" part, corresponding to the conjugacy class of Cartan subspaces with maximal $\mathfrak{p}$-part (the maximal split Cartan subspaces) and the "most discrete" part (sometimes called the fundamental series), corresponding to the conjugacy class of Cartan subspaces with maximal $\mathfrak{k}$-part (the fundamental Cartan subspaces). The series corresponding to the remaining conjugacy classes of Cartan subspaces would then be called "the intermediate series." If the fundamental Cartan subspaces are compact, then the "most discrete" part is in fact the **discrete series**, that is, the irreducible subrepresentations of $\mathcal{L}$.

In fact, this analogy with the group case holds rather precisely, as we shall explain below. In Section 4 we discuss discrete series and in Section 5 the most continuous series. In Sections 6–7 we then discuss the Plancherel and Paley-Wiener theorems for G/H.

4. The Discrete Series

The basic existence theorem is the following. We retain the terminology and notation from above. Let $L^2_d(G/H) \subset L^2(G/H)$ be the closed linear span of the irreducible subrepresentations of $\mathcal{L}$.

Theorem 1 ([33], [58]). *Let G/H be a semisimple symmetric space. Then the discrete series space $L^2_d(G/H)$ is nonzero if and only if*

$$\operatorname{rank}(G/H) = \operatorname{rank}(K/K \cap H). \tag{6}$$

The condition (6) means that G/H has a compact Cartan subspace. An equivalent more geometric formulation is that it has a compact maximally flat subsymmetric space.

In the group case this result reduces to Harish-Chandra's theorem, that the existence of discrete series is equivalent to the existence of a compact Cartan subgroup, cf. [39]. In fact the proof in [33] of the existence part of the theorem is different from Harish-Chandra's proof for the group case; see also [49], where the symmetric space viewpoint has been adapted in the proof for the group case.

We shall now discuss Problems (a), (b) and (c) for the discrete series. Assume (as we may by the above theorem) that (6) holds, and let $\mathfrak{t}$ be a compact Cartan subspace of $\mathfrak{q}$. Let Σ be the root system of $\mathfrak{t}_\mathbb{C}$ in $\mathfrak{g}_\mathbb{C}$ and Σ_c the subsystem of $\mathfrak{t}_\mathbb{C}$ in $\mathfrak{k}_\mathbb{C}$. Let W and W_c be the corresponding reflection groups.

A rough classification of the discrete series is obtained by means of the commutative algebra $\mathbb{D}(G/H)$. Recall that the characters of $\mathbb{D}(G/H)$ are parametrized by $\mathfrak{t}^*_\mathbb{C}/W$ via the Harish-Chandra isomorphism $\chi\colon \mathbb{D}(G/H) \to S(\mathfrak{t})^W$. Let $\mathcal{E}_\lambda(G/H)$ denote the joint eigenspace for $\mathbb{D}(G/H)$ in $C^\infty(G/H)$ corresponding to the character χ_λ, where $\lambda \in \mathfrak{t}^*_\mathbb{C}$. Then $\mathcal{E}_{w\lambda}(G/H) = \mathcal{E}_\lambda(G/H)$ for all $w \in W$. Since $\mathbb{D}(G/H)$ is commutative and its symmetric elements act as essentially selfadjoint operators on $L^2(G/H)$, there is a joint spectral resolution of $L^2(G/H)$ for this algebra. The resulting decomposition is G-invariant because of the invariance of the elements in $\mathbb{D}(G/H)$. It follows (see [2]) that $L^2_d(G/H)$ admits an orthogonal G-invariant decomposition

$$L^2_d(G/H) = \widehat{\bigoplus_\lambda} L^2_\lambda(G/H),$$

where $L^2_\lambda(G/H)$ is the closure in $L^2(G/H)$ of $L^2(G/H) \cap \mathcal{E}_\lambda(G/H)$, and where the sum extends over the W-orbits in the set of those $\lambda \in \mathfrak{t}^*_\mathbb{C}$ for which $L^2_\lambda(G/H)$ is nontrivial. In order to parametrize the discrete series we must then determine this set of λ's, and for each λ therein the irreducible subrepresentations of $L^2_\lambda(G/H)$.

Let $\Lambda \subset i\mathfrak{t}^*$ denote the set of elements $\lambda \in i\mathfrak{t}^*$ satisfying the following conditions (i)-(iii):

(i) $\langle \lambda, \alpha \rangle \neq 0$ for all $\alpha \in \Sigma$.

Given that (i) holds, let

$$\Sigma^+ = \{\alpha \in \Sigma \mid \langle \lambda, \alpha \rangle > 0\}. \tag{7}$$

Then this is a positive system for Σ. Put $\Sigma_c^+ = \Sigma^+ \cap \Sigma_c$, and let ρ and ρ_c be defined as half the sum of the Σ^+-roots and Σ_c^+-roots, respectively, counted with multiplicities.

(ii) $\lambda + \rho$ is a weight for T_H, i.e., $e^{\lambda+\rho}$ is well defined on T_H. Here T_H denotes the torus in G/H corresponding to $\mathfrak{t}$ (that is, $T_H = T/(T \cap H)$ where $T = \exp \mathfrak{t}$).

(iii) $\langle \lambda - \rho, \beta \rangle \geq 0$ for each compact simple root β in Σ^+.

That β is **compact** means that the root space $\mathfrak{g}_{\mathbb{C}}^{\beta}$ is contained in $\mathfrak{k}_{\mathbb{C}}$. Notice that (ii) implies that Λ is a discrete subset of $i\mathfrak{t}^*$.

Under the assumption that λ is in Λ there is a rather simple construction (which we shall outline below) of a $\mathfrak{g}$-invariant subspace $\mathcal{U}_{\lambda,K}$ of the space $C^\infty(K; G/H)$ of K-finite functions in $C^\infty(G/H)$, and it can be shown that this subspace $\mathcal{U}_{\lambda,K}$ is contained in $L^2_\lambda(G/H)$. Let $\mathcal{U}_\lambda$ denote the closure of $\mathcal{U}_{\lambda,K}$ in $L^2(G/H)$; then $\mathcal{U}_\lambda$ is a closed invariant subspace of $L^2_\lambda(G/H)$. Let π_λ denote the representation of G on $\mathcal{U}_\lambda$.

For "large" $\lambda \in \Lambda$, or more precisely if $\langle \lambda + \rho - 2\rho_c, \alpha \rangle \geq 0$ for all $\alpha \in \Sigma_c^+$, it can be shown by elementary methods that $\mathcal{U}_\lambda \neq \{0\}$. For the remaining λ's one has to add a more technical assumption in order to ensure that $\mathcal{U}_\lambda \neq \{0\}$. We shall not state this condition here. (The condition is stated in [53] together with a proof of its necessity for the nonvanishing of $\mathcal{U}_\lambda$; the sufficiency is claimed but not proven in the paper.)

Theorem 2 ([58], [67]). *The discrete series space $L^2_d(G/H)$ is spanned by the $\mathcal{U}_\lambda$'s with $\lambda \in \Lambda$. Moreover for each $\lambda \in \Lambda$ either the representation π_λ is irreducible or $\mathcal{U}_\lambda$ is zero, and if $\lambda, \lambda' \in \Lambda$ we have $\mathcal{U}_{\lambda'} = \mathcal{U}_\lambda$ if and only if $\lambda' = w\lambda$ for some $w \in W_c$.*

It follows that if $\lambda \in \mathfrak{t}_{\mathbb{C}}^*$ then $L^2_\lambda(G/H)$ is the sum of those $\mathcal{U}_{w\lambda}$ for which $w \in W$ and $w\lambda \in \Lambda$. In particular $L^2_\lambda(G/H)$ has at most as many components as the order of the quotient W/W_c.

With this result, Problem (a) is almost solved as regards to the discrete series. It is conjectured that $\pi_{\lambda'}$ is unitarily equivalent to π_λ if and only if $\mathcal{U}_{\lambda'} = \mathcal{U}_\lambda$, or equivalently, in view of the above, that each discrete series has multiplicity one in the Plancherel formula. The conjecture is proved for all classical groups G, and is only open for a few exceptional cases for very special values of λ (see [19]).

Evaluation at the base point in G/H gives rise to an H-fixed distribution vector η_λ for $\mathcal{U}_\lambda$, for which it is easily seen that we have

$$f^\wedge(\pi_\lambda)(\eta_\lambda) = \mathrm{P}_\lambda f, \qquad f \in C_c^\infty(G/H),$$

where P_λ is the orthogonal projection of $L^2(G/H)$ onto $\mathcal{U}_\lambda$. It follows that if we take $\mathcal{V}^o_{\pi_\lambda} = \mathbb{C}\,\eta_\lambda$ and use on it the Hilbert space structure obtained from the identification with $\mathbb{C}$ in which $\eta_\lambda = 1$, then $d\mu(\pi_\lambda) = 1$. In other words, the Plancherel measure restricts to the counting measure on the discrete series. This provides the solution to Problems (b) and (c) for the discrete series.

At this point it is however interesting to note the following. Though the discrete series has been parametrized as above, it seems to be an open problem to determine an explicit expression for the spherical distribution $\xi_\lambda \colon f \mapsto \langle f^\wedge(\pi_\lambda)\eta_\lambda \mid \eta_\lambda \rangle$ on G/H associated to η_λ (or equivalently for the projection operator P_λ, which is given by convolution with ξ_λ). In the group case one knows that ξ_λ is given by $d_\lambda \Theta_\lambda$, where d_λ is the formal degree and Θ_λ the character of π_λ (see [38, §5]), but there is no obvious generalization of this formula.

We shall not try to describe the proofs of the above theorems. However as the construction of $\mathcal{U}_{\lambda,K}$ can be described by quite elementary methods we would like to indicate it.

Let the notation be as above, and recall the decomposition (1) of $\mathfrak{g}$. Let $\mathfrak{g}^d$ be the real form of $\mathfrak{g}_{\mathbb{C}}$ given by

$$\mathfrak{g}^d = \mathfrak{h} \cap \mathfrak{k} \oplus i(\mathfrak{h} \cap \mathfrak{p}) \oplus i(\mathfrak{q} \cap \mathfrak{k}) \oplus \mathfrak{q} \cap \mathfrak{p},$$

where i is the imaginary unit. Assume (again for simplicity of exposition) that G is a real form of a linear complex Lie group $G_{\mathbb{C}}$, and let G^d be the real form of $G_{\mathbb{C}}$ whose Lie algebra is $\mathfrak{g}^d$. Then the subgroup $K^d = G^d \cap H_{\mathbb{C}}$ is a maximal compact subgroup. The Riemannian symmetric space G^d/K^d is called the **noncompact Riemannian form** of G/H. The subgroup $H^d = G^d \cap K_{\mathbb{C}}$ of G^d is a (not necessarily compact) real form of $K_{\mathbb{C}}$. Let $(G \cap G^d)_e$ denote the identity component of $G \cap G^d$. Then both G and G^d are contained in the set $K_{\mathbb{C}}(G \cap G^d)_e H_{\mathbb{C}}$. The K-finite functions on G/H extend naturally to left $K_{\mathbb{C}}$-finite and right $H_{\mathbb{C}}$-invariant functions on this set (and so do the H^d-finite functions on G^d/K^d, provided the H^d-action admits a holomorphic extension to $K_{\mathbb{C}}$). We call this **partial holomorphic extension**. Let $C^\infty(K; G/H)$ and $C^\infty(H^d; G^d/K^d)$ be the spaces of K-finite and H^d-finite smooth functions on G/H and G^d/K^d, respectively. There is a natural action of $\mathfrak{g}_{\mathbb{C}}$ on both of these spaces.

Theorem 3 ([33]). *Partial holomorphic extension defines a $\mathfrak{g}_{\mathbb{C}}$-equivariant linear injection $f \to f^r$ of $C^\infty(K; G/H)$ into $C^\infty(H^d; G^d/K^d)$, the image of which is the set of functions in $C^\infty(H^d; G^d/K^d)$ for which the H^d-action extends holomorphically to $K_{\mathbb{C}}$. Moreover, f is a joint eigenfunction for $\mathbb{D}(G/H)$ if and only if f^r is a joint eigenfunction for $\mathbb{D}(G^d/K^d)$.*

As an example it is quite easily seen in the group case that $G^d = (G_1)_{\mathbb{C}}$, $H^d = (K_1)_{\mathbb{C}}$ and $K^d = U_1$, where K_1 is a maximal compact subgroup in G_1 and U_1 a compact real form of $(G_1)_{\mathbb{C}}$ containing K_1.

The construction of G^d/K^d and Theorem 3 hold independently of assumption (6). However, the latter assumption is crucial for the following construction.

Since G^d/K^d is a Riemannian symmetric space the joint eigenfunctions for the algebra $\mathbb{D}(G^d/K^d)$ can be described by means of the so-called "generalized Poisson transform." This is defined as follows. It follows from the fact that $\mathfrak{t}$ is a maximal Abelian subspace of $\mathfrak{q}$ that $\mathfrak{t}^r = i\mathfrak{t}$ is a maximal Abelian split subspace for $\mathfrak{g}^d$. Hence there is an Iwasawa decomposition

$$G^d = K^d T^r N^d \tag{8}$$

of G^d with $T^r = \exp \mathfrak{t}^r$ that corresponds to a given Σ^+. Let $P^d = M^d T^r N^d$ be the corresponding minimal parabolic subgroup in G^d, and for $\lambda \in \mathfrak{t}^*_{\mathbb{C}}$ let $\mathcal{D}'_\lambda = \mathcal{D}'_\lambda(G^d/P^d)$ be the space of $(\lambda - \rho)$-homogeneous distributions on G^d/P^d, that is, the space of generalized functions f on G^d satisfying

$$f(gman) = a^{\lambda-\rho} f(g), \qquad g \in G^d,\ m \in M^d,\ a \in T^r,\ n \in N^d.$$

The group G^d acts from the left on this space. The **Poisson transform** $\mathcal{P}_\lambda : \mathcal{D}'_\lambda \to C^\infty(G^d/H^d)$ is defined by

$$\mathcal{P}_\lambda f(x) = \int_{K^d} f(xk)\, dk = \int_{K^d} p_\lambda(x,k) f(k)\, dk, \qquad x \in G^d.$$

Here the **Poisson kernel** $p_\lambda \in C^\infty(G^d \times K^d)$ is defined by $p_\lambda(x,k) = a^{-\lambda-\rho}$, where $a \in T^r$ is the T^r-part of $x^{-1}k$ in the decomposition (8). It is known that $\mathcal{P}_\lambda$ is a G^d-equivariant injective transformation into a joint eigenspace for $\mathbb{D}(G^d/K^d)$ in $C^\infty(G^d/K^d)$ if Σ^+ is given by (7); see e.g. [48, §II.3.4].

The nonvanishing of $\mathcal{U}_\lambda$ for "large" $\lambda \in \Lambda$ follows by a simple construction of an element ψ_λ in $\mathcal{U}_{\lambda,K}$ involving the following formula and Theorem 3:

$$\psi_\lambda^r(x) = \int_{K\cap H} p_\lambda(x,k)\,dk, \qquad x \in G^d;$$

see [33] or [34].

Let $\mathcal{D}'_{\lambda,H^d}$ be the set of H^d-finite elements in $\mathcal{D}'_\lambda$, and let $\mathcal{D}'_{\lambda,H^d}(H^dP^d)$ denote the subset of elements supported on the H^d-orbit H^dP^d in G^d/P^d (which is closed, according to [52] or [63, Proposition 7.1.8]). Let now $\lambda \in \Lambda$. Then condition (ii) implies that the H^d-finite action on $\mathcal{D}'_{\lambda,H^d}(H^dP^d)$ extends to a holomorphic $K_{\mathbb{C}}$-action. The space $\mathcal{U}_{\lambda,K}$ is defined by

$$\mathcal{U}_{\lambda,K} = \{f \in C^\infty(K;G/H) \mid f^r \in \mathcal{P}_\lambda(\mathcal{D}'_{\lambda,H^d}(H^dP^d))\}.$$

The proof that $\mathcal{U}_{\lambda,K} \subset L^2_\lambda(G/H)$ can be found in [58] (see also [9, Theorem 19.1]).

5. The Most Continuous Part of $L^2(G/H)$

In this section we discuss Problems (a), (b), and (c) of Section 3.2 for the "most continuous part" of $L^2(G/H)$ (to be defined below). The main references are [11] and [13].

5.1. The Fourier transform

Let notation be as in Section 2. In [11] and [13] the assumptions on G/H are somewhat more general, but for the sake of exposition we shall not discuss this point further. The representations $\pi_{\xi,\lambda}$ that occur in the most continuous part of $L^2(G/H)$ are constructed as follows. Let $P = MAN$ be a parabolic subgroup of G, with the indicated Langlands decomposition, satisfying $\sigma\theta P = P$ and being minimal with respect to this condition. Then M and A are σ-stable. Let $\mathfrak{a}_{\mathrm{q}} = \mathfrak{a}\cap\mathfrak{q}$, where $\mathfrak{a}$ is the Lie algebra of A. Then it follows that $\mathfrak{a}_{\mathrm{q}}$ is a maximal Abelian subspace of $\mathfrak{p}\cap\mathfrak{q}$, and that the Levi part MA of P is the centralizer of $\mathfrak{a}_{\mathrm{q}}$ in G. Let $(\xi,\mathcal{H}_\xi) \in M^\wedge_{\mathrm{fu}}$, the set of (equivalence classes of) finite-dimensional irreducible unitary representations of M, and let $\lambda \in i\mathfrak{a}^*$. We require that $\lambda \in i\mathfrak{a}^*_{\mathrm{q}}$, that is, that λ vanish on $\mathfrak{a}\cap\mathfrak{h}$. Then by definition $\pi_{\xi,\lambda}$ is the induced representation $\pi_{P,\xi,\lambda} = \mathrm{Ind}^G_{P=MAN}(\xi\otimes e^\lambda\otimes 1)$ (the **principal series** for G/H), that is, the representation space $\mathcal{H}_{\xi,\lambda}$ consists of (classes of) $\mathcal{H}_\xi$-valued measurable functions f on G, square integrable on K and satisfying

$$f(gman) = a^{-\lambda-\rho}\xi(m)^{-1}f(g), \qquad g\in G,\ m\in M,\ a\in A,\ n\in N, \tag{9}$$

and G acts from the left. Here $\rho = \frac{1}{2}\operatorname{Tr}\mathrm{Ad}_{\mathfrak{n}} \in \mathfrak{a}^*_{\mathrm{q}}$. (The convention in (9) differs from the above cited references: The induction takes place on the opposite side.) The representations $\pi_{\xi,\lambda}$ are irreducible for almost all $\lambda \in i\mathfrak{a}^*_{\mathrm{q}}$ by Bruhat's theorem (see [6, Theorem 2.6]).

The Plancherel decomposition for the most continuous part of $L^2(G/H)$ is obtained by realizing the abstract Fourier transform explicitly for the principal series.

This realization is then a partial isometry of $L^2(G/H)$ onto the direct integral

$$\int_{\xi,\lambda}^{\oplus} m_\xi \, \pi_{\xi,\lambda} \, d\mu(\xi,\lambda). \tag{10}$$

The multiplicities m_ξ (which happen to be independent of λ) and the measure $d\mu(\xi,\lambda)$ are explicitly described below. The most continuous part of $L^2(G/H)$, denoted $L^2_{\mathrm{mc}}(G/H)$, is then by definition the orthocomplement of the kernel of this partial isometry. Its Plancherel decomposition is exactly given by (10).

In order to realize the Fourier transform we must first discuss the space $\mathcal{V}_{\xi,\lambda} = (\mathcal{H}_{\xi,\lambda}^{-\infty})^H$. Let $\mathcal{W} \subset N_K(\mathfrak{a}_{\mathrm{q}})$ be a fixed set of elements such that $w \mapsto HwP$ parametrizes the open $H \times P$ orbits on G. (It is known (see [62] or [52]) that any set of representatives for the double quotient $N_{K\cap H}(\mathfrak{a}_{\mathrm{q}})\backslash N_K(\mathfrak{a}_{\mathrm{q}})/Z_K(\mathfrak{a}_{\mathrm{q}})$ can be used as $\mathcal{W}$ – in particular, $\mathcal{W}$ is finite.) Viewing an element $f \in \mathcal{H}_{\xi,\lambda}^{-\infty}$ as an $\mathcal{H}_\xi$-valued distribution on G, satisfying appropriate conditions of homogeneity for the right action of P, we see easily that if f is H-invariant then f must restrict to a smooth function on each open $H \times P$ orbit. Hence it makes sense to evaluate f at the elements of $\mathcal{W}$, and in fact its restriction to the open orbit HwP will be uniquely determined from the value at w. We denote this value by $\mathrm{ev}_w(f)$. It is easily seen that ev_w maps $\mathcal{V}_{\xi,\lambda}$ into the space $\mathcal{H}_\xi^{w^{-1}(M\cap H)w}$ of $w^{-1}(M\cap H)w$-fixed elements in $\mathcal{H}_\xi$. (Note that $w^{-1}Mw = M$, but $w^{-1}Hw$ may differ from H.) Let $V(\xi)$ denote the formal direct sum

$$V(\xi) = \bigoplus_{w\in\mathcal{W}} \mathcal{H}_\xi^{w^{-1}(M\cap H)w}, \tag{11}$$

provided with the direct-sum inner product. (Thus, by definition the summands are mutually orthogonal, even though this may not be the case in $\mathcal{H}_\xi$.) Furthermore, let

$$\mathrm{ev}\colon \mathcal{V}_{\xi,\lambda} \to V(\xi)$$

denote the direct sum of the maps ev_w. The construction of the induced representations $\pi_{\xi,\lambda}$ and of the map ev makes sense for $\lambda \in \mathfrak{a}_{\mathrm{q}\mathbb{C}}^*$, the complex linear dual of $\mathfrak{a}_{\mathrm{q}}$, even though the representations need not be unitary for λ outside $i\mathfrak{a}_{\mathrm{q}}^*$. We now have

Theorem 4 ([3]). *The map* ev *is bijective for generic* $\lambda \in \mathfrak{a}_{\mathrm{q}\mathbb{C}}^*$.

In this context "generic" means "outside a countable union of complex hyperplanes." For generic λ, let

$$j(\xi,\lambda)\colon V(\xi) \to \mathcal{V}_{\xi,\lambda}$$

be the inverse of ev. Then by definition we have for $\eta \in V(\xi)$ that the restriction of the distribution $j(\xi,\lambda)(\eta)$ to the open $H\times P$ orbit HwP, $w \in \mathcal{W}$, is the smooth $\mathcal{H}_\xi$-valued function given by

$$j(\xi,\lambda)(\eta)(hwman) = a^{-\lambda-\rho}\xi(m^{-1})\eta_w. \tag{12}$$

Here η_w denotes the w-component of η, viewed as an element of $\mathcal{H}_\xi$. Notice that if G/H is a Riemannian symmetric space, so that $H = K$, then we have $G = HP$ by the Iwasawa decomposition. Hence we can take $\mathcal{W} = \{e\}$, and since $M \subset K = H$

we have $V(\xi) = \{0\}$ unless ξ is the trivial representation $\mathbf{1}$, in which case $V(\mathbf{1}) = \mathbb{C}$. Then $j(\mathbf{1}, \lambda)$ is completely determined by (12); in fact we have

$$j(\mathbf{1}, \lambda)(x) = e^{-(\lambda+\rho)H(x)},$$

where $H \colon G \to \mathfrak{a}$ is the Iwasawa projection (since $V(\mathbf{1}) = \mathbb{C}$, we can omit η). Thus the kernel $p_\lambda(x, k) = j(\mathbf{1}, \lambda)(x^{-1}k)$ on $G/K \times K$ is the generalized Poisson kernel. For general G/H we can supplement (12) as follows: If $\mathrm{Re}\langle \lambda + \rho, \alpha \rangle < 0$ for all α in the set Σ^+ of positive roots (the $\mathfrak{a}$-roots of $\mathfrak{n} = \mathrm{Lie}(N)$), then $j(\xi, \lambda)(\eta)$ is the continuous function on G given by (12) on HwP for all $w \in \mathcal{W}$ and vanishing on the complement of these sets. (The condition on λ ensures the continuity.) For elements λ outside the above region the distribution $j(\xi, \lambda)$ can be obtained from the above by meromorphic continuation. (See [59], [56], [3]; these results have been generalized to other principal series representations in [21], [24].)

Having constructed the H-invariant distribution vectors $j(\xi, \lambda)\eta$ as above, we can now attempt to write down a Fourier transform for the principal series. For $f \in C_c^\infty(G/H)$ we consider the map

$$(\xi, \lambda) \mapsto f^\wedge(\pi_{\xi,\lambda}) j(\xi, \lambda) = \pi_{\xi,\lambda}(f)\, j(\xi, \lambda) \in \mathcal{H}_{\xi,\lambda}^\infty \otimes V(\xi)^*. \tag{13}$$

In the Riemannian case this is exactly the Fourier transform, as defined by Helgason (see [43]). However when G/H is not Riemannian a new phenomenon may occur: by the above definitions (13) is a meromorphic function in λ, which may have singularities on the set $i\mathfrak{a}_\mathrm{q}^*$ of interest for the Plancherel decomposition, and thus it may not make sense for some singular $\lambda \in i\mathfrak{a}_\mathrm{q}^*$. This unpleasantness is overcome by a suitable normalization of $j(\xi, \lambda)$ that removes the singularities. The normalization is carried out by means of the standard intertwining operators $A(\bar{P}, P, \xi, \lambda)$ from $\pi_{P,\xi,\lambda}$ to $\pi_{\bar{P},\xi,\lambda}$, where $\bar{P}$ is the parabolic subgroup opposite to P. Let

$$j^\circ(\xi, \lambda) = A(\bar{P}, P, \xi, \lambda)^{-1} j(\bar{P}, \xi, \lambda),$$

where $j(\bar{P}, \xi, \lambda)$ is constructed as $j(\xi, \lambda)$ above, but with P replaced by $\bar{P}$. Since the intertwining operator $A(\bar{P}, P, \xi, \lambda)$ is bijective for generic λ, it follows that

$$j^\circ(\xi, \lambda) \colon V(\xi) \to \mathcal{V}_{\xi,\lambda}$$

is again a bijection, for generic λ. Moreover, we now have

Theorem 5 ([11]). *The meromorphic function $\lambda \mapsto j^\circ(\xi, \lambda)$ is regular on $i\mathfrak{a}_\mathrm{q}^*$.*

We can now define the **Fourier transform** $f \mapsto f^\wedge$ for the principal series properly by (13), but with j replaced by j°:

$$f^\wedge(\xi, \lambda) = \pi_{\xi,\lambda}(f)\, j^\circ(\xi, \lambda) \in \mathcal{H}_{\xi,\lambda}^\infty \otimes V(\xi)^*.$$

Notice that when G/H is Riemannian the normalization makes our Fourier transform different from that of Helgason – in this case the normalization amounts to a division by Harish-Chandra's c-function $\mathbf{c}(\lambda)$. See [10] for the determination of j° in the group case.

We can now give the solution to Problem (b) for this part of $L^2(G/H)$: We take $\mathcal{V}_{\xi,\lambda}^o = \mathcal{V}_{\xi,\lambda}$, and give it the Hilbert space structure that makes $j^\circ(\xi, \lambda)$ an isometry. The solution to Problem (c) is as follows. Let $\mathcal{H}$ be the Hilbert space given by

$$\mathcal{H} = \int_{\xi,\lambda}^\oplus \mathcal{H}_{\xi,\lambda} \otimes V(\xi)^* \, d\mu(\xi, \lambda), \tag{14}$$

with the measure $d\mu(\xi,\lambda) = \dim(\xi)\, d\lambda$, where $d\lambda$ is Lebesgue measure on $i\mathfrak{a}_{\mathrm{q}}^*$ (suitably normalized). Here ξ runs over $M_{\mathrm{fu}}^\wedge$ (but of course the ξ's with $V(\xi)$ trivial do not contribute), and λ runs over an open chamber $i\mathfrak{a}_{\mathrm{q}}^{*+}$ in $i\mathfrak{a}_{\mathrm{q}}^*$ for the Weyl group $W_{\mathrm{q}} = N_K(\mathfrak{a}_{\mathrm{q}})/Z_K(\mathfrak{a}_{\mathrm{q}})$.

Theorem 6 ([13]). *Let $f \in C_c^\infty(G/H)$. Then $f^\wedge \in \mathcal{H}$ and $\|f^\wedge\| \leq \|f\|_2$. Moreover, the map $f \mapsto f^\wedge$ extends to an equivariant partial isometry $\mathfrak{F}$ of $L^2(G/H)$ onto $\mathcal{H}$. In particular, the multiplicity of $\pi_{\xi,\lambda}$ is $m_\xi = \dim V(\xi)$ for almost all λ.*

We define the **most continuous part** $L^2_{\mathrm{mc}}(G/H)$ of $L^2(G/H)$ as the orthocomplement of the kernel of $\mathfrak{F}$. That $\mathfrak{F}$ is a **partial isometry** means by definition that it restricts to an isometry of $L^2_{\mathrm{mc}}(G/H)$ onto $\mathcal{H}$. In [13] it is shown that $L^2_{\mathrm{mc}}(G/H)$ is "large" in $L^2(G/H)$ in a certain sense – in particular its orthocomplement (the kernel of $\mathfrak{F}$) has trivial intersection with $C_c^\infty(G/H)$. (Thus $f \mapsto f^\wedge$ is injective, even though the extension $\mathfrak{F}$ need not be.)

Moreover, if G/H has split rank one, that is if $\dim \mathfrak{a}_{\mathrm{q}} = 1$, then there are at most two conjugacy classes of Cartan subspaces, and hence one expects from the analogy with the group case as mentioned earlier that only the corresponding two "series" of representations will be present. Indeed this is the case; it is shown in [13] that the kernel of $\mathfrak{F}$ decomposes discretely when the split rank is one. Thus, in this case the Plancherel decomposition of $L^2(G/H)$ can be determined from Theorem 6 together with the description of the discrete series (see Section 4 above), except for the explicit determination of the Hilbert space structure on $\mathcal{V}_\pi^o$ for the discrete series representations π.

On the other hand, when G/H is Riemannian, then $\mathfrak{F}$ is injective and Theorem 6 gives the complete Plancherel decomposition of $L^2(G/H)$. (In the formulation of Harish-Chandra and Helgason the Plancherel measure is $|\mathbf{c}(\lambda)|^{-2}\, d\lambda$, but here the factor $|\mathbf{c}(\lambda)|^{-2}$ disappears because of the normalization of j°.)

A further discussion of the multiplicities m_π can be found in [10].

5.2. The spherical Fourier transform

The isomorphism of (14) onto $L^2_{\mathrm{mc}}(G/H)$ (the "inverse Fourier transform") can be given more explicitly when one restricts to K-finite functions. In this subsection we shall discuss this restriction, which happens to be crucial in the proofs of Theorems 5 and 6.

5.2.1. Eisenstein integrals. Let (μ, V_μ) be a fixed, irreducible unitary representation of K. Taking μ-components in (14) we have

$$\mathcal{H}^\mu = \int_{\xi,\lambda}^\oplus \mathcal{H}^\mu_{\xi,\lambda} \otimes V(\xi)^*\, d\mu(\xi,\lambda). \tag{15}$$

Moreover, by Frobenius reciprocity we have

$$\mathcal{H}^\mu_{\xi,\lambda} \simeq \mathrm{Hom}_{M\cap K}(V_\mu, \mathcal{H}_\xi) \otimes V_\mu \tag{16}$$

as K-modules (where K acts on the second component in the tensor product), for all $\xi \in M_{\mathrm{fu}}^\wedge, \lambda \in \mathfrak{a}_{\mathrm{q}\mathbb{C}}^*$. Note that since each representation $\xi \in M_{\mathrm{fu}}^\wedge$ is trivial on the noncompact part of M, we have that $\xi|_{M\cap K}$ is irreducible, and that $\mathrm{Hom}_{M\cap K}(V_\mu, \mathcal{H}_\xi)$ is nontrivial if and only if this restriction occurs as a subrepresentation of $\mu|_{M\cap K}$. We use the notation $\xi \uparrow \mu$ to indicate this occurrence; it happens only for finitely many ξ. Thus if we pass to K-types, the integral over ξ in (15) becomes a finite sum, hence more manageable. In analogy with the earlier definition of the space $V(\xi)$ we now define the space $\mathcal{V}(\mu)$ to be the formal direct sum

$$\mathcal{V}(\mu) = \bigoplus_{w \in \mathcal{W}} V_\mu^{w^{-1}(K \cap M \cap H)w}.$$

It is easily seen from the above that

$$\mathcal{V}(\mu) \simeq \bigoplus_{\xi \uparrow \mu} \mathrm{Hom}_{M \cap K}(\mathcal{H}_\xi, V_\mu) \otimes V(\xi). \tag{17}$$

Hence in view of (16) we have

$$\mathcal{V}(\mu)^* \otimes V_\mu \simeq \bigoplus_{\xi \uparrow \mu} \mathcal{H}^\mu_{\xi,\lambda} \otimes V(\xi)^* \tag{18}$$

for all $\lambda \in \mathfrak{a}^*_{\mathrm{q}\mathbb{C}}$. From (15) and (18) we finally obtain

$$\mathcal{H}^\mu \simeq \int_\lambda^\oplus \mathcal{V}(\mu)^* \otimes V_\mu \, d\lambda \simeq L^2(i\mathfrak{a}_\mathrm{q}^{*+}) \otimes \mathcal{V}(\mu)^* \otimes V_\mu. \tag{19}$$

This isomorphism indicates that the Fourier transform, when restricted to K-finite functions of type μ, can be considered as a map into the $\mathcal{V}(\mu)^* \otimes V_\mu$-valued functions on $i\mathfrak{a}^*_\mathrm{q}$.

Instead of working with K-finite scalar-valued functions on G/H, it is convenient to consider "μ-spherical" functions f on G/H, that is, V_μ-valued functions satisfying

$$f(kx) = \mu(k) f(x), \qquad k \in K,\, x \in G/H.$$

If $L^2(G/H;\mu)$ denotes the space of square integrable such functions, then by contraction we have a K-equivariant isomorphism

$$\gamma_\mu \colon L^2(G/H; \mu^\vee) \otimes V_\mu \xrightarrow{\sim} L^2(G/H)^\mu. \tag{20}$$

Again K acts on the second component in the tensor product. The map $\dim(\mu)\gamma_\mu$ is an isometry. Notice that when passing from K-finite functions to spherical functions one must also pass from μ to its contragradient $\mu^\vee$. Since $\mathcal{V}(\mu)^* = \mathcal{V}(\mu^\vee)$ we are led to the search, for each μ, of a Fourier transform that is a partial isometry of $L^2(G/H;\mu)$ onto $L^2(i\mathfrak{a}_\mathrm{q}^{*+}) \otimes \mathcal{V}(\mu)$. Going through the above isomorphisms in detail, we are led to the following construction culminating in (26), which essentially is the "projection" of the construction of $f \mapsto f^\wedge$ to functions of type μ.

For $\psi \in \mathcal{V}(\mu)$ and $\lambda \in \mathfrak{a}^*_{\mathrm{q}\mathbb{C}}$ with $\mathrm{Re}\langle \lambda + \rho, \alpha \rangle < 0$ for all $\alpha \in \Sigma^+$, let $\widetilde{\psi}_\lambda$ be the V_μ-valued function on G defined by

$$\widetilde{\psi}_\lambda(x) = \begin{cases} a^{-\lambda-\rho} \mu(m^{-1}) \psi_w & \text{if } x = hwman \in Hw(M \cap K)AN,\ w \in \mathcal{W}, \\ 0 & \text{if } x \notin \cup_{w \in \mathcal{W}} HwP, \end{cases}$$

where ψ_w denotes the w-component of ψ. (Note that $M = w^{-1}(M \cap H)w(M \cap K)$, and hence $Hw(M \cap K)AN = HwMAN$.) It can be shown that $\widetilde{\psi}_\lambda$ is continuous as a function of x, and has a distribution-valued meromorphic continuation in $\lambda \in \mathfrak{a}^*_{\mathrm{q}\mathbb{C}}$. Let $E_\mu(\psi, \lambda)$ be the μ-spherical function on G/H defined by

$$E_\mu(\psi, \lambda)(x) = \int_K \mu(k) \widetilde{\psi}_\lambda(x^{-1}k)\, dk.$$

It can be seen that the vector components of $E_\mu(\psi, \lambda)$ are linear combinations of generalized matrix coefficients formed by the $j(\xi, \lambda)\eta$ for $\eta \in V(\xi)$ and $\xi \uparrow \mu^\vee$, using K-finite vectors of type μ; in particular, $E_\mu(\psi, \lambda)$ is a smooth function on G/H, even when $\widetilde{\psi}_\lambda$ is only a distribution. We call these functions **Eisenstein integrals** for G/H. When G/H is Riemannian and μ is the trivial K-type $\mathbf{1}$, the construction produces the spherical functions

$$\varphi_\lambda(x) = \int_K e^{-(\lambda+\rho)H(x^{-1}k)}\,dk, \tag{21}$$

and for other K-types we get the generalized spherical functions of [45]. In the group case the Eisenstein integrals defined in this manner coincide, up to normalization, with Harish-Chandra's Eisenstein integrals associated to the minimal parabolic subgroup.

The spherical functions are eigenfunctions for the invariant differential operators on G/K – in analogy we have

$$DE_\mu(\psi, \lambda) = E_\mu(\chi_\mu(D, \lambda)\psi, \lambda) \tag{22}$$

for all $D \in \mathbb{D}(G/H)$. Here $\chi_\mu(D)$ is an $\mathrm{End}(\mathcal{V}(\mu))$-valued polynomial in λ. Just as is the case for the spherical functions, one can derive an asymptotic expansion from this "eigenequation." Here we have to recall the "KAH"-decomposition of G,

$$G = \mathrm{cl} \bigcup_{w\in\mathcal{W}} KA_{\mathrm{q}}^{+}w^{-1}H, \tag{23}$$

where A_{q}^{+} is the exponential of the positive chamber in $\mathfrak{a}_{\mathrm{q}}$ corresponding to Σ^+, and where the union inside the closure operator cl is disjoint. Since the Eisenstein integrals are K-spherical, we have to consider their behavior on $A_{\mathrm{q}}^{+}w^{-1}$, for all $w \in \mathcal{W}$. Notice that when G/H is Riemannian there is only one "direction" to control, since the KAH-decomposition then specializes to the Cartan decomposition $G = \mathrm{cl}\, KA^+K$. The expansion is essentially as follows (see [4]):

$$E_\mu(\psi, \lambda)(aw^{-1}) = \sum_{s\in W_{\mathrm{q}}} a^{s\lambda-\rho}[C(s, \lambda)\psi]_w + \text{ lower order terms in } a, \tag{24}$$

for $a \in A_{\mathrm{q}}^{+}$ and $w \in \mathcal{W}$, where W_{q} is as defined above Theorem 6 and the "c-function" $\lambda \mapsto C(s, \lambda)$ is a meromorphic function on $\mathfrak{a}_{\mathrm{q}\mathbb{C}}^{*}$ with values in $\mathrm{End}(\mathcal{V}(\mu))$. It follows easily from the μ-sphericality that we have

$$E_\mu(\psi, \lambda)(aw^{-1}) \in V_\mu^{w^{-1}(K\cap M\cap H)w}$$

for $a \in A_{\mathrm{q}}$. The expansion converges for $a \in A_{\mathrm{q}}^{+}$; the "lower-order terms" involve powers of the form $a^{s\lambda-\rho-\nu}$ where ν is a sum of positive roots.

The expression (24) is analyzed in [12], where it is shown that it takes the form

$$E_\mu(\psi, \lambda)(aw^{-1}) = \sum_{s\in W_{\mathrm{q}}} \Phi_w(s\lambda, a)[C(s, \lambda)\psi]_w \tag{25}$$

for each $w \in \mathcal{W}$, where $\Phi_w(\lambda, \cdot) \in \mathrm{End}(V_\mu^{w^{-1}(K\cap H\cap M)w})$ is given on A_{q}^{+} by a converging power series with $a^{\lambda-\rho}$ as its leading term.

5.2.2. The Fourier transform. It would now be natural to define the Fourier transform $\mathcal{F}_\mu f$ of a function $f \in C_c^\infty(G/H;\mu)$, the space of compactly supported and smooth μ-spherical functions on G/H, as the $\mathcal{V}(\mu)$-valued function φ on $\mathfrak{a}_{\mathrm{q}\mathbb{C}}^*$ given by

$$\langle \varphi(\lambda) \mid \psi \rangle = \int_{G/H} \langle f(x) \mid E_\mu(\psi, -\bar{\lambda})(x) \rangle \, dx, \qquad \psi \in \mathcal{V}(\mu),$$

where the inner products $\langle \cdot \mid \cdot \rangle$ are the sesquilinear Hilbert space inner products on $\mathcal{V}(\mu)$ and V_μ, respectively. Via the isomorphisms in (19) and (20) this would essentially correspond to the Fourier transform in (13). However, as with $j(\xi,\lambda)$ we have the problem that $E_\mu(\psi,\lambda)$, which is meromorphic in λ, may have singularities on $i\mathfrak{a}_\mathrm{q}^*$. Again we have to carry out a normalization: the **normalized Eisenstein integral** is defined by

$$E_\mu^\circ(\psi,\lambda) = E_\mu(C(1,\lambda)^{-1}\psi,\lambda).$$

In other words, the Eisenstein integral is normalized by its asymptotics, so that we have $E_\mu^\circ(\psi,\lambda)(aw^{-1}) \sim a^{\lambda-\rho}\psi_w$ for $a \in A_\mathrm{q}^+$, $w \in \mathcal{W}$, and $\operatorname{Re}\lambda$ strictly dominant. It can be shown that this normalization corresponds to the one on $j(\xi,\lambda)$, in the sense that the vector components of $E_\mu^\circ(\psi,\lambda)$ are linear combinations of matrix coefficients formed by the $j^\circ(\xi,\lambda)\eta$ for $\eta \in V(\xi)$ and $\xi \uparrow \mu^\vee$, using K-finite vectors of type μ. Moreover, it can be shown that the statement of Theorem 5 is equivalent with the following "K-finite version":

Theorem 7 ([11]). *The meromorphic function $\lambda \mapsto E_\mu^\circ(\psi,\lambda)$ is regular on $i\mathfrak{a}_\mathrm{q}^*$, for every $\mu \in K^\wedge$ and $\psi \in \mathcal{V}(\mu)$.*

A proof of Theorem 7, different from the original proof in [11] and valid for the generalized principal series as well, is given in [7]. With the result of Theorem 7 in mind we define the **μ-spherical Fourier transform** $\mathcal{F}_\mu f$ as above, but with E_μ replaced by E_μ°, that is, by

$$\langle \mathcal{F}_\mu f(\lambda) \mid \psi \rangle = \int_{G/H} \langle f(x) \mid E_\mu^\circ(\psi, -\bar{\lambda})(x) \rangle \, dx, \qquad \psi \in \mathcal{V}(\mu). \tag{26}$$

Then $\mathcal{F}_\mu f$ corresponds to $f^\wedge$ via the isomorphisms in (18) and (20).

When G/H is Riemannian and $\mu = \mathbf{1}$, the normalization again amounts to division by $\mathbf{c}(\lambda)$, and thus $\mathcal{F}_\mu f$ is in this case related to the spherical Fourier transform of f as follows:

$$\mathcal{F}_\mu f(\lambda) = \mathbf{c}(-\lambda)^{-1} \int_{G/K} f(x)\varphi_{-\lambda}(x)\, dx,$$

where φ_λ is the elementary spherical function in (21). If G/H is Riemannian and μ is nontrivial there is a similar relation, also involving $\mathbf{c}(\lambda)^{-1}$, to the Fourier transform in [45].

If $C^\circ(s,\lambda) = C(s,\lambda)C(1,\lambda)^{-1}$, then we have from (24)-(25)

$$\begin{aligned} E_\mu^\circ(\psi,\lambda)(aw^{-1}) &= \sum_{s\in W_\mathrm{q}} a^{s\lambda-\rho}[C^\circ(s,\lambda)\psi]_w + \text{ lower order terms in } a \\ &= \sum_{s\in W_\mathrm{q}} \Phi_w(s\lambda,a)[C^\circ(s,\lambda)\psi]_w. \end{aligned} \tag{27}$$

The following theorem generalizes results of Helgason and Harish-Chandra (the Maass-Selberg relations) for the Riemannian case and the group case, respectively. See [44, Theorem 6.6], [40, Lemma 17.6].

Theorem 8 ([4], [5]). *For every $s \in W_{\mathrm{q}}$ we have the following identity of meromorphic functions:*

$$C^\circ(s,\lambda)C^\circ(s,-\bar{\lambda})^* = I_{\mathcal{V}(\mu)}, \qquad \lambda \in \mathfrak{a}^*_{\mathrm{q}\mathbb{C}}.$$

In particular, for $\lambda \in i\mathfrak{a}^*_{\mathrm{q}}$, the endomorphism $C^\circ(s,\lambda)$ of $\mathcal{V}(\mu)$ is unitary.

Notice that by Riemann's boundedness theorem it follows from the above result that the meromorphic function $\lambda \mapsto C^\circ(s,\lambda)$ has no singularities on $i\mathfrak{a}^*_{\mathrm{q}}$. Therefore the possible singularities of $E^\circ_\mu(\psi,\lambda)$ must occur in the lower order terms of (27). This observation plays a crucial role in the proof of Theorem 7.

On G/K the spherical functions satisfy the functional equation $\varphi_{s\lambda} = \varphi_\lambda$, for all $s \in W_{\mathrm{q}}$. The analog for the normalized Eisenstein integral on G/H is

$$E^\circ_\mu(C^\circ(s,\lambda)\psi, s\lambda) = E^\circ_\mu(\psi,\lambda). \tag{28}$$

See [4, Proposition 16.4]. For the group case, see also [40, Lemma 17.2].

Though $E^\circ_\mu(\psi,\lambda)$ by Theorem 7 is regular on $i\mathfrak{a}^*_{\mathrm{q}}$, it will in general have singularities elsewhere on $\mathfrak{a}^*_{\mathrm{q}\mathbb{C}}$. It is remarkable, though, that in a certain direction only finitely many singularities occur. To be more precise, one has the following. Let

$$(\mathfrak{a}^*_{\mathrm{q}\mathbb{C}})_+ = \{\lambda \in \mathfrak{a}^*_{\mathrm{q}\mathbb{C}} \mid \mathrm{Re}\langle\lambda,\alpha\rangle \geq 0,\ \alpha \in \Sigma^+\},$$

and put $(\mathfrak{a}^*_{\mathrm{q}\mathbb{C}})_- = -(\mathfrak{a}^*_{\mathrm{q}\mathbb{C}})_+$.

Theorem 9 ([4]). *There exists a polynomial π' on $\mathfrak{a}^*_{\mathrm{q}\mathbb{C}}$ that is a product of linear factors of the form $\lambda \mapsto \langle\lambda,\alpha\rangle+$constant, with α a root, such that $\pi'(\lambda)E^\circ_\mu(\psi,\lambda)$ is holomorphic on a neighborhood of $(\mathfrak{a}^*_{\mathrm{q}\mathbb{C}})_+$.*

Notice that π' depends on the K-type μ. Notice also that when G/H is Riemannian we actually have that $E^\circ_\mu(\psi,\lambda)$ itself is holomorphic on $(\mathfrak{a}^*_{\mathrm{q}\mathbb{C}})_+$. Indeed, the spherical functions are everywhere holomorphic, and the normalizing divisor $\mathbf{c}(\lambda)$ has no zeros on this set. Thus, for this case one can take $\pi' = 1$.

It follows from Theorem 9 and (26) that if we put

$$\pi(\lambda) = \overline{\pi'(-\bar{\lambda})} \tag{29}$$

then $\lambda \mapsto \pi(\lambda)\mathcal{F}_\mu f(\lambda)$ is holomorphic on a neighborhood of $(\mathfrak{a}^*_{\mathrm{q}\mathbb{C}})_-$.

5.2.3. Wave packets. For the μ-spherical Fourier transform a "partial inversion formula" is given in [13] as follows. For a $\mathcal{V}(\mu)$-valued function φ on $i\mathfrak{a}^*_{\mathrm{q}}$ of suitable decay one can form a **wave packet**, which is the superposition of normalized Eisenstein integrals, with amplitudes given by φ, i.e.,

$$\mathcal{J}_\mu\varphi(x) = \int_{i\mathfrak{a}^*_{\mathrm{q}}} E^\circ_\mu(\varphi(\lambda),\lambda)(x)\,d\lambda, \qquad x \in G/H. \tag{30}$$

It is easily seen that the transform $\mathcal{J}_\mu$ is the transpose of $\mathcal{F}_\mu$. For the Euclidean Fourier transform (and more generally for the spherical Fourier transform on a Riemannian symmetric space) this transform is also the inverse of $\mathcal{F}_\mu$; the inversion formula states that $\mathcal{J}_\mu\mathcal{F}_\mu$ is the identity operator when measures are suitably normalized. In the non-Riemannian generality of G/H this cannot be expected, because of the possible presence of discrete series. However we do have

Theorem 10 ([13]). *There exists an invariant differential operator D on G/H, depending on μ, that satisfyies the following:*

(i) *As an operator on $C_c^\infty(G/H)$, D is injective and symmetric.*
(ii) $\mathcal{J}_\mu \mathcal{F}_\mu f = f$ *for all* $f \in D(C_c^\infty(G/H;\mu))$.

From (22) one can derive that $\mathcal{J}_\mu \mathcal{F}_\mu D = \mathcal{J}_\mu \chi_\mu(D) \mathcal{F}_\mu = D \mathcal{J}_\mu \mathcal{F}_\mu$. Hence it follows from (ii) that $D(\mathcal{J}_\mu \mathcal{F}_\mu f - f) = 0$ for all $f \in C_c^\infty(G/H;\mu)$. Nevertheless, one cannot then conclude from (i) that in fact $\mathcal{J}_\mu \mathcal{F}_\mu f = f$ because $\mathcal{J}_\mu \mathcal{F}_\mu f$ is not compactly supported in general. The presence of D is important; for example, D annihilates all the discrete series in $L^2(G/H;\mu)$.

The proof of Theorem 10 is very much inspired by Rosenberg's proof (see [61] or [47, Ch. IV, §7]) of the inversion formula for the spherical Fourier transform on G/K, in which case one can take $D = 1$. A key step in both proofs is the use of a "shift argument," originally used by Helgason for the proof of the Paley-Wiener theorem, where the integration in $\mathcal{J}_\mu$ (after use of (27)) is moved away from $i\mathfrak{a}_{\mathrm{q}}^*$ in the direction of $(\mathfrak{a}_{\mathrm{q}\mathbb{C}}^*)_-$, using Cauchy's theorem. It can be seen that one meets only a finite number of singular hyperplanes in this shift. The purpose of the operator D is to remove these singularities so that no residues are present. (Among other things this means that π should be a divisor in $\chi_\mu(D)$.) The shift allows one to conclude that $\mathcal{J}_\mu \mathcal{F}_\mu D f$ is compactly supported whenever f is, which is an important step in the proof of the theorem.

Theorem 10 is crucial in the proof of Theorem 6. Via the isomorphism (20) one obtains with $\mathcal{J}_{\mu^\vee}$ an explicit formula for the restriction to $\mathcal{H}^\mu$ of the isomorphism of $\mathcal{H}$ onto $L^2_{\mathrm{mc}}(G/H)$.

6. The Plancherel Formula for $L^2(G/H)$. The Intermediate Series

In a more recent development than what was described above, both the Plancherel formula for the full space $L^2(G/H)$ and the Paley-Wiener theorem have been obtained. Both of these results were announced in the seminar at the Mittag-Leffler Institute in November, 1995.

The Plancherel theorem was announced by Delorme; the proof has appeared in [29]. (In 1986 Oshima (see [57, p. 604]) announced that he had obtained a Plancherel formula, but the details have not appeared.)

The Paley-Wiener theorem was announced by the first and last named author of the present paper. They also announced that their proof implies the Plancherel formula for spaces with one conjugacy class of Cartan subspaces, and that in general their proof implies the Plancherel formula under the hypothesis that the identity of Theorem 8 (the Maass-Selberg relations) is valid for generalized Eisenstein integrals (see below). The validity of this hypothesis, which also plays a main role in Delorme's work, has been established by Carmona and Delorme in [25]. The details of the work of van den Ban and Schlichtkrull will appear in [16].

The theory of Eisenstein integrals that was developed in the previous section for the most continuous part of $L^2(G/H)$ can be generalized to the intermediate series as well. This has been done in a series of papers by Delorme and others, [21], [24], [23], [7], [27], [28], [25]. In the above we referred already to the generalization (in [25]) of the identity in Theorem 8 (the Maass-Selberg relations) to these intermediate series. The proof is based on the method of truncation, which was introduced in this context by Delorme in [28]. As a consequence of the generalization of Theorem 8, the regularity in Theorem 7 is extended (also in [25]) to the (generalized) Eisenstein integrals corresponding to the intermediate series. These results are of significance in both of the mentioned approaches to the Plancherel formula. Another important ingredient in [29] (but not in [16]) is an a priori characterization of the support of the Plancherel measure (cf. [24, Appendix C]), which in turn is derived from a result of Bernstein [18].

In [16] the Plancherel formula is derived from an inversion formula for the Fourier transform $\mathcal{F}_\mu$ that was defined on $C_c^\infty(G/H;\mu)$ in (26). This inversion formula is based on the "shift argument" that was described after Theorem 10. Without the presence of the operator D one obtains by this shift an expression involving generalized residues. It is these residues that give rise to the intermediate and the discrete series. At this point the theory of semisimple symmetric spaces resembles (in fact was inspired by) the theory of automorphic forms. The method of using residues to obtain lower-dimensional spectrum plays an important role in the fundamental works of Selberg ([64], [65]) and Langlands ([50], [51]). See also [54]. The use of residues by Langlands is in a multi-variable setting; it is of an inductive nature, involving composition of a sequence of residues along singular hyperplanes that have a common intersection. (A residue along a singular hyperplane is essentially a residue in one transversal variable.) One of the complications in the theory is that a priori the occurring compositions of sequences of one-variable residues may depend on the orders of the sequences. Arthur [1] makes a similar inductive use of one-variable residues in his proof of the Paley-Wiener theorem for a semisimple Lie group.

In [14] the residue calculus needed for analysis on semisimple symmetric spaces is developed. The treatment is entirely in terms of root systems, without reference to analysis on symmetric spaces. Again residues come about as compositions of sequences of one variable residues, but the problem of dependence on the order is circumvented by using an idea of Heckman and Opdam [41]. The result is a definition of residue operators attached to any intersection of affine hyperplanes parallel to root hyperplanes.

To be somewhat more specific, let $(G/H)_+ \subset G/H$ be the dense open subset

$$(G/H)_+ = \bigcup_{w\in\mathcal{W}} KA_{\mathfrak{q}}^+ w^{-1}H,$$

(see (23)), and define a $\operatorname{Hom}(\mathcal{V}(\mu), V_\mu)$-valued function $E_+(\lambda,\,\cdot\,)$ on $(G/H)_+$ by

$$E_+(\lambda, kaw^{-1})\psi = \mu(k)\Phi_w(\lambda, a)\psi_w,$$

(see (25)) for $\lambda \in \mathfrak{a}_{\mathfrak{q}\mathbb{C}}^*$ generic, $k \in K$, $a \in A_{\mathfrak{q}}^+$, $w \in \mathcal{W}$, and $\psi \in \mathcal{V}(\mu)$. Then (27) takes the form

$$E_\mu^\circ(\psi,\lambda)(x) = \sum_{s\in W_{\mathfrak{q}}} E_+(\lambda, x)C^\circ(s,\lambda)\psi, \qquad x \in (G/H)_+. \tag{31}$$

We define, for $f \in C_c^\infty(G/H;\mu)$, $x \in (G/H)_+$, and $\eta \in \mathfrak{a}_{\mathfrak{q}}^*$ generic,

$$\mathcal{T}_\mu^\eta \mathcal{F}_\mu f(x) = |W_{\mathfrak{q}}| \int_{\eta + i\mathfrak{a}_{\mathfrak{q}}^*} E_+(\lambda, x)\mathcal{F}_\mu f(\lambda)\, d\lambda \in V_\mu; \tag{32}$$

it can be shown that this integral converges and defines a smooth function on $(G/H)_+$. The previously mentioned shift argument involves two steps. The first step is the identification of the wave packet $\mathcal{J}_\mu \mathcal{F}_\mu f$ with $\mathcal{T}_\mu^\eta \mathcal{F}_\mu f$ for $\eta = 0$ (or, if this is a singular value, with a certain limit). This is done simply by insertion of (31) into the integral (30) that defines the wave packet. The second step is the actual shift. In the integral (32) η is shifted from 0 towards infinity in the antidominant direction. During this shift a finite number of singular hyperplanes is passed, and some generalized residues are created. For η sufficiently antidominant all the singular hyperplanes have been passed, and $\mathcal{T}_\mu^\eta \mathcal{F}_\mu f$ is then independent of η. We call it (that is, $\mathcal{T}_\mu^\eta \mathcal{F}_\mu f$ for η sufficiently antidominant) a **pseudo wave packet** and denote it by $\mathcal{T}_\mu \mathcal{F}_\mu f$. It is a smooth μ-spherical function on $(G/H)_+$, and it can be shown by taking the limit $\eta \to \infty$ that it vanishes outside a subset of $(G/H)_+$ with compact closure in G/H.

We can now state the inversion formula for the Fourier transform $\mathcal{F}_\mu$.

Theorem 11 ([16]). *Let $f \in C_c^\infty(G/H;\mu)$. Then*

$$\mathcal{T}_\mu \mathcal{F}_\mu f(x) = f(x)$$

for all $x \in (G/H)_+$.

Theorem 11 is established by induction on $\dim \mathfrak{a}_{\mathrm{q}}$. The shift argument described earlier results in a formula expressing the difference $\mathcal{T}_\mu \mathcal{F}_\mu f - \mathcal{J}_\mu \mathcal{F}_\mu f$ of the pseudo wave packet and the wave packet as a sum of integrals of generalized residues. These residual integrals are by their construction given only as smooth functions on $(G/H)_+$; a crucial step is to extend them to smooth functions on G/H. (In fact, the residual integrals are not individually extended; only certain finite combinations extend.)

Let us indicate how the inversion formula and the smooth extension are obtained in the simplest case, when $\dim \mathfrak{a}_{\mathrm{q}} = 1$. (In this case the result in fact follows already from the theory developed in [13].) The residual integrals, by which the pseudo wave packet $\mathcal{T}_\mu \mathcal{F}_\mu f$ differs from the wave packet $\mathcal{J}_\mu \mathcal{F}_\mu f$, are in this case just ordinary residues. If D is as in Theorem 10, then the effect of D is exactly to annihilate these residues, and hence $D\mathcal{T}_\mu \mathcal{F}_\mu f = D\mathcal{J}_\mu \mathcal{F}_\mu f = Df$ by Theorem 10 (ii). Thus the difference $\mathcal{T}_\mu \mathcal{F}_\mu f - f$, which is defined on $(G/H)_+$, is annihilated by D. Being also K-finite this difference is then an analytic function on $(G/H)_+$. However, since both $\mathcal{T}_\mu \mathcal{F}_\mu f$ and f are compactly supported they agree on a nonempty open set, hence everywhere. In other words, the desired inversion formula holds. Moreover the sum of the residues, which we have now identified with $f - \mathcal{J}_\mu \mathcal{F}_\mu f$, extends smoothly to G/H.

The latter conclusion is the starting point for the inductive step that gives the proof for $\dim \mathfrak{a}_{\mathrm{q}} = 2$. In this case there occur two kinds of residual integrals: those along one-dimensional singular hyperplanes, and point residues, which are taken where the singular lines meet. Using some results from [15] and the smooth extension for $\dim \mathfrak{a}_{\mathrm{q}} = 1$, one obtains the smooth extension for the sum of the residual line integrals. The argument for the inversion formula and the smooth extension of the sum of the point residues is now similar to the argument outlined above for $\dim \mathfrak{a}_{\mathrm{q}} = 1$.

The inversion formula in Theorem 11 is the key to the Plancherel formula. According to the proof outlined above, this inversion formula can be written in the form

$$f = \mathcal{J}_\mu \mathcal{F}_\mu f + \text{residual contributions}.$$

What remains for the Plancherel formula is essentially to identify these residual contributions in terms of the intermediate series and the most discrete series. The residues are taken along the singular hyperplanes of the functions involved, and at the intersections of these hyperplanes "higher-order" residues occur. The residues of the highest order are the point residues; it is the sum of these point residues that eventually becomes identified as the projection of f to the discrete series. (In particular, if the discrete series is absent this means that the point residues cancel out.) First, however, the residues of lower order are identified in terms of generalized principal series representations induced from proper parabolic subgroups. It is here that the proof uses Carmona's and Delorme's generalization [25] of Theorem 7. In particular, it follows that these lower-dimensional residual integrals define Schwartz functions. Hence, as a consequence of the inversion formula, the sum of the point residues is also a Schwartz function. Since this is a finite sum of $\mathbb{D}(G/H)$-finite functions, one can conclude that it belongs to the discrete series.

7. A Paley-Wiener Theorem for G/H

Let π' be the minimal polynomial satisfying the conclusion of Theorem 9, and as before let π be given by (29). We define the **pre-Paley-Wiener space**, $\mathcal{M}_\mu$ as the space of $\mathcal{V}(\mu)$-valued meromorphic functions φ on $\mathfrak{a}^*_{q\mathbb{C}}$ satisfying the following conditions:

(i) $\varphi(s\lambda) = C^\circ(s,\lambda)\varphi(\lambda)$ *for all* $s \in W_q$ *and* $\lambda \in \mathfrak{a}^*_{q\mathbb{C}}$.
(ii) $\pi(\lambda)\varphi(\lambda)$ *is holomorphic on a neighborhood of* $(\mathfrak{a}^*_{q\mathbb{C}})_-$.
(iii) *There exist a constant* $R > 0$ *and, for every* $n \in \mathbb{N}$, *a constant* $C > 0$ *such that*

$$\|\pi(\lambda)\varphi(\lambda)\| \le C(1+|\lambda|)^{-n} e^{R|\mathrm{Re}\,\lambda|}$$

for all $\lambda \in (\mathfrak{a}^*_{q\mathbb{C}})_-$.

It can be seen that $\mathcal{F}_\mu$ maps $C_c^\infty(G/H;\mu)$ into $\mathcal{M}_\mu$. (Properties (i) and (ii) are straightforward consequences of (28) and Theorems 8 and 9, whereas (iii) requires a more difficult estimate for $E^\circ_\mu(\psi,\lambda)$.) It follows from the Paley-Wiener theorem of Helgason and Gangolli (see [47, Ch. IV, §7]), that when G/H is Riemannian and μ is the trivial K-type then $\mathcal{F}_\mu$ is a surjection onto the pre-Paley-Wiener space, as defined above for this special case. However in general one has to require further conditions on a function $\varphi \in \mathcal{M}_\mu$ before it belongs to $\mathcal{F}_\mu(C_c^\infty(G/H;\mu))$. Briefly put, the extra condition is that any existing relation between the normalized Eisenstein integrals and their derivatives (with respect to λ) should be reflected by a similar condition on φ. More precisely, we require that:

For all finite collections of $\partial_1,\dots,\partial_k \in S(\mathfrak{a}^*_q)$ *(that is, constant coefficient differential operators on* $\mathfrak{a}^*_q$*),* $\psi_1,\dots,\psi_k \in \mathcal{V}(\mu)$, *and* $\lambda_1,\dots,\lambda_k \in (\mathfrak{a}^*_{q\mathbb{C}})_-$ *for which the relation*

$$\sum_{i=1}^{k} \partial_i \big[\pi(\lambda)\,\langle \psi \mid E^\circ_\mu(\psi_i,-\bar\lambda)(x)\rangle\big]_{\lambda=\lambda_i} = 0 \tag{33}$$

holds for every $\psi \in \mathcal{V}(\mu)$ *and all* $x \in G/H$, *we have also the relation*

$$\sum_{i=1}^{k} \partial_i \big[\, \pi(\lambda) \, \langle \varphi(\lambda) \mid \psi_i \rangle \big]_{\lambda=\lambda_i} = 0. \tag{34}$$

The space of functions $\varphi \in \mathcal{M}_\mu$ satisfying this requirement is denoted PW_μ. It is clear from the definition (26) of $\mathcal{F}_\mu f$ that it belongs to this space for $f \in C_c^\infty(G/H;\mu)$.

Theorem 12 ([13], [16]). *The* μ*-spherical Fourier transform* $\mathcal{F}_\mu$ *is a bijection of* $C_c^\infty(G/H;\mu)$ *onto the Paley-Wiener space* PW_μ.

The injectivity of $\mathcal{F}_\mu$ is an immediate corollary of Theorem 10: If $\mathcal{F}_\mu f = 0$, then $\mathcal{F}_\mu D f = \chi_\mu(D)\mathcal{F}_\mu f = 0$. Hence $Df = 0$ by (ii), and hence $f = 0$ by (i). The injectivity of $f \mapsto f^\wedge$ asserted earlier (below Theorem 6) is a consequence, by density of the K-finite functions in $C_c^\infty(G/H)$.

The proof of the surjectivity is based on the residue calculus that was described in the previous section. More precisely, given a function $\varphi \in \mathcal{M}_\mu$, one forms a pseudo wave packet from it as in (32), specifically

$$\mathcal{T}_\mu\varphi(x) = |W_{\mathfrak{q}}| \int_{\eta + i\mathfrak{a}_{\mathfrak{q}}^*} E_+(\lambda, x)\varphi(\lambda)\, d\lambda, \qquad x \in (G/H)_+,$$

where η is sufficiently antidominant. As before, one shows that $\mathcal{T}_\mu\varphi$ is supported on a subset of $(G/H)_+$ with compact closure in G/H. The surjectivity of $\mathcal{F}_\mu$ is then a consequence of the following result.

Theorem 13 ([16]). *Assume that* $\varphi \in \mathrm{PW}_\mu$. *Then the pseudo wave packet* $\mathcal{T}_\mu\varphi$ *extends to a smooth function on* G/H *belonging to* $C_c^\infty(G/H;\mu)$. *Moreover,*

$$\mathcal{F}_\mu \mathcal{T}_\mu \varphi = \varphi. \tag{35}$$

The proof of this result is based on the same shift that was applied in the proof of Theorem 11. By this shift one expresses the pseudo wave packet $\mathcal{T}_\mu\varphi$ as the sum of the wave packet $\mathcal{J}_\mu\varphi$ and a residual part. Let us again outline the argument for the case when $\dim \mathfrak{a}_{\mathfrak{q}} = 1$ (in which case it is already given in [13]). By a clever idea introduced by Campoli [22] for the split rank one group case and also used by Arthur [1], there exists a function $f \in C_c^\infty(G/H;\mu)$ whose Fourier transform agrees with φ to some specified order of derivatives at the finitely many locations where residues are taken. Hence the residual part of $\mathcal{T}_\mu\varphi$ is identical with the residual part of $\mathcal{T}_\mu\mathcal{F}_\mu f$, which was shown to extend smoothly in the proof of Theorem 11. Since also $\mathcal{J}_\mu\varphi$ is smooth on G/H we conclude that $\mathcal{T}_\mu\varphi$ extends smoothly on G/H. As mentioned before Theorem 13, the support of $\mathcal{T}_\mu\varphi$ is compact; hence $\mathcal{T}_\mu\varphi \in C_c^\infty(G/H;\mu)$. In particular, it makes sense to form the Fourier transform in (35). It follows from part of the proof of Theorem 10 that (35) holds when both sides are multiplied by the polynomial $\mu(D)$ (see [13, Lemma 21.10]). Hence it holds also without this polynomial in front (as an identity between meromorphic functions).

For the Riemannian symmetric spaces the surjectivity of $\mathcal{F}_\mu$ (with an arbitrary K-type μ) is a consequence of the Paley-Wiener theorem in [45], and for the group G itself, considered as a symmetric space, it follows from [1], as mentioned earlier. Though inspired by [1], the proof outlined above differs from Arthur's treatise in several important respects. First of all, Arthur appeals to Harish-Chandra's Plancherel theorem, whereas here the idea is to prove both the Plancherel theorem and the Paley-Wiener theorem from the same kind of reasoning. In this respect the present proof is in the same spirit as that of Helgason and Rosenberg for G/K. Secondly, Arthur uses in the inductive argument a lifting theorem due to Casselman (see [1, Theorem II.4.1]). The use of this result (the proof of which seems still to be unpublished) is here replaced by the application of the theory of asymptotic families in [15].

A partial Paley-Wiener theorem for G/H was earlier obtained in [30]. The result, that a certain natural subspace of PW_μ is contained in the range of the Fourier transform, was obtained by means of Theorem 3. This, as well as an application of the Paley-Wiener theorem to construct multipliers, is explained in [8].

References

1. J. Arthur, A Paley-Wiener theorem for real reductive groups, *Acta Math.* **150** (1983), 1–89.
2. E. P. van den Ban, Invariant differential operators on a semisimple symmetric space and finite multiplicities in a Plancherel formula, *Arkiv Mat.* **25** (1987), 175–187.
3. E. P. van den Ban, The principal series for a reductive symmetric space I. H-fixed distribution vectors, *Annales Scient. Ecole Norm. Sup.* **4**, **21** (1988), 359–412.
4. E. P. van den Ban, The principal series for a reductive symmetric space II. Eisenstein integrals, *J. Func. Anal.* **109** (1992), 331–441.
5. E. P. van den Ban, The action of intertwining operators on H-fixed generalized vectors in the minimal principal series of a reductive symmetric space, *Indag. Math.* (to appear).
6. E. P. van den Ban, Induced representations and the Langlands classification, *these* PROCEEDINGS, pp. 123–155.
7. E. P. van den Ban, J. Carmona, and P. Delorme, Paquets d'ondes dans l'espace de Schwartz d'un espace symétrique réductif, *J. Func. Anal.* **139** (1996), 225–243.
8. E. P. van den Ban, M. Flensted-Jensen and H. Schlichtkrull, Basic harmonic analysis on pseudo-Riemannian symmetric spaces, *Noncompact Lie Groups and Some of Their Applications, Proceedings of the San Antonio NATO Workshop, 1993* (E. A. Tanner and R. Wilson, eds.), vol. C 429, Kluwer, NATO ASI Series, 1994, pp. 69–101.
9. E. P. van den Ban and H. Schlichtkrull, Asymptotic expansions and boundary values of eigenfunctions on Riemannian symmetric spaces, *J. Reine Angew. Math.* **380** (1987), 108–165.
10. E. P. van den Ban and H. Schlichtkrull, Multiplicities in the Plancherel decomposition for a semisimple symmetric space, *Representation Theory of Groups and Algebras*, Contemporary Mathematics, vol. 145, 1993, pp. 163–180.
11. E. P. van den Ban and H. Schlichtkrull, Fourier transforms on a semisimple symmetric space, *Invent. Math.* (to appear).
12. E. P. van den Ban and H. Schlichtkrull, Expansions for Eisenstein integrals on semisimple symmetric spaces, *Arkiv Mat.* (to appear).
13. E. P. van den Ban and H. Schlichtkrull, The most continuous part of the Plancherel decomposition for a reductive symmetric space, *Annals of Math.* **145** (1997), 267–364.
14. E. P. van den Ban and H. Schlichtkrull, A residue calculus for root systems, preprint, 1996.
15. E. P. van den Ban and H. Schlichtkrull, Analytic families of eigenfunctions on a reductive symmetric space, in preparation.
16. E. P. van den Ban and H. Schlichtkrull, The Paley-Wiener theorem and the Plancherel decomposition for a reductive symmetric space, in preparation.
17. M. Berger, Les espaces symétriques non compacts, *Annales Scient. Ecole Norm. Sup.* **74** (1957), 85–177.
18. J. N. Bernstein, On the support of the Plancherel measure, *J. Geom. Phys.* **5** (1988), 663–710.
19. F. Bien, *$\mathcal{D}$-modules and spherical representations*, Princeton University Press, Princeton, 1990.
20. N. Bopp and P. Harinck, Formule de Plancherel pour $\mathrm{GL}(n,\mathbf{R})/\mathrm{U}(p,q)$, *J. Reine Angew. Math.* **428** (1992), 45–95.

21. J.-L. Brylinski and P. Delorme, Vecteurs distributions H-invariants pour les séries principales géneralisées d'espaces symétriques réductifs et prolongement méromorphe d'integrales d'Eisenstein, *Invent. Math.* (1992), 619–664.
22. O. Campoli, *The complex Fourier transform for rank-1 semisimple Lie groups*, Ph.D. Thesis, Rutgers University, New Brunswick, NJ, 1976.
23. J. Carmona, Terme constant des fonctions tempérées sur un espace symétrique réductif, preprint, Luminy, 1994.
24. J. Carmona and P. Delorme, Base méromorphe de vecteurs distributions H-invariants pour les séries principales géneralisées d'espaces symétriques réductifs. Equation fonctionelle, *J. Func. Anal.* **122** (1994), 152–221.
25. J. Carmona and P. Delorme, Transformation Fourier pour les espaces symétriques réductifs, preprint, 1996.
26. E. Cartan, Sur la détermination d'un système orthogonal complet dans un espace de Riemann symétrique clos, *Rend. Circ. Mat. Palermo* **53** (1929), 217–252.
27. P. Delorme, Intégrales d'Eisenstein pour les espaces symétriques réductifs: Temperance, majorations. Petite matrice B, *J. Func. Anal.* **136** (1996), 422–509.
28. P. Delorme, Troncature pour les espaces symétriques réductifs, preprint, 1996.
29. P. Delorme, Formule de Plancherel pour les espaces symétriques réductifs, preprint, 1996.
30. P. Delorme and M. Flensted-Jensen, Towards a Paley-Wiener theorem for semisimple symmetric spaces, *Acta Math.* **167** (1991), 127–151.
31. G. van Dijk and M. Poel, The Plancherel formula for the pseudo-Riemannian space $\mathrm{GL}(n,\mathbf{R})/\mathrm{GL}(n-1,\mathbf{R})$, *Compositio Math.* **58** (1986), 371–397.
32. J. Faraut, Distributions sphériques sur les espaces hyperboliques, *J. Math. Pures Appl.* **58** (1979), 369–444.
33. M. Flensted-Jensen, Discrete series for semisimple symmetric spaces, *Annals of Math.* **111** (1980), 253–311.
34. M. Flensted-Jensen, *Analysis on Non-Riemannian Symmetric Spaces*, Regional Conference Series in Mathematics, vol. 61, American Mathematical Society, Providence, 1986.
35. P. Harinck, Fonctions généralisées sphériques induites sur $G_{\mathbb{C}}/G_{\mathbb{R}}$ et applications, *J. Func. Anal.* **103** (1992), 104–127.
36. P. Harinck, Inversion des intégrales orbitales et formule de Plancherel pour $G_{\mathbb{C}}/G_{\mathbb{R}}$, *C. R. Acad. Sci. Paris* **320** (1995), 1295–1298.
37. P. Harinck, Fonctions orbitales sur G_C/G_R. Formule d'inversion des intégrales orbitales et formule de Plancherel, preprint, 1996.
38. Harish-Chandra, Representations of semisimple Lie groups VI, *Amer. J. Math.* **78** (1956), 564–628.
39. Harish-Chandra, Discrete series for semisimple Lie groups, I, II, *Acta Math.* **113** (1965), 241–318; ; **116** (1966), 1–111.
40. Harish-Chandra, Harmonic analysis on real reductive groups, III. The Maass-Selberg relations, *Annals of Math.* **104** (1976), 117–201.
41. G. Heckman and E. Opdam, Yang's system of particles and Hecke algebras, *Annals of Math.* **145** (1997), 139–173.
42. G. Heckman and H. Schlichtkrull, *Harmonic Analysis and Special Functions on Symmetric Spaces*, Perspectives in Mathematics, vol. 16, Academic Press, Boston, 1994.
43. S. Helgason, A duality for symmetric spaces with applications to group representations, *Advances in Math.* **5** (1970), 1–154.
44. S. Helgason, The surjectivity of invariant differential operators on symmetric spaces I, *Annals of Math.* **98** (1973), 451–479.
45. S. Helgason, A duality for symmetric spaces with applications to group representations, II. Differential equations and eigenspace representations, *Advances in Math.* **22** (1976), 187–219.
46. S. Helgason, *Differential Geometry, Lie Groups, and Symmetric Spaces*, Academic Press, New York, 1978.
47. S. Helgason, *Groups and Geometric Analysis*, Academic Press, Orlando, 1984.
48. S. Helgason, *Geometric Analysis on Symmetric Spaces*, Amer. Math. Soc., 1994.
49. A. W. Knapp, *Representation Theory of Semisimple Groups: An Overview Based on Examples*, Princeton University Press, Princeton, 1986.
50. R. P. Langlands, Eisenstein series, *Algebraic Groups and Discontinuous Subgroups*, Proc. Symp. Pure Math., vol. 9, American Mathematical Society, Providence, 1966, pp. 235–252.

51. R. P. Langlands, *On the Functional Equations Satisfied by Eisenstein Series*, Lecture Notes in Mathematics, vol. 544, Springer-Verlag, Berlin, 1976.
52. T. Matsuki, The orbits of affine symmetric spaces under the action of minimal parabolic subgroups, *J. Math. Soc. Japan* **31** (1979), 331–357.
53. T. Matsuki, A description of discrete series for semisimple symmetric spaces II, *Advanced Studies in Pure Math.* **14** (1988), 531–540.
54. C. Moeglin and J.-L. Waldspurger, *Spectral Decomposition and Eisenstein Series*, Cambridge University Press, Cambridge, 1995.
55. V. F. Molchanov, Plancherel's formula for pseudo-Riemannian symmetric spaces of rank 1, *Soviet Math. Doklady* **34** (1987), 323–326.
56. G. Ólafsson, Fourier and Poisson transformation associated to a semisimple symmetric space, *Invent. Math.* **90** (1987), 605–629.
57. T. Oshima, A realization of semisimple symmetric spaces and construction of boundary value maps, *Advances Studies in Pure Math.* **14** (1988), 603–650.
58. T. Oshima and T. Matsuki, A description of discrete series for semisimple symmetric spaces, *Advanced Studies in Pure Math.* **4** (1984), 331–390.
59. T. Oshima and J. Sekiguchi, Eigenspaces of invariant differential operators on an affine symmetric space, *Invent. Math.* **57** (1980), 1–81.
60. R. Penney, Abstract Plancherel theorems and a Frobenius reciprocity theorem, *J. Func. Anal.* **18** (1975), 177–190.
61. J. Rosenberg, A quick proof of Harish-Chandra's Plancherel theorem for spherical functions on a semisimple Lie group, *Proc. Amer. Math. Soc.* **63** (1977), 143–149.
62. W. Rossmann, The structure of semisimple symmetric spaces, *Canad. J. Math.* **31** (1979), 157–180.
63. H. Schlichtkrull, *Hyperfunctions and Harmonic Analysis on Symmetric Spaces*, Birkhäuser, Boston, 1984.
64. A. Selberg, Harmonic analysis and discontinuous groups in weakly symmetric Riemannian spaces with applications to Dirichlet series, *J. Indian Math. Soc.* **20** (1956), 47–87.
65. A. Selberg, Discontinuous groups and harmonic analysis, *Proceedings of the International Congress of Mathematicians 1962*, Institut Mittag-Leffler, Djursholm, Sweden, 1963, pp. 177–189.
66. M. Sugiura, Fourier series of smooth functions on compact Lie groups, *Osaka Math. J.* **8** (1971), 33–47.
67. D. A. Vogan, Irreducibility of discrete series representations for semisimple symmetric spaces, *Advanced Studies in Pure Math.* **14** (1988), 191–221.
68. N. R. Wallach, *Real Reductive Groups, I, II*, Academic Press, Boston, 1988, 1992.

Department of Mathematics, University of Utrecht, P. O. Box 80010, 3508 TA Utrecht, The Netherlands
E-mail address: ban@math.ruu.nl

Department of Mathematics and Physics, The Royal Veterinary and Agricultural University, Thorvaldsensvej 40, DK 1871 Frederiksberg C, Denmark
E-mail address: mfj@dina.kvl.dk

Department of Mathematics, University of Copenhagen, Universitetsparken 5, DK 2100 Copenhagen Ø, Denmark
E-mail address: schlicht@math.ku.dk

Proceedings of Symposia in Pure Mathematics
Volume **61** (1997), pp. 219–243

Cohomology and Group Representations

David A. Vogan, Jr.

Contents

This article is based on three lectures ostensibly devoted to "cohomological induction," a method for constructing unitary representations of reductive Lie groups. In fact the lectures concerned mostly more elementary cohomological notions, beginning with de Rham cohomology of compact manifolds. When the manifolds are related to Lie groups, de Rham cohomology is related to Lie algebra cohomology. In this way questions about de Rham cohomology can sometimes be translated into questions about cohomological properties of group representations. Cohomological induction appears at the very end, as a way to construct representations having these cohomological properties.

I am grateful to the organizers for the opportunity to participate in this conference. Tony Knapp's notes are responsible for whatever connection exists between this article and the original lectures.

1. Cohomology of Locally Symmetric Spaces

Suppose G is a connected real reductive algebraic group, and $K \subset G$ is a maximal compact subgroup. The homogeneous space G/K is a Riemannian symmetric space; it is diffeomorphic to $\mathbb{R}^n$. Suppose now that $\Gamma \subset G$ is a torsion-free discrete subgroup. Then Γ acts freely on G/K on the left, so that the double coset space

$$X = \Gamma\backslash G/K \tag{1.1)(a}$$

1991 *Mathematics Subject Classification.* Primary 22E46, 17B20.
Supported in part by NSF grant DMS-9402994.

is a smooth manifold (in fact a Riemannian locally symmetric space). Since G/K is simply connected, it is the universal cover of X; so

$$\pi_1(X) \simeq \Gamma. \tag{1.1)(b}$$

But even more is true. Because G/K is contractible, X is a "$K(\Gamma, 1)$," an Eilenberg-MacLane space. It may be thought of as a kind of geometric incarnation of the discrete group Γ. According to the original definition of the cohomology of the group Γ, we have

$$H^i(\Gamma, \mathbb{C}) \simeq H^i(X, \mathbb{C}). \tag{1.1)(c}$$

If G/K is a Hermitian symmetric space, then it is a complex Stein manifold. The complex structure is inherited by X. If Γ is cocompact in G, then X has in a natural way the structure of a projective algebraic variety; it is a Shimura variety. (Actually the most interesting Shimura varieties arise from noncocompact arithmetic subgroups Γ, by compactification of X.) A great deal is known about the cohomology of Shimura varieties; some background may be found in [9]. From the point of view of the Langlands program, however, the most basic example of a Riemannian locally symmetric space has $G = GL(n, \mathbb{R})$ and $K = O(n)$. In that case X is not a complex manifold (unless $n = 2$), and there seem to be few ideas about what kind of special extra structure X might carry.

At any rate, we want to study the cohomology of X using the de Rham theorem. The de Rham complex has differential

$$d : (\text{complex-valued } p\text{-forms on } X) \to (\text{complex-valued } p+1\text{-forms on } X).$$

Its cohomology groups are $H^p(X, \mathbb{C})$. We want to study this complex in group-theoretic terms. We begin by replacing X by a homogeneous space G/H. The first case to look at is G itself. A p-form on G is a section of $\wedge^p(T^*G)$. Because G is a Lie group, T^*G can be trivialized by left-invariant forms. This leads to a trivialization of p-forms, as follows. Think of the Lie algebra $\mathfrak{g}$ as consisting of the left-invariant vector fields on G. If ω is a p-form on G and $X_1, \ldots, X_p \in \mathfrak{g}$ are left-invariant vector fields, then

$$\omega(X_1, \ldots, X_p) \in C^\infty(G). \tag{1.2)(a}$$

This construction provides an identification

$$(p\text{-forms on } G) \simeq \operatorname{Hom}_{\mathbb{R}}(\wedge^p \mathfrak{g}, C^\infty(G)). \tag{1.2)(b}$$

(We have been a little vague about the coefficients: for complex-valued p-forms one must use complex-valued smooth functions, and for real-valued forms real-valued smooth functions.)

The next problem is to compute the differential. If ω is a p-form on a smooth manifold M and $X_0, \ldots, X_p$ are vector fields, then

$$\begin{aligned} d\omega(X_0, \ldots, X_p) = &\sum_{i=0}^{p} (-1)^i X_i \cdot \omega(X_0, \ldots, \widehat{X_i}, \ldots, X_p) \\ &+ \sum_{i<j} (-1)^{i+j} \omega([X_i, X_j], X_0, \ldots, \widehat{X_i}, \ldots, \widehat{X_j}, \ldots, X_p). \end{aligned} \tag{1.3}$$

(See for example [18], Proposition 2.25(f).) Here in the first sum the vector field X_i acts on the smooth function $\omega(X_0, \ldots, \widehat{X_i}, \ldots, X_p)$. This formula is well suited to the identification (1.2)(b) of forms on G, because the left-invariant forms are closed under Lie bracket. The resulting formula for d on $\mathrm{Hom}_{\mathbb{R}}(\wedge^p\mathfrak{g}, C^\infty(G))$ involves just two things: the action of $\mathfrak{g}$ on $C^\infty(G)$ by differentiation on the right, and the Lie bracket on $\mathfrak{g}$.

Now suppose $H \subset G$ is a closed subgroup. We want to identify p-forms on G/H as "special" p-forms on G. There is a submersion $\pi : G \to G/H$. The corresponding pullback is an inclusion

$$\pi^* : (p\text{-forms on } G/H) \hookrightarrow (p\text{-forms on } G). \tag{1.4}$$

Pullback of forms by smooth maps always commutes with d, so that this is an inclusion of complexes. It is not difficult to identify the image.

Proposition 1.5. *In the setting of* (1.4), *a p-form $\omega \in \mathrm{Hom}_{\mathbb{R}}(\wedge^p\mathfrak{g}, C^\infty(G))$ comes from G/H if and only if*

1. $\omega(X, \dots) = 0$ *whenever* $X \in \mathfrak{h}$, *and*
2. $\omega \in \mathrm{Hom}_H(\wedge^p\mathfrak{g}, C^\infty(G))$. *Here H acts on $\wedge^p\mathfrak{g}$ by the adjoint action, and on $C^\infty(G)$ by right translation.*

Consequently there is an identification

$$(p\text{-forms on } G/H) \simeq \mathrm{Hom}_H(\wedge^p\mathfrak{g}/\mathfrak{h}, C^\infty(G)).$$

If Γ is a discrete subgroup of G acting freely and properly discontinuously on G/H (on the left), then there is an identification

$$(p\text{-forms on } \Gamma\backslash G/H) \simeq \mathrm{Hom}_H(\wedge^p\mathfrak{g}/\mathfrak{h}, C^\infty(\Gamma\backslash G)).$$

In all cases the formula for d is (1.3): it involves the action of $\mathfrak{g}$ on $C^\infty(G)$ or $C^\infty(\Gamma\backslash G)$ by differentiation on the right. The formula for the complex involves also the right translation action of H on $C^\infty(G)$ or $C^\infty(\Gamma\backslash G)$. In order to apply representation theory to this picture, we will try to decompose $C^\infty(\Gamma\backslash G)$ into pieces invariant under these two right actions, and then study the contribution of each piece separately to $H^p(\Gamma\backslash G/H, \mathbb{C})$. Here is a natural formal setting for this study.

Definition 1.6. A **pair** is a tuple $(\mathfrak{g}, H)$ where $\mathfrak{g}$ is a finite-dimensional real Lie algebra and H is a Lie group with $\mathfrak{h} \subset \mathfrak{g}$. We also assume given an action Ad of H on $\mathfrak{g}$ by Lie algebra automorphisms, compatible with the adjoint action of H on $\mathfrak{h}$.

If G is a Lie group with Lie algebra $\mathfrak{g}$ and H is a Lie subgroup of G, then $(\mathfrak{g}, H)$ is in a natural way a pair. For us the most important example will be the pairs $(\mathfrak{g}, K)$ with G a reductive Lie group and K a maximal compact subgroup.

Definition 1.7. Suppose $(\mathfrak{g}, H)$ is a pair. A $(\mathfrak{g}, H)$**-module** is a complex vector space V endowed with representations of $\mathfrak{g}$ and H, subject to the following conditions.

1. The action of H on V is **locally finite**. That is, each $v \in V$ belongs to a finite-dimensional H-invariant subspace V_1, and the representation of H on V_1 is smooth.
2. The differential of the action of H (which makes sense by condition (1)) is equal to the restriction to $\mathfrak{h}$ of the action of $\mathfrak{g}$.
3. If $X \in \mathfrak{g}$, $h \in H$, and $v \in V$, then $h \cdot (X \cdot v) = (\mathrm{Ad}(h)X) \cdot (h \cdot v)$.

Example 1.8. Suppose H is a closed subgroup of a Lie group G. There are representations of $\mathfrak{g}$ and H on $C^\infty(G)$, by differentiation and translation on the right. These satisfy condition (3) in Definition 1.7, and even a version of (2). (One needs to impose an appropriate topology on $C^\infty(G)$ to make sense of the limit appearing in the definition of derivative.) But condition (1) fails unless H is finite. We can circumvent the problem in the following way. Write ρ for the action of G on $C^\infty(G)$ by right translation:

$$(\rho(g)f)(x) = f(xg) \qquad (g, x \in G).$$

Now define

$$C^\infty(G)_H = \{f \in C^\infty(G) \mid \dim(\langle \rho(h)f \mid h \in H\rangle) < \infty\}.$$

Here $\langle \rho(h)f\rangle$ is the space spanned by all right translates of f by elements of H. The subspace $C^\infty(G)_H$ is preserved by the action of $\mathfrak{g}$, and obviously it satisfies (1) of Definition 1.7. Consequently $C^\infty(G)_H$ is a $(\mathfrak{g}, H)$-module. If Γ is any subgroup of G, then the space $C^\infty(\Gamma\backslash G)_H$ of functions invariant by Γ on the left is a $(\mathfrak{g}, H)$-submodule.

Example 1.9. Suppose G is a reductive group, K is a maximal compact subgroup, and $(\pi, \mathcal{H}_\pi)$ is a continuous representation on a Hilbert space. Write $\mathcal{H}_\pi^\infty$ for the space of smooth vectors of π. This is a dense subspace of $\mathcal{H}_\pi$ invariant under the action of G, and it carries a natural representation of the Lie algebra $\mathfrak{g}$. By analogy with the preceding example, we can define

$$\mathcal{H}_{\pi,K}^\infty = \{v \in \mathcal{H}_\pi^\infty \mid \dim(\langle \pi(k)v \mid k \in K\rangle) < \infty\}$$

the space of K-finite smooth vectors of π. This space is invariant under the action of $\mathfrak{g}$ (although not under the action of G), and is therefore a $(\mathfrak{g}, K)$-module, called the **Harish-Chandra module of** π. Because K is compact, it is easy to check that $\mathcal{H}_{\pi,K}^\infty$ is dense in $\mathcal{H}_\pi$.

The construction in the preceding example makes sense for any compact subgroup of any Lie group. What makes it particularly interesting when G is reductive and K is maximal compact are theorems of Harish-Chandra, which say that when π is irreducible and unitary, then $\mathcal{H}_{\pi,K}^\infty$ is algebraically irreducible (as a $(\mathfrak{g}, K)$-module) and determines π.

Definition 1.10. Suppose $(\mathfrak{g}, H)$ is a pair, and V is a $(\mathfrak{g}, H)$-module (Definitions 1.6 and 1.7). The adjoint action Ad of H on $\mathfrak{g}$ preserves $\mathfrak{h}$, and therefore descends to $\mathfrak{g}/\mathfrak{h}$. We can therefore define

$$\Omega^p(\mathfrak{g}, H; V) = \mathrm{Hom}_H(\wedge^p(\mathfrak{g}/\mathfrak{h}), V),$$

the V-valued p-forms for $(\mathfrak{g}, H)$. We want to define a differential making this a complex. For $\omega \in \Omega^p(\mathfrak{g}, H; V)$, we define $d\omega$ by

$$\begin{aligned} d\omega(X_0, \dots, X_p) = &\sum_{i=0}^{p} (-1)^i X_i \cdot \omega(X_0, \dots, \widehat{X_i}, \dots, X_p) \\ &+ \sum_{i<j} (-1)^{i+j} \omega([X_i, X_j], X_0, \dots, \widehat{X_i}, \dots, \widehat{X_j}, \dots, X_p) \end{aligned}$$

whenever $X_i \in \mathfrak{g}$. The action in the first sum is given by the representation of $\mathfrak{g}$ on the range of ω. It is not difficult to check that $d\omega \in \Omega^{p+1}(\mathfrak{g}, H; V)$; and the fact that $d^2 = 0$ follows from the Jacobi identity for $\mathfrak{g}$. We may therefore define

$$H^p(\mathfrak{g}, H; V) = (\ker d \text{ on } \Omega^p(\mathfrak{g}, H; V))/(\operatorname{im} d \text{ on } \Omega^{p-1}(\mathfrak{g}, H; V)),$$

the **relative Lie algebra cohomology of** $(\mathfrak{g}, H)$ **with coefficients in** V.

Proposition 1.11. *Suppose H is a closed subgroup of the Lie group G, and that $\Gamma \subset G$ is a discrete group acting freely and properly discontinuously on G/H (so that $\Gamma\backslash G/H$ is a manifold). Define $C^\infty(\Gamma\backslash G)_H$ as in Example* 1.8. *Then there is a natural isomorphism*

$$H^p(\Gamma\backslash G/H, \mathbb{C}) \simeq H^p(\mathfrak{g}, H; C^\infty(\Gamma\backslash G)_H).$$

This is a formal consequence of the de Rham theorem, Proposition 1.5, and the definitions.

To make further progress along the lines suggested at the beginning of these notes, we need to decompose $C^\infty(\Gamma\backslash G)_H$ as a $(\mathfrak{g}, H)$-module. The simplest results are available when G is reductive, K is maximal compact, and Γ is cocompact and torsion free. In that case the unitary representation of G on $L^2(\Gamma\backslash G)$ is a Hilbert space direct sum of irreducible representations having finite multiplicity:

$$L^2(\Gamma\backslash G) \simeq \widehat{\bigoplus_{\pi\in\widehat{G}}} m_\pi \mathcal{H}_\pi. \tag{1.12)(a}$$

Here m_π is a nonnegative integer, the **multiplicity** of π in $L^2(\Gamma\backslash G)$. (Often it can be identified as the dimension of some classical space of automorphic functions.) For example, if $(\tau, \mathbb{C})$ is the trivial representation of G, then m_τ is the dimension of the space of G-invariant functions in $L^2(\Gamma\backslash G)$. Obviously the only G-invariant functions are constant; and since these belong to L^2 (since $\Gamma\backslash G$ is assumed to be compact), we get

$$m_\tau = 1 \qquad (\tau = \text{trivial representation of } G). \tag{1.12)(b}$$

In order to apply Proposition 1.11 we need to understand not just the decomposition of L^2 but the more subtle decomposition of $C^\infty(\Gamma\backslash G)$. It turns out that the smooth vectors in each $\mathcal{H}_\pi$ map (by (1.12)(a)) to smooth functions on $\Gamma\backslash G$; so there are inclusions

$$\bigoplus_{\pi\in\widehat{G}} m_\pi \mathcal{H}_\pi^\infty \hookrightarrow C^\infty(\Gamma\backslash G), \tag{1.12)(c}$$

$$\bigoplus_{\pi\in\widehat{G}} m_\pi \mathcal{H}_{\pi,K}^\infty \hookrightarrow C^\infty(\Gamma\backslash G)_K. \tag{1.12)(d}$$

At least in the case of (1.12)(d), one can describe exactly how the sum on the left must be completed to give an isomorphism. This leads to the following fundamental result of Matsushima.

Theorem 1.13 ([11]; see [2], Theorem VII.3.2). *Suppose G is a real reductive algebraic group, K is a maximal compact subgroup, and Γ is a torsion-free cocompact subgroup. Use the notation of* (1.12). *The inclusion of* (1.12)(d) *and the* *isomorphism of Proposition* 1.11 *induce an isomorphism*

$$\bigoplus_{\pi\in\widehat{G}} m_\pi H^p(\mathfrak{g},K;\mathcal{H}^\infty_{\pi,K}) \simeq H^p(\Gamma\backslash G/K,\mathbb{C}).$$

Matsushima's theorem accomplishes in this setting the goal of disassembling the cohomology of the space $\Gamma\backslash G/K$ into contributions of irreducible representations. In the next section we will begin to examine those individual contributions.

2. Cohomology of Irreducible Representations: the Trivial Representation

If we recall Harish-Chandra's theorem that the space of smooth K-finite vectors in an irreducible unitary representation of a reductive group G is an algebraically irreducible $(\mathfrak{g},K)$-module, then Theorem 1.13 suggests

Problem 2.1. Determine the set of irreducible $(\mathfrak{g},K)$-modules V for which $H^*(\mathfrak{g},K;V)\neq 0$, and compute the cohomology in those cases.

This problem can be completely solved when $\operatorname{rank} G = \operatorname{rank} K$, and quite a bit is known about it in general. There are only finitely many inequivalent V for which the cohomology is nonzero, and it is not terribly difficult to list the candidates. (In this connection an old result of David Wigner (see [2], Theorem I.4.1) says that the cohomology can be nonzero only if the center of the enveloping algebra acts in V as in the trivial representation. This already reduces matters to a finite set of candidates.) Actually computing the cohomology is more difficult, and involves the full strength of the ideas around the Kazhdan-Lusztig conjectures: $\mathcal{D}$-modules, the Beilinson-Bernstein localization theory, and perverse sheaves.

Fortunately for us, Problem 2.1 is not quite the right question. The answer simplifies enormously if we change it to

Problem 2.2. Determine the set of irreducible unitary $(\mathfrak{g},K)$-modules V for which $H^*(\mathfrak{g},K;V)\neq 0$, and compute the cohomology in those cases.

To see what kind of answer we can expect, we begin with an example. Suppose

$$G=U(p,q),\qquad K=U(p)\times U(q). \tag{2.3)(a}$$

This means that G is the group of complex-linear transformations of $\mathbb{C}^{p+q}$ preserving the Hermitian form

$$|z_1|^2+\cdots+|z_p|^2-|z_{p+1}|^2-\cdots-|z_{p+q}|^2. \tag{2.3)(b}$$

Theorem 2.4 ([17]). *In the setting of* (2.3), *the set of irreducible unitary* $(\mathfrak{g},K)$*-modules* V *with* $H^*(\mathfrak{g},K;V)\neq 0$ *is in one-to-one correspondence with all expressions*

$$\begin{array}{ll} & p_i+q_i\neq 0 \\ p=p_1+\cdots+p_r & \\ & p_i=0\Rightarrow q_i=1 \\ q=q_1+\cdots+q_r & \\ & q_i=0\Rightarrow p_i=1. \end{array}$$

In that case, there is a dimension shift R *(depending on the* p_i *and* q_i*) so that* $H^*(\mathfrak{g},K;V)$ *may be computed in terms of the cohomology of a compact symmetric space:*

$$H^m(\mathfrak{g},K;V)\simeq H^{m-R}\Big(\prod_{i=1}^r U(p_i+q_i)/(U(p_i)\times U(q_i)),\mathbb{C}\Big).$$

Here $U(p_i+q_i)/(U(p_i)\times U(q_i))$ is the Grassmanian of p_i-planes in $\mathbb{C}^{p_i+q_i}$.

The trivial representation $V=\mathbb{C}$ corresponds in this parametrization to the case $r=1$; that is, to the expressions $p=p_1$ and $q=q_1$. The dimension shift R is zero, as we will see in Theorem 2.10 below.

We will eventually give a similarly precise and explicit result for any G. For the rest of this section, we will concentrate on the problem of computing the cohomology groups of the Grassmann varieties appearing in the theorem. We begin with a closer look at the complex of Definition 1.10.

Definition 2.5. The pair $(\mathfrak{g},H)$ (Definition 1.6) is said to be **symmetric** if we are given an involutive automorphism σ of $\mathfrak{g}$ such that σ commutes with $\mathrm{Ad}(H)$, and $\mathfrak{g}^\sigma=\mathfrak{h}$. In this case we write $\mathfrak{q}$ for the -1 eigenspace of σ, so that

$$\mathfrak{g}=\mathfrak{h}\oplus\mathfrak{q},\qquad \mathrm{Ad}(H)(\mathfrak{q})\subset\mathfrak{q}.$$

The fact that σ is a Lie algebra automorphism means that

$$[\mathfrak{h},\mathfrak{h}]\subset\mathfrak{h},\qquad [\mathfrak{h},\mathfrak{q}]\subset\mathfrak{q},\qquad [\mathfrak{q},\mathfrak{q}]\subset\mathfrak{h}.$$

Two examples will be important for us: the pairs $(\mathfrak{g},K)$ with G reductive and K maximal compact; and the pairs $(\mathfrak{a},1)$ with $\mathfrak{a}$ an abelian Lie algebra.

Proposition 2.6. *If $(\mathfrak{g},H)$ is a symmetric pair, then*

$$H^p(\mathfrak{g},H;\mathbb{C})=\Omega^p(\mathfrak{g},H;\mathbb{C})=\mathrm{Hom}_H(\wedge^p(\mathfrak{g}/\mathfrak{h}),\mathbb{C})=\mathrm{Hom}_H(\wedge^p\mathfrak{q},\mathbb{C})$$

(Definition 1.10). That is, the differential in this complex is zero.

PROOF. Suppose that $\omega\in\mathrm{Hom}_H(\wedge^p\mathfrak{q},\mathbb{C})$. We want to show that $d\omega=0$. So suppose $X_0,\dots,X_p\in\mathfrak{q}$. Then

$$\begin{aligned} d\omega(X_0,\dots,X_p)=&\sum_{i=0}^{p}(-1)^iX_i\cdot\omega(X_0,\dots,\widehat{X_i},\dots,X_p)\\ &+\sum_{i<j}(-1)^{i+j}\omega([X_i,X_j],X_0,\dots,\widehat{X_i},\dots,\widehat{X_j},\dots,X_p)\end{aligned}$$

The terms in the first sum are all zero since $\mathfrak{g}$ acts trivially on $\mathbb{C}$. The last display in Definition 2.5 show that all the brackets $[X_i,X_j]$ belong to $\mathfrak{h}$; so the terms in the second sum are zero as well. □

Corollary 2.7. *If $\mathfrak{g}$ is an abelian Lie algebra, then $H^p(\mathfrak{g};\mathbb{C})=\mathrm{Hom}(\wedge^p\mathfrak{g},\mathbb{C})$.*

Definition 2.8. Suppose G is a reductive Lie group and K is a maximal compact subgroup. Write $\mathfrak{g}=\mathfrak{k}+\mathfrak{p}$ for the corresponding Cartan decomposition; thus $\mathfrak{p}$ is the -1 eigenspace of a Cartan involution. It follows from the bracket relations in Definition 2.5 that

$$\mathfrak{u}=\mathfrak{k}+i\mathfrak{p}\subset\mathfrak{g}_{\mathbb{C}}$$

is a real form of $\mathfrak{g}_{\mathbb{C}}$. A **compact dual** for G is a connected compact group U endowed with a subgroup isomorphic to (and denoted) K, with the property that the Lie algebra of U is isomorphic to $\mathfrak{u}$ in a K-equivariant way.

Example 2.9. Suppose $G = SO(2,1)$, the group of linear transformations of $\mathbb{R}^3$ preserving the quadratic form $x^2 + y^2 - z^2$, and having determinant one. For a maximal compact subgroup we can take $S(O(2) \times O(1)) \simeq O(2)$; the isomorphism sends a matrix $A \in O(2)$ to

$$\begin{pmatrix} A & 0 \\ 0 & \det A \end{pmatrix} \in S(O(2) \times O(1)).$$

(Here one of the zeros is a 2×1 matrix, and the other is 1×2.) The complexification of G is the group $G_{\mathbb{C}}$ of linear transformations of $\mathbb{C}^3$ preserving the same quadratic form and having determinant 1. Inside $\mathbb{C}^3$ there is another real form $V = \mathbb{R}^2 + i\mathbb{R}$; the quadratic form on $\mathbb{C}^3$ restricts to a positive definite real form on V. The subgroup U of $G_{\mathbb{C}}$ preserving V is isomorphic to $SO(3)$, and it contains K. It is easy to see that U is a compact dual for $SO(2,1)$. Notice that the homogeneous space U/K is $\mathbb{RP}^2$. Consequently

$$H^p(U/K;\mathbb{C}) = \begin{cases} \mathbb{C}, & \text{if } p = 0; \\ 0, & \text{otherwise.} \end{cases}$$

The method of the preceding example is rather general.

Theorem 2.10. *Suppose G is the group of real points of a reductive algebraic group with $G_{\mathbb{C}}$ connected, and K is a maximal compact subgroup of G. Let U be a maximal compact subgroup of $G_{\mathbb{C}}$ containing K. Then U is a compact dual of G. There are natural isomorphisms*

$$\begin{aligned} H^p(\mathfrak{g}, K;\mathbb{C}) &\simeq \operatorname{Hom}_K(\wedge^p \mathfrak{p}, \mathbb{C}) \simeq i^p \operatorname{Hom}_K(\wedge^p(i\mathfrak{p}), \mathbb{C}) \\ &\simeq H^p(\mathfrak{u}, K;\mathbb{C}) \simeq H^p(U/K;\mathbb{C}). \end{aligned}$$

REMARK. Notice that this result shows how to compute the relative Lie algebra cohomology with coefficients in the trivial representation as the cohomology of a natural compact compact manifold (in fact a compact symmetric space).

PROOF. Write θ for the Cartan involution of G fixing K. We can always realize G as a subgroup of $GL(n,\mathbb{R})$ in such a way that the θ acts by inverse transpose: $\theta g = {}^t g^{-1}$. Once this is done, $G_{\mathbb{C}}$ becomes a subgroup of $GL(n,\mathbb{C})$, and the complex conjugation action defining the real form is just conjugation of matrices. The complexification of θ is still inverse transpose, which is a holomorphic automorphism of order two commuting with complex conjugation. We may therefore define a new real form σ of $G_{\mathbb{C}}$ by $\sigma g = {}^t\overline{g}^{-1}$. The group U of real points is just $G_{\mathbb{C}} \cap U(n)$, which is compact; so U must be a compact real form of G. By construction U contains K, and it is easy to check that the Lie algebra is $\mathfrak{k} + i\mathfrak{p}$. So U is a compact dual of G. All the isomorphisms in the theorem follow from Proposition 2.6 except for the very last one. For that, Proposition 1.5 shows that $\operatorname{Hom}_K(\wedge^p(i\mathfrak{p}), \mathbb{C})$ may be identified with the space of p-forms on U invariant under left translation. Because U is connected, the action of U by left translation on $H^p(U/K, \mathbb{C})$ is trivial. It follows that every cohomology class is represented by a U-invariant p-form, and the isomorphism we want follows. □

A complete description of the cohomology groups of the space U/K in Theorem 2.10 may be found (at least for connected K) in [3], as Theorem V on page 465.

The method of the next example applies to the Hermitian symmetric cases; but other ideas are required in general.

Example 2.11. Suppose $G = U(p,q)$, $K = U(p) \times U(q)$, and $U = U(p+q)$. Write $n = p+q$. Then U/K is the Grassmann variety of p-planes in $\mathbb{C}^n$. The group $G_{\mathbb{C}}$ may be identified with $GL(n,\mathbb{C})$; so $\mathfrak{g}_{\mathbb{C}}$ consists of all $n \times n$ matrices. We have

$$\mathfrak{k}_{\mathbb{C}} = \mathfrak{gl}(p,\mathbb{C}) \times \mathfrak{gl}(q,\mathbb{C}), \tag{2.12)(a}$$

$$\mathfrak{p}_{\mathbb{C}} = \left\{ \begin{pmatrix} 0 & A \\ B & 0 \end{pmatrix} \,\middle|\, A \in M_{p\times q}(\mathbb{C}),\ B \in M_{q\times p}(\mathbb{C}) \right\}. \tag{2.12)(b}$$

Consequently

$$\mathfrak{p}_{\mathbb{C}} \simeq \mathrm{Hom}_{\mathbb{C}}(\mathbb{C}^p, \mathbb{C}^q) \oplus \mathrm{Hom}_{\mathbb{C}}(\mathbb{C}^q, \mathbb{C}^p) = \mathfrak{p}_{\mathbb{C}}^- \oplus \mathfrak{p}_{\mathbb{C}}^+; \tag{2.12)(c}$$

the last equality is a definition. The spaces $\mathfrak{p}_{\mathbb{C}}^{\pm}$ are the holomorphic and antiholomorphic tangent spaces for the complex structures on G/K and U/K associated with the Hermitian symmetric structures. We will also use the fact that the standard invariant bilinear form $\langle X, Y\rangle = \mathrm{tr}\, XY$ on $\mathfrak{gl}(n,\mathbb{C})$ restricts to an identification $\mathfrak{p}_{\mathbb{C}}^- \simeq (\mathfrak{p}_{\mathbb{C}}^+)^*$. Consequently

$$\wedge^m \mathfrak{p}_{\mathbb{C}} \simeq \bigoplus_{a+b=m} (\wedge^a \mathfrak{p}_{\mathbb{C}}^+) \otimes (\wedge^b \mathfrak{p}_{\mathbb{C}}^-) \simeq \bigoplus_{a+b=m} \mathrm{Hom}\left(\wedge^a \mathfrak{p}_{\mathbb{C}}^-, \wedge^b \mathfrak{p}_{\mathbb{C}}^-\right). \tag{2.12)(d}$$

This bigrading is related to the Hodge structure on the cohomology of U/K and $\Gamma\backslash G/K$. Inserting this description in Theorem 2.10, we find

$$H^m(\mathfrak{g}, K; \mathbb{C}) \simeq \mathrm{Hom}_K(\wedge^m \mathfrak{p}_{\mathbb{C}}, \mathbb{C}) \simeq \bigoplus_{a+b=m} \mathrm{Hom}_K\left(\wedge^a \mathfrak{p}_{\mathbb{C}}^+, \wedge^b \mathfrak{p}_{\mathbb{C}}^+\right). \tag{2.12)(e}$$

To continue, we need to understand $\wedge^a \mathfrak{p}_{\mathbb{C}}^+$ as a representation of $K = U(p) \times U(q)$; or, equivalently, as a representation of $K_{\mathbb{C}} = GL(p,\mathbb{C}) \times GL(q,\mathbb{C})$.

For that, we consider the parabolic subgroup of $GL(n,\mathbb{C})$

$$Q = \left\{ \begin{pmatrix} A & B \\ 0 & C \end{pmatrix} \,\middle|\, A \in GL(p,\mathbb{C}),\ B \in M_{p\times q}(\mathbb{C}),\ C \in GL(q,\mathbb{C}) \right\}. \tag{2.12)(f}$$

Then Q has a Levi decomposition $Q = LN$, with $L = K_{\mathbb{C}}$ and $Lie(N) = \mathfrak{p}_{\mathbb{C}}^+$. Because N is abelian, Corollary 2.7 implies

$$\left(\wedge^a \mathfrak{p}_{\mathbb{C}}^+\right)^* = H^a(\mathfrak{n}; \mathbb{C}). \tag{2.12)(g}$$

The last cohomology group is computed by Kostant's version of the Bott-Borel-Weil theorem:

Theorem 2.13 ([7]). *Suppose $Q = LN$ is a parabolic subgroup of a complex reductive Lie group G, and that F is an irreducible finite-dimensional representation of G. Then $H^*(\mathfrak{n}; F)$ is a sum of inequivalent irreducible representations of L, parametrized by the quotient of Weyl groups $W(G)/W(L)$. The number of summands in degree a is the number of elements of $W(G)$ of length a that are minimal representatives for their $W(L)$ cosets.*

The statement is explained more completely in [7]; a special case is discussed in section 2 of [14]. In order to apply Kostant's theorem to our present situation (with $F = \mathbb{C}$), we just need to compute the Weyl group elements in question. Here $W(G) = S_n$, the symmetric group of all permutations of $\{1, \ldots, n\}$, and $W(L)$ is the natural subgroup $S_p \times S_q$. A permutation σ is minimal in its $W(L)$ coset if and only if

$$\sigma(1) < \cdots < \sigma(p), \qquad \sigma(p+1) < \cdots < \sigma(p+q). \tag{2.14)(a}$$

Suppose that is the case; we want to know the length of σ. For k between 1 and p, define integers d_k between 0 and q by the requirements

$$\begin{cases} d_k = 0 & & \text{if } \sigma(k) < \sigma(p+1); \\ d_k = d & (0 < d < q) & \text{if } \sigma(p+d) < \sigma(k) < \sigma(p+d+1); \text{ and} \\ d_k = q & & \text{if } \sigma(p+q) < \sigma(k). \end{cases} \tag{2.14)(b}$$

Then it is easy to check that

$$0 \le d_1 \le d_2 \le \cdots \le d_p \le q, \qquad \sum d_k = \ell(\sigma). \tag{2.14)(c}$$

Conversely, each sequence $\{d_k\}$ satisfying the inequalities in (2.14)(c) corresponds to a unique permutation σ as (2.14)(a). Combining these calculations, Theorem 2.13, and (2.12)(g), we get

Corollary 2.15. *The exterior algebra $\wedge\mathfrak{p}^+_{\mathbb{C}}$ is a direct sum of inequivalent representations of $K_{\mathbb{C}}$. The number of representations appearing in degree a is equal to the number of sequences of integers*

$$0 \le d_1 \le d_2 \le \cdots \le d_p \le q, \qquad \sum d_k = a.$$

The total number appearing in all degrees is $\binom{n}{p}$.

Applying the formula in (2.12)(e) now gives

Corollary 2.16. *Suppose $G = U(p,q)$ and $K = U(p) \times U(q)$. Then the cohomology $H^*(\mathfrak{g}, K; \mathbb{C})$ is nonzero only in even degrees. More precisely, the dimension of $H^{2a}(\mathfrak{g}, K; \mathbb{C})$ is equal to the number of sequences of integers*

$$0 \le d_1 \le d_2 \le \cdots \le d_p \le q, \qquad \sum d_k = a.$$

The total dimension of the cohomology (and the Euler characteristic) is equal to $\binom{n}{p}$.

The formula of Corollary 2.16 shows that the cohomology occurs in degrees ranging from 0 to $2pq$, and that it has dimension 1 in those extreme degrees. This is consistent with Theorem 2.10, since U/K is a compact complex manifold of dimension pq.

Corollary 2.16 and Theorem 2.10 together compute completely the cohomology groups appearing in Theorem 2.4.

3. Cohomology of Irreducible Representations: the Discrete Series

We saw in Corollary 2.16 that the cohomology of the trivial representation is quite complicated. It is therefore natural to fear that the cohomology of something as complicated as a discrete series representation will be completely incomprehensible. This is not the case, and that fact is significant. The point is that discrete series representations are in many senses among the "atoms" of the representation theory of reductive groups. The trivial representation (in the Langlands classification, or in the theory of Eisenstein series) appears as a residue from the reduciblity of a certain principal series representation; it can be properly understood only in the context of a fairly complete understanding of that reducibility, and of all the other pieces involved in it. Once this point of view is thoroughly grasped, what is amazing is that one can give any kind of closed formula for the cohomology of the trivial representation, and that such formulas were given twenty years before the invention of intersection cohomology.

For this section, we will assume that G is a connected reductive group having a compact Cartan subgroup

$$T \subset K \subset G. \tag{3.1)(a}$$

We follow roughly the notation of [14], section 5. We fix therefore a system of positive roots Φ^+ for T in $\mathfrak{g}_{\mathbb{C}}$, and write

$$\rho = \tfrac{1}{2} \sum_{\alpha \in \Phi^+} \alpha. \tag{3.1)(b}$$

We will use the trivial weight $0 \in \Lambda$ for T; this has the required property that $0 + \rho$ is dominant and regular for Φ^+. We define

$$\pi(\Phi^+) = \text{discrete series representation with character } \Theta_\rho. \tag{3.1)(c}$$

This is the representation with Harish-Chandra parameter ρ. (Wigner's result mentioned after Problem 2.1 guarantees that discrete series representations with other Harish-Chandra parameters cannot have nonvanishing cohomology; this fact can also be deduced from a calculation like the one given for Theorem 3.2 below.) We will write

$$X(\Phi^+) = \text{Harish-Chandra module of } \pi(\Phi^+). \tag{3.1)(d}$$

Finally, recall from [14], section 3 that Φ^+ is the disjoint union of the compact and noncompact positive roots:

$$\Phi^+ = \Phi_c^+ \cup \Phi_n^+. \tag{3.1)(e}$$

Define

$$R = |\Phi_n^+| = \tfrac{1}{2} \dim G/K. \tag{3.1)(f}$$

Theorem 3.2. *With notation as in* (3.1), *the cohomology of the discrete series representation is given by*

$$H^p(\mathfrak{g}, K; X(\Phi^+)) = \begin{cases} 0 & \text{if } p \neq R \\ \mathbb{C} & \text{if } p = R. \end{cases}$$

PROOF. We try to compute the $X(\Phi^+)$-valued p-forms for $(\mathfrak{g}, K)$ (Definition 1.10). Suppose μ is the highest weight of a representation of K occurring in both $\wedge^p \mathfrak{p}_{\mathbb{C}}$ and in $X(\Phi^+)$. The first requirement means that μ must be a sum of p distinct noncompact roots, so that

$$\mu = \beta_1 + \cdots + \beta_r - \beta_1' - \cdots - \beta_s'.$$

Here $\{\beta_1, \ldots, \beta_r\}$ and $\{\beta_1', \ldots, \beta_s'\}$ are subsets of Φ_n^+, and $r + s = p$. On the other hand, the Corollary to Theorem 1 of section 5 in [14] says that the second requirement means μ is of the form

$$\mu = 2\rho_n + \sum_{\gamma \in \Phi^+} c_\gamma \gamma,$$

with c_γ a nonnegative integer. Consequently

$$\beta_1 + \cdots + \beta_r - \beta'_1 - \cdots - \beta'_s = 2\rho_n + \sum_{\gamma \in \Phi^+} c_\gamma \gamma.$$

Each positive root has strictly positive inner product with ρ. Taking the inner product of both sides with ρ, we conclude that

$$r = |\Phi_n^+|, \quad s = 0, \quad c_\gamma = 0.$$

In particular, $p = r + s = |\Phi_n^+|$, and $\mu = 2\rho_n$.

It follows first of all that $\Omega^p = 0$ for $p \neq R$. For $p = R$, the only representation of K common to $\wedge^R \mathfrak{p}_\mathbb{C}$ and $X(\Phi^+)$ is the one of highest weight $2\rho_n$. This has multiplicity one in $X(\Phi^+)$ by [14], and multiplicity one in $\wedge^R \mathfrak{p}_\mathbb{C}$ by an easy computation. So $\dim \Omega^R = 1$. Since all the other forms are zero, the differentials in the complex must be zero; and the theorem follows. □

If G/K is Hermitian symmetric, the "Hodge type" of the cohomology class of $X(\Phi^+)$ is equal to (a, b), where

$$a = |\Phi_n^+ \cap (\text{roots in } \mathfrak{p}_\mathbb{C}^+)|, \quad b = |\Phi_n^+ \cap (\text{roots in } \mathfrak{p}_\mathbb{C}^-)|.$$

4. Introduction to Cohomologically Induced Representations

In this section we will introduce a family of representations "interpolating" between the trivial representation and the discrete series representations $X(\Phi^+)$. We work with a connected real reductive group G in Harish-Chandra's class ([4], section 3). (Allowing G to be disconnected but still in Harish-Chandra's class complicates the notation slightly, but does not introduce any essential new difficulties.) We fix a maximal compact subgroup $K \subset G$, and write θ for the corresponding Cartan involution. Just as in Definition 2.8, the Cartan decomposition is written $\mathfrak{g} = \mathfrak{k} + \mathfrak{p}$.

Definition 4.1. A **θ-stable parabolic subalgebra of $\mathfrak{g}$** is a parabolic subalgebra $\mathfrak{q} \subset \mathfrak{g}_\mathbb{C}$ such that

1. $\theta\mathfrak{q} = \mathfrak{q}$, and
2. $\overline{\mathfrak{q}} \cap \mathfrak{q} = \mathfrak{l}_\mathbb{C}$ is a Levi subalgebra of $\mathfrak{q}$.

Here the bar refers to complex conjugation with respect to the real form $\mathfrak{g}$ of $\mathfrak{g}_\mathbb{C}$. Necessarily the Levi subalgebra $\mathfrak{l}_\mathbb{C}$ is defined over $\mathbb{R}$; the real subalgebra $\mathfrak{l}$ is θ-stable, and is in fact the normalizer of $\mathfrak{q}$ in $\mathfrak{g}$. We define the **Levi subgroup of $\mathfrak{q}$** by

$$L = \{g \in G \mid \mathrm{Ad}(g)(\mathfrak{q}) \subset \mathfrak{q}\}.$$

Notice that we refer to $\mathfrak{q}$ as a θ-stable parabolic subalgebra of $\mathfrak{g}$ even though it is actually a subalgebra of $\mathfrak{g}_\mathbb{C}$.

Proposition 4.2 ([6], Chapter V). *Suppose $\mathfrak{q}$ is a θ-stable parabolic subalgebra of $\mathfrak{g}$ with Levi subgroup L. Then*

1. *L is a connected real reductive group of the same rank as G.*
2. *L is preserved by θ, and the restriction of θ to L is a Cartan involution.*
3. *L contains a maximal torus $T \subset K$.*

We will be interested in θ-stable parabolics up to conjugation by K. Proposition 4.2 shows that we may therefore study those containing a fixed maximal torus in K. Here is a construction that gives all of them.

Construction 4.3. Fix a maximal torus $T \subset K$. Recall that the centralizer H of T in G is a Cartan subgroup. It has Cartan decomposition $H = TA$, with $\mathfrak{a}$ the centralizer of T in $\mathfrak{p}$. Write $\Phi_c \subset i\mathfrak{t}_0^*$ for the set of roots of T in $\mathfrak{k}_{\mathbb{C}}$, so that

$$\mathfrak{k}_{\mathbb{C}} = \mathfrak{t}_{\mathbb{C}} + \sum_{\alpha \in \Phi_c} \mathfrak{k}_{\mathbb{C},\alpha}.$$

Similarly, write $\Phi_n \subset i\mathfrak{t}_0^*$ for the set of nonzero weights of T on $\mathfrak{p}_{\mathbb{C}}$, so that

$$\mathfrak{p}_{\mathbb{C}} = \mathfrak{a}_{\mathbb{C}} + \sum_{\beta \in \Phi_n} \mathfrak{p}_{\mathbb{C},\beta}.$$

We write $\Phi = \Phi_c \cup \Phi_n$, a subset of $i\mathfrak{t}_0^*$ with multiplicities. Actually it is convenient to abuse notation slightly to allow an element of Φ to remember whether it came from Φ_c or Φ_n. A formal way to do this is to regard an element of Φ as a character of the group generated by T and θ; θ acts by $+1$ on elements of Φ_c, and by -1 on elements of Φ_n.

Now fix a system of positive roots Φ_c^+ for T in $\mathfrak{k}_{\mathbb{C}}$. Fix a weight $\lambda \in i\mathfrak{t}_0^*$ that is dominant for K; that is, so that

$$\langle \lambda, \alpha \rangle \geq 0 \qquad (\alpha \in \Phi_c^+).$$

We define the **θ-stable parabolic associated to λ** by

$$\mathfrak{q}(\lambda) = \mathfrak{h}_{\mathbb{C}} + \sum_{\substack{\gamma \in \Phi \\ \langle \lambda, \gamma \rangle \geq 0}} \mathfrak{g}_{\mathbb{C},\gamma}.$$

The corresponding Levi subalgebra is

$$\mathfrak{l}(\lambda)_{\mathbb{C}} = \mathfrak{h}_{\mathbb{C}} + \sum_{\substack{\gamma \in \Phi \\ \langle \lambda, \gamma \rangle = 0}} \mathfrak{g}_{\mathbb{C},\gamma}.$$

The Levi subgroup $L(\lambda)$ may be described as follows. Extend λ to a complex-linear functional on all of $\mathfrak{g}$, by making it zero on each weight space $\mathfrak{g}_{\mathbb{C},\gamma}$ (for $\gamma \in \Phi$). Then λ takes purely imaginary values on $\mathfrak{g}_0$. The group $L(\lambda)$ is just the stabilizer of λ in the coadjoint action:

$$L(\lambda) = \{g \in G \mid \mathrm{Ad}^*(g)(\lambda) = \lambda\}.$$

Proposition 4.4. *Every θ-stable parabolic subalgebra of $\mathfrak{g}$ arises by Construction* 4.3. *In particular,*

1. *there are only finitely many K-conjugacy classes of θ-stable parabolic subalgebras; and*
2. *the Levi subgroups of θ-stable parabolic subalgebras are precisely the isotropy groups for the coadjoint action of G at elements of $\mathfrak{k}^*$.*

This is a fairly easy consequence of Proposition 4.2. The coadjoint orbits passing through $\mathfrak{k}^*$ are called **elliptic**; so the homogeneous spaces $G/L(\lambda)$ are precisely the elliptic coadjoint orbits.

Example 4.5. Suppose again that $G = U(p,q)$, $K = U(p) \times U(q)$. The Cartan involution is conjugation by the diagonal matrix whose first p entries are $+1$ and whose last q entries are -1. Write $n = p+q$, so that $G_{\mathbb{C}} \simeq GL(n,\mathbb{C})$ as in Example 2.11. Suppose we are given an r-tuple of pairs (p_i, q_i) of nonnegative integers, so that

$$\sum p_i = p, \qquad \sum q_i = q, \qquad p_i + q_i \neq 0.$$

(These conditions are slightly weaker than the ones in Theorem 2.4.) We can rearrange the coordinates in $\mathbb{C}^n$ so that our Hermitian form has p_1 plus signs, followed by q_1 minus signs, followed by p_2 plus signs, and so on:

$$|z_1|^2 + \cdots |z_{p_1}|^2 - |z_{p_1+1}|^2 - \cdots |z_{p_1+q_1}|^2 + \cdots$$

In this new realization, the Cartan involution is still conjugation by a diagonal matrix with entries ± 1. Now let $\mathfrak{q}$ be the block-upper-triangular parabolic subalgebra of $\mathfrak{gl}(n,\mathbb{C})$ with blocks of sizes $p_1 + q_1,\ p_2 + q_2, \ldots$ along the diagonal. Then $\mathfrak{q}$ is a θ-stable parabolic subalgebra. The corresponding Levi subgroup consists of diagonal blocks; it is

$$L = U(p_1, q_1) \times \cdots \times U(p_r, q_r).$$

It is not difficult to see that these are all the θ-stable parabolic subalgebras in $\mathfrak{g}$, up to conjugation by K; and in fact no two of these are conjugate.

Here is the main theorem.

Theorem 4.6. *Suppose G is a connected real reductive Lie group in Harish-Chandra's class, and $\mathfrak{q}$ is a θ-stable parabolic subalgebra of $\mathfrak{g}$ with Levi subgroup L (Definition 4.1). Write $\mathfrak{u}$ for the nil radical of $\mathfrak{q}$, and define*

$$R = \dim \mathfrak{u} \cap \mathfrak{p}_{\mathbb{C}}.$$

1. *Attached to $\mathfrak{q}$ there is an irreducible unitary representation $\pi(\mathfrak{q})$ of G. Up to equivalence, $\pi(\mathfrak{q})$ depends only on the K-conjugacy class of $\mathfrak{q}$.*
2. *Write $X(\mathfrak{q})$ for the Harish-Chandra module of $\pi(\mathfrak{q})$. Then*

$$H^p(\mathfrak{g}, K; X(\mathfrak{q})) \simeq H^{p-R}(\mathfrak{l}, L \cap K; \mathbb{C}).$$

3. *Suppose π is an irreducible unitary representation of G with Harish-Chandra module X, and that $H^*(\mathfrak{g}, K; X) \neq 0$. Then there is a θ-stable parabolic subalgebra $\mathfrak{q}$ of $\mathfrak{g}$ so that $\pi \simeq \pi(\mathfrak{q})$.*

We will say a little bit about the proof of this theorem in sections 5 and 6. Here are some remarks. In the setting of Construction 4.3, a θ-stable Borel subalgebra containing $\mathfrak{t}$ is the same as a choice Φ^+ of a system of positive roots for Φ. When in addition $\operatorname{rank} G = \operatorname{rank} K$, we have already defined a representation $\pi(\Phi^+)$ attached to such a positive system: it is a discrete series representation. In this case $L = T = L \cap K$, so that the formula in Theorem 4.6 for the cohomology agrees with the formula in Theorem 3.2.

If $\mathfrak{q} = \mathfrak{g}_{\mathbb{C}}$, then $L = G$. We take $\pi(\mathfrak{g}_{\mathbb{C}})$ to be the trivial representation of G; then the formula in Theorem 4.6 for the cohomology is a tautology.

If $G = U(p,q)$, then Theorem 4.6 can be combined with Example 4.5 and Theorem 2.10 to give something very close to Theorem 2.4. The differences arise because the list of representations in Theorem 4.6 has a few repetitions. These have been edited out of the list in Theorem 2.4. (To get inequivalent representations in

Theorem 4.6, one should impose the additional requirement on $\mathfrak{q}$ that L have no nonabelian compact simple factors. This is done in Theorem 2.4 by the last two conditions on the p_i and q_i.)

The representations $\pi(\mathfrak{q})$ were first constructed in general (as possibly nonunitary representations) by Parthasarathy in [12]. It seems very likely that he was aware of their connection with Lie algebra cohomology. At any rate the calculation of cohomology in Theorem 4.6 is (as we will see in the next section) not very difficult. Part 3 of the theorem was proved in [17], using powerful partial results of Kumaresan from [10]. The last part of the theorem, that $\pi(\mathfrak{q})$ is actually unitary, was proved in [16].

5. Cohomologically Induced Representations: Characterization and Cohomology

In this section we will give a characterization of the representations $\pi(\mathfrak{q})$ in Theorem 4.6, and use it to compute their cohomology. The main ingredient is a certain representation of K constructed from the θ-stable parabolic $\mathfrak{q}$. In order to describe this representation, it is helpful to have a slight reformulation of Construction 4.3. In the notation of that construction, the bilinear form defines an isomorphism $i\mathfrak{t}_0^* \simeq i\mathfrak{t}_0$. Let $H_\lambda \in i\mathfrak{t}_0$ be the element corresponding to λ. Explicitly, this means

$$\gamma(H_\lambda) = \langle \lambda, \gamma \rangle \qquad (\gamma \in i\mathfrak{t}_0^*). \tag{5.1)(a}$$

The θ-stable parabolic associated to λ (Construction 4.3) is then

$$\mathfrak{q}(\lambda) = \mathfrak{q}(H_\lambda) = \mathfrak{h}_{\mathbb{C}} + \sum_{\substack{\gamma\in\Phi \\ \gamma(H_\lambda)\geq 0}} \mathfrak{g}_{\mathbb{C},\gamma}. \tag{5.1)(b}$$

Similarly, its Levi subgroup is

$$L(\lambda) = L(H_\lambda) = \{g \in G \mid \mathrm{Ad}(g)(H_\lambda) = H_\lambda\}. \tag{5.1)(c}$$

Define

$$2\rho(\mathfrak{u}\cap\mathfrak{p}) = \sum_{\substack{\gamma\in\Phi_n \\ \gamma(H_\lambda)>0}} \gamma \quad \in \quad \widehat{T} \subset i\mathfrak{t}_0^*, \tag{5.1)(d}$$

the sum of the roots of T in $\mathfrak{u}\cap\mathfrak{p}$.

Proposition 5.2. *In the setting of* (5.1), *write* $R = \dim \mathfrak{u}\cap\mathfrak{p}_{\mathbb{C}}$ *as in Theorem* 4.6. *The largest eigenvalue of* $\mathrm{Ad}(H_\lambda)$ *on* $\wedge\mathfrak{p}_{\mathbb{C}}$ *is equal to* $2\rho(\mathfrak{u}\cap\mathfrak{p})(H_\lambda)$. *The corresponding eigenspace is isomorphic to*

$$\wedge^R(\mathfrak{u}\cap\mathfrak{p}_{\mathbb{C}}) \otimes \wedge(\mathfrak{l}_{\mathbb{C}}\cap\mathfrak{p}_{\mathbb{C}}).$$

The adjoint action of $\mathfrak{u}\cap\mathfrak{k}_{\mathbb{C}}$ *is trivial on this space.*

PROOF. The triangular decomposition $\mathfrak{p}_{\mathbb{C}} = \mathfrak{u}\cap\mathfrak{p}_{\mathbb{C}} + \mathfrak{l}_{\mathbb{C}}\cap\mathfrak{p}_{\mathbb{C}} + \bar{\mathfrak{u}}\cap\mathfrak{p}_{\mathbb{C}}$ gives rise to a decomposition of the exterior algebra

$$\wedge\mathfrak{p}_{\mathbb{C}} = (\wedge(\mathfrak{u}\cap\mathfrak{p}_{\mathbb{C}})) \otimes (\wedge(\mathfrak{l}_{\mathbb{C}}\cap\mathfrak{p}_{\mathbb{C}})) \otimes (\wedge(\bar{\mathfrak{u}}\cap\mathfrak{p}_{\mathbb{C}}))$$

Any weight of T appearing is a sum of weights from the three factors. According to (5.1), $\mathrm{Ad}(H_\lambda)$ has positive eigenvalues on the first factor, zero eigenvalues on the second, and negative eigenvalues on the third. This proves everything but the last claim. For that, (5.1) implies also that $\mathrm{Ad}(\mathfrak{u}\cap\mathfrak{k}_{\mathbb{C}})$ acts to raise the eigenvalues of $\mathrm{Ad}(H_\lambda)$. □

For the next result, we need to fix a set of positive roots of T in $\mathfrak{l}_{\mathbb{C}}\cap\mathfrak{k}_{\mathbb{C}}$; this allows us to speak of highest weights for representations of $L\cap K$. Adjoining to this the set of roots of T in $\mathfrak{u}\cap\mathfrak{k}_{\mathbb{C}}$ gives a set of positive roots of T in $\mathfrak{k}_{\mathbb{C}}$, and so allows us to speak of highest weights for representations of K.

Corollary 5.3. *Let μ_L be the highest weight of a representation δ_L of $L\cap K$ appearing in $\wedge(\mathfrak{l}_{\mathbb{C}}\cap\mathfrak{p}_{\mathbb{C}})$.*

1. *There is a unique representation δ of K of highest weight $\mu=\mu_L+2\rho(\mathfrak{u}\cap\mathfrak{p}_{\mathbb{C}})$.*
2. *There is a natural isomorphism*

$$\mathrm{Hom}_K(V_\delta,\wedge^p\mathfrak{p}_{\mathbb{C}})\simeq\mathrm{Hom}_{L\cap K}(V_{\delta_L},\wedge^{p-R}(\mathfrak{l}_{\mathbb{C}}\cap\mathfrak{p}_{\mathbb{C}})).$$

3. *Suppose γ is a nonempty sum of roots in $\mathfrak{u}$. Then the representation of K of highest weight $\mu+\gamma$ does not occur in $\wedge\mathfrak{p}_{\mathbb{C}}$.*

PROOF. Suppose τ_L is any irreducible representation of $L\cap K$ of highest weight γ, and W is a representation of K. Then the Cartan-Weyl theory tells us that there is at most one representation (τ,V_τ) of K of highest weight γ; and

$$\mathrm{Hom}_K(V_\tau,W)\simeq\mathrm{Hom}_{L\cap K}(V_{\tau_L},W^{\mathfrak{u}\cap\mathfrak{k}_{\mathbb{C}}})\subset\mathrm{Hom}_{L\cap K}(V_{\tau_L},W). \tag{5.4)(a}$$

If τ does not exist, then the same formula is true with $V_\tau=0$. We apply (5.4)(a) to $\tau_L=\delta_L\otimes\wedge^R(\mathfrak{u}\cap\mathfrak{p}_{\mathbb{C}})$. Evidently the element H_λ of $\mathfrak{l}_{\mathbb{C}}\cap\mathfrak{k}_{\mathbb{C}}$ acts on τ_L by the scalar $2\rho(\mathfrak{u}\cap\mathfrak{p})(H_\lambda)$. Proposition 5.2 therefore allows us to conclude that

$$\mathrm{Hom}_{L\cap K}(V_{\tau_L},\wedge^p\mathfrak{p}_{\mathbb{C}})\simeq\mathrm{Hom}_{L\cap K}(V_{\delta_L},\wedge^{p-R}(\mathfrak{l}_{\mathbb{C}}\cap\mathfrak{p}_{\mathbb{C}})). \tag{5.4)(b}$$

Furthermore any $L\cap K$-map on the left must automatically take values in the $(\mathfrak{u}\cap\mathfrak{k}_{\mathbb{C}})$-invariants. Now (5.4) gives conclusion (2) of the corollary. The right side of (5.4)(b) is nonzero (for some p) by the assumption on δ_L; so V_τ cannot be zero, and conclusion (1) follows. For conclusion (3), we apply (5.4) again with τ_L equal to the representation of $L\cap K$ of highest weight $\mu+\gamma$. By (5.1), H_λ acts on τ_L by the scalar

$$\mu(H_\lambda)+\gamma(H_\lambda)>\mu(H_\lambda)=2\rho(\mathfrak{u}\cap\mathfrak{p})(H_\lambda).$$

This eigenvalue does not occur in $\wedge\mathfrak{p}_{\mathbb{C}}$; so (5.4)(a) implies that V_τ cannot occur in $\wedge\mathfrak{p}_{\mathbb{C}}$. □

Corollary 5.5. *In the setting of* (5.1), *there is a unique irreducible representation $\delta(\mathfrak{q})$ of K of highest weight $2\rho(\mathfrak{u}\cap\mathfrak{p}_{\mathbb{C}})$. We have*

$$\mathrm{Hom}_K(V_{\delta(\mathfrak{q})},\wedge^p\mathfrak{p}_{\mathbb{C}})\simeq\mathrm{Hom}_{L\cap K}(\mathbb{C},\wedge^{p-R}(\mathfrak{l}_{\mathbb{C}}\cap\mathfrak{p}_{\mathbb{C}})).$$

This is just Corollary 5.3 with δ_L equal to the trivial representation of $L\cap K$. Here is a characterization of the representations in Theorem 4.6.

Theorem 5.6 ([17], Proposition 6.1). *Suppose $\mathfrak{q}$ is a θ-stable parabolic subalgebra of $\mathfrak{g}$, and $\delta(\mathfrak{q})$ is the representation of K described in Corollary* 5.5. *Then there is a unique irreducible unitary representation $\pi(\mathfrak{q})$ of G with the following properties:*

1. *The restriction of $\pi(\mathfrak{q})$ to K contains $\delta(\mathfrak{q})$ exactly once.*

2. *Every representation of* K *appearing in* $\pi(\mathfrak{q})$ *has highest weight* $2\rho(\mathfrak{u}\cap\mathfrak{p}_{\mathbb{C}})+\gamma$, *with* γ *a sum of roots of* T *in* $\mathfrak{u}$.
3. *The Casimir operator (a central element of the universal enveloping algebra) acts by* 0 *in* $\pi(\mathfrak{q})$.

Only the uniqueness part of this statement is proved in [17]; the existence appears in [16]. We will discuss the construction of $\pi(\mathfrak{q})$ in section 6. Assuming that we have constructed this representation, let us see how to calculate the Lie algebra cohomology. As in Theorem 4.6, we write $X(\mathfrak{q})$ for the Harish-Chandra module. According to Definition 1.10, this is calculated by a complex

$$\Omega^p(\mathfrak{g}, K; X(\mathfrak{q})) = \operatorname{Hom}_K(\wedge^p\mathfrak{p}_{\mathbb{C}}, X(\mathfrak{q})). \tag{5.7)(a}$$

According to Corollary 5.3 and Theorem 5.6, the only representation of K occurring in both $X(\mathfrak{q})$ and $\wedge\mathfrak{p}_{\mathbb{C}}$ is $\delta(\mathfrak{q})$. Corollary 5.5 then gives

$$\Omega^p(\mathfrak{g}, K; X(\mathfrak{q})) \simeq \operatorname{Hom}_K(\wedge^p\mathfrak{p}_{\mathbb{C}}, V_{\delta(\mathfrak{q})}) \simeq \operatorname{Hom}_{L\cap K}(\wedge^{p-R}(\mathfrak{l}_{\mathbb{C}}\cap\mathfrak{p}_{\mathbb{C}}), \mathbb{C}). \tag{5.7)(b}$$

Consequently

$$\Omega^p(\mathfrak{g}, K; X(\mathfrak{q})) \simeq \Omega^{p-R}(\mathfrak{l}, L\cap K; \mathbb{C}). \tag{5.7)(c}$$

We have seen in Proposition 2.6 that the differential in the second complex is zero. The same is true of the first:

Proposition 5.8 ([2], Proposition II.3.1). *Suppose that* X *is the Harish-Chandra module of a unitary representation of* G, *and that the Casimir operator acts by zero on* X. *Then the differential in* $\Omega^p(\mathfrak{g}, K; X)$ *is zero; so*

$$H^p(\mathfrak{g}, K; X) \simeq \operatorname{Hom}_K(\wedge^p\mathfrak{p}_{\mathbb{C}}, X).$$

In light of Proposition 5.8, the formula (5.7)(c) immediately implies the cohomology formula in Theorem 4.6.

6. Cohomologically Induced Representations: Construction

In this section we will say a little about the construction of a unitary representation $\pi(\mathfrak{q})$ satisfying the conditions in Theorem 5.6. There are a number of ways to construct a Harish-Chandra module satisfying conditions (1)–(3) of Theorem 5.6, beginning with Parthasarathy's method in [12]. The only method known for constructing a *unitary* representation is algebraic in nature, and is based on ideas of Zuckerman. It is the subject of [6]; we will say almost nothing about it. Instead we will discuss a more analytic construction suggested by Kostant in [8], and elaborated by Schmid in [13]. The tools are those of complex analysis; so we begin with some general remarks about that.

Proposition 6.1. *Suppose* G *is a Lie group and* H *is a closed subgroup. Write* $\mathfrak{h}\subset\mathfrak{g}$ *for their Lie algebras. Then* G*-invariant complex structures on the homogeneous space* G/H *are in one-to-one correspondence with complex Lie subalgebras* $\mathfrak{q}\subset\mathfrak{g}_{\mathbb{C}}$, *having the following two properties.*

1. *We have* $\mathfrak{q}\cap\overline{\mathfrak{q}} = \mathfrak{h}_{\mathbb{C}}$, *and* $\mathfrak{q}+\overline{\mathfrak{q}} = \mathfrak{g}_{\mathbb{C}}$.
2. *The complexified adjoint action of* H *on* $\mathfrak{g}_{\mathbb{C}}$ *preserves* $\mathfrak{q}$.

SKETCH OF PROOF. This is well-known and (almost) elementary. Suppose we are given a $\mathfrak{q}$ satisfying these two conditions. The first condition (together with the fact that $\mathfrak{q}$ is a complex subspace of $\mathfrak{g}_{\mathbb{C}}$) means that $\mathfrak{q}$ defines a complex structure on the tangent space $\mathfrak{g}/\mathfrak{h}$ to G/H at eH. Next, we use the action of G to move this complex structure to all the other tangent spaces; the second condition guarantees that this is well-defined. In this way we get a G-invariant almost complex structure on G/H. The fact that $\mathfrak{q}$ is a Lie subalgebra means that this almost complex structure is integrable. By the Newlander-Nirenberg theorem (this is the not-so-elementary part of the argument) an integrable almost complex structure is a complex structure. The converse is similar (but entirely elementary). □

Notice that $\mathfrak{q}$ and H are almost a pair in the sense of Definition 1.6. The only change is that $\mathfrak{q}$ is a complex Lie algebra instead of a real one. (We could define a **complex pair** accordingly, but we will spare the reader.) In any case it is more or less clear what a $(\mathfrak{q}, H)$-module ought to be, by analogy with Definition 1.7; we simply require the representation of $\mathfrak{q}$ to be complex-linear instead of real-linear.

It is well-known that the G-equivariant complex vector bundles on G/H are parametrized naturally by the finite-dimensional complex representations of H. Here is the analogous result for holomorphic bundles.

Proposition 6.2. *Suppose G is a Lie group and H is a closed subgroup. Suppose that we are given a G-invariant complex structures on the homogeneous space G/H corresponding to the complex Lie algebra $\mathfrak{q} \subset \mathfrak{g}_{\mathbb{C}}$ (Proposition 6.1). Then the G-equivariant holomorphic vector bundles on G/H are naturally parametrized by the finite-dimensional $(\mathfrak{q}, H)$-modules (Definition 1.7). This parametrization sends a vector bundle $\mathcal{V}$ to the fiber $V = \mathcal{V}_{eH}$.*

We omit the proof. If V is a finite-dimensional $(\mathfrak{q}, H)$-module, then the corresponding holomorphic vector bundle on G/H is written $\mathcal{V} = G \times_{\mathfrak{q},H} V$.

If $\mathcal{V}$ is a holomorphic vector bundle on a complex manifold X, then one can define Dolbeault cohomology groups $H^{0,p}(X, \mathcal{V})$. (The definition uses a certain differential $\overline{\partial}$ on $(0,p)$-forms with values in $\mathcal{V}$. It is formally quite similar to the de Rham d on ordinary forms.) For $p = 0$, the Dolbeault cohomology is the space of all holomorphic sections of $\mathcal{V}$. If X is a Stein manifold, the higher cohomology groups are all zero. The Dolbeault theorem asserts that $H^{0,p}(X, \mathcal{V})$ is isomorphic to the pth Čech cohomology of X with coefficients in the sheaf $\mathcal{O}(\mathcal{V})$ of germs of holomorphic sections of $\mathcal{V}$.

If now $\mathcal{V}$ is a G-equivariant holomorphic vector bundle on G/H, then there is a natural action of G on the Dolbeault complex, and so on the cohomology groups $H^{0,p}(G/H, \mathcal{V})$. In this way we get a representation of G on $H^{0,p}(G/H, \mathcal{V})$. The representations we want to discuss are of this form.

Suppose now that we are in the setting of Definition 4.1, so that $\mathfrak{q}$ is a θ-stable parabolic subalgebra of $\mathfrak{g}$ with Levi subgroup L. By Definition 4.1 and Proposition 6.1, $\mathfrak{q}$ defines a G-invariant holomorphic structure on G/L. It is not difficult to see that $\mathfrak{q} \cap \mathfrak{k}_{\mathbb{C}}$ defines a K-invariant holomorphic structure on $K/L \cap K$, and the natural inclusion

$$K/(L \cap K) \to G/L \tag{6.3)(a}$$

is a holomorphic embedding. We now introduce a holomorphic line bundle on G/L. Write $\mathfrak{u}$ for the nil radical of $\mathfrak{q}$, so that we have a Levi decomposition

$$\mathfrak{q} = \mathfrak{l}_{\mathbb{C}} \oplus \mathfrak{u}. \tag{6.3)(b}$$

This decomposition is invariant under L. Under $L \cap K$ we have a further decomposition

$$\mathfrak{u} = (\mathfrak{u} \cap \mathfrak{k}_{\mathbb{C}}) \oplus (\mathfrak{u} \cap \mathfrak{p}_{\mathbb{C}}). \tag{6.3(c)}$$

We write

$$R = \dim(\mathfrak{u} \cap \mathfrak{p}_{\mathbb{C}}), \quad S = \dim(\mathfrak{u} \cap \mathfrak{k}_{\mathbb{C}}). \tag{6.3(d)}$$

Then one sees easily that

$$S = \dim_{\mathbb{C}} K/(L \cap K), \qquad R + S = \dim_{\mathbb{C}} G/L. \tag{6.3(e)}$$

Example 6.4. This example has G disconnected, and so does not quite meet our hypotheses; but it is nevertheless attractive. Let G be the general linear group $GL(2n, \mathbb{R})$, and let X be the Grassmann variety of n-dimensional complex planes in $\mathbb{C}^{2n}$. This is a compact complex manifold of complex dimension n^2; indeed it is a projective algebraic variety. The complex group $G_{\mathbb{C}} = GL(2n, \mathbb{C})$ acts transitively on X. The isotropy group at the standard copy of $\mathbb{C}^n \subset \mathbb{C}^{2n}$ is

$$Q = \left\{ \begin{pmatrix} A & B \\ 0 & C \end{pmatrix} \,\middle|\, A, C \in GL(n, \mathbb{C}),\ B \in M_{n \times n}(\mathbb{C}) \right\};$$

so $X \simeq G_{\mathbb{C}}/Q$.

Now G acts on X, but the action is not transitive. Here is a way to understand the orbits. Suppose V is an n-plane in $\mathbb{C}^{2n}$. Then $\overline{V}$ (the set of vectors obtained from V by conjugating coordinate by coordinate) is another n-plane; so $V \cap \overline{V} = W_{\mathbb{C}}$ is a subspace of $\mathbb{C}^{2n}$ defined over $\mathbb{R}$; that is, it is the complexification of a subspace W of $\mathbb{R}^{2n}$. Similarly, $V + \overline{V} = U_{\mathbb{C}}$ is the complexification of a subspace $U \supset W$ of $\mathbb{R}^{2n}$. Write d for the dimension of W; evidently $0 \leq d \leq n$. The spaces U, V, and W have the following properties.

$$W \subset U \subset \mathbb{R}^{2n}, \qquad \dim W = d, \qquad \dim U = 2n - d; \tag{6.4(a)}$$

$$W_{\mathbb{C}} \subset V \subset U_{\mathbb{C}}; \tag{6.4(b)}$$

$$V/W_{\mathbb{C}} \text{ defines a complex structure } J_{U/W} \text{ on } U/W. \tag{6.4(c)}$$

(Explicitly, $V/W_{\mathbb{C}}$ is the $+i$ eigenspace of the complexification of $J_{U/W}$.)

Conversely, suppose $W \subset U$ are subspaces of $\mathbb{R}^{2n}$, of dimensions d and $2n - d$ respectively; and suppose we are given a complex structure $J_{U/W}$ on U/W. Then the complex structure corresponds to a complex subspace $V' \subset (U/W)_{\mathbb{C}}$ of dimension $n - d$. The preimage V of V' in $W_{\mathbb{C}}$ is an n-dimensional subspace, and it gives rise to W and U by the construction above. In this way we find a bijection between the collection of n-planes in $\mathbb{C}^{2n}$ and the collection of triples $(W, U, J_{U/W})$ satisfying (6.4)(a)–(c).

In terms of this description, it is easy to understand the orbits of $G = GL(2n, \mathbb{R})$ acting on X. The dimension d of W is obviously constant on orbits. Write X_d for the set of all triples $(W, U, J_{U/W})$ as above with $\dim W = d$. It is easy to see that G is transitive on pairs of subspaces $W \subset U$ of dimensions d and $2n - d$; and that the isotropy group at (W, U) maps onto $GL(W/U)$. This last group acts transitively on the complex structures on W/U; so we conclude that G acts transitively on X_d. In particular, there are exactly $n + 1$ orbits. Only one of these is open; it is X_0, which is just the space of all complex structures on $\mathbb{R}^{2n}$. For a base point in X_0 we may take some standard complex structure $\mathbb{R}^{2n} \simeq \mathbb{C}^n$; the isotropy group is evidently $GL(n,\mathbb{C})$, so that

$$GL(2n,\mathbb{R})/GL(n,\mathbb{C}) \simeq X_0 = \{\text{complex structures on } \mathbb{R}^{2n}\} \subset X \simeq G_{\mathbb{C}}/Q.$$

Because the standard complex structure J on $\mathbb{R}^{2n}$ is given by a skew-symmetric matrix (consisting of n diagonal blocks $\begin{pmatrix} 0 & 1 \\ -1 & 0 \end{pmatrix}$), the group $L = GL(n,\mathbb{C})$ is the Levi factor of a θ-stable parabolic subalgebra. Consequently X_0 is one of the spaces considered in (6.3). The compact subvariety $K/(L \cap K) = O(2n)/U(n)$ is easy to identify in this case: it consists of all complex structures J on $\mathbb{R}^{2n}$ which preserve the inner product. We compute $S = \dim_{\mathbb{C}} K/(L \cap K) = (n^2 - n)/2$ and $R = (n^2 + n)/2$.

We turn now to a consideration of Dolbeault cohomology groups on the spaces G/L. As we indicated before (6.3), the higher cohomology groups vanish in the case of a Stein manifold. Now a compact complex submanifold of a Stein manifold is necessarily finite; but G/L has the compact complex submanifold $K/(L \cap K)$, which has complex dimension S. Schmid and Wolf have shown that G/L comes as close to being a Stein manifold as this subvariety will allow. Here is a precise statement.

Theorem 6.5 ([15]). *G/L is $(S+1)$-complete in the sense of Andreotti and Grauert.*

What this means is that G/L admits an exhaustion function (a nonnegative smooth function ϕ with $\phi^{-1}([0,N])$ compact for all N) such that the Levi form of ϕ has at most S nonpositive eigenvalues at each point of G/L. The Levi form is a Hermitian form on the holomorphic tangent bundle constructed from second partial derivatives of ϕ. In holomorphic local coordinates, its matrix is $\partial^2\phi/\partial z_i \partial \overline{z}_j$.

Corollary 6.6 ([1], page 250). *If $\mathcal{S}$ is any coherent sheaf on G/L, then $H^p(G/L,\mathcal{S}) = 0$ for $p > S$. In particular, the Dolbeault cohomology $H^{0,p}(G/L,\mathcal{V})$ with coefficients in a holomorphic vector bundle $\mathcal{V}$ vanishes for $p > S$.*

We can now introduce the line bundle on G/L that we will be working with.

Definition 6.7. Suppose $\mathfrak{q}$ is a θ-stable parabolic subalgebra for G, with Levi factor L. Use the notation of (6.3). Consider the one-dimensional $(\mathfrak{q}, L)$-module

$$L_{2\rho(\mathfrak{u})} = \wedge^{R+S}(\mathfrak{g}_{\mathbb{C}}/\mathfrak{q})^* \simeq \wedge^{R+S}\mathfrak{u} \simeq \wedge^{R+S}(\mathfrak{q}/\mathfrak{l}_{\mathbb{C}}) \tag{6.7)(a}$$

The first description exhibits $L_{2\rho(\mathfrak{u})}$ as the fiber at eL of the top exterior power of the holomorphic cotangent bundle of G/L. The corresponding holomorphic line bundle

$$\mathcal{L}_{2\rho(\mathfrak{u})} = G \times_{\mathfrak{q},L} L_{2\rho(\mathfrak{u})} \tag{6.7)(b}$$

on G/L is therefore the canonical bundle. As a $(\mathfrak{q} \cap \mathfrak{k}_{\mathbb{C}}, L \cap K)$-module, $L_{2\rho(\mathfrak{u})}$ has a factorization

$$L_{2\rho(\mathfrak{u})} = \wedge^R(\mathfrak{u} \cap \mathfrak{p}_{\mathbb{C}}) \otimes \wedge^S(\mathfrak{u} \cap \mathfrak{k}_{\mathbb{C}}); \tag{6.7)(c}$$

the factors are denoted $L_{2\rho(\mathfrak{u}\cap\mathfrak{p}_{\mathbb{C}})}$ and $L_{2\rho(\mathfrak{u}\cap\mathfrak{k}_{\mathbb{C}})}$ respectively. They induce holomorphic line bundles

$$\mathcal{L}_{2\rho(\mathfrak{u}\cap\mathfrak{p}_{\mathbb{C}})} = K \times_{\mathfrak{q}\cap\mathfrak{k}_{\mathbb{C}}, L\cap K} L_{2\rho(\mathfrak{u}\cap\mathfrak{p}_{\mathbb{C}})} \tag{6.7)(d}$$

and similarly $\mathcal{L}_{2\rho(\mathfrak{u}\cap\mathfrak{k}_{\mathbb{C}})}$ on $K/(L \cap K)$. This last is the canonical bundle for $K/(L \cap K)$.

Finally, we define $\widetilde{\pi}(\mathfrak{q})$ to be the representation of G on the Dolbeault cohomology space

$$\widetilde{\mathcal{H}}(\mathfrak{q}) = H^{0,S}(G/L, \mathcal{L}_{2\rho(\mathfrak{u})}). \tag{6.7)(e}$$

Notice that Corollary 6.6 guarantees that this is the highest degree in which the cohomology can be nonzero.

The definition needs some remarks. First, the representation space is usually infinite-dimensional. We therefore need a topology on it to make any sensible statements. The natural topology comes from the Dolbeault complex. The $(0, S)$-forms with values in $\mathcal{L}_{2\rho(\mathfrak{u})}$ carry a natural C^∞ topology, and the closed forms constitute a closed subspace. The exact forms, however, do not obviously constitute a closed subspace; so the quotient topology on $H^{0,S}$ is not obviously Hausdorff. Wong has shown in [19] that the exact forms actually are closed, so that the topology is Hausdorff.

Theorem 4.6 asks for a unitary representation on a Hilbert space. The space $\widetilde{\mathcal{H}}(\mathfrak{q})$ is not a Hilbert space unless it is finite-dimensional; so $\widetilde{\pi}(\mathfrak{q})$ cannot be exactly the representation we are looking for. Wong also shows in [19] that $\widetilde{\pi}(\mathfrak{q})$ is infinitesimally equivalent to a representation constructed algebraically by Zuckerman; and this representation was already known from [16] to be unitary. We will have no more to say about the details of this (successful) approach to proving Theorem 4.6, concentrating instead on ideas of Schmid for analyzing $\widetilde{\pi}(\mathfrak{q})$. These ideas are taken from his dissertation, which was published in [13]. We choose them because they are easier to understand, and because they motivate many arguments in the algebraic theory.

Theorem 5.6 suggests that we ought to find some connection between $\widetilde{\pi}(\mathfrak{q})$ and the representation $\delta(\mathfrak{q})$ of K (Corollary 5.5). The first step is provided by the following result.

Lemma 6.8. *In the setting of Definition* 6.7, *the representation of K on the Dolbeault cohomology of the line bundle $\mathcal{L}_{2\rho(\mathfrak{u})}$ is the irreducible representation $\delta(\mathfrak{q})$ described in Corollary* 5.5*:*

$$V_{\delta(\mathfrak{q})} \simeq H^{0,S}(K/(L \cap K), \mathcal{L}_{2\rho(\mathfrak{u})}).$$

PROOF. Write W for the cohomology group in the lemma. All such cohomology groups (with coefficients in irreducible equivariant vector bundles) are computed by the Bott-Borel-Weil theorem. But in this case we can manage with even less. Recall that S is the complex dimension of $K/(L \cap K)$, and that the line bundle factors as $\mathcal{L}_{2\rho(\mathfrak{u}\cap\mathfrak{p}_\mathbb{C})} \otimes \mathcal{L}_{2\rho(\mathfrak{u}\cap\mathfrak{k}_\mathbb{C})}$. The second factor is the canonical bundle of $K/(L \cap K)$. The Serre Duality Theorem provides an isomorphism

$$W^* \simeq H^{0,0}(K/(L \cap K), \mathcal{L}^*_{2\rho(\mathfrak{u}\cap\mathfrak{p}_\mathbb{C})})$$

We observed after Proposition 6.2 that the group on the right is the space of holomorphic sections of the line bundle. According to the Borel-Weil theorem, W^* is therefore the irreducible representation of K of lowest weight $-2\rho(\mathfrak{u} \cap \mathfrak{p}_\mathbb{C})$. By Corollary 5.5, $W^* \simeq V^*_{\delta(\mathfrak{q})}$. The lemma follows. □

To go further, we need some additional notation. In the setting of Definition 6.7, let us write $\mathcal{O}^G$ for the sheaf of germs of holomorphic sections of $\mathcal{L}_{2\rho(\mathfrak{u})}$ on G/L, and $\mathcal{O}^K$ for the corresponding sheaf on $K/(L \cap K)$. We may also regard $\mathcal{O}^K$

as a sheaf on G/L supported on $K/(L \cap K)$. According to Definition 6.7, Lemma 6.8, and the remarks after Proposition 6.2, the Čech cohomology groups of these sheaves in degree S are

$$H^S(G/L, \mathcal{O}^G) \simeq \widetilde{\mathcal{H}}(\mathfrak{q}) \tag{6.9)(a}$$

$$H^S(G/L, \mathcal{O}^K) \simeq H^S(K/(L \cap K), \mathcal{O}^K) \simeq V(\mathfrak{q}). \tag{6.9)(b}$$

So we are looking for a connection between the sheaves $\mathcal{O}^G$ and $\mathcal{O}^K$ on G/L. This is provided by the restriction map: any holomorphic germ on G/L has a restriction to $K/(L \cap K)$. The restriction map is surjective (on sheaves of germs), since any holomorphic germ on $K/(L \cap K)$ has an extension to a germ on G/L. Its kernel is the sheaf $\mathcal{V}^1$ of germs of holomorphic sections of $\mathcal{L}_{2\rho(\mathfrak{u})}$ on G/L that vanish on $K/(L \cap K)$. We therefore have a short exact sequence of sheaves on G/L

$$0 \to \mathcal{V}^1 \to \mathcal{O}^G \to \mathcal{O}^K \to 0. \tag{6.9)(c}$$

These are all coherent sheaves; so the vanishing theorem of Corollary 6.6 applies. The long exact sequence in sheaf cohomology attached to (6.9)(c) therefore ends in degree S; in light of (6.9)(a) and (6.9)(b), the last terms are

$$\cdots \to H^S(G/L, \mathcal{V}^1) \to \widetilde{\mathcal{H}}(\mathfrak{q}) \to V(\mathfrak{q}) \to 0. \tag{6.9)(d}$$

As an immediate consequence, we deduce that

$$\delta(\mathfrak{q}) \text{ occurs in } \widetilde{\pi}(\mathfrak{q}). \tag{6.9)(e}$$

This is a (small) step in the direction of Theorem 5.6. To continue, we need to understand the representations of K appearing in the cohomology of $\mathcal{V}^1$. Schmid's method for doing so is to introduce the sheaves

$$\mathcal{V}^n = \text{germs of sections of } \mathcal{L}_{2\rho(\mathfrak{u})} \text{ vanishing to } n\text{th order on } K/(L \cap K) \tag{6.9)(f}$$

on G/L. So for example

$$\mathcal{V}^0 = \mathcal{O}^G, \qquad \mathcal{V}^0/\mathcal{V}^1 = \mathcal{O}^K.$$

The next result is a generalization of Lemma 6.8.

Lemma 6.10 ([13], (4.3)). *Suppose we are in the setting of Definition 6.7; use the notation of (6.9). Then for all $n \geq 0$, the quotient sheaf $\mathcal{V}^n/\mathcal{V}^{n+1}$ is supported on $K/(L \cap K)$. It may be described as follows. Write $\mathcal{N}$ for the holomorphic normal bundle of $K/(L \cap K)$ in G/L, and $\mathcal{N}^*$ for the dual bundle. Explicitly,*

$$\mathcal{N}^* \simeq K \times_{\mathfrak{q} \cap \mathfrak{k}_\mathbb{C}, L \cap K} (\mathfrak{g}_\mathbb{C}/(\mathfrak{q} + \mathfrak{k}_\mathbb{C}))^* \simeq K \times_{\mathfrak{q} \cap \mathfrak{k}_\mathbb{C}, L \cap K} (\mathfrak{u} \cap \mathfrak{p}_\mathbb{C}).$$

Write $S^n(\mathcal{N}^*)$ for the nth symmetric power of $\mathcal{N}^*$, and $\mathcal{O}(\mathcal{W})$ for the sheaf of germs of holomorphic sections of a vector bundle $\mathcal{W}$. Then

$$\mathcal{V}^n/\mathcal{V}^{n+1} \simeq \mathcal{O}(S^n(\mathcal{N}^*) \otimes \mathcal{L}_{2\rho(\mathfrak{u})}).$$

In particular, every cohomology group of $\mathcal{V}^n/\mathcal{V}^{n+1}$ is a finite-dimensional representation of K.

If δ is an irreducible representation of K appearing in $H^S(G/L, \mathcal{V}^n/\mathcal{V}^{n+1})$, then the highest weight of δ must be of the form $2\rho(\mathfrak{u} \cap \mathfrak{p}_\mathbb{C}) + \gamma$, with γ a sum of n roots of T in $\mathfrak{u} \cap \mathfrak{p}_\mathbb{C}$.

The first part of the lemma amounts to a coordinate-free treatment of Taylor expansions; it can be done with $K/(L \cap K) \subset G/L$ replaced by any closed complex submanifold of a complex manifold. The second part is a generalization of Lemma 6.8, and can be proved in a similar way. We omit the details.

Corollary 6.11. *Suppose we are in the setting of Definition* 6.7; *use the notation of* (6.9). *The quotient sheaf* $\mathcal{O}^G/\mathcal{V}^{n+1}$ *is supported on* $K/(L \cap K)$, *and has finite-dimensional cohomology sheaves. Consider the short exact sequence*

$$0 \to \mathcal{V}^{n+1} \to \mathcal{O}^G \to \mathcal{O}^G/\mathcal{V}^{n+1} \to 0.$$

The corresponding long exact sequence in cohomology ends in degree S, *and the last terms are*

$$\cdots \to H^S(G/L, \mathcal{V}^{n+1}) \to \widetilde{\mathcal{H}}(\mathfrak{q}) \to H^S(K/(L \cap K), \mathcal{O}^G/\mathcal{V}^{n+1}) \to 0.$$

Any irreducible representation of K *appearing in this last group must have highest weight* $2\rho(\mathfrak{u} \cap \mathfrak{p}_{\mathbb{C}}) + \gamma$, *with* γ *a sum of at most* n *roots of* T *in* $\mathfrak{u} \cap \mathfrak{p}_{\mathbb{C}}$.

This follows from Lemma 6.10 just as we deduced (6.9)(d) above.

Let us see where we stand. For each nonnegative integer n, we define a subspace of $\widetilde{\mathcal{H}}(\mathfrak{q})$ by

$$\begin{aligned} \widetilde{\mathcal{H}}(\mathfrak{q})^n &= \text{kernel of the map } \widetilde{\mathcal{H}}(\mathfrak{q}) \to H^S(K/(L \cap K), \mathcal{O}^G/\mathcal{V}^{n+1}) \\ &= \text{image of the map } H^S(G/L, \mathcal{V}^{n+1}) \to \widetilde{\mathcal{H}}(\mathfrak{q}) \end{aligned} \qquad (6.12)(a)$$

Because of the first description, $\widetilde{\mathcal{H}}(\mathfrak{q})^n$ is a closed K-invariant subspace of $\widetilde{\mathcal{H}}(\mathfrak{q})$. It is also clear from the definitions that there are containments

$$\widetilde{\mathcal{H}}(\mathfrak{q})^n \subset \widetilde{\mathcal{H}}(\mathfrak{q})^m \qquad (n \geq m). \qquad (6.12)(b)$$

The Lie algebra $\mathfrak{g}$ acts on Dolbeault cocycles by first-order differential operators. It is plausible to think that such operators should decrease order of vanishing along a subvariety by at most one. This is true, and is proved in [13], Lemma 6.8:

$$\widetilde{\pi}(\mathfrak{q})(X)\widetilde{\mathcal{H}}(\mathfrak{q})^{n+1} \subset \widetilde{\mathcal{H}}(\mathfrak{q})^n \qquad (X \in \mathfrak{g}). \qquad (6.12)(c)$$

Now define

$$\widetilde{\mathcal{H}}(\mathfrak{q})^\infty = \bigcap_n \widetilde{\mathcal{H}}(\mathfrak{q})^n. \qquad (6.12)(d)$$

This is a closed, K-invariant, $\mathfrak{g}$-invariant subspace of $\widetilde{\mathcal{H}}(\mathfrak{q})$. Here is what one can prove fairly easily using these ideas.

Theorem 6.13. *Suppose we are in the setting of Definition* 6.7; *use the notation of* (6.12). *Then the* $(\mathfrak{g}, K)$*-module of* K*-finite vectors in* $\widetilde{\mathcal{H}}(\mathfrak{q})/\widetilde{\mathcal{H}}(\mathfrak{q})^\infty$ *satisfies the three conditions in Theorem* 5.6. *More precisely:*

1. *The restriction to* K *contains* $\delta(\mathfrak{q})$ *exactly once.*
2. *Every representation of* K *appearing has highest weight* $2\rho(\mathfrak{u} \cap \mathfrak{p}_{\mathbb{C}}) + \gamma$, *with* γ *a sum of roots of* T *in* $\mathfrak{u} \cap \mathfrak{p}_{\mathbb{C}}$.
3. *The Casimir operator acts by* 0 *(even on all of* $\widetilde{\mathcal{H}}(\mathfrak{q})$*).*

This is all more or less clear from Corollary 6.11 and (6.12), except for the assertion about the Casimir operator. That is a routine calculation analogous to the calculation of infinitesimal characters for induced representations (see for example [5], Proposition 8.22). (The Casimir acts on cohomology classes by differentiation on the left. Since it is central, we may as well differentiate on the right. But cohomology classes satisfy some differential equations on the right, and these allow us to show that the Casimir action is zero.) We omit the details.

In this way we can construct at least a nonunitary representation satisfying the requirements of Theorem 5.6. We conclude with a few more remarks about its relationship to $\widetilde{\pi}(\mathfrak{q})$. Suppose first that $S = 0$, so that $\widetilde{\mathcal{H}}(\mathfrak{q})$ is the space of holomorphic sections of a line bundle on G/L. The subspace $\widetilde{\mathcal{H}}(\mathfrak{q})^n$ consists of sections vanishing to order n at the point $K/(L \cap K)$. Since a nonzero holomorphic function cannot vanish to infinite order at a point, we see that $\widetilde{\mathcal{H}}(\mathfrak{q})^\infty = 0$.

In general (when $S \neq 0$), $\widetilde{\mathcal{H}}(\mathfrak{q})^n$ may be identified with Čech cohomology classes admitting representatives involving holomorphic functions that vanish to order n along $K/(L \cap K)$. It follows that $\widetilde{\mathcal{H}}(\mathfrak{q})^\infty$ corresponds to Čech cohomology classes admitting for every n representatives involving holomorphic functions that vanish to order n along $K/(L \cap K)$. Of course we will have to choose different representatives for different values of n, but there is no general argument to rule out the existence of nonzero classes. On the other hand, Schmid's beautiful analysis of $\widetilde{\mathcal{H}}(\mathfrak{q})/\widetilde{\mathcal{H}}(\mathfrak{q})^\infty$ (roughly outlined in Lemma 6.10, Corollary 6.11, and Theorem 6.13) certainly gives reason to hope that $\widetilde{\mathcal{H}}(\mathfrak{q})^\infty = 0$. This is true, and is part of the result of Wong already mentioned:

Theorem 6.14 ([19]). *Suppose we are in the setting of Definition* 6.7; *use the notation of* (6.12). *Then* $\widetilde{\mathcal{H}}(\mathfrak{q})^\infty = 0$. *Consequently* $\widetilde{\mathcal{H}}(\mathfrak{q})$ *is a smooth Fréchet representation of* G *whose* $(\mathfrak{g}, K)$*-module satisfies the conditions* (1)–(3) *of Theorems* 5.6 *or* 6.13.

References

[1] A. Andreotti and H. Grauert, Théorèmes de finitude pour la cohomologie des espaces complexes, *Bull. Soc. Math. France* **90** (1962), 193–259.

[2] A. Borel and N. Wallach, *Continuous Cohomology, Discrete Subgroups, and Representations of Reductive Groups*, Princeton University Press, Princeton, New Jersey, 1980.

[3] W. Greub, S. Halperin, and R. Vanstone, *Connections, Curvature, and Cohomology. Volume III: Cohomology of principal bundles and homogeneous spaces*, Pure and Applied Mathematics, Vol. 47–III, Academic Press (Harcourt Brace Jovanovich, Publishers), New York-London, 1976.

[4] Harish-Chandra, Harmonic analysis on reductive groups I. The theory of the constant term, *J. Func. Anal.* **19** (1975), 104–204.

[5] A. Knapp, *Representation Theory of Real Semisimple Groups: an Overview Based on Examples*, Princeton University Press, Princeton, New Jersey, 1986.

[6] A. Knapp and D. Vogan, *Cohomological Induction and Unitary Representations*, Princeton University Press, Princeton, New Jersey, 1995.

[7] B. Kostant, Lie algebra cohomology and the generalized Borel-Weil theorem, *Annals of Math.* **74** (1961), 329–387.

[8] B. Kostant, Orbits, symplectic structures, and representation theory, *Proceedings of the United States-Japan Seminar in Differential Geometry, Kyoto, Japan, 1965*, Nippon Hyoronsha, Tokyo, 1966, p. 71.

[9] R. Kottwitz, Shimura varieties and λ-adic representations, *Automorphic Forms, Shimura Varieties, and L-functions* (L. Clozel and J. Milne, eds.), Perspectives in Mathematics **10**, vol. I, Academic Press, San Diego, 1990, pp. 161–209.

[10] S. Kumaresan, On the canonical $\mathfrak{k}$-types in the irreducible unitary $\mathfrak{g}$-modules with nonzero relative cohomology, *Invent. Math.* **59** (1980), 1–11.

[11] Y. Matsushima, On Betti numbers of compact locally symmetric Riemannian manifolds, *Jour. Diff. Geom.* **1** (1967), 99–109.

[12] R. Parthasarathy, A generalization of the Enright-Varadarajan modules, *Compositio Math.* **36** (1978), 53–73.

[13] W. Schmid, Homogeneous complex manifolds and representations of semisimple Lie groups, *Representation Theory and Harmonic Analysis on Semisimple Lie Groups* (P. Sally and D. Vogan, eds.), Mathematical Surveys and Monographs **31**, American Mathematical Society, Providence, Rhode Island, 1989, pp. 223–286.
[14] W. Schmid, Discrete series, *these* PROCEEDINGS, pp. 83–113.
[15] W. Schmid and J. Wolf, A vanishing theorem for open orbits on complex flag manifolds, *Proc. Amer. Math. Soc.* **92** (1984), 461–464.
[16] D. Vogan, Unitarizability of certain series of representations, *Annals of Math.* **120** (1984), 141–187.
[17] D. Vogan and G. Zuckerman, Unitary representations with nonzero cohomology, *Compositio Math.* **53** (1984), 51–90.
[18] F. Warner, *Foundations of Differentiable Manifolds and Lie Groups*, Scott, Foresman and Company, Glenview, Illinois, 1971.
[19] H. Wong, *Dolbeault cohomologies and Zuckerman modules associated with finite rank representations*, Ph.D. dissertation, Harvard University, 1991; Dolbeault cohomological realization of Zuckerman modules associated with finite rank representations, *J. Func. Anal.* **129** (1995), 428–454.

DEPARTMENT OF MATHEMATICS, MASSACHUSETTS INSTITUTE OF TECHNOLOGY, CAMBRIDGE, MASSACHUSETTS 02139, U.S.A.
E-mail address: dav@math.mit.edu

Proceedings of Symposia in Pure Mathematics
Volume **61** (1997), pp. 245–302

Introduction to the Langlands Program

A. W. Knapp

This article is an introduction to automorphic forms on the adeles of a linear reductive group over a number field. The first half is a summary of aspects of local and global class field theory, with emphasis on the local Weil group, the L functions of Artin and Hecke, and the role of Artin reciprocity in relating the two kinds of L functions. The first half serves as background for the second half, which discusses some structure theory for reductive groups, the definitions of automorphic and cusp forms, the Langlands L group, L functions, functoriality, and some conjectures. Much of the material in the second half may be regarded as a brief introduction to the Langlands program. There are ten sections:

1. Local Fields and Their Weil Groups

This section contains a summary of information about local fields and their Weil groups. Four general references for this material are [Frö], [La], [Ta3], and [We4].

By a **local field** is meant any nondiscrete locally compact topological field. Let F be a local field. If α is a nonzero element of F, then multiplication by α is an automorphism of the additive group of F and hence carries additive Haar measure to a multiple of itself. This multiple is denoted $|\alpha|_F$, and it satisfies

$$\int_F h(\alpha^{-1}x)\,dx = \int_F h(x)\,d(\alpha x) = |\alpha|_F \int_F h(x)\,dx,$$

1991 *Mathematics Subject Classification.* Primary 11F70, 11R39, 11S37, 22E55.
This article is based partly on lectures by Don Blasius in Edinburgh. The author is grateful to Jonathan Rogawski and David Vogan for offering a number of suggestions about the exposition, to Dinakar Ramakrishnan for supplying a proof of Theorem 8.8 and giving permission to include it here, and to Hervé Jacquet for answering many questions.

where dx is an additive Haar measure. We refer to $|\alpha|_F$ as the **module** of α. By convention, $|0|_F = 0$. The function $\alpha \mapsto |\alpha|_F$ is continuous on F and is a homomorphism of the multiplicative group $F^\times$ into the multiplicative group $\mathbb{R}_+^\times$ of positive reals.

The local field F is said to be **nonarchimedean** if $|\alpha|_F$ satisfies the **ultrametric inequality**

$$|\alpha+\beta| \le \max\{|\alpha|, |\beta|\}.$$

Otherwise F is **archimedean**.

A classification of local fields appears in [We4]. There are only two archimedean local fields, $\mathbb{R}$ and $\mathbb{C}$. In the nonarchimedean case, the set of nonzero values of $|\cdot|_F$ is a discrete subgroup of $\mathbb{R}_+^\times$. The nonarchimedean local fields divide into two kinds. Those of characteristic 0 turn out to be the **p-adic fields**, namely the finite extensions of the fields $\mathbb{Q}_p$ of p-adic numbers for each prime number p. Those of characteristic not 0 turn out to be the fields of Laurent series in one variable (finite in negative powers) over the various finite fields. In this article we consider only local fields of characteristic 0.

Let us review various constructions of $\mathbb{Q}_p$. One way to define $\mathbb{Q}_p$ is as the completion of $\mathbb{Q}$ in the metric $d(x,y) = |x-y|_p$, where $|ap^n/b|_p = p^{-n}$ if a and b are integers prime to p. The metric extends to $\mathbb{Q}_p$, giving it a locally compact topology, and the field operations extend as well. It is easy to see that $|\alpha|_{\mathbb{Q}_p}$, which we abbreviate as $|\alpha|_p$, is just the distance between α and 0 in the completed metric.

Another construction begins with the definition of the maximal compact ring $\mathbb{Z}_p$ of p-adic integers, which is taken to be the inverse limit

$$\mathbb{Z}_p = \varprojlim_n \mathbb{Z}/(p^n).$$

The ring $\mathbb{Z}_p$ is an integral domain having a unique maximal ideal, namely $p\mathbb{Z}_p$. The ideals $p^n\mathbb{Z}_p$ form a neighborhood basis of 0, and $\{\alpha+p^n\mathbb{Z}_p\}$ is a neighborhood basis about $\alpha \in \mathbb{Z}_p$. Then $\mathbb{Q}_p$ may be described algebraically as the field of quotients of $\mathbb{Z}_p$ or as $\mathbb{Z}_p \otimes \mathbb{Q}$ or as $\mathbb{Z}[p^{-1}] = \bigcup_{n\ge 1} p^{-n}\mathbb{Z}_p$. The last of these descriptions provides the topology; each $p^{-n}\mathbb{Z}_p$ is to be open and homeomorphic with $\mathbb{Z}_p$. The inclusion $\mathbb{Z}_p \subset \mathbb{Q}_p$ makes $\mathbb{Z}_p$ a compact subring of the local field $\mathbb{Q}_p$. Here $\mathbb{Z}_p$ is precisely the set of all $\alpha \in \mathbb{Q}_p$ with $|\alpha|_p \le 1$, and the maximal ideal $p\mathbb{Z}_p$ is the set of all $\alpha \in \mathbb{Q}_p$ with $|\alpha|_p < 1$.

Let K be a finite extension of $\mathbb{Q}_p$ with $[K : \mathbb{Q}_p] = n$. If K is decomposed as a direct sum of n one-dimensional $\mathbb{Q}_p$ vector spaces, then the decomposition automatically respects the topology. The set

$$\mathcal{O}_K = \{\alpha \in K \bigm| |\alpha|_K \le 1\}$$

is a compact open subring of K called the **ring of integers** of K. It is the unique maximal compact subring of K and is equal to the integral closure of $\mathbb{Z}_p$ in K. The group of units $\mathcal{O}_K^\times$ of $\mathcal{O}_K$ is the set of elements α with $|\alpha|_K = 1$. The ring $\mathcal{O}_K$ has the following properties:

1) $\mathcal{O}_K$ has a unique maximal ideal $\mathfrak{m}_K$, namely the set of elements α with $|\alpha|_K < 1$. The ideal $\mathfrak{m}_K$ is principal, having any element of maximal module as generator. Such an element will typically be denoted ϖ_K and is a **prime element**. Every nonzero ideal of $\mathcal{O}_K$ is principal and is a power of $\mathfrak{m}_K$ (with the corresponding power of ϖ_K as generator).
2) $\mathcal{O}_K$ is the inverse limit of $\mathcal{O}_K/(\varpi_K^n)$ on n.

3) $k_K = \mathcal{O}_K/\mathfrak{m}_K$ is a finite field k_K, the **residue field**. The inclusion of $\mathbb{Z}_p$ into $\mathcal{O}_K$ induces a map of $\mathbb{Z}_p/p\mathbb{Z}_p$ into k_K, and thus $\mathbb{Z}_p/p\mathbb{Z}_p$ may be regarded as the prime field of k_K. We write the number of elements in k_K as $q = p^f$ and call f the **residue degree** of K over $\mathbb{Q}_p$. The module of ϖ_K is q^{-1}.
4) By (1), the ideal $p\mathcal{O}_K$ of $\mathcal{O}_K$ is of the form $\mathfrak{m}_K^e = (\varpi_K^e)$ for some integer e; e is called the **ramification degree** of K over $\mathbb{Q}_p$.
5) $[K : \mathbb{Q}_p] = ef$ since $p^{-n} = |p|_K = |\varpi_K|_K^e = q^{-e}$.

Now let $K \supset F$ be two finite extensions of $\mathbb{Q}_p$, and write q_K, f_K, e_K and q_F, f_F, e_F for the respective integers q, f, e. Also let k_K and k_F be the residue fields. Then we have an inclusion $k_F \subset k_K$. We define $f = [k_K : k_F]$, and we let e be the integer such that $\varpi_F\mathcal{O}_K = \mathfrak{m}_K^e$. The integers f and e are the **residue degree** and **ramification degree** of K over F. Then

$$f = f_K/f_F, \quad e = e_K/e_F, \quad \text{and} \quad [K : F] = ef.$$

Fix finite extensions $K \supset F$ of $\mathbb{Q}_p$, and suppose that K/F is Galois. Any F automorphism of K is module-preserving and hence maps $\mathcal{O}_K$ to itself and $\mathfrak{m}_K$ to itself. It therefore induces a automorphism of the quotient, which is the residue field k_K, and this automorphism fixes the residue field k_F of F. The result is a homomorphism $\mathrm{Gal}(K/F) \to \mathrm{Gal}(k_K/k_F)$. Since $\mathrm{Gal}(k_K/k_F)$ is cyclic with generator $x \mapsto x^{q_F}$, the image of a member of $\mathrm{Gal}(K/F)$ is necessarily of the form $x \mapsto x^{q_F^n}$ for some integer n.

Theorem 1.1. *If K/F is a Galois extension of finite extensions of $\mathbb{Q}_p$ and if k_K and k_F are the respective residue fields, then* $\mathrm{Gal}(K/F)$ *maps onto* $\mathrm{Gal}(k_K/k_F)$.

REFERENCE. See [La, p. 15]. For a formulation of this result without the assumption that K/F is Galois, see [Frö, p. 26].

In the setting of Theorem 1.1, we obtain an exact sequence

$$1 \longrightarrow I_{K/F} \longrightarrow \mathrm{Gal}(K/F) \longrightarrow \mathrm{Gal}(k_K/k_F) \longrightarrow 0, \tag{1.1}$$

where $I_{K/F}$ is the kernel, which is called the **inertia group** of K over F. We shall be interested in the effect on (1.1) of letting K swell to $\overline{F}$.

For the moment let us drop the assumption that K/F is Galois. Let f and e be the residue and ramification degrees of K over F. We say that the extension K/F is **unramified** if $e = 1$. In the Galois case, the group $\mathrm{Gal}(k_K/k_F)$ has order f, and the exactness of (1.1) implies that $I_{K/F}$ has order e; thus K/F is unramified if and only if $\mathrm{Gal}(K/F) \to \mathrm{Gal}(k_K/k_F)$ is an isomorphism.

Theorem 1.2. *Let F be a finite extension of $\mathbb{Q}_p$, and let k be a finite extension of the residue field k_F. Then there exists an unramified finite extension K of F with residue field $k_K \cong k$. Such a field K is unique up to F isomorphism and is Galois over F.*

REFERENCE. [Frö, p. 26] or [Se2, p. 54].

Consequently for each $f \geq 1$, there is, up to F isomorphism, a unique unramified extension $K = K_f$ of F of degree f, and K/F is Galois. By Theorem 1.1, $\mathrm{Gal}(K/F) \cong \mathrm{Gal}(k_K/k_F)$ is a cyclic group of order $q = p^f$ whose generator Fr is the lift of the generator $x \mapsto x^{|k_F|}$ of $\mathrm{Gal}(k_K/k_F)$. The element Fr of $\mathrm{Gal}(K/F)$ is called the **Frobenius element** and is characterized among members of $\mathrm{Gal}(K/F)$ by the congruence $\mathrm{Fr}(x) \equiv x^{|k_F|} \bmod \mathfrak{m}_K$.

Fix an algebraic closure $\overline{F}$ of F, and regard each $K = K_f$ as contained in $\overline{F}$. When two residue degrees f and f' have the property that f divides f', the multiplicative property of residue degrees and ramification degrees implies that $K_f \subset K_{f'}$. The result, as f varies, is a directed system of subfields of $\overline{F}$. The union of these subfields is called the **maximal unramified extension** of F and is denoted F_{ur}.

Every field map over F of a subfield of $\overline{F}$ into $\overline{F}$ extends to an automorphism of $\overline{F}$, and it follows that every member of $\mathrm{Gal}(F_{\mathrm{ur}}/F)$ extends to a member of $\mathrm{Gal}(\overline{F}/F)$. In other words, there is an exact sequence

$$1 \longrightarrow I_F \longrightarrow \mathrm{Gal}(\overline{F}/F) \longrightarrow \mathrm{Gal}(F_{\mathrm{ur}}/F) \longrightarrow 0 \tag{1.2}$$

in which I_F is the kernel of π. The group I_F is called the **inertia group** of F (or of $\overline{F}$ over F).

Now let us take the inverse limit of (1.1), letting K swell to $\overline{F}$. Then we obtain a homomorphism of $\mathrm{Gal}(\overline{F}/F)$ into $\mathrm{Gal}(\overline{k}_F/k_F)$, where $\overline{k}_F$ is the algebraic closure of k_F. Then we have the following result.

Theorem 1.3. *Every finite extension of F in F_{ur} is unramified. The natural homomorphism of* $\mathrm{Gal}(\overline{F}/F)$ *into* $\mathrm{Gal}(\overline{k}_F/k_F)$ *descends via* (1.2) *to a topological isomorphism of* $\mathrm{Gal}(F_{\mathrm{ur}}/F)$ *onto* $\mathrm{Gal}(\overline{k}_F/k_F)$.

REFERENCE. [Frö, p. 28] or [Se2, p. 54–55].

Let us identify $\mathrm{Gal}(\overline{k}_F/k_F)$, so that we can make (1.2) more explicit. The field $\overline{k}_F$ is the union of its finite subfields. These form a directed system under divisibility of degree, and the degree can be any positive integer. Therefore $\mathrm{Gal}(\overline{k}_F/k_F)$ is the inverse limit of cyclic groups $\mathbb{Z}/n\mathbb{Z}$, indexed according to divisibility of the indices n. The resulting compact group is denoted $\widehat{\mathbb{Z}}$ and is isomorphic to $\prod_p \mathbb{Z}_p$. We regard the Frobenius element $x \mapsto x^{|k_F|}$ in $\mathrm{Gal}(\overline{k}_F/k_F)$ as the integer $+1$, and the subgroup $\mathbb{Z}$ of $\widehat{\mathbb{Z}}$ generated by $+1$ is dense in $\widehat{\mathbb{Z}}$. The isomorphism of Theorem 1.3 tells us that $\mathrm{Gal}(F_{\mathrm{ur}}/F) \cong \widehat{\mathbb{Z}}$, and we let Fr be the (Frobenius) element of $\mathrm{Gal}(F_{\mathrm{ur}}/F)$ that corresponds to $+1$ in $\widehat{\mathbb{Z}}$. The rewritten form of (1.2) is then

$$1 \longrightarrow I_F \longrightarrow \mathrm{Gal}(\overline{F}/F) \xrightarrow{\ \pi\ } \widehat{\mathbb{Z}} \longrightarrow 0 \tag{1.3}$$

where π is defined to be restriction from $\overline{F}$ to F_{ur} composed with the isomorphism to $\widehat{\mathbb{Z}}$.

The Langlands theory of L groups and L functions makes extensive use of the Weil group of a local field. Let us define this group, sketch some of its properties, and formulate the theorems of local class field theory in terms of it. A reference for Weil groups is [Ta3].

We continue with F as a finite extension of $\mathbb{Q}_p$ and with $\overline{F}$ as an algebraic closure of F; the archimedean case is postponed to the next section. Let $G_F = \mathrm{Gal}(\overline{F}/F)$. We write $\widehat{\mathbb{Z}} = \mathrm{Gal}(F_{\mathrm{ur}}/F)$, and we let $\mathbb{Z}$ be the infinite cyclic subgroup of $\widehat{\mathbb{Z}}$ generated by the Frobenius element Fr. With π as in (1.3), the **Weil group** of F is defined as an abstract group by $W_F = \pi^{-1}(\mathbb{Z}) \subset \mathrm{Gal}(\overline{F}/F)$. The Weil form of the exact sequence (1.3) is then

$$1 \longrightarrow I_F \longrightarrow W_F \xrightarrow{\pi} \mathbb{Z} \longrightarrow 0. \tag{1.4}$$

The relative topology from (1.3) gives $\mathbb{Z}$ an unusual topology, but we change matters to give $\mathbb{Z}$ the discrete topology. Correspondingly we retopologize W_F so that π is continuous and I_F is homeomorphic with $\pi^{-1}(\{0\})$.

To understand π and W_F better, let K/F be a finite Galois extension. Then G_F maps onto $\mathrm{Gal}(K/F)$ with kernel G_K, and in turn $\mathrm{Gal}(K/F)$ maps onto $\mathrm{Gal}(k_K/k_F)$ by Theorem 1.1. The effect is to associate to any member σ of G_F an integer n and the automorphism $x \mapsto x^{q_F^n}$, with n depending on K and defined modulo the residue degree of K/F. The inverse limit of the resulting tuple of n's, as K varies, defines the member $\pi(\sigma)$ of $\widehat{Z}$. For $\pi(\sigma)$ to be in $\mathbb{Z}$, the condition is that the inverse limit can be regarded as a single integer n. That is, the members of the Weil group W_F are those members of G_F that induce on the algebraic closure $\overline{k}_F$ the automorphism $x \mapsto x^{q_F^n}$ for some integer n.

If E/F is a finite extension, then Galois theory says that G_E is the subgroup of G_F fixing E. Hence W_E is a subgroup of G_F. The next theorem identifies this subgroup.

Theorem 1.4. *If E/F is a finite extension, then $W_E = G_E \cap W_F$.*

PROOF. We may regard k_E as containing k_F with $\overline{k}_E = \overline{k}_F$. The subgroup $G_E \cap W_F$ consists of the members of G_F that induce on $\overline{k}_F$ the automorphism $x \mapsto x^{q_F^n}$ for some integer n and that fix k_E. Then $x^{q_F^n} = x$ for all $x \in k_E$, and it follows that $[k_E : k_F]$ divides n. If we let $a = n/[k_E : k_F]$, then $q_F^n = q_F^{[k_E:k_F]a} = q_E^a$. Hence $G_E \cap W_F \subset W_E$. The reverse inclusion is trivial, and the theorem follows.

Since $\mathbb{Z}$ is dense in $\widehat{\mathbb{Z}}$, it follows that W_F is dense in G_F. If E/F is a finite extension, then G_F/G_E is a finite set, and the image of W_F in it is dense. Therefore W_F maps onto G_F/G_E. Because of Theorem 1.4, we obtain a bijection

$$W_F/W_E \xrightarrow{\sim} G_F/G_E \cong \mathrm{Hom}_F(E, \overline{F}), \tag{1.5}$$

where $\mathrm{Hom}_F(E, \overline{F})$ is the set of indicated field maps. If E/F is Galois, then $\mathrm{Hom}_F(E, \overline{F}) = \mathrm{Gal}(E/F)$ and (1.5) is a group isomorphism.

Let us mention an alternate definition of W_F. If G is a topological group, we let G^c be the closure of the commutator subgroup and we define $G^{\mathrm{ab}} = G/G^c$. The closed subgroup G_F^c of $G_F = \mathrm{Gal}(\overline{F}/F)$ corresponds to a subfield F^{ab} of $\overline{F}$ called the **maximal abelian extension** of F. Its Galois group $\mathrm{Gal}(F^{\mathrm{ab}}/F)$ is just G_F^{ab}.

Let K be a finite Galois extension of F lying in $\overline{F}$, and form K^{ab}. Since $\mathrm{Gal}(F_{\mathrm{ur}}/F) = \widehat{\mathbb{Z}}$ is abelian, we have $K^{\mathrm{ab}} \supset F^{\mathrm{ab}} \supset F_{\mathrm{ur}}$. Therefore we have surjective maps induced by restriction:

$$\mathrm{Gal}(\overline{F}/F) \to \mathrm{Gal}(K^{\mathrm{ab}}/F) \to \mathrm{Gal}(F_{\mathrm{ur}}/F) = \widehat{\mathbb{Z}}. \tag{1.6}$$

The inverse image of $\mathbb{Z} \subset \widehat{\mathbb{Z}}$ in $\mathrm{Gal}(\overline{F}/F)$ was defined to be W_F, and we define the intermediate inverse image of $\mathbb{Z}$ in $\mathrm{Gal}(K^{\mathrm{ab}}/F)$ to be $W_{K/F}$. This construction carries with it surjective maps $W_F \to W_{K/F}$, and these are compatible as K varies. In addition, any element of $\overline{F}$ is in some finite Galois extension K of F and therefore is also in K^{ab}. Hence no nontrivial element of W_F can restrict to the identity on every $W_{K/F}$, and it follows that W_F is the inverse limit

$$W_F = \varprojlim_K W_{K/F}. \tag{1.7}$$

2. Local Class Field Theory

In this section we shall state the main results from local class field theory and translate them into statements about Weil groups. For most of this section, we let F be a finite extension of $\mathbb{Q}_p$ and $\overline{F}$ be an algebraic closure of F. References for the material in this section are [Ne], [Se1], [Se2], and [Ta3].

Let K be a finite Galois extension of F lying in $\overline{F}$ and having $[K:F]=n$, and let $G_{K/F} = \mathrm{Gal}(K/F)$. It is known that $H^2(G_{K/F}, K^\times)$ is cyclic of order n with a canonical generator $u_{K/F}$. This can be proved rather quickly with the aid of the theory of the Brauer group [Se1, p. 137], and also a direct cohomological proof is possible [Se1, p. 130].

Theorem 2.1. *If K/F is a finite Galois extension, then the canonical generator $u_{K/F}$ of $H^2(G_{K/F}, K^\times)$ defines (by means of "cup product") an isomorphism of $G_{K/F}^{\mathrm{ab}}$ onto $F^\times/N_{K/F}(K^\times)$, where $N_{K/F}(\,\cdot\,)$ denotes the norm map.*

REFERENCE. [Se1, p. 140].

The inverse $\theta_{K/F} : F^\times/N_{K/F}(K^\times) \to G_{K/F}^{\mathrm{ab}}$ of the isomorphism in Theorem 2.1 is called the **local reciprocity map** of K/F. When $G_{K/F}$ is abelian, then $G_{K/F}^{\mathrm{ab}}$ equals $G_{K/F}$ and $\theta_{K/F}$ is an isomorphism of $F^\times/N_{K/F}(K^\times)$ onto $G_{K/F}$.

If $x \in F^\times$ lies in the coset $\bar{x}$ of $F^\times/N_{K/F}(K^\times)$, then we write $\theta_{K/F}(\bar{x}) = (x, K/F)$.

The symbols $(x, K/F)$ define homomorphisms $F^\times \to G_{K/F}$ that are compatible [Se1, p. 140] when $K' \supset K \supset F$ and K'/F is finite abelian. Taking the inverse limit, we obtain a homomorphism $\theta_F : F^\times \to G_{F^{\mathrm{ab}}/F}$ or $\theta_F : F^\times \to G_F^{\mathrm{ab}}$.

In the exact sequence (1.3), the homomorphism π has an abelian image and therefore descends to a homomorphism π^{ab} defined on $G_F^{\mathrm{ab}} = G_{F^{\mathrm{ab}}/F}$. Accordingly we replace (1.3) by

$$1 \longrightarrow I_{F^{\mathrm{ab}}/F} \longrightarrow G_F^{\mathrm{ab}} \xrightarrow{\pi^{\mathrm{ab}}} \widehat{\mathbb{Z}} \longrightarrow 0, \tag{2.1}$$

where $\widehat{\mathbb{Z}} = G_{F_{\mathrm{ur}}/F}$ and where $I_{F^{\mathrm{ab}}/F}$ is the kernel of π^{ab}. By Galois theory we may interpret $I_{F^{\mathrm{ab}}/F}$ as $G_{F^{\mathrm{ab}}/F_{\mathrm{ur}}}$.

We shall compute $\pi^{\mathrm{ab}} \circ \theta_F : F^\times \to \widehat{\mathbb{Z}}$. For any $x \in F^\times$, the number $|x|_F$ is a power of q^{-1}, and we define $v(x)$ to be that power.

Theorem 2.2. *If K/F is a finite unramified extension and if* Fr *in $G_{K/F}$ denotes the Frobenius element, then $(x, K/F) = \mathrm{Fr}^{v(x)}$ for all $x \in F^\times$.*

REFERENCE. [Se1, p. 141].

Corollary 2.3. *For any $x \in F^\times$, $\pi^{\mathrm{ab}}(\theta_F(x)) = v(x)$ as a member of $\widehat{\mathbb{Z}}$.*

Any inverse image in G_F of the element $\theta_F(x)$ of G_F^{ab} therefore lies in the Weil group W_F.

We shall prove below in Lemma 2 that the homomorphism $W_F^{\mathrm{ab}} \to G_F^{\mathrm{ab}}$ induced by $W_F \to G_F$ is one-one, and then we may regard $(\pi^{\mathrm{ab}})^{-1}(\mathbb{Z})$ as W_F^{ab}. Thus (2.1) gives us an exact sequence

$$1 \longrightarrow G_{F^{\mathrm{ab}}/F_{\mathrm{ur}}} \longrightarrow W_F^{\mathrm{ab}} \xrightarrow{\pi^{\mathrm{ab}}} \mathbb{Z} \longrightarrow 0. \tag{2.2}$$

Lemma 1. $W_F^{\mathrm{c}} = G_F^{\mathrm{c}}$.

PROOF. Certainly $W_F^{\mathrm{c}} \subset G_F^{\mathrm{c}}$, and it is $G_F^{\mathrm{c}} \subset W_F^{\mathrm{c}}$ that needs proof. Let x be a member of W_F with $\pi(x) = 1 \in \mathbb{Z}$. Then W_F is the semidirect product of $\{x^n\}_{n=-\infty}^{\infty}$ with I_F. So W_F^{c} is the closure of the subgroup of I_F generated by all commutators of I_F and all elements $x^n i x^{-n} i^{-1}$ with $i \in I_F$.

Let Y be the smallest closed subgroup of G_F containing x. Since Y is compact abelian and $\mathbb{Z}$ is dense in $\widehat{\mathbb{Z}}$, $\pi(Y) = \widehat{\mathbb{Z}}$. If $g \in G_F$ is given, choose $y \in Y$ with $\pi(g) = \pi(y)$. Then $\pi(gy^{-1}) = 0$ shows that gy^{-1} is in I_F, and the identity $g = (gy^{-1})y$ therefore shows that every element of G_F is the product of an element of I_F and an element of Y. So G_F^{c} is the closure of the subgroup of I_F generated by all commutators of I_F and all elements $yiy^{-1}i^{-1}$ with $y \in Y$ and $i \in I_F$. The commutators of I_F are in W_F^{c}, and the element $yiy^{-1}i^{-1}$ is the limit of elements $x^n i x^{-n} i^{-1}$ with n varying through a suitable sequence. Hence $yiy^{-1}i^{-1}$ is in W_F^{c}, and $G_F^{\mathrm{c}} \subset W_F^{\mathrm{c}}$.

Lemma 2. *The homomorphism $W_F^{\mathrm{ab}} \to G_F^{\mathrm{ab}}$ induced by $W_F \to G_F$ is one-one.*

PROOF. We need to prove that $W_F \cap G_F^{\mathrm{c}} = W_F^{\mathrm{c}}$. Since $G_F^{\mathrm{c}} \subset I_F \subset W_F$, we need $G_F^{\mathrm{c}} = W_F^{\mathrm{c}}$. But this is just what Lemma 1 gives.

With Lemma 2 proved, (2.2) now follows. Corollary 2.3 implies that the restriction of θ_F to $\mathcal{O}_F^\times$ is in the kernel of π^{ab} on W_F^{ab}. Hence it is in the image of $G_{F^{\mathrm{ab}}/F_{\mathrm{ur}}}$. We can put this information and the full strength of Corollary 2.3 together in a diagram with exact rows and commutative squares

$$\begin{array}{ccccccccc} 1 & \longrightarrow & \mathcal{O}_F^\times & \longrightarrow & F^\times & \xrightarrow{v} & \mathbb{Z} & \longrightarrow & 0 \\ & & \downarrow{\scriptstyle \theta_F} & & \downarrow{\scriptstyle \theta_F} & & \downarrow{\scriptstyle 1} & & \\ 1 & \longrightarrow & G_{F^{\mathrm{ab}}/F_{\mathrm{ur}}} & \longrightarrow & W_F^{\mathrm{ab}} & \xrightarrow{\pi^{\mathrm{ab}}} & \mathbb{Z} & \longrightarrow & 0. \end{array} \tag{2.3}$$

Let us now state in its classical form the Existence Theorem of local class field theory.

Theorem 2.4. *The map $K \to F^\times/N_{K/F}(K^\times)$ is a bijection of the set of finite abelian Galois extensions K of F onto the set of open subgroups of $F^\times$ of finite index.*

REFERENCE. [Se1, p. 143].

Use of the Weil group allows us to restate this result more simply.

Corollary 2.5. *The local reciprocity map θ_F is a topological isomorphism of $F^\times$ onto W_F^{ab}.*

PROOF. If K/F is a finite abelian extension within F^{ab}, then Theorem 2.1 shows that the composition of θ_F followed by the quotient map $G_F^{\mathrm{ab}} \to G_{K/F}$ carries $F^\times$ onto $G_{K/F}$. Letting K vary, we see that θ_F carries $F^\times$ onto a dense subgroup of G^{ab}. Since $\mathcal{O}_F^\times$ is compact, it follows from (2.3) that θ_F carries $\mathcal{O}_F^\times$ onto $G_{F^{\mathrm{ab}}/F_{\mathrm{ur}}}$. A second application of (2.3) shows that θ_F carries $F^\times$ onto W_F^{ab}. The kernel of θ_F is $\bigcap_K N_{K/F}(K^\times)$, the intersection being taken over all finite extensions K of F lying in F^{ab}. If i and j are any integers ≥ 0, then the set $\{(\varpi_F^i)^n(1+\mathcal{O}_F^j) \mid n \in \mathbb{Z}\}$ is an open subgroup of $F^\times$ of finite index. By Theorem 2.4 it is $N_{K/F}(K^\times)$ for some finite abelian K/F. Hence

$$\bigcap_K N_{K/F}(K^\times) \subset \bigcap_{i,j} \{(\varpi_F^i)^n(1+\mathcal{O}_F^j) \mid n \in \mathbb{Z}\} = \{1\},$$

and θ_F is one-one. Consequently $\theta_F : F^\times \to W_F^{\mathrm{ab}}$ is a group isomorphism. Since $\theta_F : \mathcal{O}_F^\times \to G_{F^{\mathrm{ab}}/F_{\mathrm{ur}}}$ is continuous and $\mathcal{O}_F^\times$ is compact, $\theta_F : \mathcal{O}_F^\times \to G_{F^{\mathrm{ab}}/F_{\mathrm{ur}}}$ is a homeomorphism. Then it follows that $\theta_F : F^\times \to W_F^{\mathrm{ab}}$ is a homeomorphism.

REMARKS.

1) In Corollary 2.5, θ_F carries a prime element of $\mathcal{O}_F$ to an element of W_F^{ab} that acts as a Frobenius automorphism in every unramified extension. Thus $\theta_F(x)(y) = y^{|x|_F^{-1}}$ for y in any unramified extension. Some authors adjust a sign somewhere to make $\theta_F(x)(y) = y^{|x|_F}$; see [Ta3, p. 6] for a discussion of this point. For these authors the later definitions of L functions are likely to be what we, with our traditional definitions, would call the L function of the contragredient.

2) Theorem 2.4, which is the difficult result in local class field theory, is essentially equivalent with Corollary 2.5. A proof that Corollary 2.5 implies Theorem 2.4 may be based on [Se1, p. 144].

3) Corollary 2.5 implies that the (continuous) one-dimensional representations of $F^\times$ are parametrized by the continuous homomorphisms of W_F into $\mathbb{C}^\times$. This is a point of departure for conjectures of Langlands about parametrizing irreducible representations of linear reductive groups over F. We return to this matter in §8.

Let us return to the group $W_{K/F}$ defined from (1.6), where K is a finite Galois extension of F lying in $\overline{F}$. The kernel of the map $W_F \to W_{K/F}$ in (1.7), in view of (1.6), is the set of all $w \in W_F$ that act as Galois elements by 1 on K^{ab}. They are in particular members of G_F. Being 1 on K, they are in G_K. Being 1 on K^{ab}, they are in G_K^c, which equals W_K^c by Lemma 1. As a result we have

$$W_{K/F} \cong W_F/W_K^c. \tag{2.4}$$

We form the exact sequence

$$1 \longrightarrow W_K/W_K^c \longrightarrow W_F/W_K^c \longrightarrow W_F/W_K \longrightarrow 0. \tag{2.5}$$

The quotient W_K/W_K^c is just $W_{K/K}$ by (2.4), and this is by definition the subgroup of G_K^{ab} inducing an integral power of the Frobenius. By Lemma 2 and the derivation of (2.2), we can identify this subgroup with W_K^{ab}. Applying Corollary 2.5 and substituting into (2.5) from (1.5) amd (2.4), we obtain an exact sequence

$$1 \longrightarrow K^\times \longrightarrow W_{K/F} \longrightarrow G_{K/F} \longrightarrow 1. \tag{2.6}$$

Such an exact sequence yields by standard cohomology of groups a member of $H^2(G_{K/F}, K^\times)$. Tracking down the isomorphisms that led to (2.6) allows one to identify this cohomology element.

Theorem 2.6. *The cohomology class of the exact sequence* (2.6) *in the group* $H^2(G_{K/F}, K^\times)$ *is exactly the canonical generator* $u_{K/F}$.

REFERENCE. [Ta3, pp. 4–5].

Theorem 2.6 allows a fairly explicit understanding of $W_{K/F}$ for nonarchimedean local fields.

For the archimedean local fields $\mathbb{R}$ and $\mathbb{C}$, we turn Theorem 2.6 around and use it as a *definition* of the Weil group. In the case of $\mathbb{R}$, $H^2(G_{\mathbb{C}/\mathbb{R}}, \mathbb{C}^\times)$ is cyclic of order 2, and $W_{\mathbb{C}/\mathbb{R}}$ is defined correspondingly to be a group that fits into a nonsplit exact sequence

$$1 \longrightarrow \mathbb{C}^\times \longrightarrow W_{\mathbb{C}/\mathbb{R}} \longrightarrow G_{\mathbb{C}/\mathbb{R}} \longrightarrow 1.$$

Specifically we take $W_{\mathbb{C}/\mathbb{R}} = \mathbb{C}^\times \cup j\mathbb{C}^\times$, where j acts on $\mathbb{C}^\times$ by complex conjugation and where $j^2 = -1 \in \mathbb{C}^\times$. In view of (1.7), we make the definition

$$W_{\mathbb{R}} = W_{\mathbb{C}/\mathbb{R}} = \mathbb{C}^\times \cup j\mathbb{C}^\times.$$

Similarly we are led to define $W_{\mathbb{C}} = W_{\mathbb{C}/\mathbb{C}} = \mathbb{C}^\times$. With these definitions we readily check that Theorem 2.1 and Corollary 2.5 remain valid for $\mathbb{R}$ and $\mathbb{C}$.

3. Adeles and Ideles

Adeles occur in the study of "global fields," which are of two kinds. The global fields of characteristic 0 are the **number fields**, the finite extensions of $\mathbb{Q}$. The global fields of characteristic nonzero are the finite extensions of the formal rational functions over a finite field. We shall limit our discussion to number fields. General references for adeles are [Cas], [La], and [We4].

The idea with adeles is to study number-theoretic questions about a number field by first studying congruences. For example, to study the factorization of a monic polynomial with integer coefficients, we first study the factorization modulo each prime. In addition, we consider any limitations imposed by treating the polynomial as having real coefficients. Thus, in the case of $\mathbb{Q}$, we use a structure that incorporates congruences modulo each prime (as well as powers of the prime), together with information about $\mathbb{R}$. The ring of adeles $\mathbb{A}_{\mathbb{Q}}$ is the structure in question. We defer to §5 the way it carries information about factorization of polynomials.

In the case of $\mathbb{Q}$, let $P = \{\infty\} \cup \{\text{primes}\}$. For $v \in P$, the field $\mathbb{Q}_v$ is to be the field of p-adic numbers if v is a prime p, and it is to be $\mathbb{R}$ if $v = \infty$. Let $S \subset P$ be a finite set containing ∞, and define

$$\mathbb{A}_{\mathbb{Q}}(S) = \Big(\prod_{v \in S} \mathbb{Q}_v\Big) \times \Big(\prod_{v \notin S} \mathbb{Z}_v\Big).$$

With the product topology, $\mathbb{A}_{\mathbb{Q}}(S)$ is a locally compact commutative topological ring. If $S_1 \subset S_2$, then $\mathbb{A}_{\mathbb{Q}}(S_1) \subset \mathbb{A}_{\mathbb{Q}}(S_2)$. The directed system of inclusions allows us to define $\mathbb{A}_{\mathbb{Q}}$ as the direct limit

$$\mathbb{A}_{\mathbb{Q}} = \varinjlim_{S} \mathbb{A}_{\mathbb{Q}}(S) = \bigcup_S \mathbb{A}_{\mathbb{Q}}(S).$$

The direct-limit topology makes each $\mathbb{A}_{\mathbb{Q}}(S)$ be open in $\mathbb{A}_{\mathbb{Q}}$, the relative topology being the locally compact topology above. Then $\mathbb{A}_{\mathbb{Q}}$ is a locally compact commutative topological ring known as the **adeles** of $\mathbb{Q}$.

Elements of $\mathbb{A}_{\mathbb{Q}}$ may be regarded as tuples

$$x = (x_\infty, x_2, x_3, x_5, \dots, x_v, \dots) = \prod_v x_v$$

with **almost all** (i.e., all but finitely many) x_v having $|x_v|_v \leq 1$. (Here $|\cdot|_\infty$ denotes the module for the local field $\mathbb{R}$, which is just the usual absolute value.) Often one writes simply $x = (x_v)$.

The adeles are the result of a construction called restricted direct product. Suppose that I is a nonempty index set, that X_i is a locally compact Hausdorff space for each $i \in I$, and that a compact open subset K_i of X_i is specified for all i outside a finite subset S_∞ of I. If S is any finite subset of I containing S_∞, we can define

$$A(S) = \Big(\prod_{i \in S} X_i\Big) \times \Big(\prod_{i \notin S} K_i\Big), \tag{3.1}$$

and $A(S)$ will be locally compact Hausdorff. The direct limit A of the $A(S)$'s as S increases is called the **restricted direct product** of the X_i relative to the K_i. The space A is locally compact Hausdorff, and each $A(S)$ is open in it. An element of $\prod_{i \in I} x_i$ of the Cartesian product $\prod_{i \in I} X_i$ is in A if and only if x_i is in K_i for almost all i.

In practice, X_i is usually a locally compact group and K_i is a compact open subgroup. Then A is a locally compact group. In the case of $\mathbb{A}_{\mathbb{Q}}$, the finite set S_∞ is $\{\infty\}$, each X_i is a locally compact ring (namely $\mathbb{Q}_p$ or $\mathbb{R}$), and K_i is a compact open subring (namely $\mathbb{Z}_p$); thus $\mathbb{A}_{\mathbb{Q}}$ is a locally compact ring.

For a general number field F (possibly $\mathbb{Q}$ itself), we construct the ring $\mathbb{A}_F$ of adeles of F as follows. A **completion** of F is a pair (λ, K), where K is a local field and $\lambda : F \to K$ is a field map with dense image. Two completions (λ, K) and (λ', K') are equivalent is there is a topological isomorphism $\rho : K \to K'$ such that $\rho \circ \lambda = \lambda'$. A **place** is an equivalence class of completions. Places are typically denoted v, and a representative of the corresponding local field is denoted F_v. When $F = \mathbb{Q}$, the only places are those coming from embedding $\mathbb{Q}$ in $\mathbb{R}$ and in each $\mathbb{Q}_p$ for p prime.

An isomorphism that exhibits two completions as corresponding to the same place preserves the module. Consequently restriction of the module to F gives a well defined function $|\cdot|_v$ on F.

Suppose that F'/F is an extension of number fields, and let w be a place of F'. Regard w as a field map $w : F' \to F'_w$. It is not hard to see that the closure $\overline{w(F)}$ of $w(F)$ is a local field, that F'_w is a finite algebraic extension of $\overline{w(F)}$, and that the restriction of w to F determines a place v of F. In this case we say that v is the place of F that lies below w, and that w lies above v. We write $w \mid v$.

Theorem 3.1. *Let F'/F be an extension of number fields, and let v be a place of k. Then there exists a place of F' lying above v, and there are only finitely many such places.*

REFERENCE. [We4, p. 45].

We can apply Theorem 3.1 to construct $\mathbb{A}_F$. The X_i are the various F_v. All of these are p-adic fields except those with v lying above the place ∞ of $\mathbb{Q}$, and Theorem 3.1 says that there are only finitely many such places. We take S_∞ to be the set of places lying above ∞, and we let the K_i's be the rings of integers $\mathcal{O}_v$ in F_v. Then $\mathbb{A}_F$ is the restricted direct product of the F_v relative to the K_v, and it is a locally compact commutative ring.

We can get some insight into the places of a number field F by treating F as an extension of $\mathbb{Q}$ and considering all places lying over a place of $\mathbb{Q}$ (a prime or ∞). By the theory of semisimple algebras, the algebra $F \otimes \mathbb{Q}_v$ over $\mathbb{Q}_v$ is a finite direct sum of fields, each of which is a finite extension of $\mathbb{Q}_v$. Fix an algebraic closure $\overline{\mathbb{Q}}_v$, and consider the set of field maps $\mathrm{Hom}_{\mathbb{Q}}(F, \overline{\mathbb{Q}}_v)$. The group $D_v = \mathrm{Gal}(\overline{\mathbb{Q}}_v/\mathbb{Q}_v)$ acts on this set of maps by acting on the values of each map. Let $P_v(F) = D_v \backslash \mathrm{Hom}_{\mathbb{Q}}(F, \overline{\mathbb{Q}}_v)$. Then

$$F \otimes_{\mathbb{Q}} \mathbb{Q}_v \cong \bigoplus_{w \in P_v(F)} F_w,$$

with the right side involving each place lying over v just once.

Let us consider this decomposition for the infinite places, those lying above ∞ of $\mathbb{Q}$. We shall write F_∞ for the algebra $F \otimes_{\mathbb{Q}} \mathbb{R}$. This $\mathbb{R}$ algebra is of the form $\mathbb{R}^{r_1} \times \mathbb{C}^{r_2}$ for some integers $r_1 \geq 0$ and $r_2 \geq 0$ satisfying $r_1 + 2r_2 = [F : \mathbb{Q}]$. Since $D_\infty = \mathrm{Gal}(\mathbb{C}/\mathbb{R})$ consists of 1 and complex conjugation, $P_\infty(F)$ consists of the set $\mathrm{Hom}_{\mathbb{Q}}(F, \mathbb{C})$ of embeddings of F into $\mathbb{C}$, with two embeddings identified when they are complex conjugates of one another. Thus r_1 is the number of embeddings into $\mathbb{R}$, and r_2 is the number of complex-conjugate pairs of nonreal embeddings of F into $\mathbb{C}$.

Next let us consider the finite places. Recall that the ring of integers $\mathcal{O}$ in F consists of all elements satisfying a monic polynomial equation with $\mathbb{Z}$ coefficients. The places lying over the primes of $\mathbb{Q}$ are related to the nontrivial prime ideals of $\mathcal{O}$. Let v be a place of F lying above a prime p for $\mathbb{Q}$. It is easy to see that the mapping $F \to F_v$ carries $\mathcal{O}$ into the ring of integers $\mathcal{O}_v$ of F_v. Let $\mathfrak{m}_v$ be the maximal ideal of $\mathcal{O}_v$, and let P_v be the inverse image in $\mathcal{O}$ of $\mathfrak{m}_v$. Then P_v is a prime ideal of $\mathcal{O}$. That is, every finite place of F leads to a prime ideal of $\mathcal{O}$.

This fact admits a converse. Before stating the converse, we recall that the nonzero ideals of $\mathcal{O}$ admit unique factorizations as products of prime ideals, the exponents of the prime ideals being integers ≥ 0. The notion of ideal can be extended to fractional ideal; a **fractional ideal** is just a set of the form $n^{-1}I$ for an ideal I and some nonzero $n \in \mathbb{Z}$. It is not hard to see that the nonzero fractional ideals form a group. Consequently the nonzero fractional ideals admit unique factorizations as products of prime ideals, the exponents of the prime ideals being integers that are not necessarily ≥ 0.

Theorem 3.2. *Let P be a nontrivial prime ideal of $\mathcal{O}$, and let $q = |\mathcal{O}/P|$. For each $x \in F^\times$, let $v(x)$ be the power of P that appears in the factorization of the principal fractional ideal $(x) = x\mathcal{O}$, and define $|x|_P = q^{-v(x)}$. Then $|\cdot|_P$ defines a metric on F, and the completion of F in this metric is a local field whose module is the continuous extension of $|\cdot|_P$. The result is a place of F, and the prime ideal of $\mathcal{O}$ associated to this place is just P.*

Since the factorization of (x) in the theorem is finite, it follows that $|x|_v$ is different from 1 for only finitely many places v. One consequence of this fact is that any $x \in F$ embeds diagonally as an element of $\mathbb{A}_F$. We shall make constant use of this embedding. It is tempting to write $\operatorname{diag} F$ for the image, but this notation soon becomes unwieldy and it is customary to denote the image simply by F. Briefly the subset F of $\mathbb{A}_F$ always means the *diagonally embedded* version of F unless the contrary is stated.

A second consequence of the fact that $|x|_v \neq 1$ for only finitely many v is that $\prod_v |x|_v$ is well defined for each $x \in F^\times$. The next theorem tells the value of this product.

Theorem 3.3 (Artin product formula). *$\prod_v |x|_v = 1$ for all $x \in F^\times$.*

REFERENCE. [La, p. 99].

Theorem 3.4. *The image of F in $\mathbb{A}_F$ is discrete, and the quotient group $\mathbb{A}_F/F$ is compact.*

REFERENCE. [La, p. 139].

The construction of the ideles of the number field F is a second use of the notion of a restricted direct product. The index set I is the set $\{v\}$ of places, the factors X_i are the multiplicative groups $F_v^\times$, the subset S_∞ is the set of infinite places, and the K_i are the groups of units $\mathcal{O}_v^\times$. The restricted direct product is a locally compact abelian group denoted $\mathbb{A}_F^\times$ and called the **ideles** of F.

If x is in $F^\times$, then the tuple consisting of x in every place is an idele as a consequence of the discussion before Theorem 3.3. Hence $F^\times$ embeds diagonally in $\mathbb{A}_F^\times$. We denote the image simply by $F^\times$, understanding that $F^\times$ is *diagonally embedded* unless the contrary is stated.

If one tracks down the definitions, the topology on $\mathbb{A}_F^\times$ is not the relative topology from $\mathbb{A}_F$ but is finer. Actually $\mathbb{A}_F^\times$ gets the relative topology from

$$\{(x, y) \in \mathbb{A}_F \times \mathbb{A}_F \mid xy = 1\}.$$

Since $F^\times$ is discrete in $\mathbb{A}_F$ (Theorem 3.4) and since the topology on $\mathbb{A}_F^\times$ is finer than the relative topology on $\mathbb{A}_F$, $F^\times$ is discrete in $\mathbb{A}_F^\times$.

Being a locally compact abelian group, $\mathbb{A}_F$ has a Haar measure $\mu_{\mathbb{A}_F}$, and this can evidently be taken to be the product of the Haar measure on each F_v if the latter are normalized at almost every place to assign mass one to $\mathcal{O}_v$. Multiplication by any $r \in \mathbb{A}_F^\times$ is an automorphism of $\mathbb{A}_F$ and hence carries $\mu_{\mathbb{A}_F}$ to a multiple of itself. We write

$$d\mu_{\mathbb{A}_F}(rx) = |r|_{\mathbb{A}_F}\, d\mu_{\mathbb{A}_F}(x)$$

and call $|r|_{\mathbb{A}_F}$ the **module** of r. Arguing first with r equal to 1 in all but one place and then passing to general r, we see that $|r|_{\mathbb{A}_F} = \prod_v |r_v|_v$. Theorem 3.3 therefore implies that any element x of $F^\times$ has $|x|_{\mathbb{A}_F} = 1$.

Let $(\mathbb{A}_F^\times)^1$ be the subgroup of elements of $\mathbb{A}_F^\times$ of module 1, i.e., the kernel of the homomorphism $|\cdot|_{\mathbb{A}_F} : \mathbb{A}_F^\times \to \mathbb{R}_+^\times$. We have just seen that $F^\times$ lies in this subgroup, and we saw above that $F^\times$ is discrete. This proves the easy half of the following theorem.

Theorem 3.5. *The diagonal embedding $F^\times \to \mathbb{A}_F^\times$ carries $F^\times$ to a discrete subgroup of $(\mathbb{A}_F^\times)^1$, and the quotient $(\mathbb{A}_F^\times)^1/F^\times$ is compact.*

REMARKS. The compactness of the quotient is closely related to the Dirichlet Unit Theorem. For a proof of the result and a discussion of the connection, see [Cas, pp. 70–73] and [La, pp. 142–146]. For a direct argument in a more general context, see [We4, p. 76].

The relationship between $\mathbb{A}_F^\times$ and $(\mathbb{A}_F^\times)^1$ is that

$$\mathbb{A}_F^\times = (\mathbb{A}_F^\times)^1 \times R \tag{3.2}$$

for a noncanonical subgroup R isomorphic to $\mathbb{R}_+^\times$. In fact, fix an infinite place v. Then $F_v^\times$ contains a subgroup $\mathbb{R}_+^\times$, and we let R be the image of this subgroup in $\mathbb{A}_F^\times$ under inclusion into the vth place. It is clear that the module map $|\cdot|_{\mathbb{A}_F}$ carries R one-one onto $\mathbb{R}_+^\times$. Thus (3.2) follows.

From (3.2) and Theorem 3.5, it follows that

$$\mathbb{A}_F^\times/H \text{ is compact} \tag{3.3}$$

for any subgroup H of $\mathbb{A}_F^\times$ containing $RF^\times$. Recall that $F_\infty = F \otimes_{\mathbb{Q}} \mathbb{R}$ gives the archimedean component of $\mathbb{A}_F$ (with 1 in all finite places). Then $F_\infty^\times$ gives the archimedean component of $\mathbb{A}_F^\times$. Since $F_\infty^\times$ contains R, (3.3) implies that $\mathbb{A}_F^\times/(F_\infty^\times F^\times)$ is compact. Let U_1 be the ring $\prod_{v \text{ finite}} \mathcal{O}_v$, so that $K_1 = U_1^\times$ is just $\prod_{v \text{ finite}} \mathcal{O}_v^\times$.

Corollary 3.6. *Suppose that K is any open compact subgroup of K_1. Then the set $X_K = F^\times\backslash\mathbb{A}_F^\times/(F_\infty^\times K)$ is finite.*

REMARK. When $K = K_1$, use of the relationship discussed below between ideles and fractional ideals allows one to identify X_{K_1} with the ideal class group of F. Thus the corollary gives an adelic proof of the finiteness of the class number.

PROOF. The above considerations show that $\mathbb{A}_F^\times/(F_\infty^\times F^\times)$ is compact, and hence X_K is compact. On the other hand, $F_\infty^\times K$ is open in $\mathbb{A}_F^\times$. Thus $\mathbb{A}_F^\times/(F_\infty^\times K)$ is discrete, and consequently X_K is discrete.

Historically, ideles were introduced before adeles. Chevalley's purpose in introducing ideles was to extend class field theory to infinite abelian extensions. But, as is indicated in the introduction of [Ch], the theory of ideles served also the purpose of reinterpreting results about fractional ideals and related notions. We give some details about this point in order to prepare for Artin reciprocity, which will be discussed in the next section. For each $x = (x_v) \in \mathbb{A}_F^\times$, we create a fractional ideal as follows. Each finite place v corresponds to some prime ideal $\wp$ in $\mathcal{O}$. With $q_v = |\mathcal{O}_v/\mathfrak{m}_v|$ and $|x_v|_v = q_v^{-v(x_v)}$, associate to x the fractional ideal $\prod_\wp \wp^{v(x_v)}$. If $x = (x_0, x_0, \ldots)$ is in $F^\times$, then this definition reproduces the fractional ideal (x_0) in its factored form; hence there is no ambiguity if we refer to this fractional ideal as (x) in every case. If S is a finite set of places containing S_∞ and if x is an idele, we define $(x)^S = \prod_{\wp \notin S} \wp^{v(x_v)}$.

Artin reciprocity initially involves a homomorphism of fractional ideals into an abelian Galois group, and we shall want to lift this homomorphism to a homomorphism of the group of ideles. We shall impose a continuity condition. Let S be a finite set of places containing S_∞, and let I^S be the subgroup of fractional ideals whose $\wp$ factor is (1) for every finite $\wp \in S$. If G is an abelian topological group, we say that a homomorphism $\varphi : I^S \to G$ is **admissible** if, for each neighborhood N of 1 in G, there exists $\delta > 0$ such that $\varphi((a)^S)$ is in N whenever a is a member of $F^\times$ such that $|a-1|_v < \delta$ for all $v \in S$ (including all infinite places). To reinterpret admissible φ's in terms of ideles, we use the following **Weak Approximation Theorem**.

Theorem 3.7. *Let $|\cdot|_n$, $1 \le n \le N$ be distinct places of F, finite or infinite. If x_n, $1 \le n \le N$, are members of F and if $\epsilon > 0$ is given, then there exists $\xi \in F^\times$ with $|\xi - x_n|_n < \epsilon$ for $1 \le n \le N$.*

REFERENCE. [Cas, p. 48].

Theorem 3.8. *Let S be a finite set of places containing S_∞, and let G be a compact abelian group (usually finite abelian or S^1). If $\varphi : I^S \to G$ is an admissible homomorphism, then there exists a unique homomorphism $\tilde{\varphi} : \mathbb{A}_F^\times \to G$ such that*

(a) *$\tilde{\varphi}$ is continuous*
(b) *$\tilde{\varphi}$ is 1 on $F^\times$*
(c) *$\tilde{\varphi}(x) = \varphi((x)^S)$ for all $x = (x_v) \in \mathbb{A}_F^\times$ such that $x_v = 1$ for all $v \in S$ (including all infinite places).*

REFERENCE. [Ta2, pp. 169–170].

PROOF OF UNIQUENESS. Given $x \in \mathbb{A}_F^\times$, choose, by Theorem 3.7, a sequence of elements a_n of $F^\times$ such that $a_n \to x^{-1}$ at the places of S. For $y \in \mathbb{A}_F^\times$, let y^S be the idele with all entries in places of S set equal to 1, and let y_S be the idele with all entries in places outside S set equal to 1. For every n, we have

$$\begin{aligned} \tilde{\varphi}(x) &= \tilde{\varphi}(a_n x) && \text{by (b)} \\ &= \tilde{\varphi}((a_n x)_S)\tilde{\varphi}((a_n x)^S) && \text{since } \tilde{\varphi} \text{ is a homomorphism} \\ &= \tilde{\varphi}((a_n x)_S)\varphi((a_n x)^S) && \text{by (c).} \end{aligned}$$

The first factor on the right side tends to 1 by (a). Thus

$$\tilde{\varphi}(x) = \lim_{n \to \infty} \varphi((a_n x)^S).$$

4. Artin Reciprocity

A general reference for Artin reciprocity is [Ta2]. Let K/F be a Galois extension of number fields, and let $G = \mathrm{Gal}(K/F)$ be the Galois group.

Let v be a finite place of F, and let $\wp$ be the corresponding prime ideal of $\mathcal{O}_F$. The ideal $\wp\mathcal{O}_K$ of $\mathcal{O}_K$ has a factorization into prime ideals of $\mathcal{O}_K$, say

$$\wp\mathcal{O}_K = P_1^{e_1} \cdots P_g^{e_g}$$

with the P_j distinct and with all $e_j > 0$. Here the P_j's are exactly the prime ideals of $\mathcal{O}_K$ that contain $\wp$, and each P_j has $P_j \cap \mathcal{O}_F = \wp$. We say that the P_j **lie above** $\wp$. From the correspondence of finite places to prime ideals, it is easy to see that the places w_j corresponding to the P_j's are exactly the places that lie above v.

Theorem 4.1. *In the factorization* $\wp\mathcal{O}_K = P_1^{e_1}\cdots P_g^{e_g}$ *into prime ideals in* $\mathcal{O}_K$, *the Galois group* G *permutes* $P_1,\ldots,P_g$, *and the action of* G *on the* g-*element set is transitive. Consequently* $e_1 = \cdots = e_g$. *Moreover, if* $|\mathcal{O}_F/\wp| = q$ *and* $|\mathcal{O}_K/P_j| = q^{f_j}$, *then* $f_1 = \cdots = f_g$. *If* e *denotes the common value of the* e_j *and if* f *denotes the common value of the* f_j, *then*

$$efg = [K : F].$$

The prime ideal $\wp$ of $\mathcal{O}_F$ is said to **ramify** in K if $e > 1$. Ramification is an exceptional occurrence: If v lies above the place p of $\mathbb{Q}$, then ramification of $\wp$ in K implies that p divides the absolute discriminant of K. In particular, only finitely many prime ideals of $\mathcal{O}_F$ ramify in K.

In the situation of the theorem, let P be one of $P_1,\ldots,P_g$. Let w and v be the places of K and F corresponding to P and $\wp$, so that w lies above v. Write K_P and $F_\wp$ for the completions. Define

$$G_P = \{\sigma \in G \mid \sigma(P) = P\}.$$

This group is called the **decomposition group** relative to P. The members of G_P acts as isometries of K in the norm $|\cdot|_P$. Consequently σ extends to an automorphism of K_P, and we see that we can think in terms of an inclusion

$$G_P \hookrightarrow \mathrm{Gal}(K_P/F_\wp). \tag{4.1}$$

Theorem 4.2. *The embedding of* G_P *in* (4.1) *is onto* $\mathrm{Gal}(K_P/F_\wp)$, *and* $K_P/F_\wp$ *is a Galois extension of local fields.*

We observed before Theorem 1.1 that each member of $\mathrm{Gal}(K_P/F_\wp)$ acts by an isometry and consequently induces an automorphism of $\mathrm{Gal}(K_P/F_\wp)$ into $\mathrm{Gal}(k'_P/k_\wp)$, where k'_P and $k_\wp$ are the respective residue fields. Theorem 1.1 says that the resulting homomorphism is onto. Thus we can rewrite the exact sequence (1.1) in this context as

$$1 \longrightarrow I_P \longrightarrow G_P \longrightarrow \mathrm{Gal}(k'_P/k_\wp) \longrightarrow 0, \tag{4.2}$$

where the inertia group I_P is defined to be the kernel. With e, f, and g as in Theorem 4.1, we know that $efg = [K : F]$. Since G acts transitively on $P_1,\ldots,P_g$ by Theorem 4.1, the isotropy subgroup G_P at P has $efg/g = ef$ elements. By Theorem 4.2, $|\mathrm{Gal}(K_P/F_\wp)| = ef$. Our definitions make $|\mathrm{Gal}(k'_P/k_\wp)| = f$, and therefore $|I_P| = e$.

In other words, if $\wp$ is unramified in K (as is the case for almost all prime ideals of F), the extension $K_P/F_\wp$ is unramified in the sense of §1. In this case, $\mathrm{Gal}(K_P/F_\wp)$ contains a well defined Frobenius element, as in the definition following Theorem 1.2. If P is replaced by another prime ideal P' lying over $\wp$, then G_P is conjugate to $G_{P'}$ by an element of G carrying P to P', and this conjugacy carries the Frobenius element to the Frobenius element. In terms of $\wp$ as a given piece of data, the Frobenius element is then any element of a certain conjugacy class of G.

Artin reciprocity deals with the situation that $G = \mathrm{Gal}(K/F)$ is abelian. In this case when $\wp$ is a prime ideal of $\mathcal{O}_F$ that is unramified in K, the conjugacy class of Frobenius elements reduces to a single element, and we can unambiguously denote the Frobenius element by the notation

$$\left[\frac{K/F}{\wp}\right] \in G.$$

Let S be the finite set of all infinite places of F and all finite places of F that ramify in K. If $X = \prod_j \wp_j^{n_j}$ is the factorization of a fractional ideal of F into primes, we recall that $X^S = \prod_{j \notin S} \wp_j^{n_j}$. Then we can define

$$\left[\frac{K/F}{X^S}\right] = \prod_{j \notin S} \left[\frac{K/F}{\wp_j}\right]^{n_j}.$$

The resulting homomorphism of I^S into G is called the **Artin symbol** of K/F.

Theorem 4.3 (Artin reciprocity, first form). *Let K/F be a finite abelian Galois extension of degree n, and let S be the finite set of all infinite places of F and all finite places of F that ramify in K. If $a \in F^\times$ is such that a is in $(F_v^\times)^n$ for every $v \in S$ and if (a) denotes the fractional ideal $a\mathcal{O}_F$, then*

$$\left[\frac{K/F}{(a)^S}\right] = 1.$$

REFERENCE. [Ta2, p. 167].

This first form of Artin reciprocity is the weakest of three forms that we shall consider. However, it is already strong enough so that with a little computation it implies quadratic reciprocity [Cas-Fr, pp. 348–350]. It also implies a more general m^{th} power reciprocity theorem due to Kummer.

We shall now sharpen the statement of Artin reciprocity so as to be able to bring Theorem 3.8 to bear.

Theorem 4.4 (Artin reciprocity, second form). *Let K/F be a finite abelian Galois extension, and let S be the finite set of all infinite places of F and all finite places of F that ramify in K. There exists $\delta > 0$ such that whenever $a \in F^\times$ has $|a-1|_v < \delta$ for all $v \in S$, then*

$$\left[\frac{K/F}{(a)^S}\right] = 1.$$

REFERENCE. [Ta2, p. 167].

Theorem 4.4 implies Theorem 4.3 by a simple argument [Ta2, p. 167] using the Weak Approximation Theorem (Theorem 3.7).

In the terminology at the end of §3, Theorem 4.4 says that the homomorphism $I^S \to G$ defined by the Artin symbol is admissible. By Theorem 3.8, the Artin symbol lifts uniquely to a continuous homomorphism of the **idele class group of** F,

$$C_F = \mathbb{A}_F^\times / F^\times, \tag{4.3}$$

into G. Let us call this homomorphism the **Artin map** of K/F and denote it by $\theta_{K/F} : C_F \to G$.

Theorem 4.5. *Let K/F be a finite abelian Galois extension, let v be a place of F, and let w be a place of K lying above v. If $i_v : F_v^\times \to C_F$ denotes the composition of inclusion of F_v into the vth place of $\mathbb{A}_F^\times$ followed by the quotient map to C_F, then the Artin map and the local reciprocity map are related by $\theta_{K/F} \circ i_v = \theta_{K_w/F_v}$ as homomorphisms $F_v^\times \to G$.*

REFERENCE. [Ta2, p. 175].

Theorems 4.4 and 4.5 are the main facts about the Artin symbol and Artin map that we need in the next section. For completeness we include a little more information at this time. It is apparent from Theorem 4.5 and the precise statement of local class field theory in Theorem 2.4 that there has to be a sharper statement of Artin reciprocity than in Theorem 4.4. Here is such a result.

Theorem 4.6 (Artin reciprocity, third form). *Let K/F be a finite abelian Galois extension, and let S be the finite set of all infinite places of F and all finite places of F that ramify in K. If an element $a \in F^\times$ is a norm from K_w for all w lying over places of S, then*

$$\left[\frac{K/F}{(a)^S}\right] = 1.$$

REFERENCE. [Ta2, p. 176].

Now we think of K as varying. Namely we fix an algebraic closure $\overline{F}$ of F and consider finite abelian Galois extensions K of F lying in $\overline{F}$. Let G_F be the Galois group of $\overline{F}$ over F, and let $G_F^{\mathrm{ab}} = G_F/G^c$ be the Galois group of the maximal abelian extension of F. If K is a finite abelian extension of F, then the Artin map $\theta_{K/F}$ is a continuous homomorphism of C_F into $G_{K/F} = \mathrm{Gal}(K/F)$. These homomorphisms have a compatibility property that allows us to lift them to a single continuous homomorphism with values in the inverse limit, namely $\theta_F : C_F \to G_F^{\mathrm{ab}}$. We call θ_F the **Artin map** of F. It follows from the various compatibility properties that Theorem 4.5 can be restated in this notation as

$$\theta_F \circ i_v = \theta_{F_v} \tag{4.4}$$

as homomorphisms $F_v^\times \to G_F^{\mathrm{ab}}$.

Lemma. *The Artin map θ_F carries C_F onto G_F^{ab}.*

SKETCH OF PROOF. By (3.2), $\mathbb{A}_F^\times = R \times (\mathbb{A}_F^\times)^1$. Since $F^\times$ lies in $(\mathbb{A}_F^\times)^1$, we obtain $C_F \cong R \times C_F^1$, where $C_F^1 = (\mathbb{A}_F^\times)^1/F^\times$. Since R is connected and G_F^{ab} is totally disconnected, $\theta_F(R) = 1$. Thus $\theta_F(C_F) = \theta_F(C_F^1)$, and this is compact by Theorem 3.5. On the other hand, one shows that the composition of θ_F and passage to any finite quotient of G_F^{ab} carries C_F onto the finite quotient. Hence $\theta_F(C_F)$ is dense in G_F^{ab}.

In practice, Artin reciprocity is proved at the same time as the Existence Theorem of global class field theory, whose statement is given in Theorem 4.7 below. If v is a place of F and w is a place of K lying above v, then the norm map $N_{K_w/F_v} : K_w^\times \to F_v^\times$ is well defined. We set $N_{K/F} : \mathbb{A}_K^\times \to \mathbb{A}_F^\times$ equal to the coordinate-by-coordinate product $N_{K/F} = \prod_w N_{K_w/F_v}^\times$. It is clear that $N_{K/F}$ carries diagonally embedded $K^\times$ to diagonally embedded $F^\times$ and therefore descends to a homomorphism $N_{K/F} : C_K \to C_F$.

Theorem 4.7. *The map $K \mapsto C_F/N_{K/F}(C_K)$ is a bijection of the set of finite abelian Galois extensions K of F onto the set of open subgroups of C_F of finite index. The field corresponding to a subgroup B is the fixed field of the subgroup $\theta_F(B)$ of G_F^{ab}.*

REFERENCES. [Ta2, p. 172] and [Ar-Ta, p. 70].

Corollary 4.8. *The kernel of the Artin map θ_F of C_F onto G_F^{ab} is the identity component $(C_F)_0$ of C_F.*

REFERENCE. [Ta2, p. 173].

We can summarize Corollary 4.8 as saying that the sequence

$$1 \longrightarrow (C_F)_0 \longrightarrow C_F \xrightarrow{\theta_F} G_F^{\mathrm{ab}} \longrightarrow 0$$

is exact.

5. Artin L Functions

At the beginning of §3, we mentioned that the ring of adeles carries information about the factorization of polynomials, and we shall elaborate on this assertion now. The Artin L functions to be introduced in this section encode this information as explicit functions of a complex variable given by product formulas. Artin reciprocity enables one to recognize certain Artin L functions as arising in another way that shows that they have nice analytic properties.

Example 1. For the polynomial $R(X) = X^2 + 1$, we ask how $R(X)$ reduces modulo p for primes $p \neq 2$. Before giving the well known answer, let us encode the problem in a generating function. Put

$$\left(\frac{-1}{p}\right) = \begin{cases} +1 & \text{if } X^2+1 \text{ factors completely modulo } p \\ -1 & \text{if } X^2+1 \text{ is irreducible modulo } p, \end{cases}$$

and define

$$L(s) = \prod_{p\neq 2} \frac{1}{1-\left(\frac{-1}{p}\right)p^{-s}}.$$

This certainly converges for $\operatorname{Re} s > 1$. The well known answer to our question amounts to giving the pattern for $\left(\frac{-1}{p}\right)$, which is

$$\left(\frac{-1}{p}\right) = \begin{cases} +1 & \text{if } p = 4k+1 \\ -1 & \text{if } p = 4k-1. \end{cases}$$

This is the simplest case of quadratic reciprocity. The point to observe is that the pattern is described by finitely many linear congruences. If we define

$$\chi(n) = \begin{cases} +1 & \text{if } n \equiv 1 \bmod 4 \\ -1 & \text{if } n \equiv 3 \bmod 4 \\ 0 & \text{if } n \text{ even}, \end{cases}$$

then

$$L(s) = \prod_{p} \frac{1}{1-\chi(p)p^{-s}}.$$

From the property $\chi(mm') = \chi(m)\chi(m')$, we obtain

$$L(s) = \sum_{n=1}^{\infty} \frac{\chi(n)}{n^s}.$$

In this form, $L(s)$ becomes more manageable. This series is absolutely convergent for $\operatorname{Re} s > 1$. Use of summation by parts shows that $L(s)$ converges for $\operatorname{Re} s > 0$, and an elementary argument shows that $L(s)$ continues to an entire function. It is not hard to see that $L(1) \neq 0$, and from this fact it follows that there are infinitely many primes $p = 4k + 1$ and infinitely many primes $p = 4k - 1$. In other words, interesting information about primes has been encoded in $L(s)$ at a spot on the boundary of the region where $L(s)$ converges absolutely. Finding the pattern for $\dfrac{1}{1 - \left(\frac{-1}{p}\right)}$ enables us to extract this information.

Example 2 ([Buhl] and [Lgl6]). Let us consider the polynomial

$$R(X) = X^5 + 10X^3 - 10X^2 + 35X - 18.$$

This has discriminant $2^6 5^8 11^2$, and the question is to find the pattern of how $R(X)$ reduces modulo p for $p \neq 2, 5, 11$. For example, we can readily find by computer that $R(X)$ is irreducible modulo p for $p = 7, 13, 19, 29, 43, 47, 59, \ldots$. Similarly we find that $R(X)$ splits completely for $p = 2063, 2213, 2953, 3631, \ldots$. What is the pattern? These sequences of primes are not related to linear congruences, and the Langlands theory gives conjectures that describe the pattern. Let F be the splitting field of $R(X)$ over $\mathbb{Q}$, and let G be the Galois group. Since the discriminant is a square, $G \subset A_5$. The group G contains a Frobenius element Fr_p for each $p \neq 2, 5, 11$, and this element is the lift to G_p of a generator of the Galois group of $R(X) \bmod p$. Modulo $p = 7$, $R(X)$ is irreducible; so G has an element of order 5. Modulo $p = 3$, $R(X)$ is the product of two linear factors and an irreducible cubic; so G has an element of order 3. Since A_5 is generated by any two elements of respective orders 5 and 3, we conclude that $G = A_5$.

For any $p \neq 2, 5, 11$, the Galois group of $R(X) \bmod p$ tells us a great deal about the factorization of $R(X) \bmod p$. The generator of this group is a Frobenius element, which can be any element in a particular conjugacy class of G. In the case of A_5, the order of an element determines the conjugacy class of the element in A_5 unless the order is 5, for which there are two conjugacy classes. The order of the Frobenius element is f. Order 4 does not occur in A_5, and thus $f = 1$, 2, 3, or 5. If $f = 5$, $R(X) \bmod p$ is irreducible. If $f = 3$, $R(X) \bmod p$ is the product of two distinct linear factors and an irreducible cubic. If $f = 2$, the element of G has to be the product of two 2-cycles; thus $R(X) \bmod p$ has to be the product of a linear factor and two distinct irreducible quadratic factors. Finally if $f = 1$, then $R(X) \bmod p$ splits into five distinct linear factors. The value of f determines the conjugacy class of Frobenius elements in A_5 completely unless $f = 5$. For $f = 5$, there are two conjugacy classes; see [Buhl, p. 53] for how to distinguish them.

In order to encode the full information about the conjugacy classes of the Frobenius elements in one or more generating functions, we can proceed as follows. Let σ be a finite-dimensional representation of G over $\mathbb{C}$. Then the generating function is

$$L(s, \sigma) = \prod_{p \neq 2,5,11} \frac{1}{\det(1 - p^{-s}\sigma(\mathrm{Fr}_p))},.$$

except that suitable factors for $p = 2, 5, 11$ need to be included. The goal is to recognize this function in another form and thereby to find the pattern of the coefficients. This is carried out in [Buhl].

Let K/F be a Galois extension of number fields, with Galois group G, and let $\sigma : G \to \mathrm{Aut}_{\mathbb{C}}(V)$ be a finite-dimensional complex representation of G. (As always, we build continuity into the definition of "representation.") The **Artin L function** is defined to be

$$L(s,\sigma) = L(s,\sigma,K/F) = \prod_{\wp} L_{\wp}(s,\sigma),$$

the product being taken over the nontrivial prime ideals $\wp$ in $\mathcal{O}_F$. Here s is a complex variable.

Fix $\wp$, put $q = |\mathcal{O}_F/\wp|$, and let P be a prime ideal in $\mathcal{O}_K$ lying over $\wp$. The definition of $L_{\wp}(s,\sigma)$ is a little simpler if $\wp$ is unramified in K, and we consider that case first. Then there is a well defined Frobenius element Fr_P in G_P, and we put[1]

$$L_{\wp}(s,\sigma) = \det(1 - \sigma(\mathrm{Fr}_P)q^{-s})^{-1} \qquad \text{for } \mathrm{Re}\, s > 0. \tag{5.1a}$$

Let the eigenvalues of $\sigma(\mathrm{Fr}_P)$ be $\varepsilon_1, \dots, \varepsilon_{\dim V}$; these are roots of unity since Fr_P has finite order. Then

$$L_{\wp}(s,\sigma) = \prod_{i=1}^{\dim V} (1 - \varepsilon_i q^{-s})^{-1}.$$

As P varies, Fr_P moves in a conjugacy class of G. The eigenvalues of $\sigma(\mathrm{Fr}_P)$ do not change, and the second formula for $L_{\wp}(s,\sigma)$ shows that the function depends only on $\wp$.

Now suppose $\wp$ is allowed to be ramified in K, so that the inertia group I_P is nontrivial. Let V^{I_P} be the subspace of V on which $\sigma(I_P)$ acts as the identity; this would be all of V in the unramified case. Then $\sigma(G_P)$ preserves this space. If Fr_P is one lift to G_P of the canonical generator of the Galois group of the residue field extension, then the most general lift is $\mathrm{Fr}_P i_P$ with $i_P \in I_P$. Thus $\sigma(\mathrm{Fr}_P)$ is unambiguous as a linear transformation on V^{I_P}, and we define[2]

$$L_{\wp}(s,\sigma) = \det(1 - \sigma(\mathrm{Fr}_P)|_{V^{I_P}} q^{-s})^{-1} \qquad \text{for } \mathrm{Re}\, s > 0. \tag{5.1b}$$

Again we can rewrite this using eigenvalues, and we see that the result is independent of P.

Each Artin L function converges for $\mathrm{Re}\, s > 1$. Artin L functions have the following additional properties (see [Hei, pp. 222–223], [La, pp. 236–239], and [Mar, p. 9]):

1) $L(s, \sigma_1 \oplus \sigma_2, K/F) = L(s,\sigma_1,K/F)L(s,\sigma_2,K/F)$.
2) Suppose that $F \subset E \subset K$ and that E is Galois over F. Let $H = \mathrm{Gal}(K/E)$, a normal subgroup of $G = \mathrm{Gal}(K/F)$. If $\tilde{\sigma}$ is a representation of G lifted from a representation σ of G/H, then

$$L(s,\tilde{\sigma},K/F) = L(s,\sigma,E/F).$$

Consequently an Artin L function depends only on s and a continuous finite-dimensional representation of $\mathrm{Gal}(\overline{F}/F)$.

[1]Concerning the choice of Fr_P or its inverse in this formula (i.e., σ or its contragredient), see Remark 1 after Corollary 2.5.

[2]See the footnote with (5.1a).

3) Suppose that $F \subset E \subset K$ with E/F possibly not Galois. Let σ_0 be a representation of $G_0 = \mathrm{Gal}(K/E)$. Then

$$L(s, \mathrm{ind}_{G_0}^{G}(\sigma_0), K/F) = L(s, \sigma_0, K/E).$$

A **quasicharacter** is a continuous homomorphism into $\mathbb{C}^\times$. Let F be a number field. By a **Grossencharacter** of F is meant a quasicharacter of C_F.

Example 1. Fix a positive integer m. A **Dirichlet character** modulo m is the lift χ to $\mathbb{Z}$ of a character of the multiplicative group $(\mathbb{Z}/m\mathbb{Z})^\times$, with $\chi(a)$ set equal to 0 if a and m are not relatively prime. Fix such a χ. With $F = \mathbb{Q}$, let S consist of ∞ and the primes dividing m. In the notation of §3, define a homomorphism $\varphi : I^S \to S^1$ by $\varphi((a/b)^S) = \chi(a)/\chi(b)$ whenever a and b are integers relatively prime to m. If p is a prime dividing m and p^c is the exact power of p dividing m, then $|\frac{a}{b} - 1|_p < p^{-c}$ implies that a and b are congruent modulo p^c. Hence if $|\frac{a}{b} - 1|_p < m^{-1}$ for all p dividing m, then a and b are congruent modulo m, and it follows that $\chi(a) = \chi(b)$. In other words φ is admissible in the sense of §3. By Theorem 3.8 there exists a unique unitary Grossencharacter $\tilde{\varphi}$ of $\mathbb{Q}$ such that

$$\tilde{\varphi}(1, \dots, 1, p, 1, \dots) = \chi(p)$$

for all primes p not dividing m; here $(1, \dots, 1, p, 1, \dots)$ denotes the idele that is p in the pth place and is 1 elsewhere.

Example 2. If ω is any (continuous) character of G_F^{ab}, then the composition $\omega \circ \theta_F$ with the Artin map of F is a Grossencharacter. Since ω has to factor through a finite quotient of G_F^{ab}, it is the same to consider compositions $\omega_0 \circ \theta_{K/F}$, where K is a finite abelian Galois extension of F and ω_0 is a character of $\mathrm{Gal}(K/F)$.

Let φ be a Grossencharacter of F. Following Hecke in spirit, we shall associate an L function $L(s, \varphi)$ to φ. For each finite place v with corresponding prime ideal $\wp$ of $\mathcal{O}_F$, we can restrict φ to the coordinate $F_\wp^\times$, obtaining a quasicharacter $\varphi_\wp : F_\wp^\times \to \mathbb{C}^\times$. Let $\mathcal{O}_\wp$ be the ring of integers in $F_\wp$. We say that $\varphi_\wp$ is **ramified** if $\varphi_\wp|_{\mathcal{O}_\wp^\times}$ is nontrivial. For each $\wp$, let $\varpi_\wp$ be a prime element in $\mathcal{O}_\wp$. Then the definition is[3]

$$L(s, \varphi) = \prod_{\substack{\wp \text{ unramified} \\ \text{for } \varphi}} L_\wp(s, \varphi) = \prod_{\substack{\wp \text{ unramified} \\ \text{for } \varphi}} \left(1 - \frac{\varphi_\wp(\varpi_\wp)}{|\mathcal{O}_F/\wp|^s}\right)^{-1}. \tag{5.2}$$

The functions $L(s, \varphi)$ have nice analytic properties. They have meromorphic continuations to $\mathbb{C}$ and satisfy a functional equation relating the values at s and $1 - s$. The only possible pole is at $s = 1$ and is at most simple; there is no pole if φ is nontrivial on $(\mathbb{A}_F^\times)^1$. These results are essentially due to Hecke. Later Tate [Ta1] found an important different proof that uses local-global methods. For an exposition of Tate's work and a higher-dimensional generalization, see Jacquet's lecture [Ja2].

Sometimes authors include in $L(s, \varphi)$ extra factors for the infinite places that involve a gamma function and powers of certain numbers. See [Kna2] for a description of these. Shortly we shall use this kind of completed L function, writing Λ for it. Inclusion of factors for the infinite places affects the poles of $L(s, \varphi)$ only slightly and makes the functional equation much simpler.

[3] See the footnote with (5.1a).

Theorem 5.1. *If K/F is a finite abelian Galois extension of number fields and ω is a character of* $\mathrm{Gal}(K/F)$, *then the Artin L function $L(s,\omega,K/F)$ equals the Hecke L function $L(s,\varphi)$ of the Grossencharacter $\varphi=\omega\circ\theta_{K/F}$.*

PROOF. Let H be the kernel of ω in $G=\mathrm{Gal}(K/F)$, and let E be the fixed field of H in K. Then E/F is a finite abelian Galois extension of F, and ω descends to a one-one character ω_0 of $G/H=\mathrm{Gal}(E/F)$. By property (2) of Artin L functions, we have $L(s,\omega,K/F)=L(s,\omega_0,E/F)$. Also $\varphi=\omega_0\circ\varphi_{E/F}$. Thus it is enough to prove that $L(s,\omega_0,E/F)=L(s,\omega_0\circ\theta_{E/F})$. We do so factor by factor.

Let $\wp$ be a nontrivial prime ideal in $\mathcal{O}_F$. We show that $\wp$ is unramified in E if and only if $\wp$ is unramified for φ. With S as the set of infinite places and places that ramify in E, first suppose that $\wp$ is not in S. If $x_\wp$ is in $\mathcal{O}_\wp^\times$, then the proof of Theorem 3.8 (with $a_n=1$ for all n) shows that

$$\varphi_\wp(x_\wp)=\varphi(\dots,1,x_\wp,1,\dots)=\varphi((\{x_\wp\})^S)=\omega_0\left[\frac{E/F}{(1)}\right]=1.$$

Hence $\wp$ is unramified for φ.

Conversely suppose $\wp$ is unramified for $\varphi=\omega_0\circ\theta_{E/F}$. This means that $\varphi_\wp=\omega_0\circ\theta_{E/F}\circ i_\wp$ is 1 on $\mathcal{O}_\wp^\times$. By Theorem 4.5, $\omega_0\circ\theta_{E_P/F_\wp}$is 1 on $\mathcal{O}_\wp^\times$. Since ω_0 is one-one, $\theta_{E_P/F_\wp}$ is 1 on $\mathcal{O}_\wp^\times$. Referring to (2.3) and Corollary 2.5, we see that the inertia group I_P is trivial. Thus $\wp$ is not in S.

Now suppose that $\wp$ is ramified (in both senses). Then the $\wp$ factor of $L(s,\varphi)$ is 1 by definition. Meanwhile the $\wp$ factor of $L(s,\omega_0,E/F)$ is $(1-\omega_0(\mathrm{Fr}_P)|_{\mathbb{C}^{I_P}}q^{-s})^{-1}$. Since ω_0 is one-one and I_P is nontrivial, $\mathbb{C}^{I_P}=0$. Thus the $\wp$ factor for each L function is 1 in the ramified case.

Finally suppose that $\wp$ is unramified (in both senses). Theorem 4.5 gives

$$\varphi_\wp=\varphi\circ i_\wp=\omega\circ\theta_{K/F}\circ i_\wp=\omega_0\circ\theta_{E/F}\circ i_\wp=\omega_0\circ\theta_{E_P/F_\wp}. \tag{5.3}$$

Then
$$L_\wp(s,\omega_0,E/F)=(1-\omega_0(\mathrm{Fr}_P)q^{-s})^{-1}$$
by definition, while

$$\begin{aligned}
L_\wp(s,\varphi)&=(1-\varphi_\wp(\varpi_\wp)q^{-s})^{-1} &&\text{by definition}\\
&=(1-\omega_0\theta_{E_P/F_\wp}(\varpi_\wp)q^{-s})^{-1} &&\text{by (5.3)}\\
&=(1-\omega_0(\mathrm{Fr}_P)q^{-s})^{-1} &&\text{by Theorem 2.2.}
\end{aligned}$$

Hence the $\wp$ factors for the two L functions match in the unramified case.

Corollary 5.2. *If K/F is a finite abelian Galois extension of number fields and ω is a character of* $\mathrm{Gal}(K/F)$, *then $L(s,\omega,K/F)$ extends to be entire in $\mathbb{C}$. For the trivial character $\omega=1$, $L(s,1,K/F)$ extends to be meromorphic in $\mathbb{C}$ with a simple pole at $s=1$.*

Of course, it is immediate from the properties of Artin L functions that many more such functions are entire. For example, let K/F be any finite Galois extension. Then the Artin L function for a representation of $\mathrm{Gal}(K/F)$ induced from a nontrivial one-dimensional representation is entire. Thus for any **monomial group**, i.e., any finite group whose irreducible representations are all induced from one-dimensional representations of subgroups, the Artin L function of any nontrivial irreducible representation is entire. Dihedral groups are examples of monomial groups.

Artin Conjecture. *Let K/F be any finite Galois extension of number fields, and let σ be a nontrivial irreducible representation of* $\mathrm{Gal}(K/F)$. *Then $L(s,\sigma,K/F)$ extends to be entire in $\mathbb{C}$.*

The Artin L function, which we defined to include only factors (5.1) from the finite places, has a natural completion by adjoining some gamma factors for the infinite places (see [Kna2, (3.6) and (4.6)], and we shall denote the completed L function by $\Lambda(s,\sigma,K/F)$. It is actually conjectured that $\Lambda(s,\sigma,K/F)$ is entire if σ is irreducible and nontrivial.

Despite the fact that the Artin Conjecture is not known, Brauer's Induction Theorem says that an Artin L function has a continuation to all of $\mathbb{C}$ that is at least meromorphic and that the continued function satisfies the same kind of functional equation as a Hecke L function. A more detailed statement of Brauer's result is as follows.

Theorem 5.3. *The group character of any complex finite-dimensional representation of a finite group is an integer combination of group characters of representations induced from one-dimensional representations of subgroups. Consequently any Artin L function has a continuation to all of $\mathbb{C}$ that is at least meromorphic. Moreover each Artin L function satisfies a functional equation of the form*

$$\Lambda(s,\sigma,K/F)=\varepsilon(s,\sigma,K/F)\Lambda(1-s,\sigma^\vee,K/F), \tag{5.4}$$

where $\sigma^\vee$ is the contragredient and $\varepsilon(s,\sigma,K/F)$ is entire and nonvanishing.

The Langlands theory proposes addressing the Artin Conjecture by introducing L functions that generalize those of Hecke and by showing that Artin L functions are always of this kind. Aspects of this theory occupy much of the remainder of this article. For another exposition of this kind, see [Gelb2].

6. Linear Reductive Algebraic Groups

Let F be a number field. Informally a **linear algebraic group** of n-by-n matrices over F is a group G of n-by-n matrices defined by polynomial equations in n^2 matrix variables with coefficients in F. For a precise definition, see [Bo2], [Bo3], or [Wel].

Clearing fractions in the defining equations of G, we may assume that the coefficients are all in $\mathcal{O}_F$. If R is a torsion-free commutative ring containing $\mathcal{O}_F$ as a subring, then the group of R points of G is well defined, independently of how we cleared fractions, and we denote this group by $G(R)$. The group G will be said to be **unipotent** if $G(\mathbb{C})$ consists entirely of unipotent matrices, while G is **reductive** if $\{1\}$ is the only connected unipotent normal subgroup of $G(\mathbb{C})$. The reader may wish to think of a reductive G as being GL_n or SL_n, and little will be lost for current purposes by doing so.

A simple example of a linear algebraic group is the affine line, whose R points are the elements of R. To work with this algebraic group as a linear group, we may view it as the group of all matrices $\begin{pmatrix} 1 & * \\ 0 & 1 \end{pmatrix}$. Another example is the multiplicative group, whose R points are the elements of $R^\times$; this group is also called GL_1. The affine line is unipotent, and the multiplicative group is reductive.

If v is a place of F, we can topologize $GL_n(F_v)$ as an open subset of n^2-dimensional space, and the result is a locally compact group. Then $G(F_v)$ is given the relative topology from $GL_n(F_v)$ and is a locally compact group. If v is finite, then $G(\mathcal{O}_v)$ is a compact subgroup of $G(F_v)$.

For each finite set S of places containing the set S_∞ of infinite places, the group $G(\mathbb{A}_F(S))$ is well defined since the ring $\mathbb{A}_F(S)$ contains diagonally embedded $\mathcal{O}_F$. The group $G(\mathbb{A}_F(S))$ is nothing more than the direct product of all $G(F_v)$ for $v \in S$ and all $G(\mathcal{O}_v)$ for $v \notin S$. It is locally compact. Its topology may be described alternatively as the relative topology from $GL_n(\mathbb{A}_F(S))$.

Similarly, the group $G(\mathbb{A}_F)$ of $\mathbb{A}_F$ points of G is well defined since $\mathbb{A}_F$ contains diagonally embedded F. When G is the affine line, $G(\mathbb{A}_F)$ is just the group of adeles of F; when G is the multiplicative group, $G(\mathbb{A}_F)$ is the group of ideles of F. In every case, a member of $G(\mathbb{A}_F)$ may be regarded as a tuple of matrices indexed by the places of F, the matrix in the vth place being in $G(F_v)$, with almost all such matrices lying in $G(\mathcal{O}_v)$.

As is the case with the ideles, the topology on $G(\mathbb{A}_F)$ is not necessarily the relative topology from n-by-n matrices over $\mathbb{A}_F$. Instead the topology is the restricted direct product topology of the $G(F_v)$ relative to the $G(\mathcal{O}_v)$. In other words, it is the direct limit topology from the subgroups $G(\mathbb{A}_F(S))$, which are to be open. Alternatively we can topologize $G(\mathbb{A}_F)$ by embedding it in matrices $\begin{pmatrix} x & 0 \\ 0 & (\det x)^{-1} \end{pmatrix}$ of size $n+1$ and giving it the relative topology from $(n+1)^2$-dimensional space over $\mathbb{A}_F$. With these definitions, $G(\mathbb{A}_F)$ gets the relative topology from $GL_n(\mathbb{A}_F)$.

Because of this second way of realizing the topology, it follows that the group G(diagonally embedded F), which equals diagonally embedded $G(F)$, is a discrete subgroup of $G(\mathbb{A}_F)$. We write $G(F)$ for this subgroup. The first theorem generalizes Corollary 3.6.

Theorem 6.1. *If G is a reductive linear algebraic group, then the number of double cosets in $G(F)\backslash G(\mathbb{A}_F)/G(\mathbb{A}_F(S_\infty))$ is finite. For $G = GL_n$, the number of double cosets is the class number of F. For $G = SL_n$, the number of double cosets is* 1.

References.

1) In a classical setting with $G(\mathbb{Z}) \subset G(\mathbb{Q})$, this theorem is due to Borel and Harish-Chandra [Bo-HC]. The result in the classical setting says in part that if G is a semisimple linear algebraic group over $\mathbb{Q}$, then $G(\mathbb{Z})\backslash G(\mathbb{R})$ has finite volume. The result in an adelic setting appears in Borel [Bo1, p. 19]. The relationship between the two settings will be described below.

2) If the group $G(\mathbb{A}_F(S_\infty))$ in the statement of the theorem is replaced by an open subgroup of finite index, then it is clear that the number of double cosets remains finite.

More is true than is asserted in the theorem. The hypothesis that G is reductive is unnecessary. In any event, let $G_\infty = G(F_\infty)$ be the archimedean component of $G(\mathbb{A}_F)$ (equal to 1 at every finite place). We say that G has the **strong approximation property** if $G(F)G_\infty$ is dense in $G(\mathbb{A}_F)$. In this case the number of double cosets in $G(F)\backslash G(\mathbb{A}_F)/G(\mathbb{A}_F(S_\infty))$ is automatically 1. The (unipotent) affine line has the strong approximation property [Cas, p. 67], and so does the reductive SL_n. For more discussion of this property, see [Kne].

In the classical setting for automorphic forms, one works with the quotient $G(\mathbb{Z})\backslash G(\mathbb{R})$. Following [Bo-Ja, p. 195], let us see how this quotient space is related to the double coset decomposition in Theorem 6.1. We decompose $G(\mathbb{A}_F(S_\infty)) =$

$G_\infty \times K_1$, where $G_\infty = G(F_\infty)$ is the archimedean component and where $K_1 = G(U_1) = G\big(\prod_{v \text{ finite}} \mathcal{O}_v\big)$ is the nonarchimedean component (equal to 1 at every infinite place). Let K be any open subgroup of finite index in K_1, and use Theorem 6.1 to write

$$G(\mathbb{A}_F) = \coprod_{c \in C} G(F) c G_\infty K \tag{6.1}$$

as a disjoint union, for some finite subset C of $G(\mathbb{A}_F)$. Without loss of generality, we may assume that the members of C all have component 1 at the infinite place. For $c \in C$, define $G_c = G_\infty c K c^{-1}$ and $\Gamma_c = G_c \cap G(F)$. The group G_c is open in $G(\mathbb{A}_F)$, and the discreteness of $G(F)$ in $G(\mathbb{A}_F)$ implies that Γ_c is a discrete subgroup of G_c. Since cKc^{-1} is compact, we may use projection on the infinite places to identify Γ_c with a discrete subgroup of G_∞. If f is a right K invariant function on $G(\mathbb{A}_F)$ and if c is in C, let f_c be the function $x \mapsto f(cx)$ on G_∞. Then we readily check that the map

$$f \mapsto \{f_c\}_{c \in C} \tag{6.2}$$

is a bijection of the space of functions on $G(F)\backslash G(\mathbb{A}_F)/K$ with the space of functions on $\coprod_{c \in C} (\Gamma_c \backslash G_\infty)$. Thus we obtain an identification

$$G(F)\backslash G(\mathbb{A}_F)/K = \coprod_{c \in C} (\Gamma_c \backslash G_\infty). \tag{6.3}$$

Formula (6.3) is especially simple in cases where $C = \{1\}$. Examples, all with $K = K_1$, are when G is the affine line and F is arbitrary, when $G = SL_n$ and F is arbitrary, and when $G = GL_n$ and $F = \mathbb{Q}$. For these examples, (6.3) becomes

$$F\backslash \mathbb{A}_F / \prod_{\wp} \mathcal{O}_{F_\wp} = \mathcal{O}_F \backslash F_\infty, \tag{6.4a}$$

$$SL_n(F)\backslash SL_n(\mathbb{A}_F) / \prod_{\wp} SL_n(\mathcal{O}_{F_\wp}) = SL_n(\mathcal{O}_F)\backslash SL_n(F_\infty) \tag{6.4b}$$

and

$$GL_n(\mathbb{Q})\backslash GL_n(\mathbb{A}_\mathbb{Q}) / \prod_{p} GL_n(\mathbb{Z}_p) = GL_n(\mathbb{Z})\backslash GL_n(\mathbb{R}). \tag{6.4c}$$

The right side of (6.3) is more concrete than the left side, but part of the action is lost in working with the right side rather than with the adeles. For instance, in the adelic picture of (6.4c), each of the groups $GL_n(\mathbb{Q}_p)$ acts on $GL_n(\mathbb{Q})\backslash GL_n(\mathbb{A}_\mathbb{Q})$, and the corresponding action on this space by functions on $GL_n(\mathbb{Q}_p)$ biinvariant under $GL_n(\mathbb{Z}_p)$ descends to an action on the left side of (6.4c). This action is hidden, however, in the realization as $GL_n(\mathbb{Z})\backslash GL_n(\mathbb{R})$.

Let us describe the functor "restriction of the ground field," which has the property of reducing aspects of the theory over the number field F to the theory over $\mathbb{Q}$. We follow [Wel]. Let $d = [F : \mathbb{Q}]$, and let $\sigma_1, \ldots, \sigma_d$ be the distinct field maps of F into $\overline{\mathbb{Q}}$ fixing $\mathbb{Q}$. Let V be an affine variety over F. A pair (W, p), in which W is an affine variety over $\mathbb{Q}$ and $p : W \to V$ is an algebraic map defined over F, is said to be a variety obtained from V by **restriction of the ground field** from F to $\mathbb{Q}$ if the map

$$(p^{\sigma_1}, \ldots, p^{\sigma_d}) : W \to \prod_{j=1}^{d} V^{\sigma_j},$$

which is defined over $\overline{\mathbb{Q}}$, is an isomorphism of varieties. Such a pair (W, p) exists and is unique up to a canonical isomorphism over $\mathbb{Q}$. It is customary to drop the p from the notation and write $W = R_{F/\mathbb{Q}}V$, regarding $R_{F/\mathbb{Q}}$ as a functor from affine varieties over F to affine varieties over $\mathbb{Q}$. Restriction of the ground field has the key property that $(R_{F/\mathbb{Q}}V)(A) = V(F \otimes_{\mathbb{Q}} A)$ for any $\mathbb{Q}$ algebra A.

Examples.

1) If V is the affine line over F, then W is the vector space V regarded as a $\mathbb{Q}$ vector space of dimension d. To put this example in the above context, let $\{\alpha_j\}_{j=1}^d$ be a basis of F over $\mathbb{Q}$, and define $p(x_1, \dots, x_d) = \sum_j \alpha_j x_j$. Then $p^\sigma(x_1, \dots, x_d) = \sum_j \alpha_j^\sigma x_j$, and

$$(p^{\sigma_1}, \dots, p^{\sigma_d})(x_1, \dots, x_d) = \big(\sum_j \alpha_j^{\sigma_1} x_1, \dots, \sum_j \alpha_j^{\sigma_d} x_j\big).$$

The fact that $(p^{\sigma_1}, \dots, p^{\sigma_d})$ is an isomorphism follows from the fact that $\det\{\alpha_j^{\sigma_i}\} \neq 0$.

2) The previous example may be extended to an n-dimensional affine space V over F in obvious fashion, with

$$p(x_{11}, \dots, x_{1d}, x_{21}, \dots, x_{nd}) = \big(\sum_{j=1}^d \alpha_j x_{1j}, \dots, \sum_{j=1}^d \alpha_j x_{nj}\big).$$

More generally if V is the variety in n-dimensional affine space defined by some polynomials $P(X_1, \dots, X_n)$, we can define a variety W in the nd-dimensional affine space W by rewriting

$$P\big(\sum_{j=1}^d \alpha_j x_{1j}, \dots, \sum_{j=1}^d \alpha_j x_{nj}\big) = \alpha_1 p_1(x_{11}, \dots, x_{nd}) + \cdots + \alpha_d p_d(x_{11}, \dots, x_{nd})$$

and replacing P by $p_1, \dots, p_d$.

Restriction of the ground field is a functor of linear algebraic groups. Specifically if $V = G$ is a linear algebraic group over F, then the pair $(R_{F/\mathbb{Q}}, p)$ can be taken to consist of a linear algebraic group over $\mathbb{Q}$ and a homomorphism over F, and the linear algebraic group structure is unique up to a canonical isomorphism of algebraic groups. The map $(p^{\sigma_1}, \dots, p^{\sigma_d}) : R_{F/\mathbb{Q}}G \to \prod_{j=1}^d G^{\sigma_j}$ is then an isomorphism over $\overline{\mathbb{Q}}$ of algebraic groups. Because of the formula $(R_{F/\mathbb{Q}}G)(A) = G(F \otimes_{\mathbb{Q}} A)$, the isomorphism $(p^{\sigma_1}, \dots, p^{\sigma_d})$ induces isomorphisms

$$\begin{aligned} (R_{F/\mathbb{Q}}G)(\overline{\mathbb{Q}}) &\cong G(\overline{Q}) \times \cdots \times G(\overline{Q}), \\ (R_{F/\mathbb{Q}}G)(\mathbb{Q}) &\cong G(F), \\ (R_{F/\mathbb{Q}}G)(\mathbb{R}) &\cong G_\infty, \\ (R_{F/\mathbb{Q}}G)(\mathbb{A}_{\mathbb{Q}}) &\cong G(\mathbb{A}_F). \end{aligned} \tag{6.5}$$

Similarly we can define $R_{E/F}G$ as a reductive group over F whenever E/F is an extension of number fields and G is a reductive group over E.

With G defined over F, let $X^*(G)_F$ be the set of all F rational homomorphisms of G into GL_1. If χ is in $X^*(G)_F$, then χ extends at each place to a continuous homomorphism $\chi_v : G(F_v) \to F_v^\times$. Let $\chi_{\mathbb{A}_F} : G(\mathbb{A}_F) \to \mathbb{A}_F^\times$ be the product of the χ_v. Then $|\chi_{\mathbb{A}_F}|_{\mathbb{A}_F}$ is a homomorphism of $G(\mathbb{A}_F)$ into $\mathbb{R}_+^\times$. Define $G(\mathbb{A}_F)^1 = \bigcap_{\chi \in X^*(G)_F} \ker |\chi_{\mathbb{A}_F}|_{\mathbb{A}_F}$.

For example, if $G = GL_n$, then $X^*(G)_F$ consists of the integral powers of the determinant, and $|\det_{\mathbb{A}_F}(g_v)|_{\mathbb{A}_F} = \prod_v |\det g_v|_v$. Hence $G(\mathbb{A}_F)^1$ consists of the n-by-n matrices over $\mathbb{A}_F$ for which the module of the determinant is 1. In the special case that $n = 1$, this reduces to the group $(\mathbb{A}_F^\times)^1$ that appears in Theorem 3.5. The following theorem generalizes Theorem 3.5.

Theorem 6.2. *Suppose that $G(\mathbb{C})$ is connected. Then the group $G(F)$ lies in $G(\mathbb{A}_F)^1$, and the quotient space $G(F)\backslash G(\mathbb{A}_F)^1$ has finite volume. Moreover, $G(F)\backslash G(\mathbb{A}_F)^1$ is compact if and only if every unipotent element of $G(F)$ belongs to the radical of $G(F)$.*

REFERENCES. [Bo-HC] and [Bo1, p. 22].

An example of a nonabelian G for which compactness of $G(F)\backslash G(\mathbb{A}_F)^1$ follows from this theorem is the multiplicative group $D^\times$ of a finite-dimensional division algebra D of F with center F. For a direct proof of the compactness in this case, see [Gf-Gr-P] or [We4, p. 74]. To prove the compactness from Theorem 6.2, we note that a unipotent element of $G(F)$ is of the form $1 + x$ with x nilpotent. Since D is closed under addition, x must be in $D(F)$. Since x is nilpotent, x is not invertible. Thus $x = 0$, and the theorem applies.

7. Automorphic Forms and Automorphic Representations

Historically the theory of automorphic forms began with modular and cusp forms for the group $SL_2(\mathbb{Z})$. For a detailed discussion of such forms, see [Kna1] or [Shi]. Briefly if $g = \begin{pmatrix} a & b \\ c & d \end{pmatrix}$ is in $GL(2, \mathbb{R})$, let $g(z) = \frac{az+b}{cz+d}$ for all nonreal complex numbers z, and define

$$j(g, z) = (cz + d)(\det g)^{1/2}.$$

In our discussion, $j(g, z)$ will occur only in the form $j(g, z)^2$, and we may consequently use either determination of the square root of $(\det g)^{1/2}$.

A **modular form** of weight k (an even integer) for $SL_2(\mathbb{Z})$ is an analytic function $f(z)$ on the upper half plane such that

$$f(\gamma(z)) = j(\gamma, z)^k f(z) \qquad \text{for all } \gamma \in SL_2(\mathbb{Z})$$

and such that f is analytic at ∞ in the following sense: The function f, being analytic and periodic under $z \mapsto z + 1$, has an expansion $f(z) = \sum_{n=-\infty}^{\infty} c_n e^{2\pi i n z}$, and the condition of analyticity at ∞ is that $c_n = 0$ for $n < 0$. The condition at ∞ for a modular form can alternatively be formulated as the **slow-growth condition**

$$|f(x + iy)| \le C y^N \qquad \text{for some } C \text{ and } N \text{ as } y \to +\infty.$$

A **cusp form** is a modular form that vanishes at ∞ in the sense that $c_0 = 0$. The additional condition that a modular form is a cusp form can be reformulated as the vanishing of an integral:

$$\int_0^1 f(x + iy)\, dx = 0 \qquad \text{for some or equivalently every } y > 0.$$

A cusp form satisfies the **rapid-decrease condition** that for each N

$$|f(x+iy)| \le Cy^{-N} \qquad \text{for some } C \text{ as } y \to +\infty.$$

A cusp form of weight k can be expanded as $f(z) = \sum_{n=1}^{\infty} c_n e^{2\pi i n z}$, and the associated L function of f, defined by $L(s,f) = \sum_{n=1}^{\infty} c_n/n^s$, satisfies a functional equation relating the values at s and $k-s$. Hecke introduced what we now call **Hecke operators** on the space of cusp forms of weight k. A computation using contour integrals shows that this space is finite-dimensional. The Hecke operators commute and are simultaneously diagonable. The eigenfunctions all have $c_1 \neq 0$, and if c_1 is normalized to be 1 for an eigenfunction, then the corresponding L function has an Euler product expansion, the product being taken over all primes.

Gelfand and Fomin were the first to notice that cusp forms could be realized as smooth vectors in representations of a certain ambient Lie group. It is possible to arrange for modular and cusp forms to lift both to $SL_2(\mathbb{Z})\backslash SL_2(\mathbb{R})$ and to $GL_2(\mathbb{Z})\backslash GL_2(\mathbb{R})$, and we shall indicate both liftings.

The details for the lifting to $SL_2(\mathbb{Z})\backslash SL_2(\mathbb{R})$ are in [Gelb1] and also [Bump]. Given a modular form f as above, we define

$$\phi_{f,\infty}(g) = f(g(i))j(g,i)^{-k} \tag{7.1}$$

for $g \in SL_2(\mathbb{R})$. Then $\phi_\infty = \phi_{f,\infty}$ has the properties that

(i) $\phi_\infty(\gamma g) = \phi_\infty(g)$ for all $\gamma \in SL_2(\mathbb{Z})$

(ii) $\phi_\infty(gr(\theta)) = e^{-ik\theta}\phi_\infty(g)$ for all $r(\theta) = \begin{pmatrix} \cos\theta & -\sin\theta \\ \sin\theta & \cos\theta \end{pmatrix}$

(iii) $\phi_\infty(g)$ satisfies the slow-growth condition that

$$\left|\phi_\infty\left(\begin{pmatrix} 1 & x \\ 0 & 1 \end{pmatrix}\begin{pmatrix} y^{1/2} & 0 \\ 0 & y^{-1/2} \end{pmatrix} r(\theta)\right)\right| \le Cy^N \qquad \text{for some } C \text{ and } N \text{ as } y \to +\infty.$$

(iv) $\Omega\phi_\infty = -\dfrac{k}{2}\left(\dfrac{k}{2}-1\right)\phi_\infty$ for a suitable normalization of the Casimir operator Ω of $SL_2(\mathbb{R})$

(v) if f is a cusp form, then ϕ_∞ is **cuspidal** in the sense that

$$\int_0^1 \phi_\infty\left(\begin{pmatrix} 1 & x \\ 0 & 1 \end{pmatrix} g\right) dx = 0 \qquad \text{for all } g.$$

For the lifting to $GL_2(\mathbb{Z})\backslash GL_2(\mathbb{R})$, we start by extending f to all nonreal complex numbers by putting $f(-z) = f(z)$. Then we define $\phi_{f,\infty}(g)$ by (7.1) for $g \in GL_2(\mathbb{R})$. The invariance property in (i) extends to be valid for $\gamma \in GL_2(\mathbb{Z})$, properties (ii) through (v) are unchanged, and there is one new property:

(vi) $\phi_\infty(zg) = \phi_\infty(g)$ for all g in the center of $GL_2(\mathbb{R})$.

It is natural to expect at first that $SL_2(\mathbb{R})$ is the better group to use for the above lifting, but it has become customary to lift to $GL_2(\mathbb{R})$ or the positive-determinant subgroup rather than $SL_2(\mathbb{R})$ in order to be able to incorporate Hecke operators more conveniently into the theory.

There are some other classical theories of automorphic forms that can be lifted to Lie groups in the same way. The theory of Maass forms [Maa] concerns certain nonholomorphic functions on the upper half plane, and these lift to $GL_2(\mathbb{Z})\backslash GL_2(\mathbb{R})$.

A theory [We3] begun by Hecke for analytic functions on the upper half plane transforming under the group

$$\Gamma_0(N) = \left\{ \left(\begin{smallmatrix} a & b \\ c & d \end{smallmatrix}\right) \in SL_2(\mathbb{Z}) \,\middle|\, N \text{ divides } c \right\}$$

leads to functions on $GL_2(\mathbb{R})$ invariant under the group generated by $\Gamma_0(N)$ and $\left(\begin{smallmatrix} 1 & 0 \\ 0 & -1 \end{smallmatrix}\right)$. In addition, the theory of Hilbert modular forms [Ga] leads to quotients of products of several copies of $GL_2(\mathbb{R})$, and the theory of Siegel modular forms [Si] leads to quotients of real symplectic groups.

In each case the theory can be reinterpreted in an adelic setting by means of (6.3). For example, with classical modular and cusp forms with respect to $SL_2(\mathbb{Z})$, the isomorphism (6.4c) tells us that f should be lifted to

$$\phi_f(g) = f(g_\infty(i))j(g_\infty, i)^{-k} \qquad \text{for } g \in GL_2(\mathbb{A}_\mathbb{Q})$$

if $g = \gamma g_\infty k_1$ is the decomposition of $g \in GL_2(\mathbb{A}_\mathbb{Q})$ according to (6.4c) reformulated as

$$GL_2(\mathbb{A}_\mathbb{Q}) = GL_2(\mathbb{Q})GL_2(\mathbb{R}) \prod_p GL_n(\mathbb{Z}_p).$$

The function $\phi = \phi_f$ on $GL_2(\mathbb{A}_\mathbb{Q})$ has the properties that

(i) $\phi(\gamma g) = \phi(g)$ for all $\gamma \in GL_2(\mathbb{Q})$

(ii) $\phi(gk_1) = \phi(g)$ for all $k_1 \in \prod_p GL_2(\mathbb{Z}_p)$

(iii) $\phi(gr_\infty(\theta)) = e^{-ik\theta}\phi(g)$ for all $r_\infty(\theta) = \begin{pmatrix} \cos\theta & -\sin\theta \\ \sin\theta & \cos\theta \end{pmatrix}$ at the infinite place

(iv) as a function of the variable in the infinite place, ϕ satisfies the equation $\Omega\phi = -\frac{k}{2}\left(\frac{k}{2} - 1\right)\phi$ for a suitable normalization of the Casimir operator Ω of $SL_2(\mathbb{R})$

(v) $\phi(zg) = \phi(g)$ for all scalar z in $GL_2(\mathbb{A}_\mathbb{Q})$

(vi) $\phi(g)$ satisfies the following slow-growth condition: for each $c > 0$ and compact subset ω of $GL_2(\mathbb{A}_\mathbb{Q})$, there exist constants C and N such that

$$\phi\left(\left(\begin{smallmatrix} a & 0 \\ 0 & 1 \end{smallmatrix}\right) g\right) \leq C|a|^N$$

for all $g \in \omega$ and $a \in \mathbb{A}_\mathbb{Q}^\times$ with $|a|_{\mathbb{A}_\mathbb{Q}} > c$

(vii) if f is a cusp form, then ϕ is **cuspidal** in the sense that

$$\int_{\mathbb{Q}\backslash\mathbb{A}_\mathbb{Q}} \phi\left(\left(\begin{smallmatrix} 1 & x \\ 0 & 1 \end{smallmatrix}\right) g\right) dx = 0 \qquad \text{for all } g \in GL_2(\mathbb{A}_\mathbb{Q}).$$

The group $SL_2(\mathbb{Z})$, relative to which f satisfies an invariance property, is captured by the compact group in (ii). The relevant identity is

$$GL_2(\mathbb{Q}) \cap \left(GL_2(\mathbb{R}) \times \prod_p GL_2(\mathbb{Z}_p)\right) = GL_2(\mathbb{Z}).$$

For Hecke's theory with $\Gamma_0(N)$, the corresponding compact group that appears in (ii) is $\prod_p K_p'$, where

$$K_p' = \left\{ \left(\begin{smallmatrix} a & b \\ c & d \end{smallmatrix}\right) \in GL_2(\mathbb{Z}_p) \,\middle|\, c/N \in \mathbb{Z}_p \right\}.$$

The group K'_p coincides with $GL_2(\mathbb{Z}_p)$ for all p prime to N, and the relevant identity is

$$GL_2(\mathbb{Q}) \cap \big(GL_2(\mathbb{R}) \times \prod_p K'_p\big) = \Gamma_0(N) \cup \left(\begin{smallmatrix} 1 & 0 \\ 0 & -1 \end{smallmatrix}\right) \Gamma_0(N).$$

This adelic setting is what we generalize to arbitrary reductive groups, following [Bo-Ja]. Let F be a number field, let $\mathbb{A} = \mathbb{A}_F$ be the adeles of F, and let G be a reductive group over F such that $G(\mathbb{C})$ is connected. Let Z be a maximal F split torus of the center of G. Let $G_\infty = G(F_\infty)$ be the archimedean component of $G(\mathbb{A})$, and let $G(\mathbb{A}_f)$ be the nonarchimedean component, so that

$$G(\mathbb{A}) = G_\infty \times G(\mathbb{A}_f). \tag{7.2}$$

Let

$$\begin{aligned} K_\infty &= \text{a maximal compact subgroup of the Lie group } G_\infty \\ \mathfrak{g} &= \text{complexification of the (real) Lie algebra of } G_\infty \\ U(\mathfrak{g}) &= \text{universal enveloping algebra of } \mathfrak{g} \\ Z(\mathfrak{g}) &= \text{center of } U(\mathfrak{g}). \end{aligned}$$

Let K_1 be the open compact subgroup $G\big(\prod_{v \text{ finite}} \mathcal{O}_v\big)$ of $G(\mathbb{A}_f)$.

A complex-valued function f on $G(\mathbb{A})$ is **smooth** if it is continuous and, when viewed as a function of two arguments (x, y) as in (7.2), it is smooth in x for each fixed y and is locally constant of compact support in y for each fixed x. Let ρ be a finite-dimensional representation of K_∞, let $J \subset Z(\mathfrak{g})$ be an ideal of finite codimension, and let K be an open subgroup of K_1. A smooth function f on $G(\mathbb{A})$ is **automorphic** relative to (ρ, J, K) if

(i) $f(\gamma g) = f(g)$ for all $\gamma \in G(F)$
(ii) $f(gk) = f(g)$ for all $k \in K$
(iii) the span of the right translates of f by members of K_∞ is finite-dimensional, and every irreducible constituent of this representation of K_∞ is a constituent of ρ
(iv) the ideal J, acting in the G_∞ variable of (7.2), has $Jf = 0$
(v) for each $y \in G(\mathbb{A}_f)$, the function $x \mapsto f(xy)$ on G_∞ satisfies a certain slow-growth condition.

We shall not make (v) any more precise, but instead we refer to [Bo-Ja]; Theorem 7.3 below will clarify condition (v) in the principal case of interest. The set of automorphic functions relative to (ρ, J, K) will be denoted $\mathcal{A}(\rho, J, K)$.

When $G = GL_1$, any Grossencharacter gives an example of an automorphic form relative to a suitable triple.

Theorem 7.1. *For every (ρ, J, K), $\mathcal{A}(\rho, J, K)$ is finite-dimensional.*

This theorem is fundamental. It is due to Harish-Chandra; see [HC1] and [HC2, p. 8], where it is proved in the setting of the right side of (6.3). The translation into the current setting (which is the left side of (6.3)) is in [Bo-Ja, p. 195].

Theorem 7.2. *Let a smooth function f on $G(\mathbb{A})$ satisfy* (i) *through* (iv) *above and also*

$$f(zx) = \chi(z) f(x) \qquad \textit{for all } z \in Z(\mathbb{A}) \textit{ and } x \in G(\mathbb{A}) \tag{7.3}$$

for some (unitary) character of $Z(F)\backslash Z(\mathbb{A})$*, so that* $|f|$ *may be regarded as a function on* $(Z(\mathbb{A})G(F))\backslash G(\mathbb{A})$*. If* $|f|$ *is in* $L^p((Z(\mathbb{A})G(F))\backslash G(\mathbb{A}))$ *for some* $p \geq 1$*, then* f *satisfies condition* (v) *and hence is an automorphic form.*

REFERENCE. See [Bo-Ja, pp. 191 and 195]. The proof makes use of Theorem 6.2 and [HC2].

A **cusp form** is an automorphic form f such that (7.3) holds for some unitary character χ of $Z(F)\backslash Z(\mathbb{A})$ and such that

$$\int_{N(F)\backslash N(\mathbb{A})} f(ng)\, dn = 0 \tag{7.4}$$

for the unipotent radical N of every proper parabolic subgroup of G and for all $g \in G(\mathbb{A})$. For $G = GL_1$, the condition (7.4) is empty, and therefore all unitary Grossencharacters are cusp forms for GL_1. The classical analytic cusp forms relative to $SL_2(\mathbb{Z})$ yield cusp forms for $G = GL_2$ in the sense of (7.4), and so do Hecke's cusp forms relative to the subgroup $\Gamma_0(N)$ of $SL_2(\mathbb{Z})$. For general G, let ${}^0\mathcal{A}(\rho, J, K)$ be the space of cusp forms relative to (ρ, J, K).

Theorem 7.3. *Let a smooth function* f *on* $G(\mathbb{A})$ *satisfy* (i) *through* (iv) *above, as well as the cuspidal condition* (7.4) *and the condition* (7.3) *for some (unitary) character of* $Z(F)\backslash Z(\mathbb{A})$*. Then the following conditions are equivalent:*

(i) f *satisfies* (v) *and hence is a cusp form*
(ii) f *is bounded*
(iii) $|f|$ *is in* $L^2((Z(\mathbb{A})G(F))\backslash G(\mathbb{A}))$.

Actually there is a sharper result: A smooth function f satisfying (i) through (iv) of an automorphic form, as well as the slow-growth condition in (v) and the condition (7.3) for some unitary χ, is automatically rapidly decreasing as soon as it satisfies the cuspidal condition (7.4). We already noted this fact for classical cusp forms relative to $SL_2(\mathbb{Z})$. The result for general G requires precise definitions of "slow growth" and "rapidly decreasing," which we omit.

We want to define the notion of an automorphic representation of $G(\mathbb{A})$. Put $\mathcal{A} = \mathcal{A}(\rho, J, K)$. The idea is that an automorphic representation is any irreducible subquotient of $\mathcal{A}$, but the trouble is that $\mathcal{A}$ need not be mapped to itself under right translation by $G(\mathbb{A})$. Specifically, K_∞ finiteness need not be preserved under right translation by G_∞.

The idea is to make $\mathcal{A}$ into a module for an algebra $\mathcal{H}$ (the **Hecke algebra**) that reflects the action by $G(F_v)$ for each finite place v and reflects the action by $U(\mathfrak{g})$ and K_∞ at the infinite places. We summarize the construction, which is given in more detail in [Fl]. For each finite place v, let $\mathcal{H}_v$ be the algebra (under convolution) of all complex-valued locally constant functions of compact support on $G(F_v)$. Haar measure on $G(F_v)$ is to be normalized so that $G(\mathcal{O}_v)$ has measure 1, and then the characteristic function I_v of $G(\mathcal{O}_v)$ is an idempotent in $\mathcal{H}_v$. The algebra $\mathcal{H}_v$ contains a directed system of further idempotents, namely the normalized characteristic function of each open subgroup of $G(\mathcal{O}_v)$. An $\mathcal{H}_v$ module is **approximately unital** if, for each member of the module, all idempotents corresponding to sufficiently small open subgroups act as the identity. It is fairly easy to see that **smooth** representations of $G(F_v)$, i.e., those in which each vector is fixed by some compact open subgroup, correspond exactly to approximately unital

$\mathcal{H}_v$ modules. Such a representation is called **admissible** if the set of vectors fixed by any compact open subgroup is finite-dimensional.

There is a natural way of forming a restricted tensor product of the algebras $\mathcal{H}_v$ with respect to the idempotents I_v. The resulting algebra $\mathcal{H}_f$ is the part of $\mathcal{H}$ corresponding to the finite places of F and is generated by product functions that equal I_v at almost every place. A tuple of local idempotents, one for each $\mathcal{H}_v$ with almost all of them being I_v, yields another idempotent in $\mathcal{H}_f$, and the idempotents obtained in this way are indexed by a directed set. A right $\mathcal{H}_f$ module is **smooth** if each member of the module is fixed by all idempotents corresponding to members of the directed set that are sufficiently large. The module is **admissible** if the set of vectors fixed by any of these idempotents is finite-dimensional.

Next we let $\mathcal{H}_\infty$ be the convolution algebra of all K_∞ finite distributions on G_∞ that are supported on K_∞. This algebra is studied extensively in [Kna-Vo]. It contains a directed family of idempotents as follows. Let dk denote normalized Haar measure on K_∞. For each class of irreducible representations τ of K_∞, let χ_τ be the character and let d_τ be the degree. The directed family of idempotents is indexed by all finite subsets of τ's, the idempotent corresponding to a given set being the sum of $d_\tau \chi_\tau$ for all τ in the set. A right $\mathcal{H}_\infty$ module is **approximately unital** if, for each member of the module, all sufficiently large idempotents act as the identity. It is shown in [Kna-Vo, pp. 75 and 90] that $(\mathfrak{g}, K_\infty)$ modules correspond exactly to approximately unital $\mathcal{H}_\infty$ modules. Such a module is **admissible** if the set of vectors fixed by any of these idempotents is finite-dimensional, i.e., if each K_∞ type has finite multiplicity.

We define $\mathcal{H} = \mathcal{H}_\infty \otimes \mathcal{H}_f$. Smoothness and admissibility of right $\mathcal{H}$ modules are defined using idempotents that are pure tensors from $\mathcal{H}_\infty$ and $\mathcal{H}_f$. Then $\mathcal{A}$ is a smooth right $\mathcal{H}$ module. An **automorphic representation** of $\mathcal{H}$ is any irreducible subquotient of $\mathcal{A}$. Similarly if we put ${}^0\mathcal{A} = {}^0\mathcal{A}(\rho, J, K)$, then a **cuspidal automorphic representation** of $\mathcal{H}$ is any irreducible subquotient of ${}^0\mathcal{A}$.

If f is an automorphic form, then it is immediate from Theorem 7.1 that $f * \mathcal{H}$ is a smooth admissible $\mathcal{H}$ module. It follows that every automorphic representation of $\mathcal{H}$ is smooth and admissible. Such representations are commonly called automorphic representations of $G(\mathbb{A})$ even though not all of $G(\mathbb{A})$ really acts.

More generally a topologically irreducible $G(\mathbb{A})$ module is said to be **automorphic** if its underlying space of smooth vectors is an automorphic representation of $\mathcal{H}$. According to [Fl, Theorem 4], if χ is any (unitary) character of $Z(F)\backslash Z(\mathbb{A})$, then any $G(\mathbb{A})$ invariant irreducible closed subspace of

$$\begin{aligned} &L^2(G(F)\backslash G(\mathbb{A}))_\chi \\ &= \big\{f \,\big|\, |f| \in L^2((Z(\mathbb{A})G(F))\backslash G(\mathbb{A})) \text{ and } f(zx) = \chi(z)f(x) \text{ for } z \in Z(\mathbb{A}), x \in G(\mathbb{A})\big\} \end{aligned}$$

is automorphic in this sense.

Theorem 7.4. *The subspace of cuspidal functions in $L^2(G(F)\backslash G(\mathbb{A}))_\chi$ decomposes discretely with finite multiplicities. Consequently whenever f is a cusp form, $f * \mathcal{H}$ is a finite direct sum of cuspidal automorphic representations.*

REFERENCE. This theorem is due to Gelfand and Piatetski-Shapiro [Gelf-Pi]. See [Gelb1, p. 33] for a discussion when $G = GL_2$, and see [HC2, p. 9] for the general case.

REMARK. It follows from the theorem that cuspidal automorphic representations are unitarizable. That is, they are the underlying smooth representations for irreducible unitary representations of $G(\mathbb{A})$.

Hecke's cusp forms discussed earlier in the section lead to cusp forms in the adelic sense, by Theorem 7.3. Those whose L function has an Euler product expansion lead to adelic cusp forms that generate single (irreducible) cuspidal automorphic representations.

8. Langlands Theory for GL_n

In this section we describe what the Langlands theory proposes for GL_n. The theory for $G = GL_n$ may be regarded as a special case of the general theory, which will be discussed in the next section.

Fix a number field F. The theory wants to associate to each automorphic representation of $GL_n(\mathbb{A}_F)$ an L function given initially as a product of elementary L factors, one for each place of F. The method is arranged to be a direct generalization of the way in which an L function is attached to a Grossencharacter in (5.2), the case of a unitary Grossencharacter being the special case $n = 1$. At the same time it generalizes the way in which an L function is attached to a classical cusp form, except that the starting point is the Euler product expansion and not the series expansion; this situation is an instance of the special case $n = 2$ of the Langlands theory. For general n and a given automorphic representation of $GL_n(\mathbb{A}_F)$, the Langlands theory gives a precise definition of the elementary L factors at almost every place, and their product is convergent in a half plane. The definition at the remaining finite set of places hinges on a conjecture known as the Local Langlands Conjecture, which we shall discuss shortly.

By a theorem of Flath [Fl], any irreducible smooth admissible representation π of $GL_n(\mathbb{A}_F)$ (or, more precisely, of the Hecke algebra) is a "restricted tensor product" $\pi = \bigotimes_v \pi_v$ of irreducible smooth admissible representations π_v of the respective factors $GL_n(F_v)$ of $GL_n(\mathbb{A}_F)$. To define the L function of π, it is therefore enough to describe the elementary L factor of the representation π_v of $GL_n(F_v)$.

Thus let k be a local field of characteristic 0, and let π be an irreducible representation of $GL_n(k)$. Approximately, the theory proposes attaching to π a continuous homomorphism $\varphi : W_k \to GL_n(\mathbb{C})$ with certain properties. Here W_k is the Weil group of k. Then the L factor for π is defined to be a certain elementary function attached to φ that generalizes any of the factors on the right side of (5.2).

The above description is correct if k is archimedean, but the use of the Weil group W_k is insufficient if k is nonarchimedean. For reasons that we illustrate by example later in this section, the group W_k in the nonarchimedean case is replaced by the Weil-Deligne group W'_k, which we shall define now.

Let $\theta_k : k^\times \to W_k^{\text{ab}}$ be the local reciprocity map of Corollary 2.5. Let $\|w\|$ denote the effect on $w \in W_k$ of the composition of passing from $W_k \to W_k^{\text{ab}}$, followed by passing to $k^\times$ by θ_k^{-1}, followed by passing to $\mathbb{R}_+^\times$ by $|\cdot|_k$. The **Weil-Deligne group** W'_k is the semidirect product of W_k and $\mathbb{C}$, where W_k acts on $\mathbb{C}$ by $wxw^{-1} = \|w\|x$. Thus the multiplication rule on $\mathbb{C} \times W_k$ is $(a_1, w_1)(a_2, w_2) = (a_1 + \|w_1\|a_2, w_1w_2)$.

In the nonarchimedean case, a continuous homomorphism $\varphi : W'_k \to GL_n(\mathbb{C})$ is called an **admissible homomorphism** if φ is holomorphic in the $\mathbb{C}$ variable, if $\varphi(\mathbb{C})$ consists of unipotent matrices, and if $\varphi(W_k)$ consists of semisimple matrices. Such a homomorphism amounts to specifying a pair (ρ, X), where

$\rho : W_k \to GL(n,\mathbb{C})$ is a continuous homomorphism such that $\varphi(W_k)$ consists of semisimple matrices and X is a nilpotent endomorphism of $\mathbb{C}^n$ such that $\rho(w)X\rho(w)^{-1} = \|w\|X$ for all $w \in W_k$. Let Fr be an element of W_k such that $\|\mathrm{Fr}\| = q^{-1}$, where q is the order of the residue field; the element Fr is uniquely determined modulo the inertia group I_k of (1.4). If V_X^I is the subspace of $\ker X$ fixed by $\rho(I_k)$, then the elementary L factor associated to φ is[4]

$$L(s,\varphi) = \det(1 - (\rho(\mathrm{Fr})|_{V_X^I})q^{-s})^{-1}. \tag{8.1}$$

Parenthetically let us mention a substitute for the Weil-Deligne group that one encounters in the literature. With the right definition the admissible homomorphisms into $GL_n(\mathbb{C})$ for the substitute group will correspond to the admissible homomorphisms of W_k', and ultimately no semidirect products will be involved in the definition of the substitute group. For $w \in W_k$, let h_w be the matrix $\begin{pmatrix} \|w\|^{1/2} & 0 \\ 0 & \|w\|^{-1} \end{pmatrix}$ in $SL_2(\mathbb{C})$, and identify $z \in \mathbb{C}$ with $\begin{pmatrix} 1 & z \\ 0 & 1 \end{pmatrix}$. The above action of W_k on $\mathbb{C}$ by automorphisms translates into conjugation of $\begin{pmatrix} 1 & z \\ 0 & 1 \end{pmatrix}$ by h_w. Since conjugation by h_w extends to an automorphism of all of $SL_2(\mathbb{C})$, we can identify the action of W_k on $\mathbb{C}$ with a subaction of the action of W_k on $SL_2(\mathbb{C})$ by automorphisms. The semidirect product of $SL_2(\mathbb{C})$ by W_k with respect to this action then consists of pairs (x, w) with $x \in SL_2(\mathbb{C})$ and $w \in W_k$, where $(x_1, w_1)(x_2, w_2) = (x_1 h_{w_1} x_2 h_{w_1}^{-1}, w_1 w_2)$. This group is isomorphic to the direct product $SL_2(\mathbb{C}) \times W_k$ by $(x, w) \mapsto (x h_w, w)$, and $SL_2(\mathbb{C}) \times W_k$ is then used as the substitute for the Weil-Deligne group. We shall not use this substitute, however.

In the archimedean case, an **admissible homomorphism** $\varphi : W_k \to GL_n(\mathbb{C})$ is just a continuous homomorphism such that $\varphi(W_k)$ consists of semisimple matrices. The elementary L factor associated to such a φ is a nowhere-zero meromorphic function involving Γ functions and may be found in [Kna2, p. 404].

Two admissible homomorphisms φ_1 and φ_2 are said to be equivalent if they are conjugate via $GL_n(\mathbb{C})$, i.e., if there exists $g \in GL_n(\mathbb{C})$ with $g\varphi_1(x)g^{-1} = \varphi_2(x)$ for all x in the Weil-Deligne group or Weil group, as appropriate. The set of equivalence classes of admissible homomorphisms is denoted $\Phi(GL_n(k))$.

For each local field k of characteristic 0, let $\Pi(GL_n(k))$ be the set of equivalence classes of (smooth) irreducible admissible representations of $GL_n(k)$. If k is nonarchimedean, "equivalence" here means equivalence as representations of the group (or of the Hecke algebra). If k is archimedean, "equivalence" means infinitesimal equivalence (or equivalence as representations of the Hecke algebra). Leaving aside some further definitions for the moment, we can state the Local Langlands Conjecture as follows.

Local Langlands Conjecture. *$\Pi(GL_n(k))$ is indexed by $\Phi(GL_n(k))$ in a natural way that is compatible with twisting by Grossencharacters and respects L factors and ε factors.*

For $\varphi \in \Phi(GL_n(k))$ and $\pi \in \Pi(GL_n(k))$, the "twists" $\varphi \otimes \alpha$ and $\pi \otimes \alpha$ by a Grossencharacter α are defined toward the end of this section. The requirement about twists is that if φ corresponds to π, then $\varphi \otimes \alpha$ corresponds to $\pi \otimes \alpha$ for every α.

[4]See the footnote for (5.1a).

Let us discuss the requirement on L factors and ε factors. We can define the **Langlands elementary L factor** of an irreducible π to be the elementary L factor of the corresponding φ:

$$L(s,\pi) = L(s,\varphi) \qquad \text{if } \varphi \leftrightarrow \pi, \tag{8.2}$$

with $L(s,\varphi)$ as in (8.1). In Jacquet's lecture [Ja2] may be found a completely different construction of elementary L factors attached to irreducible admissible representations (due to Godement and Jacquet [Go-Ja]), and this construction does not depend on any conjectures. We require that the two kinds of L factors match. Properties of the Godement-Jacquet L functions will be discussed below after Corollary 8.6.

We mentioned ε functions attached to Galois representations in connection with Theorem 5.3. These are products of local ε factors that we have not defined, are 1 at almost every place, and generalize the Gauss sums that occur in the functional equations of Dirichlet L functions [Kna1, p. 216]. Local ε factors depend upon additional data, but let us suppress this point. In a fashion similar to that for Galois representations, a **local ε factor** may be associated to each $\varphi \in \Phi(GL_n(k))$. The correspondence of the Local Langlands Conjecture then allows one to define a **Langlands ε factor** for $\pi \in \Pi(GL_n(k))$ by

$$\varepsilon(s,\pi) = \varepsilon(s,\varphi) \qquad \text{if } \varphi \leftrightarrow \pi.$$

Meanwhile the Godement-Jacquet construction also defines an ε factor for each π in $\Pi(GL_n(k))$. We require that the two constructions match.

The paper [Kud, p. 380] gives a full list of the requirements that the conjectural local Langlands correspondence is supposed to satisfy.

The Local Langlands Conjecture is a theorem when $n = 1$. In this case the result amounts to a restatement of Corollary 2.5. For general n and for k archimedean, the conjecture is a theorem of Langlands; see [Kna2] for an exposition. Much is known in the nonarchimedean case. See [Kud] for an exposition. The conjecture is known for $n = 2$ ([Kut] and [Tu]), for $n = 3$ ([Hen1]), and for all relatively prime n and p [Moy]. For more recent work, see [Ha]. Henniart [Hen2] has shown that there is at most one candidate for the local Langlands correspondence respecting ε factors for "pairs" in the sense of [Ja-P-S]. See [Kud, §4.2] for more detail.

A part of the correspondence is easy to understand. For definiteness, let k be nonarchimedean. Let ρ be a continuous homomorphism of W_k into the diagonal subgroup of $GL_n(\mathbb{C})$. Composing with projection to each diagonal entry, we see that ρ is completely determined by n quasicharacters $\chi_1, \ldots, \chi_n$ of $W_k^{\mathrm{ab}} \cong k^\times$. Let us write $\rho = \rho_{(\chi_1,\ldots,\chi_n)}$. If we put $X = 0$, then

$$\varphi = \varphi_{(\chi_1,\ldots,\chi_n)} = (\rho_{(\chi_1,\ldots,\chi_n)}, 0)$$

is an admissible homomorphism. Let T be the diagonal subgroup of $GL_n(k)$, and let N be the upper-triangular subgroup with 1's on the diagonal. The data $(\chi_1, \ldots, \chi_n)$ give us a quasicharacter of T, and it is natural to associate to $\varphi_{(\chi_1,\ldots,\chi_n)}$ the induced representation

$$\operatorname{ind}_{TN}^{GL_n(k)}((\chi_1, \ldots, \chi_n) \otimes 1) \tag{8.3}$$

given by normalized induction (a member of the **nonunitary principal series**). This representation may be taken to be the π that corresponds to φ when π is irreducible, and it is known that irreducibility occurs if and only if there is no pair of indices i and j such that $\chi_i\chi_j^{-1}$ equals $|\cdot|_k$.

When (8.3) is reducible, the π that is associated to $\varphi_{(\chi_1,\dots,\chi_n)}$ is a certain irreducible subquotient of (8.3) known as the **Langlands subquotient**. To describe the Langlands subquotient, we first remark that the set of irreducible subquotients of (8.3) is unchanged (apart from equivalence) when $\chi_1,\dots,\chi_n$ are permuted. In order to have the indexing $\Phi \leftrightarrow \Pi$ depend only on equivalence classes, we require that the Langlands subquotient not be affected by permutation. Next, we can introduce complex numbers $s_1,\dots,s_n$ such that $|\chi_i| = |\cdot|_k^{s_i}$. The numbers s_i are not unique, but their real parts are unique. If the real parts of $s_1,\dots,s_n$ are nonincreasing, then the Langlands subquotient is the (unique) irreducible quotient of (8.3). The result is that the Langlands subquotient is determined in every case. See [Moe2] in this volume for further discussion.

A special case of the nonunitary principal series of particular interest is the **unramified principal series**, those members of the nonunitary principal series having a nonzero fixed vector under $GL_n(\mathcal{O}_k)$. By Frobenius reciprocity these are just the representations (8.3) for which each $\chi_i(x)$ depends only on the module of x. Thus each χ_i is of the form $|\cdot|^{s_i}$ for some complex s_i. The Langlands subquotient of an unramified principal series is the irreducible subquotient containing a nonzero vector fixed by $GL_n(\mathcal{O}_k)$. All members of $\Pi(GL_n(k))$ having a nonzero vector fixed by $GL_n(\mathcal{O}_k)$ are of this form, up to equivalence. Up to equivalence, they are parametrized by orbits of the symmetric group on tuples $(s_1,\dots,s_n)$ of complex numbers modulo $2\pi i(\log q)^{-1}\mathbb{Z}$. In view of (8.2), *the elementary L factor of Langlands is well defined* in the case of a member of $\Pi(GL_n(k))$ having a nonzero vector fixed by $GL_n(\mathcal{O}_k)$. Such a member of $\Pi(GL_n(k))$ is said to be **unramified**.

Let us now discuss the need for the Weil-Deligne group W_k' rather than just the Weil group W_k. Suppose that $n = 2$. The irreducible admissible representations of $GL_2(k)$ are then of three kinds—the **supercuspidal** representations (those whose matrix coefficients are compactly supported modulo the center), the Langlands subquotients of the nonunitary principal series, and the **special representations**. In (8.3), reducibility occurs for $GL_2(k)$ exactly when $\chi_1\chi_2^{-1} = |\cdot|_k^{\pm1}$. In this case, there are two irreducible subquotients, the Langlands subquotient and one other. These "other" representations are the special representations. Their matrix coefficients are square integrable modulo the center, but these representations are not supercuspidal.

For $n = 2$, the Godement-Jacquet L factors of supercuspidal representations are 1, of special representations involve one factor with q^{-s} in it, and of Langlands subquotients of nonunitary principal series involve two factors with q^{-s} in them. For the Local Langlands Conjecture to be valid, it is necessary to arrange for some other two-dimensional representations (of W_k or a substitute) than the irreducible ones (which give 1 as L factor) and the direct sums of one-dimensional ones (which give L factors that are the product of two expressions involving q^{-s}). Englargement to the Weil-Deligne group allows the existence of indecomposable yet reducible two-dimensional representations of W_k'. For one of these representations, the space is $\mathbb{C}^2$ with a basis e_1, e_2 and with action

$$\rho(w)e_i = \|w\|^{i-1}e_i, \qquad Xe_1 = e_2, \qquad Xe_2 = 0.$$

The most general indecomposable yet reducible two-dimensional representation of W_k', up to equivalence, is the tensor product of this one with a one-dimensional representation. The equivalence classes of indecomposable two-dimensional representations are used to parametrize the special representations up to equivalence.

Use of W'_k to settle $GL_2(k)$ looks somewhat artificial, but W'_k looks more reasonable when one considers what is known about GL_n. For $GL_n(k)$ with k nonarchimidean, the irreducible admissible representations have been classified. The starting point is the supercuspidal representations, which have been classified by Bushnell-Kutzko and by Corwin. Bernstein and Zelevinski showed how to classify the irreducible admissible representations in terms of the supercuspidal representations. See [Kud] and [Moe2] for more detail and for references. The nature of the Bernstein-Zelevinski part of the classification implies that if the supercuspidal representations of $GL_n(k)$ correspond to the n-dimensional irreducible admissible homomorphisms $W_k \to GL_n(\mathbb{C})$ of the Weil group W_k for all n, then $\Pi(GL_n(k))$ is parametrized by the n-dimensional admissible homomorphisms $W'_k \to GL_n(\mathbb{C})$ of the Weil-Deligne group W'_k. (See [Ja1, (3.7)] and [Kud, p. 381].)

Now we return to the number field F. Let $\pi = \bigotimes_v \pi_v$ be an irreducible admissible representation of $GL_n(\mathbb{A}_F)$. Then we can define the **Langlands L functions** by

$$\begin{aligned} L(s,\pi) &= \prod_{\text{finite } v} L(s,\pi_v) \\ \Lambda(s,\pi) &= \prod_{\text{all } v} L(s,\pi_v), \end{aligned} \tag{8.4}$$

where the factors are given by (8.2) and ultimately (8.1). The first question is one of convergence. Built into the factorization of π as $\bigotimes_v \pi_v$ is the following addendum, which simplifies questions of convergence greatly.

Theorem 8.1. *If $\pi = \bigotimes_v \pi_v$ is an irreducible admissible representation of $GL_n(\mathbb{A}_F)$, then almost every π_v is unramified, i.e., has a nonzero fixed vector under $GL_n(\mathcal{O}_v)$ and is therefore the Langlands subquotient of an unramified principal series.*

REFERENCE. [Fl, p. 181].

Corollary 8.2. *If π is a unitarizable irreducible admissible representation of $GL_n(\mathbb{A}_F)$, then $L(s,\pi)$ converges absolutely for* Re s *sufficiently large.*

REFERENCE. This is essentially due to Langlands. See [Bo4, p. 50].

In fact, Theorem 8.1 shows that it is enough to consider the factors of π that are unramified principal series. Flath's results show that we may take each factor to be unitarizable, and then all that is needed is an estimate on the L factor for a unitarizable unramified principal series.

Corollary 8.3. *If π is a cuspidal automorphic representation of $GL_n(\mathbb{A}_F)$, then $L(s,\pi)$ converges absolutely for* Re s *sufficiently large.*

REFERENCE. This kind of result is due to Langlands [Lgl2] and predates [Fl]. The proof is immediate from the above results: The cuspidal representation is unitarizable by Theorem 7.4, and convergence follows from Corollary 8.2.

With considerably more effort, Langlands has addressed convergence of $L(s,\pi)$ for general automorphic π. We need another corollary of Theorem 8.1.

Corollary 8.4. *Let $P = MN$ be the usual Levi decomposition of a standard parabolic subgroup of GL_n. Let σ be a unitarizable irreducible admissible representation of $M(\mathbb{A}_F)$, and let π be an irreducible subquotient of* $\operatorname{ind}_{M(\mathbb{A}_F)N(\mathbb{A}_F)}^{GL_n(\mathbb{A}_F)}(\sigma \otimes 1)$. *Then $L(s, \pi)$ converges absolutely for* $\operatorname{Re} s$ *sufficiently large.*

REFERENCE. [Bo4, p. 52].

Theorem 8.5. *An irreducible admissible representation π of $GL_n(\mathbb{A}_F)$ is automorphic if and only if π is equivalent with an irreducible subquotient of*

$$\operatorname{ind}_{M(\mathbb{A}_F)N(\mathbb{A}_F)}^{GL_n(\mathbb{A}_F)}(\sigma \otimes 1)$$

for some cuspidal automorphic representation σ of $M(\mathbb{A}_F)$.

REFERENCE. [Lgl4, p. 204].

Corollary 8.6. *If π is an automorphic representation of $GL_n(\mathbb{A}_F)$, then $L(s, \pi)$ converges absolutely for* $\operatorname{Re} s$ *sufficiently large.*

REFERENCE. This result of Langlands combines the above ideas. See [Bo4, p. 52]. By Theorem 8.5, π is a constituent of a representation induced from a cuspidal representation, and the cuspidal representation is unitarizable by Theorem 7.4. Then convergence follows from Corollary 8.4.

The next question concerns the analytic properties of automorphic representations. We shall be content with the results in the cuspidal case. For the general automorphic case, see [Ja1, p. 83]. As we mentioned above, Godement and Jacquet [Go-Ja] have constructed L and Λ functions in a way that is completely different from using the Local Langlands Conjecture. Their theory is summarized in this volume in [Ja2], and conjecturally their functions, which we denote $L^{\mathrm{GJ}}(s, \pi)$ and $\Lambda^{\mathrm{GJ}}(s, \pi)$, coincide with those in (8.4). What is known is that the Godement-Jacquet L functions agree with the Langlands L functions at every place where π_v is unramified. (This condition holds for almost every place by Theorem 8.1.) The Godement-Jacquet L functions have good global analytic properties, as follows.

Theorem 8.7. *If π is a cuspidal representation of $GL_n(\mathbb{A}_F)$, the Godement-Jacquet function $\Lambda^{\mathrm{GJ}}(s, \pi)$ extends to be meromorphic in $\mathbb{C}$ with singularities given at most by simple poles at $s = 0$ and $s = 1$. Moreover, $\Lambda^{\mathrm{GJ}}(s, \pi)$ satisfies a functional equation of the form*

$$\Lambda^{\mathrm{GJ}}(s, \pi) = \varepsilon^{\mathrm{GJ}}(s, \pi)\Lambda^{\mathrm{GJ}}(1 - s, \pi^\vee),$$

where $\pi^\vee$ is the contragredient and $\varepsilon^{\mathrm{GJ}}(s, \pi)$ is a multiple of N^{-s} for some integer N. The function $\Lambda^{\mathrm{GJ}}(s, \pi)$ is actually entire unless π is a unitary character of $GL_1(\mathbb{A}_F)$ trivial on the elements of module 1.

When $n = 1$, the L functions in Theorem 8.7 (with the elementary factors from the infinite places dropped) are the ones attached to unitary Grossencharacters by Hecke and Tate, and Theorem 8.7 generalizes the work in Tate's thesis [Ta1]. When $n = 2$, these L functions (after a change of parameter) generalize the L functions attached by Hecke to cusp forms relative to $SL_2(\mathbb{Z})$; Hecke proved that his L functions are entire and satisfy a functional equation.

An important aspect of the Langlands theory for GL_n is a reciprocity conjecture that implies the Artin Conjecture (end of §5). We state this conjecture of Langlands in two forms—the first version involving almost all places and either Langlands L functions or Godement-Jacquet L functions, the second version involving all places and Godement-Jacquet L functions. For the second version we need to complete the Artin L function $L(s, \sigma)$ to a function $\Lambda(s, \sigma)$ by adjoining elementary L factors for the infinite places; these factors are listed explicitly in [Kna2, (3.6) and (4.6)].

Langlands Reciprocity Conjecture, first version. *For any irreducible representation σ of $\operatorname{Gal}(\overline{F}/F)$ in $GL_n(\mathbb{C})$, there exists a cuspidal automorphic representation π of $GL_n(\mathbb{A}_F)$ such that the Artin L function of σ agrees with the Langlands L function of π at almost every place where π is unramified.*

Langlands Reciprocity Conjecture, second version. *For any irreducible representation σ of $\operatorname{Gal}(\overline{F}/F)$ in $GL_n(\mathbb{C})$, there exists a cuspidal automorphic representation π of $GL_n(\mathbb{A}_F)$ such that the completed Artin L function $\Lambda(s, \sigma)$ of σ is identical with the Godement-Jacquet L function $\Lambda^{\mathrm{GJ}}(s, \pi)$ of π. In more detail, $L_v(s, \sigma) = L_v^{\mathrm{GJ}}(s, \pi)$ for every finite place v, and $L_\infty(s, \sigma) = L_\infty^{\mathrm{GJ}}(s, \pi)$ if L_∞ and L_∞^{GJ} denote the products of the factors for the infinite places.*

If $\pi = \bigotimes_v \pi_v$, then we have noted that the Langlands L factor for π_v agrees with the Godement-Jacquet L factor for π_v when π_v is unramified and that π_v is unramified at almost every place. Therefore the second version of the conjecture implies the first. The converse is addressed by the following.

Theorem 8.8. *The first version of the Langlands Reciprocity Conjecture for σ implies the second for σ. Consequently either version of the Langlands Reciprocity Conjecture for σ implies the Artin Conjecture (end of §5) for σ.*

We postpone the discussion of Theorem 8.8 to the end of this section except to note that the second sentence of the theorem follows by combining Theorem 8.7 and the second version of the conjecture.

Let us discuss the conjecture itself. The basic case for which this conjecture is known is Theorem 5.1, which handles $n = 1$. That theorem says that abelian Artin L functions are Hecke L functions of Grossencharacters. On the one hand, this result is essentially equivalent with Artin reciprocity. On the other hand, it establishes that the pattern of $\varphi_\wp(\pi_\wp)$, which governs the factorization of certain polynomials modulo $\wp$, is given in terms of arithmetic progressions in the abelian case. The general conjecture may therefore be regarded as a statement about reciprocity on the one hand and a statement about the pattern of factorization of polynomials modulo $\wp$ on the other hand.

It is natural to expect that the Langlands Reciprocity Conjecture holds for $\sigma = \operatorname{ind} \rho$ whenever it holds for ρ and $\operatorname{ind} \rho$ is irreducible. But it is an open problem to prove such a statement, even for ρ one-dimensional. In the special case that ρ is one-dimensional and the induced representation is two-dimensional, Jacquet and Langlands [Ja-Lgl] did manage to prove the statement, and it follows that the Langlands Reciprocity Conjecture holds when $n = 2$ for σ of "dihedral type" in the sense of [Ro2, §16].

In fundamental work [Lgl5], Langlands handled some additional cases when $n = 2$, and later Tunnell [Tu] was able to deduce an improved result using the methods of Langlands. These results of Langlands and Tunnell are the subject of the lectures [Ro2] by Rogawski. The statement is as follows.

Theorem 8.9. *If σ is a two-dimensional complex representation of* $\mathrm{Gal}(\overline{F}/F)$ *with solvable image, then the Langlands Reciprocity Conjecture holds for σ.*

For further work with $n = 2$, see [Buhl] and [Fre].

Any initial attempt at proving the Langlands Reciprocity Conjecture raises the following question: If we have irreducible admissible representations π_v for each place v of F and if $\pi = \bigotimes_v \pi_v$ is well defined, how do we tell whether π is automorphic? In other words, how is the global behavior relative to $GL_n(F)$ reflected in the system of representations π_v? Theorem 8.1 gives a necessary condition; almost all π_v must be Langlands subquotients of unramified principal series. What else can be said?

We are especially interested in the cuspidal case. If we are given an irreducible admissible $\pi = \bigotimes_v \pi_v$, the **Multiplicity One Theorem** below says that there is at most one way that π can occur as a cuspidal representation. Recall from Theorem 7.4 that the cuspidal part of $L^2(GL_n(F)\backslash GL_n(\mathbb{A}_F))_\chi$ decomposes discretely with finite multiplicities.

Theorem 8.10. *If π is an irreducible admissible representation of $GL_n(\mathbb{A}_F)$ occurring in the cuspidal part of $L^2(GL_n(F)\backslash GL_n(\mathbb{A}_F))_\chi$ for some unitary character χ, then π occurs with multiplicity one.*

REFERENCE. [Sha].

For two Grossencharacters $\chi = \bigotimes_v \chi_v$ and $\chi' = \bigotimes_v \chi'_v$, an equality $\chi_v = \chi'_v$ for almost all v implies $\chi = \chi'$. In fact, we may assume that $\chi_v = 1$ for almost all v. Applying the Weak Approximation Theorem (Theorem 3.7) and using that χ is 1 on $F^\times$ and is continuous, we see that χ equals 1 everywhere. The following **Strong Multiplicity One Theorem** generalizes this result from GL_1 to GL_n.

Theorem 8.11. *Let $\pi = \bigotimes_v \pi_v$ and $\pi' = \bigotimes_v \pi'_v$ be irreducible admissible representations of $GL_n(\mathbb{A}_F)$ occurring in the cuspidal part of $L^2(GL_n(F)\backslash GL_n(\mathbb{A}_F))_\chi$ for some unitary character χ. If π_v is equivalent with π'_v for almost all v, then $\pi = \pi'$.*

REFERENCE. [Ja-Sha, p. 553].

We do not conclude merely that π and π' are equivalent. Theorem 8.10 allows us to deduce equality from equivalence. It follows from Theorem 8.11 that the cuspidal automorphic representation π in the Langlands Reciprocity Conjecture is unique if it exists.

Thus cuspidal representations are rather rigid. But how do we tell when we have one? When $F = \mathbb{Q}$, we can phrase the question in a related way. If $L(s)$ is a Dirichlet series obtained by expanding out a product of elementary factors as in (8.1), when is $L(s)$ the L function of a cuspidal automorphic representation? Weil [We2], generalizing work of Hecke, answered the question about when a Dirichlet series $\sum_{n=1}^\infty a_n/n^s$ comes from a cusp form $\sum_{n=1}^\infty a_n e^{2\pi i n\tau}$ for some $\Gamma_0(N)$, and Jacquet-Langlands [Ja-Lgl] answered the corresponding question about general cuspidal representations of GL_2. Weil's result, known as the Weil Converse Theorem, says that if every twist $\sum_{n=1}^\infty a_n\chi(n)/n^s$ by a primitive Dirichlet character χ modulo r with r prime to N extends to an entire function bounded in vertical strips and satisfying a suitable functional equation, then $\sum_{n=1}^\infty a_n e^{2\pi i n\tau}$ is a cusp form for $\Gamma_0(N)$.

The Jacquet-Langlands result [Ja-Lgl, p. 397] below uses the Weil group W_F of the number field F, a notion treated in [Ta3] whose precise definition we do not need. We list the properties of W_F that we shall use:

1) There is a natural continuous homomorphism $W_F \to \mathrm{Gal}(\overline{F}/F)$, and it of course induces a homomorphism $W_F^{\mathrm{ab}} \to \mathrm{Gal}(\overline{F}/F)^{\mathrm{ab}}$.
2) The Artin map $F^\times\backslash\mathbb{A}_F^\times \to \mathrm{Gal}(\overline{F}/F)^{\mathrm{ab}}$ factors as

$$F^\times\backslash\mathbb{A}_F^\times \xrightarrow{\sim} W_F^{\mathrm{ab}} \longrightarrow \mathrm{Gal}(\overline{F}/F)^{\mathrm{ab}}.$$

3) There is a canonical continuous homomorphism $j_v : W_{F_v} \to W_F$ such that the diagram

$$\begin{array}{ccc} W_{F_v} & \xrightarrow{j_v} & W_F \\ \downarrow & & \downarrow \\ \mathrm{Gal}(\overline{F}_v/F_v) & \longrightarrow & \mathrm{Gal}(\overline{F}/F) \end{array}$$

commutes.
4) The isomorphisms $F^\times\backslash\mathbb{A}_F^\times \xrightarrow{\sim} W_F^{\mathrm{ab}}$ of (2) and $F_v^\times \xrightarrow{\sim} W_{F_v}^{\mathrm{ab}}$ of Corollary 2.5 are such that the diagram

$$\begin{array}{ccc} F_v^\times & \longrightarrow & F^\times\backslash\mathbb{A}_F^\times \\ \downarrow & & \downarrow \\ W_{F_v}^{\mathrm{ab}} & \longrightarrow & W_F^{\mathrm{ab}} \end{array}$$

commutes. (This strengthens Theorem 4.5.)

The Jacquet-Langlands Converse Theorem also uses Weil's generalization of Artin L functions in which a representation σ of the Galois group $\mathrm{Gal}(\overline{F}/F)$ is replaced by an admissible homomorphism of the Weil group W_F into some $GL_n(\mathbb{C})$. The local definition is in (8.1), and (3) says that it is consistent with Artin's definition. Further properties of the Weil L functions $L(s,\sigma)$ and $\Lambda(s,\sigma)$ are given in [Ja-Lgl, pp. 393–394]. Like Artin L functions, the Weil L functions are known to have meromorphic continuations and to satisfy a functional equation (5.4). Thus a hypothesis about a functional equation need not be included in the theorem.

To state the theorem we need to explain a notion of **twisting** in this context. Let σ be an n-dimensional representation of $\mathrm{Gal}(\overline{F}/F)$ or, more generally as a consequence of (1) above, of the Weil group W_F of the number field F. Let α be a Grossencharacter.

The twist $\sigma \otimes \alpha$ will be an admissible n-dimensional representation of W_F. To define $\sigma\otimes\alpha$, all we have to do is interpret α as a one-dimensional representation of W_F. This we can do since α is a one-dimensional representation of $F^\times\backslash\mathbb{A}_F^\times$ and this group is isomorphic with W_F^{ab} by (2). Thus $\sigma\otimes\alpha$ is meaningful, and so are the Weil L function $L(s,\sigma\otimes\alpha)$ and its completed version $\Lambda(s,\sigma\otimes\alpha)$. The decomposition $\alpha = \bigotimes \alpha_v$ is unambiguous in the two interpretations as a consequence of (4).

Theorem 8.12 (Jacquet-Langlands Converse Theorem). *Suppose that σ is a two-dimensional representation of W_F, and suppose that $\pi(\sigma_v)$ is the irreducible admissible representation of $GL_2(F_v)$ corresponding to $\sigma \circ j_v$. If, for every unitary Grossencharacter α, the Weil L functions $\Lambda(s,\sigma\otimes\alpha)$ and $\Lambda(s,\sigma^\vee\otimes\omega^{-1})$ are* *entire functions bounded in vertical strips, then* $\bigotimes_v \pi(\sigma_v)$ *is a cuspidal automorphic representation of* $GL_2(\mathbb{A}_F)$.

A corresponding theorem is known for GL_3. See [Pi] and [Co-Pi] for results for GL_n with $n \geq 4$, where additional hypotheses are needed. Converse theorems play a role in the work of Langlands on the Artin Conjecture; see Rogawski's lectures [Ro2]).

Let us return to Theorem 8.8. A proof for $n = 2$ appears in [Lgl5, pp. 23–24]. See also [De-S]. The result for general n appears to be a folk theorem, with no proof appearing in the literature. We are grateful to Dinakar Ramakrishnan for supplying us with the proof that follows and for giving permission that it be reproduced here.

Let σ be an n-dimensional representation of $\text{Gal}(\overline{F}/F)$, and let π be a cuspidal automorphic representation of $GL_n(\mathbb{A}_F)$, with σ and π as in the first version of the Langlands Reciprocity Conjecture. We write σ_v for the restriction of σ to $\text{Gal}(\overline{F}_v/F_v)$. The same representation π of $GL_n(\mathbb{A}_F)$ will occur in the two versions of the conjecture, as it must by Theorem 8.11, and we decompose π as $\pi = \bigotimes \pi_v$.

If τ is any finite-dimensional representation of W_{F_v}, we say that τ is **ramified** if the restriction of τ to the inertia group in (1.4) is nontrivial. If τ is unramified, τ descends to a representation of the abelian group $\mathbb{Z}$ in (1.4). A representation of $\text{Gal}(\overline{F}_v/F_v)$ yields a representation of W_{F_v} by restriction, with the inertia group unchanged, and we may thus speak of ramification for a representation of $\text{Gal}(\overline{F}_v/F_v)$.

We say that our given representation σ of $\text{Gal}(\overline{F}/F)$ is **ramified** at v if σ_v is ramified. The continuity of σ implies that σ descends to $\text{Gal}(K/F)$ for some finite Galois extension K of F. Almost every finite place of F is unramified in K, and each of these places is a place where σ_v is unramified. Therefore σ is unramified at almost every place.

The statement of the conjecture gives us a finite set of places where π is unramified but the L functions of σ and π are not known to agree. We enlarge this set to a finite set S so that S contains all finite places where σ or π is ramified, as well as all infinite places. Since the Langlands and Godement-Jacquet L factors of π agree at all places $v \notin S$, we may write our given equality as

$$L_v(s,\sigma) = L_v^{\text{GJ}}(s,\pi) \qquad \text{for } v \notin S, \tag{8.5}$$

with an Artin L factor on the left side and a Godement-Jacquet L factor on the right side. The idea is to get information by twisting σ and π by suitable unitary Grossencharacters α.

The twist $\sigma \otimes \alpha$ was discussed above. A little explanation is in order for $\pi \otimes \alpha$. The representation π is an irreducible constituent of some $L^2(GL_n(F)\backslash GL_n(\mathbb{A}))_\chi$ generated by cuspidal functions. We can regard α as a representation of the center $Z(\mathbb{A})$ trivial on $Z(F)$, and then $\pi \otimes \alpha$ is an irreducible constituent of $L^2(GL_n(F)\backslash GL_n(\mathbb{A}))_{\chi\alpha}$ generated by cuspidal functions.

If v is a finite place, let $\mathfrak{m}_{F_v}$ be the maximal ideal of $\mathcal{O}_{F_v}$. The **conductor** of a character β of $F_v^\times$ will be defined to be the integer $m \geq 0$ such that β is trivial on $1 + \mathfrak{m}_{F_v}^m$ but not on $1 + \mathfrak{m}_{F_v}^{m-1}$. (For $m = 0$, we understand the condition to be that β is trivial on $\mathcal{O}_{F_v}^\times$; for $m = 1$, we understand the condition to be that β is trivial on $1 + \mathfrak{m}_{F_v}$ but not on $\mathcal{O}_{F_v}^\times$.) By Corollary 2.5 we may regard β as a character of $W_{F_v}^{\text{ab}}$, hence as a one-dimensional representation of W_{F_v}. The condition that $m = 0$ for β is the same as the condition that this one-dimensional representation be trivial on the inertia group in (1.4); thus we may consistently define β to be **unramified** if $m = 0$.

Lemma 1. *Under the assumption that $L_v(s,\sigma) = L_v^{\mathrm{GL}}(s,\pi)$ for $v \notin S$,*

$$L_v(s, \sigma \otimes \alpha) = L_v^{\mathrm{GJ}}(s, \pi \otimes \alpha) \qquad \text{for } v \notin S \tag{8.6}$$

for any Grossencharacter α.

PROOF. First suppose that α is unramified at v. Since σ is unramified at v, (8.1) gives the formulas

$$L_v(s,\sigma) = \det(1 - \sigma_v(\mathrm{Fr})q^{-s})^{-1}, \tag{8.7a}$$

$$L_v(s, \sigma \otimes \alpha) = \det(1 - \sigma_v(\mathrm{Fr})\alpha_v(\varpi_v)q^{-s})^{-1}, \tag{8.7b}$$

where ϖ_v is a member of $\mathcal{O}_{F_v}$ of module q^{-1} and $\sigma_v(\mathrm{Fr})$ is some unitary matrix. If the eigenvalues of $\sigma_v(\mathrm{Fr})$ are $a_1, \ldots, a_n$, then

$$L_v(s,\sigma) = \prod_{j=1}^{n} (1 - a_j q^{-s})^{-1}, \tag{8.8a}$$

$$L_v(s, \sigma \otimes \alpha) = \prod_{j=1}^{n} (1 - a_j \alpha_v(\varpi_v)q^{-s})^{-1}. \tag{8.8b}$$

Since π is unramified at v, the discussion with (8.3) shows that π_v is the Langlands subquotient of some unramified

$$\mathrm{ind}_{TN}^{GL_n(F_v)}((\chi_1, \ldots, \chi_n) \otimes 1) \tag{8.9}$$

and that the associated L factor is the factor for $(\rho_{(\chi_1,\ldots,\chi_n)}, 0)$. Thus

$$L_v^{\mathrm{GJ}}(s,\pi) = \prod_{j=1}^{n} (1 - b_j q^{-s})^{-1}, \tag{8.10}$$

where $b_j = \chi_j(\varpi_v)$. The twist of (8.9) by α_v is the unramified

$$\mathrm{ind}_{TN}^{GL_n(F_v)}((\chi_1\alpha_v, \ldots, \chi_n\alpha_v)) \otimes 1), \tag{8.11}$$

and it follows from [Ja1, (3.4)] that

$$L_v^{\mathrm{GJ}}(s, \pi \otimes \alpha) = \prod_{j=1}^{n} (1 - b_j \alpha_v(\varpi_v)q^{-s})^{-1}. \tag{8.12}$$

The assumed equality of (8.8a) and (8.10) forces every symmetric polynomial in n variables to agree at $(a_1, \ldots, a_n)$ and $(b_1, \ldots, b_n)$, and hence we have an equality of unordered sets $\{a_1, \ldots, a_n\} = \{b_1, \ldots, b_n\}$. Therefore (8.8b) equals (8.12).

Now suppose that α is ramified at v. We shall show that $L_v(s, \sigma \otimes \alpha)$ and $L_v^{\mathrm{GJ}}(s, \pi \otimes \alpha)$ are both 1. In the case of σ, decompose σ_v into the direct sum of irreducibles τ. Since σ is unramified at v, each τ is unramified and thus $\tau \otimes \alpha_v$ is ramified. Let V_τ be the space on which τ acts. The invariant subspaces of V_τ under τ are the same as those under $\tau \otimes \alpha_v$, and the inertial invariants form a proper such subspace. By irreducibility the inertial invariants are 0, and $L_v(s, \tau \otimes \alpha_v) = 1$. Since $L_v(s, \sigma \otimes \alpha)$ is the product of such factors, $L_v(s, \sigma \otimes \alpha) = 1$.

In the case of π, we still have that $L_v^{\mathrm{GJ}}(s,\pi\otimes\alpha)$ is to be computed from (8.12) and that $L_v^{\mathrm{GJ}}(s,\pi\otimes\alpha)=\prod_{j=1}^n L_v^{\mathrm{GJ}}(s,\chi_j\alpha_v)$ by [Ja1, (3.4)]. But this time $\chi_j\alpha_v$ is one-dimensional and ramified. The subspace of invariants is 0, $L_v^{\mathrm{GJ}}(s,\chi_j\alpha_v)$ is 1, and $L_v^{\mathrm{GJ}}(s,\pi\otimes\alpha)=1$.

Lemma 2. *Let T and T' be finite sets of places, with T containing only finite places. Fix integers $m_u\ge 0$ for $u\in T$. Then there exists a Grossencharacter α such that*

(i) *$\alpha_v=1$ for all v in T' and*
(ii) *for each u in T the conductor of α_u is $\ge m_u$.*

REFERENCE. This is a special case of [Ar-Ta, Theorem 5, p. 103].

From the end of §7, we know that the cuspidal automorphic representation π is unitary. The results of Flath [Fl] therefore imply that every local component π_v is unitary. Although π_v is not known to be tempered (see [Moe1] and [Moe2] for "tempered"), π_v does satisfy another property—it is "generic" in a sense to be defined below.

Let N be the algebraic group of matrices $x=(x_{ij})$ with x_{ij} equal to 0 for $i>j$, equal to 1 for $i=j$, and unrestricted for $i<j$. Fix a place v and a nontrivial unitary character ψ_v of the additive group F_v, and consider a unitary character θ_v of $N(F_v)$ of the form

$$\theta_v(x)=\psi(c_1x_{12}+\cdots+c_{n-1}x_{n-1,n}).$$

We say that θ_v is **nondegenerate** if $c_1\cdots c_{n-1}\ne 0$. Let $\Pi^{(v)}$ be an irreducible unitary admissible representative of $GL_n(F_v)$, and let $V^{(v)}$ be the space of its underlying smooth representation (the space of C^∞ vectors if v is infinite, the space of vectors fixed by some idempotent of $\mathcal{H}_v$ if v is finite). The representation $\Pi^{(v)}$ is said to be **generic** if there exists a nonzero continuous linear functional λ on $V^{(v)}$ and some nondegenerate θ_v as above such that

$$\lambda(\Pi^{(v)}(ng)x)=\theta_v(n)\lambda(\Pi^{(v)}(g)x)\qquad\text{for all } n\in N(F_v),\ g\in GL_n(F_v),\ x\in V^{(v)}.$$

Lemma 3. *If $\Pi=\bigotimes_v\Pi_v$ is a cuspidal automorphic representation of $GL_n(\mathbb{A}_F)$, then each Π_v is a unitary generic representation of $GL_n(F_v)$.*

REFERENCE. [Sha, Corollary, p. 190].

PROOF OF THEOREM 8.8.

Step 1. If v is any finite place, then there exists an integer $m=m(v,\sigma)\ge 0$ such that $L_v(s,\sigma_v\otimes\beta)=1$ for every unitary character β of $F_v^\times$ of conductor $\ge m$.

In fact, since the L factor of a direct sum of representations is the product of the L factors of the summands, it is enough to prove that the local factor $L_v(s,\tau\otimes\beta)$ is 1 for an irreducible τ when m is large enough. Fix τ.

If there exists β_0 with $\tau\otimes\beta_0$ unramified (at v), then $\tau\otimes\beta_0$ factors through the abelian group $\mathbb{Z}$ in (1.4), and τ must be one-dimensional. So we can regard τ as a character of $F_v^\times$, say with conductor m_0. If β has conductor $>m_0$, then $\tau\otimes\beta$ will be ramified, and $L_v(s,\tau\otimes\beta)$ will be 1.

Otherwise $\tau\otimes\beta$ is ramified (at v) for every unitary character β of $F_v^\times$. Let τ act on V_τ. The invariant subspaces of V_τ under τ are the same as under $\tau\otimes\beta$, and the inertial invariants form a proper such subspace. By irreducibility the inertial invariants are 0, and therefore $L_v(s,\tau\otimes\beta)=1$.

Step 2. If v is any finite place, then there exists an integer $m'=m'(v,\pi)\geq 0$ such that $L_v^{\mathrm{GJ}}(s,\pi_v\otimes\beta)=1$ for every unitary character β of $F_v^\times$ of conductor $\geq m'$.

In fact, the Langlands classification ([Moe2]) shows that π_v is the Langlands quotient of a representation induced from a parabolic subgroup with the product of a unitary discrete series and a quasicharacter on the Levi factor. The representation on the Levi factor is just the tensor product of similar representations on the component subgroups $GL_{n_i}(F_v)$, and [Ja1, (3.4)] shows that the L factor of π_v is the product of the L factors of these representations on the component subgroups $GL_{n_i}(F_v)$. Thus it is enough to handle a representation of $GL_k(F_v)$ that is the product of a unitary discrete series and a quasicharacter. Work of Zelevinsky discussed in [Moe2] shows that such a representation η is the unique irreducible submodule of an induced representation of the form

$$\operatorname{ind}_P^{GL(k,F_v)}\big(((\mu\otimes\gamma)\otimes(|\cdot|^{\frac{k-1}{2}},\dots,|\cdot|^{-\frac{k-1}{2}}))\otimes 1\big).$$

Here $P=MU$ is the standard parabolic subgroup associated to the partition $(d,d,\dots,d)$ for some divisor d of n, M is the product of copies of $GL(d,F_v)$, and the representation $\mu\otimes\gamma$ of M is the product of a (unitary) supercuspidal μ and a quasicharacter γ. By Theorem 8.2 of [Ja-P-S],

$$L_v^{\mathrm{GJ}}(s,\eta\otimes\beta)=L_v^{\mathrm{GJ}}(s,\mu\otimes(\gamma\beta|\cdot|^{\frac{k}{d}-1}))$$

for any quasicharacter β of $F_v^\times$. If $d>1$, then this L factor is 1 for any β by [Ja1, (1.3.5)]. Thus we may assume that $d=1$, in which case μ is just a unitary character of $F_v^\times$ and the argument is finished as in the case of the Galois representation σ.

Step 3.

$$L_\infty(s,\sigma)L_\infty^{\mathrm{GJ}}(1-s,\pi^\vee)=\varepsilon(s)L_\infty^{\mathrm{GJ}}(s,\pi)L_\infty(1-s,\sigma^\vee)\tag{8.13}$$

for an entire nonvanishing function $\varepsilon(s)$.

In fact, let T be the subset of finite places in S, and let T' be the subset of infinite places. For each u in T, let m_u denote the maximum of the numbers $m(u,\sigma)$, $m(u,\sigma^\vee)$, $m'(u,\pi)$, and $m'(u,\pi^\vee)$ given by Steps 1 and 2. Let α be a Grossencharacter chosen by Lemma 2 for the data T, T', and $\{m_u\mid u\in T\}$. Then we have

$$\begin{aligned}L_u(s,\sigma\otimes\alpha)=L_u^{\mathrm{GJ}}(s,\pi\otimes\alpha)&=1 &&\text{for } u\in T\\ L_v(s,\sigma\otimes\alpha)&=L_v(s,\sigma) &&\text{for } v\in T'\\ L_v^{\mathrm{GJ}}(s,\pi\otimes\alpha)&=L_v^{\mathrm{GJ}}(s,\pi) &&\text{for } v\in T'.\end{aligned}\tag{8.14}$$

Similar formulas are valid for $\sigma^\vee$ and $\pi^\vee$. The global functional equations of $\Lambda(s,\sigma\otimes\alpha)$ and $\Lambda^{\mathrm{GJ}}(s,\pi\otimes\alpha)$ given in (5.4) (as generalized by Weil) and in Theorem 8.7 imply that

$$\Lambda(s,\sigma\otimes\alpha)\Lambda^{\mathrm{GJ}}(1-s,\pi^\vee\otimes\alpha^\vee)=\varepsilon(s)\Lambda^{\mathrm{GJ}}(s,\pi\otimes\alpha)\Lambda(1-s,\sigma^\vee\otimes\alpha^\vee),\tag{8.15}$$

where $\varepsilon(s)$ is entire and nonvanishing. Let L^S or $L^{\mathrm{GJ}S}$ denote a product of factors corresponding to the places not in S. Substituting from (8.14) into (8.15), we obtain

$$\begin{aligned}&L_\infty(s,\sigma)L^S(s,\sigma\otimes\alpha)L_\infty^{\mathrm{GJ}}(1-s,\pi^\vee)L^{\mathrm{GJ}S}(1-s,\pi^\vee\otimes\alpha^\vee)\\ &\qquad=\varepsilon(s)L_\infty^{\mathrm{GJ}}(s,\pi)L^{\mathrm{GJ}S}(s,\pi\otimes\alpha)L_\infty(1-s,\sigma^\vee)L^S(1-s,\sigma^\vee\otimes\alpha^\vee).\end{aligned}$$

Use of Lemma 1 allows us to cancel all the factors L^S and $L^{\mathrm{GJ}S}$ in this formula, and we arrive at (8.13).

Step 4.

$$L_\infty(s,\sigma) = L_\infty^{\mathrm{GJ}}(s,\pi) \qquad \text{and} \qquad L_\infty(1-s,\sigma^\vee) = L_\infty^{\mathrm{GJ}}(1-s,\pi^\vee).$$

In fact, inspection of the formulas in [Kna2, (3.6) and (4.6)] shows for each infinite place w that $L_w(s,\sigma_w)$ is nowhere vanishing and has no poles for $\operatorname{Re} s > 0$. Similarly $L_w(1-s,\sigma_w^\vee)$ is nowhere vanishing and has no poles for $\operatorname{Re} s < 0$.

We expect corresponding properties for π, but we get less. By [Ja1, (5.1)], each $L_w^{\mathrm{GJ}}(s,\pi_w)$ is an Artin L factor and hence is nowhere vanishing. The conclusions about poles are more difficult to prove: $L_w(s,\pi_w)$ has no poles for $\operatorname{Re} s \geq \frac{1}{2}$, and $L_w(1-s,\pi_w^\vee)$ has no poles for $\operatorname{Re} s \leq \frac{1}{2}$. This result is stated as [Ba-R, Proposition 2.1]. Its proof combines the fact that π_w is unitary and generic (as follows from Lemma 3) with a classication of (irreducible) unitary generic representations (which follows readily from Vogan's classification of irreducible unitary representations. (See [Moe2] for a qualitative discussion of Vogan's classification in this volume.)

If we rewrite (8.13) as

$$L_\infty(s,\sigma)/L_\infty^{\mathrm{GJ}}(s,\pi) = \varepsilon(s)L_\infty^{\mathrm{GJ}}(1-s,\sigma^\vee)/L_\infty^{\mathrm{GJ}}(1-s,\pi^\vee),$$

then we see that the left side has no poles or zeros for $\operatorname{Re} s \geq \frac{1}{2}$ and the right side has no poles or zeros for $\operatorname{Re} s \leq \frac{1}{2}$. Therefore both sides are entire and nonvanishing.

To complete Step 4, it is enough to show that a product of elementary L factors for archimedean places is determined by its poles. Referring to [Kna2, (3.6) and (4.6)], we see that each such elementary L factor is of the form $\pi^{-\frac{s}{2}-t}\Gamma\big(\frac{s}{2}+t\big)$ or $2(2\pi)^{-s-u}\Gamma(s+u)$ for some t or u. If we put $\Gamma_{\mathbb{R}}(s) = \pi^{-\frac{s}{2}}\Gamma\big(\frac{s}{2}\big)$, then we conclude from the well known formula $\Gamma\big(\frac{z}{2}\big)\Gamma\big(\frac{z}{2}+\frac{1}{2}\big) = \pi^{1/2}2^{-z+1}\Gamma(z)$ that

$$2(2\pi)^{-s-u}\Gamma(s+u) = \Gamma_{\mathbb{R}}(s+u)\Gamma_{\mathbb{R}}(s+u+1).$$

Hence every product of elementary L factors for archimedean places is of the form $\prod_{j=1}^r \Gamma_{\mathbb{R}}(s+t_j)$ for suitable t_j's. No factor has a zero, and the pole of $\Gamma_{\mathbb{R}}(s+t_j)$ with $\operatorname{Re} s$ largest occurs at $s=-t_j$ and has residue 2. Thus we can decompose the product functions one factor at a time by finding the pole with $\operatorname{Re} s$ as large as possible and dividing off a corresponding $\Gamma_{\mathbb{R}}(s+t_j)$.

Step 5. If v is any finite place in S, then $L_v(s,\sigma) = L_v^{\mathrm{GJ}}(s,\pi)$.

In fact, let T be the set of finite places in S other than v, and let T' be the union of $\{v\}$ and the set of infinite places. For each u in T, let m_u denote the maximum of the numbers $m(u,\sigma)$, $m(u,\sigma^\vee)$, $m'(u,\pi)$, and $m'(u,\pi^\vee)$ given by Steps 1 and 2. Let α be a Grossencharacter chosen by Lemma 2 for the data T, T', and $\{m_u \mid u \in T\}$. Arguing as in Step 3 and taking into account the result of Step 4, we are led to an equation

$$L_v(s,\sigma)/L_v^{\mathrm{GJ}}(s,\pi) = \varepsilon_1(s)L_v^{\mathrm{GJ}}(1-s,\pi^\vee)/L_v^{\mathrm{GJ}}(1-s,\sigma^\vee). \tag{8.16}$$

where $\varepsilon_1(s)$ is an entire nonvanishing function.

Using the definition of $L_v(s,\sigma_v)$ in (5.1b) and taking into account that σ is unitary, we see that $L_v(s,\sigma_v)$ is a product of factors $(1-cq_v^{-s})^{-1}$ with $|c|=1$. Hence $L_v(s,\sigma_v)$ is nowhere vanishing and has no poles for $\operatorname{Re} s > 0$. Similarly $L_v(1-s,\sigma^\vee)$ is nowhere vanishing and has no poles for $\operatorname{Re} s < 1$.

Again we expect corresponding formulas for π, but matters are not so simple. The conclusions are that $L_v^{\mathrm{GJ}}(s,\pi)$ has no poles for $\operatorname{Re} s \geq \frac{1}{2}$ and $L_v^{\mathrm{GJ}}(1-s,\pi^\vee)$ has no poles for $\operatorname{Re} s \leq \frac{1}{2}$, and again the result is stated as [Ba-R, Proposition 2.1]. Its proof uses Lemma 3 and a classification of the irreducible unitary generic representations. (A somewhat different proof may be found in [Ja-P-S].)

Arguing as in Step 4, we write (8.16) as

$$L_v(s,\sigma)/L_v^{\mathrm{GJ}}(s,\pi) = \varepsilon_1(s) L_v^{\mathrm{GJ}}(1-s,\sigma^\vee)/L_v^{\mathrm{GJ}}(1-s,\pi^\vee).$$

The restrictions on zeros and poles imply that each side is entire and nonvanishing. The left side is of the form $(1+P(q_v^{-s}))/(1+Q(q_v^{-s}))$, where P and Q are polynomials without constant term. For this expression to have neither poles nor zeros, we must have $P=Q$. Therefore $L_v(s,\sigma) = L_v^{\mathrm{GJ}}(s,\pi)$. This completes Step 5 and the proof of Theorem 8.8.

9. L Groups and General Langlands L Functions

With GL_n in place as a model, we can now describe what the Langlands theory proposes for an arbitrary linear reductive group G over a number field F. It will be assumed throughout that $G(\mathbb{C})$ is connected.

The material in this section is largely due to Langlands [Lgl2] and [Lgl3], and the presentation amounts to a summary of the exposition [Bo4]. The reader may wish to consult [Bo4] for a more precise and detailed account and [Bl-Ro] for a discussion that includes a number of examples.

The theory introduces a group ${}^LG$ that is the semidirect product of a certain complex reductive group $\widehat{G}$ and the Galois group $\operatorname{Gal}(\overline{F}/F)$, with $\widehat{G}$ normal.[5]

When $G = GL_n$, $\widehat{G}$ is $GL_n(\mathbb{C})$, and the semidirect product is a direct product; the definitions will show that the Galois group can often be ignored in this case, and we are reduced to the situation in §8.

The same construction as for the L group of $G(F)$ yields, for each place v of F, an L group for $G(F_v)$. The complex group $\widehat{G}$ is unchanged, and the Galois group $\operatorname{Gal}(\overline{F}/F)$ is cut down to the decomposition subgroup $\operatorname{Gal}(\overline{F}_v/F_v)$.

Thinking in terms of F_v, let us consider the case of an arbitrary local field k of characteristic 0 containing F. Then we can form an L group for $G(k)$. An "admissible homomorphism" is a certain kind of homomorphism φ (to be described below) of the Weil group W_k or the Weil-Deligne group W_k', according as k is archimedean or nonarchimedean, into the L group of $G(k)$ that "covers the identity mapping on $\operatorname{Gal}(\overline{k}/k)$." Here "$\varphi$ covers the identity mapping on $\operatorname{Gal}(\overline{k}/k)$" means that when φ is followed by the map of the L group to $\operatorname{Gal}(\overline{k}/k)$, the result is the usual map of W_k or W_k' into $\operatorname{Gal}(\overline{k}/k)$. Two admissible homomorphisms are equivalent if they are conjugate via $\widehat{G}$, and the set of equivalence classes is denoted $\Phi(G(k))$.

We associate an elementary L factor $L(s,\varphi,r)$ to this situation whenever r is a representation of ${}^LG$ into some $GL_n(\mathbb{C})$ that is holomorphic in the $\widehat{G}$ variable. The

[5]To handle more advanced topics such as endoscopy, some adjustment in the definition of ${}^LG$ is needed. The traditional adjustment is to replace the Galois group by a Weil group. Another possibility, discussed in [Ad-Ba-V], is to use an extension of $\widehat{G}$ by the Galois group that is not necessarily a semidirect product.

definition is simply

$$L(s, \varphi, r) = L(s, r \circ \varphi), \tag{9.1}$$

where the right side is given by (8.1). When G is GL_n and r is the standard representation, this definition reduces to the definition (8.1) for the group GL_n.

As with GL_n, we let $\Pi(G(k))$ be the set of equivalence classes of irreducible admissible representations of $G(k)$.

Local Langlands Conjecture. *$\Pi(G(k))$ is partitioned in a natural way into finite nonempty subsets Π_φ indexed by all $\varphi \in \Phi(G(k))$.*

As in the case of GL_n, the correspondence is to be consistent with a number of conditions. See [Bo4, p. 43] for details. Ideally the correspondence should be consistent with functoriality, which we shall define in §10. Consistency with functoriality implies that the sets Π_φ are not necessarily singleton sets, unlike the case of GL_n. (See §10 below.) A set Π_φ is called an L **packet**, and members of the same Π_φ are said to be L **indistinguishable**.

The conjecture is known to be true if G is a torus ([Lgl1], summarized in [Bo4, p. 41]) and if k is archimedean ([Lgl3], summarized in [Bo4, p. 46]). Cases with $G = GL_n$ for which it is true were discussed in §8. Also we shall insist that representations with a nonzero $G(\mathcal{O}_k)$ fixed vector (i.e., the Langlands subquotients of the unramified principal series) be parametrized by φ's in a particular way; this parametrization we shall discuss below.

The Local Langlands Conjecture allows us to define an elementary L factor whenever π is an irreducible admissible representation of $G(k)$ and r is a finite-dimensional holomorphic representation of ${}^LG$. The definition is simply

$$L(s, \pi, r) = L(s, \varphi, r) \qquad \text{if } \varphi \leftrightarrow \pi, \tag{9.2}$$

with $L(s, \varphi, r)$ as in (9.1). When G is GL_n and r is the standard representation, this definition reduces to the definition (8.2) for GL_n.

Before continuing, let us fill in some details in the above discussion.

We begin with a rough description of the L group ${}^LG$. The group $\widehat{G}$ is a connected complex reductive group, having the same dimension as G and having root system equal to the system of coroots $2\alpha/\langle\alpha, \alpha\rangle$ of G. The question of the exact size of the center is somewhat involved and will be described in a moment. For our purposes, it will be enough to know that

(i) if $G = GL_n$, then $\widehat{G} = GL_n(\mathbb{C})$
(ii) if $G(\mathbb{C})$ is simply connected, then $\widehat{G}$ is an adjoint group
(iii) if $G(\mathbb{C})$ is an adjoint group, then $\widehat{G}$ is simply connected.

Here are some examples:

G	$\widehat{G}$
SL_n	$PGL_n(\mathbb{C})$
PGL_n	$SL_n(\mathbb{C})$
Sp_{2n}	$SO_{2n+1}(\mathbb{C})$
SO_{2n+1}	$Sp_{2n}(\mathbb{C})$

Langlands [Lgl3] captured the exact size of the center of $\widehat{G}$ by working with weight lattices, but the treatment in [B4] makes matters axiomatic with the notion of root datum, which is described in [Sp]. A **root datum** is a 4-tuple $(X, \Delta, X^\vee, \Delta^\vee)$, where

(i) X and $X^\vee$ are free abelian groups of finite rank in duality by a pairing $X \times X^\vee \to \mathbb{Z}$ denoted $\langle \cdot , \cdot \rangle$,

(ii) Δ and $\Delta^\vee$ are reduced root systems lying in subspaces of $X \subset X \otimes \mathbb{R}$ and $X^\vee \subset X^\vee \otimes \mathbb{R}$, respectively,

(iii) Δ and $\Delta^\vee$ are in bijection by a map $\alpha \mapsto \alpha^\vee$ such that $\langle \alpha, \alpha^\vee \rangle = 2$ for all $\alpha \in \Delta$,

(iv) Δ is preserved by the maps $s_\alpha : X \to X$ given by $s_\alpha(x) = x - \langle x, \alpha^\vee \rangle \alpha$, and

(v) $\Delta^\vee$ is preserved by the maps $s_{\alpha^\vee} : X^\vee \to X^\vee$ given by $s_{\alpha^\vee}(x) = x - \langle x, \alpha \rangle \alpha^\vee$.

To our reductive group G and a maximal torus T is associated a root datum $\psi(G,T) = (X, \Delta, X^\vee, \Delta^\vee)$ as follows: X is the group of rational characters $X^*(T)$ (i.e., algebraic homomorphisms $T \to GL_1$), $X^\vee$ is the group $X_*(T)$ of one-parameter subgroups (i.e., algebraic homomorphisms $GL_1 \to T$), the form $\langle x, u \rangle$ for $x \in X^*(T)$ and $u \in X_*(T)$ is given by $x(u(t)) = t^{\langle x,u \rangle}$ for all $t \in \overline{F}^\times$, Δ is the root system of (G,T), and $\Delta^\vee$ is the system of coroots described as a subset of $X_*(T)$ in [Sp, pp. 6–7]. The theorem below says that all root data arise in this way.

Theorem 9.1. *For any root datum Ψ, there exists a connected reductive group G and a maximal torus T in G such that $\Psi = \psi(G,T)$. The pair (G,T) is unique up to isomorphism over $\overline{F}$.*

REFERENCE. [Sp, p. 9].

If $\Psi = (X, \Delta, X^\vee, \Delta^\vee)$ is a root datum, then $\Psi^\vee = (X^\vee, \Delta^\vee, X, \Delta)$ is easily seen to be a root datum. Starting from (G,T), we assemble the root datum $\psi(G,T)$ and apply Theorem 9.1 to $\psi(G,T)^\vee$. The group $\widehat{G}$ is defined to be the $\mathbb{C}$ points of the reductive group produced by the theorem. Also we define $\widehat{T}$ to be the $\mathbb{C}$ points of the maximal torus produced by the theorem.

All this discussion really involved only groups defined over $\overline{F}$. We bring in F through an action of $\mathrm{Gal}(\overline{F}/F)$. Fix a positive system Δ^+. If γ is in $\mathrm{Gal}(\overline{F}/F)$, then γ carries $T(\overline{F})$ to $T(\overline{F})$ and Δ to Δ, and there exists a member g_γ of $G(\overline{F})$ that normalizes $T(\overline{F})$ such that $g_\gamma \gamma$ carries Δ^+ to itself. Any two such elements g_γ are in the same coset relative to $T(\overline{F})$, and thus we can associate to γ a unique permutation of Δ^+. This composite element also acts as an automorphism of $X^*(T)$ and $X_*(T)$. As γ varies, we obtain compatible group actions of $\mathrm{Gal}(\overline{F}/F)$ on Δ^+, $X^*(T)$, and $X_*(T)$. Thus we obtain a group action of $\mathrm{Gal}(\overline{F}/F)$ on the dual root datum $(X_*(T), \Delta^\vee, X^*(T), \Delta)$ preserving positive roots. This action lifts uniquely to an action of $\mathrm{Gal}(\overline{F}/F)$ on $\widehat{G}$ once we choose root vectors for the simple roots. Taking all the choices into account, we find that this action is canonical up to inner automorphism by a member of $\widehat{T}$. In any event, ${}^L G$ is defined to be the semidirect product of $\mathrm{Gal}(\overline{F}/F)$ and $\widehat{G}$. Henceforth we shall usually discuss only groups G that are split over F; then the action is trivial, and ${}^L G$ is the direct product and $\widehat{G}$ and the Galois group.

Let us now return to $G(k)$ with k a local field of characteristic 0 that contains F. We shall define admissible homomorphisms more precisely, but only in the case that

$G(k)$ is quasisplit over k. An element of ${}^LG$ is said to be semisimple or unipotent according as its $\widehat{G}$ component is semisimple or unipotent. If k is nonarchimedean, an **admissible homomorphism** is a continuous homomorphism φ of the Weil-Deligne group W_k' into ${}^LG$ with the following properties:

(i) φ covers the identity mapping of $\mathrm{Gal}(\overline{k}/k)$,
(ii) φ is holomorphic in the $\mathbb{C}$ variable of W_k', and every member of $\varphi(\mathbb{C})$ is unipotent, and
(iii) every member of $\varphi(W_k)$ is semisimple.

(Another condition, involving "relevance" of parabolic subgroups, is needed if $G(k)$ is not quasisplit. See [Bo4, p. 40].) If k is archimedean, admissible homomorphisms are defined in the same way except that W_k replaces W_k' and condition (ii) is dropped.

The definitions are now complete in the case of $G(k)$, and the Local Langlands Conjecture allows us to associate to each irreducible admissible representation π of $G(k)$ and holomorphic representation r of ${}^LG$ an elementary L factor $L(s, \pi, r)$.

It is known what these L factors should be in the case of unramified principal series. To keep matters simple, let us suppose that G is split over the nonarchimedean field k. Let Q be the compact subgroup $Q = G(\mathcal{O}_k)$. We shall assume that Q is a "hyperspecial maximal compact subgroup" in the sense of the exposition [Ti, pp. 35, 55]. Examples where this condition is satisfied are where $G(\mathcal{O}_k)$ is $GL_n(\mathcal{O}_k)$ in $GL_n(k)$, $Sp_n(\mathcal{O}_k)$ in $Sp_n(k)$, or $SO_n(\mathcal{O}_k)$ in $SO(k)$; here SO_n is the group of matrices of determinant one preserving a quadratic form

$$x_1 x_{p+1} + \cdots x_p x_{2p} + x_{2p+1}^2$$

in which $p = \left[\frac{n}{2}\right]$ and the last term is absent if n is even. See [Car, p. 140] and [Bo4, pp. 38–39, 45] for further discussion of the terms "special" and "hyperspecial."

Suppose that T is a k split maximal torus chosen so that $T(k) \cap Q$ is maximal compact in $T(k)$. Let $\mathcal{H}(G(k), Q)$ be the subalgebra of bi-Q-invariant functions in the Hecke algebra of $G(k)$. This subalgebra is known to be commutative, and the characteristic function of Q is the identity. Choose a Borel subalgebra $B = TN$ containing T, normalize Haar measure dm on $N(k)$ so that $N \cap Q$ gets total measure 1, and let $\Delta(t)$ be the positive function on $T(k)$ defined by $\Delta(t) = d(tnt^{-1})/dn$. For $f \in \mathcal{H}(G(k), Q)$, define

$$Sf(t) = \Delta(t)^{1/2} \int_{N(k)} f(tn)\, dn \qquad \text{for } t \in T(k).$$

Theorem 9.2. *With the above assumptions and notation, the mapping S is an algebra isomorphism of $\mathcal{H}(G(k), Q)$ onto the subalgebra $\mathcal{H}(T(k), T(k) \cap Q)^W$ of Weyl-group invariants in $\mathcal{H}(T(k), T(k) \cap Q)$.*

REFERENCE. This theorem is due to Satake, and the mapping S is called the **Satake isomorphism**. See [Car, p. 147].

Now let π be an irreducible admissible representation with a nonzero vector ϕ fixed by Q. The space of all vectors fixed by Q is one-dimensional, and it follows that $\pi(\mathcal{H}(G(k), Q))$ operates on ϕ by scalars. The result is an algebra homomorphism $\lambda = \lambda(\pi)$ of $\mathcal{H}(G(k), Q)$ into $\mathbb{C}$. Sorting out the notation with the aid of [Car, p. 134], we see that the algebra homomorphisms of $\mathcal{H}(T(k), T(k) \cap Q)$ into $\mathbb{C}$ are simply the members of $\widehat{T}$, and it follows that we may identify an algebra homomorphism of $\mathcal{H}(T(k), T(k) \cap Q)^W$ into $\mathbb{C}$ with a W orbit in $\widehat{T}$. Thus Theorem 9.2 implies that we may regard $\lambda(\pi)$ as a W orbit in $\widehat{T}$.

We may understand this construction in other terms as follows. The Hecke algebra $\mathcal{H}(T(k), T(k) \cap Q)$ is just the $\mathbb{C}$ group ring of $\widehat{T}$, which we write as $\mathbb{C}[\widehat{T}]$. The subspace $\mathbb{C}[\widehat{T}]^W$ of Weyl-group invariants is the coordinate ring of the affine variety $\widehat{T}/W$, and then it becomes clear that a homomorphism of this algebra into $\mathbb{C}$ is simply a W orbit in $\widehat{T}$. If π, as above, is an irreducible admissible representation of $G(k)$ with a nonzero vector fixed by Q, then the point it defines in $\widehat{T}/W$ is called the **Langlands class** of π. For $G = GL_n$, this point can be represented by an n-tuple $(\varepsilon_1, \ldots, \varepsilon_n)$ up to permutation, and the local L factor that we used in §8 amounts to $L(s, \pi) = \prod_{i=1}^n (1 - \varepsilon_i q^{-s})^{-1}$ in these terms.

In the general case that we are studying, the elementary L factor associated to π and a holomorphic representation r of ${}^L G$ is taken to be

$$L(s, \pi, r) = \det(1 - r(\lambda(\pi))q^{-s})^{-1}. \tag{9.3}$$

(See [Bo4, p. 39].) In (9.3) any representative of the W orbit $\lambda(\pi)$ in $\widehat{T}$ is to be used, and the determinant is independent of the choice. To arrange for this L factor to come from an admissible homomorphism φ, choose Fr in W_k with $\|\mathrm{Fr}\| = q^{-1}$, and fix a representative of $\lambda(\pi)$ in $\widehat{T}$. The idea is to make $\varphi(\mathrm{Fr}) = (\lambda(\pi), \mathrm{Fr})$ and to make φ behave in an "unramified" fashion. To do so, define $\varepsilon : W_k' \to \mathbb{Z}$ by

$$\|\mathrm{Fr}\|^{\varepsilon(z,w)} = \|w\| \qquad \text{for } z \in \mathbb{C} \text{ and } w \in W_k,$$

and let

$$\varphi(z, w) = (\lambda(\pi), \mathrm{Fr})^{\varepsilon(z,w)}.$$

With this φ associated to π, the definition (9.2) attaches (9.3) to π and r as elementary L factor.

When $G(k)$ is not k split, the treatment of unramified principal series involves considerably more structure theory, and the Satake isomorphism is more complicated to state. For an exposition of the structure theory, see [Ti] and [Car]. For the definition of the elementary L factor associated to an irreducible admissible representation with a nonzero fixed vector under $G(\mathcal{O}_k)$, see [Bo4, pp. 39 and 44–45].

Let us return to the number field F. As was true for GL_n, the results of [Fl] show that an irreducible admissible representation π of $G(\mathbb{A}_F)$ is a restricted tensor product $\pi = \bigotimes_v \pi_v$, and moreover almost every π_v has a nonzero vector fixed under $G(\mathcal{O}_v)$. Let r be a holomorphic representation of the L group of $G(F)$. Since the L group of $G(F_v)$ may be taken to be a subgroup of $G(F)$, we obtain by restriction a holomorphic representation r of the L group of each $G(F_v)$. For the irreducible admissible representation $\pi = \bigotimes_v \pi_v$ of $G(\mathbb{A}_F)$, we can then define

$$\begin{aligned} L(s, \pi, r) &= \prod_{\text{finite } v} L(s, \pi_v, r) \\ \Lambda(s, \pi, r) &= \prod_{\text{all } v} L(s, \pi_v, r), \end{aligned} \tag{9.4}$$

with $L(s, \pi_v, r)$ as in (9.2) (and almost always as in (9.3)).

Theorem 9.3. *If π is an automorphic representation of $G(\mathbb{A}_F)$ and r is a holomorphic representation of ${}^LG$, then $L(s, \pi, r)$ converges absolutely for* $\operatorname{Re} s$ *sufficiently large.*

REFERENCE. This theorem is due to Langlands. The line of argument was given in Corollary 8.2 through Corollary 8.6.

The global analytic properties of $\Lambda(s, \pi, r)$, if any, are known only in special cases. It is expected that these functions are meromorphic and satisfy a functional equation. For more information, see [Bo4].

10. Functoriality

Functoriality refers to translating knowledge of a holomorphic homomorphism of L groups into results about automorphic representations. Much of [Lgl2] raises specific questions about this problem, and [Bo4, §§ 15–17] discusses progress as of the late 1970s. More recent progress is the subject of [Lgl7].

A homomorphism $\omega : G \to H$ between reductive Lie groups over a local or global field induces a holomorphic homomorphism $\psi : {}^LH \to {}^LG$ if the image of ω is normal [Bo4, p. 29], and moreover ψ covers the identity mapping of the Galois group. For example, the inclusion ω of $SL_2(\mathbb{R})$ into $GL_2(\mathbb{R})$ induces the natural quotient map

$$\psi : GL_2(\mathbb{C}) \times \Gamma \to PGL_2(\mathbb{C}) \times \Gamma$$

of L groups, where $\Gamma = \operatorname{Gal}(\mathbb{C}/\mathbb{R})$. An admissible homomorphism

$$\varphi : W_{\mathbb{R}} \to GL_2(\mathbb{C}) \times \Gamma$$

induces by composition the admissible homomorphism

$$\psi \circ \varphi : W_{\mathbb{R}} \to PGL_2(\mathbb{C}).$$

The resulting map $\Phi(GL_2(\mathbb{R})) \to \Phi(SL_2(\mathbb{R}))$ gives us a correspondence (not a function!) $\Pi(GL_2(\mathbb{R})) \to \Pi(SL_2(\mathbb{R}))$ since the Local Langlands Conjecture is a theorem in the archimedean case. Examining matters, we see that we associate to each irreducible admissible representation of $GL_2(\mathbb{R})$ all of its constituents on restriction to $SL_2(\mathbb{R})$. Nothing very deep is happening here, but the fact that each discrete series of $GL_2(\mathbb{R})$ and certain principal series decompose into two inequivalent pieces on restriction to $SL_2(\mathbb{R})$ forces some members of $\Pi(SL_2(\mathbb{R}))$ to have more than one element.

Let us call a holomorphic homomorphism $\psi : {}^LH \to {}^LG$ covering the identity of the Galois group an L **homomorphism**. Not every L homomorphism arises from a homomorphism $G \to H$; in fact, most do not. We can still ask

(a) in the local case whether such a ψ induces a correspondence $\Pi(H) \to \Pi(G)$ and
(b) in the global case whether such a ψ induces a correspondence of automorphic representations to automorphic representations (or cuspidal representations to cuspidal representations).

In each case we ask that the correspondence respect L functions, or as much of

L functions as is known. (Also we ask that the correspondence respect ε factors, which we have largely ignored in this article.)

To fix the ideas, let us continue with the notation of §9. Thus let F be a number field, and let G be a reductive group over F. We denote by k any local field containing F, so that $G(k)$ is defined.

First let us consider the local case. If we accept the Local Langlands Conjecture, then (a) is solved, by the same argument as in the example above: The L homomorphism $\psi : {}^LH \to {}^LG$ induces a map $\psi_* : \Phi(H) \to \Phi(G)$ (at least if G is quasisplit, so that the all the conditions on a member of $\Phi(G)$ were given in §9). In turn, ψ_* induces a correspondence $\Pi(H) \to \Pi(G)$, and the correspondence respects L factors.

From an organizational point of view, we could insist that the Local Langlands Conjecture is to be proved first (or else taken as a working hypothesis), and then functoriality is to be addressed. But [Bo4] points out some early cases in which a partial result about local functoriality was established and then used to obtain a partial result about the Local Langlands Conjecture. If local functoriality and the Local Langlands Conjecture ultimately turn out to be true, the proofs may therefore have to start from some basic information (the φ's for the unramified principal series, the L and ε factors for GL_n and the standard representation, and some other conditions) and establish local functoriality and the Local Langlands Conjecture together.

Nevertheless, to keep matters brief, we shall take local functoriality as a working hypothesis and consider the global case (b). Here is one possible statement of the problem.

Question (Global functoriality). *Let G and H be reductive groups over F with G quasisplit, and let $\psi : {}^LH \to {}^LG$ be an L homomorphism. For each place v of F, let ψ_v be the restriction of ψ to a map between the L groups of $H(F_v)$ and $G(F_v)$, let $(\psi_v)_* : \Phi(H(F_v)) \to \Phi(G(F_v))$ be the induced map on admissible homomorphisms, and let $(\psi_v)_*$ also denote the correspondence $\Pi(H(F_v)) \to \Pi(G(F_v))$ obtained from the Local Langlands Conjecture. Let $\pi = \bigotimes_v \pi_v$ be an automorphic representation of $H(\mathbb{A}_F)$.*

(i) *Does there exists a choice $\Pi_v \in (\psi_v)_*(\pi_v)$ for every v such that $\Pi = \bigotimes_v \Pi_v$ is an automorphic representation of $G(\mathbb{A}_F)$?*
(ii) *If so, and if π is cuspidal, under what conditions is Π cuspidal?*

This is an extremely deep question, even if the condition $\Pi_v \in (\psi_v)_*(\pi_v)$ is required only at almost every place. (Relaxing the requirement in this way allows one to address the question without first establishing the Local Langlands Conjecture.) Here are two illustrations, taken from [Lgl2], of just how deep it is. Let us write $\Gamma = \mathrm{Gal}(\overline{F}/F)$ and $\Gamma_v = \mathrm{Gal}(\overline{F}_v/F_v)$ with $\Gamma_v \subset \Gamma$.

Example 1. Let $G = GL_n$ and $H = \{1\}$, so that ${}^LH = \{1\} \times \Gamma$ and ${}^LG = GL_n(\mathbb{C}) \times \Gamma$. Fix an n-dimensional representation σ of Γ. The map $\psi : {}^LH \to {}^LG$ given by $\psi(1, \gamma) = (\sigma(\gamma), \gamma)$ is an L homomorphism. If v is a finite place of F, then the only admissible homomorphism for $H(F_v)$ is $\varphi_0^{(v)}(z, w) = (1, \iota(w))$, where $\iota : W_{F_v} \to \Gamma_v$ is the inclusion. Let r be the representation $1 \times \sigma$ of ${}^LH$, and let $p : W_k' \to W_k$ be the natural quotient map. Then the elementary L factor for the representation r at the finite place v is

$$\begin{aligned} L(s,1,r) &= L(s, r\circ\varphi_0^{(v)}) && \text{by (9.1)} \\ &= L(s,\sigma\circ\iota\circ p) && \qquad\qquad (10.1) \\ &= \det(1-(\sigma\circ\iota(\mathrm{Fr})|_{(\mathbb{C}^n)^I})q^{-s})^{-1} && \text{by (8.1)}, \end{aligned}$$

and the right side coincides with the Artin L factor (5.1) for r. Hence the global L function for the cuspidal representation 1 of $H(\mathbb{A}_F)=\{1\}$ and for the representation r coincides with the Artin L function of r. If the answer to (i) is affirmative, let $\Pi\in\psi_*(1)$ be an automorphic representation of $GL_n(\mathbb{A}_F)$ for which $\Pi_v\in(\psi_v)_*(1_v)$ for each v. The admissible homomorphism Φ_v for Π_v has $\Phi_v=(\psi_v)_*(\varphi_0^{(v)})=\psi_v\circ\varphi_0^{(v)}$. Let R be the standard representation of ${}^LG$ (trivial on Γ). Then we have

$$\begin{aligned} L(s,\Pi_v) = L(s,\Pi_v,R) &= L(s,\psi_v\circ\varphi_0^{(v)},R) && \text{by (9.2)} \\ &= L(s,R\circ\psi_v\circ\varphi_0^{(v)}) && \text{by (9.1)} \\ &= L(s,\sigma\circ\iota\circ p) \\ &= L(s,1,r) && \text{by (10.1)}. \end{aligned}$$

Thus an affirmative answer to (i) for this situation implies that the Artin L function for any n-dimensional representation of Γ is the L function of an automorphic representation of GL_n. An affirmative answer to (ii) implies that the latter representation is cuspidal, i.e., gives an affirmative answer to the Langlands Reciprocity Conjecture of §8.

Example 2. Let H be general, and let r be a holomorphic representation of ${}^LH$ into $GL_n(\mathbb{C})$. Put $G=GL_n$, and define $\psi(x,\gamma)=(r(x,\gamma),\gamma)$. Let π be an automorphic representation of H. If the answer to (i) is affirmative, let $\Pi\in\psi_*(\pi)$ be a corresponding automorphic representation of $GL_n(\mathbb{A}_F)$. Tracking down the definitions as in Example 1, we find that

$$L(s,\pi,r)=L(s,\Pi),$$

i.e., the L function of π and the representation r is a standard L function for GL_n. If we assume about GL_n that Langlands L functions and Godement-Jacquet L functions coincide, then it follows from a generalization of Theorem 8.7 proved in [Ja1] that $\Lambda(s,\pi,r)$ has a meromorphic continuation to $\mathbb{C}$ and satisfies a functional equation. Moreover if π is cuspidal and the answer to (ii) is affirmative, then (with some exceptions that can be sorted out) $L(s,\pi,r)$ and $\Lambda(s,\pi,r)$ are entire.

We mention two situations in which substantial progress has been made in establishing global functoriality. Both these situations are discussed in more detail in Rogawski's lectures [Ro2].

1) *Adjoint representation for* GL_2. This is an instance where Example 2 can be carried out. For automorphic representations π of GL_2, we consider L functions $L(s,\pi,\mathrm{Ad})$, where $\mathrm{Ad}: GL_2(\mathbb{C})\to GL_3(\mathbb{C})$ is the adjoint representation. As in Example 2, the goal is to exhibit these L functions as standard L functions $L(s,\Pi)$ with Π automorphic for GL_3. This amounts to establishing global functoriality when $G=GL_3$, $H=GL_2$, and $\psi: {}^LH\to{}^LG$ is given by $\psi(x,\gamma)=(\mathrm{Ad}(x),\gamma)$. The positive result here is due to Gelbart and Jacquet [Gelb-Ja].

2) *Base change for GL_2.* Let E be any finite extension of the number field F. Let $H = GL_2$ over F, and let $G = R_{E/F}(GL_2)$ be the group over F given by restriction of ground field as in (6.5). As with (6.5), for any F algebra A, $(R_{E/F})(GL_2(A)) = GL_2(E \otimes_F A)$. Thus

$$G(\overline{F}) = GL_2(E \otimes_F \overline{F}) = GL_2(\overline{F}) \times \cdots \times GL_2(\overline{F})$$

and

$$G(F) = GL_2(E \otimes_F F) = GL_2(E).$$

The L groups are

$${}^L H = GL_2(\mathbb{C}) \times \mathrm{Gal}(\overline{F}/F)$$

and

$${}^L G = (GL_2(\mathbb{C}) \times \cdots \times GL_2(\mathbb{C})) \rtimes \mathrm{Gal}(\overline{F}/F),$$

the second one being a semidirect product. The action of $\mathrm{Gal}(\overline{F}/F)$ on the product $GL_2(\mathbb{C}) \times \cdots \times GL_2(\mathbb{C})$ permutes the coordinates. This action factors through $\mathrm{Gal}(E/F)$ if E is Galois over F. The map $\psi : {}^L H \to {}^L G$ is given by the diagonal map on the identity component and by the identity map on the Galois group. Langlands [Lgl5] proved global functoriality in this setting when E/F is a cyclic Galois extension of prime degree; this is the long step in the proof of new cases of Artin's Conjecture established by Langlands. Arthur and Clozel [Ar-Cl] proved the corresponding instance of global functoriality for GL_n when E/F is cyclic Galois of prime degree.

There is also a considerable amount of more recent progress. Various newer results on the analytic properties of $L(s, \pi, r)$ are summarized in [Ra]. Rogawski [Ro1] has made an extensive study of automorphic representations of the group U_3. Here E/F is a quadratic extension of number fields, and $G = U_3$ is the associated unitary group. For the group $H = U_2 \times U_1$, there is an embedding ${}^L H \to {}^L G$, and Rogawski's work addresses functoriality for this map. Rogawski also studies functoriality for the map ${}^L G \to {}^L \widetilde{G}$, where $\widetilde{G} = R_{E/F}(G)$; this is the base change lifting from U_3 to GL_3 over E. This work is applied to arithmetic geometry in [Lgl-Ra]; the forward of [Lgl-Ra] puts a number of aspects of the Langlands program in perspective.

References

[Ad-Ba-V] Adams, J., D. Barbasch, and D. A. Vogan, *The Langlands Classification and Irreducible Characters for Real Reductive Groups*, Birkhäuser, Boston, 1992.

[Ar-Cl] Arthur, J., and L. Clozel, *Simple Algebras, Base Change, and the Advanced Theory of the Trace Formula*, Princeton University Press, Princeton, 1989.

[Ar-Ta] Artin, E., and J. Tate, *Class Field Theory*, W. A. Benjamin Inc., New York, 1967.

[Ba-R] Barthel, L., and D. Ramakrishnan, A non-vanishing result for twists of L-functions of $\mathrm{GL}(n)$, *Duke Math. J.* **74** (1994), 681–700.

[Bl-Ro] Blasius, D., and J. D. Rogawski, Zeta functions of Shimura varieties, *Motives*, Proc. Symp. Pure Math., vol. 55, Part II, American Mathematical Society, Providence, 1994, pp. 525–571.

[Bo1] Borel, A., Some finiteness properties of adele groups over number fields, *Publ. Math. I.H.E.S.* **16** (1963), 5–30.

[Bo2] Borel, A., Linear algebraic groups, *Algebraic Groups and Discontinuous Subgroups*, Proc. Symp. Pure Math., vol. 9, American Mathematical Society, Providence, 1966, pp. 3–19.

[Bo3] Borel, A., *Linear Algebraic Groups*, W. A. Benjamin, New York, 1969.

[Bo4] Borel, A., Automorphic L-functions, *Automorphic Forms, Representations, and L-Functions*, Proc. Symp. Pure Math., vol. 33, Part II, American Mathematical Society, Providence, 1979, pp. 27–61.

[Bo-HC] Borel, A., and Harish-Chandra, Arithmetic subgroups of algebraic groups, *Annals of Math.* **75** (1962), 485–535.

[Bo-Ja] Borel, A., and H. Jacquet, Automorphic forms and automorphic representations, *Automorphic Forms, Representations, and L-Functions*, Proc. Symp. Pure Math., vol. 33, Part I, American Mathematical Society, Providence, 1979, pp. 189–202.

[Buhl] Buhler, J. P., *Icosahedral Galois Representations*, Lecture Notes in Mathematics, vol. 654, Springer-Verlag, Berlin, 1978.

[Bump] Bump, D., *Automorphic Forms and Representations*, Cambridge University Press, Cambridge, 1997.

[Car] Cartier, P., Representations of $\mathfrak{p}$-adic groups: A survey, *Automorphic Forms, Representations, and L-Functions*, Proc. Symp. Pure Math., vol. 33, Part I, American Mathematical Society, Providence, 1979, pp. 111–155.

[Cas] Cassels, J. W. S., Global fields, *Algebraic Number Theory* (J. W. S. Cassels and A. Fröhlich, eds.), Academic Press, London, 1967, pp. 42–84.

[Cas-Fr] Cassels, J. W. S., and A. Fröhlich, *Algebraic Number Theory*, Academic Press, London, 1967.

[Ch] Chevalley, C., La théorie du corps de classes, *Annals of Math.* **41** (1940), 394–418.

[Co-Pi] Cogdell, J. W., and I. I. Piatetski-Shapiro, Converse theorems for GL_n, *Publ. Math. I.H.E.S.* **79** (1994), 157–214.

[De-S] Deligne, P., and J.-P. Serre, Formes modulaires de poids 1, *Annales Scient. Ecole Norm. Sup.* **7** (1974), 507–530.

[Fl] Flath, D., Decomposition of representations into tensor products, *Automorphic Forms, Representations, and L-Functions*, Proc. Symp. Pure Math., vol. 33, Part I, American Mathematical Society, Providence, 1979, pp. 179–183.

[Fre] Frey, G. (ed.), *On Artin's Conjecture for Odd 2-dimensional Representations*, Lecture Notes in Mathematics, vol. 1585, Springer-Verlag, Berlin, 1994.

[Frö] Fröhlich, A., Local fields, *Algebraic Number Theory* (J. W. S. Cassels and A. Fröhlich, eds.), Academic Press, London, 1967, pp. 1–41.

[Ga] Garrett, P. B., *Holomorphic Hilbert Modular Forms*, Wadsworth & Brooks/Cole, Pacific Grove, CA, 1990.

[Gelb1] Gelbart, S. S., *Automorphic Forms on Adele Groups*, Princeton University Press, Princeton, NJ, 1975.

[Gelb2] Gelbart, S. S., An elementary introduction to the Langlands program, *Bull. Amer. Math. Soc.* **10** (1984), 177–219.

[Gelb-Ja] Gelbart, S. S., and H. Jacquet, A relation between automorphic representations of GL(2) and GL(3), *Annales Scient. École Norm. Sup.* **11** (1978), 471–542.

[Gelf-Pi] Gelfand, I. M., and I. Piatetski-Shapiro, Automorphic forms and representation theory, *Trudy Moscov. Mat. Obšč.* **12** (1963), 389–412.

[Gf-Gr-P] Gelfand, I. M., M. I. Graev, and I. I. Piatetskii-Shapiro, *Representation Theory and Automorphic Functions*, W. B. Saunders, Philadelphia, 1969.

[Go-Ja] Godement, R., and H. Jacquet, *Zeta-functions of simple algebras*, Lecture Notes in Mathematics, vol. 260, Springer-Verlag, New York, 1972.

[Ha] Harris, M., *Supercuspidal representations in the cohomology of Drinfel'd upper half spaces; elaboration of Carayol's program*, preprint, 1994; *Invent. Math.* (to appear).

[HC1] Harish-Chandra, Automorphic forms on a semisimple Lie group, *Proc. Nat. Acad. Sci. USA* **45** (1959), 570–573.

[HC2] Harish-Chandra, *Automorphic Forms on Semisimple Lie Groups*, Lecture Notes in Mathematics, vol. 62, Springer-Verlag, Berlin, 1968.

[Hei] Heilbronn, H., Zeta-functions and L-functions, *Algebraic Number Theory* (J. W. S. Cassels and A. Fröhlich, eds.), Academic Press, London, 1967, pp. 204–230.

[Hen1] Henniart, G., La conjecture de Langlands locale pour GL(3), *Mem. Soc. Math. France* **11–12** (1984), 1–186.

[Hen2] Henniart, G., Caractérisation de la correspondence de Langlands locale par les facteurs ε de paires, *Invent. Math.* **113** (1993), 339–350.

[Ja1] Jacquet, H., Principal L functions of the linear group, *Automorphic Forms, Representations, and L-Functions*, Proc. Symp. Pure Math., vol. 33, Part II, American Mathematical Society, Providence, 1979, pp. 63–86.

[Ja2] Jacquet, H., Principal L-functions for $GL(n)$, *these* PROCEEDINGS, pp. 321–329.

[Ja-Lgl] Jacquet, H., and R. P. Langlands, *Automorphic Forms on* GL(2), Lecture Notes in Mathematics, vol. 114, Springer-Verlag, Berlin, 1970.

[Ja-P-S] Jacquet, H., I. I. Piatetski-Shapiro, and J. A. Shalika, Rankin-Selberg convolutions, *Amer. J. Math.* **105** (1983), 367–464.

[Ja-Sha1] Jacquet, H., and J. A. Shalika, On Euler products and the classification of automorphic representations I, *Amer. J. Math.* **103** (1981), 499–558.

[Ja-Sha2] Jacquet, H., and J. Shalika, Rankin-Selberg convolutions: Archimedean theory, *Festschrift in Honor of I. I. Piatetski-Shapiro on the Occasion of His Sixtieth Birthday, Part I (Ramat-Aviv, 1989)*, Israel Math. Conf. Proc., vol. 2, Weizmann, Jerusalem, 1990, pp. 125–207.

[Kna1] Knapp, A. W., *Elliptic Curves*, Princeton University Press, Princeton, 1992.

[Kna2] Knapp, A. W., Local Langlands correspondence: the archimedean case, *Motives*, Proc. Symp. Pure Math., vol. 55, Part II, American Mathematical Society, Providence, 1994, pp. 393–410.

[Kna-Vo] Knapp, A. W., and D. A. Vogan, *Cohomological Induction and Unitary Representations*, Princeton University Press, Princeton, NJ, 1995.

[Kne] Kneser, M., Strong approximation, *Algebraic Groups and Discontinuous Subgroups*, Proc. Symp. Pure Math., vol. 9, American Mathematical Society, Providence, 1966, pp. 187–196.

[Kud] Kudla, S. S., The local Langlands correspondence: the non-archimedean case, *Motives*, Proc. Symp. Pure Math., vol. 55, Part II, American Mathematical Society, Providence, 1994, pp. 365–391.

[Kut] Kutzko, P., The local Langlands conjecture for GL(2) of a local field, *Annals of Math.* **112** (1980), 381–412.

[La] Lang, S., *Algebraic Number Theory*, Springer-Verlag, New York, 1986.

[Lgl1] Langlands, R. P., *Representations of abelian algebraic groups*, duplicated notes, Yale University, New Haven, CT, 1968.

[Lgl2] Langlands, R. P., Problems in the theory of automorphic forms, *Lectures in Modern Analysis and Applications III*, Lecture Notes in Mathematics, vol. 170, Springer-Verlag, Berlin, 1970, pp. 18–61.

[Lgl3] Langlands, R. P., On the classification of irreducible representations of real algebraic groups, mimeographed notes, Institute for Advanced Study, Princeton, NJ, 1973, *Representation Theory and Harmonic Analysis on Semisimple Lie Groups* (P. J. Sally and D. A. Vogan, eds.), Math. Surveys and Monographs, vol. 31, American Mathematical Society, Providence, 1989, pp. 101–170.

[Lgl4] Langlands, R. P., On the notion of an automorphic representation, *Automorphic Forms, Representations, and L-Functions*, Proc. Symp. Pure Math., vol. 33, Part I, American Mathematical Society, Providence, 1979, pp. 203–207.

[Lgl5] Langlands, R. P., *Base Change for* GL(2), Princeton University Press, Princeton, 1980.

[Lgl6] Langlands, R. P., Representation theory: its rise and its role in number theory, *Proceedings of the Gibbs Symposium (New Haven, CT, 1989)*, American Mathematical Society, Providence, RI, 1990, pp. 181-210.

[Lgl7] Langlands, R. P., Where stands functoriality today?, *these* PROCEEDINGS, pp. 457–471.

[Lgl-Ra] Langlands, R. P., and D. Ramakrishnan (eds.), *The Zeta Functions of Picard Modular Surfaces*, Les Publications CRM, Montréal, 1992.

[Maa] Maass, H., Über eine neue Art von nicht analytischen automorphen Funktionen und die Bestimmung Dirichletscher Reihen durch Funktionalgleichungen, *Math. Annalen* **121** (1949), 141–183.

[Mar] Martinet, J., Character theory and Artin L-functions, *Algebraic Number Fields: L-Functions and Galois Properties* (A. Fröhlich, ed.), Academic Press, London, 1977, pp. 1–87.

[Moe1] Mœglin, C., Representations of GL(n) over the real field, *these* PROCEEDINGS, pp. 157–166.

[Moe2] Mœglin, C., Representations of GL(n,F) in the nonarchimedean case, *these* PROCEEDINGS, pp. 303–319.

[Moy] Moy, A., Local constants and the tame Langlands correspondence, *Amer. J. Math.* **108** (1986), 863–929.

[Ne] Neukirch, J., *Class Field Theory*, Springer-Verlag, Berlin, 1986.

[Pi] Piatetski-Shapiro, I., *Festschrift in Honor of I. I. Piatetski-Shapiro on the Occasion of His Sixtieth Birthday, Part II*, Israel Math. Conf. Proc., vol. 3, 1990, pp. 185–195.

[Ra] Ramakrishnan, D., Pure motives and automorphic forms, *Motives*, Proc. Symp. Pure Math., vol. 55, Part II, American Mathematical Society, Providence, 1994, pp. 411–446.

[Ro1] Rogawski, J. D., *Automorphic Representations of Unitary Groups in Three Variables*, Princeton University Press, Princeton, 1990.

[Ro2] Rogawski, J. D., Functoriality and the Artin conjecture, *these* PROCEEDINGS, pp. 331–353.

[Se1] Serre, J.-P., Local class field theory, *Algebraic Number Theory* (J. W. S. Cassels and A. Fröhlich, eds.), Academic Press, London, 1967, pp. 129–161.

[Se2] Serre, J.-P., *Local Fields*, Springer-Verlag, New York, 1979.

[Sha] Shalika, J. A., The multiplicity one theorem for GL_n, *Annals of Math.* **100** (1974), 171–193.

[Shi] Shimura, G., *Introduction to the Arithmetic Theory of Automorphic Functions*, Princeton University Press, Princeton, NJ, 1971.

[Si] Siegel, C. L., *Topics in Complex Function Theory*, vol. III, Wiley-Interscience, New York, 1973.

[Sp] Springer, T. A., Reductive groups, *Automorphic Forms, Representations, and L-Functions*, Proc. Symp. Pure Math., vol. 33, Part I, American Mathematical Society, Providence, 1979, pp. 3–27.

[Ta1] Tate, J. T., Fourier analysis in number fields and Hecke's zeta-functions, Ph.D. Thesis, Princeton University, 1950, *Algebraic Number Theory* (J. W. S. Cassels and A. Fröhlich, eds.), Academic Press, London, 1967, pp. 305–347.

[Ta2] Tate, J. T., Global class field theory, *Algebraic Number Theory* (J. W. S. Cassels and A. Fröhlich, eds.), Academic Press, London, 1967, pp. 162–203.

[Ta3] Tate, J., Number theoretic background, *Automorphic Forms, Representations, and L-Functions*, Proc. Symp. Pure Math., vol. 33, Part II, American Mathematical Society, Providence, 1979, pp. 3–26.

[Ti] Tits, J., Reductive groups over local fields, *Automorphic Forms, Representations, and L-Functions*, Proc. Symp. Pure Math., vol. 33, Part I, American Mathematical Society, Providence, 1979, pp. 29–69.

[Tu] Tunnell, J., On the local Langlands conjecture for GL(2), *Invent. Math.* **46** (1978), 179–200.

[We1] Weil, A., *Adeles and Algebraic Groups*, bound notes, Institute for Advanced Study, Princeton, NJ, 1961.

[We2] Weil, A., Über die Bestimmung Dirichletscher Reihen durch Funktionalgleichungen, *Math. Annalen* **168** (1967), 149–156.

[We3] Weil, A., *Dirichlet Series and Automorphic Forms*, Lecture Notes in Mathematics, vol. 189, Springer-Verlag, Berlin, 1971.

[We4] Weil, A., *Basic Number Theory*, Springer-Verlag, New York, 1973.

DEPARTMENT OF MATHEMATICS, STATE UNIVERSITY OF NEW YORK, STONY BROOK, NEW YORK 11794, U.S.A.

E-mail address: `aknapp@ccmail.sunysb.edu`

Proceedings of Symposia in Pure Mathematics
Volume **61** (1997), pp. 303–319

Representations of GL(n,F) in the Nonarchimedean Case

C. Mœglin

Introduction

In the nonarchimedean case, the representations of $\mathrm{GL}_n(F)$ are less well known than in the archimedean case, but the general theory is completely understood. For a history as of 1976, see the introduction of [B-Z]; a major development was [J-L], where $n = 2$, and a complete exposition occurs in [B-Z] itself. After that, in [Z], Zelevinsky, using [B-Z2], classified all the isomorphism classes of "smooth" (see **4**) irreducible representations of $\mathrm{GL}_n(F)$ using as basic data the "cuspidal" ones (see **6**). Using that work, Tadič [T] classified the unitary dual; the statement of the result is completely analogous to the result in the archimedean case [V] except that no cohomological induction appears. An understanding of cuspidal representations (and more) is due to Bushnell and Kutzko [Bu-K].

The Local Langlands Conjecture asserts that there is a unique bijection between the set of isomorphism classes of irreducible smooth representations of $\mathrm{GL}_n(F)$ and the set of conjugacy classes of homomorphisms from the Weil-Deligne group of F into $\mathrm{GL}_n(\mathbb{C})$ (with some admissibility conditions) preserving L and ϵ factors for pairs of representations. This Langlands conjecture is not completely proved although there has been some progress by Henniart [He1, He2] and Harris [Ha1, Ha2].

This article will survey the general theory following very closely [B-Z]. For lectures on the archimedean case, see [M]. I would like to thank A. Knapp for helpful comments.

We use the following notation throughout. The field F is a finite extension of $\mathbb{Q}_p$, v_F is the valuation of F, and O_F is the ring of integers:

$$O_F := \{x \in F \mid v_F(x) \geq 0\}.$$

We denote by $\wp$ the maximal ideal of O_F, and we fix a generator ω_F for this ideal. Let $\mathbb{N} = \{1, 2, 3, ...\}$; the symbol n will always refer to a member of this set.

1. Topological Questions

$\mathrm{GL}_n(F)$ is by definition the multiplicative subgroup of $End_F F^n$ of invertible matrices. It is endowed with the inherited topology. So we have to define the

1991 *Mathematics Subject Classification.* Primary 22E50.

topology on $End_F F^n$. For all $x \in End_F F^n$, a basis of open neighborhoods is the set $\{U_{x,z} \mid z \in \mathbb{Z}\}$ such that:

$$U_{x,z} := x + \wp^z \, End_{O_F} O_F^n.$$

It is clear that $\mathrm{GL}_n(F)$ is an open topological subgroup in $End_F F^n$. Moreover, for all $z \in \mathbb{Z}$, $U_{Id,z}$ is compact and is included in $\mathrm{GL}_n(F)$ if $z \geq 1$. This implies that each point in $\mathrm{GL}_n(F)$ has a basis of neighborhoods that are open compact subgroups; in particular, $\mathrm{GL}_n(F)$ is a locally compact topological group. As such, its Haar measure is left and right invariant (the group is unimodular); see [B-Z, 1.18].

2. Maximal Compact Subgroups

Recall that O_F is the ring of integers in F and denote:

$$\mathrm{GL}_n(O_F) := \{g \in End_{O_F} F^n \mid g^{-1} \in End_{O_F} F^n\}.$$

By Cramer's Rule, $g \in \mathrm{GL}_n(F)$ is in $\mathrm{GL}_n(O_F)$ if and only if $g \in End_{O_F}(O_F)$ and $\det g$ is a unit in F^*.

Theorem. *$GL_n(F)$ contains a unique conjugacy class of maximal compact subgroups, and each such subgroup is open. One element in this class is* $\mathrm{GL}_n(O_F)$.

Let us prove this theorem. Recall that a **lattice** L in F^n is by definition an O_F-submodule of rank n. Fix two lattices L and L' in F^n. It is known (and easy to prove) that there exist a basis $e_1, \cdots, e_n$ of L and members $z_1, \cdots, z_n$ of $\mathbb{Z}^n$ such that:

$$\omega_F^{z_1} e_1, \cdots, \omega_F^{z_n} e_n \quad \text{is a basis of } L'. \tag{1}$$

From (1), it is clear that L and L' are conjugate by an element of $\mathrm{GL}_n(F)$.

Denote by L_0 the standard lattice; this means:

$$L_0 = O_F^n \subset F^n.$$

By definition $\mathrm{GL}_n(O_F)$ is the stabilizer of L_0. Since any lattice L in F^n is conjugate to L_0, $Stab_{\mathrm{GL}_n(F)} L$ is conjugate to $\mathrm{GL}_n(O_F)$. We still have to prove:

$$\mathrm{GL}_n(O_F) \text{ is an open compact subgroup} \tag{2}$$

and

$$\begin{array}{l}\text{if } K \text{ is any compact subgroup of } \mathrm{GL}_n(F), \text{ then} \\ \text{there exists a lattice } L \text{ of } F^n \text{ such that:} \\ \qquad\qquad K \subset Stab_{\mathrm{GL}_n(F)} L.\end{array} \tag{3}$$

$\mathrm{GL}_n(O_F)$ is compact and open: In fact, $End_{O_F} O_F^n$ is compact and open by definition of the topology. Also the function $g \mapsto \det g$ is continuous. Hence $\mathrm{GL}_n(O_F)$ is the intersection of a closed subset with a compact subset and is also the intersection of two open subsets. Thus (2) follows.

To prove (3), let K be as in (3) and observe that $K \cap \mathrm{GL}_n(O_F)$ is open in K; this implies that $K/(K \cap \mathrm{GL}_n(O_F))$ is a finite set. Put:

$$L := \sum_{g \in K/(K \cap \mathrm{GL}_n(O_F))} g L_0.$$

This is a finite sum of lattices and hence is a lattice. Clearly:

$$K \subset Stab_{\mathrm{GL}_n(F)} L.$$

Thus (3) follows.

3. Bruhat, Cartan, Iwasawa Decompositions

Bruhat decomposition (same result as in the archimedean case). Fix a Borel subgroup B, for example the set of upper triangular matrices, and a maximal torus T contained in it, for example the set of diagonal matrices. Let:

$$\mathcal{W} := (Norm_{\mathrm{GL}_n(F)} T)/T.$$

Then:

$$\mathrm{GL}_n(F) = \bigcup_{w \in \mathcal{W}} B\tilde{w}B,$$

where $\tilde{w}$ is any representative of w in $Norm_{\mathrm{GL}_n(F)} T$ and the union is a disjoint union. This is proved in [Bo-T, 5.15]. In that reference, one finds the correct formulation for any reductive group, split or not.

Cartan decomposition. Define:

$$A^+ := \left\{ t(z_1, \cdots, z_n) = \begin{pmatrix} \omega_F^{z_1} & & \\ & \ddots & \\ & & \omega_F^{z_n} \end{pmatrix} \,\middle|\, (z_1, \cdots, z_n) \in \mathbb{Z}^n,\ z_1 \leq \cdots \leq z_n \right\}.$$

Then the Cartan decomposition is:

$$\mathrm{GL}_n(F) = \mathrm{GL}_n(O_F) A^+ \mathrm{GL}_n(O_F). \tag{1}$$

To prove (1), let $g \in \mathrm{GL}_n(F)$ and put $L := gL_0$, where L_0 is the standard lattice as in **2**. Then (1) follows from **2** (1) with $L' := L_0$.

Iwasawa decomposition. A parabolic subgroup of $\mathrm{GL}_n(F)$ is the stabilizer of a flag. Here a **flag** is a sequence of r ($r \in \mathbb{N}$) subspaces of F^n:

$$0 \subset X_1 \subset \cdots \subset X_{r-1} \subset X_r = F^n.$$

Let P be a parabolic subgroup of $\mathrm{GL}_n(F)$, and let K be any maximal compact subgroup of $\mathrm{GL}_n(F)$. Then:

$$\mathrm{GL}_n(F) = PK. \tag{2}$$

To prove (2), denote by $(0 = X_0 \subset X_1 \subset \cdots \subset X_{r-1} \subset X_r)$ (with $r \in \mathbb{N}$) the flag with stabilizer P, and denote by L the lattice in F^n with stabilizer K (see **2**). Then for all $i \in [1, r]$, $L_i := L \cap X_i$ is an O_F submodule of X_i of maximal rank, hence a lattice in X_i. Let $g \in \mathrm{GL}_n(F)$ and put:

$$(gL)_i := gL \cap X_i \qquad \text{for } \forall i \in [1, r].$$

Step by step, for $i \in [1, r]$, one constructs $p_i \in Aut X_i$ such that $p_i(gL)_i = L_i$. To do so, for all $i \in [1, r]$, we fix Y_i (resp. $Y(g)_i$), a supplementary subspace of X_{i-1} in X_i, in such a way that:

$$L_i = L \cap Y_i \oplus L_{i-1}, \tag{3}$$

resp.

$$(gL)_i = gL \cap Y(g)_i \oplus (gL)_{i-1}. \tag{4}$$

We assume that p_{i-1} is constructed with the right property, and we fix $u_i \in Aut(X_i)$ such that $(u_i)_{|X_{i-1}} = Id$ and $u_i(Y(g)_i) = Y_i$. Now $u_i((gL)_i \cap Y(g)_i)$ is a lattice in Y_i and hence is conjugate by $m_i \in Aut(Y_i)$ to $L_i \cap Y_i$. Extend m_i to an element of $Aut(X_i)$ by the identity on X_{i-1}; extend also p_{i-1} to an element of $Aut(X_i)$ by the identity on Y_i and put:

$$p_i := p_{i-1} m_i u_i.$$

We have:

$$p_i((gL)_{i-1}) = p_{i-1}((gL)_{i-1}) = L_{i-1},$$
$$p_i(gL \cap Y(g)_i) = m_i\big(u_i(gL \cap Y(g)_i)\big) = L_i \cap Y_i.$$

From (3) and (4), we obtain our claim $p_i((gL)_i) = L_i$. Put:

$$p := p_r.$$

Clearly p stabilizes the fixed flag; hence $p \in P$. By construction $pgL = L$; hence $pg \in K$. This finishes the proof of (2).

4. Smooth Representations in Complex Vector Spaces

Let V be a $\mathbb{C}$-vector space of finite or infinite dimension and let $\pi : \mathrm{GL}_n(F) \to Aut(V)$ be a **representation**, i.e., a homomorphism of groups.

4.1. Definition. π is **smooth** if and only if, for all $v \in V$, $Stab_{\mathrm{GL}_n(F)} v$ contains an open subgroup.

If V is finite-dimensional, smoothness is just a continuity condition. The term "smooth" is due to F. Rodier; in [B-Z] the notation is Alg. Bernstein began the study of smooth representations in general. In [J-L] the representations are admissible in the sense of **7** below.

4.2. Smooth characters of F^*. Here we assume that $n = 1$ and $\dim V = 1$. We denote by U_F the group of units in O_F. We may identify $U_F = \mathrm{GL}_1(O_F)$.

Proposition. *Let χ be a smooth character of F^*.*

(i) *Suppose that $\chi_{|U_F} \equiv 1$ (χ is said to be **unramified**). Then there exists $s \in \mathbb{C}/(2i\pi/\log q_F)\mathbb{Z}$ such that:*

$$\chi(x) = q_F^{-s\, v_F(x)} = |x|_F^s \qquad \textit{for } \forall x \in F^*.$$

(ii) *For general χ, there exists a character χ' of finite order such that $(\chi')^{-1}\chi$ is unramified. Of course, χ' is not unique.*

PROOF.

(i) By the valuation, F^*/U_F is isomorphic to $\mathbb{Z}$; this implies (i).

(ii) By smoothness, $Ker\,\chi \cap U_F$ is open in U_F and hence of finite index. So $\chi(U_F)$ is a finite subgroup of $\mathbb{C}^*$. Fix d such that $\chi(U_F)^d = \{1\}$. With this choice, χ^d is unramified. We fix $s \in \mathbb{C}$ such that:

$$\chi^d = |\cdot|_F^s.$$

Let $\chi' := \chi^{-1}|\cdot|_F^{s/d}$; then $(\chi')^d \equiv 1$ and $\chi\chi'$ is unramified. This proves (ii).

4.3. Finite-dimensional smooth representations.

Proposition. *Let* (π, V) *be a finite-dimensional smooth representation of* $\mathrm{GL}_n(F)$*; then* π *is trivial on* $SL_n(F)$ *. In particular, if* (π, V) *is irreducible, then* $\dim V = 1$ *and there exists a smooth character* χ *of* F^* *such that:*

$$\pi = \chi \circ \det .$$

In fact, let (π, V) be a smooth finite-dimensional representation of $\mathrm{GL}_n(F)$, and fix B a basis of V as a vector space. The group:

$$Ker\,\pi = \bigcap_{v \in B} Stab_{\mathrm{GL}_n(F)} v,$$

contains an open subgroup. So $Ker\,\pi$ is open and, of course, normal. Denote by U the subgroup of unipotent upper triangular matrices; $Ker\,\pi \cap U$ is an open subgroup of U normalized by the diagonal torus. So:

$$Ker\,\pi \cap U = U.$$

As any unipotent element of $\mathrm{GL}_n(F)$ is conjugate to an element of U, $Ker\,\pi$ contains the subgroup of $\mathrm{GL}_n(F)$ generated by the unipotent elements. This subgroup is precisely $SL_n(F)$.

Observe now that $\mathrm{GL}_n(F)/SL_n(F)$ is abelian. Hence $\pi(\mathrm{GL}_n(F))$ generates an abelian subalgebra in $End_{\mathbb{C}} V$. In particular there is a one-dimensional subspace of V stable under this algebra. We have just proved that if (π, V) is irreducible then $\dim V = 1$. As $SL_n(F)$ is the kernel of the determinant, the proposition follows.

4.4. Induction. Let $H \subset \mathrm{GL}_n(F)$ be a closed subgroup, and let (ρ, W) be a smooth representation of H. By a **right smooth function** from $\mathrm{GL}_n(F)$ to W, we mean a function f such that there exists a compact open subgroup K_f of $\mathrm{GL}_n(F)$ stabilizing f on the right:

$$f(gk) = f(g) \qquad \text{for } \forall g \in \mathrm{GL}_n(F) \text{ and } \forall k \in K_f.$$

The usual condition entering the definition of induced representation is:

$$f(hg) = \delta_H^{1/2} \rho(h) f(g) \qquad \text{for } \forall h \in H \text{ and } g \in \mathrm{GL}_n(F), \tag{1}$$

where $\delta_H^{1/2}$ is the positive square root of the modular function for H (we use the fact that $\mathrm{GL}_n(F)$ is unimodular); see [B-Z, 1.18].

We have two ways to define the induced representation of (ρ, W):

$$Ind(\rho, W) := \{f : \mathrm{GL}_n(F) \to W, \text{ right smooth function with property (1)}\}$$

$$ind(\rho, W) := \{f : \mathrm{GL}_n(F) \to W, \text{ locally constant function with property (1) and of compact support modulo the left action of } H\}.$$

The representation of $\mathrm{GL}_n(F)$ in each vector space is by right translations. If $H\backslash \mathrm{GL}_n(F)$ is compact, then the two definitions coincide. This is the case if H is a parabolic subgroup. To avoid confusion when more than one subgroup appears, we will write $ind_H^{\mathrm{GL}_n(F)}$ instead of ind.

For any smooth representation (τ, Y) of G, we denote by $\tau_{|H}$ the tensor product of $\delta_H^{-1/2}$ with the ordinary restriction to H of the representation. Frobenius reciprocity [B-Z, 2.28] gives us an isomorphism:

$$Hom_{\mathrm{GL}_n(F)}\big((\tau, Y), Ind(\rho, W)\big) \simeq Hom_H\big((\tau_{|H}, Y), (\rho, W)\big) \tag{2}$$

obtained by composing with:

$$f \in Ind(\rho, W) \mapsto f(Id) \in W.$$

4.5. Contragradient representation. Let (π, V) be a smooth representation of $\mathrm{GL}_n(F)$. We denote by V' the linear dual of V and put:

$$V^* := \{v' \in V', \text{ such that } Stab_{\mathrm{GL}_n(F)} v' \text{ contains an open subgroup}\}.$$

The representation π^* in V^* is the obvious one and is called the **contragradient** [B-Z, 2.13]. The following are easy to prove (see loc. cit.).

Remarks.
(i) The pairing between (π, V) and (π^*, V^*) is nondegenerate.
(ii) If H, ρ, W are as in **4.4**, then

$$ind(\rho, W)^* \simeq Ind(\rho^*, W^*).$$

(iii) Assume that V is admissible (see **7** below for the definition). Then $V \simeq (V^*)^*$, and $V \otimes V^*$ is the algebra of smooth endomorphisms of V (defined as endomorphisms that are right and left invariant by an open subgroup).

If H, ρ, W are as in **4.4**, we have also the isomorphism [B-Z, 2.29]:

$$Hom_H\big((\rho, W), ((\tau_{|H})^*, (Y_{|H})^*)\big) \simeq Hom_{\mathrm{GL}_n(F)}\big(ind(\rho, W), (\tau^*, Y^*)\big) \tag{1}$$

given by associating to λ in the first space:

$$f \in ind(\rho, W) \mapsto \Big(y \in W \mapsto \int_{H\backslash \mathrm{GL}_n(F)} \langle \lambda(f(g)), \tau(g)y\rangle\, dg\Big),$$

where dg is a relatively invariant function on the quotient space: $d(hg) = d(hg^{-1}) = \delta_H^{-1} dg$ for all $h \in H$ (see [B-Z, 1.20].

5. Jacquet Module

5.1. Definition. Let (π, V) be a smooth representation of $\mathrm{GL}_n(F)$, let U be a unipotent subgroup (defined as a group that is a successive extension of the additive group F). The group U is the union of its compact subgroups. Denote:

$$V[U] := \mathbb{C}\{\pi(u)v - v \mid u \in U, v \in V\}.$$

The space $V/V[U]$ is naturally a smooth representation of the normalizer of U in $\mathrm{GL}_n(F)$. We denote by V_U this representation tensored by the character $\delta^{-1/2}_{Norm_{\mathrm{GL}_n(F)} U}$ in the notation of **4.4**, and we call $V[U]$ the **Jacquet module**.

The twisting is done so that whenever P is a parabolic subgroup of $\mathrm{GL}_n(F)$ with unipotent radical U and (ρ, W) is a smooth representation of P trivial on U, then:

$$Hom_{\mathrm{GL}_n(F)}(V, ind(W)) \simeq Hom_{P/U}(V_U, W). \tag{1}$$

(See **4.4** (2) above and use the compactness of $P\backslash \mathrm{GL}_n(F)$.) Taking into account that U is the union of its compact subgroups, we obtain the following easy but very useful characterization of $V[U]$:

Proposition. *Let (π, V) as above. Then:*

$$v \in V[U] \iff \exists K_v \text{ an open subgroup of } U \text{ such that } \int_{K_v} \pi(k)v = 0.$$

Denote by $\lim_{\to K'} V^{K'}$ the inductive limit of the spaces of K'-fixed points when K' runs over the compact subgroups of U, where the maps are defined for all $K_1 \subset K_2$ as:

$$V^{K_1} \to V^{K_2} \qquad \text{by} \qquad v \mapsto (meas(K_2))^{-1} \int_{K_2} \pi(k)v\, dk.$$

The proposition says that we have a natural isomorphism:

$$\lim_{\to K'} V^{K'} \simeq V_U. \tag{2}$$

In particular, passing to the Jacquet module is an exact functor.

In the special case that U is the unipotent radical of a parabolic P, denote by $\overline{U}$ the unipotent radical of the opposite parabolic subgroup $\overline{P}$ (to define the **opposite parabolic subgroup**, one has to fix a Levi subgroup L in P, and $\overline{P}$ is the unique parabolic subgroup intersecting P in L; see [Bo, 14.21]). One can show in this case ([C, 4.2.5]) that if V is admissible, then:

$$(V^*)_{\overline{U}} \simeq (V_U)^*.$$

From **4.5** (1), it follows, with the above notation, that:

$$Hom_{P/U=P\cap\overline{P}}(W, V_U) \simeq Hom_{\mathrm{GL}_n(F)}(ind_{\overline{P}}^{\mathrm{GL}_n(F)}(W), V). \tag{3}$$

5.2. Some computations. For an example we shall compute the Jacquet module of an induced representation when $n = 2$. Let B be the Borel subgroup of $\mathrm{GL}_2(F)$ consisting of the upper triangular matrices, denote by U its unipotent radical, and let χ be a character of the diagonal torus T. As usual we extend χ to a character of B trivial on U. Denote by σ the nontrivial element of the Weyl group. Put:

$$I(\chi) := ind(\chi).$$

We will prove:

Proposition. *There is a natural exact sequence:*

$$0 \to \mathbb{C}_{\sigma\chi} \to I(\chi)_U \to \mathbb{C}_\chi \to 0.$$

This sequence is split if and only if $\chi \not\simeq \sigma\chi$.

PROOF. In fact, let $C_c^\infty(\cdot)$ denote the set of locally constant functions with compact support. Use the Bruhat decomposition:

$$U\backslash \mathrm{GL}_2(F) = U\backslash B \cup U\backslash B\sigma B,$$

where $U\backslash B$ is closed and $U\backslash B\sigma B$ is open. From this decomposition we obtain the following exact sequence:

$$0 \to C_c^\infty(U\backslash B\sigma B) \to C_c^\infty(U\backslash G) \to C_c^\infty(U\backslash B) \to 0.$$

Project each space of functions on the left $\delta^{1/2}\chi$-T-semi-invariant functions ($\delta^{1/2}$ is the positive square root of the modular function of B); this is possible using integration on left T-cosets. We call such a space $C^{\infty}_{c,\delta^{1/2}\chi}(\cdot)$. We form Jacquet modules by integrating over the right U-cosets and tensoring by $\delta^{-1/2}$, and we obtain the following exact sequence:

$$0 \to C^{\infty}_{c,\delta^{1/2}\chi}(U\backslash B\sigma B)_U \to C^{\infty}_{c,\delta^{1/2}\chi}(U\backslash \mathrm{GL}_2(F))_U \to C^{\infty}_{c,\delta^{1/2}\chi}(U\backslash B)_U \to 0.$$

We have:

$$C^{\infty}_{c,\delta^{1/2}\chi}(U\backslash \mathrm{GL}_2(F))_U = I(\chi)_U,$$
$$C^{\infty}_{c,\delta^{1/2}\chi}(U\backslash B)_U \simeq \mathbb{C}_{\chi},$$

and a little less obviously (the shift by $\delta^{1/2}$ in the definition is here important):

$$C^{\infty}_{c,\delta^{1/2}\chi}(U\backslash B\sigma B)_U \simeq \mathbb{C}_{\sigma\chi}.$$

This gives the exact sequence of the proposition. This sequence obviously splits if $\chi \not\simeq \sigma\chi$.

Conversely assume that $\chi = \sigma\chi$ and that the sequence splits. In view of **5.1** (1), these conditions imply:

$$\dim Hom_{\mathrm{GL}_2(F)}(I(\chi), I(\chi)) = 2,$$

and $I(\chi)$ is not irreducible. It is not elementary to prove that $I(\chi)$ is in fact irreducible, but see [J-L]. This gives the needed contradiction and completes the proof.

The semisimplification of a Jacquet module for a general induced representation has been computed and has been put in a beautiful form in [B-Z2]: Let P and P' be two parabolic subgroups of $\mathrm{GL}_n(F)$, both containing a fixed Borel subgroup B. Denote by U_P and $U_{P'}$ the unipotent radicals of P and P' and by M_P and $M_{P'}$ Levi subgroups of P and P' containing a given torus in B. Denote by $\mathcal{W}$ the Weyl group of $\mathrm{GL}_n(F)$, $\mathcal{W}^P$ (resp. $\mathcal{W}^{P'}$) the Weyl group of M_P (resp. $M_{P'}$). View $\mathcal{W}^P$ and $\mathcal{W}^{P'}$ as subgroups of $\mathcal{W}$. To avoid confusion, we use a subscript to indicate the group from which we induce and a superscript the group to which we induce.

Theorem [B-Z2]. *With the above notation, let (ρ, W) be a smooth representation of P trivial on U_P. Then the semisimplification of $\left(ind_P^{\mathrm{GL}_n(F)}(\rho, W)\right)_{U_{P'}}$ equals:*

$$\bigoplus_{w\in\mathcal{W}^{P'}\backslash\mathcal{W}/\mathcal{W}^P} ind^{M_{P'}}_{M_{P'}\cap wM_Pw^{-1}} ad(w)(W_{U_{M_P\cap w^{-1}P'w}}),$$

where $U_{M_P\cap w^{-1}P'w}$ is the unipotent radical of the parabolic subgroup $M_P \cap w^{-1}P'w$ of M_P.

6. Cuspidal Representations

A big difference between archimedean and nonarchimedean fields is the existence, in the nonarchimedean case, of representations that have no realization as a subquotient of an induced representation from a proper parabolic subgroup. Such representations are called **cuspidal** (or **supercuspidal**), and their general properties have been studied by Harish-Chandra (see **8** below). We will begin by another definition and discuss later the equivalence. The presentation is due to [B-Z].

Definition [B-Z, 3.18]. Let (π, V) be a smooth representation of $\mathrm{GL}_n(F)$. We say that (π, V) is **quasicuspidal** if and only if $V_U = 0$ for the unipotent radical of any parabolic subgroup of $\mathrm{GL}_n(F)$. The representation (π, V) is said to be **cuspidal** if and only if it is quasicuspidal, is finitely generated, and has a central character.

Denote by Z the center of $\mathrm{GL}_n(F)$.

Theorem [B-Z, 3.21]. *Let (π, V) a smooth representation of $\mathrm{GL}_n(F)$. The following conditions are equivalent:*

(1) *(π, V) is quasicuspidal;*
(2) *for every $v \in V$ and $v^* \in V^*$ the* **local coefficient** $\zeta_{v,v^*}(g)$ given by:
$$g \in \mathrm{GL}_n(F) \mapsto \langle \pi(g^{-1})v, v^* \rangle$$
has compact support modulo the center Z in the sense that $Z\backslash(Z\, supp\, \zeta_{v,v^*})$ is compact;
(3) for every open compact subgroup K' of $\mathrm{GL}_n(F)$ and for every $v \in V$, the group:
$$G_{v,K'} := \{g \in \mathrm{GL}_n(F) \mid \pi(K')\pi(g)v \neq 0\}$$
is compact modulo the center Z;
(4) (π^*, V^*) is quasi-cuspidal.

The equivalence between (1) and (2) is due to Harish-Chandra, and (3) is due to [B-Z]. For the proof, see [B-Z, 3.21]. During the proof, Bernstein and Zelevinsky prove the following lemma, which will be used in **7**:

Lemma [B-Z, proof of 2.40]. *Assertion* (2) *in the above theorem implies that for every $v \in V$ and every open compact subgroup K' of $\mathrm{GL}_n(F)$, the space:*
$$E_{v,K'} := span\{\pi(K')\pi(g)v \mid g \in \mathrm{GL}_n(F) \text{ and } v_F(\det g) = 0\}$$
is of finite dimension.

7. Admissibility

Following Harish-Chandra, we say that a smooth representation (π, V) is **admissible** if for every compact open subgroup K' the dimension of the fixed-point space $V^{K'}$ is finite. The following corollary is due to Harish-Chandra, Jacquet, and Howe. The formulation here is due to [B-Z]:

Theorem [B-Z, 3.25 and 2.41]. *Every cuspidal representation is admissible. Every irreducible representation is admissible.*

Proof. Let (π, V) be cuspidal, so that (π, V) is finitely generated and the lemma in **6** is applicable. Denote by J a finite set of generators for V, and let K' be a compact open subgroup of $\mathrm{GL}_n(F)$. Denote by G^1 the subgroup of $\mathrm{GL}_n(F)$ of matrices with determinant of valuation 0 and notice that ZG^1 is of finite index in $\mathrm{GL}_n(F)$. Fix a finite set X of representatives of the left cosets of ZG^1. Then we have:
$$V^{K'} = \sum_{v \in J,\, x \in X} \left(\pi(G^1)\pi(Z)\pi(x)v\right)^{K'} = \sum_{v \in J,\, x \in X} E_{\pi(x)v,K'},$$

in the notation of the lemma. The lemma implies that this space is finite-dimensional.

To prove the second conclusion of the theorem, suppose that (π, V) is not cuspidal but is irreducible. We first embed (π, V) in an induced representation obtained from a cuspidal representation of a proper parabolic subgroup (trivial on the unipotent radical). To do so, we use **5.1** (1), taking P minimal among parabolics with the property that $V_{U_P} \neq 0$. By the obvious property of inductivity for Jacquet modules, V_{U_P} is quasicuspidal. In addition, V_{U_P} is finitely generated. In fact, V is finitely generated, even has a single generator v. Also $\mathrm{GL}_n(F) = P\,\mathrm{GL}_n(O_F)$ (Iwasawa decomposition). Now $\mathrm{GL}_n(O_F) \cap Stab\, v$ is of finite index in $\mathrm{GL}_n(O_F)$, and the claim follows. We can then use Zorn's lemma to obtain an irreducible quotient (ρ, W) of V_{U_P}; this is cuspidal. It is now enough to prove admissibility for $ind_P^{\mathrm{GL}_n(F)}(\rho, W)$, where P is a proper parabolic and (ρ, W) is cuspidal. Let K' be as above; there exists $a \in \mathbb{N}$ big enough so that $K' \supset K_a$, where:

$$K_a = Id + \wp^a\, End_{O_F}(O_F^n, O_F^n).$$

It is therefore enough to prove that the space of K_a fixed vectors, for any $a \in \mathbb{N}$, is of finite dimension. Fix such an a, and notice that $\mathrm{GL}_n(O_F)$ contains and normalizes K_a. Use the Iwasawa decomposition to write:

$$\begin{aligned}\left(ind_P^{\mathrm{GL}_n(F)}\rho, W\right)^{K_a} &= \left(ind_{\mathrm{GL}_n(O_F)\cap P}^{\mathrm{GL}_n(O_F)}\rho, W\right)^{K_a} \\ &\hookrightarrow \bigoplus_{k\in \mathrm{GL}_n(O_F)/K_a} \left(ind_{K_a\cap P}^{K_a}\rho, W\right)^{K_a} \\ &\simeq \bigoplus_{k\in \mathrm{GL}_n(O_F)/K_a} W^{K_a\cap P}.\end{aligned}$$

The set $\mathrm{GL}_n(O_F)/K_a$ is finite and $W^{K_a \cap P}$ is finite-dimensional by admissibility of cuspidal representations. The second conclusion of the theorem follows.

The theorem obviously implies that any smooth representation of finite length is admissible. Conversely Howe has proved that any admissible and finitely generated representation is of finite length; see [B-Z, 4.1] for a proof.

Using the same method as in the proof of the theorem and the computation of the Jacquet module of induced representations (see **5.2**), one obtains that for any parabolic P with unipotent radical U_P and for any smooth representation (π, V) of $\mathrm{GL}_n(F)$ of finite length, the Jacquet module V_{U_P} is admissible (in fact of finite length). This result is due to Jacquet and can be proved without **5.2**.

8. Projectivity of Cuspidal Representations

Theorem see [B-Z, 2.44]. *Let (π, V) be a cuspidal irreducible representation of $\mathrm{GL}_n(F)$, and let (ρ, W) be any smooth representation of $\mathrm{GL}_n(F)$ with a central character. Then there exists a decomposition of W into two $\mathrm{GL}_n(F)$-invariant subspaces:*

$$W = W_\pi \oplus W[\pi],$$

such that W_π is semisimple isotypic of type π and $W[\pi]$ has no subquotient isomorphic to π.

To prove this result, Bernstein and Zelevinsky prove the theorem below. Let χ be a character of the center Z of $\mathrm{GL}_n(F)$, and denote by $C_{c,\chi}^\infty(\mathrm{GL}_n(F))$ the space of locally constant functions that are χ-semi-invariant under Z and have compact support modulo Z.

Theorem [B-Z]. *Let (π, V) be a cuspidal irreducible representation of $\mathrm{GL}_n(F)$ with central character χ^{-1}.*

(i) *The natural embedding, given by the local coefficients (see (2) in the theorem of **6**):*

$$V \otimes V^* \hookrightarrow C^\infty_{c,\chi}(\mathrm{GL}_n(F))$$

has an inverse:

$$f \in C^\infty_{c,\chi} \mapsto \Big(\pi(f) : v \in V \mapsto \int_{Z\backslash \mathrm{GL}_n(F)} f(g)\pi(g)v\,dg\Big),$$

for a suitable choice of Haar measure dg. Hence $V \otimes V^*$ is a direct factor in $C^\infty_{c,\chi}(\mathrm{GL}_n(F))$ stable for the $\mathrm{GL}_n(F)$ right and left actions.

(ii) *As a subquotient of $C^\infty_{c,\chi}(\mathrm{GL}_n(F))$, $V \otimes V^*$ appears only with multiplicity one.*

For more information on the structure of $C^\infty_{c,\chi}(\mathrm{GL}_n(F))$, see [B].

Remark. Any cuspidal irreducible representation (π, V) has a nontrivial extension by itself. In fact, form the representation ρ of $\mathrm{GL}_n(F)$ in $V \oplus V$ defined by:

$$\rho(g)(v_1 + v_2) = (\pi(g)v_1 + v_F(\det g)\pi(g)v_2) \oplus \pi(g)v_2$$

for all $g \in \mathrm{GL}_n(F)$ and $v_1 \oplus v_2 \in V \oplus V$. This representation has no central character and hence is not semisimple. It is of length 2.

9. The Subquotient Theorem

This theorem is due to Casselman. See [C].

Theorem. *Let (π, V) be an irreducible smooth representation of $\mathrm{GL}_n(F)$. Then there exist a parabolic subgroup P of $\mathrm{GL}_n(F)$ with unipotent radical U_P and a cuspidal irreducible representation (ρ, W) of P trivial on U_P such that (π, V) is isomorphic to a subquotient of the induced representation from (ρ, W). Moreover, (π, V) has the same property relative to another set of data $\{P', (\rho', W')\}$ if and only if any Levi subgroup, $M_{P'}$, of P' is conjugate to a Levi subgroup M_P of P in such a way that $\rho'_{|M_{P'}}$ is isomorphic to $\rho_{|M_P}$.*

The existence of P and (ρ, W) is clear; see the proof of the theorem in **7**. In that construction we realize (π, V) as a submodule of $ind_P^{\mathrm{GL}_n(F)}(\rho, W)$, with (ρ, W) a cuspidal representation of P/U_P. To prove the asserted uniqueness, fix such an embedding. Assume now that $\{P', (\rho', W')\}$ is another set of data such that (π, V) is isomorphic to a subquotient of $ind_{P'}^{\mathrm{GL}_n(F)}(\rho', W')$. The exactness of the Jacquet module functor implies that (ρ, W) is a subquotient of $ind(\rho', W')_{U_P}$. One can compute this space; see **5.2**. So in the notation of **5.2** (except that the $'$ has been interchanged), there exists $w \in W^P\backslash W/W^{P'}$ such that (ρ, W) is a subquotient of:

$$ind^{M_P}_{M_P\cap wM_{P'}w^{-1}} ad(w)\rho'_{U_{M_{P'}\cap w^{-1}Pw}}.$$

As (ρ, W) is cuspidal, it follows from the projectivity (see **8**) and Frobenius reciprocity that such a representation cannot be a subquotient of a proper induced representation. In other words $M_P \cap wM_{P'}w^{-1} = M_P$. Moreover cuspidality of (ρ', W') still implies that $U_{M_{P'} \cap w^{-1}Pw} = \{Id\}$; this means that $M_{P'}$ is included in a Levi subgroup of $w^{-1}Pw$. These conclusions combine to give the equality:

$$M_P = wM_{P'}w^{-1}.$$

Also ρ is a subquotient of $ad(w)\rho'$ and hence is isomorphic to $ad(w)\rho'$.

We have still to prove the converse that any set of data $\{P', (\rho', W')\}$ as in the theorem leads to an embedding. In more sophisticated terms it is enough to prove that the semisimplification of an induced representation depends only on the association class of the inducing data. To prove this statement, the simplest way is to use character theory as in the archimedean case. We omit the details.

Remarks. The last theorem in effect defines what is called the **cuspidal support** of a smooth irreducible representation, namely the set of conjugates under $\mathrm{GL}_n(F)$ of the pair $(M_P, \rho_{|M_P})$ in the theorem. In first approximation, the cuspidal support is analogous to the infinitesimal character of an Harish-Chandra module in the archimedean case. To have a classification of the equivalence classes of the smooth irreducible representations, as in the archimedean case, one has to introduce other basic objects; usually one takes the discrete series representations and the tempered representations. In **11**, we will explain how to obtain the Langlands classification with these objects.

10. Discrete Series and Tempered Representations

10.1. The definitions.

Definition. Let (π, V) be a smooth irreducible representation of $\mathrm{GL}_n(F)$. We assume that the central character of π is unitary. The representation (π, V) is said to be a **discrete series** if all its local coefficients (see **6**) are square integrable modulo the center.

To understand this definition in an algebraic context, one has to use the following formula due to Casselman ([C]) and true for any reductive group: Let (π, V) be a smooth representation of $\mathrm{GL}_n(F)$, and let P be a parabolic subgroup of $\mathrm{GL}_n(F)$. Denote by M_P a Levi subgroup of P and by π_P the representation of P in the Jacquet module V_{U_P}. Denote by $\overline{P}$ the opposite parabolic subgroup of P containing M_P and by $V_{U_{\overline{P}}}$ the corresponding Jacquet module. Recall (see **5.1**) that we have a duality between V_{U_P} and $(V^*)_{U_{\overline{P}}}$ denoted by $\langle \cdot, \cdot \rangle_P$. Fix $v \in V$, $v^* \in V^*$, and $\mathcal{V}$ a neighborhood of 1 in $\mathrm{GL}_n(F)$. Then there exists an open compact subgroup K_u of U_P such that for all a in the center of M_P with $aK_ua^{-1} \subset \mathcal{V}$:

$$\langle \pi(a)v, v^* \rangle = \delta_P^{1/2}(a) \langle \pi_P(a)v_P, v^*_{\overline{P}} \rangle_P.$$

Here δ_P is the module function of P, and v_P (resp. $v^*_{\overline{P}}$) is the image of v in V_{U_P} (resp. of v^* in $V_{U_{\overline{P}}}$).

Notice that the center of M_P acts by generalized characters on V_{U_P}; these characters are called the **exponents** of π relative to P. So an algebraic definition of discrete series is:

Definition. If (π, V) is as above, then (π, V) is a discrete series if and only if the real part of its exponents relative to any parabolic subgroup are linear combination of the positive simple roots with strictly positive coefficients.

By the inductive property of the Jacquet modules, one can replace "any parabolic subgroup" in the preceeding definition by "any parabolic subgroup minimal for the property $V_P \neq 0$."

The equivalence of the two definitions is easily seen when the cuspidal support of the representation is the conjugacy class of a character of the diagonal torus; one has only to use the Cartan decomposition. In the general case, one has to use the Cartan decomposition and **6** (2).

Definition. Let (π, V) be a smooth irreducible representation of $\mathrm{GL}_n(F)$. Then (π, V) is said to be a **tempered representation** if its exponents relative to any parabolic subgroup are linear combinations with nonnegative coefficients of the positive simple roots.

This condition is equivalent to the following: (π, V) is a subquotient of an induced representation of a discrete series of a Levi subgroup of $\mathrm{GL}_n(F)$.

Notice that the induced representation obtained with a discrete series is unitary, hence semisimple.

The direction $\Leftarrow$ of the equivalence in the definition is obvious using **5.2**. To prove the other direction, assume that the exponents of (π, V) have the right property. Fix P a parabolic subgroup such that V_{U_P} is nonzero but cuspidal (see **9**), and fix Q a parabolic subgroup containing P, with Levi subgroup M_Q, and such that V_{U_Q} contains an irreducible quotient τ_Q of M_Q with unitary central character. We assume that Q is minimal with this property. Using the exactness and the inductive property of Jacquet modules, one proves that τ_Q is a discrete series of M_Q. By Frobenius reciprocity (π, V) is a submodule of the induced representation $ind_Q^{GL_n(F)} \tau_Q$.

10.2. The classification of the discrete series for $\mathrm{GL}_n(\mathbf{F})$.

For a general reductive group, one does not know how to classify the discrete series in terms of cuspidal representations. But for $\mathrm{GL}_n(F)$ this has been done by Zelevinsky [Z]. The basic fact to be able to classify such representations is to know when the induced representation from a maximal parabolic subgroup and a cuspidal representation of such a group is reducible (as in all this paper, cuspidal does not imply unitary). For $\mathrm{GL}_n(F)$ this has been done by Bernstein and Zelevinsky [B-Z2]: let $m, m' \in \mathbb{N}$ such that $m + m' = n$. Let ρ, ρ' be irreducible cuspidal representations of $\mathrm{GL}_m(F)$ and $\mathrm{GL}_{m'}(F)$ respectively. The induced representation of $\rho \otimes \rho'$ is usually denoted $\rho \times \rho'$, and:

$$\rho \times \rho' \text{ is reducible if and only if } m = m' \text{ and } \rho' \simeq \rho \otimes |\det{}_{\mathrm{GL}_m(F)}|^{\pm 1}.$$

Let $a, b \in \mathbb{N}$ such that $n = ab$, and let ρ be an irreducible cuspidal representation of $\mathrm{GL}_b(F)$. Fix a flag of F^n:

$$X_a := F^n \supset X_{a-1} \supset \cdots \supset \cdots \supset X_i \supset \cdots \supset X_0 = 0$$

such that $\dim X_i = ib$ for all $i \in [0, a]$. Denote by P the stabilizer of this flag. The group P is a parabolic subgroup with Levi subgroup M isomorphic to a copies of $\mathrm{GL}_b(F)$. So we can define the induced representation from P to $\mathrm{GL}_n(F)$:

$$\underset{i \in [1,a]}{\times} (\rho \otimes |\det{}_{\mathrm{GL}_b(F)}|^{(a-2i+1)/2}). \tag{1}$$

Theorem [Z]. *With the above notation, the representation* (1) *admits a unique irreducible submodule, denoted* $\delta(\rho, a)$. *This submodule is a discrete series. If* (π, V) *is an irreducible discrete series of* $\mathrm{GL}_n(F)$, *then there exists* a, b, ρ *as above, unique up to isomorphism, such that:*

$$(\pi, V) \simeq \delta(a, \rho).$$

10.3. Interpretation in terms of the Local Langlands Conjecture.

Fix a, b, ρ as in **10.2**. First, one notices that the cuspidal support of $\delta(\rho, a)$ is the conjugacy classes of:

$$\big(\times_{i\in[1,a]} \mathrm{GL}_b(F), \otimes_{i\in[1,a]} \rho|\det{}_{\mathrm{GL}_b(F)}|^{(a-2i+1)/2}\big). \tag{1}$$

We assume that the Local Langlands Conjecture has been proved for $\mathrm{GL}_b(F)$ (see the introduction for references and in this volume see [Kn]). This means that to ρ we have an associated morphism (W_F is the Weil group of F):

$$\sigma_\rho : W_F \to \mathrm{GL}_b(\mathbb{C})$$

satisfying a list of properties; one of them is that the representation of W_F defined by σ_ρ is irreducible. One embeds $\mathrm{GL}_b(\mathbb{C})$ in $\mathrm{GL}_n(\mathbb{C})$ by fixing an isomorphism:

$$\mathbb{C}^n \simeq \mathbb{C}^b \otimes \mathbb{C}^a.$$

So one looks at:

$$\sigma_\rho : W_F \to \mathrm{GL}_b(\mathbb{C}) \otimes \{1\} \subset \mathrm{GL}_b(\mathbb{C}) \otimes \mathrm{GL}_a(\mathbb{C}) \to \mathrm{GL}_n(\mathbb{C}).$$

The irreducibility of σ_ρ implies:

$$Cent_{\mathrm{GL}_n(\mathbb{C})}\sigma_\rho(W_F) = \mathrm{GL}_a(\mathbb{C}).$$

This means that any morphism from $W_F \times SL_2(\mathbb{C})$ into $\mathrm{GL}_n(\mathbb{C})$ with σ_ρ as restriction to W_F is of the form $\sigma_\rho \otimes \psi_u$, where ψ_u is a homomorphism from $SL_2(\mathbb{C})$ into $\mathrm{GL}_a(\mathbb{C})$. Zelevinsky has proved:

Theorem [Z]. *There exists a bijection between the set of equivalence classes of smooth irreducible representations of* $\mathrm{GL}_n(F)$ *and the set of conjugacy classes of algebraic homomorphisms from* $W_F \times SL_2(\mathbb{C})$ *into* $\mathrm{GL}_n(\mathbb{C})$. *In this bijection, the image of* $\delta(\rho, a)$ *is the class of homomorphisms* $\sigma_\rho \otimes \psi_u$ *such that the image of a regular unipotent element in* $SL_2(\mathbb{C})$ *is a regular unipotent element in* $\mathrm{GL}_a(\mathbb{C})$.

The bijection is consistent with the properties of the Local Langlands Conjecture. See Zelevinsky [Z, sec. 10].

This shows that $W_F \times SL_2(\mathbb{C})$ is the group necessary for the Langlands functoriality, in contrast with the archimedean case where W_F is enough. This fact was discovered explicitly by Deligne probably by looking geometrically at the case of GL_2. The group $W_F \times SL_2(\mathbb{C})$ is now called the **Weil-Deligne group** of F. For another form of the Weil-Deligne group and the relationship between the two, see [Kn, sec. 8].

We have to be careful when we look at a more general group than $\mathrm{GL}_n(F)$. It is not true that for a general reductive group, cuspidal representations should correspond to homomorphisms from $W_F \times SL_2(\mathbb{C})$ into the L-group, trivial on $SL_2(\mathbb{C})$. This failure amounts to the fact that the (conjectural) lift of a cuspidal representation of a reductive group to a general linear group has no reason to be cuspidal.

10.4. The classification of the tempered representations for general reductive groups.

With obvious notational changes the results of 10.1 extend to general reductive groups. Then the classification of the tempered representations follows from the (unknown) classification of the discrete series and from the study of the reducibility of the induced representations by discrete series. This last step can be done, in principle, as in the archimedean case, using the theory of the R-group: one defines R using the intertwining oparators. This is due to Harish-Chandra and Silberger; see [S2] and [S3]. For a discussion of the R-group, see [L, sec. IV.2]. But the explicit determination of the R-group is not known in general; examples are known where R is not abelian. Also one guesses that in general there is no isomorphism between $\mathbb{C}[R]$ and $\mathbb{C}[End\,\pi]$ (π being the induced representation); instead one might have to twist $\mathbb{C}[R]$ by a cocycle. But for $\mathrm{GL}_n(F)$ we have no reducibility of this kind; this has been proved by Zelevinsky without use of the R-group:

Theorem [Z]. *The induced representation from an irreducible discrete series is irreducible.*

11. The Langlands Quotient Theorem

Theorem ([S]; see also [Co]). *Let (π, V) be an irreducible smooth representation of $GL_n(F)$. Then there exist a unique (up to conjugacy) parabolic subgroup Q with Levi subgroup M_Q and a unique isomorphism class of tempered representations ρ of M_Q with central character in the positive Weyl chamber, such that π is a quotient of the induced representation $ind_Q^{\mathrm{GL}_n(F)}\rho$. In this case π appears with multiplicity one as a subquotient of $ind_Q^{\mathrm{GL}_n(F)}\rho$, and $ind_Q^{\mathrm{GL}_n(F)}\rho$ has π as a unique irreducible quotient.*

For the proof we follow [Co]. Let P be a parabolic subgroup with Levi subgroup M_P. Let us introduce notation for the flag stabilized by P:

$$0 \subset F^{n_1} \subset \cdots \subset F^{n_t=n} \tag{1}$$

with $t \in \mathbb{N}$ and $0 < n_1 < \cdots < n_t = n$. Let λ be a character of the center of M_P. Using **4.2**, we can write λ as the product of t characters of F^* in the form $\lambda = \{\chi_i|\cdot|_F^{s_i}, i \in [1,t]\}$. Define recursively the set of numbers:

$$\ell_1 := \max\big\{i \in [1,t] \,\big|\, Re\,\big(\sum_{j\in[1,i]} s_j\big)/n_i \text{ is minimal}\big\},$$

$$\ell_2 := \max\big\{i \in [\ell_1+1,t] \,\big|\, Re\,\big(\sum_{j\in[\ell_1+1,i]} s_j\big)/n_i \text{ is minimal}\big\},$$

$$\vdots$$

Call $r-1$ the last step where ℓ_{r-1} is defined. Look at the subflag of (1) given by:

$$0 \subset F^{\ell_1} \subset \cdots \subset F^{\ell_{r-1}} \subset F^{\ell_r:=n}.$$

Denote by $P_{\lambda,-}$ the stabilizer of that flag and by $M_{\lambda,-}$ its Levi subgroup containing M_P. The restriction of λ to the center of $M_{\lambda,-}$ is denoted λ_-. (In fact, $Re\,\lambda_-$ is just the orthogonal projection of $Re\,\lambda$ on the negative Weyl chamber, but we do not want to stop to define all the words in this sentence.)

Now, let $\tilde{\pi}$ be a smooth irreducible representation of $\mathrm{GL}_n(F)$. Take P to be a parabolic subgroup minimal with respect to the property that the Jacquet module of $\tilde{\pi}$ relative to the unipotent radical U_P of P is nonzero. Decompose:

$$\tilde{\pi}_{U_P} = \bigoplus_{\lambda \in X(\tilde{\pi}, P)} \tilde{\pi}_\lambda, \tag{2}$$

where $X(\tilde{\pi}, P)$ is a suitable subset of the set of smooth characters of the center of M_P and where, for $\lambda \in X(\tilde{\pi}, P)$, $\tilde{\pi}_\lambda$ is the λ generalized eigenspace. We have defined $P_{\lambda,-}$ above, and we fix $\lambda_0 \in X(\tilde{\pi}, P)$ such that $\tilde{\pi}_{\lambda_0} \neq 0$ and $P_0 := P_{\lambda_0,-}$ is minimal with this property. Denote by U_0 the unipotent radical of P_0 and by M_0 a Levi subgroup containing M_P. As in (2), decompose:

$$\tilde{\pi}_{U_0} = \bigoplus_{\nu \in X(\tilde{\pi}, P_0)} \tilde{\pi}_\nu.$$

Put:

$$\tilde{\pi}_0 := \bigoplus_{\nu \in X(\tilde{\pi}, P_0), \nu_- = \lambda_-} \tilde{\pi}_\nu.$$

By the obvious transitivity of the Jacquet module functor, $\tilde{\pi}_0 \neq 0$. Moreover such a representation is tempered (its exponents are nonnegative combinations of positive roots). Choose an irreducible quotient τ of $\tilde{\pi}_0$ as in the proof of **7**. Using Frobenius reciprocity, we have an embedding:

$$\tilde{\pi} \hookrightarrow ind_{P_0}^{\mathrm{GL}_n(F)} \tau.$$

The regularity of λ_- as a character of the center of M_0 and the computation of **5.2** prove that τ appears as a subquotient of the Jacquet module $\big(ind_{P_0}^{\mathrm{GL}_n(F)} \tau\big)_{U_0}$ precisely with multiplicity one (for more details see [Co, Lemme 6], but the computation is a standard application of **5.2**). By exactness of the Jacquet module functor, we find that $\tilde{\pi}$ appears with multiplicity one as a subquotient of $ind_{P_0}^{\mathrm{GL}_n(F)} \tau$. Let π' be any irreducible submodule of $ind_{P_0}^{\mathrm{GL}_n(F)} \tau$. By Frobenius reciprocity, π'_{U_0} admits τ as a quotient. Then the exactness of the Jacquet module functor and the multiplicity-one property of τ yield $\pi' = \tilde{\pi}$.

Finally use duality: take $\tilde{\pi} = \pi^*$. Then we obtain π as the unique irreducible quotient of $ind_{P_0}^{\mathrm{GL}_n(F)} \tau^*$, and π appears as a subquotient of this induced representation with multiplicity one. Existence in the theorem is proved. Uniqueness is proved using the same kind of argument.

References

[B] I. N. Bernstein, Le centre de Bernstein, rédigé par P. Deligne, *Représentations des Groupes Réductifs sur un Corps Local* (I. N. Bernstein, P. Deligne, D. Kazhdan, and M.-F. Vignéras, eds.), Herrmann, coll. travaux en cours, Paris, 1984.

[B-Z] I. N. Bernshtein and A. V. Zelevinskii, Representations of the Group $GL_n(F)$ where F is a non-Archimedean local field (Russian), *Uspehi Mat. Nauk* **31** (1976), no. 3(189), 5–70; Russian Math. Surveys **31:3** (1976), 1–68.

[B-Z2] I. N. Bernstein and A. V. Zelevinsky, Induced representations of reductive $\mathfrak{p}$-adic groups I, *Annales Scient. Ecole Norm. Sup.* **10** (1977), 441–472.

[Bo] A. Borel, *Linear Algebraic Groups*, W. A. Benjamin, New York, 1969; second edition, Graduate Texts in Mathematics, Springer-Verlag, vol. 126, 1991.

[Bo-T] A. Borel and J. Tits, Groupes réductifs, *Publ. Math. I.H.E.S.* **27** (1965), 55–150.

[Bu-K] C. J. Bushnell and P. C. Kutzko, *The Admissible Dual of* GL(N) *via Compact Open Subgroups*, Annals of Mathematics Studies, vol. 129, Princeton University Press, Princeton, NJ, 1993.

[C] W. Casselman, *Introduction to the theory of admissible representations of p-adic reductive groups*, unpublished notes.

[Co] F. Courtes, *Le Théorème du Quotient de Langlands pour les Corps p-adiques*, Mémoire de DEA, Université de Paris 7, 1991.

[Ha1] M. Harris, Supercuspidal representations in the cohomology of Drinfel'd upper half spaces; elaboration of Carayol's program, *Invent. Math.*, in press.

[Ha2] M. Harris, *The local Langlands conjecture for GL(n) over a p-adic field* $n < p$, preprint, Paris, 1996.

[He1] G. Henniart, La conjecture de Langlands locale numérique pour GL(n), *Annales Scient. Ecole Norm. Sup.* **21** (1988), 497–544.

[He2] G. Henniart, Caractérisation de la correspondence de Langlands locale par les facteurs ϵ de paires, *Invent. Math.* **113** (1993), 339–350.

[J-L] Jacquet, H., and R. P. Langlands, *Automorphic Forms on* GL(2), Lecture Notes in Mathematics, vol. 114, Springer-Verlag, Berlin, 1970.

[Kn] A. W. Knapp, Introduction to the Langlands program, *these* PROCEEDINGS, pp. 245–302.

[L] R. P. Langlands, *Les Débuts d'une Formule des Traces Stable*, Publications Mathématiques, vol. 13, L'Université Paris VII, Paris, 1983.

[M] C. Moeglin, Representations of GL(n) over the real field, *these* PROCEEDINGS, pp. 157–166.

[S1] A. J. Silberger, The Langlands quotient theorem for p-adic groups, *Math. Annalen* **236** (1978), 95–104.

[S2] A. J. Silberger, The Knapp-Stein dimension theorem for p-adic groups, *Proc. Amer. Math. Soc.* **68** (1978), 243–246; **76** (1979), 169–170.

[S3] A. J. Silberger, *Introduction to Harmonic Analysis on Reductive p-adic Groups*, Mathematical Notes, vol. 23, Princeton University Press, Princeton, NJ, 1979.

[T] M. Tadić, Classification of unitary representations in irreducible representations of general linear group (non-Archimedean case), *Annales Scient. Ecole Norm. Sup.* **19** (1986), 335–382.

[V] D. A. Vogan, The unitary dual of GL(n) over an archimedean field, *Invent. Math.* **83** (1986), 449–505.

[Z] A. V. Zelevinsky, Induced representations of reductive $\mathfrak{p}$-adic groups II, *Annales Scient. Ecole Norm. Sup.* **13** (1980), 165–210.

DÉPARTEMENT DE MATHÉMATIQUES, UNIVERSITÉ DE PARIS VII, F-75 251 PARIS CEDEX 05, FRANCE

E-mail address: moeglin@math.jussieu.fr

Proceedings of Symposia in Pure Mathematics
Volume **61** (1997), pp. 321–329

Principal L-functions for $GL(n)$

Hervé Jacquet

1. Introduction

Our purpose is to explain how "Tate's thesis" ([T]) generalizes to the general linear group $GL(n)$. In this way, we hope to stress the analogy between ordinary L-functions and automorphic L-functions (see [B]). The reference [J] contains an extensive bibliography of papers on this subject. Among these papers we quote [Tam], [Ma] and [GJ]: [Tam] is one of the earliest papers on the subject; [Ma] and [GJ] contain the idea needed to treat the case of the cusp forms; and [GJ] and [J] contain a complete account of the theory.

For an introduction, we briefly review Tate's thesis in the case of $\mathbb{Q}$. Let p be a place of $\mathbb{Q}$, that is, a prime number or infinity. Consider the local field $\mathbb{Q}_p$. Thus $\mathbb{Q}_\infty = \mathbb{R}$ and, for p a prime number, $\mathbb{Q}_p$ is the field of p-adic numbers. Let χ be a character of $\mathbb{Q}_p^\times$ and Φ be a Schwartz-Bruhat function on $\mathbb{Q}_p$. The **local Zeta integral of Tate** has the form:

$$Z(\Phi, \chi, s) = \int_{\mathbb{Q}_p^\times} \Phi(x)\, \chi(x) |x|^s \, d^\times x. \tag{1}$$

Here $d^\times x$ is a Haar measure on $\mathbb{Q}_p^\times$. The integral converges for $\Re s > 0$ and extends to a meromorphic function of s. Tate introduces a **local L-factor** $L(s, \chi)$. Its main property is that the ratio

$$\frac{Z(\Phi, \chi, s)}{L(s, \chi)}$$

is an entire function of s; moreover, the ratio is identically 1 for a suitable choice of Φ. The ratio satisfies a functional equation. To formulate it we choose a nontrivial additive character ψ of $\mathbb{Q}_p$ and we denote by

$$\widehat{\Phi}(x) = \int_{\mathbb{Q}_p} \Phi(x)\psi(xy)\, dx$$

the corresponding **Fourier transform** of Φ. The Haar measure is taken to be self-dual with respect to ψ, that is,

$$\Phi(0) = \int \widehat{\Phi}(x)\, dx.$$

1991 *Mathematics Subject Classification.* Primary 11R39, 11R42, 22E55.

Then the functional equation reads:

$$\frac{Z(\widehat{\Phi}, \chi^{-1}, 1-s)}{L(1-s, \chi^{-1})} = \epsilon(s, \chi, \psi) \frac{Z(\Phi, \chi, s)}{L(s, \chi)}. \tag{2}$$

The factor ϵ that appears in the equation is an exponential function of s.

If p is finite and χ is unramified, that is, has the form

$$\chi(x) = |x|^{s_1}$$

then

$$L(s, \chi) = \frac{1}{1 - p^{-s_1 - s}}.$$

If furthermore the largest ideal on which the character ψ is trivial is the ring of p-adic integers $\mathbb{Z}_p$, then the ϵ factor is 1.

Now we go to a global situation. We let χ be a character of $\mathbb{Q}_{\mathbb{A}}^{\times}$ which is trivial on $\mathbb{Q}^{\times}$. It will be convenient to denote by $\mathbb{Q}^1$ be the set of ideles of absolute value 1. Then $\mathbb{Q}_{\mathbb{A}}^{\times}$ is the direct product of $\mathbb{R}^+$ and $\mathbb{Q}^1$. Moreover the quotient $\mathbb{Q}^1/\mathbb{Q}^{\times}$ is compact. We will assume that χ is trivial on $\mathbb{R}^+$. We consider the **global Zeta integral of Tate**

$$Z(\Phi, \chi, s) = \int_{\mathbb{Q}_{\mathbb{A}}^{\times}} \Phi(x)\chi(x)|x|^s \, d^{\times}x, \tag{3}$$

where $d^{\times}x$ is a Haar measure on $\mathbb{Q}_{\mathbb{A}}^{\times}$ and Φ is a Schwartz-Bruhat function on $\mathbb{A}$. The integral converges for $\Re s > 1$. We can write it in the form:

$$Z(\Phi, \chi, s) = \int_0^{\infty} \Big(\int_{\mathbb{Q}^1/\mathbb{Q}^{\times}} \sum_{\xi \in \mathbb{Q}^{\times}} \Phi(tz\xi) \, dz \Big) t^s \, d^{\times}t \tag{4}$$

Here $d^{\times}t = dt/t$ is a Haar measure on $\mathbb{R}^+$ and dz a suitable Haar measure on $\mathbb{Q}^1$. To continue we let ψ be a nontrivial character of $\mathbb{A}$ trivial on $\mathbb{Q}$. Such a character is a product of local ones:

$$\psi(x) = \prod_p \psi_p(x_p),$$

if $x = (x_p)$. Moreover, for almost all p, the ring of p-adic integers $\mathbb{Z}_p$ is the largest ideal on which ψ_p is trivial. The **Fourier transform** of Φ is then defined by:

$$\widehat{\Phi}(x) = \int_{\mathbb{A}} \Phi(y)\psi(xy) \, dy.$$

Once more the Haar measure dx on $\mathbb{A}$ is self-dual:

$$\int \widehat{\Phi}(x) \, dx = \Phi(0).$$

The Poisson summation formula reads:

$$\sum_{\xi \in \mathbb{Q}} \Phi(\xi) = \sum_{\xi \in \mathbb{Q}} \widehat{\Phi}(\xi).$$

We can use the Poisson summation formula to write

$$\sum_{\xi \in \mathbb{Q}^{\times}} \Phi(tz\xi) = \sum_{\xi \in \mathbb{Q}^{\times}} \widehat{\Phi}(z^{-1}t^{-1}\xi)t^{-1} + \widehat{\Phi}(0)t^{-1} - \Phi(0).$$

We break up the outer integral of (4) into the sum of the integral from 0 to 1 and the integral from 1 to infinity and use the above relation to get:

$$Z(\Phi,\chi,s)= \int_1^{\infty}\int_{\mathbb{Q}^1/\mathbb{Q}^\times}\sum_{\xi\in\mathbb{Q}^\times}\Phi(tz\xi)\,dz\,t^s\,d^\times t \tag{5}$$

$$+\int_0^1\int_{\mathbb{Q}^1/\mathbb{Q}^\times}\sum_{\xi\in\mathbb{Q}^\times}\widehat{\Phi}(t^{-1}z^{-1}\xi)\,dz\,t^{1-s}\,d^\times t \tag{6}$$

$$+\widehat{\Phi}(0)\int_{\mathbb{Q}^1/\mathbb{Q}^\times}\chi(z)\,dz\int_0^1 t^{s-1}\,d^\times t \tag{7}$$

$$-\Phi(0)\int_{\mathbb{Q}^1/\mathbb{Q}^\times}\chi(z)\,dz\int_0^1 t^s\,d^\times t. \tag{8}$$

For simplicity, we shall assume that χ is a nontrivial character. Then the two last integrals are zero. We may further change t to t^{-1} and z to z^{-1} in the second expression to arrive at:

$$Z(\Phi,\chi,s)= \int_1^{+\infty}\int_{\mathbb{Q}^1/\mathbb{Q}^\times}\sum_{\xi\in\mathbb{Q}^\times}\Phi(tz\xi)\,dz\,t^s\,d^\times t \tag{9}$$

$$+\int_1^{+\infty}\int_{\mathbb{Q}^1/\mathbb{Q}^\times}\sum_{\xi\in\mathbb{Q}^\times}\Phi(tz\xi)\,dz\,t^{1-s}\,d^\times t. \tag{10}$$

This new expression is convergent for all values of s and gives the analytic continuation of the Zeta integral as an entire function of s. In addition, we obtain the functional equation:

$$Z(\widehat{\Phi},\chi^{-1},1-s)=Z(\Phi,\chi,s). \tag{11}$$

Next Tate defines:

$$L(s,\chi)=\prod_p L(s,\chi_p). \tag{12}$$

The product is over all places p, including the infinite place. This infinite product converges absolutely for $\Re s>1$. He also defines:

$$\epsilon(s,\chi)=\prod_p \epsilon(s,\chi_p,\psi_p).$$

In this product, almost all factors are equal to 1 and the product does not depend on the choice of ψ. The properties of the global Zeta integral translate to the fact that $L(s,\chi)$ extends to an entire function of s and satisfies the functional equation:

$$L(s,\chi)=\epsilon(s,\chi)L(1-s,\chi^{-1}). \tag{13}$$

Suppose that $\chi=1$ (trivial character). Then $L(s,1)$ is the Riemann Zeta function—times the appropriate Γ-factor. The above discussion must then be slightly modified and the result is that $L(s,1)$ is meromorphic with simple poles at $s=0$ and $s=1$. In general if χ is nontrivial, it defines a primitive Dirichlet character χ_0 say and then $L(s,\chi)$ equals the Dirichlet L-function $L(s,\chi_0)$.

2. Local Theory for $GL(2)$

We pass to the case of the group $GL(2)$. We first consider the local situation. We consider a unitary irreducible representation π of $G_p = GL(2, \mathbb{Q}_p)$ on a Hilbert space H. A function of the form

$$\omega(g) = (\pi(g)u, u), \qquad u \in H, \ \|u\| = 1,$$

will be called a **matrix coefficient** of π.

In particular, suppose p is a prime number and set

$$K_p = GL(2, \mathbb{Z}_p).$$

This is a compact open subgroup of G_p. We say that π is **spherical** if it contains a (unit) vector u fixed under K. The vector is then unique up to a scalar factor and the corresponding coefficient is the spherical function ω_π attached to π. Such representations are parametrized by pairs of complex numbers (z_1, z_2). Indeed, let s_i be such that $p^{-s_i} = z_i$. Let ϕ be the function on G_p defined by

$$\phi\left[\begin{pmatrix} a_1 & x \\ 0 & a_2 \end{pmatrix} k\right] = |a_1|^{1/2+s_1}|a_2|^{-1/2+s_2}, \qquad k \in K_p.$$

Then

$$\omega_\pi(g) = \int_K \phi(kg)\, dk.$$

In fact, the complex numbers z_1, z_2 are limited by the condition $|z_i| = 1$ or $p^{-1/2} \leq \Re z_i \leq p^{1/2}$.

In general to a given representation π we can associate **local Zeta integrals** of the form:

$$Z(\Phi, \omega, s) = \int_{G_p} \Phi(g)\, \omega(g) |\det g|^{s+1/2}\, d^\times g, \tag{14}$$

where ω is a matrix coefficient of π and Φ is a Schwartz-Bruhat function on the space of 2×2 matrices. This integral converges for $\Re s$ sufficiently large and extends to a meromorphic function of s. As in the case of $GL(1)$ we introduce a factor $L(s, \pi)$. Its main property is the fact that the ratio

$$\frac{Z(\Phi, \omega, s)}{L(s, \pi)}$$

is entire. Moreover, we can choose ω_i and Φ_i, $1 \leq i \leq r$, in such a way that

$$\sum_i Z(\Phi_i, \omega_i, s) = L(s, \pi).$$

In particular, if p is a prime number and π is spherical as above, then

$$L(s, \pi) = \frac{1}{(1 - z_1 p^{-s})(1 - z_2 p^{-s})}.$$

Let Φ be the characteristic function of $M(2 \times 2, \mathbb{Q}_p)$. Since Φ is K-invariant:

$$Z(\Phi, \omega_\pi, s) = \int \Phi(g)\phi(g) |\det g|^{s+1/2}\, dg \tag{15}$$

$$= \int \Phi\left[\begin{pmatrix} a_1 & x \\ 0 & a_2 \end{pmatrix}\right] |a_1|^{s+s_1}|a_2|^{s+s_2}\, dx\, d^\times a_1\, d^\times a_2 \tag{16}$$

$$= L(s, \pi). \tag{17}$$

There is also a functional equation. We denote by $\widehat{\Phi}$ the **Fourier transform** of Φ:

$$\widehat{\Phi}(x) = \int_{M(2\times 2,\mathbb{Q}_p)} \Phi(y)\psi(\mathrm{tr}(xy))\,dy.$$

We set $\check{\omega}(g) = \omega(g^{-1})$. This is a matrix coefficient of the representation $\tilde{\pi}$ contragredient to π. Then the functional equation reads:

$$\frac{Z(\widehat{\Phi},\check{\omega},1-s)}{L(1-s,\tilde{\pi})} = \epsilon(s,\pi,\psi)\,\frac{Z(\Phi,\omega,s)}{L(s,\pi)} \tag{18}$$

Here the factor ϵ is again an exponential function of s.

3. Global Theory for $GL(2)$

We pass to the global theory for $GL(2)$. We regard $G = GL(2)$ as an algebraic group over $\mathbb{Q}$. We set $G(\mathbb{A}) = GL(2,\mathbb{A})$, $G(\mathbb{Q}) = GL(2,\mathbb{Q})$ and $G_p = GL(2,\mathbb{Q}_p)$. We denote by G^1 the set of $g \in GL(2,\mathbb{A})$ such that $|\det g| = 1$. The quotient $G(\mathbb{Q})\backslash G^1$ has finite volume. Recalling that $\mathbb{Q}_{\mathbb{A}}^{\times}$ is the direct product of $\mathbb{R}^+$ and $\mathbb{Q}^1$, we let Z^+ be the group of scalar matrices of the form

$$t = \begin{pmatrix} t & 0 \\ 0 & t \end{pmatrix}$$

with t in $\mathbb{R}^+$. Then we have a direct product $G(\mathbb{A}) = G^1 Z^+$. Whenever convenient we identify a function on G^1 with a function on $G(\mathbb{A})$ invariant under Z^+. We also denote by A the group of diagonal matrices, by P the group of upper triangular matrices and by N the group of upper triangular matrices with unit diagonal. A function f on G^1 is said to be **automorphic** if

$$f(\gamma g) = f(g)$$

for all $\gamma \in G(\mathbb{Q})$. The **constant term** of f is then the function f_N on $N(\mathbb{A})\backslash G(\mathbb{A})$ defined by:

$$f_N(g) := \int_{N(\mathbb{Q})\backslash N(\mathbb{A})} f(ng)\,dn = \int_{\mathbb{A}/\mathbb{Q}} f\left[\begin{pmatrix} 1 & x \\ 0 & 1 \end{pmatrix} g\right] dx.$$

For the purpose of this lecture we define an **automorphic representation** π of $G(\mathbb{A})$ as a unitary irreducible representation that occurs discretely in the Hilbert space

$$L^2(G(\mathbb{Q})\backslash G^1).$$

This Hilbert space contains the closed invariant subspace L_0 spanned by the cuspidal functions, that is, the functions f such that $f_N = 0$. The representation of $G(\mathbb{A})$ on L_0 decomposes discretely (with multiplicity one). The other representations which appear discretely are the ones of the form $g \mapsto \chi(\det(g))$, where χ is a character of $\mathbb{Q}^1/\mathbb{Q}^{\times}$. Thus let π be a component of L_0. One can write such a representation as an infinite tensor product $\pi = \otimes \pi_p$, where π_p is an irreducible representation of G_p. The precise meaning of such a tensor product decomposition is hard to describe briefly and may be found in [F]. For our purpose, it will suffice to say that π has matrix coefficients of the form

$$\omega(g) = \prod_p \omega_p(g_p).$$

For all p, ω_p is a matrix coefficient of π_p. In addition, for almost all p, the representation π_p is spherical and the coefficient ω_p is the corresponding spherical coefficient. Consider the **global Zeta integral**

$$Z(\Phi, s, \omega) = \int_{G(\mathbb{A})} \Phi(g)\omega(g)|\det g|^{s+1/2}\, d^\times g. \tag{19}$$

It converges when $\Re s$ is sufficiently large. We are going to see that it extends to an entire function of s and satisfies the functional equation:

$$Z(\widehat{\Phi}, \check{\omega}, 1-s) = Z(\Phi, \omega, s). \tag{20}$$

Indeed, we can write ω as

$$\omega(g) = \int_{G(\mathbb{Q})\backslash G^1} f_1(hg)\overline{f}_2(h)\, dh$$

where f_1, f_2 belong to L_0. Then

$$Z(\Phi, \omega, s) = \int_{G(\mathbb{A})} \Phi(g)\Big(\int_{G(\mathbb{Q})\backslash G^1} f_1(hg)\overline{f}_2(h)dh\Big)|\det g|^{s+1/2}\, dg.$$

Changing variables gives:

$$\int\int \Phi(h^{-1}g)f_1(g)\overline{f}_2(h)\, dh\, |\det g|^{s+1/2}\, dg$$

or

$$\int_0^{+\infty}\Big[\int\int\Big(\sum_{\xi\in G(\mathbb{Q})} \Phi(h_2^{-1}\xi t h_1)\Big) f_1(h_1)\overline{f}_2(h_2)\, dh_1\, dh_2\Big] t^{2s+1}\, d^\times t.$$

The outer integral is over $\mathbb{R}^+$ and the inner integrals over $G(\mathbb{Q})\backslash G^1$. As in the case of $GL(1)$ we break the integral further into the sum of

$$\int_1^{+\infty}\Big[\int\int\Big(\sum_{\xi\in G(\mathbb{Q})} \Phi(h_2^{-1}\xi t h_1)\Big) f_1(h_1)\overline{f}_2(h_2)\, dh_1\, dh_2\Big] |t|^{2s+1}\, d^\times t \tag{21}$$

and

$$\int_0^{1}\Big[\int\int\Big(\sum_{\xi\in G(\mathbb{Q})} \Phi(h_2^{-1}\xi t h_1)\Big) f_1(h_1)\overline{f}_2(h_2)\, dh_1\, dh_2\Big] |t|^{2s+1}\, d^\times t. \tag{22}$$

Just as in the case of $GL(1)$ the first integral converges for all s. Of course the domain of integration is not compact but this is compensated by the fact that the functions f_i are rapidly decreasing in a suitable sense. We use the Poisson summation formula to transform the integrand of the second term:

$$\sum_{\xi\in G(\mathbb{Q})} \Phi(h_2^{-1}\xi t h_1)t^{2s+1} = \sum_{\xi\in G(\mathbb{Q})} \widehat{\Phi}(h_1^{-1}\xi t^{-1} h_2)\, t^{2s-3} \tag{23}$$

$$+ \sum_{\sigma} \widehat{\Phi}(h_1^{-1}\sigma t^{-1} h_2)\, t^{2s-3} \tag{24}$$

$$- \sum_{\sigma} \Phi(h_2^{-1}\sigma t h_1)\, t^{2s+1} \tag{25}$$

$$+ \widehat{\Phi}(0)\, t^{2s-3} \tag{26}$$

$$- \Phi(0)\, t^{2s+1}. \tag{27}$$

The sums in (24) and (25) are over all rational matrices σ of rank 1. We have to integrate this against $f_1(h_1)\overline{f}_2(h_2)dh_1dh_2$. The terms containing $\Phi(0)$ and $\widehat{\Phi}(0)$ do not contribute because the functions f_i are orthogonal to the constant functions on G^1. The sum over all matrices of rank 1 in (25) can be written as

$$\sum_{\gamma_1\in P(\mathbb{Q})\backslash G(\mathbb{Q})}\sum_{\gamma_2\in P(\mathbb{Q})\backslash G(\mathbb{Q})}\sum_{\alpha\in\mathbb{Q}^\times}\Phi\left[h_1^{-1}\gamma_1^{-1}\begin{pmatrix}0&0\\0&\alpha\end{pmatrix}t\gamma_2h_2\right].$$

Integrating against the functions f_i we find:

$$\int_{P(\mathbb{Q})\backslash G^1}\int_{P(\mathbb{Q})\backslash G1}\sum_{\alpha\in\mathbb{Q}^\times}\Phi\left[h_1^{-1}\begin{pmatrix}0&0\\0&\alpha\end{pmatrix}th_2\right]f_1(h_1)\overline{f}_2(h_2)\,dh_1\,dh_2.$$

But

$$\Phi\left[h_1^{-1}\begin{pmatrix}0&0\\0&\alpha\end{pmatrix}th_2\right]$$

depends only on the class of h_i modulo $N(\mathbb{A})$. Thus this integral equals

$$\int_{A(\mathbb{Q})N(\mathbb{A}))\backslash G^1}\int_{A(\mathbb{Q})N(\mathbb{A})\backslash G^1}\sum_{\alpha\in\mathbb{Q}^\times}\Phi\left[h_1^{-1}\begin{pmatrix}0&0\\0&\alpha\end{pmatrix}th_2\right]f_{1N}(h_1)\overline{f}_{2N}(h_2)\,dh_1\,dh_2,$$

which is 0 because of the cuspidality of f_1 and f_2. Thus the expression (22) is in fact equal to

$$\int_0^1\left[\int\int\Big(\sum_{\xi\in G(\mathbb{Q})}\widehat{\Phi}(h_1^{-1}\xi t^{-1}h_2)\Big)f_1(h_1)\overline{f}_2(h_2)\,dh_1\,dh_2\right]t^{2s-3}\,d^\times t.$$

Changing t to its inverse, we finally obtain:

$$Z(\Phi,\omega,s)$$
$$=\int_1^{+\infty}\int\int\Big(\sum_{\xi\in G(\mathbb{Q})}\Phi(h_2^{-1}\xi th_1)\Big)f_1(h_1)\overline{f}_2(h_2)\,dh_1\,dh_2\,t^{2s+1}\,d^\times t \quad (28)$$
$$+\int_1^{+\infty}\int\int\Big(\sum_{\xi\in G(\mathbb{Q})}\widehat{\Phi}(h_1^{-1}\xi th_2)\Big)f_1(h_1)\overline{f}_2(h_2)\,dh_1\,dh_2\,t^{3-2s}\,d^\times t. \quad (29)$$

This expression is convergent for all s and provides the analytic continuation of our integrals. Moreover the functional equation (20) is clear.

4. Automorphic L-functions for $GL(2)$

Let π be an automorphic cuspidal representation of $GL(2)$. As in the case of $GL(1)$ we define a **global L-function**:

$$L(s,\pi):=\prod_p L(s,\pi_p). \quad (30)$$

The infinite product converges for $\Re s$ sufficiently large. We also introduce an **epsilon factor**

$$\epsilon(s,\pi)=\prod_p\epsilon(s,\pi_p,\psi_p).$$

Then the results on the Zeta integral amount to saying that $L(s,\pi)$ extends to an entire function of s and satisfies the functional equation

$$L(s,\pi) = \epsilon(s,\pi)L(1-s,\tilde{\pi}). \tag{31}$$

Apart from a translation, the L-function attached to a holomorphic (new) modular form is of the form $L(s,\pi)$ for a suitable π and the previous result is a theorem of Hecke. Note that the classical distinction between old and new forms is absorbed into the representation theory point of view. The advantage of this approach is that it stresses the analogy between abelian L-functions and L-functions attached to automorphic representations of $GL(2)$. Note that if π has the form $\pi(g) = \chi(\det g)$ then:

$$L(s,\pi) = L(s+\tfrac{1}{2},\chi)L(s-\tfrac{1}{2},\chi).$$

The theory extends to $GL(n)$ (see [B], [Tam], [GJ], [J], [Ma]). We first discuss the case of **cuspidal** automorphic representations. The condition of cuspidality is that the constant term along the unipotent radicals of all parabolic subgroups vanish. Of course, this condition is empty if $n=1$. One defines then the space L_0 of cuspidal elements in $L^2(G(\mathbb{Q})\backslash G^1)$. The representation of $G(\mathbb{A})$ on L_0 decomposes discretely (with multiplicity one). A **cuspidal automorphic representation** π is then a component of L_0. One defines an L-function $L(s,\pi)$ as an infinite product of factors $L(s,\pi_p)$. For almost all p the L-factor is the reciprocal of a polynomial of degree n in p^{-s}. The infinite product $L(s,\pi)$ converges for $\Re s$ sufficiently large and extends to an meromorphic function of s that satisfies a functional equation. In fact the function is entire, except in the case $n=1$ and $\pi=1$, the trivial character, where the function has simple poles at $s=0$ and $s=1$.

To have a more complete theory one needs to describe the space of all automorphic forms (see [BJ]). An **automorphic form** is a smooth function on $G(\mathbb{A})$ that is automorphic, i.e., is invariant under $G(\mathbb{Q})$ on the left and satisfies additional conditions. There is a condition of slow growth at infinity. The other conditions are algebraic: the function is supposed to be K-finite on the right, where K is the product of the orthogonal group and the groups $GL(n,\mathbb{Z}_p)$. Finally the function satisfies some differential equations: it is annihilated by an ideal of finite codimension in the center of the enveloping algebra of G_∞. If π is cuspidal automorphic then the K-finite elements of its space are automorphic forms in this sense. The group $G(\mathbb{A})$ (or rather a suitable convolution algebra of distributions) operates on the space of automorphic forms and the irreducible components are called **automorphic representations**. To an automorphic representation π one can still attach an L-function. However, the L-function decomposes (essentially) into a product:

$$L(s,\pi) = \prod_i L(s+s_i,\pi_i),$$

where the representations π_i are automorphic cuspidal for groups $GL(r_i)$. "Essentially" here and below means that the infinite products agree at almost all places. Note that $r_i = 1$ may occur. In particular the above L-functions may have finitely many poles.

We have already observed that the Dirichlet L-functions (for a primitive character) and the Hecke L-functions (for a new form) are automorphic. Part of the conjectures of Langlands is that all the L-functions of classical number theory are essentially automorphic, that is, are essentially of the form $L(s,\pi)$, for a suitable

π. For instance the Dedekind Zeta function of an extension of degree n of $\mathbb{Q}$ should be exactly of the form $L(s, \pi)$ where π is an automorphic representation of $GL(n)$. Likewise the Artin L-functions should be automorphic. More generally, all Diophantine L-functions should be essentially automorphic L-functions.

References

[B] A. Borel, Automorphic L-functions, *Automorphic Forms, Representations, and L-Functions* (A. Borel and W. Casselman, eds.), Proc. Symp. Pure Math., vol. 33, Part II, American Mathematical Society, Providence, 1979, pp. 27–61.

[BJ] A. Borel and H. Jacquet, Automorphic forms and automorphic representations, *Automorphic Forms, Representations, and L-Functions* (A. Borel and W. Casselman, eds.), Proc. Symp. Pure Math., vol. 33, Part I, American Mathematical Society, Providence, 1979, pp. 189–202.

[GJ] R. Godement and H. Jacquet, *Zeta-functions of simple algebras*, Lecture Notes in Mathematics, vol. 260, Springer-Verlag, New York, 1972.

[J] H. Jacquet, Principal L functions of the linear group, *Automorphic Forms, Representations, and L-Functions* (A. Borel and W. Casselman, eds.), Proc. Symp. Pure Math., vol. 33, Part II, American Mathematical Society, Providence, 1979, pp. 63–86.

[Ma] G. N. Maloletkin, Zeta functions of parabolic forms (Russian), *Math. Sb. (N.S.)* **86(128)** (1971), 622–643; English translation in *Math. USSR-Sb*, **15** (1971), 619–641.

[Tam] T. Tamagawa, On the zeta-functions of a division algebra, *Annals of Math.* **77** (1963), 387–405.

[T] J. Tate, Fourier analysis in number fields and Hecke's zeta-functions, Ph.D. Thesis, Princeton University, 1950, *Algebraic Number Theory* (J. W. S. Cassels and A. Fröhlich, eds.), Academic Press, London, 1967, pp. 305–347.

Department of Mathematics, Columbia University, New York, NY 10027–4408, U.S.A.
E-mail address: hj@math.columbia.edu

Proceedings of Symposia in Pure Mathematics
Volume **61** (1997), pp. 331–353

Functoriality and the Artin Conjecture

Jonathan D. Rogawski

Contents

This article contains an exposition of the proof of the Artin conjecture for two-dimensional Galois representations of tetrahedral and octahedral type. The proofs, given in Section 3, are carried out by applying some general theorems about cuspidal representations on $GL(2)$ and $GL(3)$ to a particular situation. As such, they provide good illustrations of how the automorphic formalism works. On the other hand, it should be noted that the arguments ultimately rely on some fortunate but accidental features of the low-dimensional situation. For an interesting mathematical and historical discussion of the base change problem and its relation to the Artin conjecture, we recommend the introduction to [L2].

1991 *Mathematics Subject Classification.* Primary 11R39, 11R42; Secondary 22E55.

The first two sections provide background material, much of which has already been covered in [Kn]. More precisely, Section 1 contains a review of Artin L-functions. In Section 2, we state some general conjectures about automorphic representations of $GL(n)$, emphasizing the analogy between irreducible Galois representations and cuspidal representations. We also state the general theorems needed for the proofs in Section 3.

My thanks are due to Tony Knapp for a careful reading of the manuscript and several helpful suggestions.

Notation. Throughout this article, F denotes a number field, $\overline{F}$ an algebraic closure of F, and $Gal(\overline{F}/F)$ the absolute Galois group of F. We view $Gal(\overline{F}/F)$ as a topological group relative to the Krull topology. All finite extensions of E will implicitly be assumed to be subfields of $\overline{F}$. We write $\mathbf{A}_F$ and I_F, respectively, for the adele ring and idele group attached to F. As usual, we identify F (resp. F^*) with its image in $\mathbf{A}_F$ (resp., I_F) under the diagonal embedding. A **Hecke character** is an idele class character, that is, a continuous homomorphism from I_F to $\mathbb{C}^*$ trivial on F^*. For each place v of F, let F_v denote the completion of F relative to v. We also fix algebraic closures $\overline{F}_v$ of F_v for each place v.

I. Artin L-Functions

1. Definitions

We use the term **Galois representation** to denote a continuous homomorphism

$$\sigma : Gal(\overline{F}/F) \longrightarrow GL(V)$$

where V is a finite-dimensional complex vector space. Recall that σ is continuous if and only if σ factors through the projection $Gal(\overline{F}/F) \to Gal(E/F)$ for some finite extension E/F. The determinant $\det(\sigma)$ of σ is the complex-valued character $x \to \det(\sigma(x))$. It is identified with a Hecke character of I_F via the Artin isomorphism of class field theory. The relevant part of class field theory is reviewed in Sections 4–5 of [Kn].

The Artin L-function $L(s,\sigma)$ attached to σ is an Euler product

$$L(s,\sigma) = \prod_v L(s,\sigma_v).$$

Here our convention is that v runs over all places of F, archimedean and non-archimedean; by contrast archimedean places are not included in the definition in Section 5 of [Kn]. To define the local factor $L(s,\sigma_v)$, choose an embedding $i'_v : \overline{F} \to \overline{F}_v$. This gives rise to an embedding of Galois groups

$$i_v : Gal(\overline{F}_v/F_v) \longrightarrow Gal(\overline{F}/F)$$

via restriction. The composition $\sigma_v = \sigma \circ i_v$ is a continuous representation of $Gal(\overline{F}_v/F_v)$. It depends on the choice of i'_v, but different choices of i'_v lead to conjugate embeddings i_v. The equivalence class of σ_v is therefore well-defined and depends only on v.

In the nonarchimedean case, let k_v and $\overline{k}_v$ denote the residue fields of F_v and $\overline{F}_v$, respectively. Then $Gal(\overline{F}_v/F_v)$ acts on $\overline{k}_v$ and we have an exact sequence

$$1 \longrightarrow I_v \longrightarrow Gal(\overline{F}_v/F_v) \longrightarrow Gal(\overline{k}_v/k_v) \longrightarrow 1,$$

where I_v is the inertia subgroup. Set $q_v = Card(k_v)$. A **Frobenius element** Fr_v is an element of $Gal(\overline{F}_v/F_v)$ whose image in $Gal(\overline{k}_v/k_v)$ is the automorphism $x \to x^{q_v}$. The action of $\sigma_v(Fr_v)$ on the subspace V^{I_v} of inertial invariants in V is independent of the choice of Fr_v and we define the local factor at v by:

$$L(s, \sigma_v) = \det(1 - q^{-s}\sigma_v(Fr_v)|V^{I_v})^{-1}$$

The representation σ is said to be **unramified** at v if $\sigma_v(I_v) = 1$. In this case, the element $\sigma_v(Fr_v)$ is independent of the choice of Fr_v. The **Frobenius class** attached to v is the *conjugacy class* $\{\sigma_v(Fr_v)\}$ of $\sigma_v(Fr_v)$ in $GL(V)$. The Frobenius class is independent of the choice of embedding i_v and thus depends only on v. Furthermore, it is a **semisimple** conjugacy class, i.e., it consists of diagonalizable elements. Indeed, since Image(σ) is a finite group, $\sigma_v(Fr_v)$ is a linear transformation of finite order, hence diagonalizable. Furthermore, its eigenvalues $z_1, \ldots, z_n$ are roots of unity. Identifying $GL(V)$ with $GL_n(\mathbb{C})$, we have

$$\sigma_v(Fr_v) \sim \begin{pmatrix} z_1 & & \\ & \ddots & \\ & & z_n \end{pmatrix}$$

where $\sim$ denotes conjugacy. In this case, the definition yields

$$L(s, \sigma_v) = \prod_{j=1}^{n} (1 - q_v^{-s} z_j)^{-1}$$

Observe that $\det(\sigma_v(Fr_v)) = z_1 \cdots z_n$. Under the local Artin isomorphism sending Fr_v to a uniformizing element $\varpi_v \in F_v^*$, $\det(\sigma_v)$ is identified with the unramified character of F_v^* defined by $x \to z^{val(x)}$, where $z = z_1 \cdots z_n$ and $val(x)$ is the v-adic valuation on F_v^*.

If v is archimedean, then $F_v \approx \mathbb{R}$ or $\mathbb{C}$. In the first case, $Gal(\overline{F}_v/F_v) \approx Gal(\mathbb{C}/\mathbb{R}) = \{1, c\}$, where c denotes complex conjugation. The eigenvalues of $\sigma_v(c)$ are ± 1. We set

$$L(s, \sigma_v) = (\pi^{-\frac{s}{2}}\Gamma(\tfrac{s}{2}))^k (\pi^{-\frac{s+1}{2}}\Gamma(\tfrac{s+1}{2}))^\ell$$

where k (resp., ℓ) are the number of $+1$ (resp., -1) eigenvalues of $\sigma_v(c)$. If $F_v \approx \mathbb{C}$, then $Gal(\overline{F}_v/F_v)$ is the trivial group and we set

$$L(s, \sigma_v) = (2(2\pi)^{-s}\Gamma(s))^n$$

where $n = \dim(\sigma)$.

With these definitions, it is clear that the correspondence $\sigma \to L(s, \sigma)$ is **additive** in the sense that

$$L(s, \sigma \oplus \tau) = L(s, \sigma)L(s, \tau)$$

for any two Galois representations σ and τ.

For any finite set of places S, the partial L-function is defined as the Euler product

$$L_S(s, \sigma) = \prod_{v \notin S} L(s, \sigma_v).$$

In particular, $L(s, \sigma) = L_S(s, \sigma)$ for $S = \phi$.

Example 1. Let σ be the trivial representation and let S be the set of archimedean places. Then $L_S(s, \sigma) = \prod_{v<\infty}(1 - q_v^{-s})^{-1}$ is the Dedekind zeta function $\zeta_F(s)$ of F.

2. The Artin Conjecture

As mentioned above, the eigenvalues of $\sigma_v(Fr_v)$ are roots of unity. It follows easily that the Euler product for $L(s,\sigma)$ converges absolutely in the half plane $\mathrm{Re}(s) > 1$. The following theorem combines results of Hecke, Artin, and Brauer.

Theorem 1. *$L(s,\sigma)$ extends analytically to a meromorphic function on the complex plane $\mathbb{C}$.*

The Artin L-function $L(s,\sigma)$ also satisfies a functional equation of the form $L(s,\sigma) = \varepsilon(s,\sigma)L(1-s,\sigma^*)$ where σ^* is the contragredient representation to σ and $\varepsilon(s,\sigma)$ is the so-called "epsilon factor" [T].

We can now state the famous

Artin Conjecture. *If σ is irreducible and nontrivial, then $L(s,\sigma)$ can be analytically continued to an entire function of s.*

3. The Frobenius Classes Determine σ

Let $S(\sigma)$ be the set of places v such that either v is archimedean or σ is ramified at v. Thus σ_v is unramified if and only if $v \notin S(\sigma)$. The continuity of σ implies that $S(\sigma)$ is a finite set. It is useful to emphasize the following point:

> *a Galois representation σ defines a family of semisimple conjugacy classes $\{\sigma_v(Fr_v)\}$ in $GL_n(\mathbb{C})$ indexed by $v \notin S(\sigma)$.*

The following basic theorem asserts that this collection determines σ uniquely.

Theorem 2. *Let σ_1 and σ_2 be Galois representations of dimension n such that $\sigma_{1v}(Fr_v) \sim \sigma_{2v}(Fr_v)$ for almost all v. Then $\sigma_1 \simeq \sigma_2$.*

PROOF. This theorem is an immediate consequence of the Tchebotarev density theorem [La], [N]. Choose an extension E/F such that both σ_1 and σ_2 factor through the projection $\pi : Gal(\overline{F}/F) \to Gal(E/F)$. The Tchebotarev density theorem implies that every conjugacy class in $Gal(E/F)$ is of the form $\{\pi\circ\sigma_v(Fr_v)\}$ for infinitely many primes v of F (in fact, it says that the density of primes for which $\{\pi\circ\sigma_v(Fr_v)\}$ is a given class c in $Gal(E/F)$ is the correct one, namely $|c|/N$ where $N = |Gal(E/F)|$). In particular, if $\sigma_{1v}(Fr_v) \sim \sigma_{2v}(Fr_v)$ for almost all v, then the characters of σ_1 and σ_2 are equal and hence are equivalent. □

We shall see below that cuspidal representations also give rise to collections of conjugacy classes in $GL_n(\mathbb{C})$ and are determined by them in an analagous fashion (Sec. 9).

4. Change of Field: Induction and Restriction

Artin L-functions behave well with respect to induction and restriction. Let E/F be a finite extension and let ρ be a Galois representation of $Gal(\overline{F}/E)$. We write $Ind_E^F(\rho)$ for the representation of $Gal(\overline{F}/F)$ **induced** from ρ. Recall that $Ind_E^F(\rho)$ is the representation of $Gal(\overline{F}/F)$ by right translation on the space of all functions $f : Gal(\overline{F}/F) \to V$ such that $f(xy) = \rho(x)f(y)$ for all $x \in Gal(\overline{F}/E)$ and $y \in Gal(\overline{F}/F)$ (here V is the space on which ρ acts).

If σ is a representation of $Gal(\overline{F}/F)$, we write σ_E for the **restriction** of σ to $Gal(\overline{F}/E)$. Let

$$R_{E/F} = Ind_E^F(1)$$

denote the representation of $Gal(\overline{F}/F)$ induced from the trivial representation of $Gal(\overline{F}/E)$.

Proposition 3. *Let E/F be a finite extension.*

1. *The formation of L-functions is invariant under induction. More precisely, if ρ is a representation of $Gal(\overline{F}/E)$ and $\sigma = Ind_E^F(\rho)$, then $L(s,\sigma) = L(s,\rho)$.*
2. *Let σ be a representation of $Gal(\overline{F}/F)$. Then $L(s,\sigma_E) = L(s,\sigma \otimes R_{E/F})$.*

PROOF. Part (1) is proved in [La] (see also [Kn], Section 5). It is equivalent to the assertion

$$L(s,\sigma_v) = \prod_{w|v} L(s,\rho_w)$$

for all places v of F. Part (2) follows (1) and the general **projection formula** for induced representations: if ρ is a representation of a group G and ρ_H is its restriction to a subgroup H of finite index, then $Ind_H^G(\rho_H) \simeq Ind_H^G(1) \otimes \rho$. In our case, this formula yields $Ind_E^F(\sigma_E) = \sigma \otimes R_{E/F}$. □

II. Cuspidal Representations

Let G denote the algebraic group $GL(n)$ and Z the subgroup of scalar matrices. For any commutative ring R with identity, $G(R) = GL_n(R)$ and $Z(R) \simeq R^*$. If v is a place of F, we write G_v for $GL_n(F_v)$.

To describe the cuspidal representations of $GL_n(\mathbf{A}_F)$, we fix a unitary Hecke character ξ of I_F and regard it as a character of $Z(F)\backslash Z(\mathbf{A}_F)$ in the obvious way:

$$\begin{pmatrix} \lambda & & \\ & \ddots & \\ & & \lambda \end{pmatrix} \longrightarrow \xi(\lambda).$$

Let $L^2(\xi)$ be the Hilbert space of measurable functions $\varphi(g)$ on $GL_n(F)\backslash GL_n(\mathbf{A}_F)$ such that $\varphi(zg) = \xi(z)\varphi(g)$ for all $z \in Z(\mathbf{A}_F)$ and

$$\int_{GL_2(F)Z(\mathbf{A}_F)\backslash GL_2(\mathbf{A}_F)} |\varphi(g)|^2 dg < \infty.$$

Then $G(\mathbf{A}_F)$ acts on $L^2(\xi)$ by right translation ρ. The center $Z(\mathbf{A}_F)$ acts on $L^2(\xi)$ via ξ.

In general, if π is an irreducible admissible representation of $G(\mathbf{A}_F)$ or G_v, the center ($Z(\mathbf{A}_F)$ or Z_v) acts by a character ω_π called the **central character** of π. We view ω_π as a character of I_F or F_v^* in the two cases. Observe that if $n = 1$, then $G = Z$ and $L^2(\xi)$ is a one-dimensional vector space consisting of multiples of the function ξ.

5. Subspace of Cuspidal Functions

To define the subspace $L_o^2(\xi)$ of cuspidal functions ([GGP], [H], [JL]) let N be a standard unipotent subgroup of G attached to a partition $n = n_1 + \cdots + n_r$. By definition, N is the subgroup of elements of G with identity matrices of size $n_j \times n_j$

along the diagonal, arbitrary entries above them and zeroes below. For example, the partition $5 = 2 + 3 + 2$ corresponds to the subgroup of matrices of the form

$$\begin{pmatrix} \multicolumn{2}{c}{\multirow{2}{*}{I_2}} & * & * & * & * & * \\ & & * & * & * & * & * \\ 0 & 0 & & & & * & * \\ 0 & 0 & & I_3 & & * & * \\ 0 & 0 & & & & * & * \\ 0 & 0 & 0 & 0 & 0 & \multicolumn{2}{c}{\multirow{2}{*}{I_2}} \\ 0 & 0 & 0 & 0 & 0 & & \end{pmatrix}$$

The **constant term** of an element $\varphi \in L^2(\xi)$ relative to N is the function

$$\varphi_N(g) = \int_{N(F)\backslash N(\mathbf{A}_F)} \varphi(ng)dn.$$

We say that φ is **cuspidal** if, for all standard unipotent subgroups other than $\{e\}$, $\varphi_N(g) = 0$ for almost all g. Let $L^2_o(\xi)$ be the subspace of cuspidal functions. It is clearly invariant under ρ. We denote the restriction of ρ to $L^2_o(\xi)$ by ρ_o. See Section 7 of [Kn] for a discussion of the historical origins of "constant term" and "cuspidal."

According to a general theorem of Gelfand-Graev-Piatetskii-Shapiro [H], ρ_o decomposes as a direct sum of irreducible unitary representations of $G(\mathbf{A}_F)$. This result holds for any reductive group in place of $GL(n)$. The irreducible constituents ρ_o are called **cuspidal representations** (with central character ξ). Note that for any π with central character ξ, the contragredient representation π^* has central character ξ^{-1}.

We denote by $\mathcal{A}_F(n,\xi)$ the set of all cuspidal representation of $G(\mathbf{A}_F)$ with central character ξ and $\mathcal{A}_F(n)$ the set of all cuspidal representations (with any unitary central character). In particular, $\mathcal{A}_F(1)$ is simply the set of unitary Hecke characters of the idele group I_F.

6. Multiplicity-One Theorem

It is a basic fact that each representation occurring in the decomposition of the representation ρ_o of $GL_n(\mathbf{A})$ appears with multiplicity one. This result is due to Jacquet-Langlands [JL] for $n = 2$ and to Shalika [S] for general n. Accordingly, we have

$$\rho_o \simeq \bigoplus_{\pi \in \mathcal{A}_F(n,\xi)} \pi.$$

This multiplicity-one result need not hold for groups other than $GL(n)$. For example, it is false for $SL(n)$ if $n > 3$ [Bl]. See [BR] for a discussion of multiplicity questions.

7. Unramified Representations

We recall some facts about unramified representations of the groups G_v. Assume that v is nonarchimedean. Recall that an irreducible admissible representation π_v of G_v is called **unramified** if the space of π_v contains a nonzero vector fixed under the maximal compact subgroup $GL_n(\mathcal{O}_v)$, where $\mathcal{O}_v$ is the ring of integers of F_v. Fortunately, the unramified representations are easily classified. Let $B \subset G$ be the standard Borel subgroup of upper-triangular matrices in G, that is, the subgroup of matrices (b_{ij}) in G such that $b_{ij} = 0$ for $i > j$. An unramified character χ of B_v is a character of the form

$$\chi\left(\begin{pmatrix} a_1 & \cdots & * \\ & \ddots & * \\ & & a_n \end{pmatrix}\right) = \prod z_j^{val(a_j)}$$

where $z_1, ..., z_n$ are nonzero complex numbers. The representations $Ind_{B_v}^{G_v}(\chi)$ for χ unramified are called **unramified principal series representations**. Here, $Ind_{B_v}^{G_v}(\chi)$ denotes normalized induction, so that $Ind_{B_v}^{G_v}(\chi)$ is unitary if χ is unitary. For $\sigma \in S_n$, let $\sigma\chi$ denote the character defined by the collection $z_{\sigma(1)}, ..., z_{\sigma(n)}$.

Theorem 4.

1. *Let χ be an unramified character of B_v.. Then $Ind_{B_v}^{G_v}(\chi)$ has a unique constitutent $\pi(\chi)$ that is unramified.*
2. *Every unramified representation is of the form $\pi(\chi)$ for some unramified χ.*
3. *Let χ_1, χ_2 be unramified characters of B_v. Then $\pi(\chi_1) = \pi(\chi_2)$ if and only $\sigma\chi_1 = \chi_2$ for some $\sigma \in S_n$.*

See [Ca] for more details about these facts.

This theorem allows us to attach a conjugacy class of semisimple elements $\{g(\pi_v)\}$ in $GL_n(\mathbb{C})$ to each unramified representation π_v. Namely, we set

$$g(\pi_v) = \begin{pmatrix} z_1 & & \\ & \ddots & \\ & & z_n \end{pmatrix}$$

if $\pi_v = \pi(\chi)$ and χ is defined as above by $z_1, ..., z_n$. The contragredient π_v^* of $\pi_v = \pi(\chi)$ is the representation $\pi(\chi^{-1})$. Therefore $g(\pi_v^*) = g(\pi_v)^{-1}$.

An unramified character of Z_v (which we identify with F_v^*) is of the form $x \to z^{val(x)}$ for some $z \in \mathbb{C}^*$. The center Z_v acts on $Ind_{B_v}^{G_v}(\chi)$ (and hence also on $\pi(\chi)$) by the character $x \to (z_1 \cdots z_n)^{val(x)}$. Therefore the central character of $\pi(\chi)$ is the unramified character corresponding to $z = \det(g(\pi_v))$. The classification can be restated as follows.

Theorem 5. *Let v be a finite place. Then there is a bijection $\pi_v \to \{g(\pi_v)\}$ between the set of isomorphism classes of unramified representations of $GL_n(F_v)$ and the set of semisimple conjugacy classes in $GL_n(\mathbb{C})$. The central character of π_v is the character of F_v^* defined by $x \to (\det(g(\pi_v)))^{val(x)}$.*

Definition 1. The conjugacy class $\{g(\pi_v)\}$ of the unramified representation π_v of $GL_n(F_v)$ is called the **Langlands class** of π_v.

Sometimes it is useful to restate the classification in terms of Galois representations. A representation σ of $Gal(\overline{F}_v/F_v)$ is called **unramified** if it factors through the projection to $Gal(\overline{k}_v/k_v)$. Such a representation is uniquely determined by the image $\sigma(Fr_v)$, which is independent of the choice of Frobenius element Fr_v. We associate to π_v the unique unramified representation

$$\rho(\pi_v) : Gal(\overline{F}_v/F_v) \longrightarrow GL_n(\mathbb{C}).$$

such that $\rho(\pi_v)(Fr_v) = g(\pi_v)$. By the previous theorem, this gives a *bijection* between the set of irreducible unramified representations of $GL_n(F_v)$ and the set of unramified n-dimensional representations of $Gal(\overline{F}_v/F_v)$ such that the image of Fr_v is semisimple. Of course, this is the easy part of the local Langlands correspondence described in Section 8 of [Kn].

8. The L-Function of a Cuspidal Representation

A cuspidal representation $\pi \in \mathcal{A}_F(n)$ can be decomposed as a restricted tensor product over all places of F

$$\pi = \bigotimes{}' \pi_v.$$

For each v, π_v is an irreducible admissible representation of G_v. We refer to [F] for the precise definitions and theorems. For *almost all* nonarchimedean places v, π_v is unramified. Let $S(\pi)$ be the complement of the set of places such that π_v is unramified. Note that $S(\pi)$ contains the archimedean places. The L-function of π is an Euler product over all places v of F :

$$L(s,\pi) = \prod_v L(s,\pi_v).$$

For $v \notin S(\pi)$, the local factor is defined by the formula

$$L(s,\pi_v) = \det(1 - q_v^{-s} g(\pi_v))^{-1} = \prod_{j=1}^{n} (1 - q_v^{-s} z_j)^{-1}.$$

The general definition of $L(s,\pi_v)$ for all places v (i.e., including $v \in S(\pi)$) is given in [J1], [GJ]. We shall not need this definition in the sequel. For any finite set of places S, the L-function $L_S(s,\pi)$ of π is defined as an Euler product

$$L_S(s,\pi) = \prod_{v \notin S} L(s,\pi_v).$$

For the correspondence between these definitions and the definitions for more general reductive groups, see Section 9 of [Kn].

9. Convergence of the Euler Product

The convergence in some half-plane of the Euler product $L(s,\pi)$ attached to $\pi \in \mathcal{A}_F(n)$ is a consequence of the *unitarity* of the local components π_v. More precisely, it can be shown there exists a real number t (depending only on n) such that for all $v \notin S(\pi)$, the eigenvalues $\{z_j\}$ of $g(\pi_v)$ satisfy $|z_j| \leq q^t$ (see [Bo] for a discussion of this point). This implies the absolute convergence of $L(s,\pi)$ in the half-plane $\mathrm{Re}(s) > t+1$. The following theorem is due to Hecke and Jacquet-Langlands [JL] for $n=2$ and to Godement-Jacquet for general n [J1], [GoJ]. For an exposition in this volume, see [J2].

Theorem 6. *Let $\pi \in \mathcal{A}_F(n)$. Then $L(s,\pi)$ has an analytic continuation to an entire function of s.*

The L-function also satisfies a function equation of the form $L(s,\pi) = \varepsilon(s,\pi)L(1-s,\pi^*)$ [J1]. As mentioned above, convergence of the Euler product defining an Artin L-function in a half-plane follows from the fact that the eigenvalues of $\sigma_v(Fr_v)$ are roots of unity and hence have absolute value one for $v \notin S(\sigma)$. For cuspidal representations of $GL(n)$, we have the so-called **generalized Ramanujan conjecture**:

Conjecture 7. *Let $\pi \in \mathcal{A}_F(n)$. Then the eigenvalues of $g(\pi_v)$ have absolute value one for all $v \notin S(\pi)$.*

Unlike the case of Galois representations, however, the eigenvalues of $g(\pi_v)$ are not roots of unity in general (cf. Sec. 11 below).

Conjecture 7 reduces to the classical Ramanujan-Petersson conjecture for modular forms when π is a cuspidal representation of $GL(2)_{/\mathbb{Q}}$ such that the component π_∞ lies in the discrete series or limit of discrete series. This was proved by Deligne [De] for π_∞ discrete series and Deligne-Serre [DS] for π_∞ limit of discrete series (cf. [R]).

10. Strong Multiplicity-One Theorem

The family of Langlands classes attached to a cuspidal representation is analogous to the family of Frobenius classes attached to a Galois representation. In particular,

> *a cuspidal representation $\pi \in \mathcal{A}_F(n)$ defines a family of semisimple conjugacy classes $\{g(\pi_v)\}$ in $GL_n(\mathbb{C})$ indexed by $v \notin S(\pi)$.*

Example 2. If $n = 1$, then π is a Hecke character of I_F. If π_v is unramified, the Langlands class $g(\pi_v)$ lies in $GL_1(\mathbb{C}) = \mathbb{C}^*$. In fact, we have $g(\pi_v) = \pi_v(\varpi_v)$ where $\varpi_v \in F_v^*$ is a prime element.

The analogue of the Theorem 2 is the so-called **strong multiplicity-one** theorem due to Jacquet-Shalika [JS].

Theorem 8 (Strong Multiplicity-One). *Let $\pi_1, \pi_2 \in \mathcal{A}_F(n)$ be cuspidal representations such that $g(\pi_{1v}) \sim g(\pi_{2v})$ for almost all v. Then $\pi_1 = \pi_2$.*

11. The Langlands-Artin Conjecture

In a foundational article published in 1970, Langlands [L1] stated a collection of conjectures known under the general heading **functoriality conjecture**. They imply, as a special case, a relation between Galois representations and cuspidal representations. While transparently simple to state, it has remarkably far-reaching ramifications. So far, it has been established only in a limited number of special cases, some of which are explained in Part III below.

Langlands-Artin Conjecture. *Let σ be an irreducible Galois representation of $Gal(\overline{F}/F)$ of dimension n. Then there exists a cuspidal representation π in $\mathcal{A}_n(F)$ such that for almost all places v, $\sigma_v(Fr_v) \sim g(\pi_v)$.*

By the strong multiplicity-one theorem, there is at most one cuspidal representation π satisfying $\sigma_v(Fr_v) \sim g(\pi_v)$ for almost all v. We write $\pi(\sigma)$ for this representation, if it exists.

Remarks.

1. Suppose that $\pi = \pi(\sigma)$ exists. We may view the determinant character $\det(\sigma)$ as a character of I_F via the Artin map. Then $\omega_\pi = \det(\sigma)$. Indeed, for almost all places v of F, the local components of ω_π and $\det(\sigma)$ are the characters $x \to (\det g(\pi_v))^{val(x)}$ and $x \to (\det \sigma_v(Fr_v))^{val(x)}$, respectively. By the strong multiplicity-one theorem for Hecke characters, we obtain $\omega_\pi = \det(\sigma)$.

2. The condition $\sigma_v(Fr_v) \sim g(\pi_v)$ is equivalent to the equality of local L-functions: $L(s, \sigma_v) = L(s, \pi_v)$.
3. The Langlands-Artin conjecture implies the Artin conjecture. This implication uses Theorem 8.8 of [Kn] and Theorem 6.
4. Suppose that σ is reducible, say $\sigma = \bigoplus_{j=1}^{N} \sigma_j$ with $\dim(\sigma_j) = n_j$. Then $L(s, \sigma) = \prod_{j=1}^{N} L(s, \sigma_j)$ and hence

$$L(s, \sigma) = \prod_{j=1}^{N} L(s, \pi(\sigma_j)).$$

The map $\sigma \longrightarrow \pi(\sigma)$ is a special case of the global Langlands correspondence, which is pieced together from the local Langlands correspondences described in Section 8 of [Kn]). We emphasize that the global correspondence is certainly not surjective, even if it exists. Indeed, if π is of the form $\pi(\sigma)$, the elements $g(\pi_v)$ must be of finite order for almost all π. To say more about the image of the map $\sigma \longrightarrow \pi(\sigma)$, recall that for archimedean v, π_v corresponds to an n-dimensional representation $\rho(\pi_v)$ of the Weil group $W_{\mathbb{C}/\mathbb{R}}$. One conjectures that if $\pi = \pi(\sigma)$, then π_v and σ_v correspond under the Langlands correspondence for *all* places v of F (and not just at the unramified places). This would imply that if v is archimedean, then the representation $\rho(\pi_v)$ is equivalent to the pullback of σ_v via the projection $W_{\mathbb{C}/\mathbb{R}} \rightarrow Gal(\mathbb{C}/\mathbb{R})$. In other words, if $\pi = \pi(\sigma)$, then the archimedean components of π are conjecturally of a very special type, corresponding to Weil group representations that factor through $Gal(\mathbb{C}/\mathbb{R})$. It is sometimes conjectured that the image of $\sigma \longrightarrow \pi(\sigma)$ is *precisely* the set of cuspidal representations whose archimedean components correspond to Weil group representations factoring through $Gal(\mathbb{C}/\mathbb{R})$. This assertion, however, appears to be independent of the general functoriality conjecture.

Consider the case of $GL(2)_{/\mathbb{Q}}$. The cuspidal representations π are divided into two classes, according as the archimedean component π_∞ is "holomorphic" of some weight $k \geq 1$ or not. If π_∞ is holomorphic, then π corresponds to a classical newform of weight $k \geq 1$. According to the Deligne-Serre theorem [DS], $\pi = \pi(\sigma)$ for some Galois representation σ if $k = 1$. If $k > 1$, the Langlands classes of π are not of finite order. In this case, π is associated to a "compatible family" of ℓ-adic representations, but it does not correspond to a complex Galois representation [De]. On the other hand, the nonholomorphic cuspidal representations of $GL(2)_{\mathbb{Q}}$ are attached to classical "Maass forms" on the upper half-plane with eigenvalue λ for the Laplacian [G]. It is conjectured that such a π is of the form $\pi(\sigma)$ for some irreducible Galois representation σ if and only if $\lambda = \frac{1}{4}$. The cuspidal π with $\lambda \neq \frac{1}{4}$ have no apparent connection with Galois theory and one even speculates that almost all of the Langlands classes of such π have transcendental eigenvalues.

12. Tensor Structure

The category of Galois representations has several operations defined on it: tensor product, induction, restriction, etc. A fundamental problem is to determine whether analogous operations exist on the set of cuspidal representations. This would follow from the Langlands-Artin conjecture for the *subset* of cuspidal representations $\pi \in \mathcal{A}_F(n)$ in the image of the map $\sigma \longrightarrow \pi(\sigma)$. The general Langlands conjectures predict that these operations exist on *all* of $\mathcal{A}_F(n)$. In this section, we describe this conjecture and some of its consequences in greater detail.

Consider the tensor product $\sigma_1 \otimes \sigma_2$ of two irreducible Galois representations. If $\sigma_1 \otimes \sigma_2$ is irreducible, then the three (conjectural) cuspidal representations $\pi(\sigma_1), \pi(\sigma_2)$, and $\pi(\sigma_1 \otimes \sigma_2)$ are related by

$$g(\pi_v(\sigma_1)) \otimes g(\pi_v(\sigma_2)) \sim g(\pi_v(\sigma_1 \otimes \sigma_2))$$

for almost all places v. Recall that $\sim$ denotes conjugacy within the general linear group. It is reasonable to think of $\pi(\sigma_1 \otimes \sigma_2)$ as a kind of product of $\pi(\sigma_1)$ and $\pi(\sigma_2)$. We shall write it as $\pi(\sigma_1) \boxtimes \pi(\sigma_2)$. In general, $\sigma_1 \otimes \sigma_2$ need not be irreducible, but will decompose as a direct sum of irreducibles, say $\sigma_1 \otimes \sigma_2 \simeq \bigoplus \tau_j$. In this case

$$g(\pi_v(\sigma_1)) \otimes g(\pi_v(\sigma_2)) \sim \bigoplus g(\pi_v(\tau_j)),$$

where the right-hand side is to be interpreted as a matrix in block diagonal form with blocks made up of the matrices $g(\pi_v(\tau_j))$. By analogy, we conjecture that the tensor product of two cuspidal representations exists, even if they are not attached to Galois representations. More precisely,

Definition 2. Let $\pi \in \mathcal{A}_F(n)$ and $\pi' \in \mathcal{A}_F(m)$. Suppose we are given cuspidal representations $\pi_j \in \mathcal{A}_F(n_j)$ for $j = 1, \ldots, r$ such that $\sum_{j=1}^r n_j = mn$. We shall write

$$\pi \boxtimes \pi' = \bigoplus_{j=1}^{r} \pi_j$$

if $g(\pi_v) \otimes g(\pi'_v) \sim \bigoplus_{j=1}^r g(\pi_{jv})$ for almost all v (where the conjugacy is understood to occur in $GL_{nm}(\mathbb{C})$).

Then we have the following conjecture, which is a piece of the general Langlands functoriality conjecture.

Conjecture 9. *The tensor product $\pi \boxtimes \pi'$ exists for any two cuspidal representations $\pi \in \mathcal{A}_F(n)$ and $\pi' \in \mathcal{A}_F(m)$.*

It is convenient to form the additive monoid

$$\mathcal{A}_F = \bigoplus \mathcal{A}_F(n),$$

consisting of formal sums $\sum_{j=1}^N \pi_j$ where $\pi_j \in \mathcal{A}_F(n_j)$ for some n_j. Define

$$\deg : \mathcal{A}_F \longrightarrow \mathbb{Z}$$

by setting $\deg(\pi) = n$ for $\pi \in \mathcal{A}_F(n)$ and extending to $\mathcal{A}_F$ by additivity. For any element $\pi = \sum_{j=1}^N \pi_j$ in $\mathcal{A}_F$ of degree n, let $\{g(\pi)\}$ denote the conjugacy class of the element $\bigoplus_{j=1}^N g(\pi_j)$ in $GL_n(\mathbb{C})$.

The tensor operation $\boxtimes$ defines a distributive multiplication on $\mathcal{A}_F$. We may also conjecture that other operations of linear algebra exist on $\mathcal{A}_F$, such as exterior or symmetric powers. Recall that a homomorphism $r : GL_n(\mathbb{C}) \to GL_m(\mathbb{C})$ is called an algebraic representation if the entries of $r(g)$ are polynomial functions of the entries in g and $\det(g)^{-1}$. More generally, the functoriality conjecture predicts the following.

Conjecture 10. *Let $r : GL_n(\mathbb{C}) \to GL_m(\mathbb{C})$ be an algebraic representation of $GL_n(\mathbb{C})$. Then for all $\pi \in \mathcal{A}_F(n)$, there exists $\pi' \in \mathcal{A}_F(m)$ such that*

$$r(g(\pi_v)) \sim g(\pi'_v)$$

for almost all v.

13. The Adjoint Lifting

We now discuss a special case in which this last conjecture is known. The group $GL_2(\mathbb{C})$ acts by conjugation on the three-dimensional space of 2×2 matrices of trace zero. This defines the three-dimensional adjoint representation,

$$Ad : GL_2(\mathbb{C}) \longrightarrow GL_3(\mathbb{C}).$$

The following theorem was proved by Gelbart-Jacquet [GJ], generalizing a method introduced by Shimura for the study of symmetric square L-functions of modular forms.

Theorem 11. *Let $\pi \in \mathcal{A}_F(2)$. Then there exists an element, denoted $Ad(\pi)$, of degree 3 in $\mathcal{A}_F$ such that $Ad(g(\pi_v)) \sim g(Ad(\pi)_v)$ for almost all v.*

In particular,

$$\text{if} \quad g(\pi_v) \sim \begin{pmatrix} a & \\ & b \end{pmatrix}, \quad \text{then} \quad g(Ad(\pi)_v) \sim \begin{pmatrix} a/b & & \\ & 1 & \\ & & b/a \end{pmatrix}.$$

We can describe precisely when $Ad(\pi)$ is cuspidal in terms of automorphic induction defined below in Section 15. The element $Ad(\pi)$ is cuspidal if and only if π is not of the form $AI_E^F(\theta)$ for some quadratic extension E/F and some Hecke character θ of E. This is not exactly the description given in [GJ]. However, the results of Labesse-Langlands [LL] show that this description is equivalent with the description given in [GJ].

14. A Theorem of Jacquet and Shalika

The analogy between cuspidal representations of $GL(n)$ and Galois representations can be used very effectively to *predict* results that ought to be true. For example, if σ and τ are n-dimensional Galois representations (or representations of any finite group) with σ irreducible, then σ is isomorphic to τ if and only if $\sigma \otimes \sigma^*$ is isomorphic to $\tau \otimes \sigma^*$, where σ^* is the contragredient to σ. Indeed, $\sigma \otimes \sigma^* \approx Hom(\sigma, \sigma)$ and $\tau \otimes \sigma^* \approx Hom(\sigma, \tau)$. The image of the identity in $Hom(\sigma, \sigma)$ under an isomorphism $\sigma \otimes \sigma^* \to \tau \otimes \sigma^*$ yields an isomorphism of σ with τ. The analogue of this result for cuspidal representations of $GL(n)$ is the following theorem [JS].

Theorem 12 (Jacquet-Shalika). *Let $\pi, \pi' \in \mathcal{A}_F$ be elements of degree n with π cuspidal. If $g(\pi_v) \otimes g(\pi_v^*) \sim g(\pi'_v) \otimes g(\pi_v^*)$ for almost all v, then $\pi = \pi'$.*

15. Induction and Restriction Revisited

Let E/F be a finite extension of degree ℓ. We now define operations that correspond to induction and restriction of Galois representations. The induction operation is called **automorphic induction** and the restriction operation is called **base change**:

$$\begin{aligned} AI_E^F &: \mathcal{A}_E \longrightarrow \mathcal{A}_F && \text{(automorphic induction)} \\ BC_{E/F} &: \mathcal{A}_F \longrightarrow \mathcal{A}_E && \text{(base change).} \end{aligned}$$

Before proceeding to the definitions, we define a Galois action on $\mathcal{A}_F(n)$ (and by extension, on $\mathcal{A}_F$). First consider what happens in the Galois case. Let η be an automorphism of E/F. For any Galois representation σ of $Gal(\overline{F}/E)$, we define $\eta(\sigma)$ by $\eta(\sigma)(x) = \sigma(\eta^{-1}x\eta)$. If Fr_w is a Frobenius element of a place w of E, then $\eta^{-1}Fr_w\eta$ is a Frobenius element of the place $\eta^{-1}(w)$. In particular, $\eta(\sigma)(Fr_w) = \sigma(Fr_{\eta^{-1}(w)})$.

We define the conjugate $\eta(\Pi)$ of a representation Π of $GL_n(\mathbf{A}_E)$ to be the representation sending g to $\Pi(\eta^{-1}(g))$, where $\eta^{-1}(g)$ the matrix obtained by applying η^{-1} to the entries of g. This definition also makes sense in the local case for representations of $GL_n(E_w)$ if η is a Galois automorphism of E_w/F_w. If Π is a cuspidal representation of $GL_n(\mathbf{A}_E)$, then $\eta(\Pi)$ is again cuspidal since it is realized on the space of functions of the form $f(\eta(x))$, where $f(x)$ belongs to the subspace of $L_o(\xi)$ on which Π is realized. Furthermore, if $\Pi = \bigotimes \Pi_w$ (product over places w of E), then $\eta(\Pi) = \bigotimes \eta(\Pi_w)$. The next lemma shows that the Langlands classes of $\eta(\Pi)$ are a permutation of the Langlands classes of Π. It follows in particular that the action of Galois on $\mathcal{A}_E(n)$ is compatible with the correspondence $\sigma \to \pi(\sigma)$ in the sense that if $\pi(\sigma)$ exists, then $\pi(\eta(\sigma))$ exists and $\pi(\eta(\sigma)) = \eta(\pi(\sigma))$.

Lemma 13. *Let η be a Galois automorphism of E and let Π be a cuspidal representation of $GL_n(\mathbf{A}_E)$. Assume that $\Pi_{\eta^{-1}(w)}$ is unramified, where $\eta^{-1}(w)$ is the conjugate of w under η. Then $\eta(\Pi_w)$ is also unramified and $g(\eta(\Pi_w)) \sim g(\Pi_{\eta^{-1}(w)})$.*

PROOF. Let $w' = \eta^{-1}(w)$. Then $val_w(x) = val_{w'}(\eta^{-1}(x))$ for $x \in E$, and η^{-1} induces isomorphisms $E_w \xrightarrow{\sim} E_{w'}$ and $G_w \xrightarrow{\sim} G_{w'}$. Let χ be a character of the Borel subgroup $B_{w'}$ and set $I' = Ind_{B_{w'}}^{G_{w'}}(\chi)$. Then $\eta(I')$ is isomorphic to the representation $I = Ind_{B_w}^{G_w}(\chi \circ \eta^{-1})$. Indeed, the map $f(x) \to f(\eta(x))$ for f in the induced space of I induces an isomorphism of $\eta(I)$ with I'. If χ is unramified and $\Pi_{w'}$ is isomorphic to the unique unramified constituent of I', then $\eta(\Pi_w)$ is isomorphic to the unramified constituent of I. Suppose that χ sends an upper-triangular matrix with diagonal entries $a_1, ..., a_n$ to $\prod z_j^{val_{w'}(a_j)}$. Then $\{z_1, ..., z_n\}$ is the set of eigenvalues of $g(\Pi_{w'})$. But then $\chi \circ \eta^{-1}$ sends an upper-triangular matrix with diagonal entries $a_1, ..., a_n$ to $\prod z_j^{val_w(a_j)}$ and therefore $g(\eta(\Pi_w)) \sim g(\Pi_{\eta^{-1}(w)})$.
□

We now define an operation of automorphic induction so as to correspond to induction of Galois representations. Let $\pi \in \mathcal{A}_E(n)$. By analogy with Artin L-functions, automorphic induction should preserve L-functions. An element $\Pi \in \mathcal{A}_F$ of degree $\deg(\Pi) = n\ell$ is said to be **automorphically induced** by π if

$$L(s, \Pi_v) = \prod_{w|v} L(s, \pi_w)$$

for almost all places v of F. The product is over places of E dividing v. There is at most one Π satisfying this condition for almost all v by the strong multiplicity-one theorem. We write $\Pi = AI_E^F(\pi)$ if such a Π exists. This definition is compatible with the Langlands correspondence in the following sense: if $\pi = \pi(\rho)$ for some Galois representation ρ of $Gal(\overline{F}/E)$, then

$$AI_E^F(\pi(\rho)) = \pi(Ind_E^F(\rho)).$$

We cannot expect Π to be cuspidal in all cases; this corresponds to the fact that $Ind_E^F(\rho)$ may be reducible even if ρ is irreducible.

Assume that v is unramified in E and that π_w is unramified for all w dividing v. In this case, Π_v is unramified and we describe $g(\Pi_v)$ explicitly as follows. It suffices to determine the eigenvalues of $g(\Pi_v)$. Set $d(w) = [E_w : F_v]$. We claim that $g_v(\Pi)$ is an element in $GL_{n\ell}(\mathbb{C})$ whose set of eigenvalues is the union over w dividing v of all $(d(w))^{th}$ roots of eigenvalues of $g(\pi_w)$. This gives a total of $n\big(\sum_{w|\ell} d(w)\big) = n\ell$ eigenvalues as required, since $\sum_{w|\ell} d(w) = \ell$. To check this assertion, fix w dividing v and set $d = d(w)$. Let $\zeta = \exp(\frac{2\pi i}{d})$ and let $\{z_{wj} : 1 \le j \le n\}$ be the set of eigenvalues of $g(\pi_w)$. Then $q_v = q_w^{\frac{1}{d}}$ and

$$L(s, \pi_w) = \prod_{j=1}^{n} (1 - z_{wj} q_w^{-s})^{-1} = \prod_{j=1}^{n} \prod_{k=1}^{d} (1 - z_{wj}^{\frac{1}{d}} \zeta^k q_v^{-s})^{-1}.$$

Therefore $\prod_{w|v} L(s, \pi_w)$ is equal to $\det(1 - q_v^{-s} g(\Pi_v))^{-1}$ with $g(\Pi_v)$ as described.

The following theorem is proved in [AC].

Theorem 14. *Let E/F be a cyclic extension. Then the automorphic induction map $AI_E^F : \mathcal{A}_E \to \mathcal{A}_F$ exists. Suppose that π is cuspidal. Then $AI_E^F(\pi)$ is cuspidal unless there exists a nontrivial element $\tau \in Gal(E/F)$ such that $\tau(\pi)$ is isomorphic to π.*

The condition for $AI_E^F(\pi)$ to be cuspidal is parallel to the condition for a representation induced from a cyclic extension to be irreducible. Indeed, if ρ is a Galois representation of $Gal(\overline{F}/E)$, then $Ind_E^F(\rho)$ be irreducible if and only if ρ is not isomorphic to any of its conjugates under $Gal(E/F)$.

Example 3. Assume E/F is quadratic and let $\theta \in \mathcal{A}_E(1)$ be a Hecke character. The existence of the automorphic representation of degree two $\pi = A_E^F(\theta)$ was proved in [JL], [ST] using the theory of the Weil representation. In the classical case $F = \mathbb{Q}$, π was constructed at the level of modular forms by Hecke and Maass. Observe that if v splits into two places w and w' in E, then

$$g(\pi_v) = \begin{pmatrix} \theta(\varpi_w) & 0 \\ 0 & \theta(\varpi_{w'}) \end{pmatrix}$$

where $\varpi_w, \varpi_{w'}$ denote prime elements at w, w'. If v is unramified in E and remains prime, then

$$g(\pi_v) = \begin{pmatrix} 0 & 1 \\ \theta(\varpi_w) & 0 \end{pmatrix}$$

where w is the unique place of E dividing v, since the eigenvalues of this matrix are the two square roots of $\theta(\varpi_w)$. Furthermore, $\pi = A_E^F(\theta)$ is cuspidal if and only if $\theta \neq \theta^\tau$, where τ is conjugation of E/F. This is equivalent to the condition that θ not be of the form $\theta' \circ N_{E/F}$.

Example 4. Let E/F is a cyclic cubic extension and let $\theta \in \mathcal{A}_E(1)$ be a Hecke character. The existence of the automorphic representation $AI_E^F(\theta)$ of degree three is due to Jacquet, Piatetski-Shapiro, and Shalika [JPS2].

16. Base Change

The operation of base change for automorphic representations corresponds to restriction of representations of Galois groups. Let E/F be a finite extension, let v be a place of F that is unramified in E, and let $Fr_v \in Gal(\overline{F}/F)$ be a Frobenius element. If w is a place of E dividing v, then $Fr_v^{d(w)}$ is a Frobenius element for w, where $d(w) = [E_w : F_v]$ is the relative degree. It follows that if σ is a Galois representation of $Gal(\overline{F}/F)$ unramified at v, then $\sigma(Fr_v)^{d(w)} \sim \sigma_E(Fr_w)$. We use this observation to define the base change lift of a cuspidal representation.

Definition 3. Let $\pi \in \mathcal{A}_F(n)$ and let $\Pi \in \mathcal{A}_E$ be an element of degree n. Then Π is said to be a **base change lift** of π if for almost all places v of F we have

$$g(\Pi_w) \sim g(\pi_v)^{d(w)}$$

for all places w of E dividing v.

By the strong multiplicity one theorem, the base change lift is unique if it exists. We denote it by π_E or $BC_{E/F}(\pi)$.

Example 5. Suppose that $n = 1$, so that π is a Hecke character of I_F. Let us check that π_E exists for any extension E/F, and that $\pi_E = \pi \circ N_{E/F}$ where $N_{E/F} : I_E \to I_F$ is the global norm map on ideles. Let ϖ be a prime element in F attached to a place v that is unramified in E. Then ϖ is also a prime element in E attached to any place w of E dividing v. By definition, $g(\pi_v) = \pi_v(\varpi)$ and $N_{E_w/F_v}(\varpi) = \varpi^{d(w)}$. Since the global norm induces the local norm maps on the idele components, we see that the Langlands class at w attached $\pi \circ N_{E/F}$ is $\pi_v(\varpi)^{d(w)}$ as required.

Example 6. Suppose that $\Pi = BC_{E/F}(\pi)$ where E/F is quadratic. Then $g(\Pi_w) \sim g(\pi_v)$ or $\sim g(\pi_v)^2$, according as v splits or remains prime in E. More generally if the degree $\ell = [E : F]$ is prime, then $g(\Pi_w) \sim g(\pi_v)$ or $\sim g(\pi_v)^\ell$, according as v splits or remains prime in E.

The base change lift is conjectured to exist in all cases, but it need not be cuspidal (just as the restriction of an irreducible Galois representation may become reducible upon restriction to a subgroup of finite index). The first pioneering work on the base change problem was done by Saito and Shintani [Sh]. A complete theory of base change for $GL(2)$ and cyclic extensions of prime degree was developed by Langlands ([L2]). It was generalized to $GL(m)$ for $m > 2$ by Arthur and Clozel ([AC]). Before stating the general theorem, we note some properties of base change that follow readily from the definition:

1. Base change lifting is transitive: let $F \subset E \subset K$ is a sequence of number fields and let $\pi \in \mathcal{A}_F(n)$. Assume that $BC_{K/F}(\pi)$ and $BC_{E/F}(\pi)$ exist. Then $BC_{K/E}(BC_{E/F}(\pi))$ exists and

$$BC_{K/F}(\pi) = BC_{K/E}(BC_{E/F}(\pi)).$$

2. Base change is compatible with twisting in the following sense: let χ be a Hecke character of I_F. Then

$$BC_{E/F}(\pi \otimes \chi) = BC_{E/F}(\pi) \otimes \chi_E,$$

where $\chi_E = \chi \circ N_{E/F}$.
3. Base change is compatible with the Langlands correspondence in the following sense: if $\pi = \pi(\sigma)$ for some Galois representation σ of $Gal(\overline{F}/F)$, then $\pi_E = \pi(\sigma_E)$.

Theorem 15. *Assume that E/F is a cyclic extension of prime degree ℓ.*

(a) (Existence) *For all automorphic representations $\pi \in \mathcal{A}_F(n)$, the base change lifting π_E exists. Furthermore, π_E cuspidal unless ℓ divides n and $\pi \otimes \omega \simeq \pi$ for some nontrivial character ω of $I_F/F^* N_E(I_E)$.*
(b) (Description of fibers) *Let π, $\pi' \in \mathcal{A}_F(n)$. Then $\pi_E = \pi'_E$ if and only if there exists a character ψ of $I_F/F^* N_E(I_E)$ such that $\pi = \pi' \otimes \psi$.*
(c) (Descent) *Let $\Pi \in \mathcal{A}_E(n)$. Then there exists $\pi \in \mathcal{A}_F(n)$ such that $\Pi = \pi_E$ if and only if $\eta(\Pi) = \Pi$ for all $\eta \in Gal(E/F)$.*

Remarks.

1. Part (a) of this theorem clearly remains true for any extension K/F that can be obtained by successive cyclic extensions of prime degree, that is, for any *solvable* extension.
2. Let $\eta \in Gal(E/F)$. Then $\eta(\Pi) = \Pi$ if and only if $\eta(\Pi_w) = \Pi_w$ for almost all places w of E by the strong multiplicity-one theorem. Assume that w divides the place v of F. If v remains prime in E, then $\eta(w) = w$ and the condition $\eta(\Pi_v) = \Pi_v$ is automatically satisfied. On the other hand, if v splits completely, we may identify the groups G_w for w dividing v. In this case, the condition $\eta(\Pi) = \Pi$ implies that the local components Π_w for w dividing v are all isomorphic. For example, if $\ell = 2$, then $\eta(\Pi) = \Pi$ if and only if $\Pi_w \approx \Pi_{w'}$ whenever w, w' are two places lying above a split prime v of F.

III. Special Cases of the Artin Conjecture

We shall consider an *irreducible* two-dimensional Galois representation

$$\rho : Gal(\overline{F}/F) \longrightarrow GL_2(\mathbb{C}).$$

As before, if L is a finite extension of F, we write ρ_L for the restriction of ρ to $Gal(\overline{F}/L)$. We say that $\pi(\rho)$ **exists** if there exists a cuspidal representation $\pi(\rho)$ of $GL_2(\mathbf{A}_F)$ satisfying the Langlands-Artin conjecture.

The group $GL_2(\mathbb{C})$ acts by conjugation (the adjoint representation) on the Lie algebra $\mathfrak{sl}_2(\mathbb{C})$ of 2×2 matrices of trace zero. Choosing a basis of $\mathfrak{sl}_2(\mathbb{C})$, we obtain a three-dimensional representation which we denote by Ad :

$$Ad : GL_2(\mathbb{C}) \longrightarrow GL_3(\mathbb{C})$$

The symmetric bilinear form $Tr(AB)$ is invariant under the adjoint action, and the image of Ad is isomorphic to the complex orthogonal group $SO_3(\mathbb{C})$ defined relative to this bilinear form. The irreducible two-dimensional representations ρ are classified according to the image of $Ad \circ \rho$ in $SO_3(\mathbb{C})$. As is well-known, a finite subgroup of $SO_3(\mathbb{C})$ is either cyclic, dihedral, or isomorphic to the one of the symmetry groups of the Platonic solids:

1. **tetrahedral group**, isomorphic to A_4
2. **octahedral group**, isomorphic to S_4
3. **icosahedral group**, isomorphic to A_5

We shall say that ρ is of cyclic, dihedral, tetrahedral, ... type if Image$(Ad \circ \rho)$ is of the corresponding type. We check below that $\pi(\rho)$ exists when ρ is of cyclic or dihedral type. Our main goal is to prove that $\pi(\rho)$ exists also if ρ is of tetrahedral or octahedral type. The Langlands-Artin conjecture is still open for icosahedral Galois representations, although it has been verified in some special cases [Bu].

17. Dihedral Representations

We first check that $\pi(\rho)$ exists if ρ is of cyclic or dihedral type. We use the following lemma.

Lemma 16. *Let $\tau : G \to GL_2(\mathbb{C})$ be an irreducible two-dimensional representation of a finite group G. Then τ is of cyclic or dihedral type if and only if τ is induced from a character χ of a subgroup H of index two. Furthermore, $\chi \neq \chi^g$ where g is any element in $G - H$.*

PROOF. The representation $Ad \circ \tau$ stabilizes a line if and only if τ is of cyclic or dihedral type since there are no irreducible three-dimensional representations of cyclic or dihedral groups. We claim that $Ad \circ \tau$ stabilizes a line if and only if τ preserves a symmetric bilinear form up to multiples (i.e., Image(τ) lies in the similitude group of a symmetric bilinear form). To check the claim, let $Sym : GL_2(\mathbb{C}) \to GL_3(\mathbb{C})$ be the representation on the symmetric tensors of degree two. It is easy to check that $Sym \approx Ad \otimes \nu$ where ν is the character $\nu(g) = \det(g)$, and therefore $Sym \circ \tau$ stabilizes a line if and only if τ^* preserves a symmetric bilinear form up to multiples. Since the map $g \to \det(g)^t g^{-1}$ is an inner automorphism of $GL_2(\mathbb{C})$, two-dimensional representations have the property that $\tau^* \approx \tau \otimes \det(\tau)^{-1}$. This shows that $Sym \circ \tau^* \approx Sym \circ \tau \otimes \det(\tau)^{-1}$, and the claim follows.

The similitude group of the standard form $x_1 y_2 + x_2 y_1$ is the normalizer $N(T)$ of the diagonal subgroup $T \subset GL_2(\mathbb{C})$. Since all symmetric bilinear forms are equivalent over $\mathbb{C}$, we conclude that $Ad(\tau)$ is cyclic or dihedral if and only if Image(τ) is conjugate to a subgroup of $N(T)$. We may assume that Image$(\tau) \subset N(T)$. Now τ is irreducible, and so Image(τ) is not contained in T. Since $[N(T) : T] = 2$, we see that $H = \{g \in G : \tau(g) \in T\}$ is a subgroup of index two in G. The restriction τ_H is isomorphic to a direct sum of two distinct characters χ_1 and χ_2 of H. Any element $g \in G - H$ must interchange the χ_1 and χ_2 eigenspaces and it follows easily that $\tau \xrightarrow{\sim} Ind_H^G \chi_j$ for $j = 1$ or 2. □

Now we can prove

Theorem 17. *Assume that ρ is of cyclic or dihedral type. Then $\pi(\rho)$ exists.*

PROOF. Applying the lemma to Galois representations, we see that if ρ is cyclic or dihedral, then there is a quadratic extension E/F such that $\rho = Ind_E^F \theta$ for some character θ of $Gal(\overline{F}/E)$. The irreducibility of ρ implies that $\theta \neq \theta^\sigma$, where σ is conjugation relative to E/F . By class field theory, θ may be identified with an element of $\mathcal{A}_E(1)$ such that $\theta \neq \theta^\sigma$. The representation $AI_E^F(\theta)$ exists and is cuspidal by Example 3 (special case of Theorem 14), and we have $\pi(\rho) = AI_E^F(\theta)$. Indeed, the Langlands classes of $AI_E^F(\theta)$ coincide with the Frobenius classes of $Ind_E^F\theta$ for almost all v. $\square$

18. Tetrahedral Representations

In this section we prove the following theorem due to Langlands [L2].

Theorem 18 (Langlands). *Assume that ρ is of tetrahedral type. Then $\pi(\rho)$ exists.*

We begin with some preliminary remarks. The group A_4 has a unique irreducible representation of dimension three $\rho_{tet} : A_4 \to GL_3(\mathbb{C})$, defined via the action of A_4 on the tetrahedron in $\mathbb{R}^3$. Let us describe this representation in more detail. The six edges of the tetrahedron break up into three pairs of opposite edges. For each pair, consider the line passing through the centers of opposite edges. The three lines obtained in this way are mutually orthogonal and may be taken as the axes in $\mathbb{R}^3$. Furthermore, they are permuted by the action of A_4, yielding a map from A_4 to S_3 whose image has order 3. This defines an exact sequence

$$1 \longrightarrow V \longrightarrow A_4 \longrightarrow \mathbb{Z}/3 \longrightarrow 1$$

where $V = \mathbb{Z}/2 \oplus \mathbb{Z}/2$. We observe that ρ_{tet} is induced from the subgroup V. Indeed, V stabilizes each of the three axes and acts on them by distinct nontrivial characters. Frobenius reciprocity implies that $\rho_{tet} \approx Ind_V^{A_4}\theta$, where θ is any one of the three nontrivial characters of V. Note that the exact sequence above defines an action of $\mathbb{Z}/3$ on V and that the three nontrivial characters are permuted transitively by this action.

Now assume that ρ is of tetrahedral type. The composition of $Ad \circ \rho$ with the projection to $\mathbb{Z}/3$ yields a surjective map $Gal(\overline{F}/F) \to \mathbb{Z}/3$ whose kernel is of the form $Gal(\overline{F}/E)$, where E/F is a cyclic cubic extension. By the remarks in the previous paragraph, $Ad \circ \rho$ is isomorphic to $Ind_E^F\theta'$, where θ' is a character of order two of $Gal(\overline{F}/E)$. Furthermore, θ' is not fixed by either of the two nontrivial elements of $Gal(E/F)$. These observations allows us to conclude that $\pi(Ad \circ \rho)$ exists and is cuspidal. Indeed, θ' may be viewed as a Hecke character of finite order of the ideles I_F. According to Example 4 (special case of Theorem 14), we may automorphically induce θ' to obtain an element $AI_E^F(\theta')$ of $\mathcal{A}_F$ of degree 3. It is cuspidal since θ' is not fixed by any nontrivial element of $Gal(E/F)$. Finally, $AI_E^F(\theta') = \pi(Ad \circ \rho)$ by the compatibility of automorphic induction with the Langlands correspondence. This proves the first statement in the following lemma.

Lemma 19. *If ρ is of tetrahedral type, then the cuspidal representations $\pi(Ad(\rho))$ and $\pi(\rho_E)$ exist.*

PROOF. The representation ρ_E is irreducible. Indeed, if it were not, then it would decompose as a direct sum of two invariant lines. These lines must be permuted by $Gal(\overline{F}/F)$ under the action of ρ since $Gal(\overline{F}/E)$ is normal in $Gal(\overline{F}/F)$. Since $[E : F] = 3$, this action would have to be trivial and ρ itself would be reducible. Therefore ρ_E is irreducible. Furthermore, $Ad(\rho_E)$ is dihedral of order 4 by construction. The existence of $\pi(\rho_E)$ follows from Theorem 17. $\square$

Next, we prove the following proposition.

Proposition 20. *Let ρ be of tetrahedral type. Suppose that $\pi \in \mathcal{A}_F(2)$ has the following three properties:*

(i) $BC_{E/F}(\pi) = \pi(\rho_E)$
(ii) $\omega_\pi = \det(\rho)$
(iii) $Ad(\pi) = \pi(Ad(\rho))$.

Then $\pi = \pi(\rho)$.

PROOF. Let v be a finite place of F outside of $S(\pi) \cup S(\rho)$. Suppose that

$$g(\pi_v) \sim \begin{pmatrix} a & \\ & b \end{pmatrix}, \qquad \rho(Fr_v) \sim \begin{pmatrix} \alpha & \\ & \beta \end{pmatrix}.$$

We must show that $g(\pi_v) \sim \rho(Fr_v)$, i.e., that $\{a, b\} = \{\alpha, \beta\}$ as unordered sets. Let w be a place of E dividing v and let $d(w) = [E_w : F_v]$. Our hypotheses give us the following information:

(a) $BC_{E/F}(\pi) = \pi(\rho_E)$ implies $g(\pi_v)^{d(w)} \sim \rho(Fr_v)^{d(w)}$
(b) $\omega_\pi = \det(\rho)$ implies $ab = \alpha\beta$ (cf. Remark 1 in Sec. 11)
(c) $Ad(\pi) = \pi(Ad(\rho))$ implies $\{a/b, 1, b/a\} = \{\alpha/\beta, 1, \beta/\alpha\}$.

If $d(w) = 1$, then (a) already gives what we want. Otherwise, $d(w) = 3$. In this case, (a) and (b) imply that we may choose the labelling so that $a = \zeta\alpha$ and $b = \zeta^2\beta$ for some cube root of unity ζ. Thus we have

$$g(\pi_v) \sim \begin{pmatrix} \zeta\alpha & \\ & \zeta^2\beta \end{pmatrix}, \qquad \rho(Fr_v) \sim \begin{pmatrix} \alpha & \\ & \beta \end{pmatrix}.$$

If $\zeta = 1$, we are done. If not, (c) implies that $\zeta^{-1}\alpha/\beta = \beta/\alpha$ and hence that $\alpha/\beta = \pm\zeta^2$. If $\alpha/\beta = \zeta^2$, both matrices have eigenvalues $\{\beta, \zeta^2\beta\}$ and we are done. It remains to rule out the possibility $\alpha/\beta = -\zeta^2$. However, if $\alpha/\beta = -\zeta^2$, then

$$Ad(\rho(Fr_v)) \sim \begin{pmatrix} -\zeta^2 & & \\ & 1 & \\ & & -\zeta \end{pmatrix},$$

and this is an element of order 6. This is not possible since A_4 does not have an element of order 6! □

To conclude the proof of Theorem 18, we must show that a cuspidal representation π satisfying the conditions of Proposition 20 exists. We have seen that $\pi(\rho_E)$ exists. Furthermore, we clearly have $\eta(\rho_E) \simeq \rho_E$ for all $\eta \in Gal(\overline{F}/F)$. The same relation $\eta(\pi(\rho_E)) \simeq \pi(\rho_E)$ therefore also holds by the compatibility mentioned in Section 16. The descent part of the base change theorem (Theorem 15) implies that there exists $\pi \in \mathcal{A}_F(2)$ such that $BC_{E/F}(\pi) = \pi(\rho_E)$. According to the description of the fibers of the base change map, π is unique up to twisting by a character of the cyclic group of order three $I_F/F^*N_{E/F}(I_E)$. Let us show that there exists a unique choice of π for which $\omega_\pi = \det(\rho)$. The relation $BC_{E/F}(\pi) = \pi(\rho_E)$ implies that $\omega_\pi \circ N_{E/F} = \det(\rho) \circ N_{E/F}$; so in any case $\omega_\pi\omega = \det(\rho)$ for some character ω of $I_F/F^*N_{E/F}(I_E)$. Since

$$\omega_{\pi\otimes\omega^2} = \omega_\pi\omega^4 = \omega_\pi\omega,$$

we may (and shall) choose π so that $\det(\pi) = \det(\rho)$.

Now set

$$\Pi_1 = Ad(\pi), \qquad \Pi_2 = \pi(Ad(\rho))$$

It remains to show that condition (iii) of Proposition 20 is satisfied, i.e., that $\Pi_1 = \Pi_2$. It will suffice, of course, to prove that $g(\Pi_{1v}) \sim g(\Pi_{2v})$ for almost all finite places v of F such π_v and ρ_v are unramified. This is obvious if v splits in E, since $g(\pi_v) \sim \rho(Fr_v)$ in that case, but there does not seem to be any elementary way to conclude that $g(\Pi_{1v}) \sim g(\Pi_{2v})$ if v remains prime. Therefore, Langlands uses the result of Jacquet-Shalika (stated as Theorem 12 above) at this stage in the argument. To apply it, we must check that

$$g_v(\Pi_1) \otimes g_v(\Pi_2^*) = g_v(\Pi_2) \otimes g_v(\Pi_2^*)$$

for almost all places v. This is clear if v splits; so assume that v remains prime in E. Then the image of a Frobenius element $Fr_v \in Gal(\overline{F}/F)$ in A_4 has order 3, and hence

$$g(\Pi_{2v}) \sim Ad(\rho)(Fr_v) \sim \begin{pmatrix} 1 & & \\ & \zeta & \\ & & \zeta^2 \end{pmatrix},$$

where $\zeta \neq 1$ is a cube root of unity. In other words, $\rho(Fr_v) \sim \begin{pmatrix} \alpha & \\ & \zeta\alpha \end{pmatrix}$ for some α. We also have $g(\Pi_{2v}^*) \sim g(\Pi_{2v})$ since $g(\Pi_{2v})$ is conjugate to its inverse. Since $g(\pi_v)^3 \sim \rho(Fr_v)^3$ and $\det(g(\pi_v)) \sim \det(\rho(Fr_v))$, we can conclude that

$$g(\pi_v) \sim \begin{pmatrix} \alpha & \\ & \zeta\alpha \end{pmatrix} \quad \text{or} \quad \begin{pmatrix} \zeta^2\alpha & \\ & \zeta^2\alpha \end{pmatrix}.$$

Therefore

$$g(\Pi_{1v}) \sim \begin{pmatrix} 1 & & \\ & \zeta & \\ & & \zeta^2 \end{pmatrix} \quad \text{or} \quad \begin{pmatrix} 1 & & \\ & 1 & \\ & & 1 \end{pmatrix}.$$

Although the second possibility would spell disaster if it really occurred, we do not have to rule it out in advance because

$$g_v(\Pi_1) \otimes g_v(\Pi_2^*) = g_v(\Pi_2) \otimes g_v(\Pi_2^*)$$

in both cases, as is easily checked. With this stroke of luck, the proof of Theorem 18 is complete! □

19. Octahedral Representations

We shall now prove that $\pi(\rho)$ exists also for octahedral representations, following the argument of J. Tunnell [Tu]. Certain octahedral cases had previously been established by Langlands [L2]. The improvement due to Tunnell was made possible by the following theorem of Jacquet, Piatetski-Shapiro, and Shalika [JPS1].

Theorem 21. *Let K/F be a nonnormal cubic extension. Then for all π in $\mathcal{A}_F(2)$, the base change lifting $B_{K/F}(\pi)$ exists and is cuspidal.*

Assuming this result, we shall prove

Theorem 22 (Langlands-Tunnell). *Let ρ be a Galois representation of octahedral type. Then $\pi(\rho)$ exists.*

Let N/F be the S_4-extension defined by the kernel of $Ad(\rho)$. The group S_4 has three 2-Sylow subgroups of order 8. The conjugation action of S_4 on this set of 3 subgroups defines a epimorphism $\varphi : S_4 \to S_3$ and exact sequence

$$1 \longrightarrow V \longrightarrow S_4 \xrightarrow{\varphi} S_3 \longrightarrow 1.$$

where $V = \mathbb{Z}/2 \oplus \mathbb{Z}/2$. Let M be the fixed field of V. We define two subfields K and E of M as follows. Let E be the quadratic extension defined by the sign character of S_4 (obtained by pull-back from the sign character of S_3). Fix a 2-Sylow subgroup H containing V, and let K/F be the (non-Galois) cubic extension fixed by H. We have the following diagram of fields:

$$\begin{array}{ccccc} & & N & & \\ & & | & & \\ & & M & & \\ & / & & \backslash & \\ K & & & & E \\ & \backslash & & / & \\ & & F & & \end{array}$$

Note that ρ_E is of tetrahedral type and ρ_K is dihedral (since $Gal(N/K)$ is isomorphic to the dihedral group D_8). Therefore $\pi(\rho_E)$ and $\pi(\rho_K)$ both exist by Theorems 22 and 17, respectively. In the next lemma, we make use of quadratic base change $BC_{E/F}$ and the cubic base change $BC_{K/F}$ whose existence is guaranteed by Theorem 21.

Lemma 23. *Suppose that π is a cuspidal representation of $GL_2(\mathbf{A}_F)$ such that $\pi_E = \pi(\rho_E)$ and $\pi_K = \pi(\rho_K)$. Then $\pi = \pi(\rho)$.*

PROOF. Let v be a place of F at which both π and ρ are unramified, and suppose that

$$g(\pi_v) \sim \begin{pmatrix} a & \\ & b \end{pmatrix}, \qquad \rho(Fr_v) \sim \begin{pmatrix} a' & \\ & b' \end{pmatrix}.$$

Of course, if v splits in E or if there exists a prime of K of relative degree one dividing v, then we have $g(\pi_v) \sim \rho(Fr_v)$. Otherwise, we may conclude that $g(\pi_v)^2 \sim \rho(Fr_v)^2$ and $g(\pi_v)^3 \sim \rho(Fr_v)^3$. If $g(\pi_v)$ and $\rho(Fr_v)$ have an eigenvalue in common, then they are conjugate. Indeed, if $a = a'$, then $b^2 = b'^2$ and $b^3 = b'^3$ and hence $b = b'$. Suppose that $g(\pi_v)$ and $\rho(Fr_v)$ are not conjugate. Then they have no eigenvalue in common, and so we assume that $a' = -a$. The relation $g(\pi_v)^3 \sim \rho(Fr_v)^3$ forces $b' = \eta a$, where $\eta^3 = 1$ but $\eta \neq 1$. This gives

$$\rho(Fr_v) \sim \begin{pmatrix} -a & \\ & \eta a \end{pmatrix} \quad \text{and} \quad Ad(\rho)(Fr_v) \sim \begin{pmatrix} -\eta^2 & & \\ & 1 & \\ & & -\eta \end{pmatrix},$$

which implies that $Ad(\rho)(Fr_v)$ has order 6. This is not possible since S_4 has no elements of order 6. We conclude that $g(\pi_v) \sim \rho(Fr_v)$, as claimed. □

To prove Theorem 22, we shall construct a π satisfying the conditions of the previous lemma. We have $\rho_E \sim \tau(\rho_E)$, where τ is conjugation of E/F since ρ_E extends to ρ, and therefore $\tau(\pi(\rho_E)) = \pi(\rho_E)$. By the base change theorem (Theorem

15), $\pi(\rho_E)$ descends to a cuspidal representation of $GL_2(\mathbf{A}_F)$ in two distinct ways. Let π_1 and π_2 be the two cuspidal representations such that $BC_{E/F}(\pi_j) = \pi(\rho_E)$. Then $\pi_1 = \pi_2 \otimes \omega_{E/F}$. It will suffice to check $BC_{K/F}(\pi_j) = \pi(\rho_K)$ for one of $j = 1$ and $j = 2$, since this π_j will satisfy the conditions of the lemma.

As observed in Lemma 19 and its proof, ρ_M is irreducible and $\pi(\rho_M)$ exists. The cuspidal representation $\pi(\rho_M)$ is in the image of the base change lifting from K since $\pi(\rho_K)$ clearly lifts to $\pi(\rho_M)$. Theorem 15 implies that $\pi(\rho_M) = BC_{M/K}(\pi')$ for precisely two cuspidal representations π' of $GL_2(\mathbf{A}_K)$ and these two representation differ by a twist by $\omega_{M/K}$. The two cuspidal representations are therefore $\pi(\rho_K)$ and $\pi(\rho_K) \otimes \omega_{M/K}$.

We claim that $BC_{K/F}(\pi_1)$ and $BC_{K/F}(\pi_2)$ also lift to $\pi(\rho_M)$. Indeed, by the transitivity of base change and the compatibility of base change with the Langlands correspondence, we have

$$BC_{M/F}(\pi_j) = BC_{M/K}(BC_{K/F}(\pi_j)) = BC_{M/K}(\pi(\rho_K)) = \pi(\rho_M).$$

Since $\pi_1 = \pi_2 \otimes \omega_{E/F}$ and $\omega_{M/K} = \omega_{E/F} \circ N_{M/E}$, we have

$$BC_{K/F}(\pi_1) = BC_{K/F}(\pi_2 \otimes \omega_{E/F}) = BC_{K/F}(\pi_2) \otimes \omega_{M/K}$$

because of the compatibility of base change with twisting. We must therefore have $BC_{K/F}(\pi_j) = \pi(\rho_K)$ for some j. □

References

[AC] J. Arthur and L. Clozel, *Simple Algebras, Base Change, and the Advanced Theory of the Trace Formula*, Annals of Math. Studies, vol. 120, Princeton University Press, Princeton, 1989.

[BR] D. Blasius and J. Rogawski, Zeta functions of Shimura varieties, *Motives* (U. Jannsen, S. Kleiman, J.-P. Serre, eds.), Proc. Symp. Pure Math., vol. 55, Part II, American Mathematical Society, Providence, 1994, pp. 525–571.

[Bl] D. Blasius, On multiplicities for SL(n), *Israel J. Math.* **88,** 1–3 (1994), 237–251.

[Bo] A. Borel, Automorphic L-functions, *Automorphic Forms, Representations, and L-Functions* (A. Borel and W. Casselman, eds.), Proc. Symp. Pure Math., vol. 33, Part II, American Mathematical Society, Providence, 1979, pp. 27–61.

[Bu] J. Buhler, *Icosahedral Galois Representations*, Lecture Notes in Mathematics, vol. 654, Springer-Verlag, Berlin, 1978.

[Ca] Cartier, P., Representations of $\mathfrak{p}$-adic groups: A survey, *Automorphic Forms, Representations, and L-Functions* (A. Borel and W. Casselman, eds.), Proc. Symp. Pure Math., vol. 33, Part I, American Mathematical Society, Providence, 1979, pp. 111–155.

[De] P. Deligne, Formes modulaires et représentations ℓ-adiques, *Séminaire Bourbaki vol. 1968/69, Exposés 347–363*, Lecture Notes in Mathematics, vol. 179, Springer-Verlag, Berlin, 1971, pp. 139–185.

[DS] P. Deligne and J.-P. Serre, Formes modulaires de poids 1, *Ann. Sci. École Norm. Sup.* (4) **7** (1974), 507–530.

[F] D. Flath, Decomposition of representations into tensor products, *Automorphic Forms, Representations, and L-Functions* (A. Borel and W. Casselman, eds.), Proc. Symp. Pure Math., vol. 33, Part I, American Mathematical Society, Providence, 1979, pp. 179–183.

[G] S. Gelbart, *Automorphic Forms on Adele Groups*, Annals of Math. Studies, vol. 83, Princeton University Press, Princeton, 1975.

[GJ] S. Gelbart and H. Jacquet, A relation between automorphic representations of GL(2) and GL(3), *Ann. Sci. École Norm. Sup.* (4) **11** (1978), 471–542.

[GGP] I. Gelfand, M. Graev, and I. Pitateskii-Shapiro, *Representation Theory and Automorphic Functions*, W. B. Saunders, Philadelphia, 1969.

[Go-J] R. Godement and H. Jacquet, *Zeta-functions of Simple Algebras*, Lecture Notes in Mathematics, vol. 260, Springer-Verlag, New York, 1972.

[H] Harish-Chandra, *Automorphic Forms on Semisimple Lie Groups*, Lecture Notes in Mathematics, vol. 62, Springer-Verlag, Berlin, 1968.

[Ja1] H. Jacquet, Principal L-functions of the linear group, *Automorphic Forms, Representations, and L-Functions* (A. Borel and W. Casselman, eds.), Proc. Symp. Pure Math., vol. 33, Part II, American Mathematical Society, Providence, 1979, pp. 63–86.

[Ja2] H. Jacquet, Principal L-functions for $GL(n)$, *these* PROCEEDINGS, pp. 321–329.

[JL] H. Jacquet, H., and R. Langlands, *Automorphic Forms on* GL(2), Lecture Notes in Mathematics, vol. 114, Springer-Verlag, Berlin, 1970.

[JPS1] H. Jacquet, I. Piatetski-Shapiro, and J. Shalika, Relèvement cubique non normal, *C. R. Acad. Sci. Paris* **292** (1981), 567–571.

[JPS2] H. Jacquet, I. Piatetski-Shapiro, and J. Shalika, Automorphic forms on GL(3). I, *Annals of Math.* **109** (1979), 169–212.

[JS] H. Jacquet and J. Shalika, On Euler products and the classification of automorphic representations I and II, *Amer. J. Math.* **103** (1981), 499–558 and 777–815.

[Kn] A. W. Knapp, Introduction to the Langlands program, *these* PROCEEDINGS, pp. 245–302.

[La] S. Lang, *Algebraic Number Theory*, Springer-Verlag, New York, 1986.

[L1] R. P. Langlands, Problems in the theory of automorphic forms, *Lectures in Modern Analysis and Applications III*, Lecture Notes in Mathematics, vol. 170, Springer-Verlag, Berlin, 1970, pp. 18–61.

[L2] Langlands, R. P., *Base Change for* GL(2), Annals of Math. Studies, vol. 96, Princeton University Press, Princeton, 1980.

[LL] J.-P. Labesse and R. Langlands, L-indistinguishability for SL(2), *Canad. J. Math.* **13** (1981), 726–785.

[N] J. Neukirch, *Algebraische Zahlentheorie*, Springer-Verlag, Berlin, 1992.

[R] J. Rogawski, Modular forms, the Ramanujan conjecture and the Jacquet-Langlands correspondence, *Discrete Groups, Expanding Graphs and Invariant Measures*, by A. Lubotzky, Birkhäuser, Basel, 1994, pp. 135–176.

[S] J. Shalika, The multiplicity one theorem for GL_n, *Annals of Math.* (2) **100** (1974), 171–193.

[ST] J. Shalika and S. Tanaka, On an explicit construction of a certain class of automorphic forms, *Amer. J. Math.* **91** (1969), 1049–1076.

[Sh] T. Shintani, On liftings of holomorphic cusp forms, *Automorphic Forms, Representations, and L-Functions* (A. Borel and W. Casselman, eds.), Proc. Symp. Pure Math., vol. 33, Part II, American Mathematical Society, Providence, 1979, pp. 97–110.

[T] J. Tate, Number theoretic background, *Automorphic Forms, Representations, and L-Functions* (A. Borel and W. Casselman, eds.), Proc. Symp. Pure Math., vol. 33, Part II, American Mathematical Society, Providence, 1979, pp. 3–26.

[Tu] J. Tunnell, Artin's conjecture for representations of octahedral type, *Bull. Amer. Math. Soc.* **5** (1981), 173–175; On the local Langlands conjecture for $GL(2)$, *Invent. Math.* **46** (1978), 179–200.

DEPARTMENT OF MATHEMATICS, THE HEBREW UNIVERSITY, GIVAT RAM, JERUSALEM, ISRAEL
E-mail address: `jonr@math.huji.ac.il`

Proceedings of Symposia in Pure Mathematics
Volume **61** (1997), pp. 355–405

Theoretical Aspects of the Trace Formula for $GL(2)$

A. W. Knapp

The Selberg-Arthur trace formula is one of the tools available for approaching the conjecture of global functoriality in the Langlands program. Global functoriality is described within this volume in [Kn2]. We start with reductive groups G and H, say over the rationals $\mathbb{Q}$ for simplicity. We assume that G is quasisplit, and we suppose that we are given an L homomorphism $\psi : {}^L H \to {}^L G$. From an automorphic representation of the adeles of H, we use ψ to construct, place-by-place from the Local Langlands Conjecture (or at almost every place without the conjecture), an irreducible representation of the adeles of G. The question of global functoriality is whether the latter representation is automorphic (or, in the case that it is defined only at almost every place, whether it can be completed to an automorphic representation). If it is automorphic, then we want to know also what conditions ensure that a cuspidal representation of the adeles of H yields a cuspidal representation of the adeles of G under this process. It is known that these questions capture various deep conjectures in classical algebraic number theory, arithmetic algebraic geometry, and representation theory and that they unify and generalize such conjectures significantly.

The trace formula for the reductive group G gives information about the multiplicity of the occurrence of an irreducible representation of the adeles of G in the cuspidal spectrum. If Z denotes the center of G, the quotient $Z(\mathbb{A})G(\mathbb{Q})\backslash G(\mathbb{A})$ is almost compact in the sense that it has finite volume.[1] If $Z(\mathbb{A})G(\mathbb{Q})\backslash G(\mathbb{A})$ is actually compact and if R denotes the right regular representation of $G(\mathbb{A})$ on $L^2(Z(\mathbb{A})G(\mathbb{Q})\backslash G(\mathbb{A}))$, then the trace formula will assert the equality of two expressions for $\operatorname{Tr}(R(\varphi))$ on this L^2 space, φ being a suitably regular function of compact support on $G(\mathbb{A})$. In the notation of [Ar4], the formula in the compact case has the shape

$$\sum_{\mathfrak{o}\in\mathcal{O}} J_{\mathfrak{o}}(\varphi) = \sum_{\chi\in\mathfrak{X}} J_{\chi}(\varphi), \tag{0.1}$$

1991 *Mathematics Subject Classification.* Primary 22E45, 22E55.

This article is based in part on lectures by Laurent Clozel and Hervé Jacquet in Edinburgh and in part on subsequent discussions with Jacquet and with Jonathan Rogawski. The author is grateful for all the help he received from these individuals and the referee.

[1]In this paper we follow the standard convention that the group of $\mathbb{Q}$ points of any subgroup refers to the diagonally embedded version of that subgroup unless the contrary is explicitly indicated.

in which the left side, called the **geometric side**, consists of terms that are integrals of φ over conjugacy classes, suitably normalized by volume factors. The right side, called the **spectral side**, is a sum of expressions $m_\pi \operatorname{Tr} \pi(\varphi)$, m_π being the multiplicity of an irreducible representation π in R.

When the quotient $Z(\mathbb{A})G(\mathbb{Q})\backslash G(\mathbb{A})$ is noncompact, $R(\varphi)$ is not of trace class on all of $L^2(Z(\mathbb{A})G(\mathbb{Q})\backslash G(\mathbb{A}))$ but is of trace class on the cuspidal part. The computation of $\operatorname{Tr}(R_{\text{cusp}}(\varphi))$ is done with a "truncation parameter" T, $0 < T < \infty$, in place, and the result has the shape

$$\operatorname{Tr}(R_{\text{cusp}}(\varphi)) = \sum_{\mathfrak{o} \in \mathcal{O}} J_{\mathfrak{o}}^T(\varphi) - \sum_{\chi \in \mathfrak{X} - \mathfrak{X}(G)} J_\chi^T(\varphi), \tag{0.2}$$

with $\operatorname{Tr}(R_{\text{cusp}}(\varphi))$ being regarded as the sum of the terms $J_\chi^T(\varphi)$ with $\chi \in \mathfrak{X}(G)$, each of which is constant in T. The ingredients in (0.2) are more complicated than in (0.1): The set $\mathcal{O}$ now involves various kinds of conjugacy classes, and the terms J_χ^T involve Eisenstein series relative to proper parabolic subgroups of G. One can pass to the limit in (0.2) as $T \to +\infty$, taking into account various cancellations, and the result can be written in the qualitative form (0.1), but the interpretation of each side as a trace is lost.

In any event the trace formula does carry in it the multiplicity of each irreducible representation of the adeles of G in the cuspidal spectrum of the L^2 space, and the formula may therefore be expected to give some information toward answering the above functoriality question. In practice it is normally a comparison of the trace formulas for G and H that gives useful information, but this point will not concern us at this time.

In this paper we shall discuss aspects of the background and derivation of the trace formula for $G = GL_2$ when the number field is $\mathbb{Q}$, including a precise statement of the result. We shall treat also the case that G is a quaternion division algebra. Another article [Kn-Ro] in this volume gives some applications of the trace formula for various groups.

Although our interest in the trace formula will ultimately be in an adelic setting, it is helpful to keep in mind a certain classical setting, because the analysis there is more transparent and suggests approaches to the analysis in the adelic setting. Historically the trace formula was introduced by Selberg in [Se1] and [Se2]. Selberg worked initially in the context of a transitive group action on a Riemannian manifold in which the space of invariant differential operators is commutative, and he considered the analysis of the space of functions transforming suitably under a discrete subgroup that acts properly discontinuously. The case of the action of $SL_2(\mathbb{R})$ on the upper half plane, with $SL_2(\mathbb{Z})$ as the discrete subgroup, was of particular interest, and we may think in terms of an analysis of

$$L^2(SL_2(\mathbb{Z})\backslash SL_2(\mathbb{R})). \tag{0.3}$$

Let $G = SL_2(\mathbb{R})$ and $\Gamma = SL_2(\mathbb{Z})$. It is an elementary fact, which we prove as Theorem 1.3 below, that the right regular representation R of G on $L^2(\Gamma\backslash G)$ splits as an orthogonal direct sum

$$L^2(\Gamma\backslash G) = L^2_{\text{cusp}}(\Gamma\backslash G) \oplus L^2_{\text{cont}}(\Gamma\backslash G) \oplus \mathbb{C},$$

where the members of $L^2_{\text{cusp}}(\Gamma\backslash G)$ are functions satisfying a cuspidal condition of the kind discussed in [Kn2, §7] and where the members of $L^2_{\text{cont}}(\Gamma\backslash G)$ are essentially generated by summing the left translates by Γ of nice functions on G that have integral 0. The space $\mathbb{C}$ is the space of constant functions. The space $L^2_{\text{cont}}(\Gamma\backslash G)$ is the continuous part of the decomposition, and the known complete analysis of this space will be given in Theorem 1.4 and §2 in terms of Eisenstein series. Our analysis specializes the ones of Langlands ([Lgl1], [Lgl2], and [HC]); a different proof appears in [Go1]. See also [Lan1].

The space $L^2_{\text{cusp}}(\Gamma\backslash G)$ is the "cuspidal part" of the decomposition. It splits into a discrete sum of irreducible representations with finite multiplicities, as is shown in Theorem 1.5 and §3. Our proof specializes the one in [Go2].

Although the cuspidal part of the decomposition at first appears less complicated than the continuous part, little is known about what specific irreducible representations occur and what multiplicities they have. That is where the trace formula comes in. If φ is in $C^\infty_{\text{com}}(G)$, then the operator

$$R(\varphi)f(x) = \int_G f(xy)\varphi(y)\,dy$$

is of trace class on $L^2_{\text{cusp}}(\Gamma\backslash G)$. The trace formula implies the equality of two expressions for the trace of $R(\varphi)$ on $L^2_{\text{cusp}}(\Gamma\backslash G)$. If

$$L^2_{\text{cusp}}(\Gamma\backslash G) = \bigoplus m_\pi \pi$$

is the decomposition into irreducible constituents with multiplicities, then one of the expressions for the trace is simply $\sum m_\pi \operatorname{Tr} \pi(\varphi)$. The other expression comes from realizing $R(\varphi)$ on $L^2_{\text{cusp}}(\Gamma\backslash G)$ as an integral operator on $\Gamma\backslash G$ and is the integral of the kernel of this operator over the diagonal; the trace works out to be a sum of terms encoding conjugacy class information about φ and spectral information about the action of $R(\varphi)$ on the noncuspidal part of $L^2(\Gamma\backslash G)$. The equality of the two expressions therefore gives information about multiplicities of irreducible representations in $L^2_{\text{cusp}}(\Gamma\backslash G)$ in terms of geometric information about G. We shall indicate in §4 what computation has to be made for the trace formula, but we shall omit an explicit statement of the formula in the context (0.3). See [He1], [He2], and [Ef] for a statement of this kind. For our purposes the trace formula is better understood in an adelic context, and we shall give in §7 a precise statement of that kind.

The trace formula in the classical setting does not lend itself to the kind of comparison of traces from different groups useful for global functoriality, but it does have some direct applications. One such is that it gives a formula for the trace of each Hecke operator on each space of classical cusp forms; the resulting theorem is called the Eichler-Selberg trace formula and is discussed in [Lan1] and [Mi, Ch. 6]. A degenerate case of this argument yields a proof of the dimension formula for spaces of classical cusp forms without appealing to the Riemann-Roch Theorem.

Let us now be more specific about the adelic context. The reductive group under study will largely be GL_2, and we regard it as defined over the rationals $\mathbb{Q}$. The places v of $\mathbb{Q}$ are ∞ and all the primes, and $\mathbb{Q}_v$ is correspondingly the field of reals $\mathbb{R}$ if $v = \infty$ and is the field of p-adics $\mathbb{Q}_p$ if v is a prime p. If the restricted direct product $\mathbb{A} = \prod_v \mathbb{Q}_v$ denotes the adeles of $\mathbb{Q}$, the problem of global functoriality typically leads one to representations of $GL_2(\mathbb{A}) = \prod_v GL_2(\mathbb{Q}_v)$ of the form $\pi = \prod_v \pi_v$ with π_v an irreducible admissible representation of $GL_2(\mathbb{Q}_v)$ for each v. Roughly speaking, π is automorphic if π is involved in analysis of the quotient $Z(\mathbb{A})GL_2(\mathbb{Q})\backslash GL_2(\mathbb{A})$, where $Z(\mathbb{A})$ denotes the subgroup of scalar matrices. More particularly, the question is likely to be whether π occurs in the cuspidal part of the discrete spectrum of

$$L^2(Z(\mathbb{A})GL_2(\mathbb{Q})\backslash GL_2(\mathbb{A})). \tag{0.4}$$

The question is therefore answered by knowing whether the multiplicity of π in the cuspidal spectrum is zero or is positive, and the trace formula gives subtle information about this multiplicity.

As is noted in [Kn2, §6], the space (0.3) is a prototype for (0.4). The functions in $L^2(Z(\mathbb{A})GL_2(\mathbb{Q})\backslash GL_2(\mathbb{A}))$ that are invariant under the right action by $\prod_p GL_2(\mathbb{Z}_p)$ may be regarded as functions in $L^2(SL_2(\mathbb{Z})\backslash SL_2(\mathbb{R}))$. Thus (0.3) may be analyzed by specializing results about (0.4) to results about (0.3). On the other hand, the techniques that are used in studying (0.3) often suggest techniques for studying (0.4).

The first people to consider the decomposition of the adelic setting (0.4) were Gelfand, Graev, and Piatetski-Shapiro in 1964, and an exposition is in [Gf-Gr-P]. Later expositions are the ones by Jacquet-Langlands [Ja-Lgl], Duflo-Labesse [Du-Lab], Gelbart [Gb1], Gelbart-Jacquet [Gb-Ja], Rogawski [Ro], and Gelbart [Gb2]. The treatment [Gb1] specializes work of Arthur [Ar1], and [Gb2] specializes later work of Arthur.

In §5 we obtain the trace formula for $L^2(Z(\mathbb{A})G(\mathbb{Q})\backslash G(\mathbb{A}))$ when G is the multiplicative group of a quaternion division algebra over $\mathbb{Q}$. This space splits discretely with finite multiplicities and is considerably easier to understand than (0.4).

In §6 we give aspects of the decomposition of (0.4) into a continuous part and a discrete part, as well as aspects of the analysis of the continuous part using adelic Eisenstein series. The same section shows how some of the concepts used in studying (0.3) are adapted to yield an analysis of (0.4). For background material on adeles and automorphic representations, see [Kn2].

Finally in §7 we discuss the trace formula in the adelic setting (0.4). We relate aspects of Arthur's proof using truncation operators [Ar3], and we state the final formula and an important special case. The seven sections of this paper are thus as follows.

Arthur has extended the theory of the trace formula well beyond GL_2. For the theorem in "$\mathbb{Q}$ rank one," see [Ar1], and for a theorem about general reductive G, see [Ar2] and [Ar3]. Labesse [Lab] gives a status report as of 1990, and Gelbart [Gb2] gives an exposition of Arthur's work.

1. Overview of Decomposition of $L^2(SL_2(\mathbb{Z})\backslash SL_2(\mathbb{R}))$

We use the following notation: $G = SL(2,\mathbb{R})$, $\Gamma = SL_2(\mathbb{Z})$, $N = \left\{\begin{pmatrix} 1 & * \\ 0 & 1 \end{pmatrix}\right\}$, $\Gamma_\infty = \Gamma \cap N$, $A = \left\{\begin{pmatrix} r & 0 \\ 0 & r^{-1} \end{pmatrix}\right\}$, and $K = SO(2)$. Let $L^2(\Gamma\backslash G)$ be the space of functions on G, up to equality almost everywhere, that are left invariant under Γ and are square integrable modulo Γ. We are interested in the decomposition of the regular representation of G on $L^2(\Gamma\backslash G)$. References are [Gb1], [Go1], [Go2], [HC], [Lgl1], and [Lgl2].

If H is a closed subgroup of G, let $\mathcal{D}(H\backslash G)$ be the space of complex-valued smooth functions on G that are compactly supported modulo H.

Lemma 1.1. *If ϕ is in $\mathcal{D}(N\backslash G)$, then the function $\widehat{\phi}$ defined by $\widehat{\phi}(g) = \sum_{\gamma\in\Gamma_\infty\backslash\Gamma}\phi(\gamma g)$ is in is in $\mathcal{D}(\Gamma\backslash G)$.*

PROOF. Since $\Gamma_\infty\backslash N$ is compact, the support of ϕ is contained in $\Gamma_\infty C$ for some compact set $C\subset G$. Thus $\phi(\gamma g)\neq 0$ only for $\gamma g\in\Gamma_\infty C$. If g ranges through a compact set, then the γ's such that $\phi(\gamma g)\neq 0$ are those in a set $\Gamma_\infty C'$ with C' compact, and these form a finite subset of $\Gamma_\infty\backslash\Gamma$. Hence only finitely many terms in the sum defining $\widehat{\phi}$ contribute on any compact set of g's, and therefore $\widehat{\phi}$ is smooth. Finally the support of $\phi(\gamma\,\cdot\,)$ is contained in $\gamma^{-1}\Gamma_\infty C$, and the support of $\widehat{\phi}$ is contained in $\Gamma\Gamma_\infty C=\Gamma C$. The latter set is compact modulo Γ.

If F is any locally square integrable function on G that is left invariant under Γ_∞, we define the **constant term** of F to be the function F_0 on G given by

$$F_0(g)=\int_{\Gamma_\infty\backslash N}F(ng)\,d\dot{n}, \tag{1.1}$$

where $d\dot{n}$ has total mass 1. Since F is locally square integrable on G, Fubini's Theorem shows that $F(\,\cdot\,g)$ is locally square integrable on N for almost every g. Since $\Gamma_\infty\backslash N$ is compact, it follows for these g's that $F(\,\cdot\,g)$ is in $L^2(\Gamma_\infty\backslash N)$ and hence also is in $L^1(\Gamma_\infty\backslash N)$. Thus F_0 is defined almost everywhere.

The name "constant term" comes from the classical theory of modular forms. If the analytic function f on the upper half plane is a classical modular form of weight k relative to $SL_2(\mathbb{Z})$, then f has a Fourier expansion $f(z)=\sum_{n=0}^{\infty}c_ne^{2\pi inz}$, and the constant term c_0 of this series is given by

$$c_0=\int_{-1/2}^{1/2}f(x+iy)\,dx.$$

When f is lifted as in [Kn2, §7] to an automorphic form ϕ on G relative to Γ by means of the formula

$$\phi(g)=f(g(i))j(g,i)^{-k}, \tag{1.2}$$

in which $j(g,z)=cz+d$ when $g=\left(\begin{smallmatrix}a&b\\c&d\end{smallmatrix}\right)$, we find that the constant term ϕ_0 in the sense of (1.1) is given by $\phi_0(g)=c_0j(g,i)^{-k}$.

Lemma 1.2. *Let ϕ be a measurable function on G left invariant under N, and let F be a measurable function on G left invariant under Γ. Define $\widehat{\phi}(g)=\sum_{\gamma\in\Gamma_\infty\backslash\Gamma}\phi(\gamma g)$. If $\widehat{|\phi|}$ and F are in $L^2(\Gamma\backslash G)$, then*

$$\langle\widehat{\phi},F\rangle_{L^2(\Gamma\backslash G)}=\langle\phi,F_0\rangle_{L^2(N\backslash G)}, \tag{1.3}$$

the indicated integruls converging.

PROOF. Formally we have

$$\langle\widehat{\phi}, F\rangle_{L^2(\Gamma\backslash G)} = \int_{\Gamma\backslash G} \sum_{\Gamma_\infty\backslash\Gamma} \phi(\gamma g)\overline{F(g)}\,d\dot{g} = \int_{\Gamma_\infty\backslash G} \phi(x)\overline{F(x)}\,d\dot{x}.$$

This computation is rigorous if ϕ and F are replaced by $|\phi|$ and $|F|$, and the hypotheses say that the left side is finite in this case. Then the right side is finite, and we see that the following continuation of the above computation is justified:

$$= \int_{N\backslash G} \int_{\Gamma_\infty\backslash N} \overline{F(ng)}\phi(g)\,d\dot{n}\,d\dot{g} = \langle\phi, F_0\rangle_{L^2(N\backslash G)}.$$

This completes the proof.

Lemma 1.1 implies that $\widehat{\phi}$ is in $L^2(\Gamma\backslash G)$ if ϕ is in $\mathcal{D}(N\backslash G)$. Using Lemma 1.2, we obtain a characterization of the closure of the subspace of all such $\widehat{\phi}$.

Theorem 1.3. *The space $L^2(\Gamma\backslash G)$ is the orthogonal direct sum of G invariant subspaces*

$$L^2(\Gamma\backslash G) = L^2_{\text{cusp}}(\Gamma\backslash G) \oplus L^2_{\text{cont}}(\Gamma\backslash G) \oplus \mathbb{C},$$

where $L^2_{\text{cusp}}(\Gamma\backslash G)$ is the subspace of functions whose constant terms are 0 almost everywhere on G, $L^2_{\text{cont}}(\Gamma\backslash G)$ is the closure of the subspace of all $\widehat{\phi}$ with $\phi \in \mathcal{D}(N\backslash G)$ of integral 0, and $\mathbb{C}$ is the space of constant functions.

PROOF. If F is in $L^2(\Gamma\backslash G)$ and ϕ is in $\mathcal{D}(N\backslash G)$, we shall use the formula (1.3) of Lemma 1.2. If F is in $L^2_{\text{cusp}}(\Gamma\backslash G)$, then $F_0 = 0$ almost everywhere, and (1.3) shows that $\widehat{\phi}$ is orthogonal to F. Conversely if $\widehat{\phi}$ is orthogonal to F for all ϕ, then (1.3) shows that F_0 is orthogonal to $\mathcal{D}(N\backslash G)$ and is 0 almost everywhere. Thus $L^2_{\text{cusp}}(\Gamma\backslash G)$ is the orthogonal complement of the closure of the subspace of all $\widehat{\phi}$.

Taking $F = 1$ in (1.3), we see that $L^2_{\text{cont}}(\Gamma\backslash G)$ is a closed invariant subspace of codimension 1 in the closure of the subspace of all $\widehat{\phi}$. Since G acts unitarily, the orthogonal complement of $L^2_{\text{cont}}(\Gamma\backslash G)$ is a G invariant one-dimensional subspace, necessarily $\mathbb{C}$. The theorem follows.

We shall now describe the representation of G on $L^2_{\text{cont}}(\Gamma\backslash G)$. The group G acts on the upper half plane by linear fractional transformations, with

$$g(z) = \frac{az+b}{cz+d} \qquad \text{if } g = \begin{pmatrix} a & b \\ c & d \end{pmatrix}. \tag{1.4}$$

Moreover,

$$\operatorname{Im} g(z) = \frac{\operatorname{Im} z}{|cz+d|^2}. \tag{1.5}$$

Let $G = NAK$ be the usual Iwasawa decomposition of G. We write the K component of $g \in G$ as $\kappa(g)$. If k is in K, then

$$\begin{pmatrix} 1 & x \\ 0 & 1 \end{pmatrix} \begin{pmatrix} y^{1/2} & 0 \\ 0 & y^{-1/2} \end{pmatrix} k(i) = x + iy. \tag{1.6}$$

Thus we can read off the N and A components of g from the real and imaginary parts of $g(i)$. We write $y(g) = \operatorname{Im} g(i)$ for the imaginary part. If $y > 0$, define

$a(y) = \begin{pmatrix} y^{1/2} & 0 \\ 0 & y^{-1/2} \end{pmatrix}$. Then $a(y(g))$ is the A component of g in the Iwasawa decomposition.

We need to normalize Haar measures. We normalize dn on N to be compatible with counting measure on Γ_∞ and the measure of total mass 1 on $\Gamma_\infty\backslash N$, we normalize da on A to correspond to $\frac{dy}{y}$ on $(0,\infty)$ when $y = y(a)$ and $a = a(y)$, and we normalize dk on K to have total mass 1. If $g = nak$ is the Iwasawa decomposition of an element $g \in G$, we define

$$dg = dn\, dk\, y(a)^{-1}\, \frac{dy}{y}. \tag{1.7}$$

Then dg is a Haar measure on G.

A function F on G or K will be called **even** if $F\left(x\left(\begin{smallmatrix} -1 & 0 \\ 0 & -1 \end{smallmatrix}\right)\right) = F(x)$, **odd** if $F\left(x\left(\begin{smallmatrix} -1 & 0 \\ 0 & -1 \end{smallmatrix}\right)\right) = -F(x)$. For s complex, let $P^{+,s}$ be the spherical principal series representation of G defined as follows: $P^{+,s}$ acts initially in the space

$$\{\text{even } F \in C^\infty(G) \mid F(nag) = y(a)^{\frac{1}{2}(1+s)} F(g)\} \tag{1.8}$$

by the right regular representation with norm squared given by $\int_K |F(k)|^2\, dk$, and then it is completed to a representation in a Hilbert space. The subspace of C^∞ vectors is exactly (1.8), and the representation is unitary if $\operatorname{Re} s = 0$. If f is a smooth even function on K, then f extends to a member f_s of (1.8) by the rule

$$f_s(nak) = y(a)^{\frac{1}{2}(1+s)} f(k).$$

For each t, $P^{+,it}$ is irreducible and is unitarily equivalent with $P^{+,-it}$. Thus there exists a unique-up-to-scalar bounded linear operator intertwining $P^{+,it}$ and $P^{+,-it}$. We denote a particular normalization of this operator by $M(t)$; $M(t)$ will be defined explicitly in (2.13), and it will be unitary with $M(-it)$ as inverse.

We shall describe a certain direct integral of the unitary representations $P^{+,it}$. The underlying Hilbert space, which is denoted $\widehat{L}^2(E)$, is the set of measurable functions

$$F : i\mathbb{R} \to \{\text{even functions in } L^2(K)\}$$

(modulo null functions) such that

$$M(it)F(it) = F(-it)$$

and such that the expression

$$\|F\|^2_{\widehat{L}^2(E)} = \frac{1}{4\pi}\int_{-\infty}^{\infty} \|F(it)\|^2_{L^2(K)}\, dt$$

is finite. We make this into a representation space for G by having $P^{+,it}$ act on $F(it)_{it}$. More concretely, if U is to be the representation, we let

$$(U(g)F)(it) = \big(P^{+,it}(g)(F(it)_{it})\big)|_K.$$

The main theorem about $L^2_{\text{cont}}(\Gamma\backslash G)$ is as follows.

Theorem 1.4. *There exists a G equivariant unitary mapping E of $L^2_{\text{cont}}(\Gamma\backslash G)$ onto $\widehat{L}^2(E)$.*

This theorem will be proved in §2 by constructing the mapping E explicitly with the aid of Eisenstein series.

We come to the representation of G on $L^2_{\text{cusp}}(\Gamma\backslash G)$. Knowledge of how this representation decomposes remains far from complete. But we can say the following.

Theorem 1.5. *$L^2_{\text{cusp}}(\Gamma\backslash G)$ is the orthogonal Hilbert-space direct sum of irreducible representations, each occurring with finite multiplicity.*

The tool for proving Theorem 1.5 is Theorem 1.6 below, which will be proved in §3. Let φ be in $\mathcal{D}(G)$, and define a bounded operator $R(\varphi)$ on $L^2(\Gamma\backslash G)$ by $R(\varphi)f(x) = \int_G f(xy)\varphi(y)\,dy$. This carries any closed G invariant subspace of $L^2(\Gamma\backslash G)$ into itself.

Theorem 1.6. *For each φ in $\mathcal{D}(G)$, the operator*

$$R(\varphi) : L^2_{\text{cusp}}(\Gamma\backslash G) \to L^2_{\text{cusp}}(\Gamma\backslash G)$$

is Hilbert-Schmidt, hence compact.

Proof that Theorem 1.6 implies Theorem 1.5. In order to obtain the discrete decomposition into irreducible closed invariant subspaces, it is enough, by Zorn's Lemma, to prove that any nonzero invariant closed subspace S of $L^2_{\text{cusp}}(\Gamma\backslash G)$ contains an irreducible invariant subspace. The operator $R(\varphi)$ is self adjoint on S if $\varphi(x^{-1}) = \overline{\varphi(x)}$, and it is nonzero if φ is nonzero and φ is supported in a sufficiently small neighborhood of the identity. By Theorem 1.6 it is compact. Therefore it has a nonzero eigenvalue λ, and that eigenvalue has finite multiplicity. Let f be a nonzero eigenvector belonging to λ, and let λ have multiplicity n. Let T be the closed invariant subspace generated by f. If T is the orthogonal sum of $n+1$ closed invariant subspaces and if $P_1, \ldots, P_{n+1}$ are the orthogonal projections, then $R(\varphi)$ has eigenvalue λ on the independent vectors $P_1 f, \ldots, P_{n+1} f$, contradiction. It follows that T decomposes fully into at most n irreducible closed invariant subspaces. Any one of these subspaces is the required irreducible subspace of S.

Thus we can write $L^2_{\text{cusp}}(\Gamma\backslash G)$ as the orthogonal Hilbert-space direct sum of irreducible subspaces. Let S be such a subspace. As in the previous paragraph, we can choose φ with $\varphi(x^{-1}) = \overline{\varphi(x)}$ so that $R(\varphi)$ is nonzero on S. Since $R(\varphi)$ is compact self adjoint on S, $R(\varphi)$ has a nonzero eigenvalue λ on a nonzero subspace of S. On each irreducible summand of $L^2_{\text{cusp}}(\Gamma\backslash G)$ that is equivalent with S, $R(\varphi)$ must act with λ as an eigenvalue on the corresponding subspace. If S occurs with infinite multiplicity, then λ occurs with infinite multiplicity as an eigenvalue of $R(\varphi)$. But this contradicts the compactness of the self adjoint operator $R(\varphi)$ on $L^2_{\text{cusp}}(\Gamma\backslash G)$.

2. Decomposition of the Continuous Part

In this section we shall prove Theorem 1.4, giving an explicit decomposition of $L^2_{\text{cont}}(\Gamma\backslash G)$ when $G = SL_2(\mathbb{R})$ and $\Gamma = SL_2(\mathbb{Z})$. We continue with notation as in §1. We shall proceed somewhat along the lines of Appendix IV of [Lgl2] and then [Gb1]. For a different argument leading to a conclusion that is stated differently, see [Go1]. The technique of proof will involve Eisenstein series, which we now introduce.

If f is an even function in $C^\infty(K)$, we recall that $f_s : G \to \mathbb{C}$ is defined for $s \in \mathbb{C}$ by

$$f_s(nak) = y(a)^{\frac{1}{2}(1+s)} f(k) \tag{2.1}$$

when $n \in N$, $a \in A$, and $k \in K$. This satisfies the functional equation

$$f_s(nag) = y(a)^{\frac{1}{2}(1+s)} f_s(g)$$

and hence is a member of the representation space for the spherical principal series $P^{+,s}$.

Fix a finite-dimensional representation τ of K, and let $W(\tau)$ denote the space of complex-valued even functions on K with the property that $k \mapsto f(k_0 k)$, for each $k_0 \in K$, is a linear combination of matrix coefficients of the constituents of τ. If f is in $W(\tau)$, the corresponding **Eisenstein series** $E(g, f, s)$ is defined formally by

$$E(g, f, s) = \sum_{\gamma \in \Gamma_\infty \backslash \Gamma} f_s(\gamma g) = \sum_{\gamma \in \Gamma_\infty \backslash \Gamma} y(\gamma g)^{\frac{1}{2}(1+s)} f(\kappa(\gamma g)) \tag{2.2}$$

for $g \in G$ and $s \in \mathbb{C}$.

We can understand $\Gamma_\infty \backslash \Gamma$ with the help of the right action of G on row vectors. In this action the orbit under Γ of the row vector $(\,0 \quad 1\,)$ is all row vectors $(\,c \quad d\,)$ with c and d integers such that $\mathrm{GCD}(c, d) = 1$. The isotropy subgroup at $(\,0 \quad 1\,)$ is Γ_∞, and thus $\Gamma_\infty \backslash \Gamma$ may be identified with the set of relatively prime pairs (c, d). Evidently if $\Gamma_\infty \gamma$ corresponds to (c, d), then c and d form the bottom row of γ.

For an example let us take $\tau = 1$ and $f = 1$. If we put $g(i) = z = x + iy$, then (1.5) shows that (2.2) becomes

$$E(g, 1, s) = \sum_{\mathrm{GCD}(c,d)=1} \frac{y^{\frac{1}{2}(1+s)}}{|cz + d|^{1+s}}. \tag{2.3a}$$

Taking into account that every nonzero (m, n) in $\mathbb{Z}^2$ is uniquely the product of a positive integer and a relatively prime pair, we obtain

$$\zeta(1 + s) E(g, 1, s) = \sum_{(m,n) \neq (0,0)} \frac{y^{\frac{1}{2}(1+s)}}{|mz + n|^{1+s}}, \tag{2.3b}$$

where $\zeta(\,\cdot\,)$ is the Riemann ζ function.

The original Eisenstein series historically were series of the form

$$\sum_{(m,n) \neq (0,0)} \frac{1}{(mz + n)^k}, \tag{2.4}$$

as well as certain variants. The series is absolutely convergent if $k > 2$. In order to make sense out of the series (2.4) when $k = 2$, Hecke considered the analytic continuation in s of expressions of the form

$$\sum_{(m,n) \neq (0,0)} \frac{1}{(mz + n)^k |mz + n|^{2s}}. \tag{2.5}$$

In [Mi] these are called "Eisenstein series with parameter s," and (2.3b) is an instance of (2.5). If we take τ to be a nontrivial character of K and reinterpret

$E(g, f, s)$ on the upper half plane by reversing the formula (1.2) for lifting modular forms to G, we obtain the other instances of (2.5).

Lemma 2.1. *$E(g, f, s)$ is absolutely convergent for $\operatorname{Re} s > 1$, and the convergence is uniform for g and s in compact sets.*

PROOF. It is enough to estimate $\sum_{\gamma\in\Gamma_\infty\backslash\Gamma} y(\gamma g)^{\frac{1}{2}(1+\operatorname{Re} s)}$. This is written explicitly in (2.3a), and the larger series in (2.3b) is known to converge for $\operatorname{Re} s > 1$.

Lemma 2.2. *For any $\varepsilon > 0$, there is a constant C_ε such that*

$$|E(g, f, s)| \le C_\varepsilon(\sup_K |f|)y(g)^{\frac{1}{2}(1+\operatorname{Re} s)}$$

whenever $y(g) \ge \frac{1}{2}$ and $1 + \varepsilon \le \operatorname{Re} s \le 1 + \varepsilon^{-1}$.

PROOF. Without loss of generality, we may take $f = 1$ on K. Write $z = x+iy = g(i)$ and $\sigma = \operatorname{Re} s$. Applying (2.3a), we see that we are to estimate

$$\sum_{\operatorname{GCD}(c,d)=1} \frac{y^{\frac{1}{2}(1+\sigma)}}{((cx + d)^2 + c^2y^2)^{\frac{1}{2}(1+\sigma)}}.$$

So it is enough to show that

$$\sum_{(c,d)\neq(0,0)} ((cx + d)^2 + c^2y^2)^{-\frac{1}{2}(1+\sigma)} \tag{2.6}$$

is bounded above for $y \ge \frac{1}{2}$ and $1 + \varepsilon \le \sigma \le 1 + \varepsilon^{-1}$.

Fix $c \neq 0$. At most two d's give $|cx + d| < 1$. The contribution to (2.6) from such pairs (c, d) is therefore $\le \sum_{c\neq0} 2c^{-(1+\sigma)}y^{-(1+\sigma)} \le C_{1,\varepsilon}y^{-(1+\sigma)}$.

For the remaining terms, we can replace $cx + d$ by the nonzero integer

$$\operatorname{sgn}(cx + d)[|cx + d|].$$

Then the contribution to (2.6) from the remaining terms is

$$\le \sum_{\substack{(c,n),\\ n\neq0}} \frac{1}{(n^2 + c^2y^2)^{\frac{1}{2}(1+\sigma)}} \le 2^{1+\sigma} \sum_{\substack{(c,n),\\ n\neq0}} \frac{1}{(4n^2 + c^2)^{\frac{1}{2}(1+\sigma)}},$$

and the result follows.

An **automorphic form** on G relative to Γ is a smooth function f with the following properties:

(a) $f(\gamma g) = f(g)$ for all $\gamma \in \Gamma$
(b) f is right K finite
(c) f is $Z(\mathfrak{g})$ finite, where $Z(\mathfrak{g})$ is the center of the universal enveloping algebra of the complexified Lie algebra of G
(d) f satisfies the slow growth condition $|f(g)| \le Cy(g)^N$ for some C and N and all g with $y(g) \ge \frac{1}{2}$.

(See [Kn2, §7] and [Gb1, p. 28].)

Proposition 2.3. *For any $f \in W(\tau)$, $E(\,\cdot\,, f, s)$ is an automorphic form on G relative to Γ if $\operatorname{Re} s > 1$.*

PROOF. Properties (a) and (b) are clear from the definitions, and (d) follows from Lemma 2.2. For (c), we observe that the function $g \mapsto f(\kappa(g))y(g)^{\frac{1}{2}(1+s)}$ is in the space of the principal series $P^{+,s}$. The Casimir operator Ω acts in $P^{+,s}$ by a scalar $c(s)$ depending on s. Since Ω is central, it acts on every term of (2.2) by $c(s)$, and it acts on $E(\cdot, f, s)$ by $c(s)$. The element Ω generates $Z(\mathfrak{g})$, and (c) follows.

Although $E(\cdot, f, s)$ is an automorphic form, it need not be in $L^2(\Gamma\backslash G)$. In fact, let us check that $E(\cdot, f, s)$ is not in $L^2(\Gamma\backslash G)$ if $f = 1$ and s is real. In this case we can see that $E(g, 1, s)$ is bounded below, as well as above, by a multiple of $y(g)^{\frac{1}{2}(1+s)}$. The invariant measure on $\Gamma\backslash G$ amounts to $y^{-2}\,dx\,dy$ on the standard fundamental domain

$$S = \left\{z \,\middle|\, \operatorname{Im} z > 0, |z| \geq 1, |\operatorname{Re} z| \leq \tfrac{1}{2}\right\}$$

for Γ, and the integral of $|E(g, 1, s)|^2$ is of the order of $\int_{y=1}^{\infty} \int_{x=-1/2}^{1/2} y^{s-1}\,dx\,dy$, which is infinite for $s > 1$ (not to mention $s > 0$).

Although an individual $E(\cdot, f, s)$ is not in L^2, it turns out that suitable averages in the s variable are in L^2. Here is the construction.

Let $\mathcal{D}(N\backslash G, \tau)$ be the subspace of all $\phi \in \mathcal{D}(N\backslash G)$ such that $k \mapsto \phi(gk)$ is in $W(\tau)$ for each $g \in G$. For $\phi \in \mathcal{D}(N\backslash G, \tau)$, define the **Fourier-Laplace transform** of ϕ by

$$\Phi(g, s) = \int_0^{\infty} \phi(a(y)^{-1}g)y^{\frac{1}{2}(1+s)}\,\frac{dy}{y}. \tag{2.7}$$

This function satisfies

$$\Phi(nag, s) = y(a)^{\frac{1}{2}(1+s)}\Phi(g, s). \tag{2.8}$$

If we write $s = \sigma + it$ and $y = e^{2x}$, then we have

$$\Phi(g, \sigma + it) = \int_{-\infty}^{\infty} 2\phi(a(e^{2x})^{-1}g)e^{x(1+\sigma+it)}\,dx,$$

and Fourier inversion gives

$$2\phi(a(e^{2x})^{-1}g)e^{x(1+\sigma)} = \frac{1}{2\pi}\int_{-\infty}^{\infty} \Phi(g, \sigma + it)e^{-ixt}\,dt.$$

Taking $x = 0$ thus shows that

$$\phi(g) = \frac{1}{4\pi}\int_{\operatorname{Re} s=\sigma} \Phi(g, s)\,d|s| = \frac{1}{4\pi}\int_{\operatorname{Re} s=\sigma} y(g)^{\frac{1}{2}(1+s)}\Phi(\kappa(g), s)\,d|s|. \tag{2.9}$$

As a function of s, $\Phi(g, s)$ is a Schwartz function of $\operatorname{Im} s$ uniformly in any vertical strip of s and any compact set of g. The restriction $\Phi|_{K\times\{s\}}$ is a member of $W(\tau)$ for each s, and we shall usually abbreviate $\Phi|_{K\times\{s\}}$ as $\Phi(s)$.

Recall from Lemma 1.1 that the function

$$\widehat{\phi}(g) = \sum_{\gamma\in\Gamma_\infty\backslash\Gamma} \phi(\gamma g)$$

is in $\mathcal{D}(\Gamma\backslash G)$. Substituting from (2.9), we obtain

$$\widehat{\phi}(g) = \frac{1}{4\pi}\sum_{\gamma\in\Gamma_\infty\backslash\Gamma}\left(\int_{\operatorname{Re} s=\sigma} y(\gamma g)^{\frac{1}{2}(1+s)}\Phi(\kappa(\gamma g), s)\right)d|s|.$$

By Lemma 2.2, $\sum_\gamma |y(\gamma g)^{\frac{1}{2}(1+s)}|$ is bounded as a function of $\operatorname{Im} s$, and $\Phi(\kappa(\gamma g), s)$ is a Schwartz function of $\operatorname{Im} s$. Therefore the expression for $\widehat{\phi}(g)$ converges with absolute values inserted, and the sum and integral may be interchanged. The result is that

$$\widehat{\phi}(g) = \frac{1}{4\pi} \int_{\operatorname{Re} s = \sigma} E(g, \Phi(s), s)\, d|s|. \tag{2.10}$$

It is in this sense that suitable averages of Eisenstein series are in $L^2(\Gamma \backslash G)$.

Now we identify the constant term of an Eisenstein series. Recall from §1 that constant terms are indicated by a subscript 0. Let w denote the matrix $w = \begin{pmatrix} 0 & -1 \\ 1 & 0 \end{pmatrix}$.

Lemma 2.4. *For* $\operatorname{Re} s > 0$ *and for even functions* $f \in C^\infty(K)$, *the integral* $\int_N f_s(wng)\, dn$ *is convergent, and the formula*

$$A(s)f(g) = \int_N f_s(wng)\, dn \qquad \text{for } g \in G \tag{2.11}$$

defines a G *intertwining operator* $A(s) : P^{+,s} \to P^{+,-s}$. *As an operator from the space of even functions in* $C^\infty(K)$ *to itself,* $A(s)$ *has the following properties:*

(a) *it varies analytically in* s
(b) *it is uniformly bounded for* $\operatorname{Re} s \geq 1 + \varepsilon$
(c) *its adjoint relative to* $L^2(K)$ *is* $A(\bar{s})$.

REFERENCE. This result is elementary, and $A(s)$ is known as a **standard intertwining operator**. See Donley's lecture [Do], Moeglin's lecture [Mo], and also [Kn1, Ch. VII].

Since $A(s)$ is a G intertwining operator, it is in particular a K intertwining operator and therefore carries $W(\tau)$ to itself.

Lemma 2.5. *As an operator from* $W(\tau)$ *to itself, the operator* $A(s)$, *initially defined for* $\operatorname{Re} s > 0$, *continues to a meromorphic function of* $s \in \mathbb{C}$. *The continued family of operators has the following properties:*

(a) *the only possible poles are at* $s = 0, -2, -4, \ldots$ *and are simple*
(b) *for* $f \in W(\tau)$, $A(s)f$ *vanishes at* $s = 1$ *if* τ *does not contain the trivial representation of* K
(c) *apart from the poles,* $A(s)$ *is of at most polynomial growth in* $\operatorname{Im} s$ *in any vertical strip*
(d) *the operator* $A(-s)A(s)$ *is a meromorphic scalar depending on* s.

REFERENCE. This result is more subtle than Lemma 2.4 but is still not difficult. See [Do], [Mo], and also [Kn1, Ch. VII].

Proposition 2.6. *If* $\operatorname{Re} s > 1$ *and if* f *is in* $W(\tau)$, *then the constant term of the Eisenstein series for* f *is given by*

$$E_0(\,\cdot\,, f, s) = 2f_s + 2(M(s)f)_{-s}, \tag{2.12}$$

where $M(s)$ *is the operator*

$$M(s) = \frac{\zeta(s)}{\zeta(1+s)} A(s), \tag{2.13}$$

$A(s)$ being the operator in Lemma 2.4 and ζ being the Riemann ζ function. Here

(a) *$M(s)$ is analytic for $\operatorname{Re} s \geq 0$ except at $s = 1$, where it has at most a simple pole*
(b) *$M(s)$ is analytic at $s = 1$ if τ does not contain the trivial representation of K*
(c) *the residue of $M(s)$ at $s = 1$ is $6/\pi$ if $\tau = 1$*
(d) *the adjoint of $M(s)$ relative to the $L^2(K)$ norm on $W(\tau)$ is $M(\bar{s})$*
(e) *apart from the possible pole at $s = 1$, $M(s)$ is of at most polynomial growth in $\operatorname{Im} s$ uniformly for $0 \leq \operatorname{Re} s \leq \sigma$*
(f) *$M(-s)M(s) = 1$ as an identity of meromorphic functions.*

REMARK. Lemma 101 of [HC] shows in (e) that $M(s)$ is actually uniformly bounded in this strip, apart from the pole.

PROOF. Let H be the diagonal subgroup of G. We have seen that the coset of γ in $\Gamma_\infty \backslash \Gamma$ is characterized by the relatively prime pair (c, d) of entries of its bottom row. If $c = 0$, we obtain the cosets of ± 1. When $c \neq 0$, $\gamma = \left(\begin{smallmatrix} a & b \\ c & d \end{smallmatrix}\right)$ may be uniquely decomposed according to $NHwN$ as

$$\gamma = \begin{pmatrix} 1 & a/c \\ 0 & 1 \end{pmatrix} \begin{pmatrix} c^{-1} & 0 \\ 0 & c \end{pmatrix} \begin{pmatrix} 0 & -1 \\ 1 & 0 \end{pmatrix} \begin{pmatrix} 1 & d/c \\ 0 & 1 \end{pmatrix}.$$

Then

$$\Gamma_\infty \gamma = \begin{pmatrix} 1 & \frac{a}{c} + \mathbb{Z} \\ 0 & 1 \end{pmatrix} \begin{pmatrix} c^{-1} & 0 \\ 0 & c \end{pmatrix} \begin{pmatrix} 0 & -1 \\ 1 & 0 \end{pmatrix} \begin{pmatrix} 1 & d/c \\ 0 & 1 \end{pmatrix},$$

and the member $\nu = \left(\begin{smallmatrix} 1 & x \\ 0 & 1 \end{smallmatrix}\right)$ of Γ_∞ has

$$\Gamma_\infty \gamma \nu = \begin{pmatrix} 1 & \frac{a}{c} + \mathbb{Z} \\ 0 & 1 \end{pmatrix} \begin{pmatrix} c^{-1} & 0 \\ 0 & c \end{pmatrix} \begin{pmatrix} 0 & -1 \\ 1 & 0 \end{pmatrix} \begin{pmatrix} 1 & \frac{d}{c} + x \\ 0 & 1 \end{pmatrix}.$$

Thus we see that all the cosets $\Gamma_\infty \gamma \nu$, as ν varies, are distinct and that the number of double cosets $\Gamma_\infty \gamma \Gamma_\infty$ corresponding to a given c is $\varphi(|c|)$, where φ is Euler's φ function.

We compute

$$E_0(g, f, s) = \int_{\Gamma_\infty \backslash N} E(ng, f, s)\, d\dot{n} = \sum_{\Gamma_\infty \backslash \Gamma} \int_{\Gamma_\infty \backslash N} f_s(\gamma n g)\, d\dot{n}$$

by separating the terms $\gamma = \pm 1$ from the terms with $\gamma \in NHwN$. If we write $\gamma = \gamma(c, d)$, this expression is

$$\begin{aligned} &= 2 \int_{\Gamma_\infty \backslash N} f_s(ng)\, d\dot{n} + \sum_{\substack{\gamma(c,d) \in \Gamma_\infty \backslash \Gamma, \\ c \neq 0}} \int_{\Gamma_\infty \backslash N} f_s(\gamma(c, d) n g)\, d\dot{n} \\ &= 2 f_s(g) + \mathrm{II}, \end{aligned}$$

where

$$\begin{aligned}
\mathrm{II} &= \sum_{\substack{c\neq 0,\\ \mathrm{GCD}(c,d)=1}} \int_{\Gamma_\infty\backslash N} f_s(\gamma(c,d)ng)\,d\dot{n} \\
&= \sum_{c\neq 0} \sum_{\substack{d \bmod c,\\ \mathrm{GCD}(c,d)=1}} \sum_{k=-\infty}^{\infty} \int_{\Gamma_\infty\backslash N} f_s(\gamma(c,d+ck)ng)\,d\dot{n} \\
&= \sum_{c\neq 0} \sum_{\substack{d \bmod c,\\ \mathrm{GCD}(c,d)=1}} \sum_{\nu\in\Gamma_\infty} \int_{\Gamma_\infty\backslash N} f_s(\gamma(c,d)\nu ng)\,d\dot{n} \\
&= \sum_{c\neq 0} \sum_{\substack{d \bmod c,\\ \mathrm{GCD}(c,d)=1}} \int_N f_s(\gamma(c,d)ng)\,dn.
\end{aligned}$$

Write $\gamma(c,d) \in NHwN$ as $\gamma(c,d) = n'(c,d)h(c)wn''(c,d)$, noting that $h(c) = \begin{pmatrix} c^{-1} & 0 \\ 0 & c \end{pmatrix}$, independently of d. Then the above expression is

$$\begin{aligned}
&= \sum_{c\neq 0} \sum_{\substack{d \bmod c,\\ \mathrm{GCD}(c,d)=1}} \int_N f_s(n'(c,d)h(c)wn''(c,d)ng)\,dn \\
&= \sum_{c\neq 0} \sum_{\substack{d \bmod c,\\ \mathrm{GCD}(c,d)=1}} \int_N f_s(h(c)wng)\,dn
\end{aligned}$$

by the change of variables $n''(c,d)n \mapsto n$. In turn this is

$$= 2\sum_{c=1}^{\infty} \varphi(c)c^{-(1+s)} \int_N f_s(wng)\,dn.$$

Easy computation using Euler products shows that $\sum_{c=1}^\infty \varphi(c)c^{-(1+s)} = \dfrac{\zeta(s)}{\zeta(1+s)}$. Therefore

$$\mathrm{II} = \frac{2\zeta(s)}{\zeta(1+s)} \int_N f_s(wng)\,dn = \frac{2\zeta(s)}{\zeta(1+s)}\,(A(s)f)_{-s}(g)$$

in the notation of Lemma 2.4, and we conclude that

$$\tfrac{1}{2}E_0(g,f,s) = f_s(g) + \frac{\zeta(s)}{\zeta(1+s)}\,(A(s)f)_{-s}(g).$$

This proves (2.12) with $M(s)$ as in (2.13).

Conclusions (a) and (b) are immediate from Lemma 2.5, and (d) is immediate from Lemma 2.4c, (2.13), and analytic continuation. Before proving (c), we need an identity. The operator $A(s)$ carries 1_s to a multiple of 1_{-s} since $A(s)$ carries $W(1) = \mathbb{C}$ to itself. To compute the multiple, we calculate

$$(A(s)1)_{-s}(1) = \int_N 1_s(wn)\,dn = \int_N y(wn)^{\frac{1}{2}(1+s)}\,dn.$$

The measure dn is to be normalized consistently with the measure of total mass 1 on $\Gamma_\infty \backslash N$ and the counting measure on Γ_∞. Thus if $n = \left(\begin{smallmatrix} 1 & x \\ 0 & 1 \end{smallmatrix}\right)$, then dn is Lebesgue measure dx. Since

$$wn(i) = \begin{pmatrix} 0 & -1 \\ 1 & 0 \end{pmatrix} \begin{pmatrix} 1 & x \\ 0 & 1 \end{pmatrix} i = \frac{-1}{i+x} = \frac{-x+i}{x^2+1},$$

we have $y(wn) = \operatorname{Im} wn(i) = (x^2+1)^{-1}$. Thus

$$(A(s)1)_{-s}(1) = \int_{-\infty}^{\infty} (x^2+1)^{-\frac{1}{2}(1+s)}\, dx.$$

Consequently a trick of Euler's yields

$$\begin{aligned} \Gamma(\tfrac{1}{2}(1+s))(x^2+1)^{-\frac{1}{2}(1+s)} &= \int_0^\infty (x^2+1)^{-\frac{1}{2}(1+s)} t^{\frac{1}{2}(1+s)} e^{-t} \frac{dt}{t} \\ &= \int_0^\infty t^{\frac{1}{2}(1+s)} e^{-t(x^2+1)} \frac{dt}{t} = \int_0^\infty t^{\frac{1}{2}(1+s)} e^{-t} e^{-tx^2} \frac{dt}{t} \end{aligned}$$

and then

$$\begin{aligned} \Gamma(\tfrac{1}{2}(1+s)) \int_{-\infty}^{\infty} (x^2+1)^{-\frac{1}{2}(1+s)}\, dx &= \int_0^\infty t^{\frac{1}{2}(1+s)} e^{-t} \Big(\int_{-\infty}^{\infty} e^{-tx^2}\, dx \Big) \frac{dt}{t} \\ &= \int_0^\infty t^{\frac{1}{2}(1+s)} e^{-t} \Big(\int_{-\infty}^{\infty} e^{-\pi r^2} \sqrt{\frac{\pi}{t}}\, dr \Big) \frac{dt}{t} \\ &= \sqrt{\pi} \int_0^\infty t^{s/2} e^{-t} \frac{dt}{t} \\ &= \sqrt{\pi}\, \Gamma(\tfrac{s}{2}). \end{aligned}$$

Hence

$$(A(s)1)_{-s}(1) = \int_{-\infty}^{\infty} (x^2+1)^{-\frac{1}{2}(1+s)}\, dx = \frac{\sqrt{\pi}\, \Gamma(\frac{s}{2})}{\Gamma(\frac{1}{2}(1+s))}. \tag{2.14}$$

To prove (c), we use (2.13) and (2.14) to write

$$(M(s)1)_{-s} = \frac{\zeta(s)}{\zeta(1+s)} (A(s)1)_{-s} = \frac{\Lambda(s)}{\Lambda(1+s)} 1_{-s}, \tag{2.15}$$

where $\Lambda(s) = \pi^{-s/2}\Gamma(\frac{s}{2})\zeta(s)$. Therefore

$$\operatorname{Res}_{s=1}\{(M(s)1)_{-s}\} = \frac{\operatorname{Res}_{s=1}\{\Lambda(s)\}}{\Lambda(2)} = \frac{\pi^{-1/2}\Gamma(\frac{1}{2})\operatorname{Res}_{s=1}\{\zeta(s)\}}{\pi/6} = \frac{6}{\pi}.$$

For (f), we combine Lemma 2.5d, (2.13), and (2.15) to obtain

$$M(-s)M(s) = \frac{\Lambda(s)}{\Lambda(1-s)} \frac{\Lambda(-s)}{\Lambda(1+s)},$$

and (f) follows from the functional equation $\Lambda(1-s) = \Lambda(s)$ of the ζ function.

Finally to prove (e), we use (2.13). Lemma 2.5c tells us that $A(s)$ is of at most polynomial growth in $\operatorname{Im} s$, apart from the pole at $s = 0$, for $0 \le \operatorname{Re} s \le \sigma$. Also $\zeta(s)$ is bounded in any vertical strip, apart from its pole. And $|\zeta(1+s)|^{-1}$ is known to be at most polynomial growth in $\operatorname{Im} s$ uniformly for $0 \le \operatorname{Re} s \le \sigma$; see [Ti, p. 44]. Thus (e) follows.

Corollary 2.7. *Let ϕ and ψ be members of $\mathcal{D}(N\backslash G,\tau)$, and let Φ and Ψ be the Fourier-Laplace transforms of ϕ and ψ. Then*

$$\langle\widehat{\phi},\widehat{\psi}\rangle_{L^2(\Gamma\backslash G)}=\frac{1}{2\pi}\int_{\operatorname{Re}s=\sigma}\big(\langle\Phi(s),\Psi(-\bar{s})\rangle_{L^2(K)}+\langle M(s)\Phi(s),\Psi(\bar{s})\rangle_{L^2(K)}\big)\,d|s|$$

for any $\sigma>1$.

PROOF. By (2.10), we have

$$\widehat{\phi}(g)=\frac{1}{4\pi}\int_{\operatorname{Re}s=\sigma}E(g,\Phi(s),s)\,d|s|.$$

Taking the constant term of both sides and applying Proposition 2.6, we obtain

$$\begin{aligned}\widehat{\phi}_0(g)&=\frac{1}{4\pi}\int_{\operatorname{Re}s=\sigma}E_0(g,\Phi(s),s)\,d|s|\\&=\frac{1}{2\pi}\int_{\operatorname{Re}s=\sigma}\big(\Phi(s)_s(g)+(M(s)\Phi(s))_{-s}(g)\big)\,d|s|.\end{aligned}$$

If we write $g=na(y)k$, then Haar measure dg decomposes as $y^{-1}\,dn\,dk\,\frac{dy}{y}$, according to (1.7). Thus the invariant measure on $N\backslash G$ is $y^{-1}\,dk\,\frac{dy}{y}$. Lemma 1.2 therefore gives

$$\begin{aligned}&\langle\widehat{\phi},\widehat{\psi}\rangle_{L^2(\Gamma\backslash G)}\\&=\langle\widehat{\phi}_0,\psi\rangle_{L^2(N\backslash G)}\\&=\frac{1}{2\pi}\int_{N\backslash G}\int_{\operatorname{Re}s=\sigma}\big(\Phi(s)_s(g)+(M(s)\Phi(s))_{-s}(g)\big)\overline{\psi(g)}\,d|s|\,d\dot{g}\\&=\frac{1}{2\pi}\int_{N\backslash G}\int_{\operatorname{Re}s=\sigma}\big(y(g)^{\frac{1}{2}(1+s)}\Phi(\kappa(g),s)+y(g)^{\frac{1}{2}(1-s)}(M(s)\Phi(s))(\kappa(g))\big)\\&\qquad\qquad\times\overline{\psi(g)}\,d|s|\,d\dot{g}\\&=\frac{1}{2\pi}\int_{\operatorname{Re}s=\sigma}\int_K\int_0^\infty\big(y^{\frac{1}{2}(1+s)}\Phi(k,s)+y^{\frac{1}{2}(1-s)}(M(s)\Phi(s))(k)\big)\\&\qquad\qquad\times\overline{\psi(a(y)k)}y^{-1}\,\frac{dy}{y}\,dk\,d|s|\\&=\frac{1}{2\pi}\int_{\operatorname{Re}s=\sigma}\big(\langle\Phi(s),\Psi(-\bar{s})\rangle_{L^2(K)}+\langle M(s)\Phi(s),\Psi(\bar{s})\rangle_{L^2(K)}\big)\,d|s|.\end{aligned}$$

This completes the proof.

Now we move the line of integration in Corollary 2.7 to $\operatorname{Re}s=0$. The integrand is meromorphic, the functions $\Phi(s)$ and $\Psi(s)$ are Schwartz functions of $\operatorname{Im}z$ uniformly in vertical strips, and the growth of $M(s)$ is controlled by Proposition 2.6e. Thus we can move the line of integration by the Cauchy Integral Formula, picking up a residue term from $s=1$. The result is

$$\begin{aligned}\langle\widehat{\phi},\widehat{\psi}\rangle_{L^2(\Gamma\backslash G)}=\frac{1}{2\pi}\int_{-\infty}^{\infty}\big(\langle\Phi(it),\Psi(it)\rangle_{L^2(K)}+\langle M(it)\Phi(it),\Psi(-it)\rangle_{L^2(K)}\big)\,dt\\+\operatorname{Res}_{s=1}\big\{\langle M(s)\Phi(s),\Psi(\bar{s})\rangle_{L^2(K)}\big\}.\end{aligned}\tag{2.16}$$

By Proposition 2.6, parts (b) and (e), the second term is

$$\operatorname{Res}_{s=1}\{\langle M(s)\Phi(s),\Psi(\bar{s})\rangle_{L^2(K)}\} = \begin{cases} \frac{6}{\pi}\langle\Phi(1),\Psi(1)\rangle_{L^2(K)} & \text{if } \tau = 1 \\ 0 & \text{if 1 is not in } \tau. \end{cases}$$

We can simplify the right side of the residue term since

$$\Phi(k,1) = \int_0^\infty \phi(a(y)^{-1}k)\,dy = \int_A \phi(ak)y(a)^{-1}\,da.$$

When $\tau = 1$, this expression is constant in k and yields $\int_{N\backslash G}\phi(g)\,d\dot{g}$. When 1 is not in τ, the integral of this expression over $k \in K$ is 0. We conclude that

$$\operatorname{Res}_{s=1}\{\langle M(s)\Phi(s),\Psi(\bar{s})\rangle_{L^2(K)}\} = \frac{6}{\pi}\Big(\int_{N\backslash G}\phi(g)\,d\dot{g}\Big)\overline{\Big(\int_{N\backslash G}\psi(g)\,d\dot{g}\Big)} \tag{2.17}$$

for all τ.

Corollary 2.8. *Let ϕ and ψ be members of $\mathcal{D}(N\backslash G,\tau)$, and let Φ and Ψ be the Fourier-Laplace transforms of ϕ and ψ. Then*

$$\begin{aligned}\langle\widehat{\phi},\widehat{\psi}\rangle_{L^2(\Gamma\backslash G)} = {} & \frac{1}{4\pi}\int_{-\infty}^{\infty}\langle\Phi(it)+M(-it)\Phi(-it),\Psi(it)+M(-it)\Psi(-it)\rangle_{L^2(K)}\,dt \\ & + \frac{6}{\pi}\Big(\int_{N\backslash G}\phi(g)\,d\dot{g}\Big)\overline{\Big(\int_{N\backslash G}\psi(g)\,d\dot{g}\Big)}.\end{aligned}$$

PROOF. Averaging the effect of leaving alone the first term on the right side of (2.16) and replacing t by $-t$, we obtain

$$\begin{aligned}\langle\widehat{\phi},\widehat{\psi}\rangle_{L^2(\Gamma\backslash G)} = {} & \frac{1}{4\pi}\int_{-\infty}^{\infty}\big(\langle\Phi(it),\Psi(it)\rangle_{L^2(K)} + \langle M(it)\Phi(it),\Psi(-it)\rangle_{L^2(K)} \\ & \qquad + \langle\Phi(-it),\Psi(-it)\rangle_{L^2(K)} + \langle M(-it)\Phi(-it),\Psi(it)\rangle_{L^2(K)}\big)\,dt \\ & + (\text{residue term}).\end{aligned} \tag{2.18}$$

It follows from Proposition 2.6, parts (d) and (f), that $M(it)$ is unitary with inverse $M(-it)$. Therefore

$$\langle M(it)\Phi(it),\Psi(-it)\rangle_{L^2(K)} = \langle\Phi(it),M(-it)\Psi(-it)\rangle_{L^2(K)}$$

and $$\langle\Phi(-it),\Psi(-it)\rangle_{L^2(K)} = \langle M(-it)\Phi(-it),M(-it)\Psi(-it)\rangle_{L^2(K)}.$$

Substituting in (2.18) for the second and third terms of the integrand, we obtain the t integral of the corollary. The residue term has been evaluated in (2.17).

Corollary 2.9. *Let ϕ be in $\mathcal{D}(N\backslash G,\tau)$, and let Φ be its Fourier-Laplace transform. Then $\widehat{\phi} = 0$ if and only if $\int_{N\backslash G}\phi(g)\,dg = 0$ and $\Phi(it) = -M(-it)\Phi(-it)$ for $-\infty < t < \infty$.*

PROOF. This is immediate from Corollary 2.8 with $\psi = \phi$.

From these results we obtain the analysis of $L^2_{\text{cont}}(\Gamma\backslash G)$. In fact, let $\mathcal{PW}(\tau)$ be the space of Fourier transforms of the space $C^\infty_{\text{com}}(i\mathbb{R}, W(\tau))$ of compactly supported smooth functions on $i\mathbb{R}$ with values in $W(\tau)$. The Fourier-Laplace transform $\phi \mapsto \Phi$ is a one-one map of $\mathcal{D}(N\backslash G, \tau)$ onto $\mathcal{PW}(\tau)$. For $\Phi \in \mathcal{PW}(\tau)$, define

$$\Phi_1(it) = \Phi(it) + M(-it)\Phi(-it).$$

The map $\Phi \mapsto \Phi_1$ is a linear map of $\mathcal{PW}(\tau)$ into the subspace $\widehat{L}^2(E,\tau)$ of functions h in $L^2(i\mathbb{R}, W(\tau))$ such that $M(it)h(it) = h(-it)$, and Corollary 2.9 says that the composition $\phi \mapsto \Phi \mapsto \Phi_1$ descends to a map $\widehat{\phi} \mapsto \Phi_1$. Let us call this descended map $\widetilde{E}_\tau$, writing it as

$$\widetilde{E}_\tau : \{\widehat{\phi} \mid \phi \in \mathcal{D}(N\backslash G, \tau)\} \to \widehat{L}^2(E,\tau).$$

By Lemma 1.2 with $F = 1$, $\widehat{\phi}$ has integral 0 over $\Gamma\backslash G$ if and only if ϕ has integral 0 over $N\backslash G$. Let us restrict $\widetilde{E}_\tau$ to a map

$$E_\tau : \{\widehat{\phi} \mid \phi \in \mathcal{D}(N\backslash G, \tau) \text{ and } \int_{N\backslash G} \phi(g)\, d\dot{g} = 0\} \to \widehat{L}^2(E,\tau). \tag{2.19a}$$

Corollary 2.9 shows that E_τ is one-one, and Corollary 2.8 shows that E_τ is actually isometric apart from a factor $1/4\pi$. Let $L^2_{\text{cont}}(\Gamma\backslash G, \tau)$ be the subspace of functions $h \in L^2_{\text{cont}}(\Gamma\backslash G)$ such that $k \mapsto h(\Gamma g k)$ is in $W(\tau)$ for all $g \in G$. Theorem 1.3 shows that E_τ extends to an isometric map

$$E_\tau : L^2_{\text{cont}}(\Gamma\backslash G, \tau) \to \widehat{L}^2(E,\tau). \tag{2.19b}$$

Meanwhile, consideration of Fourier transforms shows that $\mathcal{PW}(\tau)$ is dense in $L^2(i\mathbb{R}, W(\tau))$, and so is the subspace where $\Phi(1) = 0$ (corresponding to ϕ of integral 0). Hence the image under $\phi \mapsto \Phi \mapsto \Phi_1$ of functions of integral 0 is dense in $\widehat{L}^2(E,\tau)$. Thus the map (2.19a) has dense image. Since (2.19a) is isometric, (2.19b) is onto. We may summarize as follows.

Theorem 2.10. *Let $\phi \in \mathcal{D}(N\backslash G, \tau)$ have integral 0, let Φ be its Fourier-Laplace transform, and define*

$$\Phi_1(it) = \Phi(it) + M(-it)\Phi(-it).$$

The composition of the linear maps $\phi \mapsto \Phi \mapsto \Phi_1$ descends to a well defined linear map $\widehat{\phi} \to \Phi_1$, which extends to a bounded linear map E_τ of $L^2_{\text{cont}}(\Gamma\backslash G, \tau)$ onto $\widehat{L}^2(E,\tau)$ such that

$$\|\widehat{\phi}\|^2_{L^2(\Gamma\backslash G)} = \frac{1}{4\pi}\int_{-\infty}^{\infty} \|\Phi_1(it)\|^2_{L^2(K)}\, dt.$$

The map E_τ has an equivariance property. Since $\mathcal{D}(N\backslash G, \tau)$ is not closed under translation by G, we cannot hope for G equivariance. But we can hope for as much equivariance as τ permits. Thus let R be the right regular representation of G on $\mathcal{D}(N\backslash G)$, and define

$$R(f)\phi(x) = \int_G \phi(xg) f(g)\, dg$$

for all $f \in C^\infty_{\text{com}}(G)$ such that $k \mapsto f(k^{-1}g)$ is in $W(\tau)$ for all $g \in G$. If ϕ is in $\mathcal{D}(N\backslash G, \tau)$, then a change of variables shows that $R(f)\phi$ is in $\mathcal{D}(N\backslash G, \tau)$. Let $\Phi_\phi(g, s)$ be the Fourier-Laplace transform of ϕ. Remembering from (2.8) that $\Phi_\phi(\,\cdot\,, s)$ is in the space for $P^{+,s}$, we readily check that

$$P^{+,s}(f)\Phi_\phi(x,s) = \Phi_{R(f)\phi}(x,s).$$

Passing from Φ to Φ_1 and using the intertwining property of $M(-it)$ implicit in Lemma 2.4 and analytic continuation, we obtain, in obvious notation,

$$P^{+,it}(f)(\Phi_1)_\phi(x,it) = (\Phi_1)_{R(f)\phi}(x,it).$$

Consequently E_τ is equivariant with respect to the operation of all members f of $C^\infty_{\text{com}}(G)$ such that $k \mapsto f(k^{-1}x)$ is in $W(\tau)$ for all $x \in G$.

Now we pass to the limit, in effect taking the union over all τ. Let $\widehat{L}^2(E)$ be the set of all square integrable functions h from $i\mathbb{R}$ into the even functions on K such that $M(it)h(it) = h(-it)$. The union E of the E_τ gives us an isometric map (apart from the factor $1/4\pi$) of a dense subspace of $L^2_{\text{cont}}(\Gamma\backslash G)$ onto a dense subspace of $\widehat{L}^2(E)$, and this is equivariant with respect to all members f of $C^\infty_{\text{com}}(G)$ such that $k \mapsto f(k^{-1}x)$ is in a common $W(\tau)$ for all $x \in G$. Such f's form an approximate identity, and therefore E extends to an isometry of $L^2_{\text{cont}}(\Gamma\backslash G)$ onto $\widehat{L}^2(E)$ equivariant with respect to G. This proves Theorem 1.4.

3. Discrete Decomposition of the Cuspidal Part

In this section we shall prove Theorem 1.6, giving Godement's variation [Go2] of a proof of Langlands [Lgl2]. We continue to let $G = SL_2(\mathbb{R})$ and $\Gamma = SL_2(\mathbb{Z})$, and we use other notation as in §1. Fix φ in $\mathcal{D}(G)$. Our objective is to show that the operator $R(\varphi)f(x) = \int_G f(xy)\varphi(y)\,dy$ is Hilbert-Schmidt (hence compact) on the subspace $L^2_{\text{cusp}}(\Gamma\backslash G)$ of $L^2(\Gamma\backslash G)$. The main step is to prove the following lemma.

Lemma 3.1. *For any integer $M \geq 0$, there exists a constant $C(\varphi, M)$ such that*

$$|R(\varphi)f(g)| \leq C(\varphi, M)y(g)^{-M}\|f\|_{L^2(\Gamma\backslash G)}$$

for all $f \in L^2_{\text{cusp}}(\Gamma\backslash G)$ and for all $g \in G$ such that $g(i)$ is in the standard fundamental domain

$$S = \{z \mid \operatorname{Im} z > 0, |z| \geq 1, |\operatorname{Re} z| \leq \tfrac{1}{2}\}$$

for Γ.

Remark. We need this estimate only for $M = 0$, but the estimate for general M is no harder.

Proof. Writing

$$R(\varphi)f(x) = \int_G f(xy)\varphi(y)\,dy = \int_G f(y)\varphi(x^{-1}y)\,dy = \int_{\Gamma_\infty\backslash G} \sum_{\gamma\in\Gamma_\infty} f(y)\varphi(x^{-1}\gamma y)\,dy$$

shows that
$$R(\varphi)f(x) = \int_{\Gamma_\infty\backslash G} K_\varphi(x,y)f(y)\,dy,$$

where
$$K_\varphi(x,y) = \sum_{\gamma\in\Gamma_\infty} \varphi(x^{-1}\gamma y).$$

Define functions $n : \mathbb{R} \to N$ and $t : N \to \mathbb{R}$ by $n(t) = \begin{pmatrix} 1 & t \\ 0 & 1 \end{pmatrix}$ and $t\begin{pmatrix} 1 & x \\ 0 & 1 \end{pmatrix} = x$. The function $\varphi_{x,y}(t) = \varphi(x^{-1}n(t)y)$ is in $C^\infty_{\mathrm{com}}(\mathbb{R})$, and the Poisson summation formula gives

$$\sum_{m=-\infty}^{\infty} \varphi_{x,y}(m) = \sum_{m=-\infty}^{\infty} \widehat{\varphi}_{x,y}(m),$$

where

$$\widehat{\varphi}_{x,y}(s) = \int_{\mathbb{R}} \varphi_{x,y}(t) e^{-2\pi i t s}\, dt.$$

Thus the kernel defining $R(\varphi)$ on $L^2(\Gamma\backslash G)$ is given by

$$K_\varphi(x,y) = \sum_{m=-\infty}^{\infty} \widehat{\varphi}_{x,y}(m).$$

The contribution to $R(\varphi)f$ from $m = 0$ is the main term in the sense that we shall use the hypothesis that f is in $L^2_{\mathrm{cusp}}(\Gamma\backslash G)$ to handle it. The contribution from the other terms will be treated as an error term. The term for $m = 0$ gives

$$\begin{aligned}
\int_{\Gamma_\infty\backslash G} \widehat{\varphi}_{x,y}(0) f(y)\, dy &= \int_{\Gamma_\infty\backslash G}\int_N \varphi(x^{-1}ty) f(y)\, dt\, dy \\
&= \int_{N\backslash G}\int_{s\in\Gamma_\infty\backslash N}\int_{t\in N} \varphi(x^{-1}tsy) f(sy)\, dt\, ds\, dy \\
&= \int_{N\backslash G}\int_{t\in N}\int_{s\in\Gamma_\infty\backslash N} \varphi(x^{-1}ty) f(sy)\, ds\, dt\, dy
\end{aligned}$$

after a change of variables, and the right side is 0 since f is in $L^2_{\mathrm{cusp}}(\Gamma\backslash G)$.

Now we consider the contribution to $R(\varphi)f$ from $m \neq 0$. Let C be the support of φ, and write the Iwasawa decomposition of $x \in G$ relative to $G = NAK$ as $x = n_x a_x k_x$. Since K and C are compact, we have $KC \subset N\Omega_A K$ for some compact subset Ω_A of A. If $\varphi(x^{-1}n(t)y) \neq 0$, then $x^{-1}n(t)y$ is in C. Hence $k_x^{-1}a_x^{-1}n_x^{-1}n(t)y$ is in C, and y is in $n(-t)n_x a_x k_x C \subset Na_x N\Omega_A K \subset Na_x\Omega_A K$. In other words, $a_y = a_x\omega_A$ for some $\omega_A \in \Omega_A$. If $\widehat{\varphi}_{x,y}(m) \neq 0$, we therefore have

$$\begin{aligned}
\widehat{\varphi}_{x,y}(m) &= \int_{\mathbb{R}} \varphi(x^{-1}n(t)y) e^{-2\pi i t m}\, dt \\
&= \int_{\mathbb{R}} \varphi(k_x^{-1}a_x^{-1}n_x^{-1}n(t)n_y a_y k_y) e^{-2\pi i t m}\, dt \\
&= e^{2\pi i t(n_x^{-1}n_y)m} \int_{\mathbb{R}} \varphi(k_x^{-1}a_x^{-1}n(t)a_x\omega_A k_y) e^{-2\pi i t m}\, dt \\
&= e^{2\pi i t(n_x^{-1}n_y)m} \int_{\mathbb{R}} \varphi(k_x^{-1}n(y(x)^{-1}t)\omega_A k_y) e^{-2\pi i t m}\, dt \\
&= e^{2\pi i t(n_x^{-1}n_y)m} \int_{\mathbb{R}} \varphi(k_x^{-1}n(t)\omega_A k_y) e^{-2\pi i y(x) t m} y(x)\, dt.
\end{aligned}$$

As k and k' vary through K and a varies through Ω_A, the functions $t \mapsto \varphi(kn(t)ak')$ vary in a compact family in $\mathcal{D}(\mathbb{R})$ and therefore satisfy uniform estimates. Thus we obtain

$$|\widehat{\varphi}_{x,y}(m)| \le C_{M,\varphi}\, y(x)|y(x)m|^{-M} = C_{M,\varphi}\, y(x)^{1-M}|m|^{-M}$$

for every positive integer M and all x and y in G.

Since we have seen that the $m = 0$ term gives 0, we obtain

$$\begin{aligned}|R(\varphi)f(x)| &\leq \int_{y\in\Gamma_\infty\backslash G,\, y\in Na_x\Omega_A K} \sum_{m\neq 0} |\widehat{\varphi}_{x,y}(m)||f(y)|\,dy \\ &\leq \int_{y\in n[-\frac{1}{2},\frac{1}{2}]a_x\Omega_A K} \sum_{m\neq 0} |\widehat{\varphi}_{x,y}(m)||f(y)|\,dy \\ &\leq \int_{y\in n[-\frac{1}{2},\frac{1}{2}]a_x\Omega_A K} \sum_{m\neq 0} C_{M,\varphi} y(x)^{1-M}|m|^{-M}|f(y)|\,dy \\ &\leq C_\varphi y(x)^{1-M} \int_{y\in n[-\frac{1}{2},\frac{1}{2}]a_x\Omega_A K} |f(y)|\,dy,\end{aligned}$$

with the last inequality valid for $M \geq 2$. By the Schwarz inequality this is

$$\leq C_\varphi y(x)^{1-M}\Big(\int_{y\in n[-\frac{1}{2},\frac{1}{2}]a_x\Omega_A K} dy\Big)^{1/2}\Big(\int_{y\in n[-\frac{1}{2},\frac{1}{2}]a_x\Omega_A K} |f(y)|^2\,dy\Big)^{1/2}.$$

Since $n[-\frac{1}{2},\frac{1}{2}]$ has N measure 1 and K has total measure 1, we see that

$$\int_{y\in n[-\frac{1}{2},\frac{1}{2}]a_x\Omega_A K} dy = \int_{a\in a_x\Omega_A} y(a)^{-1}\,da = y(x)\int_{a\in\Omega_A} y(a)^{-1}\,da.$$

Also if $y(x) \geq \frac{1}{2}$, then the set $n[-\frac{1}{2},\frac{1}{2}]a_x\Omega_A K$ is covered by finitely many Γ translates of the fundamental domain S. If the number of such translates is q, then

$$\int_{y\in n[-\frac{1}{2},\frac{1}{2}]a_x\Omega_A K} |f(y)|^2\,dy \leq q\int_{y(i)\in S} |f(y)|^2\,dy = q\|f\|^2_{L^2(\Gamma\backslash G)}.$$

Putting these facts together, we find that

$$|R(\varphi)f(x)| \leq C'_\varphi y(x)^{\frac{3}{2}-M}\|f\|_{L^2(\Gamma\backslash G)} \tag{3.1}$$

if $y(x) \geq \frac{1}{2}$ and $M \geq 2$. Here C'_φ is $C_\varphi\big(q\int_{\Omega_A} y(a)^{-1}\,da\big)^{1/2}$. If $x(i)$ is in S, then $y(x) \geq \frac{1}{2}$. When $y(x) \geq \frac{1}{2}$, the inequality (3.1) for all exponents $\frac{3}{2} - M$ with $M \geq 2$ implies the inequality for all integer exponents and a constant depending on the exponent. This proves the lemma.

PROOF OF THEOREM 1.6. We take $M = 0$ in Lemma 3.1. The lemma says that, for each $g \in \Gamma\backslash G$, $f \mapsto R(\varphi)f(g)$ is a bounded linear functional on $L^2_{\text{cusp}}(\Gamma\backslash G)$. Hence there exists a function K_g in $L^2_{\text{cusp}}(\Gamma\backslash G)$ such that

$$R(\varphi)f(g) = \int_{\Gamma\backslash G} K_g(x)f(x)\,dx$$

for all $f \in L^2_{\text{cusp}}(\Gamma\backslash G)$. Moreover, $\|K_g\|_{L^2(\Gamma\backslash G)} \leq C(\varphi, 0)$ for all $g \in \Gamma\backslash G$. Put $K(g,x) = K_g(x)$. If $K(\,\cdot\,,\,\cdot\,)$ is jointly measurable, then

$$\int_{\Gamma\backslash G\times\Gamma\backslash G} |K(g,x)|^2\,dx\,dg \leq \int_{\Gamma\backslash G} C(\varphi,0)^2\,dg < \infty$$

since $\Gamma\backslash G$ has finite volume, and $R(\varphi)$ is exhibited as the restriction to $L^2_{\text{cusp}}(\Gamma\backslash G)$ of a Hilbert-Schmidt operator on $L^2(\Gamma\backslash G)$ that leaves $L^2_{\text{cusp}}(\Gamma\backslash G)$ stable. Hence $R(\varphi)$ is Hilbert-Schmidt on $L^2_{\text{cusp}}(\Gamma\backslash G)$.

To complete the proof, we need to address the joint measurability of the kernel. If X is a left invariant first-order derivative, then $X(R(\varphi)f) = -R(X\varphi)f$. Applying the lemma to $X\varphi$, we conclude that $\sup|X(R(\varphi)f)| \le C\|f\|_{L^2(\Gamma\backslash G)}$. If ε and Γg are given, it follows that $|R(\varphi)f(g') - R(\varphi)f(g)| \le \varepsilon\|f\|_{L^2(\Gamma\backslash G)}$ for all $f \in L^2_{\text{cusp}}(\Gamma\backslash G)$ and for all g' sufficiently close to g. Therefore $g \mapsto K_g$ is continuous as a map of $\Gamma\backslash G$ into $L^2_{\text{cusp}}(\Gamma\backslash G)$, and we saw above that it is bounded. It is a general fact that if M is in $L^2(\Gamma\backslash G \times \Gamma\backslash G)$ and $M_g(x) = M(g, x)$, then $g \mapsto M_g$ is in $L^2(\Gamma\backslash G, L^2(\Gamma\backslash G))$. Thus we can use $\{K_g\}$ to define a continuous linear functional on $L^2(\Gamma\backslash G \times \Gamma\backslash G)$ by

$$M \mapsto \int_{g\in\Gamma\backslash G} (M_g, K_g)_{L^2(\Gamma\backslash G)}\, d\dot{g}.$$

This linear functional must be given by the complex conjugate of a (jointly measurable) member K' of $L^2(\Gamma\backslash G \times \Gamma\backslash G)$. We can replace $K(\,\cdot\,,\,\cdot\,)$ by $K'(\,\cdot\,,\,\cdot\,)$ above, have the required joint measurability, and still have $R(\varphi)f = \int_{\Gamma\backslash G} K'(\,\cdot\,, x)f(x)\,dx$ almost everywhere for each $f \in L^2_{\text{cusp}}(\Gamma\backslash G)$.

4. Introduction to the Trace Formula

A first insight into what to look for in a trace formula comes from the compact quotient case. Let G be a unimodular Lie group, let Γ be a discrete subgroup such that $\Gamma\backslash G$ is compact, and let R be the right regular representation of G on $L^2(\Gamma\backslash G)$.

Let φ be in $C^\infty_{\text{com}}(G)$, and define $R(\varphi)f(x) = \int_G f(xy)\varphi(y)\,dy$. The computation

$$R(\varphi)f(x) = \int_G f(xy)\varphi(y)\,dy = \int_G f(y)\varphi(x^{-1}y)\,dy = \int_{\Gamma\backslash G} \sum_{\gamma\in\Gamma} f(y)\varphi(x^{-1}\gamma y)\,dy$$

shows that

$$R(\varphi)f(x) = \int_{\Gamma\backslash G} K(x,y)f(y)\,dy,$$

where $K(x,y) = \sum_{\gamma\in\Gamma}\varphi(x^{-1}\gamma y)$. This sum is locally finite, and it follows that K is in $C^\infty(\Gamma\backslash G \times \Gamma\backslash G)$. Thus we can apply the following lemma.

Lemma 4.1. *Let X be a compact C^∞ manifold, and let dx be a measure on X that is a smooth function times Lebesgue measure in each coordinate neighborhood. Let K be in $C^\infty(X\times X)$, and define a bounded operator B on $L^2(X, dx)$ by $Bf(x) = \int_X K(x,y)f(y)\,dy$. Then B is of trace class, and its trace is*

$$\operatorname{Tr} B = \int_X K(x,x)\,dx.$$

REFERENCE. [Kn1, p. 341].

By the lemma, $R(\varphi)$ is of trace class. Referring to the proof in §1 that Theorem 1.6 implies Theorem 1.5, we see that $L^2(\Gamma\backslash G)$ decomposes into the direct sum of irreducible representations of G, each occurring with finite multiplicity. Let us write

$$L^2(\Gamma\backslash G) = \bigoplus_{\pi\in\widehat{G}} m_\pi \pi. \tag{4.1}$$

The lemma also gives us a formula for the trace of $R(\varphi)$, namely

$$\operatorname{Tr} R(\varphi) = \int_{\Gamma\backslash G} K(x,x)\,dx = \int_{\Gamma\backslash G} \sum_{\gamma\in\Gamma} \varphi(x^{-1}\gamma x)\,dx. \tag{4.2}$$

We can refine the right side of (4.2) by lumping terms whose elements γ are conjugate in G. For a group U, let U^γ be the centralizer of γ in U. From each conjugacy class $\mathfrak{o}$ of elements in Γ, we select a representative. Say that γ is a representative of $\mathfrak{o}_\gamma$. Then $\mathfrak{o}_\gamma$ consists of all $\delta^{-1}\gamma\delta$, where δ varies through $\Gamma^\gamma\backslash\Gamma$. Thus

$$\begin{aligned}\int_{\Gamma\backslash G} \sum_{\gamma\in\Gamma} \varphi(x^{-1}\gamma x)\,dx &= \sum_{\mathfrak{o}_\gamma} \sum_{\delta\in\Gamma^\gamma\backslash\Gamma} \int_{\Gamma\backslash G} \varphi(x^{-1}\delta^{-1}\gamma\delta x)\,dx \\ &= \sum_{\mathfrak{o}_\gamma} \int_{\Gamma^\gamma\backslash G} \varphi(x^{-1}\gamma x)\,dx \\ &= \sum_{\mathfrak{o}_\gamma} \int_{G^\gamma\backslash G} \int_{\Gamma^\gamma\backslash G^\gamma} \varphi(x^{-1}y^{-1}\gamma y x)\,dy\,dx \\ &= \sum_{\mathfrak{o}_\gamma} \operatorname{vol}(\Gamma^\gamma\backslash G^\gamma) \int_{G^\gamma\backslash G} \varphi(x^{-1}\gamma x)\,dx. \end{aligned} \tag{4.3}$$

We arrive at the following result.

Theorem 4.2. *Let G be a unimodular Lie group, let Γ be a discrete subgroup such that $\Gamma\backslash G$ is compact, let R be the right regular representation of G on $L^2(\Gamma\backslash G)$, and let φ be in $C^\infty_{\text{com}}(G)$. Then $R(\varphi)$ is of trace class, and*

$$\operatorname{Tr} R(\varphi) = \sum_{\mathfrak{o}_\gamma} \operatorname{vol}(\Gamma^\gamma\backslash G^\gamma) \int_{G^\gamma\backslash G} \varphi(x^{-1}\gamma x)\,dx. \tag{4.4a}$$

Consequently if the decomposition of $L^2(\Gamma\backslash G)$ into irreducible representations of G is as in (4.1), *then*

$$\sum_{\pi\in\widehat{G}} m_\pi \operatorname{Tr}\pi(\varphi) = \sum_{\mathfrak{o}_\gamma} \operatorname{vol}(\Gamma^\gamma\backslash G^\gamma) \int_{G^\gamma\backslash G} \varphi(x^{-1}\gamma x)\,dx. \tag{4.4b}$$

Let us consider two examples. The first example is the case that G is compact and $\Gamma = \{1\}$. If dx is normalized to have total mass 1, then (4.4b) gives

$$\varphi(1) = \sum_{\pi\in\widehat{G}} m_\pi \operatorname{Tr}\pi(\varphi), \tag{4.5}$$

which is the Fourier inversion formula for G. It is typical of the trace formula that we can get information about the multiplicities m_π by specializing φ. Indeed, if in (4.5) we take φ to be the complex conjugate of the character of π, the Schur orthogonality relations tell us that m_π equals the degree of π.

The second example with compact quotient is the case that $G = \mathbb{R}$ and $\Gamma = \mathbb{Z}$. Assuming that the measure on $\mathbb{Z}\backslash\mathbb{R}$ has total mass one, we find that the right side of (4.4b) is just $\sum_{n=-\infty}^{\infty} \varphi(n)$, while the left side is $\sum_{n=-\infty}^{\infty} \widehat{\varphi}(n)$ if $\widehat{\varphi}(n) = \int_{\mathbb{Z}\backslash\mathbb{R}} \varphi(x)e^{-2\pi i n x}\,dx$. Formula (4.4b) is therefore the Poisson summation formula for smooth functions φ of compact support.

The example that we have been studying in this paper has $G = SL_2(\mathbb{R})$ and $\Gamma = SL_2(\mathbb{Z})$. For this case, $\Gamma\backslash G$ is noncompact and (4.4b) is not directly applicable. Indeed, we saw in §3 that $L^2(\Gamma\backslash G)$ has a continuous part to its decomposition, and $R(\varphi)$ cannot always be of trace class. What we know from Theorem 1.5 is that $R(\varphi)$ is Hilbert-Schmidt on $L^2_{\text{cusp}}(\Gamma\backslash G)$ if φ is in $C^\infty_{\text{com}}(G)$. Since the composition of two Hilbert-Schmidt operators is of trace class, $R(\varphi)$ is of trace class on $L^2_{\text{cusp}}(\Gamma\backslash G)$ if φ is a finite sum of convolutions of pairs of members of $C^\infty_{\text{com}}(G)$. A theorem of Dixmier and Malliavin [Di-Ma] says that this is always the case on a Lie group, and we arrive at the following theorem.

Theorem 4.3. *For $G = SL_2(\mathbb{R})$ and $\Gamma = SL_2(\mathbb{Z})$, $R(\varphi)$ is of trace class on $L^2_{\text{cusp}}(\Gamma\backslash G)$ if φ is in $C^\infty_{\text{com}}(G)$.*

Following the line of argument in the compact quotient case, we want to obtain a formula for $\operatorname{Tr} R(\varphi)$ on $L^2_{\text{cusp}}(\Gamma\backslash G)$ by integrating a kernel on its diagonal. Although the computation at the beginning of this section shows that $R(\varphi)$ is given by the kernel

$$K(x,y) = \sum_{\gamma\in\Gamma} \varphi(x^{-1}\gamma y),$$

this kernel reflects the action of $R(\varphi)$ on all of $L^2(\Gamma\backslash G)$. It is necessary to subtract terms to account for the contributions of $L^2_{\text{cont}}(\Gamma\backslash G)$ and the constant functions. On the constant functions, $R(\varphi)$ acts as the scalar $\int_G \varphi(x)\,dx$, and this scalar is the trace. Thus we need to know the kernel $K_{\text{cont}}(x,y)$ for the action of $R(\varphi)$ on $L^2_{\text{cont}}(\Gamma\backslash G)$.

The derivation of a formula for $K_{\text{cont}}(x,y)$ is a little complicated, and we shall carry out only the formal argument, omitting the justification for some interchanges of limits. Also we shall assume that φ is two-sided K finite. See [Gb-Ja] for more details. The argument requires knowing that there is a meromorphic continuation for an Eisenstein series $E(g,f,s)$ itself (with f in some $W(\tau)$, say), not just for its constant term. Moreover, the only poles for the continued Eisenstein series are simple and coincide with the poles of the constant term, and the continued Eisenstein series satisfies growth estimates in $\operatorname{Im} s$ in any strip $0 \le \operatorname{Re} s \le \sigma$. For a proof of these facts, see [Go1] or Appendix IV of [Lgl2]. These facts have an analog in the adelic setting (0.4), and the paper [Ja] in this volume discusses this analog.

Lemma 4.4. *Let ϕ be a K finite even function in $\mathcal{D}(N\backslash G)$, and let Φ be its Fourier-Laplace transform. Then the analytically continued Eisenstein series satisfies*

$$E(g, M(s)\Phi(s), -s) = E(g, \Phi(s), s). \tag{4.6}$$

PROOF. The constant term of the right side is $2\Phi(s)_s + 2(M(s)\Phi(s))_{-s}$ when $\operatorname{Re} s > 1$, and it is this at all points where there is no pole, by analytic continuation. Similarly the constant term of the left side is $2(M(s)\Phi(s))_{-s} + 2(M(-s)M(s)\Phi(s))_s$ when $\operatorname{Re} s < -1$, and it is this at all points where there is no pole. Since $M(-s)M(s) = 1$ by Proposition 2.6f, the two sides of (4.6) have equal constant terms.

For fixed $s = s_0$, let $b(g)$ be the difference of the two sides of (4.6). Then $b(g)$ has constant term 0, and Lemma 1.2 shows that $b(g)$ is orthogonal to any L^2 function of the form $\widehat{\phi}$. Thus $b(g)$ is orthogonal to $E(g,\Phi(s),s)$ in the region of convergence $\operatorname{Re} s > 1$ and then, by analytic continuation, for all s where there is no pole. Similarly $b(g)$ is orthogonal to $E(g, M(s)\Phi(s), -s)$ in the region of convergence $\operatorname{Re} s < -1$ and then for all s where there is no pole. Therefore $b(g)$ is orthogonal to itself, and $b(g) = 0$.

Let H be the space of even functions in $L^2(K)$. As in §2, we introduce

$$\widehat{L}^2(E) = \{F \in L^2(i\mathbb{R}, H) \mid M(it)F(it) = F(-it)\}.$$

Recall that the construction in §2 started from an even K finite $\phi \in \mathcal{D}(N\backslash G)$ of integral 0 and gave a map $\phi \to \Phi \to \Phi_1$ with $\Phi_1 \in \widehat{L}^2(E)$, and this map descended to be a well defined linear map S carrying $\widehat{\phi}$ to Φ_1. Theorem 2.10 shows that S preserves norms, in the sense that

$$\langle \widehat{\phi}, \widehat{\psi} \rangle_{L^2(\Gamma\backslash G)} = \frac{1}{4\pi} \int_{-\infty}^{\infty} \langle \Phi_1(it), \Psi_1(it) \rangle_{L^2(K)}\, dt, \tag{4.7}$$

and S has dense image in $\widehat{L}^2(E)$. Hence S completes to a unitary mapping of $L^2_{\text{cont}}(\Gamma\backslash G)$ onto $\widehat{L}^2(E)$.

Lemma 4.5. *Let $\phi \in \mathcal{D}(N\backslash G)$ be K finite of integral 0, and let h be a K finite member of the space H of even functions in $L^2(K)$. Then*

$$\langle h, S\widehat{\phi}(it) \rangle_{L^2(K)} = \frac{1}{2} \int_{\Gamma\backslash G} E(g, h, it) \overline{\widehat{\phi}(g)}\, dg.$$

PROOF. Since ϕ has integral 0 over $N\backslash G$, $\Phi(1)$ has integral 0 over K. Thus $M(s)\Phi(s)$ has no pole at $s = 1$, and $E(g, \Phi(s), s)$ has no pole at $s = 1$. For $\sigma > 1$, it follows that

$$\begin{aligned}
&\langle \widehat{\phi}, \widehat{\psi} \rangle_{L^2(\Gamma\backslash G)} \\
&= \frac{1}{4\pi} \int_{\Gamma\backslash G} \Big[\int_{\operatorname{Re} s = \sigma} E(g, \Phi(s), s) \overline{\widehat{\psi}(g)}\, d|s| \Big]\, dg \quad \text{by (2.10)} \\
&= \frac{1}{4\pi} \int_{\operatorname{Re} s = \sigma} \Big[\int_{\Gamma\backslash G} E(g, \Phi(s), s) \overline{\widehat{\psi}(g)}\, dg \Big]\, d|s| \quad \text{by interchange} \\
&= \frac{1}{4\pi} \int_{t=-\infty}^{\infty} \Big[\int_{\Gamma\backslash G} E(g, \Phi(it), it) \overline{\widehat{\psi}(g)}\, dg \Big]\, dt \quad \text{by moving the line of integration}
\end{aligned} \tag{4.8}$$

since there is no pole at $s = 1$.

In (4.8), Lemma 4.4 and the change of variables $t \to -t$ allow us to replace $\Phi(it)$ by $M(-it)\Phi(-it)$. Averaging the two results yields

$$\langle \widehat{\phi}, \widehat{\psi} \rangle_{L^2(\Gamma\backslash G)} = \frac{1}{8\pi} \int_{t=-\infty}^{\infty} \Big[\int_{\Gamma\backslash G} E(g, \Phi_1(it), it) \overline{\widehat{\psi}(g)}\, dg \Big]\, dt. \tag{4.9}$$

Comparing (4.7) and (4.9), we see that

$$\int_{-\infty}^{\infty} \langle \Phi_1(it), \Psi_1(it) \rangle_{L^2(K)}\, dt = \tfrac{1}{2} \int_{t=-\infty}^{\infty} \Big[\int_{\Gamma\backslash G} E(g, \Phi_1(it), it) \overline{\widehat{\psi}(g)}\, dg \Big]\, dt. \tag{4.10}$$

On each side of (4.10), we write the integral as a sum of integrals over $(0, \infty)$ and $(-\infty, 0)$ and in the $(-\infty, 0)$ integral replace t by $-t$ and then $\Phi_1(-it)$ by

$M(it)\Phi_1(it)$. Finally on the left side we replace $\Psi_1(-it)$ by $M(it)\Psi_1(it)$, and on the right side we substitute from Lemma 4.4. The result is that

$$\int_0^\infty \langle \Phi_1(it), \Psi_1(it)\rangle_{L^2(K)}\,dt = \tfrac{1}{2}\int_{t=0}^\infty \Big[\int_{\Gamma\backslash G} E(g,\Phi_1(it),it)\overline{\widehat{\psi}(g)}\,dg\Big]\,dt. \tag{4.11}$$

The functions $t \to \Phi_1(it)$ are dense in $L^2((0,\infty),H)$, and we can pass to the limit in the Eisenstein series if we stick to a K finite function in $L^2((0,\infty),H)$. Thus (4.11) persists if $\Phi_1(it)$ is replace by any K finite function in $L^2((0,\infty),H)$. Let us use a function of the form $c(t)h$, where h is a K finite member of H and $c(\,\cdot\,)$ is in $L^2((0,\infty),\mathbb{C})$. Then we obtain

$$\int_0^\infty c(t)\langle h, \Psi_1(it)\rangle_{L^2(K)}\,dt = \tfrac{1}{2}\int_{t=0}^\infty c(t)\Big[\int_{\Gamma\backslash G} E(g,h,it)\overline{\widehat{\psi}(g)}\,dg\Big]\,dt.$$

Since $c(t)$ is arbitrary, the integrands are equal at every point of continuity, i.e., everywhere. This proves the lemma.

Proposition 4.6. *Let $\{f_\alpha\}$ be an orthonormal basis of K finite functions in H, and let φ be two-sided K finite in $C^\infty_{\mathrm{com}}(G)$. Then $R(\varphi)$ is given on $L^2_{\mathrm{cont}}(\Gamma\backslash G)$ by the kernel*

$$K_{\mathrm{cont}}(x,y) = \frac{1}{16\pi}\sum_{\alpha,\beta}\int_{-\infty}^\infty \langle P^{+,it}(\varphi)f_\beta, f_\alpha\rangle E(x,f_\alpha,it)\overline{E(y,f_\beta,it)}\,dt,$$

PROOF. Extend the linear map S to all of $L^2(\Gamma\backslash G)$ by setting S equal to 0 on $L^2_{\mathrm{cusp}}(\Gamma\backslash G)$ and $\mathbb{C}$. For ϕ and ψ of integral 0, we have

$$\langle \widehat{\phi}, \widehat{\psi}\rangle_{L^2(\Gamma\backslash G)} = \langle S\widehat{\phi}, S\widehat{\psi}\rangle_{\widehat{L}^2(E)},$$

and it follows that S^*S is the orthogonal projection of $L^2(\Gamma\backslash G)$ on $L^2_{\mathrm{cont}}(\Gamma\backslash G)$. Since S is an intertwining operator, we have

$$S^*SR(\varphi)S^*S = S^*P^{+,\cdot}(\varphi)S,$$

where $P^{+,\cdot}$ is the representation on $\widehat{L}^2(E)$. Consequently

$$\begin{aligned}
&\langle S^*SR(\varphi)S^*S\widehat{\phi}, \widehat{\psi}\rangle_{L^2(\Gamma\backslash G)}\\
&= \langle S^*P^{+,\cdot}(\varphi)S\widehat{\phi}, \widehat{\psi}\rangle_{L^2(\Gamma\backslash G)}\\
&= \langle P^{+,\cdot}(\varphi)S\widehat{\phi}, S\widehat{\psi}\rangle_{\widehat{L}^2(E)}\\
&= \frac{1}{4\pi}\int_{-\infty}^\infty \langle P^{+,it}(\varphi)S\widehat{\phi}(it), S\widehat{\psi}(it)\rangle_{L^2(K)}\,dt\\
&= \frac{1}{4\pi}\int_{-\infty}^\infty \sum_\alpha \langle P^{+,it}(\varphi)S\widehat{\phi}(it), f_\alpha\rangle_{L^2(K)}\langle f_\alpha, S\widehat{\psi}(it)\rangle_{L^2(K)}\,dt\\
&= \frac{1}{4\pi}\int_{-\infty}^\infty \sum_\alpha \overline{\langle P^{+,it}(\varphi)^*f_\alpha, S\widehat{\phi}(it)\rangle}_{L^2(K)}\langle f_\alpha, S\widehat{\psi}(it)\rangle_{L^2(K)}\,dt\\
&= \frac{1}{16\pi}\int_{-\infty}^\infty \sum_\alpha \Big[\int_{\Gamma\backslash G}\overline{E(g,P^{+,it}(\varphi)^*f_\alpha,it)}\widehat{\phi}(g)\,dg\Big]\Big[\int_{\Gamma\backslash G}E(g',f_\alpha,it)\overline{\widehat{\psi}(g')}\,dg'\Big]\,dt
\end{aligned}$$

by Lemma 4.5

$$= \int_{\Gamma\backslash G\times\Gamma\backslash G} \Big[\frac{1}{16\pi}\int_{-\infty}^{\infty}\sum_\alpha \overline{E(g, P^{+,it}(\varphi)^* f_\alpha, it)} E(g', f_\alpha, it)\, dt\Big] \widehat{\phi}(g)\overline{\widehat{\psi}(g')}\, dg\, dg'.$$

Therefore $S^*SR(\varphi)S^*S$ is given by the kernel

$$K_{\mathrm{cont}}(g', g) = \frac{1}{16\pi}\int_{-\infty}^{\infty}\sum_\alpha \overline{E(g, P^{+,it}(\varphi)^* f_\alpha, it)} E(g', f_\alpha, it)\, dt.$$

If we expand $P^{+,it}(\varphi)^* f_\alpha = \sum_\beta \langle P^{+,it}(\varphi)^* f_\alpha, f_\beta\rangle f_\beta$, then we get the result of the proposition.

As a consequence of Proposition 4.6, the kernel of $R(\varphi)$ on $L^2_{\mathrm{cusp}}(\Gamma\backslash G)\oplus\mathbb{C}$ along the diagonal is $K(x,x) - K_{\mathrm{cont}}(x,x)$. This difference is integrable over $\Gamma\backslash G$, but the separate terms are not. Some process of truncation needs to be used to avoid $\infty - \infty$ as integral, and we shall not pursue the details in this setting. See [He2] and [Ef] for further information about the classical trace formula. Actually the mechanism of the trace formula is more understandable in the adelic setting, where the interplay between characters and conjugacy classes is fairly clear, than in the setting of $SL_2(\mathbb{Z})\backslash SL_2(\mathbb{R})$, where the complicated nature of $SL_2(\mathbb{Z})$'s conjugacy classes obscures matters. In addition, significant applications require having the formula for two different algebraic groups, and it is therefore appropriate to have a derivation that can be generalized to groups other than SL_2 or GL_2. We shall therefore proceed directly to the adelic setting.

5. Digression on Quaternion Algebras

This section is the first of three sections in which we discuss the trace formula in the setting of adeles. The base number field will be $\mathbb{Q}$, and the adeles of $\mathbb{Q}$ will be denoted $\mathbb{A}$. For background on adeles and reductive algebraic groups, see the exposition [Kn2].

Before treating $G = GL_2$, we consider the case that G' is the multiplicative group of a quaternion algebra over $\mathbb{Q}$. By definition a **quaternion algebra** over a field F is a central simple algebra over F that has dimension 4 and is not equal to the full matrix algebra $M_2(F)$. Since any central simple algebra over F is a full matrix algebra over a division algebra over F, it follows that a quaternion algebra over F is a division algebra.

Let us see how to make G' into a linear algebraic group. Thus let D be a quaternion algebra over $\mathbb{Q}$. It is known that there exist integers m and n such that m, n, and mn are not squares in $\mathbb{Q}$ and such that D has a $\mathbb{Q}$ basis $\{1, u, v, w\}$ with $w = uv$ and

$$u^2 = m, \qquad v^2 = n, \qquad w^2 = -mn.$$

Furthermore

$$uv = -vu, \qquad uw = -wu, \qquad vw = -wv.$$

We may associate 2-by-2 matrices to the members of this $\mathbb{Q}$ basis by

$$1 \leftrightarrow \begin{pmatrix} 1 & 0 \\ 0 & 1\end{pmatrix}, \qquad u \leftrightarrow \begin{pmatrix} \sqrt{m} & 0 \\ 0 & -\sqrt{m}\end{pmatrix}, \qquad v \leftrightarrow \begin{pmatrix} 0 & \sqrt{n} \\ \sqrt{n} & 0\end{pmatrix}, \qquad w \leftrightarrow \begin{pmatrix} 0 & \sqrt{mn} \\ -\sqrt{mn} & 0\end{pmatrix}.$$

These matrices may also be chosen to be defined over a quadratic extension of $\mathbb{Q}$ rather than a quartic extension, for example by taking

$$1 \leftrightarrow \begin{pmatrix} 1 & 0 \\ 0 & 1\end{pmatrix}, \qquad u \leftrightarrow \begin{pmatrix} \sqrt{m} & 0 \\ 0 & -\sqrt{m}\end{pmatrix}, \qquad v \leftrightarrow \begin{pmatrix} 0 & 1 \\ n & 0\end{pmatrix}, \qquad w \leftrightarrow \begin{pmatrix} 0 & \sqrt{m} \\ -n\sqrt{m} & 0\end{pmatrix}.$$

In either case if we identify D with its effect under left multiplication on this basis, then G' is realized as an algebraic subgroup of GL_4 defined over $\mathbb{Q}$.

The determinant of the 2-by-2 matrix corresponding to

$$x = a1 + bu + cv + dw$$

is $a^2 - b^2m - c^2n + d^2mn$, and the determinant of the 4-by-4 matrix describing left multiplication by x is the square of this expression. For $v \in \{\infty, \text{primes}\}$, we see that $D \otimes_{\mathbb{Q}} \mathbb{Q}_v \cong M_2(\mathbb{Q}_v)$ if and only if $a^2 - b^2m - c^2n + d^2mn = 0$ is solvable nontrivially in $\mathbb{Q}_v$. Exactly in this case, $G'(\mathbb{Q}_v) \cong GL_2(\mathbb{Q}_v)$ and we say that G' is **unramified** or **split** at v. If v is an odd prime p, this always happens if $p \nmid m$ and $p \nmid n$, according to Corollaries 1 and 2 of [Bv-Sh, p. 50].

Let $\mathbb{A}$ be the adeles of $\mathbb{Q}$. The center Z' of G', namely the subgroup of scalar multiples of 1, has positive dimension, and consequently the quotient space $G'(\mathbb{Q})\backslash G'(\mathbb{A})$ has infinite volume. Thus instead of studying the right regular representation of $G'(\mathbb{A})$ on $L^2(G'(\mathbb{Q})\backslash G'(\mathbb{A}))$, we begin by studying the right regular representation on $L^2(Z'(\mathbb{A})G'(\mathbb{Q})\backslash G'(\mathbb{A}))$. The quotient space $Z'(\mathbb{A})G'(\mathbb{Q})\backslash G'(\mathbb{A})$ is compact as a consequence of the general theorem quoted as Theorem 6.2 in [Kn2] or a direct calculation that may be found in [Gf-Gr-P, pp. 115–119] or [We, pp. 74–75]. Despite the fact that this quotient is not a manifold, we shall see that Theorem 4.2 is still valid for it with suitable interpretations.

We study functions on $Z'(\mathbb{A})G'(\mathbb{Q})\backslash G'(\mathbb{A})$ by studying functions on $G'(\mathbb{A})$ that are left invariant under $Z'(\mathbb{A})$ and $G'(\mathbb{Q})$. But we can investigate more of $G'(\mathbb{Q})\backslash G'(\mathbb{A})$ if we consider further functions on $G'(\mathbb{A})$. Thus for each (unitary) character ω of $Z'(\mathbb{Q})\backslash Z'(\mathbb{A})$, we define $L^2(Z'(\mathbb{A})G'(\mathbb{Q})\backslash G'(\mathbb{A}), \omega)$ to be the set of f on $G'(\mathbb{A})$ such that

$$f(z\gamma g) = \omega(z)f(g) \qquad \text{for } z \in Z'(\mathbb{A}),\ \gamma \in G'(\mathbb{Q}),\ g \in G'(\mathbb{A}) \tag{5.1}$$

and such that $|f|$ is square integrable on $Z'(\mathbb{A})G'(\mathbb{Q})\backslash G'(\mathbb{A})$. We denote by R_ω the right regular representation of $G'(\mathbb{A})$ on this space. We put $\overline{G}' = Z'\backslash G'$, so that we can identify $Z'(\mathbb{A})G'(\mathbb{Q})\backslash G'(\mathbb{A})$ with $\overline{G}'(\mathbb{Q})\backslash \overline{G}'(\mathbb{A})$.

Let us write $G'(\mathbb{A}) = G'_\infty \times G'(\mathbb{A}_f)$ for the decomposition of $G'(\mathbb{A})$ according to the infinite and finite places. Recall from §7 of [Kn2] that a complex-valued function f on $G'(\mathbb{A})$ is **smooth** if it is continuous and, when viewed as a function of two arguments $(x, y) \in G'_\infty \times G'(\mathbb{A}_f)$, it is smooth in x for each fixed y and is locally constant of compact support in y for each fixed x.

We define $C^\infty_{\text{com}}(G'(\mathbb{A}), \omega^{-1})$ to be the space of smooth functions on $G'(\mathbb{A})$ such that

$$\varphi(zg) = \omega(z)^{-1}\varphi(g) \qquad \text{for } z \in Z'(\mathbb{A}),\ g \in G'(\mathbb{A}). \tag{5.2}$$

If f is in $L^2(Z'(\mathbb{A})G'(\mathbb{Q})\backslash G'(\mathbb{A}), \omega)$ and φ is in $C^\infty_{\text{com}}(G'(\mathbb{A}), \omega^{-1})$, then the function $f(xy)\varphi(y)$ on $G'(\mathbb{A}) \times G'(\mathbb{A})$ descends to a function on $G'(\mathbb{A}) \times \overline{G}'(\mathbb{A})$, and it makes sense to consider

$$R_\omega(\varphi)f(x) = \int_{\overline{G}'(\mathbb{Q})\backslash \overline{G}'(\mathbb{A})} f(xy)\varphi(y)\,dy \tag{5.3}$$

as a member of $L^2(Z'(\mathbb{A})G'(\mathbb{Q})\backslash G'(\mathbb{A}), \omega)$. Since $\omega(Z'(\mathbb{Q})) = 1$, the function $\gamma \mapsto \varphi(x^{-1}\gamma y)$ on $G'(\mathbb{Q})$ descends to a well defined function on $\overline{G}'(\mathbb{Q})$. Thus we can imitate the computation at the beginning of §4 and write

$$\begin{aligned} R_\omega(\varphi)f(x) &= \int_{\overline{G}'(\mathbb{A})} f(xy)\varphi(y)\,dy \\ &= \int_{\overline{G}'(\mathbb{A})} f(y)\varphi(x^{-1}y)\,dy \\ &= \int_{\overline{G}'(\mathbb{Q})\backslash\overline{G}'(\mathbb{A})} \sum_{\gamma\in\overline{G}'(\mathbb{Q})} f(y)\varphi(x^{-1}\gamma y)\,dy. \end{aligned}$$

Therefore

$$R_\omega(\varphi)f(x) = \int_{\overline{G}'(\mathbb{Q})\backslash\overline{G}'(\mathbb{A})} K(x,y)f(y)\,dy, \tag{5.4}$$

where $K(x,y) = \sum_{\gamma\in\overline{G}'(\mathbb{Q})}\varphi(x^{-1}\gamma y)$. The function $K(x,y)$ is defined on $G'(\mathbb{A})\times G'(\mathbb{A})$, is left invariant under $G'(\mathbb{Q})$ in each variable, and satisfies

$$K(z_1x, z_2y) = \omega(z_1)\omega(z_2)^{-1}K(x,y) \qquad \text{for } z_1 \text{ and } z_2 \text{ in } Z'(\mathbb{A}).$$

Let K_1 be the open compact subgroup $\prod_p G'(\mathcal{O}_p)$ of $G'(\mathbb{A}_f)$. The function φ is left and right invariant under some open compact subgroup K_2 of K_1, and consequently the function $K(x,y)$ is right invariant under K_2 in each variable. The right invariance in y implies that $R_\omega(\varphi)f(x)$ depends only on the function $f^{\#}(yK_2) = \operatorname{vol}(K_2)^{-1}\int_{K_2} f(yk_2)\,dk_2$ defined on $G'(\mathbb{Q})\backslash G'(\mathbb{A})/K_2$. The right invariance in x implies that $R_\omega(\varphi)f(x) = (R_\omega(\varphi)f)^{\#}(xK_2)$. Thus we can regard $R_\omega(\varphi)$ as an operator from functions on $G'(\mathbb{Q})\backslash G'(\mathbb{A})/K_2$ transforming under ω to functions on the same space. By (6.3) of [Kn2], the compact space $Z'(\mathbb{A})G'(\mathbb{Q})\backslash G'(\mathbb{A})/K_2$ is a (possibly disconnected) manifold. If $\omega = 1$, Lemma 4.1 is directly applicable. If $\omega \neq 1$, then Lemma 4.1 is indirectly applicable with the aid of a compactly supported function h on $G'(\mathbb{Q})\backslash G'(\mathbb{A})/K_2$ such that $\int_{Z'(\mathbb{A})} h(zx)\,dz = 1$ for all $x\in G'(\mathbb{A})$. The result for any ω is as follows.

Lemma 5.1. *If φ is in $C^\infty_{\text{com}}(G'(\mathbb{A}),\omega^{-1})$, then the operator $R_\omega(\varphi)$ defined by (5.3) is of trace class on $L^2(Z'(\mathbb{A})G'(\mathbb{Q})\backslash G'(\mathbb{A}),\omega)$, and its trace is*

$$\operatorname{Tr} R_\omega(\varphi) = \int_{\overline{G}'(\mathbb{Q})\backslash\overline{G}'(\mathbb{A})} K(x,x)\,dx,$$

where $K(x,x) = \sum_{\gamma\in\overline{G}'(\mathbb{Q})}\varphi(x^{-1}\gamma x)$.

The proof that Lemma 4.1 implies Theorem 4.2 may be adjusted to show that Lemma 5.1 implies the following result, which gives the trace formula for the multiplicative group of a quaternion algebra over $\mathbb{Q}$.

Theorem 5.2. *Let G' be the multiplicative group of a quaternion algebra over $\mathbb{Q}$, let Z' be the center, let $\overline{G}' = Z'\backslash\overline{G}'$, let R_ω be the right regular representation of $G'(\mathbb{A})$ on $L^2(Z'(\mathbb{A})G'(\mathbb{Q})\backslash G'(\mathbb{A}),\omega)$, and let φ be in $C^\infty_{\text{com}}(G'(\mathbb{A}),\omega^{-1})$. Then $R_\omega(\varphi)$ is of trace class, and*

$$\operatorname{Tr} R_\omega(\varphi) = \sum_{\mathfrak{o}_\gamma} \operatorname{vol}(\overline{G}'(\mathbb{Q})^\gamma\backslash\overline{G}'(\mathbb{A})^\gamma)\int_{\overline{G}'(\mathbb{A})^\gamma\backslash\overline{G}'(\mathbb{A})}\varphi(x^{-1}\gamma x)\,dx,$$

the sum being taken over conjugacy classes in $\overline{G}'(\mathbb{Q})$. *Consequently if the decomposition of* $L^2(Z'(\mathbb{A})G'(\mathbb{Q})\backslash G'(\mathbb{A}),\omega^{-1})$ *into irreducible representations of* $G'(\mathbb{A})$ *is as in* (4.1), *then*

$$\sum_{\pi\in\widehat{G'(\mathbb{A})}} m_\pi \operatorname{Tr}\pi(\varphi) = \sum_{\mathfrak{o}_\gamma} \operatorname{vol}(\overline{G}'(\mathbb{Q})^\gamma\backslash\overline{G}'(\mathbb{A})^\gamma)\int_{\overline{G}'(\mathbb{A})^\gamma\backslash\overline{G}'(\mathbb{A})} \varphi(x^{-1}\gamma x)\,dx.$$

6. Adelic Eisenstein Series

Now we turn our attention to the group $G = GL_2$. For this group we seek an understanding of functions on $G(\mathbb{Q})\backslash G(\mathbb{A})$, where $\mathbb{A}$ denotes the adeles of $\mathbb{Q}$. References are [Gf-Gr-P], [Ja-Lgl], [Du-La], [Ar1], [Gb1], [Gb-Ja], and [Ar4]. This quotient space does not have finite volume, and some adjustment has to be made. The same difficulty arose in §5 with the multiplicative group G' of a quaternion algebra: The quotient $G'(\mathbb{Q})\backslash G'(\mathbb{A})$ has infinite volume, and we in effect chose to study only functions that could be related to $Z'(\mathbb{A})G'(\mathbb{Q})\backslash G'(\mathbb{A})$, where Z' is the center. For G', we took advantage of the fact that $Z'(\mathbb{A})G'(\mathbb{Q})\backslash G'(\mathbb{A})$ is compact.

In the literature an adjustment for G is made in either of two equivalent ways. One possible adjustment, analogous to what we did for G' in §5, is to study functions that can be related to $Z(\mathbb{A})G(\mathbb{Q})\backslash G(\mathbb{A})$, where Z is the center consisting of scalar matrices. This quotient space is not compact, but it does have finite volume, as we shall see in a moment. Specifically for each character ω of $Z(\mathbb{Q})\backslash Z(\mathbb{A})$, we define $L^2(Z(\mathbb{A})G(\mathbb{Q})\backslash G(\mathbb{A}),\omega)$ to be the set of f on $G(\mathbb{A})$ such that

$$f(z\gamma g) = \omega(z)f(g) \qquad \text{for } z\in Z(\mathbb{A}),\ \gamma\in G(\mathbb{Q}),\ g\in G(\mathbb{A}) \tag{6.1a}$$

and such that $|f|$ is square integrable on $Z(\mathbb{A})G(\mathbb{Q})\backslash G(\mathbb{A})$. We shall be interested in the right regular representation R_ω of $G(\mathbb{A})$ on this space. We put $\overline{G} = Z\backslash G$, so that we can identify $Z(\mathbb{A})G(\mathbb{Q})\backslash G(\mathbb{A})$ with $\overline{G}(\mathbb{Q})\backslash\overline{G}(\mathbb{A})$.

The other possible adjustment uses the subgroup $G^1 = G(\mathbb{A})^1$ of elements $g\in G(\mathbb{A})$ such that $|\det g|_{\mathbb{A}} = 1$. The discrete subgroup $G(\mathbb{Q})$ of $G(\mathbb{A})$ lies in G^1 by the Artin product formula (Theorem 3.3 of [Kn2]), and the quotient space $G(\mathbb{Q})\backslash G^1$ is noncompact of finite volume, by the theorem of Borel and Harish-Chandra quoted as Theorem 6.2 of [Kn2]. In this approach the objective is to understand the decomposition of the right regular representation of G^1 on $L^2(G(\mathbb{Q})\backslash G^1)$. The group G^1 has center $Z^1 = G^1\cap Z(\mathbb{A})$. If $(\mathbb{A}^\times)^1$ denotes the group of ideles of module 1, then the members of Z^1 have both diagonal entries equal to the same member of $(\mathbb{A}^\times)^1$. From Theorem 3.5 of [Kn2], we know that the abelian group $\mathbb{Q}^\times\backslash(\mathbb{A}^\times)^1$ is compact. Its characters are in one-one correspondence with the characters of Z^1 that are trivial on $Z^1\cap G(\mathbb{Q})$, hence with the irreducible representations of $Z^1G(\mathbb{Q})$ that are trivial on $G(\mathbb{Q})$. The formalism

$$L^2(G(\mathbb{Q})\backslash G^1) \cong \operatorname{ind}_{G(\mathbb{Q})}^{G^1} 1 \cong \operatorname{ind}_{Z^1G(\mathbb{Q})}^{G^1}\operatorname{ind}_{G(\mathbb{Q})}^{Z^1G(\mathbb{Q})} 1$$

therefore leads to the conclusion that $L^2(G(\mathbb{Q})\backslash G^1)$ decomposes as a Hilbert space orthogonal sum

$$L^2(G(\mathbb{Q})\backslash G^1) = \sum_{\omega_0\in(\mathbb{Q}^\times\backslash(\mathbb{A}^\times)^1)^\wedge} L^2(Z^1G(\mathbb{Q})\backslash G^1,\omega_0),$$

where ω_0 is regarded as a character of $Z^1G(\mathbb{Q})$ that is trivial on $G(\mathbb{Q})$. Here $L^2(Z^1G(\mathbb{Q}))\backslash G^1, \omega_0)$ is the set of f on G^1 such that

$$f(z\gamma g) = \omega_0(z)f(g) \qquad \text{for } z \in Z^1,\ \gamma \in G(\mathbb{Q}),\ g \in G^1 \tag{6.1b}$$

and such that $|f|$ is square integrable on $Z^1G(\mathbb{Q})\backslash G^1$. Invariant integration on $Z^1G(\mathbb{Q})\backslash G^1$ can be achieved by pulling functions back to $G(\mathbb{Q})\backslash G^1$ and integrating there, and hence $Z^1G(\mathbb{Q})\backslash G^1$ has finite volume.

The inclusion of G^1 into $G(\mathbb{A})$ yields a map of G^1 into the quotient space $Z(\mathbb{A})G(\mathbb{Q})\backslash G(\mathbb{A})$, and this is onto since every member of $G(\mathbb{A})$ is the product of a member of G^1 and a positive scalar matrix at the infinite place. The map descends to a map of $Z^1G(\mathbb{Q})\backslash G^1$ onto $Z(\mathbb{A})G(\mathbb{Q})\backslash G(\mathbb{A})$, and the result is one-one since $G^1 \cap Z(\mathbb{A}) = Z^1$. Thus we may identify

$$Z(\mathbb{A})G(\mathbb{Q})\backslash G(\mathbb{A}) \cong Z^1G(\mathbb{Q})\backslash G^1.$$

When a character is taken into account, matters are a little more complicated. Let ω be a character of $Z(\mathbb{A})$ trivial on $Z(\mathbb{Q})$, and let $L^2(Z(\mathbb{A})G(\mathbb{Q})\backslash G(\mathbb{A}), \omega)$ be as in (6.1a). By the second isomorphism theorem, $Z(\mathbb{Q})\backslash Z(\mathbb{A})$ is isomorphic to $G(\mathbb{Q})\backslash Z(\mathbb{A})G(\mathbb{Q})$, and thus ω can be regarded as a character of $Z(\mathbb{A})G(\mathbb{Q})$ trivial on $G(\mathbb{Q})$. We can restrict ω to a character ω_0 of $Z^1G(\mathbb{Q})$ that is trivial on $G(\mathbb{Q})$, and then we obtain an identification of the function spaces

$$L^2(Z(\mathbb{A})G(\mathbb{Q})\backslash G(\mathbb{A}), \omega) \cong L^2(Z^1G(\mathbb{Q})\backslash G^1, \omega_0). \tag{6.2}$$

Conversely when a character ω_0 of $Z^1G(\mathbb{Q})$ that is trivial on $G(\mathbb{Q})$ is given, we can extend ω to a unitary character ω of $Z(\mathbb{A})G(\mathbb{Q})$ that is trivial on $G(\mathbb{Q})$, and we again obtain (6.2). The complication is that the extension of ω_0 to ω is not unique.

By imposing a further condition on ω, we can get around this nonuniqueness. Let $\mathbb{Q}_\infty^+$ be the group of ideles that are trivial at all finite places and are positive at the infinite place, and let Z_∞ be the subgroup of $Z(\mathbb{A})$ whose diagonal entries are in $\mathbb{Q}_\infty^+$. Then $Z(\mathbb{A}) = Z^1 \times Z_\infty$ and $Z(\mathbb{A})G(\mathbb{Q}) = Z^1G(\mathbb{Q}) \times Z_\infty$. Hence a character ω_0 of $Z^1G(\mathbb{Q})$ trivial on $G(\mathbb{Q})$ extends uniquely to a character of $Z(\mathbb{A})G(\mathbb{Q})$ trivial on $G(\mathbb{Q})$ if we impose the condition that ω is trivial on Z_∞.

We choose to study the left side of (6.2) rather than the right side. Working with the right side would make the proof of the trace formula considerably more elegant. But as we shall see in [Kn-Ro], working with the left side makes it much easier to use the trace formula in applications. It will not simplify matters to assume that ω is trivial on Z_∞, and thus we do not assume this triviality.

Henceforth we therefore fix ω as a character of $Z(\mathbb{A})$ that is trivial on $Z(\mathbb{Q})$; by extracting the upper left entry of a scalar matrix, we may regard ω alternatively as a character of $\mathbb{Q}^\times\backslash\mathbb{A}^\times$. We consider the space $L^2(Z(\mathbb{A})G(\mathbb{Q})\backslash G(\mathbb{A}), \omega)$ and the right regular representation R_ω of $G(\mathbb{A})$ on this space.

Let $N = \left(\begin{smallmatrix} 1 & * \\ 0 & 1 \end{smallmatrix}\right)$ and $M = \left(\begin{smallmatrix} * & 0 \\ 0 & * \end{smallmatrix}\right)$ as algebraic subgroups of G, and put $P = MN$. If f is in $L^2(Z(\mathbb{A})G(\mathbb{Q})\backslash G(\mathbb{A}), \omega)$, we define the **constant term** of f (along P) to be

$$f_P(g) = \int_{\mathbb{Q}\backslash\mathbb{A}} f\left(\left(\begin{smallmatrix} 1 & x \\ 0 & 1 \end{smallmatrix}\right) g\right) dx,$$

where dx has total mass one. This function is left invariant under $N(\mathbb{A})$ and $P(\mathbb{Q})$, the latter because the Artin product formula shows that conjugation by a member of $P(\mathbb{Q})$ does not change dx. Let $L^2_{\text{cusp}}(\omega)$ be the closed subspace of functions f such that $f_P(g)$ is 0 almost everywhere. This subspace is invariant under $R_\omega(G(\mathbb{A}))$.

Theorem 6.1. *If φ is in $C^\infty_{\text{com}}(G(\mathbb{A}), \omega^{-1})$, then $R_\omega(\varphi)$ is Hilbert-Schmidt, hence compact, on $L^2_{\text{cusp}}(\omega)$.*

REFERENCE FOR SKETCH. [Gb-Ja, pp. 217–218].

Corollary 6.2. *$L^2_{\text{cusp}}(\omega)$ decomposes discretely into irreducible representations having finite multiplicity.*

PROOF. The argument is the same as the proof that Theorem 1.6 implies Theorem 1.5.

Corollary 6.3. *If φ is in $C^\infty_{\text{com}}(G(\mathbb{A}), \omega^{-1})$, then $R_\omega(\varphi)$ is of trace class on $L^2_{\text{cusp}}(\omega)$.*

PROOF. We can write $\varphi(x) = \int_{Z(\mathbb{A}} \psi(zx)\omega(z)\,dz$ for some smooth function ψ of compact support on $G(\mathbb{A})$. Then ψ is a finite sum of functions $\psi_\infty \times \psi_{\text{fin}}$, where ψ_∞ is smooth of compact support at the place ∞ and ψ_{fin} is locally constant of compact support at the finite places. Form φ_∞ and φ_{fin} from ψ_∞ and ψ_{fin} by integrating over the appropriate components of $Z(\mathbb{A})$, so that $\varphi_\infty \times \varphi_{\text{fin}}$ is in $C^\infty_{\text{com}}(G(\mathbb{A}), \omega^{-1})$. A theorem of Dixmier and Malliavin [Di-Ma] shows that each ψ_∞ is a sum of terms that are each the convolution of two compactly supported smooth functions. Also each ψ_{fin} is the convolution of ψ_{fin} with the characteristic function of some open compact subgroup. Consequently $\varphi_\infty \times \varphi_{\text{fin}}$ is the finite sum of convolutions of pairs of members of $C^\infty_{\text{com}}(G(\mathbb{A}), \omega^{-1})$. Then it follows from Theorem 6.1 that $R_\omega(\varphi)$ is a finite sum of products of two Hilbert-Schmidt operators and hence is of trace class.

The next step is to identify the orthogonal complement of the subspace $L^2_{\text{cusp}}(\omega)$ of $L^2(Z(\mathbb{A})G(\mathbb{Q})\backslash G(\mathbb{A}), \omega)$ in a fashion analogous to Theorem 1.3. The dictionary for comparing subgroups of $SL_2(\mathbb{R})$ and $G(\mathbb{A})$ is that $\Gamma \leftrightarrow G(\mathbb{Q})$, $N \leftrightarrow N(\mathbb{A})$, and $\Gamma_\infty \leftrightarrow P(\mathbb{Q})$. The condition in §§1–4 that functions be even is analogous to the condition now that functions transform under ω. The proof of Lemma 1.1 used that $\Gamma_\infty \subset N$ and that $\Gamma_\infty\backslash N$ is compact, but it would have worked as well under the condition that $\Gamma_\infty\backslash N\Gamma_\infty$ is compact. We therefore obtain an adelic analog of that lemma: If ϕ is a continuous function on $G(\mathbb{A})$ satisfying

$$\phi(zn\gamma g) = \omega(z)\phi(g) \tag{6.3}$$

for $z \in Z(\mathbb{A})$, $n \in N(\mathbb{A})$, and $\gamma \in P(\mathbb{Q})$ and having compact support modulo $N(\mathbb{A})P(\mathbb{Q})$, then

$$\widehat{\phi}(g) = \sum_{\gamma \in P(\mathbb{Q})\backslash G(\mathbb{Q})} \phi(\gamma g) \tag{6.4}$$

is a locally finite sum and defines a continuous function on $G(\mathbb{A})$ satisfying (6.1b) and having compact support modulo $Z(\mathbb{A})G(\mathbb{Q})$.

Lemma 6.4. *Let ϕ be a measurable function on $G(\mathbb{A})$ left invariant under $N(\mathbb{A})P(\mathbb{Q})$ and transforming under ω, and let F be a measurable function on $G(\mathbb{A})$ as in (6.1b). If $\widehat{|\phi|}$ is square integrable modulo $Z(\mathbb{A})G(\mathbb{Q})\backslash G(\mathbb{A})$ and if F is in $L^2(Z(\mathbb{A})G(\mathbb{Q})\backslash G(\mathbb{A}), \omega)$, then*

$$\langle \widehat{\phi}, F\rangle_{L^2(Z(\mathbb{A})G(\mathbb{Q})\backslash G(\mathbb{A}))} = \langle \phi, F_P\rangle_{L^2(Z(\mathbb{A})N(\mathbb{A})P(\mathbb{Q})\backslash G(\mathbb{A}))}, \tag{6.5}$$

the indicated integrals converging.

REMARKS. This is proved in the same way as Lemma 1.2. When an integral over $Z(\mathbb{A})P(\mathbb{Q})\backslash G(\mathbb{A})$ is written as an iterated integral over $(Z(\mathbb{A})N(\mathbb{A})P(\mathbb{Q}))\backslash G(\mathbb{A})$ and $(Z(\mathbb{A})P(\mathbb{Q}))\backslash(Z(\mathbb{A})N(\mathbb{A})P(\mathbb{Q}))$, the inner integral is rewritten over $N(\mathbb{Q})\backslash N(\mathbb{A})$ by the second isomorphism theorem. The equality (6.5) depends on normalizations of Haar measures, but we postpone this detail until after the proof of Lemma 6.7 below.

Theorem 6.5. *Within $L^2(Z(\mathbb{A})G(\mathbb{Q})\backslash G(\mathbb{A}),\omega)$, the orthogonal complement of $L^2_{\text{cusp}}(\omega)$ is the closure of the space of all $\widehat{\phi}$ with ϕ continuous on $G(\mathbb{A})$, left invariant under $N(\mathbb{A})P(\mathbb{Q})$, transforming under $Z(\mathbb{A})$ by ω, and having compact support modulo $Z(\mathbb{A})N(\mathbb{A})P(\mathbb{Q})$.*

PROOF. Same as for Theorem 1.3.

Eisenstein series are used in the analysis of this orthocomplement. Let K be the maximal compact subgroup $O_2(\mathbb{R})\times\prod_p G(\mathbb{Z}_p)$ of $G(\mathbb{A})$, so that $G(\mathbb{A})=P(\mathbb{A})K$. If an element g is decomposed as $g=\left(\begin{smallmatrix} a & x\\ 0 & b\end{smallmatrix}\right)k$ with $k\in K$, we define

$$h(g)=\left|\tfrac{a}{b}\right|_{\mathbb{A}}^{1/2}. \tag{6.6}$$

This is well defined since $h(g)=1$ for any element g of $P(\mathbb{A})\cap K$. To be able to compute with the function $h(\,\cdot\,)$, we identify $\mathbb{A}^2$ with row vectors and introduce a kind of norm on $\mathbb{A}^2$. If $v_p=(\,x_p\quad y_p\,)$ is a row vector over $\mathbb{Q}_p$, we define $\|v_p\|_p=\max\{|x_p|_p,|y_p|_p\}$. A little computation shows that $\|v_pk_p\|_p=\|v_p\|_p$ for k_p in $GL_2(\mathbb{Z}_p)$. If $v_\infty=(\,x_\infty\quad y_\infty\,)$ is a row vector over $\mathbb{R}$, we define $\|v_\infty\|_\infty=\sqrt{x_\infty^2+y_\infty^2}$. Then of course, $\|v_\infty k_\infty\|_\infty=\|v_\infty\|_\infty$ for k_∞ in $O_2(\mathbb{R})$. If $v\in\mathbb{A}^2$ is decomposed as $v=v_\infty\times\prod v_p$, we let $\|v\|_{\mathbb{A}}=\|v_\infty\|_\infty\times\prod_p\|v_p\|_p$, and this norm is preserved under right multiplication by K.

Lemma 6.6. *The function $h(\,\cdot\,)$ defined in (6.6) is given on $G(\mathbb{A})$ by*

$$h(g)^{-1}=\frac{\|(\,0\quad 1\,)\,g\|_{\mathbb{A}}}{|\det g|_{\mathbb{A}}^{1/2}}.$$

PROOF. Since K preserves norms, it is sufficient to consider $g\in P(\mathbb{A})$. If $g=\left(\begin{smallmatrix} a & x\\ 0 & b\end{smallmatrix}\right)$, then

$$\frac{\|(\,0\quad 1\,)\,g\|_{\mathbb{A}}}{|\det g|_{\mathbb{A}}^{1/2}}=\frac{\|(\,0\quad b\,)\,\|_{\mathbb{A}}}{|ab|_{\mathbb{A}}^{1/2}}=\frac{|b|_{\mathbb{A}}^{1/2}}{|a|_{\mathbb{A}}^{1/2}},$$

and the result follows.

The square $h(\,\cdot\,)^2$ is an adelic analog of the function $y(\,\cdot\,)$ in §§1–4. For example, Haar measure on $G(\mathbb{A})$ may be expressed in terms of h in analogy with (1.7). If $g=pk$ is a decomposition of an element relative to $G(\mathbb{A})=P(\mathbb{A})K$, then we have

$$dg=d_lp\,dk=h(p)^{-2}\,d_rp\,dk, \tag{6.7}$$

where d_lp and d_rp are left and right Haar measures on $P(\mathbb{A})$. Normalizations of Haar measures will be discussed in more detail after the proof of Lemma 6.7 below.

The analog of summing over $\Gamma_\infty\backslash\Gamma$ will be summing over $P(\mathbb{Q})\backslash G(\mathbb{Q})$. By the Bruhat decomposition we can take as representatives γ of the cosets $P(\mathbb{Q})\gamma$ the elements 1 and $w\begin{pmatrix}1&\xi\\0&1\end{pmatrix}$ with ξ in $\mathbb{Q}$, where $w=\begin{pmatrix}0&1\\1&0\end{pmatrix}$.

The next lemma will reduce several estimates about $h(\cdot)$ to estimates in the setting of §§1–4.

Lemma 6.7. *Let $g=\begin{pmatrix}yu&xu\\0&u\end{pmatrix}$ vary through a compact subset X of $P(\mathbb{A})$, and let γ vary through matrices of the form $\gamma=w\begin{pmatrix}1&\xi\\0&1\end{pmatrix}$ with $\xi\in\mathbb{Q}$. Write $\xi=d/c$ with* GCD$(c,d)=1$*, and write also $x=x_\infty\prod_p x_p$ and $y=y_\infty\prod_p y_p$. Then there exists a constant B such that*

$$h(\gamma g)\le\frac{B}{|cz_\infty+d|}$$

for all $g\in X$ and all $\xi\in\mathbb{Q}$, where $z_\infty=x_\infty+iy_\infty$ as a member of $\mathbb{C}$.

PROOF. We have

$$\begin{aligned}h(\gamma g)^{-1}&=\|\,(0\quad 1)\,\gamma g\|_{\mathbb{A}}=\|\,(0\quad 1)\begin{pmatrix}0&1\\1&\xi\end{pmatrix}\begin{pmatrix}yu&xu\\0&u\end{pmatrix}\|_{\mathbb{A}}\\&=\|\,(yu\quad u(x+\xi))\,\|_{\mathbb{A}}=|u|_{\mathbb{A}}\,\|\,(y\quad x+\xi)\,\|_{\mathbb{A}}\\&=|u|_{\mathbb{A}}\,\|\,(cy\quad cx+d)\,\|_{\mathbb{A}}=|u|_{\mathbb{A}}\,|cz_\infty+d|\prod_p\|\,(cy_p\quad cx_p+d)\,\|_p.\end{aligned}$$

Thus it is enough to bound

$$\prod_p\|\,(cy_p\quad cx_p+d)\,\|_p=\prod_p\max(|cy_p|_p,|cx_p+d|_p)$$

below. We do so by making repeated use of the inequality

$$\max(a_1b_1,a_2b_2)\ge\max(a_1,a_2)\min(b_1,b_2)$$

valid for positive reals. There are three cases. First suppose that $|d|_p<|c|_p|x_p|_p$. Then

$$\max(|cy_p|_p,|cx_p+d|_p)=|c|_p\max(|y_p|_p,|x_p|_p)\ge|d|_p\frac{\max(|y_p|_p,|x_p|_p)}{|x_p|_p},$$

and hence

$$\begin{aligned}\max(|cy_p|_p,|cx_p+d|_p)&\ge\max\left(|c|_p\max(|y_p|_p,|x_p|_p),|d|_p\frac{\max(|y_p|_p,|x_p|_p)}{|x_p|_p}\right)\\&\ge\max(|c|_p,|d|_p)\min\left(\max(|y_p|_p,|x_p|_p),\frac{\max(|y_p|_p,|x_p|_p)}{|x_p|_p}\right)\\&=\min\left(\max(|y_p|_p,|x_p|_p),\frac{\max(|y_p|_p,|x_p|_p)}{|x_p|_p}\right)\end{aligned}$$

since $\max(|c|_p,|d|_p)=1$. Second suppose that $|d|_p>|c|_p|x_p|_p$. Then

$$\begin{aligned}\max(|cy_p|_p,|cx_p+d|_p)&=\max(|c|_p|y_p|_p,|d|_p)\\&\ge\max(|c|_p,|d|_p)\min(|y_p|_p,1)=\min(|y_p|_p,1).\end{aligned}$$

Third suppose that $|d|_p = |c|_p|x_p|_p$. Then

$$\max(|cy_p|_p, |cx_p + d|_p) \geq |c|_p|y_p|_p = |d|_p \frac{|y_p|_p}{|x_p|_p},$$

and hence

$$\begin{aligned}\max(|cy_p|_p, |cx_p + d|_p) &\geq \max\left(|c|_p|y_p|_p, |d|_p \frac{|y_p|_p}{|x_p|_p}\right)\\ &\geq \max(|c|_p, |d|_p) \min\left(|y_p|_p, \frac{|y_p|_p}{|x_p|_p}\right)\\ &= \min\left(|y_p|_p, \frac{|y_p|_p}{|x_p|_p}\right).\end{aligned}$$

Combining the three cases, we see that

$$\max(|cy_p|_p, |cx_p + d|_p) \geq \min\left(|y_p|_p, 1, \frac{|y_p|_p}{|x_p|_p}\right). \tag{6.8}$$

We claim that the product over p of (6.8) is bounded below for $g \in X$. For a single g, this is obvious since $|x_p|_p \leq 1$ and $|y_p|_p = 1$ for all but finitely many p, so that the right side of (6.8) is 1 for all but finitely many p. If a sequence $g^{(n)} \leftrightarrow (x^{(n)}, y^{(n)})$ has the product tending to 0, we can choose a convergent subsequence, say with limit $g^{(0)} \leftrightarrow (x^{(0)}, y^{(0)})$. The convergence has to take place in one of the product spaces of which $G(\mathbb{A})$ is a union, and therefore there are only finitely many p for which we do not have $|x_p^{(n)}|_p \leq 1$ and $|y_p^{(n)}|_p = 1$ for all n. For all but finitely many p, (6.8) is therefore 1 for all n, and we have convergence for the remaining p. Thus (6.8) cannot be tending to 0, and the proof is complete.

Now let us discuss normalizations of Haar measures. Discrete groups get the counting measure, and the compact group $\mathbb{Q}\backslash\mathbb{A}$ gets the measure of total mass one. However, it will not be convenient to assume that $\mathbb{Q}^\times\backslash(\mathbb{A}^\times)^1$ has total mass one. Instead we proceed as follows: We fix any Haar measure on $\mathbb{A}^\times$ and give $\mathbb{Q}^\times\backslash\mathbb{A}^\times$ the quotient measure. The group $\mathbb{Q}_\infty^+$ of ideles that are trivial at all finite places and are positive at the infinite place is isomorphic to the group $\mathbb{R}_+^\times$ of positive reals by $t \to |t|_\mathbb{A}$, and we transport dx/x on $\mathbb{R}_+^\times$ to a Haar measure on $\mathbb{Q}_\infty^+$. Then we can use the isomorphism $\mathbb{Q}^\times\backslash\mathbb{A}^\times \cong \mathbb{Q}_\infty^+ \times \mathbb{Q}^\times\backslash(\mathbb{A}^\times)^1$ to determine a Haar measure on $\mathbb{Q}^\times\backslash(\mathbb{A}^\times)^1$.

For the parabolic $P(\mathbb{A})$, we have $P(\mathbb{A}) = N(\mathbb{A})M(\mathbb{A})$ with

$$M(\mathbb{A}) = \left\{\begin{pmatrix} u & 0 \\ 0 & v\end{pmatrix} \,\middle|\, u, v \in \mathbb{A}^\times\right\}. \tag{6.9a}$$

We identify $N(\mathbb{A})$ with $\mathbb{A}$ and define Haar measure on $N(\mathbb{A})$ accordingly. Next we identify $Z(\mathbb{A})$ with $\mathbb{A}^\times$ by $\begin{pmatrix} u & 0 \\ 0 & u\end{pmatrix} \leftrightarrow u$, and then Haar measure is determined on $Z(\mathbb{A})$. The equality $M(\mathbb{A}) = \left\{\begin{pmatrix} a & 0 \\ 0 & 1\end{pmatrix}\right\} Z(\mathbb{A})$ follows from the decomposition $\begin{pmatrix} u & 0 \\ 0 & v\end{pmatrix} = \begin{pmatrix} uv^{-1} & 0 \\ 0 & 1\end{pmatrix}\begin{pmatrix} v & 0 \\ 0 & v\end{pmatrix}$, and thus we have an isomorphism $M(\mathbb{A}) \cong \mathbb{A}^\times Z(\mathbb{A})$. This isomorphism allows us to fix Haar measure on $M(\mathbb{A})$. In the notation of (6.9a), Haar measure on $M(\mathbb{A})$ is nothing more than $du\, dv$, where du and dv indicate Haar measure on $\mathbb{A}^\times$.

Next the decomposition $P(\mathbb{A}) = N(\mathbb{A})M(\mathbb{A})$ allows us to use the measures dn on $N(\mathbb{A})$ and dm on $M(\mathbb{A})$ to determine left and right Haar measures $d_l p$ and $d_r p$ on $P(\mathbb{A})$ by

$$d_r p = dn\, dm \quad \text{and} \quad d_l(p) = d_r(p^{-1}). \tag{6.9b}$$

We pick any Haar measure on K, not insisting that it have total mass one, and then we use (6.7) to determine Haar measure on $G(\mathbb{A})$. Finally we require that invariant measures on closed subgroups and quotients are to be compatible with the measure on the whole group. In particular this requirement fixes the measures on the quotients of $G(\mathbb{A})$ in (6.5). It also fixes Haar measure on $Z(\mathbb{A})G(\mathbb{Q})$ since

$$Z(\mathbb{A})\backslash Z(\mathbb{A})G(\mathbb{Q}) \cong Z(\mathbb{Q})\backslash G(\mathbb{Q}).$$

For the remainder of this section we largely follow [Gb-Ja]. For each $s \in \mathbb{C}$, we introduce a Hilbert space $H(s)$ of functions $F : G(\mathbb{A}) \to \mathbb{C}$ with

$$F\left(\begin{pmatrix} q_1 au & x \\ 0 & q_2 bv \end{pmatrix} g\right) = h\left(\begin{pmatrix} a & 0 \\ 0 & b \end{pmatrix}\right)^{1+s} \omega(bv) F\left(\begin{pmatrix} uv^{-1} & 0 \\ 0 & 1 \end{pmatrix} g\right) \tag{6.10}$$

for q_1 and q_2 in $\mathbb{Q}^\times$, a and b in $\mathbb{Q}_\infty^+$, u and v in $(\mathbb{A}^\times)^1$, x in $\mathbb{A}$, and g in $G(\mathbb{A})$. Such functions depend on u and v only as members of $\mathbb{Q}^\times\backslash(\mathbb{A}^\times)^1$, and the norm squared is taken to be

$$\int_{(\mathbb{Q}^\times\backslash(\mathbb{A}^\times)^1)\times K} \left| F\left(\begin{pmatrix} u & 0 \\ 0 & 1 \end{pmatrix} k\right)\right|^2 du\, dk. \tag{6.11}$$

If F satisfies (6.10), then F is completely determined by its values on elements $\begin{pmatrix} u & 0 \\ 0 & 1 \end{pmatrix} k$ with $u \in (\mathbb{A}^\times)^1$ and $k \in K$ since $G(\mathbb{A}) = P(\mathbb{A})K$ and since the part $\begin{pmatrix} au & x \\ 0 & bv \end{pmatrix}$ of the matrix in (6.10) is the most general member of $P(\mathbb{A})$.

Conversely let H be the Hilbert space of all f on $(\mathbb{A}^\times)^1 \times K$ such that

(i) f is left invariant under $\mathbb{Q}^\times$ in the first variable

(ii) $f(uv, k) = f\left(u, \left(\begin{smallmatrix} v & 0 \\ 0 & 1 \end{smallmatrix}\right) k\right)$ whenever $\left(\begin{smallmatrix} v & 0 \\ 0 & 1 \end{smallmatrix}\right)$ is in $(\mathbb{A}^\times)^1 \cap K$

(iii) f is square integrable on $(\mathbb{Q}^\times\backslash(\mathbb{A}^\times)^1) \times K$.

If f is in H, then we can extend f uniquely to a function $F = f_s$ in $H(s)$ by

$$f_s\left(\begin{pmatrix} au & 0 \\ 0 & bv \end{pmatrix}\begin{pmatrix} 1 & x \\ 0 & 1 \end{pmatrix} k\right) = h\left(\begin{pmatrix} a & 0 \\ 0 & b \end{pmatrix}\right)^{1+s} \omega(bv) f(uv^{-1}, k). \tag{6.12}$$

The group $G(\mathbb{A})$ operates on $H(s)$ via the right regular representation, which we denote $P^{\omega,s}$. This representation is unitary if s is imaginary.

To postpone technical difficulties until the end, fix a finite-dimensional representation η of the compact abelian group $\mathbb{Q}^\times\backslash(\mathbb{A}^\times)^1$ and a finite-dimensional representation τ of the compact group K. Both η and τ are to be thought of as large (and therefore reducible). Let $W(\eta, \tau)$ be the subspace of $f \in H$ such that $u \mapsto f(uu_0, k_0)$, for each (u_0, k_0), is a linear combination of matrix coefficients of the constituents of η and such that $k \mapsto f(u_0, k_0 k)$, for each (u_0, k_0), is a linear combination of matrix coefficients of the constituents of τ. Let $\tilde\eta = \omega\eta^c$, where η^c denotes the contragredient of η; $\tilde\eta$ will play the role of a Weyl group transform of η. Possibly by replacing η by $\eta \oplus \tilde\eta$, we may assume that $\tilde\eta = \eta$, i.e., that $\eta = \omega\eta^c$. *We make this assumption in what follows.* It will cause us no loss of generality since our interest is in what happens as η gets large.

If f is in $W(\eta,\tau)$, the **Eisenstein series** $E(g,f,s)$ corresponding to f is defined formally by

$$E(g,f,s)=\sum_{\gamma\in P(\mathbb{Q})\backslash G(\mathbb{Q})} f_s(\gamma g) \tag{6.13}$$

for $g\in G(\mathbb{A})$ and $s\in\mathbb{C}$. If a member g of $G(\mathbb{A})$ is decomposed according to $G(\mathbb{A})=P(\mathbb{A})K$ as

$$g=pk=\begin{pmatrix} au & 0\\ 0 & bv\end{pmatrix}\begin{pmatrix} 1 & x\\ 0 & 1\end{pmatrix}k$$

with a and b in $\mathbb{Q}_\infty^+$ and with u and v in $(\mathbb{A}^\times)^1$, let us write $b(g)$, $u(g)$, $v(g)$, and $\kappa(g)$ for b, u, v, and k. Then we can rewrite (6.13) as

$$E(g,f,s)=\sum_{\gamma\in P(\mathbb{Q})\backslash G(\mathbb{Q})} h(\gamma g)^{1+s}\omega(b(\gamma g)v(\gamma g))f(u(\gamma g)v(\gamma g)^{-1},\kappa(\gamma g)).$$

The functions ω and f are bounded. By Lemma 6.7 and the convergence of the series $\sum|cz+d|^{-(1+s)}$ for $\operatorname{Re}s>1$, the series for $E(g,f,s)$ is absolutely convergent if $\operatorname{Re}s>1$, and the convergence is uniform for g and s in compact sets. By Lemmas 6.7 and 2.2, there is a constant $C(\varepsilon,b)$ such that

$$|E(g,f,s)|\le C(\varepsilon,b)(\sup_K|f|)h(g)^{1+\operatorname{Re}s}$$

whenever $h(g)\ge b$ and $1+\varepsilon\le\operatorname{Re}s\le 1+\varepsilon^{-1}$. As a function of $g\in G(\mathbb{A})$, $E(g,f,s)$ is an automorphic form on $G(\mathbb{A})$ is the sense of the definition before Theorem 7.1 of [Kn2].

Let $\mathcal{C}_\omega((Z(\mathbb{A})N(\mathbb{A})P(\mathbb{Q}))\backslash G(\mathbb{A}),(\eta,\tau))$ be the set of continuous functions on $G(\mathbb{A})$ transforming as in (6.3), having compact support modulo $Z(\mathbb{A})N(\mathbb{A})P(\mathbb{Q})$, and satisfying the condition that $\phi\left(\left(\begin{smallmatrix} r & 0\\ 0 & 1\end{smallmatrix}\right)\left(\begin{smallmatrix} u & 0\\ 0 & 1\end{smallmatrix}\right)k\right)$ is in $W(\eta,\tau)$ for fixed $r\in\mathbb{Q}_\infty^+$ and is smooth for $r\in\mathbb{Q}_\infty^+$ when u and k are fixed, with uniform estimates on the smoothness as u and k vary. We define the **Fourier-Laplace transform** of such a function ϕ by

$$\Phi(g,s)=\int_0^\infty \phi(a(y)^{-1}g)y^{\frac12(1+s)}\,\frac{dy}{y}. \tag{6.14}$$

The function $\Phi(\,\cdot\,,s)$ on $G(\mathbb{A})$ is in $H(s)$ for each s, and the restriction to the subgroup $(\mathbb{Q}^\times\backslash(\mathbb{A}^\times)^1)\times K$ is in H. We write $\Phi(s)$ for the restriction. Just as in (2.9), Fourier inversion gives

$$\begin{aligned}\phi(g)&=\frac{1}{4\pi}\int_{\operatorname{Re}s=\sigma}\Phi(g,s)\,d|s|\\ &=\frac{1}{4\pi}\int_{\operatorname{Re}s=\sigma}h(g)^{1+s}\omega(v(g))\Phi\left(\begin{pmatrix} u(g)v(g)^{-1} & 0\\ 0 & 1\end{pmatrix}\kappa(g),s\right)d|s|\end{aligned} \tag{6.15}$$

for any real σ. With $\widehat{\phi}$ defined as in (6.4), we obtain, as in (2.10),

$$\widehat{\phi}(g)=\frac{1}{4\pi}\int_{\operatorname{Re}s=\sigma}E(g,\Phi(s),s)\,d|s| \tag{6.16}$$

for $\sigma>1$.

Proposition 6.8. *If* $\operatorname{Re} s > 1$ *and if* f *is in* $W(\eta, \tau)$, *then the constant term of the Eisenstein series for* f *is given by*

$$E_P(\,\cdot\,, f, s) = f_s + (M(s)f)_{-s} \tag{6.17}$$

for an operator $M(s) : W(\eta, \tau) \to W(\eta, \tau)$ *given by*

$$(M(s)f)_{-s}(g) = \int_{N(\mathbb{A})} f_s(wng)\, dn.$$

PROOF. We start from

$$E(g, f, s) = f_s(g) + \sum_{\xi \in \mathbb{Q}} f_s\left(w \left(\begin{smallmatrix} 1 & \xi \\ 0 & 1 \end{smallmatrix}\right) g\right).$$

Replacing g by ng and integrating for $n \in N(\mathbb{Q})\backslash N(\mathbb{A})$ gives

$$\begin{aligned} E_P(f, g, s) &= f_s(g) + \int_{N(\mathbb{Q})\backslash N(\mathbb{A})} \sum_{\xi \in \mathbb{Q}} f_s\left(w \left(\begin{smallmatrix} 1 & \xi \\ 0 & 1 \end{smallmatrix}\right) ng\right) dn \\ &= f_s(g) + \int_{N(\mathbb{A})} f_s(wng)\, dn, \end{aligned}$$

as required. An easy change of variables shows that $M(s)$ carries $W(\eta, \tau)$ to itself because $\eta = \tilde{\eta}$, i.e., $\eta = \omega \eta^c$.

Corollary 6.9. *Let* ϕ *and* ψ *be members of* $\mathcal{C}_\omega((Z(\mathbb{A})N(\mathbb{A})P(\mathbb{Q}))\backslash G(\mathbb{A}), (\eta, \tau))$, *and let* Φ *and* Ψ *be the Fourier-Laplace transforms of* ϕ *and* ψ. *Then*

$$\begin{aligned} &\langle \widehat{\phi}, \widehat{\psi} \rangle_{L^2(Z(\mathbb{A})G(\mathbb{Q})\backslash G(\mathbb{A}))} \\ &\quad = \frac{1}{4\pi} \int_{\operatorname{Re} s = \sigma} \Big(\langle \Phi(s), \Psi(-\bar{s}) \rangle_{L^2((\mathbb{Q}^\times \backslash (\mathbb{A}^\times)^1) \times K)} \\ &\qquad\qquad + \langle M(s)\Phi(s), \Psi(\bar{s}) \rangle_{L^2((\mathbb{Q}^\times \backslash (\mathbb{A}^\times)^1) \times K)} \Big)\, d|s| \end{aligned}$$

for any $\sigma > 1$.

The proof of Corollary 6.9 is almost the same as for Corollary 2.7. Two comments are in order. One is that the constant $1/2\pi$ in Corollary 2.7 has become $1/4\pi$ here because the formula for the constant term of an Eisenstein series no longer involves a factor of 2. The other comment concerns normalizations of Haar measure. Suppose that x, y, r_1, and r_2 are positive reals viewed as ideles at the infinite place such that

$$\begin{pmatrix} x & 0 \\ 0 & x \end{pmatrix} \begin{pmatrix} y^{1/2} & 0 \\ 0 & y^{-1/2} \end{pmatrix} = \begin{pmatrix} r_1 & 0 \\ 0 & r_2 \end{pmatrix}.$$

A little computation with Jacobian determinants shows that $\dfrac{dx\, dy}{xy} = \dfrac{dr_1\, dr_2}{r_1 r_2}$. The right side of this identity is what was defined as Haar measure for the infinite place of $M(\mathbb{A})$, and therefore dy/y is Haar measure for the subgroup of all $a(y) = \begin{pmatrix} y^{1/2} & 0 \\ 0 & y^{-1/2} \end{pmatrix}$. Representatives of the cosets of $Z(\mathbb{A})\backslash G(\mathbb{A})$ are the matrices $\begin{pmatrix} 1 & x \\ 0 & 1 \end{pmatrix} a(y) \begin{pmatrix} u & 0 \\ 0 & 1 \end{pmatrix} k$ with $y > 0$, $u \in (\mathbb{A}^\times)^1$, $x \in \mathbb{A}$, and $k \in K$, and it follows that Haar measure on $Z(\mathbb{A})\backslash G(\mathbb{A})$ is $y^{-1}\,dx\,du\,dk\,\dfrac{dy}{y}$. The invariant measure to use on $Z(\mathbb{A})N(\mathbb{A})\backslash G(\mathbb{A})$ is therefore $y^{-1}\,du\,dk\,\dfrac{dy}{y}$.

Theorem 6.10. *The family of operators $M(s)$, initially given as an analytic family $M(s): W(\eta,\tau) \to W(\eta,\tau)$, extends to be meromorphic for $s \in \mathbb{C}$. The only possible pole for $\operatorname{Re} s \geq 0$ is at $s = 1$ and is at most simple. As a function of $\operatorname{Im} s$, $M(s)$ is uniformly at most of polynomial growth, apart from the pole, in any vertical strip $0 \leq \operatorname{Re} s < \sigma$. The continued operators satisfy $M(-s)M(s) = 1$ as a meromorphic function of s.*

REFERENCE. See [Gb-Ja] and also Jacquet's article [Ja] in this volume.

Now we move the line of integration in Corollary 6.9 to $\operatorname{Re} s = 0$, just as in §2. The integrand is meromorphic, the functions $\Phi(s)$ and $\Psi(s)$ are Schwartz functions of $\operatorname{Im} z$ uniformly in vertical strips, and the growth of $M(z)$ is controlled by Theorem 6.10. We can move the line of integration by the Cauchy Integral Formula and an easy passage to the limit, picking up a residue term from $s = 1$. The result is

$$
\begin{aligned}
&\langle \widehat{\phi}, \widehat{\psi}\rangle_{L^2(Z(\mathbb{A})G(\mathbb{Q})\backslash G(\mathbb{A}))} \\
&\quad = \frac{1}{4\pi}\int_{-\infty}^{\infty} \Big(\langle \Phi(it), \Psi(it)\rangle_{L^2((\mathbb{Q}^\times\backslash(\mathbb{A}^\times)^1)\times K)} \\
&\qquad\qquad + \langle M(it)\Phi(it), \Psi(-it)\rangle_{L^2((\mathbb{Q}^\times\backslash(\mathbb{A}^\times)^1)\times K)}\Big)\,dt \\
&\qquad + \frac{1}{2}\lim_{s\to 1}\langle (s-1)M(s)\Phi(1), \Psi(1)\rangle_{L^2((\mathbb{Q}^\times\backslash(\mathbb{A}^\times)^1)\times K)}
\end{aligned}
\tag{6.18}
$$

for ϕ and ψ as in Corollary 6.9.

Next we simplify this expression, using that $\tilde{\eta} = \eta$. The residue term, to which we return shortly, may be shown to be

$$
c\sum_{\chi^2=\omega}\langle \Phi(1), \chi\circ\det\rangle_{L^2((\mathbb{Q}^\times\backslash(\mathbb{A}^\times)^1)\times K)}\overline{\langle \Psi(1), \chi\circ\det\rangle}_{L^2((\mathbb{Q}^\times\backslash(\mathbb{A}^\times)^1)\times K)},
$$

where c is a known positive constant. (See [Gb-Ja, p. 227].) For the integral term on the right side of (6.18), the first step is to check from the definitions that $\tilde{\eta} = \eta$ implies $M(s)^* = M(\bar{s})$ for $\operatorname{Re} s > 1$. Then this relation persists for all s by analytic continuation. Since $M(-s)M(s) = 1$ by Theorem 6.10, it follows that $M(it)$ is unitary with inverse $M(-it)$. Then (6.18) may be rewritten by the techniques of Corollary 2.8 as

$$
\begin{aligned}
&\langle \widehat{\phi}, \widehat{\psi}\rangle_{L^2(Z(\mathbb{A})G(\mathbb{Q})\backslash G(\mathbb{A}))} \\
&\quad = \frac{1}{8\pi}\int_{-\infty}^{\infty} \langle \Phi(it) + M(-it)\Phi(-it), \Psi(it) + M(-it)\Psi(-it)\rangle_{L^2((\mathbb{Q}^\times\backslash(\mathbb{A}^\times)^1)\times K)}\,dt \\
&\qquad + c\sum_{\chi^2=\omega}\langle \Phi(1), \chi\circ\det\rangle_{L^2((\mathbb{Q}^\times\backslash(\mathbb{A}^\times)^1)\times K)}\overline{\langle \Psi(1), \chi\circ\det\rangle}_{L^2((\mathbb{Q}^\times\backslash(\mathbb{A}^\times)^1)\times K)}.
\end{aligned}
\tag{6.19}
$$

With this formula in place, the kind of analysis in §2, in view of Theorem 6.5, shows that $L^2_{\text{cusp}}(\omega)^\perp$ is the sum of a direct integral of the representations $H(s)$, together with a discrete contribution from the residues at $s = 1$. This is the adelic analog of Theorem 1.4. For details, see [Gb-Ja, §4]. We denote the direct integral term by $L^2_{\text{cont}}(\omega)$ and the term for the various residues by $L^2_{\text{res}}(\omega)$. The decomposition may be summarized as

$$L^2(Z(\mathbb{A})G(\mathbb{Q})\backslash G(\mathbb{A}), \omega) = L^2_{\text{cusp}}(\omega) \oplus L^2_{\text{cont}}(\omega) \oplus L^2_{\text{res}}(\omega). \tag{6.20}$$

The residues come from one-dimensional representations of $G(\mathbb{A})$, necessarily of the form $g \mapsto \chi(\det g)$. The corresponding members of $L^2_{\text{res}}(\omega)$ are the functions $f(g) = \chi(\det g)$. Since f is to be left invariant under $G(\mathbb{Q})$, χ is a character of $\mathbb{Q}^\times\backslash\mathbb{A}^\times$. Since f is to transform by ω under $Z(\mathbb{A})$, $\chi^2 = \omega$. Thus the decomposition of $L^2_{\text{res}}(\omega)$ is a Hilbert space direct sum

$$L^2_{\text{res}}(\omega) = \bigoplus_{\chi^2=\omega} \mathbb{C}_{\chi\circ\det}. \tag{6.21}$$

7. Adelic Trace Formula

We continue with notation as in §6. In the decomposition (6.20) the difficult term to understand is $L^2_{\text{cusp}}(\omega)$. The operator $R_\omega(\varphi)f(x) = \int_{\overline{G}(\mathbb{A})} f(xy)\varphi(y)\,dy$, for $\varphi \in C^\infty_{\text{com}}(G(\mathbb{A}), \omega^{-1})$, acts on $L^2(Z(\mathbb{A})G(\mathbb{Q})\backslash G(\mathbb{A}), \omega)$ and leaves $L^2_{\text{cusp}}(\omega)$ stable. It is of trace class on this subspace, by Corollary 6.3. The adelic trace formula gives an explicit expression for the trace of this operator on the subspace $L^2_{\text{cusp}}(\omega)$. The final formula is stated in [Gf-Gr-P], [Ja-Lgl], [Du-La], [Ar1], [Gb1], [Gb-Ja], and [Ar4], and we shall follow [Gb-Ja].

The idea is that $R_\omega(\varphi)$ is given by manageable integral operators on the whole space and on the subspaces $L^2_{\text{cont}}(\omega)$ and $L^2_{\text{res}}(\omega)$. Let kernels for these integral operators be $K(x,y)$, $K_{\text{cont}}(x,y)$, and $K_{\text{res}}(x,y)$. Then the operator on $L^2_{\text{cusp}}(\omega)$ must be given by the kernel

$$K_{\text{cusp}}(x,y) = K(x,y) - K_{\text{cont}}(x,y) - K_{\text{res}}(x,y), \tag{7.1}$$

and the trace in question ought to be the integral of $K_{\text{cusp}}(x,x)$ over the quotient $Z(\mathbb{A})G(\mathbb{Q})\backslash G(\mathbb{A})$.

These kernels are not uniquely defined as functions on $G(\mathbb{A}) \times G(\mathbb{A})$ without some further restriction. In the case of $K(x,y)$, the same derivation as for (5.4) leads to the formula

$$K(x,y) = \sum_{\gamma\in\overline{G}(\mathbb{Q})} \varphi(x^{-1}\gamma y). \tag{7.2}$$

Then $K(x,y)$ is left invariant in each variable under $G(\mathbb{Q})$ and satisfies

$$K(z_1x, z_2y) = \omega(z_1)\omega(z_2)^{-1}K(x,y) \qquad \text{for } z_1, z_2 \in Z(\mathbb{A}). \tag{7.3}$$

It is this condition that determines $K(x,y)$ uniquely.

Similarly to determine the kernels $K_{\text{cont}}(x,y)$ and $K_{\text{res}}(x,y)$ uniquely, we insist that they satisfy the same invariance properties as $K(x,y)$. Then $K_{\text{cont}}(x,y)$ and $K_{\text{res}}(x,y)$ can be written down fairly explicitly. The techniques for $K_{\text{cont}}(x,y)$ are the same as for Proposition 4.6. To get at $K_{\text{cont}}(x,y)$, we need to know that the Eisenstein series themselves, not just their constant terms, admit meromorphic continuations.

Theorem 7.1. *If f is in a subspace $W(\eta,\tau)$ of H, then the function $s \mapsto E(g,f,s)$, initially given as an analytic function for* $\operatorname{Re} s > 1$, *extends to be meromorphic in $\mathbb{C}$. Its constant term is given by the analytic continuation of $E_P(g,f,s)$,* *and $E(g,f,s)$ has the same poles as $E_P(g,f,s)$. Also $E(g,f,s)$ is at most of polynomial growth in* $\operatorname{Im} s$ *in any vertical strip* $0 \le \operatorname{Re} s \le \sigma$.

REFERENCE. For a discussion of this theorem, see [Ja] in this volume.

To obtain an expression for $K_{\text{cont}}(x,y)$, we proceed as in Proposition 4.6. We can choose an orthonormal basis $\{f_\alpha\}$ of H such that each f_α is in some $W(\eta,\tau)$. Theorem 7.1 shows that $E(x,f_\alpha,it)$ is meaningful. If $P^{\omega,it}$ is the unitary representation of $G(\mathbb{A})$ on $H(it)$, then calculations in [Gb-Ja, pp. 232–234] show that[2]

$$K_{\text{cont}}(x,y) = \frac{1}{8\pi} \sum_{\alpha,\beta} \int_{-\infty}^{\infty} \langle P^{\omega,it}(\varphi) f_\beta, f_\alpha\rangle E(x,f_\alpha,it)\overline{E(y,f_\beta,it)}\,dt. \tag{7.4}$$

Moreover, an easy computation with (6.21) shows that

$$K_{\text{res}}(x,y) = (\operatorname{vol}(Z(\mathbb{A})G(\mathbb{Q})\backslash G(\mathbb{A})))^{-1} \sum_{\chi^2=\omega} \chi(\det x)\overline{\chi(\det y)} \int_{\overline{G}(\mathbb{A})} \varphi(g)\chi(\det g)\,dg. \tag{7.5}$$

A direct attempt to integrate $K_{\text{cusp}}(x,x)$ with the aid of formulas (7.1) through (7.5) leads to $\infty - \infty$, and a more subtle approach is needed. Selberg [Sel] already saw the need for truncation in the classical setting (0.3), but his method was adapted to a fundamental domain for $SL_2(\mathbb{Z})$ in the upper half plane. We shall use the truncation operator of Arthur [Ar3], which does not require a fundamental domain for its definition. Expositions of this operator appear in [Gb-Ja] and [Mo-Wa].

Recall that $w = \begin{pmatrix} 0 & 1 \\ 1 & 0 \end{pmatrix}$. When w is embedded in $\mathbb{A}^\times$, we regard it as embedded diagonally.

Lemma 7.2. *For any $n \in N(\mathbb{A})$ and $g \in G(\mathbb{A})$, $h(wng) \le h(g)^{-1}$.*

PROOF. Let us write $g = n'ak$ with $n' \in N(\mathbb{A})$, a diagonal, and $k \in K$. Then $wng = wnn'ak = (waw^{-1})(wn''w^{-1})k$. It follows from Lemma 6.6 that $h\begin{pmatrix} 1 & 0 \\ x & 1 \end{pmatrix} \le 1$, and therefore

$$h(wng) = h(waw^{-1})h(wn''w^{-1}) = h(g)^{-1}h(wn''w^{-1}) \le h(g)^{-1}.$$

Corollary 7.3. *If $h(\gamma_0 g) > 1$ for some $\gamma_0 \in P(\mathbb{Q})\backslash G(\mathbb{Q})$, then $h(\gamma g) < 1$ for all other $\gamma \in P(\mathbb{Q})\backslash G(\mathbb{Q})$.*

REMARK. Since $h(p) = 1$ for $p \in P(\mathbb{Q})$ by Artin's product formula, $h(\gamma g)$ is well defined as a function of γ in $P(\mathbb{Q})\backslash G(\mathbb{Q})$.

PROOF. We may assume that $\gamma_0 = 1$. By the Bruhat decomposition, $\gamma = w\begin{pmatrix} 1 & \xi \\ 0 & 1 \end{pmatrix}$ for some $\xi \in \mathbb{Q}$. Then $\gamma = wn$ for some $n \in N(\mathbb{A})$, and Lemma 7.2 gives $h(\gamma g) = h(wng) \le h(g)^{-1} < 1$.

Fix $T \in \mathbb{R}$ with $T > 0$, and let I_T be the characteristic function of the set $[e^T, +\infty)$. For $T > 0$, the Arthur **truncation operator** Λ^T is defined on all

[2] The formula (5.20) in [Gb-Ja] for K_{cont} has a coefficient $1/4\pi$. The reason for this apparent discrepancy is that $dy = \frac{1}{2}\,dt$.

locally integrable complex-valued functions f on $G(\mathbb{A})$ that are left invariant under $G(\mathbb{Q})$ by

$$\Lambda^T f(g) = f(g) - \sum_{\gamma \in P(\mathbb{Q})\backslash G(\mathbb{Q})} f_P(\gamma g) I_T(h(\gamma g)). \tag{7.6}$$

The sum[3] in (7.6) has at most one nonzero term, by Corollary 7.3, and $f_P(\gamma g)$ depends only on the coset of γ in $P(\mathbb{Q})\backslash G(\mathbb{Q})$. Thus $\Lambda^T f(g)$ is well defined. It is clearly left invariant under $G(\mathbb{Q})$. If f is **cuspidal** in the sense that $f_P = 0$, then $\Lambda^T f = f$.

Corollary 7.4. *If $T > 0$, then $(\Lambda^T f)_P(g) = 0$ unless $I_T(h(g)) = 0$.*

PROOF. Assume that $I_T(h(g)) = 1$. Lemma 7.2 shows that $I_T(h(wng)) = 0$ for all $n \in N(\mathbb{A})$. Hence

$$\begin{aligned}\Lambda^T f(ng) &= f(ng) - f_P(ng) I_T(h(ng)) - \sum_{\xi \in \mathbb{Q}} f_P\left(w \left(\begin{smallmatrix} 1 & \xi \\ 0 & 1 \end{smallmatrix}\right) ng\right) I_T\left(h\left(w\left(\begin{smallmatrix} 1 & \xi \\ 0 & 1 \end{smallmatrix}\right) ng\right)\right) \\ &= f(ng) - f_P(g).\end{aligned}$$

Integrating over $n \in N(\mathbb{Q})\backslash N(\mathbb{A})$ therefore gives

$$(\Lambda^T f)_P(g) = f_P(g) - f_P(g) = 0.$$

Corollary 7.5. *If $T > 0$, then $\Lambda^T(\Lambda^T f) = \Lambda^T f$.*

PROOF. We have

$$\Lambda^T(\Lambda^T f)(g) = (\Lambda^T f)(g) - \sum_{\gamma} (\Lambda^T f)_P(\gamma g) I_T(h(\gamma g)).$$

If $I_T(h(\gamma g)) \neq 0$, then Corollary 7.4 shows that $(\Lambda^T f)_P(\gamma g) = 0$. Hence every term in the sum is 0.

Proposition 7.6. *If $T > 0$, then Λ^T is a Hermitian operator on the space $L^2(Z(\mathbb{A})G(\mathbb{Q})\backslash G(\mathbb{A}), \omega)$.*

REFERENCE. [Gb-Ja, p. 230] or [Ar3, pp. 91–92].

Because of Corollary 7.5 and Proposition 7.6, Λ^T is an orthogonal projection on $L^2(Z(\mathbb{A})G(\mathbb{Q})\backslash G(\mathbb{A}), \omega)$, and we know that its image contains $L^2_{\text{cusp}}(\omega)$. Note, however, that the truncation operator does not commute with the action of $G(\mathbb{A})$, and its image is not $G(\mathbb{A})$ invariant.

In order to obtain more subtle properties of the truncation operator, it is helpful to understand more of the geometry of the action of $G(\mathbb{Q})$ on $G(\mathbb{A})$. Recall that products from $N(\mathbb{A}) \times M(\mathbb{A}) \times K$ cover $G(\mathbb{A})$. Let

$$M_\infty = \{m \in M(\mathbb{A}) \mid \text{diagonal entries of } m \text{ are in } \mathbb{Q}_\infty^+\}$$

$$M^1 = \{m \in M(\mathbb{A}) \mid \text{diagonal entries of } m \text{ are in } (\mathbb{A}^\times)^1\}.$$

Here M_∞ is the direct product of $Z_\infty = M_\infty \cap \mathbb{Z}(\mathbb{A})$ and $A_\infty = \{a(y) \mid y \in \mathbb{Q}_\infty^+\}$. Then $M(\mathbb{A}) = M_\infty M^1$, and hence products from $N(\mathbb{A}) \times M_\infty \times M^1 \times K$ cover $G(\mathbb{A})$.

[3] Instead of using I_T, Arthur uses a function $\hat{\tau}_P$ and incorporates T into its argument. Arthur's notation is especially suited to higher rank groups.

A **Siegel set** $\mathcal{S}$ is a subset of $G(\mathbb{A})$ consisting of all $nrmk$ with n in a compact subset of $N(\mathbb{A})$, r in M_∞ with $h(r) \geq c > 0$, m in a compact subset of M^1, and k in K. The set $\mathcal{S}$ is the product of Z_∞ and the set $\mathcal{S}^1 = \mathcal{S} \cap G^1$. The set $\mathcal{S}^1$ may be viewed as an adelic analog of a rectangular set in the upper half plane that is unbounded above but is bounded on the other three sides.

Lemma 7.7. *Let $c' > 0$. On any compact set of elements $g \in G(\mathbb{A})$, there are only finitely many $\gamma \in P(\mathbb{Q})\backslash G(\mathbb{Q})$ such that $h(\gamma g) \geq c'$ for some g in the compact set.*

PROOF. By the Bruhat decomposition we can take as coset representatives γ the elements 1 and $w\begin{pmatrix} 1 & \xi \\ 0 & 1 \end{pmatrix}$ with $\xi \in \mathbb{Q}$. Thus it is enough to consider $h(\gamma g)$ for $\gamma = w\begin{pmatrix} 1 & \xi \\ 0 & 1 \end{pmatrix} = \begin{pmatrix} 0 & 1 \\ 1 & \xi \end{pmatrix}$. Since right translation by K does not affect $h(g)$, we may assume that g is in $P(\mathbb{A})$. Write $g = \begin{pmatrix} yu & xu \\ 0 & u \end{pmatrix}$ with $x = x_\infty \prod_p x_p$ and $y = y_\infty \prod_p y_p$, and put $\xi = d/c$ with $\mathrm{GCD}(c,d) = 1$. If $h(\gamma g) \geq c'$ for some g in the given compact set, then Lemma 6.7 gives

$$B^{-1}|cz_\infty + d| \leq h(\gamma g)^{-1} \leq c'^{-1},$$

where $z_\infty = x_\infty + iy_\infty$. This can happen for only finitely many pairs (c,d) if (x_∞, y_∞) lies in a compact subset of the upper half plane, and the lemma follows.

Proposition 7.8. *Let $\mathcal{S}$ be a Siegel set, and let $c' > 0$. Then there are only finitely many $\gamma \in P(\mathbb{Q})\backslash G(\mathbb{Q})$ such that $h(\gamma g) \geq c'$ for some $g \in \mathcal{S}$.*

PROOF. Write $\mathcal{S} = Z_\infty \mathcal{S}^1$ with $\mathcal{S}^1 \subset G^1$. The subset of $g \in \mathcal{S}^1$ with $c' \leq h(g) \leq 1$ is compact and is handled by Lemma 7.7. For $z \in Z_\infty$, we have $h(\gamma z_\infty g) = h(\gamma g)$, and therefore there are only finitely many $\gamma \in P(\mathbb{Q})\backslash G(\mathbb{Q})$ have $h(\gamma g) \geq c'$ for some $g \in \mathcal{S}$. To complete the proof, consider the subset of $g \in \mathcal{S}$ with $h(g) > 1$. For such g, Corollary 7.3 shows that $h(\gamma g) \leq 1$ whenever γ is nontrivial in $P(\mathbb{Q})\backslash G(\mathbb{Q})$.

Corollary 7.9. *Let $\mathcal{S}$ be a Siegel set. Then there are only finitely many $\gamma \in G(\mathbb{Q})$ such that $\gamma\mathcal{S}$ meets $\mathcal{S}$.*

PROOF. Say that $h(g) \geq c'$ for $g \in \mathcal{S}$. According to Proposition 7.8, the elements γ in $G(\mathbb{Q})$ for which $\gamma\mathcal{S}$ meets $\mathcal{S}$ lie in finitely many cosets of $P(\mathbb{Q})\backslash G(\mathbb{Q})$. If there are infinitely many such elements γ, then there is some $\gamma_0 \in G(\mathbb{Q})$ such that $\varepsilon_j \gamma_0 \mathcal{S}$ meets $\mathcal{S}$ for infinitely many ε_j in $P(\mathbb{Q})$.

Suppose that the coset of γ_0 is trivial. Then we may take $\gamma_0 = 1$, so that $\varepsilon_j \mathcal{S}$ meets $\mathcal{S}$ for infinitely many ε_j. Since ε_j is in G^1, $\varepsilon_j \mathcal{S}^1$ meets $\mathcal{S}^1$ for infinitely many ε_j. Since $h(\varepsilon_j s) = h(s)$ and since $\mathcal{S}^1$ is compact in all other directions, we obtain a contradiction to the discreteness of $P(\mathbb{Q})$.

Thus we may suppose that the coset of γ_0 in $P(\mathbb{Q})\backslash G(\mathbb{Q})$ is nontrivial. If $\varepsilon_j \gamma_0 \mathcal{S}$ meets $\mathcal{S}$, then $\varepsilon_j \gamma_0 \mathcal{S}^1$ meets $\mathcal{S}^1$. If s is in $\mathcal{S}^1$, then $h(\varepsilon_j \gamma_0 s) = h(\gamma_0 s)$. When $h(s) > 1$, Corollary 7.3 shows that $h(\gamma_0 s) \leq 1$. And the part of $\mathcal{S}^1$ where $h(s) \leq 1$ is compact. Hence h is bounded on $\varepsilon_j \gamma_0 \mathcal{S}^1$ uniformly in j. Since $\varepsilon_j \gamma_0 \mathcal{S}^1$ meets $\mathcal{S}^1$, the points of intersection lie in a compact subset of $\mathcal{S}^1$, and we may assume that these points of intersection $\varepsilon_j \gamma_0 s_j = s_j'$ converge, say to s_0'. Applying h shows that $h(\gamma_0 s_j) \to h(s_0')$. Let $s_j = n_j a_j k_j$ with $n_j \in N(\mathbb{A})$, $a_j \in M(\mathbb{A}) \cap G^1$, and $k_j \in K$.

Since γ_0 is nontrivial, we may assume that $\gamma_0 = wn'$ with $n' \in N(\mathbb{A})$. Then

$$\begin{aligned} h(\gamma_0 s_j) = h(wn'n_j a_j k_j) &= h((wa_j w^{-1})(wa_j^{-1}n'n_j a_j w^{-1})(wk_j)) \\ &= h(wa_j w^{-1})h(wa_j^{-1}n'n_j a_j w^{-1}) \end{aligned}$$

and hence

$$h(\gamma_0 s_j)h(s_j) = h(wa_j^{-1}n'n_j a_j w^{-1}).$$

Since $n'n_j$ is bounded within $N(\mathbb{A})$ while $h(a_j)$ is bounded below, $wa_j^{-1}n'n_j a_j w^{-1}$ lies in a compact subset of G^1. Therefore h is bounded away from 0 and $+\infty$ on it. Consequently $h(s_j)$ is bounded away from 0 and $+\infty$. We may therefore assume that s_j converges within G^1, say to s_0. Then $\lim \varepsilon_j^{-1}s_0' = \gamma_0 s_0$ exists in G^1, and ε_j converges. This is a contradiction since $P(\mathbb{Q})$ is discrete.

Proposition 7.10. *If $\mathcal{S}$ is a sufficiently large Siegel set, then $G(\mathbb{Q})\mathcal{S} = G(\mathbb{A})$.*

REMARK. Corollary 7.9 and Proposition 7.10 together show that Siegel sets for many purposes are adequate substitutes for fundamental domains for the action of $G(\mathbb{Q})$ on $G(\mathbb{A})$. For a generalization to all reductive groups, see [Bo].

PROOF. It is known [Lan2, p. 140] that $\mathcal{D} = [-\frac{1}{2}, \frac{1}{2}) \times \prod_p \mathbb{Z}_p$ is a fundamental domain for $\mathbb{Q}\backslash\mathbb{A}$. Then $\tilde{\mathcal{D}} = \left\{\left(\begin{smallmatrix} 1 & x \\ 0 & 1 \end{smallmatrix}\right) \,\middle|\, x \in \mathcal{D}\right\}$ is a fundamental domain for $N(\mathbb{Q})\backslash N(\mathbb{A})$. Let $\mathcal{C}_0$ be the compact subset $\{1\} \times \prod_p \mathbb{Z}_p^\times$ of $(\mathbb{A}^\times)^1$; the set $\mathcal{C}_0$ has the property that $\mathbb{Q}^\times \mathcal{C}_0 = (\mathbb{A}^\times)^1$. Let $\mathcal{C}$ be the subset of M^1 whose diagonal entries are in $\mathcal{C}_0$, and define

$$\mathcal{S} = \tilde{\mathcal{D}} \times Z_\infty \times \{a(y) \in A_\infty \mid y \geq \tfrac{\sqrt{3}}{2}\} \times \mathcal{C} \times K.$$

Given $g \in G(\mathbb{A})$, we are to left-translate g into $\mathcal{S}$ via $G(\mathbb{Q})$. Lemma 7.7 shows that we may assume that $h(\gamma g) \leq h(g)$ for all $\gamma \in G(\mathbb{Q})$. Write $g = nak$ with $n \in N(\mathbb{A})$, $a \in M(\mathbb{A})$, and $k \in K$. Left translating by a member of $M(\mathbb{Q})$, we may assume that a is in $M_\infty \mathcal{C}$. Left translating further by $N(\mathbb{Q})$, we may assume that n is in $\tilde{\mathcal{D}}$.

We have

$$h(wnak) = h((waw^{-1})(wa^{-1}naw^{-1})wk) = h(waw^{-1})h(wa^{-1}naw^{-1}),$$

and therefore

$$h(wa^{-1}naw^{-1}) = h(wg)h(g) \leq h(g)^2. \tag{7.7}$$

We can decompose n and a according to the infinite and finite places as $n = \left(\begin{smallmatrix} 1 & x_\infty x_{\text{fin}} \\ 0 & 1 \end{smallmatrix}\right)$ and $a = a(y)a_{\text{fin}}$. Taking into account the form of $\tilde{\mathcal{D}}$ and $\mathcal{C}$, we see that $wa^{-1}naw^{-1}$ is $\left(\begin{smallmatrix} 1 & 0 \\ y^{-1}x_\infty & 1 \end{smallmatrix}\right)$ at the place ∞ and is $\left(\begin{smallmatrix} 1 & 0 \\ x' & 1 \end{smallmatrix}\right)$ with $x' \in \prod \mathbb{Z}_p$ at the finite places. By Lemma 6.6

$$h(wa^{-1}naw^{-1}) = (1 + y^{-2}x_\infty^{-2})^{-1/2} = \frac{y}{\sqrt{y^2 + x_\infty^2}}. \tag{7.8}$$

Since $h(g) = h(a) = h(a(y)) = y^{1/2}$, comparison of (7.7) and (7.8) shows that $y/\sqrt{y^2 + x_\infty^2} \leq y$, i.e., $y^2 + x_\infty^2 \geq 1$. Since $|x_\infty| \leq \frac{1}{2}$, $y^2 \geq \frac{3}{4}$. Thus our particular left translate of g via $G(\mathbb{Q})$ is in $\mathcal{S}$.

Now let us return to Λ^T. Suppose that $T \geq T_0 > 0$. If Ω is a compact subset of $G(\mathbb{A})$, then only finitely many terms in the sum for $\Lambda^T f(g)$ can be nonzero for some $g \in \Omega$, by Lemma 7.7. Taking T large enough, we can make $I_T(h(\gamma g)) = 0$ for each such term. Thus we obtain the following result.

Proposition 7.11. *$\Lambda^T f$ converges to f uniformly on compact subsets of $G(\mathbb{A})$.*

Under some mild restrictions on f, $\Lambda^T f$ is small at infinity in a certain sense. To make this idea precise, we shall use Siegel sets. If $\mathcal{S}$ is a Siegel set, again write $\mathcal{S} = Z_\infty \mathcal{S}^1$ with $\mathcal{S}^1 \subset G^1$. Then the part of $\mathcal{S}^1$ where $h(g) \leq 1$ is compact. For $h(g) > 1$, Corollary 7.3 shows that

$$\Lambda^T f(g) = f(g) - f_P(g) I_T(h(g)).$$

Thus if the $\mathcal{S}^1$ component of a member g of $\mathcal{S}$ is far enough out, we obtain

$$\Lambda^T f(g) = f(g) - f_P(g).$$

To get an idea why this difference is small in favorable circumstances, suppose that $f = F * \psi$ with F bounded and left invariant under $G(\mathbb{Q})$ and with ψ continuous of compact support on $G(\mathbb{A})$. Then

$$f(g) - f_P(g) = \int_{G(\mathbb{A})} \int_{N(\mathbb{Q})\backslash N(\mathbb{A})} F(x)[\psi(x^{-1}g) - \psi(x^{-1}ng)]\, dn\, dx. \tag{7.9}$$

It is easy to check that as g tends to ∞ through $\mathcal{S}$, $g^{-1}ng$ tends uniformly to 1 for n in any compact subset of $N(\mathbb{A})$. Therefore

$$\int_{G(\mathbb{A})} |\psi(x^{-1}g) - \psi(x^{-1}ng)|\, dx = \int_{G(\mathbb{A})} |\psi^{-1}(x^{-1}) - \psi(x^{-1}g^{-1}ng)|\, dx$$

tends to 0 as the $\mathcal{S}^1$ component of $g \in \mathcal{S}$ tends to ∞, and (7.9) has limit 0.

Let us state a general result. A function f on $G(\mathbb{A})$ that is left invariant under $G(\mathbb{Q})$ is said to be **slowly increasing** if, for each Siegel set $\mathcal{S}$, there are constants C and N such that

$$|f(g)| \leq Ch(g)^N \qquad \text{for all } g \in \mathcal{S}. \tag{7.10}$$

Because of Proposition 7.10, this condition controls the global growth of f at infinity for $G(\mathbb{A})$. The function f is said to be **rapidly decreasing** if, for each Siegel set $\mathcal{S}$ and integer $-N$, there is a constant C such that

$$|f(g)| \leq Ch(g)^{-N} \qquad \text{for all } g \in \mathcal{S}. \tag{7.11}$$

Let $G(\mathbb{A}_{\text{fin}})$ be the part of $G(\mathbb{A})$ corresponding to the finite places, and let K_0 be an open compact subgroup of $G(\mathbb{A}_{\text{fin}})$. If the above function f is right invariant under K_0, then f may be viewed as a function on the space $G(\mathbb{Q})\backslash G(\mathbb{A})/K_0$, which is a smooth manifold. Let us say that f is smooth if this descended function is smooth.

Proposition 7.12. *Let K_0 be an open subgroup of $G(\mathbb{A}_{\text{fin}})$. Suppose that f is a function on $G(\mathbb{A})$ that is left invariant under $G(\mathbb{Q})$, right invariant under K_0, and smooth. If f and all its left invariant derivatives are slowly increasing, then $\Lambda^T f$ is rapidly decreasing.*

REFERENCE. [Ar3, Lemma 1.4].

Finally we can return to the formula (7.1) for $K_{\text{cusp}}(x,y)$. We follow [Gb-Ja]. Let P_{cusp} be the orthogonal projection on $L^2_{\text{cusp}}(\Omega)$. It is not hard to see that $K_{\text{cusp}}(x,y)$ is in $L^2_{\text{cusp}}(\omega^{-1})$ as a function of the second variable. When we therefore apply the truncation operator Λ_2^T in the second variable, we obtain

$$K_{\text{cusp}}(x,y) = \Lambda_2^T K(x,y) - \Lambda_2^T K_{\text{cont}}(x,y) - \Lambda_2^T K_{\text{res}}(x,y).$$

It turns out that each term on the right side is now integrable over the diagonal and that

$$\operatorname{Tr}(P_{\text{cusp}} R_\omega(\varphi) P_{\text{cusp}}) = \int \Lambda_2^T K(x,x)\,dx - \int \Lambda_2^T K_{\text{cont}}(x,x)\,dx - \int \Lambda_2^T K_{\text{res}}(x,x)\,dx \tag{7.12}$$

with the integrals extending over $\overline{G}(\mathbb{Q})\backslash\overline{G}(\mathbb{A})$.

In place of (7.2) we have the formula

$$\begin{aligned}\Lambda^T K(x,x) &\\ = \sum_{\gamma\in\overline{G}(\mathbb{Q})} \varphi(x^{-1}\gamma x) &- \sum_{\xi\in P(\mathbb{Q})\backslash G(\mathbb{Q})} \int_{N(\mathbb{Q})\backslash N(\mathbb{A})} \Big(\sum_{\gamma\in\overline{G}(\mathbb{Q})} \varphi(x^{-1}\gamma n\xi x) I_T(h(\xi x))\Big)\,dn.\end{aligned}$$

For T large enough, the right side may be shown to be

$$= \sum_{\gamma\in\overline{G}(\mathbb{Q})} \varphi(x^{-1}\gamma x) - \sum_{\xi\in P(\mathbb{Q})\backslash G(\mathbb{Q})} \int_{N(\mathbb{A})} \sum_{\mu\in\overline{M}(\mathbb{Q})} \varphi(x^{-1}\xi^{-1}\mu n\xi x) I_T(h(\xi x))\,dn. \tag{7.13}$$

We group these terms according to the type of γ or μ. We say that γ is **elliptic** if its eigenvalues are not in $\mathbb{Q}$, **hyperbolic regular** if its eigenvalues are distinct rationals, **singular** if γ is the product of a scalar matrix and a unipotent matrix. From γ elliptic we get

$$\sum_{\substack{\gamma \text{ elliptic} \\ \text{in } \overline{G}(\mathbb{Q})}} \varphi(x^{-1}\gamma x).$$

From γ and μ hyperbolic regular, we get

$$\sum_{\substack{\gamma \text{ hyperbolic} \\ \text{regular in } \overline{G}(\mathbb{Q})}} \varphi(x^{-1}\gamma x) - \sum_{\xi\in P(\mathbb{Q})\backslash G(\mathbb{Q})} \int_{N(\mathbb{A})} \sum_{\substack{\mu\in\overline{M}(\mathbb{Q}), \\ \mu\neq 1}} \varphi(x^{-1}\xi^{-1}\mu n\xi x) I_T(h(\xi x))\,dn.$$

From γ and μ singular we get

$$\sum_{\substack{\gamma\in\overline{G}(\mathbb{Q}), \\ \text{unipotent}}} \varphi(x^{-1}\gamma x) - \sum_{\xi\in P(\mathbb{Q})\backslash G(\mathbb{Q})} \int_{N(\mathbb{A})} \varphi(x^{-1}\xi^{-1}n\xi x) I_T(h(\xi x))\,dn.$$

The term with the elliptic elements is handled just as in (4.3): From each conjugacy class $\mathfrak{o}$ of elliptic elements in $\overline{G}(\mathbb{Q})$, we select a representative. Say that γ is a representative of $\mathfrak{o}_\gamma$. Then $\mathfrak{o}_\gamma$ consists of all $\delta^{-1}\gamma\delta$, where δ varies through $\overline{G}(\mathbb{Q})^\gamma\backslash\overline{G}(\mathbb{Q})$. Upon integration over $x \in \overline{G}(\mathbb{Q})\backslash\overline{G}(\mathbb{A})$, the term with the elliptic elements gives

$$\int_{\overline{G}(\mathbb{Q})\backslash\overline{G}(\mathbb{A})} \sum_{\gamma \text{ elliptic}} \varphi(x^{-1}\gamma x)\, dx = \sum_{\mathfrak{o}_\gamma \text{ elliptic}} \operatorname{vol}(\overline{G}(\mathbb{Q})^\gamma\backslash\overline{G}(\mathbb{A})^\gamma) \int_{\overline{G}(\mathbb{A})^\gamma\backslash\overline{G}(\mathbb{A})} \varphi(x^{-1}\gamma x)\, dx.$$

A more complicated computation shows that the contribution to the x integral from the hyperbolic regular elements is of the form

$$\sum_{\substack{\mathfrak{o}_\gamma \text{ hyperbolic} \\ \text{regular}}} \operatorname{vol}(\overline{G}(\mathbb{Q})^\gamma\backslash\overline{G}(\mathbb{A})^\gamma) \int_{\overline{G}(\mathbb{A})^\gamma\backslash\overline{G}(\mathbb{A})} \varphi(x^{-1}\gamma x)(-\log h(wx))\, dx + (T)(\text{constant}).$$

Without loss of generality, we can take $\gamma = \left(\begin{smallmatrix} q & 0 \\ 0 & 1 \end{smallmatrix}\right)$ with $q \in \mathbb{Q}^\times$. Then $\overline{G}(\mathbb{A})^\gamma = \overline{M}(\mathbb{A})$ and $\overline{G}(\mathbb{Q})^\gamma = \overline{M}(\mathbb{Q})$, so that $\overline{M}(\mathbb{Q})\backslash\overline{M}(\mathbb{A}) \cong \mathbb{Q}^\times\backslash(\mathbb{A}^\times)^1$. Since $\left(\begin{smallmatrix} q & 0 \\ 0 & 1 \end{smallmatrix}\right)$ and $\left(\begin{smallmatrix} 1 & 0 \\ 0 & q \end{smallmatrix}\right)$ lie in the same conjugacy class when projected to $\overline{G}(\mathbb{A})$, indexing the γ's by q counts each class twice. Thus the part of the above expression not involving T simplifies to

$$= \tfrac{1}{2}\operatorname{vol}(\mathbb{Q}^\times\backslash(\mathbb{A}^\times)^1) \sum_{\substack{q\in\mathbb{Q}^\times, \\ q\neq 1}} \int_{K\times N} \varphi\left(k^{-1}n^{-1}\left(\begin{smallmatrix} q & 0 \\ 0 & 1 \end{smallmatrix}\right) nk\right)(-\log h(wnk))\, dn\, dk.$$

The term with $\gamma = 1$ is just $\varphi(1)$, and the integral is $\operatorname{vol}(\overline{G}(\mathbb{Q})\backslash\overline{G}(\mathbb{A}))\varphi(1)$. There is also a contribution from the terms with γ unipotent but not the identity; this result has a constant term and a T term, but we shall not write these terms out.

This much deals with the integral of $\Lambda_2^T K(x,x)$. Next we consider the integral of $\Lambda_2^T K_{\text{cont}}(x,x)$. Referring to (7.4), we see that we should compute the inner product of a truncated Eisenstein series with an untruncated Eisenstein series—or, what comes to the same thing, of two truncated Eisenstein series.

Proposition 7.13. *For f_1 and f_2 in $W(\eta,\tau)$,*

$$\begin{aligned}\langle \Lambda^T E(\,\cdot\,, f_1, s), \Lambda^T E(\,\cdot\,, f_2, -\bar{s})\rangle &= 4\langle f_1, f_2\rangle T + 2\langle M(-s)M'(s)f_1, f_2\rangle \\ &\quad + \frac{1}{s}\{\langle f_1, M(-\bar{s})f_2\rangle e^{sT} - \langle M(s)f_1, f_2\rangle e^{-sT}\}.\end{aligned}$$

REMARK. The proof will show the importance of the particular form of Arthur's truncation operator.

SKETCH OF PROOF. For $\operatorname{Re} s > 1$, we start from the identities $E(g,f,s) = \sum_\gamma f_s(\gamma g)$ and $E_P(\,\cdot\,, f, s) = f_s + (M(s)f)_{-s}$, the latter given by Proposition 6.8. Then we have

$$\begin{aligned}\Lambda^T E(g,f,s) &= E(g,f,s) - \sum_\gamma E_P(\gamma g, f, s) I_T(h(\gamma g)) \\ &= \sum_\gamma f_s(\gamma g)(1 - I_T(h(\gamma g))) - \sum_\gamma (M(s)f)_{-s}(\gamma g) I_T(h(\gamma g)).\end{aligned}$$

Let $\operatorname{Re} s_1 > \operatorname{Re} s_2 > 1$. Substituting from above and writing f and f' in place of f_1 and f_2 to simplify the notation, we obtain

$$\begin{aligned}
&\langle \Lambda^T E(\,\cdot\,, f, s_1), E(\,\cdot\,, f', \bar{s}_2)\rangle_{L^2(\overline{G}(\mathbb{Q})\backslash\overline{G}(\mathbb{A}))} \\
&\quad = \int_{\overline{G}(\mathbb{Q})\backslash\overline{G}(\mathbb{A})} \sum_{\gamma\in P(\mathbb{Q})\backslash G(\mathbb{Q})} \big(f_{s_1}(\gamma g)(1 - I_T(h(\gamma g)))(M(s_1)f)_{-s_1}(\gamma g) I_T(h(\gamma g))\big) \\
&\qquad\qquad\qquad\qquad \times \overline{E(g, f', \bar{s}_2)}\, dg \\
&\quad = \int_{\overline{P}(\mathbb{Q})\backslash\overline{G}(\mathbb{A})} \big(f_{s_1}(g)(1 - I_T(h(g))) - (M(s_1)f)_{-s_1}(g) I_T(h(g))\big)\overline{E(g, f', \bar{s}_2)}\, dg \\
&\quad = \int_{\overline{M}(\mathbb{Q})N(\mathbb{A})\backslash\overline{G}(\mathbb{A})} \big(f_{s_1}(g)(1-I_T(h(g))) - (M(s_1)f)_{-s_1}(g) I_T(h(g))\big)\overline{E_P(g, f', \bar{s}_2)}\, dg \\
&\qquad\qquad\qquad\qquad\qquad\qquad\qquad\qquad\qquad\qquad \text{by Lemma 6.4} \\
&\quad = \int_{\overline{M}(\mathbb{Q})N(\mathbb{A})\backslash\overline{G}(\mathbb{A})} \big(f_{s_1}(g)(1 - I_T(h(g))) - (M(s_1)f)_{-s_1}(g) I_T(h(g))\big) \\
&\qquad\qquad\qquad \times \overline{\big(f'_{\bar{s}_2}(g) + (M(\bar{s}_2)f')_{-\bar{s}_2}(g)\big)}\, dg.
\end{aligned}$$

Now we substitute for g, reducing each function by its transformation rules to a function on $(\mathbb{A}^\times)^1 \times K$. The set of integration reduces to $A_\infty \times ((\mathbb{A}^\times)^1 \times K)$. The A_∞ integration can be done explicitly, and the $(\mathbb{A}^\times)^1 \times K$ integration gives inner products in the Hilbert space H. The result of this computation, initially valid for $\operatorname{Re} s_1 > \operatorname{Re} s_2 > 1$, extends by analytic continuation to be valid for all s_1 and s_2 where there is no singularity. We then put $s_2 = s$ and $s_1 = s + h$. Taking the limit as h tends to 0, we obtain the formula of the proposition.

We return to (7.4). Interchanging the order of integration yields

$$\begin{aligned}
&\int_{\overline{G}(\mathbb{Q})\backslash\overline{G}(\mathbb{A})} \Lambda_2^T K_{\mathrm{cont}}(x, x)\, dx \\
&\quad = \frac{1}{8\pi}\sum_{\alpha,\beta}\int_{-\infty}^{\infty} \langle P^{\omega,it}(\varphi) f_\beta, f_\alpha\rangle \Big[\int_{\overline{G}(\mathbb{Q})\backslash\overline{G}(\mathbb{A})} E(x, f_\alpha, it)\overline{\Lambda^T E(x, f_\beta, it)}\, dx\Big]\, dt.
\end{aligned}$$

The Hermitian property of Λ^T in Proposition 7.6 extends to this integral, and we can substitute from Proposition 7.13. Easy computation gives

$$\begin{aligned}
&\int_{\overline{G}(\mathbb{Q})\backslash\overline{G}(\mathbb{A})} \Lambda_2^T K_{\mathrm{cont}}(x, x)\, dx \\
&\qquad = \frac{1}{4\pi}\int_{-\infty}^{\infty} \operatorname{Tr}(M(-it)M'(it)P^{\omega,it}(\varphi))\, dt - \frac{1}{4}\operatorname{Tr}(M(0)\pi_0(\varphi)) \\
&\qquad\quad + (T)(\text{constant}) + (\text{term tending to 0 as } T \to \infty).
\end{aligned}$$

Finally the integral of $\Lambda_2^T K_{\mathrm{res}}(x, x)$ is just

$$\int_{\overline{G}(\mathbb{Q})\backslash\overline{G}(\mathbb{A})} \Lambda_2^T K_{\mathrm{res}}(x, x)\, dx \to \int_{\overline{G}(\mathbb{Q})\backslash\overline{G}(\mathbb{A})} K_{\mathrm{res}}(x, x)\, dx = \sum_{\chi^2=\omega}\int_{\overline{G}(\mathbb{A})} \varphi(x)\chi(\det x)\, dx.$$

If we substitute all these results into (7.12), we obtain an equality for all T. Some terms have a coefficient T, and these all cancel (but not in an obvious way). The other terms tend to a finite limit as T tends to ∞. In the limit as T tends to ∞, we obtain the adelic form of the trace formula, as follows.

Theorem 7.14. *If φ is in $C^\infty_{\mathrm{com}}(G(\mathbb{A}),\omega^{-1})$, then*

$$\mathrm{Tr}(P_{\mathrm{cusp}}R_\omega(\varphi)P_{\mathrm{cusp}})$$

$$\text{(i)} \qquad = \mathrm{vol}(\overline{G}(\mathbb{Q})\backslash\overline{G}(\mathbb{A}))\varphi(1)$$

$$\text{(ii)} \qquad + \sum_{\mathfrak{o}_\gamma \text{ elliptic}} \mathrm{vol}(\overline{G}(\mathbb{Q})^\gamma\backslash\overline{G}(\mathbb{A})^\gamma) \int_{\overline{G}(\mathbb{A})^\gamma\backslash\overline{G}(\mathbb{A})} \varphi(x^{-1}\gamma x)\,dx$$

$$\text{(iii)} \qquad + \text{f.p.} \int_{\overline{N}(\mathbb{A})\backslash\overline{G}(\mathbb{A})} \varphi\Big(x^{-1}\left(\begin{smallmatrix}1&1\\0&1\end{smallmatrix}\right)x\Big)\,dx$$

$$\text{(iv)} \qquad + \tfrac{1}{2}\mathrm{vol}(\mathbb{Q}^\times\backslash(\mathbb{A}^\times)^1)\sum_{\substack{q\in\mathbb{Q}^\times,\\ q\neq 1}} \int_{K\times N(\mathbb{A})} \varphi\left(k^{-1}n^{-1}\left(\begin{smallmatrix}q&0\\0&1\end{smallmatrix}\right)nk\right)(-\log h(wnk))\,dn\,dk$$

$$\text{(v)} \qquad + \frac{1}{4\pi}\int_{-\infty}^{\infty} \mathrm{Tr}(M(-it)M'(it)P^{\omega,it}(\varphi))\,dt$$

$$\text{(vi)} \qquad - \tfrac{1}{4}\mathrm{Tr}(M(0)P^{\omega,0}(\varphi))$$

$$\text{(vii)} \qquad - \sum_{\chi^2=\omega}\int_{\overline{G}(\mathbb{A})} \varphi(x)\chi(\det x)\,dx,$$

where the f.p. *term is computed as the value at $s=1$ of*

$$\left\{\int_{\mathbb{A}^\times\times K} \varphi\Big(k^{-1}\left(\begin{smallmatrix}1&a\\0&1\end{smallmatrix}\right)k\Big)|a|^s_{\mathbb{A}}\,d^\times a\,dk - (\text{principal part at } s=1)\right\}$$

when Haar measures are normalized as in (6.7) *and the remarks following Lemma* 6.7.

On the right side of the trace formula above, the terms arise as follows. The first four come from $K(x,x)$, the next two come from $K_{\mathrm{cont}}(x,x)$, and the last one comes from $K_{\mathrm{res}}(x,x)$. The first four we may regard as **geometric terms**, and the others are **spectral terms**. Of the first four, (i) is from $\gamma=1$, (ii) is from elliptic γ, (iii) is from nontrivial unipotent γ, and (iv) is from hyperbolic regular γ.

There is an important special case in which the formula simplifies considerably.

Corollary 7.15. *Suppose that $\varphi\in C^\infty_{\mathrm{com}}(G(\mathbb{A}),\omega^{-1})$ decomposes into a product $\varphi(g)=\prod_v\varphi_v(g_v)$. If there are two places v such that*

$$\int_{M(\mathbb{Q}_v)\backslash G(\mathbb{Q}_v)} \varphi_v\left(x^{-1}\left(\begin{smallmatrix}\alpha&0\\0&\beta\end{smallmatrix}\right)x\right)\,dx = 0$$

whenever α and β are distinct members of $\mathbb{Q}_v$, then terms (iii) *through* (vi) *vanish in the trace formula, so that*

$$\begin{aligned}\mathrm{Tr}(P_{\mathrm{cusp}}R_\omega(\varphi)P_{\mathrm{cusp}}) &= \mathrm{vol}(\overline{G}(\mathbb{Q})\backslash\overline{G}(\mathbb{A}))\varphi(1)\\ &\quad + \sum_{\mathfrak{o}_\gamma \text{ elliptic}} \mathrm{vol}(\overline{G}(\mathbb{Q})^\gamma\backslash\overline{G}(\mathbb{A})^\gamma)\int_{\overline{G}(\mathbb{A})^\gamma\backslash\overline{G}(\mathbb{A})}\varphi(x^{-1}\gamma x)\,dx\\ &\quad - \sum_{\chi^2=\omega}\int_{\overline{G}(\mathbb{A})}\varphi(x)\chi(\det x)\,dx.\end{aligned}$$

REFERENCE. [Gb-Ja, §7].

PROOF OF VANISHING OF (iv). Without loss of generality, we may take Haar measure on K and $N(\mathbb{A})$ to be products of Haar measures from each place.

Let v_1 and v_2 be the places in question, let $\gamma = \begin{pmatrix} q & 0 \\ 0 & 1 \end{pmatrix}$, and write $k = \prod_v k_v$ and $n = \prod_v n_v$. Lemma 6.6 shows that $h(wnk)$ is a product $\prod_v h(wn_v k_v)$. Then

$$\begin{aligned} &\int_{K\times N(\mathbb{A})} \varphi(k^{-1}n^{-1}\gamma nk)\log h(wnk)\,dn\,dk \\ &\quad= \int_{K\times N(\mathbb{A})} \prod_v \Big(\varphi_v(k_v^{-1}n_v^{-1}\gamma n_v k_v)\Big)\Big(\sum_v \log h(wn_v k_v)\Big)\prod_v dn_v\,dk_v \\ &\quad= \sum_u \Big(\prod_{v\neq u}\int_{K_v\times N(\mathbb{Q}_v)} \varphi_v(k_v^{-1}n_v^{-1}\gamma n_v k_v)\,dn_v\,dk_v\Big) \\ &\qquad\times\Big(\int_{K_u\times N(\mathbb{Q}_u)} \varphi_u(k_u^{-1}n_u^{-1}\gamma n_u k_u)\log h(wn_u k_u)\,dn_u\,dk_u\Big). \end{aligned}$$

Consider the uth term of the sum on the right side. In the product over $v \neq u$, either v_1 or v_2 must be one of the v's, and then the corresponding factor is 0 because of the hypothesis. Hence the uth term is 0, and this happens for each u.

References

[Ar1] Arthur, J., The Selberg trace formula for groups of F-rank one, *Annals of Math.* **100** (1974), 326–385.

[Ar2] Arthur, J., A trace formula for reductive groups I: terms associated to classes in $G(\mathbf{Q})$, *Duke Math. J.* **45** (1978), 911–952.

[Ar3] Arthur, J., A trace formula for reductive groups II: applications of a truncation operator, *Compositio Math.* **40** (1980), 87–121.

[Ar4] Arthur, J., The trace formula in invariant form, *Annals of Math.* **114** (1981), 1–74.

[Ar5] Arthur, J., On the inner product of truncated Eisenstein series, *Duke Math. J.* **49** (1982), 35–70.

[Bo] Borel, A., Some finiteness properties of adele groups over number fields, *Publ. Math. I.H.E.S.* **16** (1963), 5–30.

[Bv-Sh] Borevich, Z. I., and I. R. Shafarevich, *Number Theory*, Academic Press, New York, 1966.

[Di-Ma] Dixmier, J., and P. Malliavin, Factorisations de fonctions et de vecteurs indéfiniment différentiables, *Bull. des Sci. Math.* **102** (1978), 305–330.

[Do] Donley, R. W., Irreducible representations of SL(2,R), *these* PROCEEDINGS, pp. 51–59.

[Du-La] Duflo, M., and J.-P. Labesse, Sur la formule des traces de Selberg, *Annales Sci. École Norm. Sup.* **4** (1971), 193–284.

[Ef] Efrat, I. Y., *The Selberg Trace Formula for* $\mathrm{PSL}(2,\mathbb{R})^n$, Memoirs Amer. Math. Soc., vol. 65, no. 359, American Mathematical Society, Providence, 1987.

[Gb1] Gelbart, S. S., *Automorphic Forms on Adele Groups*, Princeton University Press, Princeton, NJ, 1975.

[Gb2] Gelbart, S., *Lectures on the Arthur-Selberg Trace Formula*, American Mathematical Society, Providence, RI, 1996.

[Gb-Ja] Gelbart, S., and H. Jacquet, Forms of GL(2) from the analytic point of view, *Automorphic Forms, Representations, and L-Functions*, Proc. Symp. Pure Math., vol. 33, Part I, American Mathematical Society, Providence, 1979, pp. 213–251.

[Gf-Gr-P] Gelfand, I. M., M. I. Graev, and I. I. Pitateskii-Shapiro, *Representation Theory and Automorphic Functions*, W. B. Saunders, Philadelphia, 1969.

[Go1] Godement, R., Decomposition of $L^2(G/\Gamma)$, *Algebraic Groups and Discontinuous Subgroups*, Proc. Symp. Pure Math., vol. 9, American Mathematical Society, Providence, 1966, pp. 211–224.

[Go2] Godement, R., The spectral decomposition of cusp-forms, *Algebraic Groups and Discontinuous Subgroups*, Proc. Symp. Pure Math., vol. 9, American Mathematical Society, Providence, 1966, pp. 225–234.

[HC] Harish-Chandra, *Automorphic Forms on Semisimple Lie Groups*, Lecture Notes in Mathematics, vol. 62, Springer-Verlag, Berlin, 1968.

[He1] Hejhal, D. A., *The Selberg Trace Formula for* PSL(2, $\mathbb{R}$)*, vol. 1*, Lecture Notes in Mathematics, vol. 548, Springer-Verlag, Berlin, 1976.

[He2] Hejhal, D. A., *The Selberg Trace Formula for* PSL(2, $\mathbb{R}$)*, vol. 2*, Lecture Notes in Mathematics, vol. 1001, Springer-Verlag, Berlin, 1983.

[Ja] Jacquet, H., Note on the analytic continuation of Eisenstein series: An appendix to the previous paper, *these* PROCEEDINGS, pp. 407–412.

[Ja-Lgl] Jacquet, H., and R. P. Langlands, *Automorphic Forms on* GL(2), Lecture Notes in Mathematics, vol. 114, Springer-Verlag, Berlin, 1970.

[Kn1] A. W. Knapp, *Representation Theory of Semisimple Groups: An Overview Based on Examples*, Princeton University Press, Princeton, N.J., 1986.

[Kn2] Knapp, A. W., Introduction to the Langlands program, *these* PROCEEDINGS, pp. 245–302.

[Kn-Ro] Knapp, A. W., and J. D. Rogawski, Applications of the trace formula, *these* PROCEEDINGS, pp. 413–431.

[Lab] Labesse, J.-P., The present state of the trace formula, *Automorphic Forms, Shimura Varieties, and L-Functions*, Vol. I, Academic Press, Boston, 1990, pp. 211–226.

[Lan1] Lang, S., $SL_2(R)$, Addison-Wesley, Reading, Mass., 1975; Springer-Verlag, New York, 1985.

[Lan2] Lang, S., *Algebraic Number Theory*, Springer-Verlag, New York, 1986.

[Lgl1] Langlands, R. P., Eisenstein series, *Algebraic Groups and Discontinuous Subgroups*, Proc. Symp. Pure Math., vol. 9, American Mathematical Society, Providence, 1966, pp. 235–252.

[Lgl2] Langlands, R. P., *On the Functional Equations Satisfied by Eisenstein Series*, Lecture Notes in Mathematics, vol. 544, Springer-Verlag, Berlin, 1976.

[Mi] Miyake, T., *Modular Forms*, Springer-Verlag, Berlin, 1989.

[Mo] Moeglin, C., Representations of GL(n) over the real field, *these* PROCEEDINGS, pp. 157–166.

[Mo-Wa] Moeglin, C., and J.-L. Waldspurger, *Décomposition Spectrale et Séries d'Eisenstein*, Birkhäuser, Basel, 1994.

[Ro] Rogawski, J. D., Modular forms, the Ramanujan conjecture and the Jacquet-Langlands correspondence, *Discrete Groups, Expanding Graphs and Invariant Measures*, by A. Lubotzky, Birkhäuser, Basel, 1994, pp. 135–176.

[Se1] Selberg, A., Harmonic analysis and discontinuous groups in weakly symmetric Riemannian spaces with applications to Dirichlet series, *J. Indian Math. Soc.* **20** (1956), 47–87.

[Se2] Selberg, A., Discontinuous groups and harmonic analysis, *Proceedings of the International Congress of Mathematicians 1962*, Institut Mittag-Leffler, Djursholm, Sweden, 1963, pp. 177–189.

[Ti] Titchmarsh, E. C., *The Theory of the Riemann Zeta-Function*, Oxford University Press, Oxford, 1951.

[We] Weil, A., *Basic Number Theory*, Springer-Verlag, New York, 1973.

DEPARTMENT OF MATHEMATICS, STATE UNIVERSITY OF NEW YORK, STONY BROOK, NEW YORK 11794, U.S.A.

E-mail address: `aknapp@ccmail.sunysb.edu`

Proceedings of Symposia in Pure Mathematics
Volume **61** (1997), pp. 407–412

Note on the Analytic Continuation of Eisenstein Series: An Appendix to the Previous Paper

Hervé Jacquet

1. Setting and Theorem

The previous paper [Kn] omitted any proof of the theorem that any Eisenstein series has an analytic continuation to a global meromorphic function. In a classical setting including $SL_2(\mathbb{Z})\backslash SL_2(\mathbb{R})$, the method of proof is due to Selberg. In the case of $SL_2(\mathbb{Z})\backslash SL_2(\mathbb{R})$, one can use the Poisson Summation Formula to obtain a very simple proof (see [Go]). However, this proof does not generalize. Langlands [Lgl] generalized Selberg's method to handle quotients $\Gamma\backslash G$, where Γ is an arithmetic subgroup of a reductive group G. The case of $SL_2(\mathbb{Z})\backslash SL_2(\mathbb{R})$ is treated separately in an appendix. Successive simplifications of the original argument of Selberg can be found in the work of Hejhal [He1] and [He2] and the thesis of Efrat [Ef]. I understand that these simplifications are also based, at least in part, on ideas of Selberg.

Recast in an adelic setting, the general Langlands proof appears in [Mo-Wa], which credits [Co]. The present paper is intended to facilitate the reading of [Mo-Wa] and to discuss the special case that is needed in [Kn]. I stress that the proof sketched in this note is based on Selberg's ideas. I am very grateful to Hejhal for explaining these ideas to me. Moreover, the thesis of Efrat was very useful to me. The setting for the special case is that the base field is $\mathbb{Q}$, $\mathbb{A}$ is the adeles of $\mathbb{Q}$, G is GL_2, and Z is the center of G. The paper [Kn] considers functions on $G(\mathbb{A})$ transforming by a character ω under $Z(\mathbb{A})$. Leaving aside some definitions that we shall review below, we can restate Theorem 7.1 of [Kn] as follows:

Theorem. *If f is in a subspace $W(\eta,\tau)$ of H, then the Eisenstein series $s \mapsto E(g,f,s)$, initially given as an analytic function for $\Re s > 1$, extends to be meromorphic in $\mathbb{C}$. Its constant term $E_P(g,f,s)$ is given by the analytic continuation of the constant term for $\Re s > 1$, and $E(g,f,s)$ has the same poles as $E_P(g,f,s)$. Also $E(g,f,s)$ is at most of polynomial growth in $\Im s$ in any vertical strip $0 \leq \operatorname{Re} s \leq \sigma$.*

The proof uses the theory of resolvents of compact Hermitian operators. With $G = GL_2$ and the base field equal to $\mathbb{Q}$, we discuss only the simplest kind of f, which has $\omega = 1$ and is right invariant under the maximal compact subgroup

---

1991 *Mathematics Subject Classification.* Primary 22E45, 22E55.

$K = O_2(\mathbb{R}) \times \prod_p G(\mathbb{Z}_p)$ of $G(\mathbb{A})$. It is in this case that the ideas are clearest. Then f will be uniquely determined up to a constant, and η and τ will both be trivial.

Put $\overline{G} = Z\backslash G$. We denote by L^2 and L^{loc} the spaces of square integrable and locally square integrable functions on $G(\mathbb{Q})\backslash G(\mathbb{A})$ that are invariant under $Z(\mathbb{A})$. Whenever convenient, we may identify these spaces with spaces of functions on $\overline{G}(\mathbb{Q})\backslash\overline{G}(\mathbb{A})$.

Following the notation of [Kn], we define a function $h(\,\cdot\,)$ on $G(\mathbb{A})$ by

$$h(g) = \left|\tfrac{a}{b}\right|_{\mathbb{A}}^{1/2}.$$

if

$$g = \begin{pmatrix} a & x \\ 0 & b \end{pmatrix} k.$$

The function f that we shall study is

$$f(g,s) = h(g)^{1+s}. \tag{1}$$

Let P be the upper triangular group. The function f satisfies the transformation law

$$f(pgk) = h(p)^{1+s} f(g,s) \qquad \text{for } p \in P(\mathbb{A}) \text{ and } k \in K. \tag{2}$$

Since $G(\mathbb{A}) = P(\mathbb{A})K$, f is uniquely determined up to a constant by the transformation law (2).

Then, for $\Re s > 1$, the **Eisenstein series** of our function f is defined by the convergent series:

$$E(g,s) = \sum_{\gamma \in P(\mathbb{Q})\backslash G(\mathbb{Q})} f(\gamma g, s),$$

where we have suppressed f in the notation on the left side. We will often use the notation $E(s)$ for the function of s with values in the space of functions on $G(\mathbb{A})$ defined by

$$E(s)(g) = E(g,s).$$

As in Proposition 6.8 of [Kn], the constant term E_P of the Eisenstein series is equal to

$$E_P(g,s) = f(g,s) + m(s)f(g,-s), \tag{3}$$

where $m(s)$ is a certain scalar function holomorphic for $\Re s > 1$. (It is the operator $M(s)$ in that proposition in the special case that $\eta = 1$ and $\tau = 1$.)

Our goal is to show that the functions $E(g,s)$ and $m(s)$ extend to global meromorphic functions of s satisfying the following functional equations:

$$m(s)E(g,s) = E(g,-s) \qquad \text{and} \qquad m(s)m(-s) = 1.$$

The analytic continuation of $m(s)$ and the second functional equation are actually special cases of the result quoted as Theorem 6.10 of [Kn], but we shall see that these conclusions follow also from the argument we give.

2. Analytic Continuation

Suppose that u is a smooth function of compact support on $G(\mathbb{A})$ bi-invariant under K. In fact, we will take for u the product of a smooth function of compact support u_∞ on $G(\mathbb{R})$ and the characteristic functions u_p of the maximal compact subgroups $GL_2(\mathbb{Z}_p)$. We denote by $R(u)$ the convolution of a function ϕ and the function u on the right:

$$R(u)\phi(g) = \int_{G(\mathbb{A})} \phi(gx)u(x)dx.$$

Applying $R(u)$ to f as in (1) and using (2), we see that

$$R(u)f(s) = \widehat{u}(s)f(s), \tag{4}$$

where $\widehat{u}(s) = \int_{G(\mathbb{A})} h(x)^{1+s}u(x)\,dx$. The function $\widehat{u}(s)$ is certainly entire, and it has the properties

$$\widehat{u}(s) = \widehat{u}(-s) \tag{5a}$$

and

$$\widehat{u}(s) \text{ is not constant unless it is } 0. \tag{5b}$$

To see (5) we decompose $P = TN$ with T diagonal and N upper triangular. Decomposing Haar measure and using the invariance under K, we obtain

$$\widehat{u}(s) = \int_{T(\mathbb{R})} h(t)^{-1+s}\Big[\int_{N(\mathbb{R})} u_\infty(tn)\,dn\Big]dt.$$

Now the function

$$t \mapsto h(t)^{-1}\int_{N(\mathbb{R})} u_\infty(tn)\,dn$$

is Weyl group invariant (see [Ca, p. 147] or [La]) and has compact support on $T(\mathbb{R})$. Moreover, it is an arbitrary function with these properties. Otherwise said, $\widehat{u}$ has the form

$$\widehat{u}(s) = \int_0^\infty x^s v(s)\,dx, \tag{6}$$

where v is smooth of compact support on $\mathbb{R}^+$ and verifies $v(x) = v(x^{-1})$. Moreover, $\widehat{u}$ is an arbitrary function with these properties.

Let us recall the definition and properties of the Arthur **truncation operator** given in [Ar] and discussed in this volume in [Kn, §7]. If ϕ is a function on $G(\mathbb{Q})\backslash G(\mathbb{A})$ and $T > 0$, we denote by $\Lambda^T\phi$ the function defined by:

$$\Lambda^T\phi(g) = \phi(g) - \sum_{\gamma\in P(\mathbb{Q})\backslash G(\mathbb{Q})} \phi_P(\gamma g)I_T(h(\gamma g)),$$

where I_T is the characteristic function of the set $[T, +\infty)$. The main properties of Λ^T are

$$\Lambda^T\Lambda^T = \Lambda^T. \tag{7a}$$

$$\Lambda^T R(u)\Lambda^T \text{ is compact on } L^2. \tag{7b}$$

(See the lemma in IV.2.6 of [Mo-Wa] for (7b); also the proof of Theorem 1.6 in [Kn] can readily be modified to give the assertion.)

To write down the effect of Λ^T on the Eisenstein series, we introduce an auxiliary series $E^T(g, s)$ defined by:

$$E^T(g, s) = \sum_{\gamma\in P(\mathbb{Q})\backslash G(\mathbb{Q})} f(\gamma g, s)I_T(h(\gamma g)).$$

It is proved in [Kn, §7] that the series has at most one nonzero term, and hence it is entire in s. Moreover, we readily check that for every s there is g such that $E^T(g,s) \neq 0$. Finally $E^T(g,s)$ is in L^2 for $\Re s << 0$. Then when we apply the truncation operator to $E(g,s)$, (1) and (3) immediately give

$$\Lambda^T E(s) = E(s) - E^T(s) - m(s)E^T(-s). \tag{8}$$

Summing left translates of (4) over $\gamma \in P(\mathbb{Q})\backslash G(\mathbb{Q})$ shows that the Eisenstein series is a solution of the following homogeneous eigenvalue problem:

$$R(u)E(s) = \widehat{u}(s)E(s) \qquad \text{for } \Re s > 0. \tag{9}$$

For $\Re s > 0$ the truncated Eisenstein series takes its values in L^2 and is the solution of an inhomogeneous eigenvalue problem. To formulate it we introduce the function

$$F^T(s) = \Lambda^T\big(R(u)E^T(s)\big)$$

and observe that $s \mapsto F^T(s)$ is an entire function of s with values in the space L^2. Direct calculation gives

$$\Lambda^T E^T(s) = 0. \tag{10}$$

Therefore

$$\begin{aligned}
&\big(\Lambda^T R(u)\Lambda^T - \widehat{u}(s)\big)\Lambda^T E(s) \\
&\quad = \big(\Lambda^T R(u) - \widehat{u}(s)\Lambda^T\big)\Lambda^T E(s) && \text{by (7a)} \\
&\quad = \big(\Lambda^T R(u) - \widehat{u}(s)\Lambda^T\big)\big(E(s) - E^T(s) - m(s)E^T(-s)\big) && \text{by (8)} \\
&\quad = -\Lambda^T R(u)E^T(s) - m(s)\Lambda^T R(u)E^T(-s) && \text{by (9) and (10),}
\end{aligned}$$

and the nonhomogeneous eigenvalue problem is

$$\big(\Lambda^T R(u)\Lambda^T - \widehat{u}(s)\big)\big(\Lambda^T E(s)\big) = -F^T(s) - m(s)F^T(-s).$$

The adjoint of the operator $\Lambda^T R(u)\Lambda^T$ is $\Lambda^T R(u^*)\Lambda^T$, where $u^*(x) = \overline{u(x^{-1})}$. Now let us assume that $u \neq 0$ and $u = u^*$. By (7b), $\Lambda^T R(u)\Lambda^T$ is compact Hermitian on L^2. We shall use the following lemma about resolvents.

Lemma. *Let L be a compact Hermitian operator, and let $\sigma(L)$ be the union of its spectrum and* 0. *Then the resolvent $(\lambda - L)^{-1}$, defined as a holomorphic function on $\mathbb{C} - \sigma(L)$, extends to a meromorphic function on $\mathbb{C}^\times$.*

Assuming that $\widehat{u}$ is not identically 0, let $V(u)$ be the discrete subset of $s \in \mathbb{C}$ where $\widehat{u}(s) = 0$. From the lemma (see [Mo-Wa] for details) it follows that there exists a unique meromorphic function $e^T(g,s)$, with domain the complement of $V(u)$, such that

$$\Lambda^T R(u)\Lambda^T e^T(s) - \widehat{u}(s)e^T(s) = -F^T(s). \tag{11}$$

The only possible poles of $e^T(g,s)$ are at points s where $\widehat{u}(s)$ is in the spectrum of $\Lambda^T R(u)\Lambda^T$.

From (7a) and the uniqueness of a solution of (11), it follows that

$$\Lambda^T e^T(s) = e^T(s). \tag{12}$$

Likewise, for $\Re s > 0$, from (5a) and (7a) and (11), the uniqueness of a solution of (11) when the right side is $-F^T(s) - m(s)F^T(s)$ implies that

$$\Lambda^T E(s) = e^T(s) + m(s)e^T(-s).$$

Using (8), we can rewrite this last relation in the form

$$E(s) = \widetilde{E}(s) + m(s)\widetilde{E}(-s), \tag{13}$$

where we have set

$$\widetilde{E}(s) = E^T(s) + e^T(s). \tag{14}$$

By construction e^T is meromorphic in the complement of $V(u)$ with values in L^2. On the other hand, the auxiliary series $E^T(s)$ is entire is s and locally bounded in g, hence takes its values in L^{loc}. Thus by (13) and (14), if we know the analytic continuation of $m(s)$ we obtain the analytic continuation of $E(s)$, at least as a function with values in L^{loc}. From this one can derive the analytic continuation of $E(s)$ as a smooth function. Moreover, for $\Re s << 0$, the auxiliary series $E^T(s)$ is square integrable. We conclude that on the intersection of the complement of $V(u)$ and a left half-plane, the function $\widetilde{E}(s)$ is a meromorphic function with values in L^2.

To obtain the analytic continuation of $m(s)$, we recall from (9) that E is a solution of the homogeneous eigenvalue problem

$$R(u)E(s) = \widehat{u}(s)E(s).$$

Using the previous relations we find that

$$\big(R(u)\widetilde{E}(s) - \widehat{u}(s)\widetilde{E}(s)\big) + m(s)\big(R(u)\widetilde{E}(-s) - \widehat{u}(s)\widetilde{E}(-s)\big) = 0.$$

This equation can be used to show that $m(s)$ has an analytic continuation to the complement of $V(u)$ provided

$$R(u)\widetilde{E}(s) - \widehat{u}(s)\widetilde{E}(s)$$

is not identically zero.

To see that it is not identically zero, we argue by contradiction. Then at any point where $\widetilde{E}(s)$ is defined and square integrable the eigenvalue $\widehat{u}(s)$ must be real. Thus $\widehat{u}(s)$ must be real-valued on some open set, which is impossible since $\widehat{u}(s)$ is not constant. We conclude that $m(s)$ is meromorphic in the complement of $V(u)$. It follows that $E(s)$ has a meromorphic extension to the complement of $V(u)$. By letting u vary, we conclude that both m and E have meromorphic extensions to $\mathbb{C}$.

References

[Ar] J. Arthur, A trace formula for reductive groups II: applications of a truncation operator, *Compositio Math.* **40** (1980), 87–121.

[Ca] Cartier, P., Representations of $\mathfrak{p}$-adic groups: A survey, *Automorphic Forms, Representations, and L-Functions*, Proc. Symp. Pure Math., vol. 33, Part I, American Mathematical Society, Providence, 1979, pp. 111–155.

[Co] Y. Colin de Verdières, Une nouvelle démonstration du prolongement méromorphe des séries d'Eisenstein, *C. R. Acad. Sci. Paris* **293** (1981), 361–363.

[Ef] I. Y. Efrat, *The Selberg Trace Formula for* $\mathrm{PSL}(2,\mathbb{R})^n$, Memoirs Amer. Math. Soc., vol. 65, no. 359, American Mathematical Society, Providence, 1987.

[Go] R. Godement, Decomposition of $L^2(G/\Gamma)$, *Algebraic Groups and Discontinuous Subgroups*, Proc. Symp. Pure Math., vol. 9, American Mathematical Society, Providence, 1966, pp. 211–224.

[He1] D. A. Hejhal, *The Selberg Trace Formula for* $\mathrm{PSL}(2,\mathbb{R})$*, vol. 1*, Lecture Notes in Mathematics, vol. 548, Springer-Verlag, Berlin, 1976.

[He2] D. A. Hejhal, *The Selberg Trace Formula for* $\mathrm{PSL}(2,\mathbb{R})$*, vol. 2*, Lecture Notes in Mathematics, vol. 1001, Springer Verlag, Berlin, 1983.

[Kn] A. W. Knapp, Theoretical aspects of the trace formula for $GL(2)$, *these* PROCEEDINGS, pp. 355–405.

[La] S. Lang, $\mathbf{SL}_2(\mathbf{R})$, Addison-Wesley, Reading, Mass., 1975; second edition, Springer-Verlag, New York, 1985.

[Lgl] R. P. Langlands, *On the Functional Equations Satisfied by Eisenstein Series*, Lecture Notes in Mathematics, vol. 544, Springer-Verlag, Berlin, 1976.

[Mo-Wa] C. Moeglin and J.-L. Waldspurger, *Décomposition Spectrale et Séries d'Eisenstein*, Birkhäuser, Basel, 1994.

DEPARTMENT OF MATHEMATICS, COLUMBIA UNIVERSITY, NEW YORK, NY 10027–4408, U.S.A.
E-mail address: hj@math.columbia.edu

Proceedings of Symposia in Pure Mathematics
Volume **61** (1997), pp. 413–431

Applications of the Trace Formula

A. W. Knapp and J. D. Rogawski

This paper is an introduction to some ways that the trace formula can be applied to proving global functoriality. We rely heavily on the ideas and techniques in [Kn1], [Kn2], and [Ro4] in this volume.

To address functoriality with the help of the trace formula, one compares the trace formulas for two different groups. In particular the trace formula for a single group will not be enough, and immediately the analysis has to be done in an adelic setting. The idea is to show that the trace formulas for two different groups are equal when applied respectively to suitably matched functions. The actual matching is a problem in local harmonic analysis, often quite difficult and as yet not solved in general. The equality of the trace formulas and some global analysis allow one to prove cases of functoriality of automorphic representations. We discuss three examples in the first three sections—the Jacquet-Langlands correspondence, existence of automorphic induction in a special case, and aspects of base change.

A feature of this kind of application is that it is often some variant of the trace formula that has to be used. In our applications we use the actual trace formula in the first example, a trace formula that incorporates an additional operator in the second example, and a "twisted trace formula" in the third example. The paper [Ja] in this volume describes a "relative trace formula" in this context, and [Ar2] describes the need for a "stable trace formula" and a consideration of "endoscopy." See [Ar-Cl], [Bl-Ro], [Lgl-Ra], [Ra], and [Ro2] for some advanced applications.

For completeness we mention that the trace formula has another kind of application that is not closely related to functoriality. In this case one makes an absolute use of the trace formula for a single group. We discuss briefly three examples in the last section—the Eichler-Selberg trace formula, a property of the multiplicative groups of quaternion algebras, and a consequence of the Eichler-Shimura congruence relation. The first and third of these involve GL_2 and are often discussed in the classical setting without adeles. However, we shall use adeles throughout.

The sections of the paper are as follows:

1991 *Mathematics Subject Classification.* Primary 11R39, 11R42, 22E45, 22E55.

We use the following notation. The base field will be a number field F, $\overline{F}$ will be a fixed algebraic closure of F, and $\mathbb{A}$ will denote the adeles of F. If an extension field of F appears, we shall use subscripts on $\mathbb{A}$ to distinguish the adeles of the various fields. In each application, G will denote some particular reductive linear group over F, and Z will denote the center. We fix a character ω of $Z(\mathbb{Q})\backslash Z(\mathbb{A})$ and denote by $L^2(G,\omega)$ the space of functions f on $G(\mathbb{A})$ that are left invariant under $G(\mathbb{Q})$, transform by $f(zg) = \omega(z)f(g)$ under $Z(\mathbb{A})$, and have $|f|$ square integrable on $Z(\mathbb{A})G(\mathbb{Q})\backslash G(\mathbb{A})$. Let R be the right regular representation of $G(\mathbb{A})$ on $L^2(G,\omega)$. If v is a place of F, we write G_v and Z_v for $G(F_v)$ and $Z(F_v)$.

We write $\overline{G} = Z\backslash G$. Let $C^\infty_{\text{com}}(G(\mathbb{A}),\omega^{-1})$ be the space of functions φ on $G(\mathbb{A})$ that transform by $\varphi(zg) = \omega(z)^{-1}\varphi(g)$ for $z \in Z(\mathbb{A})$ and are smooth and have compact support modulo $Z(\mathbb{A})$. For $\varphi \in C^\infty_{\text{com}}(G(\mathbb{A}),\omega^{-1})$ and $f \in L^2(G,\omega)$, we set

$$R(\varphi)f(x) = \int_{\overline{G}(\mathbb{A})} f(xy)\varphi(y)\,dy;$$

the integrand is well defined as a function of $y \in \overline{G}(\mathbb{A})$.

1. Jacquet-Langlands Correspondence

For the Jacquet-Langlands correspondence let $G = GL_2$ over F, and let G' be the multiplicative group of a **quaternion algebra** D over F, i.e., a four-dimensional central division algebra over F. Quaternion algebras and the algebraic group G' are discussed in this volume in [Kn2, §5]. See also [Ro3] and [Vi].

In particular the quotient $Z'(\mathbb{A})G'(\mathbb{Q})\backslash G'(\mathbb{A})$ is compact. Let S be the finite set of places v where $G'_v = G'(F_v)$ is **ramified**, i.e., where $D \otimes_F F_v$ remains a division algebra. It is known that S is not empty and contains an even number of places. At the remaining places we say that G'_v is **split**.

The L groups of G and G' in the sense of [Kn1, §9] are both $GL_2(\mathbb{C}) \times \Gamma$, where $\Gamma = \text{Gal}(\overline{F}/F)$. The case of global functoriality that is addressed by the Jacquet-Langlands correspondence is the identity map ${}^LG' \to {}^LG$.

Let us consider this map at the local level, where an example will be helpful. When $F = \mathbb{Q}$, G'_∞ is either $GL_2(\mathbb{R})$ or the multiplicative group $\mathbb{H}^\times$ of the ordinary quaternions, and in either case the L group is the direct product of $GL_2(\mathbb{C})$ and the two-element Galois group $\text{Gal}(\mathbb{C}/\mathbb{R})$. However, the admissible homomorphisms of the Weil group $W_\mathbb{R}$ into the two L groups are different. In the first case the definition is as in [Kn2, §9], but in the second case, [Kn2, §9] noted that another condition needs to be satisfied. That condition involves "relevance" of parabolic subgroups of ${}^LG'$ and is discussed in [Bo, pp. 32, 40]. In this way the local Langlands correspondence for the quaternion case leads only to the familiar representations of the compact-mod-center group $G'_\infty = SU(2) \times \mathbb{R}^\times_+$, while the local Langlands correspondence for $G_\infty = GL_2(\mathbb{R})$ leads both to discrete series and to principal series. At the local level, functoriality in the quaternion case matches the familiar representations of $SU(2) \times \mathbb{R}^\times_+$ with the discrete series of $GL_2(\mathbb{R})$. See Remark 1 after Theorem 1a below for more detail.

Imagine piecing together local identifications of this kind for each place and obtaining an identification of irreducible admissible representations of $G'(\mathbb{A})$ with some of those of $G(\mathbb{A})$. The assertion of global functoriality is that automorphic representations correspond to automorphic representations.

Let us state precise results as theorems. We begin with the local case. If k is a local field of characteristic 0 and A is a quaternion division algebra over k, then there exist a number field F and a division algebra D over F such that $F_v = k$ and $D(k) = A$. Because of this circumstance, there is no loss of generality in the local case in stating our results for F_v and a ramified G'_v.

In this situation consider regular semisimple elements t'_v and t_v in G'_v and G_v. We write $t'_v \sim t_v$ if t'_v and t_v "have the same eigenvalues," i.e., if the traces and determinants of t'_v as a 2-by-2 matrix as in [Kn2, §5] match those for t_v. The relation $t'_v \sim t_v$ defines a bijection between noncentral conjugacy classes in G'_v and the noncentral semisimple conjugacy classes in G_v that are not hyperbolic, i.e., do not have eigenvalues in F_v.

We can now define the **local Jacquet-Langlands correspondence**. With G'_v ramified, let π'_v and π_v be irreducible admissible representations of G'_v and G_v, respectively, and let $\theta_{\pi'_v}$ and θ_{π_v} be their global characters. We say that π'_v and π_v **correspond**, written $\pi'_v \leftrightarrow \pi_v$, if $\theta_{\pi'_v}(t') = -\theta_{\pi_v}(t)$ for all regular semisimple elements $t' \in G'_v$ and $t \in G_v$ such that $t'_v \sim t_v$.

PRELIMINARY REMARKS.

(1) Let us say that an irreducible admissible representation π_v of G_v is **elliptic** if its character θ_{π_v} does not vanish identically on the elliptic set (an **elliptic** element for this G being one with eigenvalues not in F). It is clear that π_v can correspond to some π'_v only if π_v is elliptic.

(2) If $\pi'_v \leftrightarrow \pi_v$, then the central characters of π'_v and π_v match on the centers $F_v^\times$ of G'_v and G_v.

(3) Since π'_v is finite-dimensional, it is determined by its global character $\theta_{\pi'_v}$. It follows that a given π_v corresponds to at most one π'_v. It is also true, but less obvious, that π'_v corresponds to at most one π_v. This follows from the fact that an elliptic representation of $GL_2(F_v)$ is determined by the restriction of its character to the elliptic set, which in turn follows from orthogonality relations for the global characters on the elliptic set ([Ja-Lgl]).

Theorem 1a. *The local Jacquet-Langlands correspondence is a bijection between the set $\Pi(G'_v)$ of classes of irreducible admissible representations of G'_v and the set $\Pi^2(G_v)$ of classes of discrete series representations of G_v.*

REMARKS.

(1) If $F_v = \mathbb{R}$ and G'_v is ramified, then $G'_v = SU(2) \times \mathbb{R}_+^\times$, while $G_v = GL_2(\mathbb{R})$ is the product of the group $SL_2^\pm(\mathbb{R})$ of determinant ± 1 by $\mathbb{R}_+^\times$. By the unitary trick each irreducible finite-dimensional representation π'_v of $SU(2)$ complexifies to a holomorphic representation of $SL_2(\mathbb{C})$ and then restricts to an irreducible finite-dimensional representation of $SL_2(\mathbb{R})$. To such a finite-dimensional representation V_n of $SL_2(\mathbb{R})$ of dimension n, there are two discrete series representations $\mathcal{D}^+_{n+1}$ and $\mathcal{D}^-_{n+1}$ such that V_n, $\mathcal{D}^+_{n+1}$, and $\mathcal{D}^-_{n+1}$ are the irreducible subquotients of a principal series representation of $SL_2(\mathbb{R})$. The global character of the principal series representation is 0 on the usual compact Cartan subgroup, and thus the character of V_n is the negative of the sum of the characters of $\mathcal{D}^+_{n+1}$ and $\mathcal{D}^-_{n+1}$ on this compact Cartan subgroup. Induction of either of $\mathcal{D}^+_{n+1}$ or $\mathcal{D}^-_{n+1}$ to $SL_2^\pm(\mathbb{R})$ gives a discrete series representation $\mathcal{D}_{n+1}$ whose restriction to $SL_2(\mathbb{R})$ is $\mathcal{D}^+_{n+1} \oplus \mathcal{D}^-_{n+1}$. Then $V_n \mapsto \mathcal{D}_{n+1}$ is the local Jacquet-Langlands correspondence in this case.

(2) Even though the one-dimensional representations of G'_v and G_v in the theorem can be naturally identified, they do not match under the local Jacquet-Langlands correspondence. In fact, the characters of naturally identified one-dimensional representations of G'_v and G_v agree on the elliptic set, instead of being negatives of each other. When v is nonarchimedean and π'_v is one-dimensional for G'_v of the form $\chi \circ \det$, the corresponding representation π_v of G_v is the special representation associated to χ.

(3) The local Jacquet-Langlands correspondence was obtained originally in [Ja-Lgl], using properties of the Weil representation, and an exposition appears in [Gb1]. We shall proceed quite differently, bypassing the Weil representation and deducing the local result Theorem 1a from an intermediate step in the proof of a global result, Theorem 1b below. In doing so, we follow [Gb-Ja] and [Ro3].

(4) Theorem 1a addresses GL_2. An analogous result for GL_3 was obtained by Flath in his thesis, and a generalization to GL_n was obtained in [DKV] and [Ro1]. The Weil representation no longer plays a role.

Let us turn to the global case. If v is a place where G'_v is split, we say that π'_v and π_v correspond, written $\pi'_v \leftrightarrow \pi_v$, if they are equivalent under any fixed isomorphism $G'_v \cong G_v$. This property is independent of the chosen isomorphism since every automorphism of $GL_2(F_v)$ is inner.

We can identify $Z'(F)\backslash Z'(\mathbb{A})$ canonically with $F^\times\backslash\mathbb{A}^\times$, and we can also identify $Z(F)\backslash Z(\mathbb{A})$ canonically with $F^\times\backslash\mathbb{A}^\times$. If the central character ω' for G' corresponds to the central character ω for G under these identifications, let us say that ω' and ω **match**.

Let $\mathcal{A}(G')$ be the set of classes of automorphic representations of $G'(\mathbb{A})$ with a unitary central character, and let $\mathcal{A}_*(G')$ be the set of those that are not one-dimensional. Also let $\mathcal{A}_0^S(G)$ be the set of cuspidal representations $\pi = \bigotimes \pi_v$ of $G(\mathbb{A})$ with a unitary central character and with π_v in the discrete series for every $v \in S$. The **global Jacquet-Langlands correspondence** is the map $\pi' \mapsto \pi$ in the following theorem.

Theorem 1b. *Let $\pi' = \bigotimes \pi'_v$ be in $\mathcal{A}_*(G)$, and let $\pi = \bigotimes \pi_v$ be the irreducible admissible representation of $G(\mathbb{A})$ such that $\pi'_v \leftrightarrow \pi_v$ for all v. Then π is in $\mathcal{A}_0^S(G)$. Furthermore the map $\pi' \mapsto \pi$ is a bijection between $\mathcal{A}_*(G')$ and $\mathcal{A}_0^S(G)$.*

Corollary (of proof). *Each π' in $\mathcal{A}_*(G')$ occurs with multiplicity one.*

This theorem appeared originally in [Ja-Lgl], and that book acknowledges the influence of earlier work by Shimizu; Theorem 1b has its origins also in the work of Eichler. The proof shows that the global Jacquet-Langlands correspondence preserves multiplicities, and we know that multiplicity one holds for G; thus multiplicity one holds for G', as asserted in the corollary. Note that if π' and π correspond as in Theorem 1b, then the definitions force the central character ω' of π' to match the central character ω of π.

The line of proof of Theorems 1a and 1b is to establish a global identity of traces by means of the trace formulas for G' and G and to derive both theorems from this identity. The global identity says that there is a relationship between R' on $L^2(G', \omega')$ and R on the discrete part of $L^2(G, \omega)$, i.e., the sum of the cuspidal part and the residual part, provided ω' and ω match. If $\varphi' \in C^\infty_{\text{com}}(G'(\mathbb{A}), \omega^{-1})$ and $\varphi \in C^\infty_{\text{com}}(G(\mathbb{A}), \omega^{-1})$ are paired suitably (written $\varphi' \leftrightarrow \varphi$), the relationship is that

$$\operatorname{Tr}(R'(\varphi')) = \operatorname{Tr}(P_{\text{cusp}} R(\varphi) P_{\text{cusp}}) + \operatorname{Tr}(P_{\text{res}} R(\varphi) P_{\text{res}}), \tag{1.1}$$

where P_{cusp} and P_{res} are orthogonal projections.

Let us define the notion of "matching orbital integrals." First we consider this notion locally. Recall that $\overline{G}' = Z'\backslash G'$ and $\overline{G} = Z\backslash G$. Fix v, and let Haar measures on $\overline{G}'_v$ and $\overline{G}_v$ be given. We say that φ'_v and φ_v have **matching orbital integrals** if the relevant conditions from the following list are satisfied:

(1) When v is not in S, $\varphi'_v = \varphi_v$ after G'_v has been identified with G_v.

(2) When v is in S and elliptic elements t' and t are given with $t' \sim t$, we can identify the centralizers ${G'_v}^{t'}$ and ${G_v}^t$ of t' and t and then we can compatibly normalize the Haar measures on the quotients ${\overline{G}'_v}^{t'}\backslash\overline{G}'_v$ and ${\overline{G}_v}^t\backslash\overline{G}_v$. In this situation we require

$$\int_{{\overline{G}'_v}^{t'}\backslash\overline{G}'_v} \varphi'_v(x^{-1}t'x)\,dx = \int_{{\overline{G}_v}^{t}\backslash\overline{G}_v} \varphi_v(x^{-1}tx)\,dx \qquad (t' \sim t). \tag{1.2a}$$

(3) When v is in S and a hyperbolic regular element t is given in G_v, then

$$\int_{{\overline{G}_v}^{t}\backslash\overline{G}_v} \varphi_v(x^{-1}tx)\,dx = 0. \tag{1.2b}$$

The local theorem about the existence of matching orbital integrals is that if v is in S and φ'_v is given, then there exists φ_v with matching orbital integrals in the sense of (2) and (3) above. Moreover if φ_v is given satisfying (3), then there exists φ'_v such that (2) holds. A proof of this result for nonarchimedean v may be found in Section 2 of [Ro1]. For archimedean v, see [She].

For the global case we first fix Haar measures. The book [Ja-Lgl] describes a canonical way of defining Haar measure on G'_v once Haar measure on G_v is given, and we use this normalization for every v. Then we say that $\varphi' = \prod_v \varphi'_v$ and $\varphi = \prod_v \varphi_v$ have **matching orbital integrals** (and we write $\varphi' \leftrightarrow \varphi$) if φ'_v and φ_v have matching orbital integrals for every v.

Under the assumption that $\varphi' \leftrightarrow \varphi$, let us see why (1.1) follows. We shall use the trace formulas for G' and G. Let M be the diagonal subgroup of G. From (1.2b) we have

$$\int_{M_v\backslash G_v} \varphi_v\Big(x^{-1}\Big(\begin{smallmatrix}\alpha & 0\\ 0 & \beta\end{smallmatrix}\Big)x\Big)\,dx = 0 \qquad \text{for } v \in S \tag{1.3}$$

whenever α and β are distinct members of F_v. Since S contains at least two places, Corollary 7.15 of [Kn2] says that several of the terms in the trace formula for G vanish. More precisely, under the condition (1.3), Theorem 5.2 and Corollary 7.15 of [Kn2] give

$$\operatorname{Tr}(R'(\varphi')) = \sum_{\mathfrak{o}_{\gamma'}} \operatorname{vol}(\overline{G}'(F)^{\gamma'}\backslash\overline{G}'(\mathbb{A})^{\gamma'}) \int_{\overline{G}'(\mathbb{A})^{\gamma'}\backslash\overline{G}'(\mathbb{A})} \varphi'(x^{-1}\gamma'x)\,dx \tag{1.4a}$$

and

$$\begin{aligned}\operatorname{Tr}(P_{\text{cusp}}R(\varphi)P_{\text{cusp}}) &+ \operatorname{Tr}(P_{\text{res}}R(\varphi)P_{\text{res}})\\ &= \operatorname{vol}(\overline{G}(F)\backslash\overline{G}(\mathbb{A}))\varphi(1) \qquad\qquad (1.4\text{b})\\ &\quad + \sum_{\substack{\mathfrak{o}_\gamma\\ \text{elliptic}}} \operatorname{vol}(\overline{G}(F)^{\gamma}\backslash\overline{G}(\mathbb{A})^{\gamma}) \int_{\overline{G}(\mathbb{A})^{\gamma}\backslash\overline{G}(\mathbb{A})} \varphi(x^{-1}\gamma x)\,dx,\end{aligned}$$

so that the two sides of (1.1) do resemble each other.

The right side of (1.4a) and the last term of (1.4b) are sums over $\{\gamma'\}$ or $\{\gamma\}$ of products over v of expressions

$$m_{\gamma',v}\int_{\overline{G}'_{v}{}^{\gamma'}\backslash\overline{G}'_v}\varphi'_v(x^{-1}\gamma' x)\,dx \qquad \text{and} \qquad m_{\gamma,v}\int_{\overline{G}_v{}^{\gamma}\backslash\overline{G}_v}\varphi_v(x^{-1}\gamma x)\,dx,$$

where $m_{\gamma',v}$ and $m_{\gamma,v}$ are certain local volume factors. These local volume factors have been arranged to match when $\gamma'\sim\gamma$, and the orbital integrals match in this case by (1) and (2). Each γ' is $\sim\gamma$ for some γ. We have arranged by (3) that any γ that is not $\sim\gamma'$ for some γ' has orbital integral 0; indeed this follows from the fact that a quadratic extension E embeds in the quaternion algebra if and only if no prime of S splits in E.

In proving (1.1), we are left with the term in (1.4a) having $\gamma'=1$ and the first term on the right side of (1.4b). One shows that if global Haar measures are built from our matching local ones, then

$$\operatorname{vol}(\overline{G}'(F)\backslash\overline{G}'(\mathbb{A}))=\operatorname{vol}(\overline{G}(F)\backslash\overline{G}(\mathbb{A})).$$

Moreover Harish-Chandra's formula for recovering the value of a function at 1 from its elliptic orbital integrals may be used to show that

$$\varphi'_v(1)=\begin{cases}-\varphi_v(1) & \text{for } v\in S\\ \varphi_v(1) & \text{for } v\notin S.\end{cases}$$

Since S contains an even number of places,

$$\varphi'(1)=\varphi(1).$$

Thus the remaining terms on the right sides of (1.4a) and (1.4b) match, and (1.1) follows.

Letting $n(\pi')$ and $n(\pi)$ denote multiplicities, we obtain an equality

$$\sum n(\pi')\operatorname{Tr}(\pi'(\varphi'))=\sum n(\pi)\operatorname{Tr}(\pi(\varphi))$$

when $\varphi'\leftrightarrow\varphi$, where the sum on the left is over π' in $L^2(G',\omega')$ and the sum on the right is over cuspidal and residual π in $L^2(G,\omega)$. The known multiplicity one for G means that $n(\pi)$ is 1 when it is nonzero. Thus we can rewrite this equation as

$$\sum n(\pi')\operatorname{Tr}(\pi'(\varphi'))=\sum \operatorname{Tr}(\pi(\varphi)). \tag{1.5}$$

Digression. Before indicating how Theorems 1a and 1b follow from (1.5), let us make an observation about a relationship between character identities and orbital integrals. From the Weyl integration formulas for integration over G'_v and G_v and from the fact that irreducible characters are given by locally integrable functions, the trace of an irreducible admissible representation applied to a function is an integral of orbital integrals of the function. Thus if Theorem 1a is known (i.e., if π_v has been constructed so as to be related to π'_v as in the theorem) and if $\varphi'\leftrightarrow\varphi$, then it follows that

$$\operatorname{Tr}(\pi'_v(\varphi'_v))=\begin{cases}-\operatorname{Tr}(\pi_v(\varphi_v)) & \text{for } v\in S\\ \operatorname{Tr}(\pi_v(\varphi_v)) & \text{for } v\notin S.\end{cases}$$

Since S contains an even number of places, we obtain

$$\mathrm{Tr}(\pi'(\varphi')) = \mathrm{Tr}(\pi(\varphi)).$$

Then it is a fairly simple matter to use (1.5) to see that π is automorphic. See [Gb-Ja, p. 248] for details. This is the approach of [Ja-Lgl], the local representations having been studied with the aid of the Weil representation.

Our task is different from the task in the above digression. We are to deduce both Theorem 1a and Theorem 1b from (1.5). The details are in [Gb-Ja, pp. 249–250] and in [Ro1], and our discussion will be brief.

In (1.5) let us repeat the term $\mathrm{Tr}(\pi'(\varphi'))$ according to the multiplicity of π'. Taking into account that $\varphi' = \prod_v \varphi'_v$ and $\varphi = \prod_v \varphi_v$, we can rewrite (1.5) as

$$\sum \prod_v \mathrm{Tr}(\pi'_v(\varphi'_v)) = \sum \prod_v \mathrm{Tr}(\pi_v(\varphi_v)). \tag{1.6}$$

Lemma 1. *For each $w \notin S$, fix an irreducible unitary representation τ_w of G_w. If φ'_v and φ_v have matching orbital integrals for $v \in S$, then*

$$\sum \prod_{v\in S} \mathrm{Tr}(\pi'_v(\varphi'_v)) = \sum \prod_{v\in S} \mathrm{Tr}(\pi_v(\varphi_v)), \tag{1.7}$$

where the sum on the left is over π' in $L^2(G',\omega')$ with $\pi'_w \cong \tau_w$ for all $w \notin S$ and where the sum on the right is over cuspidal and residual π in $L^2(G,\omega)$ with $\pi_w \cong \tau_w$ for all $w \notin S$. The sum on the right has at most one term.

The formula in Lemma 1 follows by applying to (1.6) a notion of generalized linear independence established in [Lab-L]. The sum on the right in (1.7) has at most one term as a consequence of strong multiplicity one for G.

Lemma 2 (Weak Jacquet-Langlands correspondence). *If $\pi' = \bigotimes \pi'_v$ occurs in $L^2(G',\omega')$, then there exists a unique $\pi = \bigotimes \pi_v$ that is cuspidal or residual in $L^2(G,\omega)$ and has $\pi'_v \cong \pi_v$ for all $v \notin S$. Moreover, π_v is in the discrete series of G_v for all $v \in S$. Conversely if $\pi = \bigotimes \pi_v$ is cuspidal in $L^2(G,\omega)$ is such that π_v is in the discrete series of G_v for all $v \in S$, then there exists at least one $\pi' = \bigotimes \pi'_v$ in $L^2(G',\omega')$ such that $\pi'_v \cong \pi_v$ for all $v \notin S$.*

Sketch of proof. Uniqueness is by strong multiplicity one. For existence, suppose on the contrary that there is no π. Applying Lemma 1 with $\tau_w = \pi'_w$ for all $w \notin S$, we see that

$$\sum \prod_{v\in S} \mathrm{Tr}(\pi'_v(\varphi'_v)) = 0,$$

where the sum is over σ' in $L^2(G',\omega')$ with $\sigma'_w \cong \pi'_w$ for all $w \notin S$. Generalized linear independence of characters [Lab-L] shows that the sum is empty, in contradiction to the fact that the sum must contain a term for each occurrence of π' in $L^2(G',\omega')$. The converse statement is proved in the same way, starting from the assumption that

$$\sum \prod_{v\in S} \mathrm{Tr}(\pi_v(\varphi_v)) = 0.$$

Lemma 2 gives the map $\pi' \to \pi$ of Theorem 1b. Let us now turn our attention to Theorem 1a. Fix π and take $\tau_w = \pi_w$ for $w \notin S$, so that the term for π appears on the right side of (1.7). The first step is to show that the sum on the left side of (1.7) is a finite sum; the argument uses ε factors and appears in Lemma 5.14 of [Ro1]. Then using linear independence of characters, we can omit from the product any place v such that π_v is a special representation (since we already know that this corresponds to a one-dimensional representation of G'_v). We may also cancel the factors for archimedean $v \in S$, since we know the local Jacquet-Langlands correspondence for archimedean places. Thus we are left with a subset $S' \subset S$ of places where each π_v is supercuspidal. Then it follows that

$$\sum \prod_{v\in S'} \operatorname{Tr}(\pi'_v(t')) = \prod_{v\in S'} \operatorname{Tr}(\pi_v(t)) \qquad \text{when } t' \sim t.$$

On the left side we have finitely many characters of the compact group $\prod_{v\in S} G'_v$. The L^2 norm of the right side on the elliptic set is 1 by the orthogonality relations for supercuspidal characters on the elliptic set ([Ro1], Lemma 5.3). We conclude that there is only one summand π' on the left side, i.e.,

$$\prod_{v\in S'} \operatorname{Tr}(\pi'_v(t')) = \prod_{v\in S'} \operatorname{Tr}(\pi_v(t)) \qquad \text{for some } \pi'.$$

Hence for each $v \in S'$ there is a scalar λ_v such that $\operatorname{Tr}(\pi'_v(t')) = \lambda_v \operatorname{Tr}(\pi_v(t))$. Again by the orthogonality relations, $|\lambda_v| = 1$. To prove that $\lambda_v = -1$, we let t' tend to 1. By continuity, $\operatorname{Tr}(\pi'_v(t'))$ tends to $\operatorname{Tr}(\pi'_v(1)) = \dim(\pi'_v)$. By an argument in [Ro1], the supercuspidal π_v has the properties that $\operatorname{Tr}(\pi_v(t))$ is continuous at $t = 1$ and $\operatorname{Tr}(\pi_v(1)) = -d(\pi_v)$, where $d(\pi_v)$ is the formal degree. Thus we obtain $\dim(\pi'_v) = -\lambda_v d(\pi_v)$. Since $\dim(\pi'_v)$ and $d(\pi_v)$ are positive, we conclude that $\lambda_v = -1$.

This completes the construction of the correspondence in Theorem 1a for all π'_v that occur as the vth component of some automorphic π' for G'. But every π'_v arises in this way. In fact, Lemma 1.5 of [Ro1] shows that we have only to apply the trace formula to a function $\varphi' = \prod \varphi'_w$ of suitably small support with φ'_v equal to a matrix coefficient of π'_v.

The rest of the argument for Theorems 1a and 1b is relatively easy, and we omit the details.

2. Automorphic Induction

Automorphic induction is described in [Ro4, §15]. The point of departure is the question whether induction of Galois representations has a counterpart for automorphic representations. Let us suppose that E/F is a finite extension of degree d of number fields. For convenience we assume that E/F is Galois. If σ is an m-dimensional representation of $G_E = \operatorname{Gal}(\overline{F}/E)$, we form $\Sigma = \operatorname{ind}_{G_E}^{G_F} \sigma$ as a representation of $\operatorname{Gal}(\overline{F}/F)$ of degree $n = md$. The L functions of σ and Σ are equal. If the Local Langlands Conjecture holds at every place, then σ parametrizes an irreducible admissible representation π of $GL_m(\mathbb{A}_E)$ whose L function equals the L function of σ. Suppose π is automorphic. Then Σ parametrizes an irreducible admissible representation Π of $GL_n(\mathbb{A}_F)$ with the same L function, and the question is whether Π is automorphic.

This situation is an instance of global functoriality. In fact, let

$$H = R_{E/F}(GL_m/E) \qquad \text{and} \qquad G = GL_n/F,$$

where $R_{E/F}$ indicates restriction of scalars (§§6 and 10 of [Kn1]). Here $H(F) = GL_m(E)$ and $H(\mathbb{A}_F) = GL_m(\mathbb{A}_E)$, and we have

$$
\begin{aligned}
{}^L H &= (GL_m(\mathbb{C}) \times \cdots \times GL_m(\mathbb{C})) \rtimes G_F \qquad (d \text{ factors of } GL_m(\mathbb{C})),\\
{}^L G &= GL_n(\mathbb{C}) \times G_F,
\end{aligned}
$$

where G_F acts on ${}^L H$ by permuting the factors. The group G_F acts through $\mathrm{Gal}(E/F)$, and $\mathrm{Gal}(E/F)$ may be regarded as a subgroup of the permutation group on d letters. Regard the permutation for $\tau \in G_F$ as a d-by-d matrix, and then replace each entry by an m-by-m zero or identity matrix. In this way we can identify τ with an n-by-n matrix, which we denote τ also. The map ${}^L H \to {}^L G$ given by

$$
((a_1, \ldots, a_d), \tau) \mapsto \left(\begin{pmatrix} a_1 & & \\ & \ddots & \\ & & a_d \end{pmatrix} \tau,\ \tau \right), \qquad a_j \in GL_m(\mathbb{C}), \tag{2.1}
$$

is a homomorphism, and the global functoriality question is whether it corresponds to a map of automorphic representations.

As in [Ro4, §15] we can rephrase the question in such a way that it is not necessary to assume the Local Langlands Conjecture. In fact, the given irreducible admissible representation π of $H(\mathbb{A}_F) = GL_m(\mathbb{A}_E)$ has a tensor product decomposition $\pi = \bigotimes_w \pi_w$ in which π_w is an unramified principal series for almost every place w of E. For the unexceptional places w, the L function $L(s, \pi_w)$ is well defined and determines π_w up to isomorphism. Similar considerations apply to a representation $\Pi = \bigotimes_v \Pi_v$ of $G(\mathbb{A}_F) = GL_n(\mathbb{A}_F)$. Suppose π is automorphic. We define Π to be **automorphically induced** by π if Π is automorphic and

$$
L(s, \Pi_v) = \prod_{w|v} L(s, \pi_w)
$$

for almost all places v of F. These conditions determine Π uniquely by strong multiplicity one, since the local L factors of Π have already determined Π_v at almost every place v. When Π is automorphically induced by π, we write $\Pi = AI_E^F(\pi)$.

Theorem 2. *Let E/F be a cyclic extension of degree d. If π is a cuspidal representation of $GL_m(\mathbb{A}_E)$, then $\Pi = AI_E^F(\pi)$ exists as an automorphic representation of $GL_{md}(\mathbb{A}_F)$. Moreover, Π is cuspidal unless there is a nontrivial element $\tau \in \mathrm{Gal}(E/F)$ such that $\tau(\pi)$ is isomorphic to π.*

In this generality Theorem 2 is due to Arthur and Clozel [Ar-Cl]. We confine our comments to the special case that $d = 2$ and $m = 1$, which is the case that was handled originally. We shall write θ for the given automorphic representation π of $GL_1(\mathbb{A}_E)$ (a unitary Grossencharacter of E) and $\Pi = \Pi_\theta$ for the automorphic representation of $GL_2(\mathbb{A}_F)$ that is to be constructed. We assume that $\theta \neq \theta^\sigma$, where σ is the nontrivial element of $\mathrm{Gal}(E/F)$, and we are to construct $\Pi = \Pi_\theta$ and show that it is cuspidal automorphic.

Four methods are known for establishing this special case:

(1) Use theta series to construct Π as a physical subspace of the appropriate L^2 space. This method is due to Hecke, Maass, and Shalika-Tanaka.

(2) Use the Jacquet-Langlands converse theorem for GL_2. See Theorem 8.12 of [Kn1] for the statement. This is the method of Jacquet and Langlands [Ja-Lgl].
(3) Derive the result at the same time as establishing base change. This is the approach of Arthur and Clozel [Ar-Cl] and uses the trace formula.
(4) Use the trace formula in a different way. This is a method of Langlands [Lgl3].

We follow a variant of method (4). (See [Lab1].)

Thus let E/F be a quadratic extension. By class field theory (particularly Theorem 4.7 in [Kn1]), there exists a nontrivial character α of $F^\times\backslash\mathbb{A}_F^\times$ such that $\ker\alpha = N_{E/F}(E^\times\backslash\mathbb{A}_E^\times)$.

We are given a unitary Grossencharacter θ for E (i.e., a character of $E^\times\backslash\mathbb{A}_E^\times$) that is not fixed by the nontrivial element σ of $\mathrm{Gal}(E/F)$, and we are to construct an automorphic representation $AI_E^F(\theta)$ of $GL_2(\mathbb{A}_F)$.

We define an operator $A: L^2(G,\omega)\to L^2(G,\omega)$ by $Af(x) = f(x)\alpha(\det x)$. Note here that $\det x$ is a square and hence lies in $\ker\alpha = N_{E/F}(E^\times\backslash\mathbb{A}_E^\times)$ if x is in $Z(\mathbb{A}_F)$. We readily check that A intertwines R and $R\otimes(\alpha\circ\det)$ in the sense that

$$A(R\otimes(\alpha\circ\det))(x)f = R(x)Af \qquad \text{for } f\in L^2(G,\omega).$$

Moreover, A has order two and sends cuspidal functions to cuspidal functions. By strong multiplicity one for GL_2, the cuspidal irreducible representations in $L^2(G,\omega)$ have multiplicity one. If we think of the effect of A on the irreducible summands Π of the cuspidal part of $L^2(G,\omega)$, then we see that there are two possibilities:

(i) $\Pi\cong\Pi\otimes\alpha$, and then the space V_Π of Π is carried to itself by A
(ii) $\Pi\not\cong\Pi\otimes\alpha$, and then A has in effect a matrix on the direct sum of the two spaces of the block form $\begin{pmatrix} 0 & * \\ * & 0\end{pmatrix}$.

The same thing remains true of $P_{\mathrm{cusp}}R(x)P_{\mathrm{cusp}}A$ for every $x\in G(\mathbb{A}_F)$ and then also of $P_{\mathrm{cusp}}R(\varphi)P_{\mathrm{cusp}}A$ for every $\varphi\in C^\infty_{\mathrm{com}}(G(\mathbb{A}),\omega^{-1})$. The latter is of trace class since $P_{\mathrm{cusp}}R(\varphi)P_{\mathrm{cusp}}$ is of trace class, and we see that

$$\mathrm{Tr}(P_{\mathrm{cusp}}R(\varphi)P_{\mathrm{cusp}}A) = \sum_{\substack{\Pi\cong\Pi\otimes\alpha,\\ \Pi\text{ cuspidal}}} \mathrm{Tr}(\Pi(\varphi)A_\Pi), \tag{2.2}$$

where A_Π is the restriction of A to V_Π.

We shall want to use a variant of the trace formula to compute this trace in a different way. Part of the philosophy of applying the trace formula is that the main terms are the orbital integrals of the elliptic elements in $G(F)$ or $\overline{G}(F)$ (an **elliptic** element for this G again being one with eigenvalues not in F). In order to isolate the elliptic terms, we write the unprojected operator as an integral operator and compute with the kernel as if the space of integration were compact. In the present case the operator $R(\varphi)A$ is an integral operator with kernel

$$\sum_{\gamma\in\overline{G}(F)} \varphi(x^{-1}\gamma y)\alpha(\det y).$$

If the space of integration were compact, this operator would be of trace class, and a familiar computation shows that the trace of this operator would be

$$= \sum_{\gamma\in\overline{G}(F)} \int_{\overline{G}(F)\backslash\overline{G}(\mathbb{A}_F)} \varphi(x^{-1}\gamma x)\alpha(\det x)\,dx$$
$$= \sum_{\mathfrak{o}_\gamma} \int_{\overline{G}(F)^\gamma\backslash\overline{G}(\mathbb{A}_F)} \varphi(x^{-1}\gamma x)\alpha(\det x)\,dx,$$

where $\mathfrak{o}_\gamma$ consists of all $\delta^{-1}\gamma\delta$ with δ varying through $\overline{G}(F)^\gamma\backslash\overline{G}(F)$. Thus our formula for the trace on the cuspidal part is

$$\operatorname{Tr}(P_{\mathrm{cusp}}R(\varphi)P_{\mathrm{cusp}}A) = \sum_{\substack{\mathfrak{o}_\gamma,\\ \gamma \text{ elliptic}}} \int_{\overline{G}(F)^\gamma\backslash\overline{G}(\mathbb{A}_F)} \varphi(x^{-1}\gamma x)\alpha(\det x)\,dx + (\text{other terms}). \tag{2.3}$$

Define

$$F_\varphi(\gamma) = \int_{\overline{G}(\mathbb{A}_F)^\gamma\backslash\overline{G}(\mathbb{A}_F)} \varphi(x^{-1}\gamma x)\alpha(\det x)\,dx. \tag{2.4}$$

This function makes sense as a function on the elliptic subset of $G(F)$ that satisfies

$$F_\varphi(g\gamma g^{-1}) = \alpha(\det g)F_\varphi(\gamma). \tag{2.5}$$

We shall relate the elliptic elements $\gamma \in G(F)$ for which $F_\varphi(\gamma) \neq 0$ to an embedded copy of $R_{E/F}(E^\times)$ in G. If $\gamma \in G(F)$ is elliptic, its eigenvalues are not in $F^\times$ but lie in some quadratic extension $K = F(\sqrt{m})$. Left multiplication by members of K on the F basis $\{1, \sqrt{m}\}$ exhibits K as the matrices $\left(\begin{smallmatrix} a & mb \\ b & a\end{smallmatrix}\right)$ with coefficients in F. Then $R_{K/F}(K^\times)$ is exhibited as an explicit algebraic subgroup H_K of G, and γ is conjugate to a member of H_K.

Let us observe that if there exists y in the centralizer $G(\mathbb{A}_F)^\gamma$ with $\alpha(\det y) \neq 1$, then $F_\varphi(\gamma) = 0$. This is immediate from (2.4).

For γ elliptic, the centralizer of γ in G is isomorphic to H_K for some quadratic extension K/F. Then $\det G(\mathbb{A}_F)^\gamma = N_{K/F}(K^\times\backslash\mathbb{A}_K^\times)$. For α to be trivial on this, we must have $N_{K/F}(K^\times\backslash\mathbb{A}_K^\times) \subset \ker\alpha$. But class field theory says that the norm group determines the quadratic extension, and therefore $K \cong E$. The group H_E is just our given algebraic group H up to canonical isomorphism, and $H(F) = E^\times$. The only members of $E^\times$ that have eigenvalues in F are the scalar matrices, i.e., the members of $Z(F) \cong F^\times$.

This proves the desired relationship: What we conclude from the above is that an elliptic $\gamma \in G(F)$ has $F_\varphi(\gamma) = 0$ unless the member γ is conjugate to a member of $E^\times$ and is not scalar. The $\mathfrak{o}_\gamma$'s in question are thus parametrized by the nontrivial cosets of the quotient $F^\times\backslash E^\times$. Two elements e and e^σ parametrize the same $\mathfrak{o}_\gamma$, but there is no other redundance. Combining this conclusion with (2.3), we obtain

$$\operatorname{Tr}(P_{\mathrm{cusp}}R(\varphi)P_{\mathrm{cusp}}A) = \tfrac{1}{2}\operatorname{vol}((\mathbb{A}_F^\times E^\times)\backslash\mathbb{A}_E^\times) \sum_{\gamma\in(F^\times\backslash E^\times)-F^\times} F_\varphi(\gamma) + (\text{other terms}). \tag{2.6}$$

The first term on the right side of (2.6) looks something like one side of the Poission summation formula for the group $(\mathbb{A}_F^\times E^\times)\backslash\mathbb{A}_E^\times$ and its cocompact discrete subgroup $F^\times\backslash E^\times$, except that one term is missing. The Poisson summation formula is what the trace formula comes down to in the abelian case, and thus we have an indication of where we are headed.

Let us now refine matters by taking the "other terms" into account. Qualitatively the trace formula for G, when modified to handle $R(\varphi)A$, says that

$$(\text{spectral terms}) = (\text{geometric terms})$$

and more specifically that

$$\begin{pmatrix}\text{cuspidal}\\ \text{terms}\end{pmatrix} + \begin{pmatrix}\text{continuous}\\ \text{terms}\end{pmatrix} + \begin{pmatrix}\text{one-dimensional}\\ \text{terms}\end{pmatrix} + \begin{pmatrix}\text{residual}\\ \text{terms}\end{pmatrix}$$
$$= \begin{pmatrix}\text{central}\\ \text{terms}\end{pmatrix} + \begin{pmatrix}\text{elliptic}\\ \text{terms}\end{pmatrix} + \begin{pmatrix}\text{hyperbolic}\\ \text{terms}\end{pmatrix} + \begin{pmatrix}\text{unipotent}\\ \text{terms}\end{pmatrix}.$$

Here (2.2) gives

$$\begin{pmatrix}\text{cuspidal}\\ \text{terms}\end{pmatrix} = \operatorname{Tr}(P_{\text{cusp}} R(\varphi) P_{\text{cusp}} A) = \sum_{\substack{\Pi \cong \Pi \otimes \alpha,\\ \Pi \text{ cuspidal}}} \operatorname{Tr}(\Pi(\varphi) A_\Pi),$$

and we have just seen that

$$\begin{pmatrix}\text{elliptic}\\ \text{terms}\end{pmatrix} = \tfrac{1}{2}\operatorname{vol}((\mathbb{A}_F^\times E^\times)\backslash \mathbb{A}_E^\times) \sum_{\gamma \in (F^\times \backslash E^\times) - F^\times} F_\varphi(\gamma).$$

An argument like the one above that proves some vanishing for $F_\varphi(\gamma)$ shows that the central terms are 0, and similar considerations show that the hyperbolic terms are 0. A one-dimensional representation is never equal to its own nontrivial twist, and hence the one-dimensional terms are 0. Further argument shows that the continuous terms are 0. Thus our refined version of (2.6) is

$$\sum_{\substack{\Pi \cong \Pi \otimes \alpha,\\ \Pi \text{ cuspidal}}} \operatorname{Tr}(\Pi(\varphi) A_\Pi) + \begin{pmatrix}\text{residual}\\ \text{terms}\end{pmatrix}$$
$$= \tfrac{1}{2}\operatorname{vol}((\mathbb{A}_F^\times E^\times)\backslash \mathbb{A}_E^\times) \sum_{\gamma \in (F^\times \backslash E^\times) - F^\times} F_\varphi(\gamma) + \begin{pmatrix}\text{unipotent}\\ \text{terms}\end{pmatrix}. \tag{2.7}$$

We would like to be able to apply the Poisson summation formula to the first term on the right side of (2.7). Thus we extend F_φ to a function on the set of regular elements in $\mathbb{A}_E^\times$ (i.e., those elements having no component central), using formula (2.4). It turns out (see [Lab-L]) that F_φ extends to a smooth function on $\mathbb{A}_E^\times$. However, its value on a central element is not given by (2.4) but rather by the unipotent terms of (2.7). Application of the Poisson summation formula therefore gives

$$\sum_{\substack{\Pi \cong \Pi \otimes \alpha,\\ \Pi \text{ cuspidal}}} \operatorname{Tr}(\Pi(\varphi) A_\Pi) + \begin{pmatrix}\text{residual}\\ \text{terms}\end{pmatrix} = \tfrac{1}{2}\operatorname{vol}((\mathbb{A}_F^\times E^\times)\backslash \mathbb{A}_E^\times) \sum_{\chi \in ((\mathbb{A}_F^\times E^\times \backslash \mathbb{A}_E^\times)^\wedge} \widehat{F_\varphi}(\chi). \tag{2.8}$$

Let $\{\chi, \chi^\sigma\}$ be given with $\chi \neq \chi^\sigma$. The same style of argument as at the end of §1 allows us to find a term on the left side corresponding to some Π such that this term equals the sum of the terms for $\{\chi, \chi^\sigma\}$ on the right side. Then Π is automorphically induced from some Grossencharacter depending on χ. Actually the Grossencharacter is of the form $\theta = \chi\rho$ for a certain ρ independent of χ. We omit the details. (Cf. [Lap-R] for a similar argument in a slightly different context.)

The proof yields a bonus. Comparison of what terms are left after the above matching shows that the cuspidal Π's with $\Pi \cong \Pi \otimes \alpha$ are exactly the cuspidal Π's that are automorphically induced. This result is due to [Lab-L].

3. Base Change

Base change is described in [Ro4, §16]. For an extension E/F, the point of departure is the question whether restriction of Galois representations from G_F to G_E has a counterpart for automorphic representations. It is explained in [Kn1, §10] how this situation is an instance of global functoriality. Namely let

$$H = GL_n/F \qquad \text{and} \qquad G = R_{E/F}(GL_n/E).$$

Here $G(F) = GL_n(E)$ and $G(\mathbb{A}_F) = GL_n(\mathbb{A}_E)$, and we have

$$\begin{aligned} {}^L H &= GL_2(\mathbb{C}) \times \mathrm{Gal}(\overline{F}/F), \\ {}^L G &= (GL_n(\mathbb{C}) \times \cdots \times GL_n(\mathbb{C})) \rtimes \mathrm{Gal}(\overline{F}/F), \end{aligned}$$

with the Galois group operating by permutations on the factors in the second case. The map ${}^L H \to {}^L G$ is given by the diagonal map on the identity component and by the identity map on the Galois group.

Making the above notions precise in a direct way requires the Local Langlands Conjecture. But as in [Ro4, §16] we can rephrase the question in such a way that it is not necessary to assume the Local Langlands Conjecture: Write the given automorphic representation π of $H(\mathbb{A}_F) = GL_n(\mathbb{A}_F)$ as $\pi = \bigotimes_v \pi_v$ with π_v equal to an unramified principal series for almost every place v of F. An irreducible admissible representation $\Pi = \bigotimes_w \Pi_w$ of $G(\mathbb{A})_F = GL_n(\mathbb{A}_E)$ is said to be a **base change lift** of π if it is automorphic and if, for almost all v and all w dividing v, π_v is isomorphic to the Langlands subquotient of an unramified principal series induced from the character χ and Π_w is isomorphic to the Langlands subquotient of an unramified principal series induced from the character $\chi^{[E_w:F_v]}$. These conditions determine Π uniquely by strong multiplicity one. When Π and π are related in this way, we write $\Pi = BC_{E/F}(\pi)$.

Theorem 3. *Let E/F be a cyclic extension of prime degree l. If π is a cuspidal representation of $GL_n(\mathbb{A}_F)$, then $\Pi = BC_{E/F}(\pi)$ exists as an automorphic representation of $GL_n(\mathbb{A}_E)$. Moreover, Π is cuspidal unless l divides n and $\pi \otimes \alpha \cong \pi$ for some nontrivial character α of $F^\times \backslash \mathbb{A}_F^\times$ that is trivial on $N_{E/F}(E^\times \backslash \mathbb{A}_E^\times)$.*

In this generality Theorem 3 is due to Arthur and Clozel [Ar-Cl]. See [Ro4] for a more complete statement. We confine our comments to the special case that $n = 2$, which was the case handled originally by Saito, Shintani, and Langlands [Lgl2]. For simplicity we shall assume $l \neq 2$. Thus write $H = GL_2/F$ and $G = R_{E/F}(GL_2/E)$, so that $H(\mathbb{A}_F) = GL_2(\mathbb{A}_F)$ and $G(\mathbb{A}_F) = GL_2(\mathbb{A}_E)$. The comparison of traces that leads to our special case of Theorem 3 involves the usual trace formula for $GL_2(\mathbb{A}_F)$ and a "twisted trace formula" for $GL_2(\mathbb{A}_E)$.

To describe this variant, we introduce the notion of **twisted conjugacy** of elements of $GL_2(E)$. By assumption, $\mathrm{Gal}(E/F)$ is cyclic of prime order l. Let σ be a generator. Define g and g' in $GL_2(E)$ to be σ **conjugate** if there exists $x \in GL_2(E)$

such that $x^{-1}g\sigma(x) = g'$. The associated norm map $N : GL_2(E) \to GL_2(E)$ is defined by $N(g) = g\sigma(g)\cdots\sigma^{l-1}(g)$. It is easy to check that

(1) $N(x^{-1}g\sigma(x)) = x^{-1}N(g)x$, and hence the $GL_2(E)$ conjugacy class of $N(g)$ equals N of the σ conjugacy class of g,
(2) $\sigma(N(g)) = g^{-1}N(g)g$,
(3) the $GL_2(E)$ conjugacy class of $N(g)$ always meets $GL_2(F)$ in a unique $GL_2(F)$ conjugacy class.

Fact (2) shows that the trace and determinant of $N(g)$ belong to F, and (3) follows from this. Define

$$\eta : \{\sigma \text{ conjugacy classes in } GL_2(E)\} \to \{\text{conjugacy classes in } GL_2(F)\}$$

by $\delta \mapsto \gamma$ if $\gamma \in GL_2(F)$ is $GL_2(E)$ conjugate to $N(\delta)$. This map can be shown to be injective, and we write $\gamma \in \eta(\delta)$. If δ is in $GL_2(E)$, the σ centralizer of δ is an algebraic subgroup G_σ^δ of G whose F points are

$$G_\sigma^\delta(F) = \{x \in GL_2(E) \mid x^{-1}\delta\sigma(x) = \delta\}.$$

Let Z be the group of 2-by-2 scalar matrices, and put

$$Z_E(\mathbb{A}_F) = Z(F)N_{E/F}(Z(\mathbb{A}_E)).$$

Fix a character ω of $Z_E(\mathbb{A}_F)$ trivial on $Z(F)$ and let ω_E be the character $z \mapsto \omega(N_{E/F}(z))$ of $Z(\mathbb{A}_E)$. The relevant L^2 spaces will be L_E^2, built from G, and L_F^2, built from H:

$$L_E^2 = \Big\{f(zg) = \omega_E(z)f(g) \text{ for all } z \in Z(\mathbb{A}_E) \;\Big|\; \int_{Z(\mathbb{A}_E)GL_2(E)\backslash GL_2(\mathbb{A}_E)} |f|^2 < \infty\Big\},$$
$$L_F^2 = \Big\{f(zg) = \omega(z)f(g) \text{ for all } z \in Z_E(\mathbb{A}_F) \;\Big|\; \int_{Z_E(\mathbb{A}_F)GL_2(F)\backslash GL_2(\mathbb{A}_F)} |f|^2 < \infty\Big\}.$$

Let R_E and R_F be the right regular representations of $GL_2(\mathbb{A}_E)$ and $GL_2(\mathbb{A}_F)$, respectively, on these L^2 spaces, and define an operator A_σ on the first such space by $(A_\sigma F)(g) = F(\sigma^{-1}g)$. We are interested in the trace of $P_{\text{cusp}}R_E(\varphi)P_{\text{cusp}}A_\sigma$.

As in §2 we write down the kernel of the unprojected operator $R_E(\varphi)A_\sigma$, which is

$$K(x,y) = \sum_{\delta\in\overline{GL}_2(E)} \varphi(x^{-1}\delta\sigma(y))$$

if we denote $Z\backslash GL_2$ by $\overline{GL}_2$. This operator is not of trace class, but we write down its (divergent) integral over the diagonal anyway, retaining the elliptic terms. The result is

$$\operatorname{Tr}(P_{\text{cusp}}R_E(\varphi)P_{\text{cusp}}A_\sigma) = \sum_{\substack{\{\delta\},\\ \sigma \text{ conjugacy class,}\\ N\delta \text{ elliptic}}} m_\delta^\sigma \int_{G_\sigma^\delta(\mathbb{A}_F)\backslash GL_2(\mathbb{A}_E)} \varphi(g^{-1}\delta\sigma(g))\,dg + (\text{other terms}). \quad (3.1)$$

We want to compare (3.1) with

$$\operatorname{Tr}(P_{\text{cusp}}R_F(f)P_{\text{cusp}}) = \sum_{\substack{\{\gamma\},\\ \gamma \text{ elliptic}}} m_\gamma \int_{GL_2^\gamma(\mathbb{A}_F)\backslash GL_2(\mathbb{A}_F)} f(g^{-1}\gamma g)\, dg + (\text{still other terms}). \tag{3.2}$$

The functions φ and f that we shall consider will be on $G(\mathbb{A}_E)$ and $G(\mathbb{A}_F)$, respectively, transforming under the central subgroups $Z(\mathbb{A}_E)$ and $Z_E(\mathbb{A}_F)$ oppositely to the members of the L^2 spaces, and being smooth of compact support modulo the central subgroups. We shall assume that φ and f are given by products of functions φ_v and f_v corresponding to the places of F such that

(i) φ_v is a function on $GL_2(E_v)$, smooth and compactly supported modulo $Z(E_v)$

(ii) $\varphi(zg) = \omega_E^{-1}(z)\varphi_v(g)$ for $z \in Z(E_v)$

and

(i′) f_v is a function on $GL_2(F_v)$, smooth and compactly supported modulo $Z(F_v)$

(ii′) $f(zg) = \omega^{-1}(z) f_v(g)$ for $z \in N_{E_v/F_v}(Z(E_v))$.

The idea is to show that

$$l \operatorname{Tr}(P_{\text{cusp}}R_E(\varphi)P_{\text{cusp}}A_\sigma) = \operatorname{Tr}(P_{\text{cusp}}R_F(f)P_{\text{cusp}}) \tag{3.3}$$

whenever φ and f are suitably compatible. In the first place one shows that $l\, m_\delta^\sigma = m_\gamma$ if $\gamma \in \eta(\delta)$; this is a Tamagawa-number problem that we shall not discuss.

In the second place one is to show that every φ gives a compatible f. The meaning of compatibility is that we can associate $\varphi_v \to f_v$ for all v in such a way that, for all regular semisimple $\gamma \in GL_2(F)$,

$$\int_{GL_2^\gamma(\mathbb{A}_F)\backslash GL_2(\mathbb{A}_F)} f(g^{-1}\gamma g)\, dg$$

is 0 if γ is not a norm from $GL_2(E)$ and is

$$\int_{G_\sigma^\delta(\mathbb{A}_F)\backslash GL_2(\mathbb{A}_E)} \varphi(g^{-1}\delta\sigma(g))\, dg$$

if $\gamma \in \eta(\delta)$. Early work with this notion is due to Saito and Shintani.

Showing that every φ gives an f is a manageable problem in local analysis. But even when we solve this problem, we have not solved the base change problem. In applying the trace formulas, we have to choose φ_v and f_v at almost every place to be the identity in the appropriate bi-K-invariant Hecke algebras (K being a standard maximal compact subgroup). Thus we have to know for almost every place that if φ_v is the normalized characteristic function of the maximal compact subgroup of $GL_2(E_v)$, then f_v can be taken to be the normalized characteristic function of the maximal compact subgroup of $GL_2(F_v)$.

This is a special case of a result that is mostly conjectural and is known as the **fundamental lemma**. Its importance is emphasized in the lectures [Lgl4] of Langlands. See Kottwitz [Ko1] for the fundamental lemma in the case of base change for GL_2. For generalizations see [Ko2], [Ar-Cl], [Lab2], [Wa1], and [Ha].

The result is that the main terms on the right sides of (3.1) and (3.2) are equal (apart from the factor of l). The hard part is to show that the contributions from the other terms cancel, so that (3.3) results. This is done in [Lgl2, §11]. The proof that (3.3) implies our case of Theorem 3 is summarized on [Lgl, p. 20], and the details are carried out in [Lgl2, §11].

4. Applications Involving a Single Group

We discuss briefly in this section three applications of the trace formula of a single group.

4.1. Eichler-Selberg trace formula. This theorem in its classical form is derived in [Ei2], and an exposition appears in [Mi]. We use the standard terminology of [Shi2]. Let $N>1$ be an integer, and let $\Gamma_0(N)$ be the subgroup of $SL_2(\mathbb{Z})$ of matrices whose lower left entry is divisible by N. Let $S_k(N)$ be the space of (analytic) cusp forms of weight k relative to $\Gamma_0(N)$. This space is known to be finite-dimensional. If p is a prime not dividing N, then the Hecke operator T_p is defined on, among other things, the space $S_k(N)$. The Eichler-Selberg trace formula computes the trace of this operator. If p is put equal to 1 in the formula, the operator T_p becomes the identity, and the formula gives an expression for the dimension of $S_k(N)$. (This dimension formula may also be derived from the Riemann-Roch Theorem.)

The Eichler-Selberg trace formula may also be derived from the (adelic) trace formula for GL_2, with the character ω of the center taken to be 1. This kind of proof has been carried out in [Du-La], although it is not immediately obvious how to correlate the terms of the GL_2 trace formula in [Du-La] with the terms in Theorem 7.14 of [Kn2]. Let us take $N=1$ for simplicity. The space $S_k(1)$ is spanned by simultaneous eigenfunctions of all the Hecke operators, and each such eigenfunction gives rise to a cuspidal automorphic representation $\pi=\pi_\infty\otimes\bigotimes_p\pi_p$ whose component π_∞ at the infinite place is a discrete series with extreme weight $\pm k$. Let $\varphi=\varphi_\infty\times\prod_p\varphi_p$ be defined as follows. For finite p, φ_p modulo center is just the characteristic function of the usual maximal compact subgroup of $GL_2(\mathbb{Q}_p)$. For ∞, φ_∞ is the matrix coefficient $(\pi_\infty(\,\cdot\,)v,v)$, where v has extreme weight. The idea is that the trace formula for GL_2 is to be applied to this φ.

The difficulty is that this φ does not have compact support modulo center. At this point one can apply the main result of [Lab3], which creates a compactly supported function behaving like the function φ above and having constant term 0. Then the GL_2 trace formula is applicable, and the formula for the trace of the Hecke operator follows.

For a generalization of this theorem, see [Ar1].

4.2. An application to quaternion algebras. Let F be a number field, and let E/F be a quadratic extension. By class field theory there exists a nontrivial character α of $F^\times\backslash\mathbb{A}_F^\times$ such that $\ker\alpha=N_{E/F}(E^\times\backslash\mathbb{A}_E^\times)$. Let D be a quaternion (division) algebra over F, let G' be the multiplicative group, and suppose that E does not embed in $D(F)$. Then the theorem is that $\pi'\not\cong\pi'\otimes\alpha$ for every automorphic representation occurring in $L^2(G',\omega)$.

The argument runs parallel to that for Theorem 2. Let R be the right regular representation of $G'(\mathbb{A}_F)$ on $L^2(G',\omega)$. Define an operator $A:L^2(G',\omega)\to L^2(G',\omega)$

by $Af(x) = f(x)\alpha(\det x)$. Then A satisfies the intertwining relation

$$A(R \otimes (\alpha \circ \det))(x)f = R(x)Af \qquad \text{for } f \in L^2(G', \omega).$$

By the Corollary to Theorem 1b, the automorphic representations in $L^2(G', \omega)$ have multiplicity one. Arguing as in §2, we obtain

$$\operatorname{Tr}(R(\varphi)A) = \sum_{\substack{\pi' \cong \pi' \otimes \alpha, \\ \pi' \text{ automorphic}}} \operatorname{Tr}(\pi'(\varphi)A_{\pi'}),$$

where $A_{\pi'}$ is the restriction of A to the space of π'.

The operator $R(\varphi)A$ is an integral operator with kernel

$$\sum_{\gamma \in \overline{G}'(F)} \varphi(x^{-1}\gamma y)\alpha(\det y).$$

Since $Z'(\mathbb{A}_F)G'(F)\backslash G'(\mathbb{A}_F)$ is compact, $R(\varphi)A$ is of trace class, and its trace is

$$\operatorname{Tr}(R(\varphi)A) = \sum_{\mathfrak{o}_\gamma} \int_{\overline{G}'(F)^\gamma \backslash \overline{G}'(\mathbb{A}_F)} \varphi(x^{-1}\gamma x)\alpha(\det x)\, dx,$$

where $\mathfrak{o}_\gamma$ consists of all $\delta^{-1}\gamma\delta$ with δ varying through $\overline{G}'(F)^\gamma \backslash \overline{G}'(F)$.

Define

$$F_\varphi(\gamma) = \int_{\overline{G}'(\mathbb{A}_F)^\gamma \backslash \overline{G}'(\mathbb{A}_F)} \varphi(x^{-1}\gamma x)\alpha(\det x)\, dx$$

for $\gamma \in \overline{G}'(F)$. If $\gamma \in \overline{G}'(F)$ is 1, this is 0. Otherwise $\gamma \in \overline{G}'(F)$ lies in a unique quadratic extension K within $D(F)$, and the centralizer consists of the nonzero elements of this quadratic extension. Since K cannot be E by hypothesis, an argument in §2 shows that $F_\varphi(\gamma) = 0$ in this case. Consequently $\operatorname{Tr}(R(\varphi)A) = 0$. Then

$$\sum_{\substack{\pi' \cong \pi' \otimes \alpha, \\ \pi' \text{ automorphic}}} \operatorname{Tr}(\pi'(\varphi)A_{\pi'}) = 0.$$

A generalized independence argument allows us to conclude that the sum is empty, and the theorem follows.

4.3. A consequence of the Eichler-Shimura congruence relation. This theorem in its classical form is derived in [Ei1] and [Shi1] and explained in [Sw-Bi]. It yields a formula for the L function of $X_0(N)$, the modular curve attached to the group $\Gamma_0(N)$ defined in §4.1, as a product of L functions of cuspidal representations of weight 2. This product formula, place by place, is a consequence of the trace formula for GL_2. See [Ih]. For a generalization to a group other than GL_2, see [Lgl-Ra]. For an exposition see [Bl-Ro].

References

[Ar1] Arthur, J., The L^2-Lefschetz numbers of Hecke operators, *Invent. Math.* **97** (1981), 257–290.

[Ar2] Arthur, J., Stability and endoscopy: Informal motivation, *these* PROCEEDINGS, pp. 443–442.

[Ar-Cl] Arthur, J., and L. Clozel, *Simple Algebras, Base Change, and the Advanced Theory of the Trace Formula*, Princeton University Press, Princeton, 1989.

[Bl-Ro] Blasius, D., and J. Rogawski, Zeta functions of Shimura varieties, *Motives* (U. Jannsen, S. Kleiman, J.-P. Serre, eds.), Proc. Symp. Pure Math., vol. 55, Part II, American Mathematical Society, Providence, 1994, pp. 525–571.

[DKV] Deligne, P., D. Kazhdan, and M.-F. Vigneras, Représentations des Algèbres Centrales Simples p-adiques, *Représentations des Groupes Réductifs sur un Corps Local* (J.-N. Bernstein, P. Deligne, D. Kazhdan, and M.-F. Vigneras, eds.), Travaux en Cours, Hermann, Paris, 1984, pp. 33–117.

[Du-La] Duflo, M., and J.-P. Labesse, Sur la formule des traces de Selberg, *Annales Sci. École Norm. Sup.* **4** (1971), 193–284.

[Ei1] Eichler, M., Quaternäre quadratische Formen und die Riemannsche Vermutung fr̈ die Kongruenzzetafunktionen, *Arch. der Math.* **5** (1954), 355–366.

[Ei2] Eichler, M., Eine Verallgemeinerung der Abelschen Integrale, *Math. Zeitschr.* **67** (1957), 267–298.

[Gb1] Gelbart, S. S., *Automorphic Forms on Adele Groups*, Princeton University Press, Princeton, NJ, 1975.

[Gb-Ja] Gelbart, S., and H. Jacquet, Forms of GL(2) from the analytic point of view, *Automorphic Forms, Representations, and L-Functions* (A. Borel and W. Casselman, eds.), Proc. Symp. Pure Math., vol. 33, Part I, American Mathematical Society, Providence, 1979, pp. 213–251.

[Ha] Hales, T. C., On the fundamental lemma for standard endoscopy: reduction to unit elements, *Canad. J. Math.* **47** (1995), 974–994.

[Ih] Ihara, Y., Hecke polynomials as congruence ζ functions in elliptic modular case, *Annals of Math.* **85** (1967), 267–295.

[Ja] Jacquet, H., Automorphic spectrum of symmetric spaces, *these* PROCEEDINGS, pp. 443–455.

[Ja-Lgl] Jacquet, H., and R. P. Langlands, *Automorphic Forms on* GL(2), Lecture Notes in Mathematics, vol. 114, Springer-Verlag, Berlin, 1970.

[Kn1] Knapp, A. W., Introduction to the Langlands program, *these* PROCEEDINGS, pp. 245–302.

[Kn2] Knapp, A. W., Theoretical aspects of the trace formula for $GL(2)$, *these* PROCEEDINGS, pp. 355–405.

[Ko1] Kottwitz, R., Orbital integrals and base change, *Automorphic Forms, Representations, and L-Functions*, Proc. Symp. Pure Math., vol. 33, Part II, American Mathematical Society, Providence, 1979, pp. 111–113.

[Ko2] Kottwitz, R. E., Base change for unit elements of Hecke algebras, *Compositio Math.* **60** (1986), 237–250.

[Lab1] Labesse, J.-P., L-indistinguishable representations and trace formula for $SL(2)$, *Lie Groups and Their Representations: Summer School of the Bolyai János Mathematical Society (Budapest 1971)* (I. M. Gelfand, ed.), John Wiley & Sons, New York, 1975, pp. 331–338.

[Lab2] Labesse, J.-P., Fonctions élémentaires et lemme fondamental pour le changement de base stable, *Duke Math. J.* **61** (1990), 519–530.

[Lab3] Labesse, J.-P., Pseudo-coefficients trés cuspidaux et K théorie, *Math. Annalen* **291** (1991), 607–616.

[Lab-L] Labesse, J.-P., and R. P. Langlands, L-indistinguishability for $SL(2)$, *Canad. J. Math.* **31** (1979), 726–785.

[Lap-R] Lapid, E., and J. Rogawski, On twists of cuspidal representations of $GL(2)$, *Forum Math.* (to appear).

[Lgl1] Langlands, R. P., *On the Functional Equations Satisfied by Eisenstein Series*, Lecture Notes in Mathematics, vol. 544, Springer-Verlag, Berlin, 1976.

[Lgl2] Langlands, R. P., *Base Change for* GL(2), Princeton University Press, Princeton, 1980.

[Lgl3] Langlands, R. P., *Les Débuts d'une Formule des Traces Stable*, Publications Mathématiques, vol. 13, L'Université Paris VII, Paris, 1983.

[Lgl4] Langlands, R. P., Where stands functoriality today?, *these* PROCEEDINGS, pp. 457–471.

[Lgl-Ra] Langlands, R. P., and D. Ramakrishnan (eds.), *The Zeta Functions of Picard Modular Surfaces*, Les Publications CRM, Montréal, 1992.

[Mi] Miyake, T., *Modular Forms*, Springer-Verlag, Berlin, 1989.

[Mo-Wa] Moeglin, C., and J.-L. Waldspurger, *Décomposition Spectrale et Séries d'Eisenstein*, Birkhäuser Verlag, Basel, 1994.

[Ra] Ramakrishnan, D., Pure motives and automorphic forms, *Motives* (U. Jannsen, S. Kleiman, J.-P. Serre, eds.), Proc. Symp. Pure Math., vol. 55, Part II, American Mathematical Society, Providence, 1994, pp. 411–446.

[Ro1] Rogawski, J. D., Representations of $GL(n)$ and division algebras over a p-adic field, *Duke Math. J.* **50** (1983), 161–196.

[Ro2] Rogawski, J. D., *Automorphic Representations of Unitary Groups in Three Variables*, Princeton University Press, Princeton, 1990.

[Ro3] Rogawski, J. D., Modular forms, the Ramanujan conjecture and the Jacquet-Langlands correspondence, *Discrete Groups, Expanding Graphs and Invariant Measures*, by A. Lubotzky, Birkhäuser, Basel, 1994, pp. 135–176.

[Ro4] Rogawski, J. D., Functoriality and the Artin conjecture, *these* PROCEEDINGS, pp. 331–353.

[She] Shelstad, D., Orbital integrals for $GL_2(\mathbf{R})$, *Automorphic Forms, Representations, and L-Functions*, Proc. Symp. Pure Math., vol. 33, Part I, American Mathematical Society, Providence, 1979, pp. 107–110.

[Shi1] G. Shimura, Correspondances modulaires et les fonctions ζ de courbes algébriques, *J. Math. Soc. Japan* **10** (1958), 1–28.

[Shi2] G. Shimura, *Introduction to the Arithmetic Theory of Automorphic Functions*, Princeton University Press, Princeton, NJ, 1971.

[Sw-Bi] Swinnerton-Dyer, H. P. F., and B. J. Birch, Elliptic curves and modular functions, *Modular Functions of One Variable IV*, Lecture Notes in Mathematics, vol. 476, Springer-Verlag, Berlin, 1975, pp. 2–32.

[Vi] Vigneras, M.-F., *Arithmétique des Algèbres de Quaternions*, Lecture Notes in Math., vol. 800, Springer-Verlag, Berlin, 1980.

[Wa1] Waldspurger, J.-L., Sur les intégrales orbitales tordues pour les groupes linéaires: un lemme fondamental, *Canad. J. Math.* **43** (1991), 852–896.

[Wa2] Waldspurger, J.-L., Homogénéité de certains distributions sur les groupes p-adiques, *Publ. Math. I.H.E.S.* **81** (1995), 25–72.

DEPARTMENT OF MATHEMATICS, STATE UNIVERSITY OF NEW YORK, STONY BROOK, NEW YORK 11794, U.S.A.

E-mail address: aknapp@ccmail.sunysb.edu

DEPARTMENT OF MATHEMATICS, THE HEBREW UNIVERSITY, GIVAT RAM, JERUSALEM, ISRAEL

E-mail address: jonr@math.huji.ac.il

Proceedings of Symposia in Pure Mathematics
Volume **61** (1997), pp. 433–442

Stability and Endoscopy: Informal Motivation

James Arthur

The purpose of this note is described in the title. It is an elementary introduction to some of the basic ideas of stability and endoscopy. We shall not discuss the techniques of the theory, which among other things entail a sophisticated use of Galois cohomology. Our aim is rather to persuade a reader that the theory was created in response to some very natural problems in harmonic analysis. The article is intended for people who are starting (or even just thinking of starting) to learn the subject.

Langlands was actually led to the theory of endoscopy by questions in algebraic geometry, particularly Shimura varieties [17, §1]. However, he quickly realized that the questions had remarkable implications for harmonic analysis. It is in this context that we will discuss the basic ideas.

We begin with a simple form of the trace formula. Suppose that G is a reductive algebraic group defined over a number field F. The adeles $\mathbb{A}$ of F are a locally compact ring in which F embeds as a discrete subring, and the group of F-rational points $G(F)$ embeds as a discrete subgroup of the locally compact group $G(\mathbb{A})$ of adelic points. We shall be concerned with the case that G is anisotropic, or equivalently, that the quotient space $G(F)\backslash G(\mathbb{A})$ is compact. It is then known that the regular representation

$$\big(R(y)\phi\big)(x) = \phi(xy), \qquad \phi \in L^2\big(G(F)\backslash G(\mathbb{A})\big),\ x, y \in G(\mathbb{A}),$$

of $G(\mathbb{A})$ on the Hilbert space $L^2\big(G(F)\backslash G(\mathbb{A})\big)$ (with the right $G(\mathbb{A})$-invariant measure on $G(F)\backslash G(\mathbb{A})$) decomposes discretely. More precisely, we can write

$$R = \bigoplus_{\pi} m_\pi \pi(f),$$

a direct sum over π in the set $\Pi\big(G(\mathbb{A})\big)$ of irreducible representations of $G\big((\mathbb{A})\big)$, with finite multiplicities $m_\pi \in \mathbb{N} \cup \{0\}$. (If G is not anisotropic, there is a subrepresentation R_{disc} of R that decomposes in this way, at least modulo the split component of the center of G.)

1991 *Mathematics Subject Classification.* Primary 11R39, 22E55.
Supported in part by a research grant from NSERC.

Selberg's original formula gives the trace of the convolution operator

$$R(f) = \bigoplus_{\pi} m_\pi \pi(f)$$

obtained by integrating R against a test function f in $C_c^\infty\big(G(\mathbb{A})\big)$. (See [21], [2], [8].) On the one hand, the trace of $R(f)$ is a discrete sum

$$I_{\text{disc}}(f) = \sum_{\pi \in \Pi(G(\mathbb{A}))} m_\pi \operatorname{tr}\big(\pi(f)\big) \tag{1}$$

of irreducible characters. The trace formula asserts that the trace of $R(f)$ can also be written as a linear combination

$$I_{\text{ell}}(f) = \sum_{\gamma \in \Gamma(G(F))} a^G(\gamma) I_G(\gamma, f) \tag{2}$$

of **orbital integrals**

$$I_G(\gamma, f) = \int_{G_\gamma(\mathbb{A}) \backslash G(\mathbb{A})} f(x^{-1}\gamma x)\,dx$$

of f. Here, $\Gamma\big(G(F)\big)$ stands for the set of conjugacy classes in $G(F)$, $G_\gamma(\,\cdot\,)$ denotes the centralizer of γ in $G(\,\cdot\,)$, and the coefficients are given by

$$a^G(\gamma) = \operatorname{vol}\big(G_\gamma(F) \backslash G_\gamma(\mathbb{A})\big).$$

The trace formula for compact quotient is thus the identity of the two expansions $I_{\text{ell}}(f)$ and $I_{\text{disc}}(f)$. These two expressions are called, respectively, the **geometric side** and the **spectral side** of the trace formula. (The general trace formula ([1], [2]) is considerably more complicated. If G is not anisotropic, $I_{\text{ell}}(f)$ and $I_{\text{disc}}(f)$ are merely the simplest of a number of such expansions on each side, parametrized by conjugacy classes of Levi subgroups of G.)

We should recall that $C_c^\infty\big(G(\mathbb{A})\big)$ is the vector space spanned by complex-valued functions

$$f = f_\infty \cdot f_{\text{fin}} = \prod_{v \in \infty} f_v \cdot \prod_{v \text{ finite}} f_v,$$

in which the Archimedean component f_∞ lies in the usual space of smooth functions of compact support. The non-Archimedean component f_{fin} is required to be a locally constant function of compact support on the group $G(\mathbb{A}_{\text{fin}})$ of finite adelic points. This condition on the non-Archimedean component implies in particular that for almost all v, f_v is the characteristic function of a hyperspecial maximal compact subgroup of $G(F_v)$ [26, §1.10, §3.1, §3.10].

If f equals $\prod_v f_v$, the global orbital integral is automatically a product

$$I_G(\gamma, f) = \prod_v I_G(\gamma, f_v) \tag{3}$$

of local orbital integrals

$$I_G(\gamma, f_v) = |D^G(\gamma)|_v^{1/2} \int_{G_\gamma(F_v) \backslash G(F_v)} f_v(x_v^{-1}\gamma x_v)\,dx_v.$$

(The Weyl discriminant

$$D^G(\gamma) = \det\big(1 - \mathrm{Ad}(\gamma)\big)_{\mathfrak{g}/\mathfrak{g}_\gamma}$$

is inserted only for general convenience. It does not appear globally because of the product formula on F^*.) On the spectral side, any irreducible admissible representation is a restricted tensor product

$$\pi = \bigotimes_v \pi_v, \qquad \pi_v \in \Pi\big(G(F_v)\big),$$

of irreducible representations of the local groups [3], and

$$\mathrm{tr}\big(\pi(f)\big) = \prod_v \mathrm{tr}\big(\pi_v(f_v)\big).$$

Automorphic representations are interesting because the components π_v are believed to carry fundamental arithmetic information. The data that parametrize the local sets $\Pi\big(G(F_v)\big)$ are very interesting in themselves, but what is especially important is the global information that implicitly relates the local data for the different components of any π with $m(\pi)$ positive. One hopes to study such information through the trace formula.

A major goal is to prove precise reciprocity laws relating $m(\pi)$ and $m(\pi')$, for representations π and π' of different groups G and G'. The most general pairs (G, G') for which such reciprocity laws should exist are given by Langlands' functoriality conjecture [14], [19]. The general functoriality conjecture is extremely deep, and will undoubtedly need more than just the trace formula for its ultimate resolution. However, there are a significant number of cases for which the trace formula seems ideally suited. It is for these cases that the theory of endoscopy has been designed.

Any discussion of these matters has to begin with the basic case solved by Jacquet and Langlands in 1968 [6, §17]. (See also [4], [8].) In this case, G is the multiplicative group of a quaternion algebra over F (which is actually only anisotropic modulo the center), and G' equals $GL(2)$. The basic idea is not hard to describe. The characteristic polynomial for G' and its analogue for G determine a canonical bijection from $\Gamma\big(G(F)\big)$ to a subset of $\Gamma\big(G'(F)\big)$. Indeed, the center of $G'(F)$ is bijective with F^*, while the conjugacy classes of noncentral elements in $G'(F)$ lie in disjoint subsets parametrized naturally by certain quadratic extensions of F. The characteristic polynomial gives an identical parametrization for a subset of the conjugacy classes in $G'(F) = GL(2, F)$. Thus, there is a canonical injection from the set of terms on the geometric side of the trace formula for G to a subset of the terms for G'. Jacquet and Langlands define a correspondence

$$f = \prod_v f_v \longrightarrow f' = \prod_v f'_v$$

from $C_c^\infty\big(G(\mathbb{A})\big)$ to $C_c^\infty\big(G'(\mathbb{A})\big)$ such that

$$I_G(\gamma, f) = I_{G'}(\gamma', f')$$

if γ' is the image of γ, and such that $I_{G'}(\gamma', f') = 0$ if γ' is not the image of any γ. It is known that

$$a^G(\gamma) - \mathrm{vol}\big(G_\gamma(F)\backslash G_\gamma(\mathbb{A})\big) = \mathrm{vol}\big(G'_{\gamma'}(F)\backslash G'(\mathbb{A})\big) = a^{G'}(\gamma'),$$

and also that the supplementary parabolic terms in the trace formula of G' vanish for the function f'. The geometric sides of the two trace formulas are therefore equal.

Once the two geometric sides have been cancelled, one can easily imagine being able to exploit the resulting equality of spectral sides. The correspondence of functions $f_v \to f'_v$ is defined locally. Moreover, at all places outside a finite set S, $G(F_v)$ is isomorphic with $GL(2, F_v) = G'(F_v)$. At these places, f'_v can simply be taken to be f_v. One can then fix the function $f_S = \prod_{v \in S} f_v$, and regard the difference of the two spectral sides as a linear form on the space spanned by the functions $f^S = \prod_{v \notin S} f_v$. In particular, if $\pi = \pi_S \pi^S \in \Pi\big(G(\mathbb{A})\big)$ is a representation with $m(\pi) \neq 0$, there will have to be a term for $G' = GL(2)$ to match the functional

$$f^S \longrightarrow m(\pi)\mathrm{tr}\big(\pi_S(f_S)\big)\,\mathrm{tr}\big(\pi^S(f^S)\big).$$

Combining this argument with the theorem of strong multiplicity one for $GL(2)$ (a general form of which is [7, Theorem 4.4]), one obtains a correspondence $\pi \to \pi'$ such that $\pi_v = \pi'_v$ for every $v \notin S$, and such that $m(\pi)$ equals $m(\pi')$.

The indirectness of the basic argument is part of its charm. The multiplicities $m(\pi)$ and $m(\pi')$ on the two groups are defined quite abstractly, in terms of the traces of two operators. They cannot be compared directly. The trace formulas convert information wrapped up in the multiplicities into concrete linear combinations of orbital integrals. However, these geometric terms become too complicated as f varies (with increasing support, for example) to be of great use for any isolated group. What really drives the argument is local harmonic analysis. It establishes that the geometric terms for G and G', complicated though each may be in isolation, match each other and cancel.

Langlands realized about 1970 that there would be a serious obstruction to extending the argument to other groups. The characteristic polynomial is behind the transfer of conjugacy classes from G to G', and the coefficients of the characteristic polynomial do have analogues for general G. For example, one can take any set of generators for the algebra of G-invariant polynomials on G. These objects can certainly be used to transfer conjugacy classes in G to classes in suitably related groups G'. However, invariant polynomials measure only *geometric* conjugacy classes, that is, conjugacy classes in a group of points over an algebraically closed field. For most G other than a general linear group, there exist nonconjugate elements in $G(F)$ that are conjugate over an algebraic closure $G(\overline{F})$. A similar phenomenon holds for the local groups $G(F_v)$. For example, in the case of $G = SL(2)$ and $F_v = \mathbb{R}$, the elements $\begin{pmatrix} \cos\theta & \sin\theta \\ -\sin\theta & \cos\theta \end{pmatrix}$ and $\begin{pmatrix} \cos\theta & -\sin\theta \\ \sin\theta & \cos\theta \end{pmatrix}$ are conjugate over $G(\mathbb{C})$, but not over $G(\mathbb{R})$. This phenomenon clearly complicates the problem of transferring conjugacy classes.

We are really thinking only of semisimple conjugacy classes here, since we do not want to deal with subtleties of geometric invariant theory. In fact, to focus on the essential problem, it is best to consider only elements γ that are **strongly regular**, which is to say that G_γ is a torus. The strongly regular elements form an open dense subset in any of the local groups $G(F_v)$. Langlands called two strongly regular elements in $G(F_v)$ **stably conjugate** if they were conjugate over $G(\overline{F}_v)$. Stable conjugacy is then an equivalence relation that is weaker than conjugacy. Any stable conjugacy class is a finite union of ordinary conjugacy classes. Now, we are accustomed to thinking of conjugacy classes as being dual to irreducible characters. In the present context, one can argue plausibly that the strongly regular conjugacy classes in $G(F_v)$ are the dual analogues of irreducible *tempered* characters on $G(F_v)$. The relation of stable conjugacy ought then to determine a parallel relation on the set of tempered characters. Langlands quickly realized that in the case $F_v = \mathbb{R}$, there was already a good candidate for such a relation in the work of Harish-Chandra.

One of Harish-Chandra's great achievements was the classification of the discrete series for real groups [5], [22]. Discrete series are of course the basic building blocks of arbitrary tempered representations. We recall that Harish-Chandra's classification consists of a parametrization and character formula that are remarkably similar to those established by Weyl in the special case of compact groups. However, there were two new aspects to Harish-Chandra's generalization. First of all, $G(\mathbb{R})$ can have several conjugacy classes of maximal tori (Cartan subgroups); the basic character formula applies only to a maximal torus $T(\mathbb{R})$ that is compact. Secondly, the real Weyl group $W_{\mathbb{R}}(G,T)$ induced by elements of $G(\mathbb{R})$ is generally smaller than the complex Weyl group $W_{\mathbb{C}}(G,T)$ induced by elements in $G(\mathbb{C})$. For example, if $G = \mathrm{Sp}(2n)$, then $W_{\mathbb{C}}(G,T)$ is isomorphic to a semidirect product $(\mathbb{Z}/2\mathbb{Z})^n \rtimes S_n$, while $W_{\mathbb{R}}(G,T)$ corresponds to the subgroup S_n. The discrete series are parametrized by $W_{\mathbb{R}}(G,T)$-orbits of regular characters on $T(\mathbb{R})$, and not the $W_{\mathbb{C}}(G,T)$-orbits that determine the finite dimensional representations of Weyl. In particular, the discrete series occur naturally in finite packets, each of which is bijective with the set $W_{\mathbb{R}}(G,T)\backslash W_{\mathbb{C}}(G,T)$ of cosets. Thinking of the L-functions he had defined earlier [14], Langlands called the relationship defined by this packet structure **L-equivalence**, and he used it as the foundation for a classification of all the irreducible representations of $G(\mathbb{R})$ [15]. (Knapp and Zuckerman [10] later determined the precise structure of the packets for representations outside the discrete series.) Shelstad then completed the theory for real groups [23], [24], [25], by showing among other things that the relationship of L-equivalence on the irreducible tempered characters was indeed dual, in a very precise sense, to the relationship of stable conjugacy on the strongly regular conjugacy classes.

Returning to the trace formula, we could formulate the first question that might come to mind as follows. Is the distribution

$$f \longrightarrow I_{\mathrm{ell}}(f), \qquad f \in C_c^{\infty}\big(G(\mathbb{A})\big),$$

defined by the geometric side stable? In other words, does it depend only on the stable orbital integrals

$$S_G(\sigma_v, f_v) = \sum_{\gamma_v \in \sigma_v} I_G(\gamma_v, f_v) \tag{4}$$

of the constituents f_v of f? The elements σ_v stand for strongly regular stable conjugacy classes in $G(F_v)$, and γ_v is summed over the conjugacy classes in a stable conjugacy class. At first glance, the answer might seem to be yes. Stable conjugacy can be defined for rational elements $\gamma \in G(F)$, and the volume $a^G(\gamma)$ ought to depend only on the stable class of γ. This would allow us to group the terms in $I_{\mathrm{ell}}(f)$ as sums

$$\sum_{\gamma \subset \sigma} I_G(\gamma, f)$$

of global orbital integrals, over the rational conjugacy classes γ in a rational stable class σ. (There will be some elements γ here that are not strongly regular, but this is really a side issue. Our assumption that G is anisotropic insures that the elements γ are at least semisimple.) If we look more closely, however, we find that the answer to the question is no. We have asked that the distribution be stable in *each* function f_v. In particular, if $\sigma_S = \prod_{v \in S} \sigma_v$ is any finite product of local (strongly regular) stable conjugacy classes, with a rational representative σ, then each ordinary conjugacy class $\gamma_S = \prod_{v \in S} \gamma_v$ in σ_S would also have to have a rational representative γ. There are simply not enough rational conjugacy classes in general for this to happen. Contrary to our first impression, then, the distribution $I_{\mathrm{ell}}(f)$ is not generally stable in f.

Thus, the initial observations of Langlands about stable conjugacy had immediate implications for two of the pillars of representation theory: Harish-Chandra's classification of discrete series and Selberg's trace formula. In the first case, there was the problem of constructing a relation on the irreducible tempered representations dual to stable conjugacy. In the case of the trace formula, the problem could be formulated as follows. Express $I_{\mathrm{ell}}(f)$ as the sum of a canonical stable distribution $S^G_{\mathrm{ell}}(f)$ and an explicit error term. The first group to be investigated was $SL(2)$. Labesse and Langlands [13] solved the problem for the anisotropic inner forms of this group (as well as for $SL(2)$ itself), and showed that the solution had remarkable implications for the spectral decomposition. In the general case, Langlands [18] was also able to solve the problem, under the assumption of two conjectures in local harmonic analysis.

Let us describe the main features of Langlands' general solution. The stable part was constructed first, and the error term was then expressed explicitly in terms of the stable parts $S^{G'}_{\mathrm{ell}}$ of trace formulas for groups G' of dimension smaller than G. The groups G', together with the quasi-split inner form G^* of G, are now known as the **elliptic endoscopic groups** for G. They are a family of quasi-split groups whose dual groups ([15, §2], [11, §1]) are of the form

$$\widehat{G}' = \widehat{G}_s = \mathrm{Cent}(\widehat{G}, s)^0.$$

The elements s range over semisimple points in the dual group $\widehat{G}$ of G, and are taken up to translation by the center of $\widehat{G}$ and up to conjugation by $\widehat{G}$. (See [16], [11, §7] and [20, §1.2].) Suppose for example that G is an inner form of a split adjoint group. Then $\widehat{G}$ is simply connected, and the centralizer of s in $\widehat{G}$ is already connected. The elliptic endoscopic groups in this case are the ones for which $\widehat{G}'$ is contained in no proper Levi subgroup of $\widehat{G}$. Thus, if G is an orthogonal group $SO(2n+1)$, $\widehat{G}$ equals $\mathrm{Sp}(2n, \mathbb{C})$, and s can be taken from among the elements

$$\begin{pmatrix} I_r & 0 & 0 \\ 0 & -I_{2n-2r} & 0 \\ 0 & 0 & I_r \end{pmatrix}, \qquad 0 \leq r \leq \left[\frac{n}{2}\right].$$

The corresponding group $\widehat{G}' = \widehat{G}_s$ is $\mathrm{Sp}(2r, \mathbb{C}) \times \mathrm{Sp}(2n-2r, \mathbb{C})$, and G' is the split group $SO(2r+1) \times SO(2n-2r+1)$. If G is more general, it is necessary to work with the full L-group

$${}^L G = \widehat{G} \rtimes \mathrm{Gal}(\overline{F}/F).$$

In this case there are further groups G' that are constructed by letting $\mathrm{Gal}(\overline{F}/F)$ act by outer automorphisms on $\widehat{G}' = \widehat{G}_s$ through the nonconnected components of the centralizer of s in $\widehat{G}$.

Langlands' stabilization of $I_{\mathrm{ell}}(f)$ was based on a hypothetical transfer

$$f = \prod_v f_v \longrightarrow f' = \prod_v f'_v \tag{5}$$

of functions on $G(\mathbb{A})$ to functions on any endoscopic group $G'(\mathbb{A})$. Later refinement has given a very precise form to the conjecture. In [20], Langlands and Shelstad constructed local transfer factors, which are explicit complex-valued functions

$$\Delta_G(\sigma'_v, \gamma_v)$$

of a stable conjugacy class σ'_v in $G'(F_v)$ and a strongly regular conjugacy class γ_v in $G(F_v)$. They vanish unless σ'_v maps (in a natural sense) to the stable conjugacy class of γ_v. The transfer factors then assume the role of the kernel in a transform

$$f_v \longrightarrow f'_v(\sigma'_v) = \sum_{\gamma_v} \Delta_G(\sigma'_v, \gamma_v) I_G(\gamma_v, f_v), \qquad f_v \in C_c^\infty\big(G(F_v)\big). \tag{6}$$

The conjecture is that for any $f_v \in C_c^\infty\big(G(F_v)\big)$, there is a function $f'_v \in C_c^\infty\big(G'(F_v)\big)$ whose stable orbital integrals are given by the values of the transform. That is,

$$f'_v(\sigma'_v) = S_{G'}(\sigma'_v, f'_v), \tag{7}$$

for any σ'_v. There is also a supplementary conjecture, known as the **fundamental lemma**, which applies to the unramified places v of G and G'. The assertion is that if f_v is the characteristic function of a hyperspecial maximal compact subgroup of $G(F_v)$, then f'_v can be taken to be the characteristic function of a hyperspecial maximal compact subgroup of $G'(\mathbb{A})$. Together, the two conjectures imply that there is a transfer correspondence from functions $f = \prod_v f_v$ in $C_c^\infty\big(G(\mathbb{A})\big)$ to functions $f' = \prod_v f'_v$ in $C_c^\infty\big(G'(\mathbb{A})\big)$. (Actually, there is a general problem of embedding ${}^LG'$ into ${}^LG$, which sometimes necessitates replacing G' by a certain central extension $\tilde{G}'$. We shall ignore this complication.)

Given the two local conjectures, Langlands' stabilization of $I_{\mathrm{ell}}(f)$ takes the form of an endoscopic expansion

$$I_{\mathrm{ell}}(f) = \sum_{G'} \iota(G, G') S_{\mathrm{ell}}^{G'}(f'), \tag{8}$$

with explicit coefficients $\iota(G, G')$. The distributions on the right are to be regarded as *stable* trace formulas for the elliptic endoscopic groups G'. They are linear combinations

$$S_{\mathrm{ell}}^{G'}(f') = \sum_{\sigma'} b^{G'}(\sigma') S_{G'}(\sigma', f') \tag{9}$$

over stable conjugacy classes σ' in $G'(F)$, with explicitly defined coefficients $b^{G'}(\sigma')$, of global stable orbital integrals

$$S_{G'}(\sigma', f') = \prod_v S_{G'}(\sigma', f'_v). \tag{10}$$

In terms of the original problem, the summand with G' equal to the quasi-split inner form G^* of G (that is, with $s = 1$) is to be regarded as the stable part of $I_{\text{ell}}(f)$, while the rest of the expansion constitutes the error term. Langlands actually dealt only with the strongly regular terms in the original trace formula. To be able to ignore the remaining singular terms, one would have to restrict f by, for example, taking f_v to be supported on the strongly regular set in $G(F_v)$ at some v. Kottwitz [12] was later able to deal with singular terms in $I_{\text{ell}}(f)$.

In the original basic case that G is the multiplicative group of a quaternion algebra, the right hand side of (8) has only one term, which corresponds to the quasi-split inner form $G' = G^* = GL(2)$ of G. The identity then leads to the correspondence $\pi \to \pi'$ of automorphic representations. It is harder to interpret the general case. The original trace formula does tell us that $I_{\text{ell}}(f)$ equals the spectral expansion $I_{\text{disc}}(f)$ defined by the trace of $R(f)$. The identity (8) suggests that $I_{\text{disc}}(f)$ is the sum of a stable part and an error term given precisely in terms of smaller endoscopic groups. That is,

$$I_{\text{disc}}(f) = \sum_{G'} \iota(G, G') S_{\text{disc}}^{G'}(f'). \tag{11}$$

This by itself does not provide a general correspondence of automorphic representations from G to any of the groups G', but it is nonetheless a striking conclusion. Very little is known about the multiplicities $m(\pi)$, especially regarding their stability properties. The identity (11) would give a precise obstruction to the distribution

$$f \longrightarrow I_{\text{disc}}(f) = \text{tr}\big(R(f)\big), \qquad f \in C_c^\infty\big(G(\mathbb{A})\big),$$

being stable, in terms of spectral information on smaller groups. A general distribution on $C_c^\infty\big(G(\mathbb{A})\big)$ could fail to be stable independently at each v in any given finite set S. The general obstruction would have to be measured by many terms, parametrized by products

$$G'_S = \prod_{v \in S} G'_v$$

of *local* endoscopic groups. The products G'_S that are the diagonal image of global endoscopic groups G' are sparse in the set of all products.

We shall conclude with a word on the role of the general trace formula. The formula (8), for anisotropic G, does not immediately imply (11). The problem is that there is no direct formula like (9) (taken in conjunction with (10) and (4)) to define the terms $S_{\text{disc}}^{G'}(f')$. These terms must instead be defined by induction on the dimension of G'. However, this really requires an analogue of (11) for the quasi-split form G^* of G. The inductive definition would take the form

$$S_{\text{disc}}^{G^*}(f) = I_{\text{disc}}^{G^*}(f) - \sum_{G' \neq G^*} \iota(G^*, G') S_{\text{disc}}^{G'}(f'), \qquad f \in C_c^\infty\big(G^*(\mathbb{A})\big),$$

and the conclusion to be drawn from (8) (or rather its analogue for G^*) is simply that $S_{\text{disc}}^{G^*}(f)$ is stable. But G^* is quasi-split, not anisotropic; so we have strayed from our original assumption on G. We begin to see that it is rather unnatural to restrict G to being anisotropic, even if we only want to study a simple version of the trace formula. It would be better to keep G arbitrary, and to restrict f so that the geometric side reduces to the form (2). The spectral part $I_{\text{disc}}(f)$ would still have more terms than just the characters $m(\pi)\text{tr}\big(\pi(f)\big)$. (See [1, (4.3) and Theorem 7.1].) However the extra terms are very interesting, and are in any case part of the story. In fact, there are compelling reasons to want to stabilize the full trace formula, with functions f that are unrestricted, even though there are many more terms on each side, and more problems to be solved.

References

1. J. Arthur, The invariant trace formula II. Global theory, *J. Amer. Math. Soc.* **1** (1988), 501–554.
2. J. Arthur, The trace formula and Hecke operators, *Number Theory, Trace Formulas and Discrete Groups, Symposium in Honor of Atle Selberg, Oslo, Norway, July 14–21, 1987* (K. E. Aubert, E. Bombieri, and D. Goldfeld, eds.), Academic Press, Boston, 1989, pp. 11–27.
3. D. Flath, Decomposition of representations into tensor products, *Automorphic Forms, Representations, and L-Functions*, Proc. Symp. Pure Math., vol. 33, Part I, American Mathematical Society, Providence, 1979, pp. 179–183.
4. S. Gelbart and H. Jacquet, Forms of GL(2) from the analytic point of view, *Automorphic Forms, Representations, and L-Functions*, Proc. Symp. Pure Math., vol. 33, Part I, American Mathematical Society, Providence, 1979, pp. 213–251.
5. Harish-Chandra, Discrete series for semisimple Lie groups. II. Explicit determination of the characters, *Acta. Math.* **116**, 1–111.
6. H. Jacquet and R. P. Langlands, *Automorphic Forms on* GL(2), Lecture Notes in Mathematics, vol. 114, Springer-Verlag, Berlin, 1970.
7. H. Jacquet and J. Shalika, On Euler products and the classification of automorphic representations, II, *Amer. J. Math.* **103** (1981), 777–815.
8. A. Knapp, Theoretical aspects of the trace formula for $GL(2)$, *these* PROCEEDINGS, pp. 355–405.
9. A. Knapp and J. Rogawski, Applications of the trace formula, *these* PROCEEDINGS, pp. 413–431.
10. A. Knapp and G. Zuckerman, Classification of irreducible tempered representations of semisimple Lie groups, *Annals of Math.* **116** (1982), 389–501.
11. R. Kottwitz, Stable trace formula: cuspidal tempered terms, *Duke Math. J.* **51** (1984), 611–650.
12. R. Kottwitz, Stable trace formula: elliptic singular terms, *Math. Annalen* **275** (1986), 365–399.
13. J.-P. Labesse and R. P. Langlands, L-indistinguishability for $SL(2)$, *Canad. J. Math.* **31** (1979), 726–785.
14. R. P. Langlands, Problems in the theory of automorphic forms, *Lectures in Modern Analysis and Applications III*, Lecture Notes in Mathematics, vol. 170, Springer-Verlag, Berlin, 1970, pp. 18–61.
15. R. P. Langlands, On the classification of irreducible representations of real algebraic groups, mimeographed notes, Institute for Advanced Study, Princeton, NJ, 1973, *Representation Theory and Harmonic Analysis on Semisimple Lie Groups* (P. J. Sally and D. A. Vogan, eds.), Math. Surveys and Monographs, vol. 31, American Mathematical Society, Providence, 1989, pp. 101–170.
16. R. P. Langlands, Stable conjugacy: definitions and lemmas, *Canad. J. Math.* **31** (1979), 700–725.
17. R. P. Langlands, On the zeta-functions of some simple Shimura varieties, *Canad. J. Math.* **31** (1979), 1121–1216.
18. R. P. Langlands, *Les Débuts d'une Formule des Traces Stable*, Publications Mathématiques, vol. 13, L'Université Paris VII, Paris, 1983.
19. R. P. Langlands, Where stands functoriality today?, *these* PROCEEDINGS, pp. 457–471.
20. R. P. Langlands and D. Shelstad, On the definition of transfer factors, *Math. Annalen* **278** (1987), 219–271.
21. A. Selberg, Harmonic analysis and discontinuous groups in weakly symmetric Riemannian spaces with applications to Dirichlet series, *J. Indian Math. Soc.* **20** (1956), 47–87.
22. W. Schmid, Discrete series, *these* PROCEEDINGS, pp. 83–113.
23. D. Shelstad, Orbital integrals and a family of groups attached to a real reductive group, *Annales Scient. Ecole Norm. Sup.* **12** (1979), 1–31.
24. D. Shelstad, Embeddings of L-groups, *Canad. J. Math.* **33** (1981), 513–558.
25. D. Shelstad, L-indistinguishability for real groups, *Math. Annalen* **259** (1982), 385–430.

26. J. Tits, Reductive groups over local fields, *Automorphic Forms, Representations, and L-Functions*, Proc. Symp. Pure Math., vol. 33, Part I, American Mathematical Society, Providence, 1979, pp. 29–69.

DEPARTMENT OF MATHEMATICS, UNIVERSITY OF TORONTO, TORONTO, ONTARIO M5S 3G3, CANADA

Proceedings of Symposia in Pure Mathematics
Volume **61** (1997), pp. 443–455

Automorphic Spectrum of Symmetric Spaces

Hervé Jacquet

1. Introduction

Let F be a number field, $F_{\mathbb{A}}$ its ring of adeles, and G a reductive group defined over F. Suppose for now that G is semisimple rather than reductive. Let f be a smooth function of compact support on $G(F_{\mathbb{A}})$. We set:

$$K_f(x,y) := \sum_{\gamma \in G(F)} f(x^{-1}\gamma y). \tag{1}$$

Then we have a **spectral decomposition** of the kernel K_f:

$$K_f(x,y) = \sum_{\chi} K_{f,\chi}(x,y). \tag{2}$$

In this sum χ is a discrete automorphic representation of the Levi factor M_χ of a parabolic subgroup $P_\chi = M_\chi U_\chi$ of G. If $M_\chi = G$ then χ is a discrete component of $L^2(G(F)\backslash G(F_{\mathbb{A}}))$ and

$$K_{f,\chi}(x,y) = \sum_i \left(\rho(f)\phi_i\right)(x)\,\overline{\phi_i}(y), \tag{3}$$

where ϕ_i is an orthonormal basis of the space $V(\chi)$ of χ and we have set:

$$\left(\rho(f)\right)\phi(x) = \int_{G(F_{\mathbb{A}})} f(g)\phi(xg)dg.$$

In general $K_{f,\chi}$ has an expression of the form:

$$K_{f,\chi}(x,y) = \int \sum_i E(x,\rho(f)\phi_i,u)\overline{E(y,\phi_i,u)}\,du, \tag{4}$$

where $E(x,\phi,u)$ is an Eisenstein series for the parabolic subgroup P_χ built out of the datum χ, and u is integrated over a suitable Euclidean space. The goal of the theory of Eisenstein series is to obtain formula (2). See [aK] in this volume for the example $G = GL(2)$.

Potentially, the formula (2) contains all the information about the discrete terms, that is, the terms for which $M_\chi = G$. Classically, to obtain this information

---

1991 *Mathematics Subject Classification.* Primary 22E55, 43A85.

effectively, one sets $x = y$ in the formula and then integrates over $G(F)\backslash G(F_{\mathbb{A}})$. This leads to divergent integrals, which need to be regularized by the "truncation process." The formula obtained by this process is the absolute **trace formula**.

When $G(F)\backslash G(F_{\mathbb{A}})$ is compact, the integrals converge and the trace formula identity is then the equality of two expressions. The first expression is:

$$\sum_{\xi} \mathrm{vol}\,(H_\xi(F)\backslash H_\xi(F_{\mathbb{A}})) \int_{H_\xi(F_{\mathbb{A}})\backslash G(F_{\mathbb{A}})} \Phi(h^{-1}\xi h)\, dh. \tag{5}$$

The sum is over a set of representatives for the conjugacy classes of $G(F)$. For each representative ξ we denote by H_ξ the centralizer of ξ in G. The second expression is a sum over all discrete components χ of $L^2(G(F)\backslash G(F_{\mathbb{A}}))$:

$$\sum_{\chi} \mathrm{Tr}\,\chi(f). \tag{6}$$

For an introduction to the trace formula for general G, see the expository articles [A1], [A2], [L2], as well as the references therein and Chapter 10 of [L1].

In this short note we propose a generalization of this setup for the trace formula to a relative situation.

2. The Spectrum

Let F and G be as above with G connected. Let $g \mapsto g^*$ be an anti-automorphism of order 2 of G and S be a connected component of the subvariety T of G defined by

$$T = \{s \in G \mid s^* = s\}.$$

Examples.

1) Let E/F be a quadratic extension of number fields and G the group $GL(n, E)$ regarded as an algebraic group over F. We denote by $s \mapsto \overline{s}$ the effect of the nontrivial element of the Galois group on a matrix s and set $s^* = {}^t\overline{s}$. Then the corresponding variety T is connected. It is the variety of invertible Hermitian matrices.

2) Let G be the group $GL(2n)$ regarded as an algebraic group over F, and let the anti-automorphism be $g \mapsto g^{-1}$. We consider the variety S of matrices conjugate to the matrix

$$\begin{pmatrix} 1_n & 0 \\ 0 & -1_n \end{pmatrix}$$

This is a connected component of the variety T of matrices s such that $s = s^{-1}$.

In our general set-up, $G(F)$ acts on $S(F)$ on the right by $(s, g) \mapsto g^*sg$. For now, assume that the action of $G(F)$ on $S(F)$ is transitive. Near the end of this section we will explain how to deal with the case where the group $G(F)$ is not transitive on $S(F)$. At any rate, in Example 2 the group $G(F)$ is indeed transitive. Let ϵ be a point of $S(F)$ and H the stabilizer of ϵ in G. Suppose that Φ is a smooth function of compact support on $S(F_{\mathbb{A}})$. Then we can define a function K_Φ on $G(F_{\mathbb{A}})$ by:

$$K_\Phi(g) := \sum_{\xi \in S(F)} \Phi(g^*\xi g). \tag{7}$$

Clearly, it satisfies

$$K_\Phi(\gamma g) = K_\Phi(g), \qquad \gamma \in G(F).$$

The space $\mathcal{K}$ spanned by the functions $K_\Phi(g)$ is invariant under right translations by $G(F_\mathbb{A})$. It is natural to try to decompose the representation of $G(F_\mathbb{A})$ on $\mathcal{K}$ into a continuous sum of irreducible representations.

To make this more precise, we consider first the case where G is semisimple and $H(F)\backslash H(F_\mathbb{A})$ is compact. We can find the **spectral decomposition** of the space $\mathcal{K}$ as follows. Given Φ, we can choose a smooth function of compact support f on $G(F_\mathbb{A})$ such that

$$\Phi(s) = \int_{H(F)\backslash H(F_\mathbb{A})} f(hg)\, dh \qquad \text{if } s = g^*\epsilon g.$$

We now consider the kernel K_f associated to f (see (1)) and its spectral expression (2). Then we can write

$$K_\Phi(g) = \sum_\chi K_{\Phi,\chi}(g), \tag{8}$$

where we have set:

$$K_{\Phi,\chi}(g) := \int_{H(F)\backslash H(F_\mathbb{A})} K_{f,\chi}(h,g)\, dh.$$

As in section 1, χ is a discrete automorphic representation of the Levi-factor M_χ of a parabolic subgroup $P_\chi = M_\chi U_\chi$ of G.

In particular, if $M_\chi = G$, then

$$K_{\Phi,\chi}(g) = \sum_i P(\rho(f)\phi_i)\, \overline{\phi_i}(g),$$

where we have set

$$P(\phi) := \int_{H(F)\backslash H(F_\mathbb{A})} \phi_i(h)\, dh.$$

Thus $K_{\Phi,\chi} \neq 0$ if and only there is $\phi \in V(\chi)$ such that $P(\phi) \neq 0$. We shall say then that χ **belongs** to the discrete automorphic spectrum of S, or that it is **distinguished** by H. Note that the linear form P is invariant under $H(F_\mathbb{A})$. Thus, at each place v of F, there is a linear form $P_v \neq 0$ on the space of χ_v such that $P_v(\pi_v(h_v)u) = P_v(u)$ for each $h_v \in H_v$ and each vector u in the space of π_v. We may express this by saying that π_v is **distinguished** by H_v. In general π_v will not be a discrete component of $L^2(H_v\backslash G_v)$.

If $M_\chi \neq G$ then

$$K_{\Phi,\chi}(s) = \sum_i \Big(\int_{H(F)\backslash H(F_\mathbb{A})} E(h,\rho(f)\phi_i,u)\, dh\Big)\, \overline{E(g,\phi_i,u)}\, du$$

if $s = g^*\epsilon g$.

Formula (8) is the **spectral decomposition** of the space $\mathcal{K}$. It is the relative analogue (in the case at hand) of formula (2).

When G is reductive but possibly not semisimple, the integrals are to be taken over a quotient $ZH(F)\backslash H(F_\mathbb{A})$, where Z is a suitable central subgroup of $G(F_\mathbb{A})$ contained in $H(F_\mathbb{A})$. If also the quotient $ZH(F)\backslash H(F_\mathbb{A})$ is not compact, then in order to carry through the above computations one needs to regularize the integrals in terms of the **truncation operator**. In such a process one will obtain a decomposition of the form (2). In particular, there will be discrete terms and continuous terms. Some discrete terms may appear that were not in the original spectral formula (2).

Just as in the absolute case the formula (8) contains potentially all the information about the spectrum. In order to obtain this information effectively, it is natural to try to imitate the construction of the absolute trace formula. Let us assume again that G is semisimple and $H(F)\backslash H(F_{\mathbb{A}})$ is compact. We consider the integral:

$$\int_{H(F)\backslash H(F_{\mathbb{A}})} K_\Phi(h)\, dh.$$

It can be computed in two ways. Using the definition of K_Φ, we obtain on the one hand:

$$\sum_\xi \mathrm{vol}(H_\xi(F)\backslash H_\xi(F_{\mathbb{A}})) \int_{H_\xi(F_{\mathbb{A}})\backslash H(F_{\mathbb{A}})} \Phi(h^*\xi h)\, dh. \tag{9}$$

The sum is over a set of representatives for the orbits of H on $S(F)$. On the other hand, for each representative ξ we denote by H_ξ the stabilizer of ξ. Using the spectral expansion (8), we get:

$$\sum_\chi \int K_{\Phi,\chi}(h)\, dh. \tag{10}$$

In particular if $M_\chi = G$ then

$$\int K_{\Phi,\chi}(h)\, dh = \sum_i P(\rho(f)\phi_i)\, \overline{P(\phi_i)}.$$

The right hand side is the spherical character attached to the representation χ and the linear form P. The equality of (9) and (10) constitutes the **relative trace formula**. When the quotient is not compact, we can expect all the complications of the absolute trace formula.

Another complication is due to the fact that the action of $G(F)$ on $S(F)$ need not be transitive. We then choose a set of representatives Y for the orbits of $G(F)$ on $S(F)$. For each $\epsilon \in Y$ we denote by H_ϵ the stabilizer of ϵ in G. To a function Φ we now associate a family of functions f_ϵ with $\epsilon \in Y$ such that

$$\Phi(s) = \int_{H_\epsilon(F)\backslash H_\epsilon(F_{\mathbb{A}})} f_\epsilon(hg)\, dh \qquad \text{if } s = g^*\epsilon g.$$

Actually to make this definition precise we make it locally and assume that S satisfies the Hasse principle: two points of $S(F)$ are conjugate under $G(F)$ if and only if at every place v they are conjugate under $G(F_v)$. It remains to define $K_{\Phi,\chi}$. If all the quotients $H_\epsilon(F)\backslash H_\epsilon(F_{\mathbb{A}})$ were compact, we would set:

$$K_{\Phi,\chi}(g) := \sum_{\epsilon\in Y} \int_{H_\epsilon(F)\backslash H_\epsilon(F_{\mathbb{A}})} K_{f_\epsilon,\chi}(h_\epsilon, g)\, dh_\epsilon$$

However, at least in Example 1 the quotient is not compact for most of the ϵ, and so we have to use truncation again. Finally to obtain the relative trace formula in the case at hand we need to choose $\epsilon_0 \in Y$ and consider the integral:

$$\int_{H_{\epsilon_0}(F)\backslash H_{\epsilon_0}(F_{\mathbb{A}})} K_\Phi(h)\, dh.$$

We discuss two applications of the previous notions, both fairly typical. One application is to the positivity of an L-function. The other application is to the principle of functoriality.

3. An Example

Just as in the absolute trace formula, one expects a notion of "instability" and "endoscopy"(see [A2], [LL], and [L2] for an introduction to these notions). We give an example here and an application. However our formulation is not explicitly in terms of endoscopy. Let E/F be a quadratic extension of number fields and η the corresponding idele class character. Let χ be an automorphic cuspidal representation of $GL(2, F_{\mathbb{A}})$ with trivial central character. Let $L(s, \chi)$ be the corresponding L-function. We normalize our notation so that the center of symmetry is at $\frac{1}{2}$. The positivity of $L(\frac{1}{2}, \chi)$ for holomorphic forms has been studied extensively (see [hK] for a brief history). In [KS] this positivity is established for Maass forms. We present here the proof given in [G1] of the following related fact:

Theorem. $L(\frac{1}{2}, \chi)L(\frac{1}{2}, \chi \otimes \eta) \geq 0.$

For now, some restrictive assumptions must be made, but the principle of the proof is general.

Let G be the group $GL(2)$ regarded as an algebraic group over F. We consider the anti-automorphism

$$\tau(g) := \begin{pmatrix} 1 & 0 \\ 0 & -1 \end{pmatrix} g^{-1} \begin{pmatrix} 1 & 0 \\ 0 & -1 \end{pmatrix}.$$

We consider the variety S of matrices $s \in G$ with $\det s = 1$ and $\tau(s) = s$. The stabilizer in G of the identity I in S is the group A of diagonal matrices. The group A operates by conjugation on S with the center Z of G operating trivially. Elements of S have the form:

$$s = \begin{pmatrix} a & b \\ c & a \end{pmatrix} \qquad \text{with } a^2 - bc = 1.$$

An element s is regular semisimple if $a^2 \neq 1$ (see [JR2]). If an element s is not regular semisimple then it has a Jordan decomposition $s = s_1 n$ where $s_1 = \pm I$ and $n \in S$ is unipotent. The regular semisimple orbits are parametrized by $\alpha \in F - \{\pm 1\}$. To obtain a set of representatives, for each α we simply choose in any way β and γ such that $\beta\gamma = \alpha^2 - 1$ and associate to α the matrix

$$\sigma_\alpha = \begin{pmatrix} \alpha & \beta \\ \gamma & \alpha \end{pmatrix}.$$

The other orbits are $\{I\}, \{-I\}$, and the orbit of the elements of the form $s_1 n_i$, where $s_1 = \pm I$, $i = 1, 2$, and

$$n_1 = \begin{pmatrix} 1 & 1 \\ 0 & 1 \end{pmatrix}, \qquad n_2 = \begin{pmatrix} 1 & 0 \\ 1 & 1 \end{pmatrix}.$$

Recall that an integral of the form

$$\int \phi(a) |\det a|^s \, da$$

where ϕ is in the space $V(\chi)$ of χ is a holomorphic multiple of $L(s,\chi)$ and that $L(s,\chi)$ is equal to an integral of this form for a suitable $\phi \in V(\chi)$ (see [JL]). Thus χ is distinguished by A if and only if $L(\frac{1}{2},\chi) \neq 0$. Similarly, an integral of the form

$$\int \phi(a)\eta(\det a)|\det a|^s\, da$$

is an holomorphic multiple of $L(s,\chi\otimes\eta)$. Thus $L(\frac{1}{2},\chi\otimes\eta) \neq 0$ if and only if there is $\phi \in V(\chi)$ such that

$$\int \phi(a)\eta(\det a)\, da \neq 0.$$

We are led to define the notion of a representation **distinguished** by a pair of the form (H,η), where H is as above and η is a character of $H(F_{\mathbb{A}})$ trivial on $H(F)$. In the situation at hand, χ is distinguished by the pair $(A,\eta\circ\det)$ if and only if

$$L(\tfrac{1}{2},\pi\otimes\chi) \neq 0.$$

If Φ is a smooth function of compact support on $S(F_{\mathbb{A}})$ we consider the integral:

$$\int_{Z(F_{\mathbb{A}})A(F)\backslash A(F_{\mathbb{A}})} K_\Phi(a)\eta(\det a)\, da.$$

This integral is divergent. Because we have a very simple situation, we can make do with something simpler than truncation. The integral is weakly convergent in the following sense. It can be computed as the sum:

$$\sum_{\alpha\neq\pm1}\Omega[\Phi,\eta;\alpha] + \sum_{\pm,\, i\in\{1,2\}}\Omega[\Phi,\eta;\pm n_i];$$

here for each α we set

$$\Omega[\Phi,\eta;\alpha] := \int_{A(F_{\mathbb{A}})/Z(F_{\mathbb{A}})} \Phi(a\sigma_\alpha a^{-1})\eta(a)\, da.$$

This integral is absolutely convergent. The remaining terms in the sum are improper integrals. For example:

$$\Omega[\Phi,\eta;n_1] := \int_{F_{\mathbb{A}}^\times} \Phi\left[\begin{pmatrix} 1 & a \\ 0 & 1\end{pmatrix}\right]\eta(a)\, d^\times a.$$

This integral can be viewed as the analytic continuation to the point $s=0$ of the following **Tate integral**, which converges for $\Re s > 1$:

$$\int_{F_{\mathbb{A}}^\times} \Phi\left[\begin{pmatrix} 1 & a \\ 0 & 1\end{pmatrix}\right]|a|^s\eta(a)\, d^\times a.$$

Note that the semisimple elements $\pm I$ do not contribute to the formula.

We now identify the multiplicative group of E with the torus $T \subset GL(2,E)$ of matrices of the form:

$$t = \begin{pmatrix} a & 0 \\ 0 & \overline{a}\end{pmatrix}.$$

As before, $a \mapsto \overline{a}$ is the conjugation in E with respect to F. We consider all (classes of) inner forms of $GL(2,F)$ that contain the torus T. A convenient way to describe the classes is as follows. Choose once and for all a set X of representatives for the cosets of $N_{E/F}(E^\times)$ in $F^\times$. For each $\epsilon \in X$ let G_ϵ be the multiplicative group of the algebra of matrices of the form:

$$g = \begin{pmatrix} a & b\epsilon \\ \bar{b} & \bar{a} \end{pmatrix}.$$

We consider the symmetric space S_ϵ of $s \in G_\epsilon$ such that $\tau(s) = s$ and $\det s = 1$. Thus S_ϵ is the space of matrices of the form

$$s = \begin{pmatrix} a & b\epsilon \\ \bar{b} & a \end{pmatrix}, \qquad a \in F,\ b \in E,\ a^2 - 1 = b\bar{b}\epsilon.$$

The stabilizer of I is the group T. It operates by conjugation and the center Z of G_ϵ operates trivially. Regular semisimple elements are those for which $a^2 \neq 1$. The remaining elements are $\pm I$. Note that the disjoint union over $\epsilon \in X$ of the regular semisimple orbits is parametrized by the set $F - \{\pm 1\}$. Indeed, for each α in this set we can write $\alpha^2 - 1 = \beta\bar{\beta}\epsilon_\alpha$ with $\epsilon_\alpha \in X$. We choose a β satisfying this condition, and then the representative is the matrix

$$\tau_\alpha = \begin{pmatrix} \alpha & \beta\epsilon_\alpha \\ \bar{\beta} & \alpha \end{pmatrix}.$$

Next, we choose for each ϵ a function Φ_ϵ on S_ϵ and consider the integral:

$$\int_{T(F_{\mathbb{A}})/Z(F_{\mathbb{A}})} K_{\Phi_\epsilon}(t)dt.$$

Then

$$\sum_{\epsilon \in X} \int_{T(F_{\mathbb{A}})/Z(F_{\mathbb{A}})} K_{\Phi_\epsilon}(t)\, dt$$
$$= \sum_{\alpha \neq \pm 1} \Omega(\Phi_{\epsilon_\alpha}; \alpha) + \sum_{\epsilon, \pm} \Phi_\epsilon(\pm I) \operatorname{vol}\left(T(F_{\mathbb{A}})/T(F)Z(F_{\mathbb{A}})\right),$$

where we have set:

$$\Omega(\Phi_\epsilon; \alpha) := \int_{T(\mathbb{A})/Z(F_{\mathbb{A}})} \Phi_\epsilon(t\tau_\alpha t^{-1})\, dt.$$

We now compare the formulae we just obtained. For given (Φ) there is a family (Φ_ϵ) (and conversely) such that, for any $\alpha \neq \pm 1$,

$$\Omega(\Phi, \eta; \alpha) = \Omega(\Phi_{\epsilon_\alpha}, \alpha). \tag{11}$$

We say that (Φ) and the family (Φ_ϵ) have **matching orbital integrals**. Then

$$\int_{A(F_{\mathbb{A}})/Z(F_{\mathbb{A}})A(F)} K_\Phi(a)\eta(a)\, da = \sum_{\epsilon \in X} \int_{T(F_{\mathbb{A}})/Z(F_{\mathbb{A}})T(F)} K_{\Phi_\epsilon}(t)\, dt. \tag{12}$$

Note in view of (11) that the terms attached to the regular semisimple elements agree. As a matter of fact there is a natural bijection between the classes of regular semisimple elements in $S(F)$ and the disjoint union of the classes of regular semisimple elements in the spaces $S_\epsilon(F)$. The bijection associates to σ_α the class of τ_α, or, more intrinsically, the class of σ is associated to the class of τ if σ and τ are conjugate in $GL(2, E)$. Thus the content of (12) is the assertion that the singular terms coming from the groups G_ϵ and the unipotent terms coming from $GL(2)$ agree.

Equating the spectral sides from the relative trace formulae, we obtain the following result. Assume that Φ and the family of the Φ_ϵ have matching orbital integrals. Suppose that Φ corresponds to f on $GL(2)$ and Φ_ϵ corresponds to f_ϵ on G_ϵ. Let χ be an automorphic cuspidal representation of $GL(2, F_{\mathbb{A}})$ (with trivial central character). Then its contribution to the relative trace formula for $GL(2)$ is

$$\sum_{\phi_i \in V(\chi)} \int \rho(f)\phi_i(a)da \overline{\int \phi_i(a)\eta(\det a)\, da}.$$

As before, the sum is over an orthonormal basis of $V(\chi)$. In particular, the above contribution is nonzero for a suitable choice of f if and only if

$$L(\tfrac{1}{2}, \chi)L(\tfrac{1}{2}, \chi \otimes \eta) \neq 0. \tag{13}$$

To such a χ there corresponds for each ϵ in a certain subset $X(\chi)$ of X a representation χ_ϵ of the group G_ϵ. We recall that the relation is that, at all places v where $G_{\epsilon v} \simeq G_v$, we have $\chi_{\epsilon v} = \chi_v$. Using a standard argument, we can then derive the equality (see [J1]):

$$\sum_{\phi_i \in V(\chi)} \int \rho(f)\phi_i(a)\, da \overline{\int \phi_i(a)\eta(\det a)\, da}$$
$$= \sum_{\epsilon \in X(\chi)} \sum_{\phi_i \in V(\chi_\epsilon)} \int \rho(f_\epsilon)\phi_i(t)\, dt \overline{\int \phi_i(t)\, dt}.$$

In fact, for a given χ, there is at most one ϵ such that the representation χ_ϵ is distinguished by T, that is, contributes to the identity above. Now the right hand side is a distribution of positive type: if for each ϵ the function f_ϵ has the form $f_\epsilon = f_1 * f_1^*$, where we set $f_1^*(g) = \overline{f(g^{-1})}$, then the right hand side is ≥ 0. In [G1] it is shown that one can choose the data in such a way that the right hand side is ≥ 0 and the left hand side is equal to a positive multiple of the product (13). We conclude that the product (13) is ≥ 0. This proves the theorem.

4. Relation with the Principle of Functoriality

Let again E/F be a quadratic extension of number fields and η the corresponding idele class character. For any automorphic cuspidal representation π of $GL(n, F_{\mathbb{A}})$, there is an automorphic (in general cuspidal) automorphic representation Π of $GL(n, E_{\mathbb{A}})$ that is a base change of π, in the sense that

$$L(s, \pi)L(s, \pi \otimes \eta) = L(s, \Pi).$$

In the case $n = 1$ π is an idele class character of F and $\Pi = \pi \circ \text{norm}$. For $n = 2$ (resp. $n \geq 2$) this is a special case of the result of [L1] (resp. [AC]).

Conjecture. *An automorphic representation Π of $GL(n, E_{\mathbb{A}})$ is a base change if and only if there is a unitary group H such that Π is distinguished by H. Furthermore, this group H can be taken to be quasisplit.*

In one direction, we can use an argument of [HLR] to show that an automorphic representation Π distinguished by a unitary group is invariant under the Galois group and therefore is a base change by [AC]. Indeed, the linear form P is invariant under $H(F_{\mathbb{A}})$. Thus if v is a place of F that splits into v_1 and v_2 in E, then on the tensor product $\pi_{v_1} \otimes \pi_{v_2}$ there is a linear form P_v that is invariant under H_v, i.e., verifies for every g and every vector u:

$$P\big(\pi_{v_1}(g) \otimes \pi_{v_2}({}^tg^{-1})u\big) = P(u).$$

Thus π_{v_1} is contragredient to the representation $g \mapsto \pi_{v_2}({}^tg^{-1})$. Since the automorphism $g \mapsto {}^tg^{-1}$ takes a representation to the contragredient representation, we see that $\pi_{v_1} = \pi_{v_2}$. Now, suppose v is a place of F inert and unramified in E, and let w be the corresponding place of E. If the representation Π_w is unramified, then it is invariant under the Galois conjugation. Thus the representation $g \mapsto \Pi(g)$ and the representation $g \mapsto \Pi(\overline{g})$ have the same components at almost all places and are necessarily equivalent (strong multiplicity one). Our assertion follows.

It remains to prove that a representation that is a base change is distinguished by a unitary group. Now to explain our approach to this converse statement, we begin by reviewing the Kuznietsov trace formula [nK]. First we consider certain orbital integrals. We fix a nontrivial additive character ψ of F if F is local and of $F_{\mathbb{A}}/F$ if F is a number field. Let A be the group of diagonal matrices in $GL(n)$, $W = W(G)$ the Weyl group of A identified with the group of permutation matrices in $GL(r, F)$, and N the group of upper triangular matrices with unit diagonal. We define an algebraic group morphism from N to F by:

$$\theta_0(u) = \sum_i u_{i,i+1}$$

and set $\theta_\psi(u) = \psi(\theta_0(u))$. We often write θ for θ_ψ. Recall that the elements of the form wa with $w \in W$ and $a \in A(F)$ form a set of representatives for the action of $N(F) \times N(F)$ on $G(F)$ defined by:

$$s \overset{(n_1,n_2)}{\mapsto} {}^tn_1 s n_2.$$

We say that wa is **relevant** if $\theta_0(n_1n_2) = 1$ when (n_1, n_2) fixes wa. If wa is relevant, then there is a standard parabolic subgroup P_w (i.e., P_w contains N) with standard Levi factor M_w (i.e., M_w contains A) such that w is the longest element of $W \cap M_w$. We then denote by A_w the center of M_w. The element a belongs to A_w. Conversely, all wa obtained in this way are relevant. We denote by $R(G)$ the set of w of the above form. If $w \in R(G)$ and $M = M_w$, we also write $w = w_M$. In particular, w_G is the longest element of $W(G)$. If F is a local (resp. global field) and Φ is a smooth function of compact support on $GL(n, F)$ (resp. $GL(n, F_{\mathbb{A}})$), for each relevant element wa in $GL(n, F)$ we consider the "Kloosterman integral":

$$I(wa, \Phi) = \int \Phi({}^tn_1 wan_2)\theta(n_1n_2)\, d(n_1, n_2).$$

The integral is taken over the quotient of $N(F) \times N(F)$ (resp. $N(F_{\mathbb{A}}) \times N(F_{\mathbb{A}})$) by the stabilizer of wa.

The Kuznietsov trace formula is obtained in the following way. Let f a smooth function of compact support on $G(F_{\mathbb{A}})$ and set as before:

$$K_f(x, y) = \sum_{\gamma \in GL(n,F)} f(x^{-1}\gamma y).$$

Then

$$\int K_f({}^t n_1, n_2)\theta(n_1^{-1}n_2)\,dn_1\,dn_2 = \sum_{w,\alpha} I(w\alpha, f).$$

The integral on the left is over $(N(F)\backslash N(F_{\mathbb{A}})) \times (N(F)\backslash N(F_{\mathbb{A}}))$. The sum on the right is over all relevant elements. To obtain a useful formula, we must fix a character ω of the center $Z \simeq F^\times$ and integrate over the center to obtain the identity:

$$\int \Big(\int_{Z(F_{\mathbb{A}})/Z(F)} K({}^t n_1, n_2 z)\omega(z)\,dz\Big)\theta(n_1^{-1}n_2)\,dn_1\,dn_2$$
$$= \sum_{w,\alpha} \int_{Z(F_{\mathbb{A}})} I(w\alpha z, f)\omega(z)\,dz. \tag{14}$$

The sum is now over all $\alpha \in A_w(F)/Z(F)$. We have then

$$\sum_{w,\alpha} \int_{Z(F_{\mathbb{A}})} I(w\alpha z, f)\omega(z)\,dz = \sum_{\chi} \int K_\chi({}^t n_1, n_2 z)\theta(n_1^{-1}n_2)\omega(z)\,dz\,dn_1\,dn_2.$$

Here the sum is over all χ that are cuspidal automorphic representations of a Levi subgroup M_χ and whose central character agrees with ω^{-1} on the center of G.

We denote by S the set of invertible Hermitian matrices in $GL(n, E)$. The group $N(E)$ operates on $S(F)$ by:

$$s \overset{n}{\mapsto} {}^t\overline{n} s n.$$

We can use this action to define the relevant orbits of $N(E)$ on $S(F)$. As before, the elements of the form wa with $w \in R(G)$ and $a \in A_w(F)$ form a set of representatives for the relevant orbits. We can then define global (resp. local) **orbital integrals** by:

$$J(wa, \Phi) = \int \Phi({}^t\overline{n} wan)\theta(n\overline{n})\,dn.$$

Note that $n\overline{n}$ is the product of an element of $N(F_{\mathbb{A}})$ (resp. $N(F)$) and an element of the derived group of $N(E_{\mathbb{A}})$ (resp. $N(E)$) so that $\theta(n\overline{n})$ is well defined. Now let Φ be as before a smooth function of compact support on $S(F_{\mathbb{A}})$. Then we consider the integral

$$\int_{N(E)\backslash N(E_{\mathbb{A}})} K_\Phi(n)\theta(n\overline{n})\,dn.$$

It is equal to

$$\sum_{w,\alpha} J(w\alpha, \Phi).$$

Moreover as before we have:

$$\int_{N(E)\backslash N(E_{\mathbb{A}})} \Big(\int_{Z(F_{\mathbb{A}})/Z(F)} K_\Phi(zn)\omega(z)\,dz\Big)\theta(n\overline{n})\,dn$$
$$= \sum_{w,\alpha} \int_{Z(F_{\mathbb{A}})} J(w\alpha z, \Phi)\omega(z)\,dz. \tag{15}$$

We now compare the formulae. We need to compare the local orbital integrals. Consider a place v of F inert in E. We then have a local quadratic extension E_w/F_v

and an additive character ψ_v of F_v. We conjecture the existence of transfer factors $\gamma(wa, \psi_v)$ such that for any f there is Φ (and conversely) with:

$$I(wa, f) = \gamma(wa, \psi_v) J(wa, \Phi).$$

We shall say then that Φ and f have **matching orbital integrals**. In particular, if the residual characteristic of F_v is odd and the order of ψ_v is 0 then the characteristic function f of $GL(n, \mathcal{O}_v)$ and the characteristic function Φ of $S(F) \cap GL(2, \mathcal{O}_w)$ should have matching orbital integrals (**fundamental lemma**). Moreover, there should be a similar statement for general Hecke functions f. In addition, there is an elementary matching at split places, with trivial transfer factors. Now let $z \in Z(F_{\mathbb{A}})$. Then the product over all inert places v of F of the transfer factors should have the form:

$$\prod_v \gamma(w\alpha z_v, \psi_v) = t(z) \qquad \text{for } \alpha \in A_w(F),\ z \in F_{\mathbb{A}},$$

where t is either the trivial character or the quadratic character η.

Now assume that Φ and f have matching orbital integrals. Then:

$$\int K_\Phi(n)\theta(\overline{n}n)\, dn\, \omega(z)\, dz = \int K_f({}^t n_1, n_2 z)\theta(n_1^{-1} n_2)\, dn_1\, dn_2\, t\omega(z)\, dz.$$

Next we compare the spectral sides. Let Σ be the set of projective equivalence classes of rational Hermitian matrices. Choose a set of representatives. As before we can associate to Φ a collection of functions f_ϵ, $\epsilon \in \Sigma$ on $GL(n, E_{\mathbb{A}})$. Then the equality of the spectral sides reads:

$$\sum_\Pi \sum_\epsilon \int_{\tilde{H}_\epsilon(F)\backslash \tilde{H}_\epsilon(F_{\mathbb{A}})} K_{f_\epsilon,\pi}(h, n)\omega(\lambda(h))\, dh\, \theta(n\overline{n})\, dn$$
$$= \sum_\pi K_{f,\pi}({}^t n_1, n_2 z)\theta(n_1^{-1} n_2)\, dn_1\, dn_2\, t\omega(z)\, dz. \tag{16}$$

Here the sum is over a set of representatives for the projectives equivalence classes of Hermitian matrices. For each representative ϵ we denote by $\tilde{H}_\epsilon$ the corresponding similitude group and by λ the similitude ratio. From this identity the conjecture would follow. (See [JY1], [JY2], [JY3], [J5].)

In general, representations distinguished by a group H should have a simple characterization in terms of the principle of functoriality (see [JLR]). For instance if $H = GL(n, F)$ then the representations should be base change from the unitary group (see [yF1]). On the other hand, if $H = O(n)$ then the representations distinguished by H should be the representations associated to a representation of the two-fold cover ([FK]); see [J2]. For the case of the symplectic group see [JR1].

References

[A1] J. Arthur, The trace formula and Hecke operators, *Number Theory, Trace Formulas and Discrete Groups, Symposium in Honor of Atle Selberg, Oslo, Norway, July 14–21, 1987* (K. E. Aubert, E. Bombieri, and D. Goldfeld, eds.), Academic Press, Boston, 1989, pp. 11–27.

[A2] J. Arthur, Stability and endoscopy: Informal motivation, *these* PROCEEDINGS, pp. 433–442.

[AC] J. Arthur and L. Clozel, *Simple Algebras, Base Change, and the Advanced Theory of the Trace Formula*, Annals of Mathematics Studies, vol. 120, Princeton University Press, Princeton, 1989.

[BFG] D. Bump, S. Friedberg, and D. Goldfeld, Poincaré series and Kloosterman sums, *The Selberg Trace Formula and Related Topics*, Contemporary Math., vol. 53, Amer. Math. Soc., Providence, 1986, pp. 39–49.

[C] J. Cogdell and I. I. Piatetski-Shapiro, *The Arithmetic and Spectral Analysis of Poincaré Series*, Perspectives in Mathematics, Academic Press, Boston, 1990.

[D-H] J.-M. Deshouillers and H. Iwaniec, Kloosterman sums and Fourier coefficients of cusp forms, *Invent. Math.* **70** (1982/83), 219–288.

[D-R-S] W. Duke, Z. Rudnick, and P. Sarnak, Density of integer points on affine homogeneous varieties, *Duke Math. J.* **71** (1993), 143–179.

[sF] S. Friedberg, Poincaré series for $GL(n)$: Fourier expansions, Kloosterman sums, and algebreo-geometric estimates, *Math. Zeitschrift* **196** (1987), 165–188.

[yF1] Y. Z. Flicker, On distinguished representations, *J. Reine Angew. Math.* **418** (1991), 139–172.

[yF2] Y. Z. Flicker, Cyclic automorphic forms on a unitary group, preprint.

[FH] Y. Z. Flicker and J. L. Hakim, Quaternionic distinguished representations, *Amer. J. Math.* **116** (1994), 683–736.

[FJ] S. Friedberg and H. Jacquet, Linear periods, *J. Reine Angew. Math.* **443** (1993), 91–139.

[FK] Y. Z. Flicker and D. Kazhdan, Metaplectic correspondence, *Publ. Math. I.H.E.S.* **64** (1986), 53–110.

[dG] D. Goldfeld, Kloosterman zeta functions for $\mathrm{GL}(n, \mathbf{Z})$, *Proc. Intern. Congress Math. Berkeley 1986*, vol. 1, American Mathematical Society, Providence, 1987, pp. 417–424.

[G1] J. Guo, On the positivity of central critical values of automorphic L-functions for GL(2), *Duke Math. J.* **83** (1996), 157–190.

[G2] J. Guo, On a Generalization of a result of Waldspurger, *Canad. J. Math.* **48** (1996), 105–142.

[G3] J. Guo, Uniqueness of generalized Waldspurger model for $GL(2n)$, preprint.

[H] J. Hakim, Distinguished p-adic representations, *Duke Math. J.* **62** (1991), 1–22.

[HLR] G. Harder, R. P. Langlands, and M. Rapoport, Algebraische Zyklen auf Hilbert-Blumenthal-Flachen, *J. Reine Angew. Math.* **366** (1986), 53–120.

[HR] M. Heumos and S . Rallis, Symplectic-Whittaker models for GL_n, *Pacific J. Math.* **146** (1990), 247–279.

[I] H. Iwaniec, On Waldspurger's Theorem, *Acta Arith.* **49** (1987), 205–212.

[J1] H. Jacquet, Sur un résultat de Waldspurger, *Annales Scient. Ecole Norm. Sup.* **19** (1986), 185–229.

[J2] H. Jacquet, On the non vanishing of some L-functions, *Proc. Indian Acad. Sci. (Math. Sci.)* **97** (1987), 117–155.

[J3] H. Jacquet, Représentations distinguées pour le groupe orthogonal, *C. R. Acad. Sci. Paris Sér. I Math.* **312** (1991), 957–961.

[J4] H. Jacquet, Relative Kloosterman integrals for GL(3), II, *Canad. J. Math.* **44** (1992), 1220–1240.

[J5] H. Jacquet, The continuous spectrum of the relative trace formula for GL(3) over a quadratic extension, *Israel J. Math.* **89** (1995), 1–59.

[JL] H. Jacquet and R. P. Langlands, *Automorphic Forms on* GL(2), Lecture Notes in Mathematics, vol. 114, Springer-Verlag, Berlin, 1970.

[JLR] H. Jacquet, K. Lai, and S. Rallis, A trace formula for symmetric spaces, *Duke Math. J.* **70** (1993), 305–372.

[JR1] H. Jacquet and S. Rallis, Symplectic periods, *J. Reine Angew. Math.* **423** (1992), 175–197.

[JR2] H. Jacquet and S. Rallis, Uniqueness of linear periods, *Compositio Math.* **102** (1996), 65–123.

[JY1] H. Jacquet and Y. Ye, Une remarque sur le changement de base quadratique, *C. R. Acad. Sci. Paris Sér. I Math.* **311** (1990), 671–676.

[JY2] H. Jacquet and Y. Ye, Relative Kloosterman integrals for GL(3), *Bull. Soc. Math. France* **120** (1992), 263–295.

[JY3] H. Jacquet and Y. Ye, Distinguished representations and quadratic base change for $GL(3)$, *Trans. Amer. Math. Soc.* **348** (1996), 913–939.

[aK] A. W. Knapp, Theoretical aspects of the trace formula for $GL(2)$, *these* PROCEEDINGS, pp. 355–405.

[kK] Kamal Khuri-Makdisi, On the Fourier coefficients of nonholomorphic Hilbert modular forms of half-integral weight, *Duke Math. J.* **84** (1996), 399–452.

[nK] N. V. Kuznietsov, The Petersson conjecture for cusp forms of weight zero and the Linnik conjecture. Sums of Kloosterman sums, *Math. Sbornik (N.S.)* **111**, (153, no. 3) (1980), 334–383.

[wK] W. Kohnen, Fourier coeffcients of modular forms of half-integral weight, *Math. Annalen* **271** (1985), 237–268.

[KS] S. Katok and P. Sarnak, Heegner points, cycles and Maass forms, *Israel J. Math.* **84** (1993), 193–227.

[KZ] W. Kohnen and D. Zagier, Values of L-series of modular forms at the center of the critical strip, *Invent. Math.* **64** (1981), 175–198.

[L1] R. P. Langlands, *Base Change for* GL(2), Princeton University Press, Princeton, 1980.

[L2] R. P. Langlands, Eisenstein series, the trace formula, and the modern theory of automorphic forms, *Number Theory, Trace Formulas and Discrete Groups, Symposium in Honor of Atle Selberg, Oslo, Norway, July 14–21, 1987* (K. E. Aubert, E. Bombieri, and D. Goldfeld, eds.), Academic Press, Boston, 1989, pp. 125–155.

[L3] R. P. Langlands, *Les Débuts d'une Formule des Traces Stable*, Publications Mathématiques, vol. 13, L'Université Paris VII, Paris, 1983.

[LL] J. P. Labesse and R. P. Langlands, L-indistinguishability for $SL(2)$, *Canad. J. Math.* **31** (1979), 726–785.

[M1] Zhengyu Mao, Relative Kloosterman integrals for the unitary group, *C. R. Acad. Sci. Paris Sér. I Math.* **315** (1992), 381–386.

[M2] Zhengyu Mao, Sur les sommes de Salié relatives, *C. R. Acad. Sci. Paris Sér. I Math.* **316** (1993), 1257–1262.

[M3] Zhengyu Mao, Relative Kloosterman integrals for GL(3), *Canad. J. Math.* **45** (1993), 1211–1230.

[RR] C. Rader and S. Rallis, Spherical characters on $\mathfrak{p}$-adic symmetric spaces, *Amer. J. Math.* **118** (1996), 91–178.

[S] G. Shimura, On the Fourier coefficients of Hilbert modular forms of half-integral weight, *Duke Math. J.* **71** (1993), 501–557.

[gS] G. Stevens, Poincaré series on GL(r) and Kloosterman sums, *Math. Annalen* **277** (1987), 25–51.

[pS] P. Sarnak, Diophantine problems and linear groups, *Proc. Intern. Congress Math., Kyoto 1990*, vol. 1, Math. Soc. Japan, Springer-Verlag, Tokyo, 1991, pp. 459–471.

[tS] T. A. Springer, Some results on algebraic groups with involutions, *Algebraic Groups and Related Topics*, Advanced Studies in Pure Mathematics, vol. 6, Kinokuniya Company Ltd., Tokyo, 1985, pp. 525–543.

[W] J. L. Waldspurger, Sur les coefficients de Fourier des formes modulaires de poids demi-entier, *J. Math. Pures Appl.* **60** (1981), 375–484.

[Y1] Y. Ye, Kloosterman integrals and base change for GL(2), *J. Reine Angew. Math.* **400** (1989), 57–121.

[Y2] Y. Ye, The fundamental lemma of a relative trace formula for GL(3), *Compositio Math.* **89** (1993), 121–162.

[Y3] Y. Ye, An integral transform and its applications, *Math. Annalen* **300** (1994), 405–417.

[Z] D. Zagier, Modular forms associated to real quadratic fields, *Invent. Math.* **30** (1975), 1–46.

DEPARTMENT OF MATHEMATICS, COLUMBIA UNIVERSITY, NEW YORK, NY 10027–4408, U.S.A.
E-mail address: `hj@math.columbia.edu`

Proceedings of Symposia in Pure Mathematics
Volume **61** (1997), pp. 457–471

Where Stands Functoriality Today?

Robert P. Langlands

1. Introduction

The notion of functoriality arose from the spectral analysis of automorphic forms but its definition was informed by two major theories: the theory of class fields as created by Hilbert, Takagi, and Artin and others; and the representation theory of semisimple Lie groups in the form given to it by Harish-Chandra. In both theories the statements are deep and general and the proofs difficult, highly structured, and incisive. Historical antecedents and contemporary influences aside, both were in large part created by the power of one or two mathematicians. Whether for intrinsic reasons or because of the impotence of the mathematicians who have attempted to solve its problems, the fate of functoriality has been different, and the theory of automorphic forms remains in 1997 as it was in 1967: a diffuse, disordered subject driven as much by the availability of techniques as by any high esthetic purpose.

Preoccupied with other matters I have drifted away from the field, so that I certainly have no remedy to offer. None the less, having had two occasions[1] to address the question posed in the title, I have tried to understand something of the techniques that have led to progress on the questions central to functoriality, their successes and their limitations, as well as the new circumstances in the theory of automorphic forms: problems and notions that once seemed peripheral to me and whose importance I failed to appreciate are now central, either because of their intrinsic importance or because of their accessibility.

Partly as an encouragement to younger, fresher mathematicians to take up the problem of functoriality, for that is one of the purposes of this school, but also partly as an idle reflection as to what I myself might undertake if I returned to it, I would like to respond to the title in broad terms, personal and certainly diffident and uncertain. My own mathematical experience and observation strongly suggest that progress is almost always the result of sustained awareness of the principal issues supplemented by some specific, concrete insight: begged, borrowed, or stolen or, happiest of all, distilled in one's own alembic. I offer no insights.

Initially there were two principal issues: functoriality itself, the relation between automorphic forms on different groups; and the identification of *motivic* L-functions,

1991 *Mathematics Subject Classification.* Primary 11R39, 22E55.

[1]In Edinburgh and at a conference to celebrate the 250th anniversary of Princeton University organized by its department of mathematics.

thus those associated to algebraic varieties over number fields, of which the zeta function, Artin L-functions, and the Hasse-Weil L-functions are the primitive examples, with *automorphic* L-functions, of which the zeta function and Hecke L-functions – of all types – are the first examples. The first issue arose and could be broached in the context of nonabelian harmonic analysis: representation theory and the trace formula. The second arose elsewhere but could also be broached in this context. Three key names[2] here are: Arthur for the trace formula; Kottwitz for the study of the Hasse-Weil zeta-functions of Shimura varieties; and Waldspurger for the solution (in part) of some important attendant problems, transfer and the fundamental lemma, in local harmonic analyis. The methods developed so far are difficult and important and probably essential for the construction of the complete theory but they have limitations, and I see no reason to believe that they alone will suffice.

In parallel to the notions and problems of functoriality arose a theory of special values of motivic L-functions, thus a collection of problems, and methods for dealing with some of them. Whereas progress in functoriality, as a part of the theory of automorphic forms, has been largely analytic, exploiting the more abstract consequences of abelian class-field theory but developing very few *arithmetic* arguments, some of the most incisive progress in the theory of special values has been purely arithmetic. Relations between the analytic theory and the arithmetical have often been uneasy. In Wiles's proof of the Fermat theorem they are fused: although the burden of the proof is borne by the arithmetic, an essential initial element is provided by the analysis. This is a clue that it would be reckless to ignore.

The notion of functoriality arose as a strategy – modeled on that of Artin for dealing with abelian Artin L-functions – for establishing the analytic continuation of all automorphic L-functions by showing that each was equal to a special kind of autormorphic L-function attached to $GL(n)$, a *standard* L-function. Some of these automorphic L-functions can be handled by other methods. Since functoriality is not available in general, one is obliged to resort to them, either to establish critical cases of functoriality[3] or to deal with finer properties of the automorphic L-functions: location of poles; properties of special values. These methods, in which the principal ingredients are theta functions and the Rankin-Selberg integral, are exploited in an unstructured, catch-as-catch-can way; it would be useful, especially in view of the increasing importance of automorphic L-functions and automorphic forms in the analytic theory of numbers,[4] to have a coherent notion of their function.

The analytic theory of automorphic forms can be developed over function fields as well as over number fields, but here, thanks largely to Drinfeld, some constructive, arithmetic elements were introduced very early. It is instructive to compare the proofs of the very little hard evidence we have that a nonabelian class-field theory exists, namely base change for $GL(2)$ or, more generally, $GL(n)$ with respect to cyclic extensions, with the proofs of class-field theory. In class-field theory the

[2]There are of course many more, but my purpose is to provide a brief guide to the subject, not to the literature. Names are mentioned to provide a little color and as implicit suggestions for further reading, nothing more.

[3]For example, the existence, established by Tunnell, of an automorphic form attached to the most general octahedral representation.

[4]Since I shall have occasion to discuss none of this, I mention two recent, and quite different, reviews, one by Duke in the Notices of the AMS of February, 1997, and one by Bump, Friedberg, and Hoffstein in the April, 1996 issue of the Bulletin of the AMS.

principal proofs are preceded by the construction and analysis of basic arithmetic objects, Kummer extensions. For base change in general, the little arithmetic information needed is simply taken from the abelian theory; the proofs themselves are entirely analytic. It is unlikely that a general nonabelian theory will be constructed so cheaply. It appears that over function fields the arithmetic – or diophantine – information is already implicitly at hand in the reductive group, it being possible to approximate the set $G(F)\backslash G(\mathbb{A})$ by points on an algebraic variety. His reflections on function fields also led Drinfeld to purely geometric formulations – in which the constant field of the function field is replaced by the field of complex numbers – of functoriality or rather of the duality in terms of which it is expressed. What influence the geometric study will have on the arithmetic problems is far from clear; on the other hand the very existence of the geometric notions and problems is further evidence that the dual group, in terms of which functoriality is defined, is a natural and not a factitious construct.

2. Basic Analytic Theory

In order to make the notions as accessible as possible, I shall begin by recalling in somewhat metaphorical terms the analytic theory[5] of automorphic forms on $GL(n)$ and then pass to the simplest kind of automorphic L-functions and the notion of functoriality. Although the theory of automorphic forms applies to all reductive groups and draws many insights and many techniques from this generality, for many, maybe all, serious applications to the theory of numbers, this generality is superfluous; only $GL(n)$ matters. On the other hand, it is unlikely that the theory on $GL(n)$, $n > 2$, would ever have been broached, had it not been suggested by experience with the symplectic and other groups.

The analytic theory is a spectral theory, for *families* of commuting operators, some of them differential operators and some of them difference operators (Hecke operators). At the analytic level the difficulties connected with the difference operators are of less importance, so that they are suppressed at this stage.[6]

Basic objects I. The first is the group $GL(n)$, but over two rings: the field $\mathbb{Q}$ of rational numbers and the ring $\mathbb{A}$ of adeles, which will not be defined here. The notion of an adele is just a formal expression of the importance of congruences for number theory. The basic spaces are $GL(n,\mathbb{Q})\backslash GL(n,\mathbb{A})/K$, in which K is the product of a compact subgroup of $GL(n,\mathbb{A})$ with a central subgroup of the same group. What are these spaces in reality? Examples:

(1) $n = 1$ — multiplicative group of $\mathbb{Z}/n\mathbb{Z}$;
(2) $n = 2$ — the upper half-plane divided by the subgroup of matrices in $SL(2,\mathbb{Z})$ that are congruent to I modulo an integer N.

[5]There are several expositions of this theory available; I leave the choice among them to the reader.

[6]In few contexts except the related one of the representation theory of real reductive groups has there been, I had thought, a successful, nontrivial theory of families of commuting differential operators. It is likely the rigidity of the group structure that permits the construction of the theory for automorphic forms. It has, however, recently been pointed out to me by Siddhartha Sahi that the spectral theory of real reductive groups now appears to be only an instance of more general theories, to which he suggests Macdonald's report in the Séminaire Bourbaki (exp. 797, 1995) as an introduction.

Notice that both these spaces carry the structure of algebraic varieties: the first is of dimension zero; the second is of dimension one – a *modular* curve of level N. This ceases to be so for $n > 2$.

Basic objects II. These are, first of all, the space of square-integrable functions on the space

$$GL(n,\mathbb{Q})\backslash GL(n,\mathbb{A})$$

or what, for our purposes, amounts to the same thing, on the family of spaces

$$GL(n,\mathbb{Q})\backslash GL(n,\mathbb{A})/K,$$

and its decomposition under the action of $GL(n,\mathbb{A})$. This is the spectral theory. The basic ingredients of the spectral theory are the cusp forms on $GL(m,\mathbb{A})$, $m = 1, 2 \ldots$. They, or perhaps better, if due attention is paid to the symmetries of the theory, the irreducible subspaces of cusp forms, are to be thought of as the elementary particles. I observe right off the bat that for each m there are infinitely many and that, although some are related to other objects, there is no question whatsoever of classifying them.

The spectrum on $GL(n,\mathbb{A})$ is obtained by choosing $m_1, \ldots, m_r$ such that $\sum m_i = n$ and putting particles $\pi_1, \ldots, \pi_r$, one for each m_i, together, moving with various velocities (in one-dimension!). Observe that particles of different types can sometimes fuse to form *bound* particles. This is rare and the possibilities are dealt with by Arthur's conjecture, which is very precise and proved in only a few instances. It is closely related to Ramanujan's conjecture.

The basic objects π, thus collections of elementary particles, fused or not, and with relative motion, also have a *local* structure – at each prime number and over the real numbers – that mimics the *global* adelic structure. For all but a finite number of primes these local objects have no internal structure; hence a global elementary particle undergoes complete fission at almost all primes and becomes just n objects with no internal structure (thus by definition, globally or locally, attached to the constant function on $GL(1)$) moving about with different velocities $s_1, s_2, \ldots, s_n$. (To push the metaphor to unwarranted extremes, they obey Bose-Einstein statistics; so the order does not matter!) What matters is in fact the conjugacy class of the matrix

$$A_p(\pi) = \begin{pmatrix} p^{is_1} & 0 & \ldots & 0 & 0 \\ 0 & p^{is_2} & \ldots & 0 & 0 \\ \vdots & \vdots & \ddots & \vdots & \vdots \\ 0 & 0 & \ldots & p^{is_{n-1}} & 0 \\ 0 & 0 & \ldots & 0 & p^{is_n} \end{pmatrix}.$$

Attach an incomplete L-function to π by

$$\text{(A)} \qquad L(s,\pi) = \prod{}' \frac{1}{\det(I - A_p(\pi)/p^s)}.$$

The product can be completed at the primes where local objects have an internal structure, thus in number-theoretical language where there is ramification, and also at the infinite prime. Classical methods, developed by many people but always resting ultimately on the imbedding of $GL(n)$ in the additive group of $n \times n$ matrices and the use of Fourier analysis, allow one to extend these functions (completed or not) to all of the complex plane as meromorphic functions with a very limited number of poles. For reasons to be explained, I refer to these L-functions as the *standard* L-functions. (This terminology has unfortunately been corrupted.)

The particle metaphor, in which what is otherwise referred to as an *automorphic representation* or *automorphic form* is viewed as a collection of particles with structure moving about at various speeds, fits well the assemblage of an automorphic representation ρ on $GL(l)$ and another σ on $GL(m)$ to construct an automorphic representation $\pi = \rho \oplus \sigma$ on $GL(n)$, $n = l + m$. This is an operation that is well understood in the context of the spectral theory. It is the theory of Eisenstein series. There is another operation that is not well understood, although its existence is conjectured as a part of functoriality. If we have an automorphic repesentation ρ of $GL(l)$ and one σ of $GL(m)$ and if this second is just a collection of m structureless particles moving about at velocities $s_1, \ldots, s_m$ then we can construct the automorphic representation $\rho \otimes \sigma$ of $GL(n)$, $n = lm$, as the sum of $\rho(s_1), \ldots, \rho(s_m)$. The representation $\rho(s)$ is just ρ with a modified velocity and is constructed by multiplying ρ by the one-dimensional representation $|\det|^{is}$. There is reason to believe that for all ρ and all σ an automorphic representation that deserves, because of its local properties, to be denoted $\rho \otimes \sigma$ can be constructed.

3. Functoriality

Functoriality is a succinct hypothesis that is easy to state once the basic notions are at hand and whose consequences are immediate and serious: the analytic continuation of Artin's L-functions and the general form of Ramanujan's conjecture.

As remarked it is believed that it is possible to construct $\rho \otimes \sigma$ even if σ has internal structure. This would be a basic form of functoriality. What is the general form and how does one arrive at it? Notice, before turning to functoriality, that the basic categories in which it is possible both to add objects of degrees l and m to arrive at one of degree $l + m$ and to multiply them to arrive at one of degree lm are the categories defined by the finite-dimensional representations of a given group (finite, compact, or algebraic). An automorphic representation is, however, almost always an infinite-dimensional representation of a very large group, and the pertinent degree has nothing to do with its dimension.

So functoriality refers to objects and categories that have not yet appeared in this review. We can study automorphic representations π on any (reductive) algebraic group G defined over $\mathbb{Q}$ (even over number fields and function fields). The principal groups are $GL(n)$. To any G we associate a complex group ${}^LG$, its L-group, by a procedure that is now fairly well known, and that I do not describe in any further detail here, except in the context of examples. For a given π, in general an automorphic repesentation of $G(\mathbb{A})$, one can associate to almost all primes p a conjugacy class $A_p(\pi)$ in ${}^LG$. If ρ is a *finite-dimensional holomorphic representation* of the L-group ${}^LG$, we can introduce the, perhaps incomplete, L-function

$$L(s, \pi, \rho) = \prod_p{}' \frac{1}{\det(I - \rho(A_p(\pi)/p^s))}. \tag{B}$$

The study of Eisenstein series yields a large number of Euler products with meromorphic continuation that can be written in this form. Since the L-group of $GL(n)$ is $GL(n, \mathbb{C})$ or at least has $GL(n, \mathbb{C})$ as a quotient, according to the context in which it is appropriate to work, the products (A) are a special case of (B).

Once the Euler products are so expressed, it is very strongly suggested that they can be continued not just for those ρ that arise from Eisenstein series, but for all ρ. Class-field theory, in the form given to it by Emil Artin, immediately suggests a strategy. Given π, an automorphic representation of the group $G(\mathbb{A})$, and ρ, an n-dimensional holomorphic representation of the complex group ${}^LG$, show that there is an automorphic representation Π of $GL(n,\mathbb{A})$ such that

$$\{A_p(\Pi)\} = \{\rho(A_p(\pi))\}$$

for almost all p. Then

$$L(s,\pi,\rho) = L(s,\Pi),$$

and, as already observed, the analytic continuation of the right-hand side is essentially a matter of Poisson summation.

This is the first form of functoriality that suggested itself. Applying it to the trivial group $\{1\}$ over a finite algebraic extension of $\mathbb{Q}$ viewed as a group over $\mathbb{Q}$ by restriction of scalars, so that the (somewhat protean) L-group becomes the Galois group $\mathrm{Gal}(K/\mathbb{Q})$ of a finite extension, I immediately realized that it would yield the nonabelian class-field theory for which Artin had searched unsuccessfully.

An extension of the first notion of functoriality that suggests itself quickly is that if G and G' are two groups and

$$\phi : {}^LG \to {}^LG',$$

then there is a corresponding map of automorphic forms $\pi \to \pi'$ such that $A_p(\pi')$ and $\phi(A_p(\pi))$ are conjugate for all almost all p. It is this non-formal, deep-lying functoriality to which the title refers.

Some progress, even substantial progress, has been made on functoriality since the idea was introduced in the late sixties. None the less we still do not understand it in any serious sense. Since functoriality, once established, entails immediately the analytic continuation of all the functions $L(s,\pi,\rho)$, and all L-functions with Euler products appearing in the theory of automorphic forms are – as yet, but probably for good – of this type, there would seem to be little point in pursuing seriously the various methods introduced for dealing with this or that special case:

(1) Fourier coefficients of Eisenstein series: Langlands, Shahidi;
(2) Hecke's first method;
(3) integration against Eisenstein series: Rankin, Selberg, ... ;
(4) integration against a product of theta functions and Eisenstein series: Shimura, ...

On the other hand, for the moment these methods sometimes yield very important information that is not otherwise available, so that it would be premature to discard them. I am not sure what their ultimate role will be.

4. The Trace Formula

Functoriality itself can, for the moment, only be established in a very limited number of cases. One technique is to use the trace formula, thus the spectral theory, some local information, and some form – to use once again a colorful language – of the uncertainty principle. The local information – endoscopy, transfer, and the fundamental lemma – has been hard won, by Waldspurger and many others, and is still incomplete. I myself thought about these matters for ten years without making any serious headway. The global arguments, in part an elaborate induction, are being carried out largely by Arthur and demand the massive deployment of technically difficult analysis in a highly structured conceptual context.

In a number of important cases (infinitely many but still of a very special nature) a homomorphism

$$\phi : {}^LG \to {}^LG'$$

is accompanied by a natural map from conjugacy classes in a semi-direct $G' \rtimes \Sigma$, where Σ is a finite cyclic group to conjugacy classes in G. Note the transition from the L-groups to the groups themselves. Then the map $\pi \to \phi(\pi)$ predicted by functoriality is reflected in character identities, and thus in the trace formula.

These character identities usually manifest themselves in relatively simple, easily understood or well understood, contexts – in particular: finite-dimensional representations of G and $G' \rtimes \Sigma$; representations of real groups; and unramified representations. New identities would be greatly appreciated. Once the basic identity is recognized, the difficulty is, first, to establish sufficiently many of the identities in local harmonic analysis suggested by it, and then to apply the trace formula to the two groups $G(\mathbb{A})$ and $G'(\mathbb{A}) \rtimes \Sigma$ simultaneously in order to compare their spectra. Note that the trace formula compares *geometric* information, thus integration over conjugacy classes in $G(\mathbb{Q})$ or $G'(\mathbb{Q}) \rtimes \Sigma$, on one side with spectral information on the other. The uncertainty principle implies that if most terms on one side vanish then all terms on both sides vanish. Various components of the spectrum appear in different ways, and it was, I believe, Arthur's reflections on the identities entailed between the terms corresponding to bound states that led him to the conjecture to which I alluded earlier. I cannot give it here. It is very delicate, and very important, and is established in some cases, but I am not certain that anyone, even Arthur himself, has fully understood its implications. Some investigators have overlooked it to their later embarrassment.

A basic ingredient in Arthur's conjecture is the *generalized* Ramanujan conjecture that affirms that if π is a cusp form for $GL(m)$ then the conjugacy classes $A_p(\pi)$ have eigenvalues of absolute value one. The generalized Ramanujan conjecture is easily shown to be a consequence of functoriality, and partial results for functoriality lead to partial results for the conjecture.

Arthur himself has focussed on two cases: either the group $G' \rtimes \Sigma$ is defined by a group G' given as $GL(n)$ over a cyclic extension of the ground field; or Σ is of order two and its nontrivial element acts as the involution $A \to {}^tA^{-1}$ on $GL(n)$ over the ground field. The first, cyclic base change, is of importance in connection with Artin's conjecture; the second gives the transfer of automorphic forms associated to the standard homomorphisms of the L-groups of orthogonal and symplectic groups into $GL(n)$. There are other possibilities,[7] still largely neglected. Moreover, the relative trace formula of Jacquet perhaps deserves the attention of an ambitious, talented youth. At the same time it is manifest that although it is unlikely that the final goals can be reached without the trace formula and the methods developed by Arthur, they are inaccessible with identities between traces alone. It is natural to suppose that inequalities between traces, combined perhaps with a deeper group-theoretical analysis (even at the level of finite groups), would be the appropriate tool, but no-one has been able to do anything with this.

The two cases of functoriality not yet established that lie nearest at hand are base change for icosahedral extensions and the existence of automorphic forms on

[7]Rogawski's study of the unitary group in three variables is an example.

$GL(n)$, $n > 4$, attached to automorphic forms on $GL(2)$ via the homomorphisms of dual groups $GL(2, \mathbb{C}) \to GL(n, \mathbb{C})$ defined by representations on symmetric tensors. These are two basic outstanding problems of the subject. A solution of the second would yield all forms of Ramanujan's conjecture for $GL(2)$. The Ramanujan-Petersson conjecture for holomorphic forms has of course been solved through the last of the Weil conjectures, whose proof was seriously influenced by ideas from the theory of automorphic forms, but the analogous problem for Maaß forms as well as the Selberg conjecture remain open.

I add that there are two other methods to establish functoriality in certain cases: converse theorems and, through Howe's conjecture, the oscillator representation. Converse theorems, although at the present stage occasionally extremely useful, sometimes indispensable, are, if class-field theory is taken as our paradigm, philosophically perverse, putting the cart before the horse. The oscillator representation, thus theta series, yields automorphic representations that are, in the sense of Arthur's conjecture, quite degenerate. Although the connection between theta series and functoriality is quite delicate, and therefore quite fascinating, I am not convinced that it is basic. None the less theta series, and therefore automorphic forms of half-integral weight, remain for me a troubling philosophical puzzle. They are obviously important; they are not unrelated to functoriality; yet the notions of functoriality cannot accommodate them.

Philosophically perverse or not, converse theorems have been very useful and there are eminent mathematicians who attach a great deal of importance to them; so it is amusing to see how to express them in the context of our metaphor. As observed, the constant function on $GL(1, \mathbb{A})$ is to be thought of as the particle without structure. The scattering matrix for the interaction of this structureless particle with an elementary particle π on $GL(n, \mathbb{A})$ is a quotient of standard L-functions

$$\Lambda(s, \pi)/\Lambda(s+1, \widetilde{\pi}).$$

Here the letter Λ is used rather than L to indicate that the function is a product over all primes, including those at infinity, so that $\Lambda(s, \pi)$ is the product of $L(s, \pi)$ with some Gamma functions and some elementary functions. Thus the standard L-functions for π describe the interaction of π with a structureless particle.

To describe the interaction of an elementary particle π on $GL(n, A)$ with another elementary particle π' on $GL(m, A)$ one needs the L-functions associated to the tensor-product representation ρ of the L-group $GL(n, \mathbb{C}) \times GL(m, \mathbb{C})$ of $GL(n, \mathbb{A}) \times GL(m, \mathbb{A})$ in $GL(mn, \mathbb{C})$. Usually one writes

$$\text{(C)} \qquad L(s, \pi \times \pi', \rho) = L(s, \pi \times \pi').$$

The usual converse theorems assert that if for each m, $1 \leq m < n$, there is a family of functions associated to all π' on $GL(m, \mathbb{A})$ with the analytic properties of the collection (C) then these families are defined by an elementary particle π. Expressed with our metaphor, the assertion is that if elementary particles in degree less than n are scattered by an unknown object that seems to be an elementary particle of degree n then it is indeed such an elementary particle.

An obvious question is whether the Riemann hypothesis is satisfied for the standard L-functions $L(s, \pi)$, and thus for all L-functions. It is generally supposed that this is so, although so far as I know the general hypothesis has not been examined numerically. The intimate connection, already observed, between the L-functions $L(s, \pi)$, in particular the zeta function itself which corresponds to the trivial representation of $GL(1,\mathbb{A})$, and the Eisenstein series appearing in the spectral theory has suggested, first, I believe, to Selberg, that the spectral theory might be used to establish the Riemann hypothesis. Selberg crushed any simple hopes by constructing, in the context of $GL(2)$ (or $SL(2)$), nonarithmetic groups for which the analogue of the hypothesis is false. But the zeta function appears in the spectral analysis of all $GL(n)$, $n = 1, \ldots$, and these are very rigid objects. Moreover, Ribet, in a paper to which we shall come, and others have successfully applied information about automorphic forms on $GL(n)$, $n > m$, to the arithmetic study of automorphic functions attached to $GL(m)$, and it is not utterly far-fetched to imagine that something similar could be done analytically. (For different implementations of such a strategy, see the review of Bump, Friedberg, and Hoffstein.) I myself have no idea, but I do remark that if π is a cusp form on $GL(m)$ then $L(s,\pi)$ appears in the spectral analysis of all $GL(mn)$, $n = 1, 2, \ldots$, so that there is at least a certain coherency in the suggestion.

5. Hasse-Weil L-functions

As already observed, the principle of functoriality is more or less the suggestion that the collection of automorphic representations behaves like the collection of (finite-dimensional) representations of a group. For automorphic representations themselves, the putative group is large and elusive, so that it is best not to think much about it. The corresponding local objects can, however, be given concrete meaning. For example, representations of a real reductive group $G(\mathbb{R})$ can be classified by homomorphisms into the L-group ${}^LG$ of an extension

$$\mathbb{C}^\times \to W \to \{1, \sigma\}.$$

The element σ is to be thought of as complex conjugation so that $\sigma z \sigma^{-1} = \bar{z}$, $z \in \mathbb{C}^\times$. In addition, the appropriate choice for σ^2, an element of $\mathbb{C}^\times$, is -1.

Then, for example, the representations of $GL(n,\mathbb{R})$ are classified, in an appropriate sense, by the continuous homomorphisms of W into the L-group of $GL(n)$ which is nothing but $GL(n,\mathbb{C})$. Only homomorphisms such that the image of $\mathbb{C}^\times$ is diagonalizable are allowed. The diagonal elements are then characters of $\mathbb{C}^\times$ and thus of the form

$$z \longrightarrow z^r (z\bar{z})^s,$$

where r is an integer and s a complex number. A very important class of homomorphisms, and thus of representations of $GL(n,\mathbb{R})$, is determined by demanding that in these diagonal characters the complex number s always be an integer, so that the character is of the form

$$z \longrightarrow z^p \bar{z}^q.$$

The presence of the two integers p and q suggests a connection with Hodge theory; so I say that representations of $GL(n,\mathbb{R})$ of this kind are of *Hodge* type.

We can then distinguish a very important class of automorphic representations π, those whose component at infinity is of Hodge type. (I have not stressed it, but an automorphic representation π of any group $G(\mathbb{A})$ can be expressed as a tensor product $\bigotimes \pi_v$, where v runs over the places, finite and infinite of $\mathbb{Q}$, thus $v = \infty, 2, 3, 5, \ldots$.) We recall, in addition, that using the Weil zeta-functions attached to varieties over finite fields it is possible to attach to a projective variety V over $\mathbb{Q}$ (more generally over finite extensons of $\mathbb{Q}$ and function fields), a global zeta function, the Hasse-Weil zeta-function and that this zeta function can be expressed as a quotient of products of Euler products

$$\frac{\prod_{i=2j+1,0\leq j<\dim(V)} L^{(i)}(s,V)}{\prod_{i=2j,0\leq j\leq\dim(V)} L^{(i)}(s,V)}. \tag{D}$$

It is generally supposed that each factor $L^{(i)}(V,s)$ that appears here, either in the numerator or in the denominator, is equal to a standard automorphic L-function $L(s,\pi)$ for a suitable $\pi = \pi(i,V)$. This π will necessarily be of Hodge type at infinity.

The group-theoretical procedure used by Artin to factor the zeta function of a finite extension of a number field into a product of Artin L-functions can be extended to a factorization of the functions $L^{(i)}(s,V)$ although the properties of this factorization are far from understood. There are very few varieties for which the analytic continuation of (D) can be established, and then only by proving first that each factor is equal to an $L(s,\pi)$. For varieties of dimension zero, the function $L^{(0)}(s,V)$ can be dealt with, but its factors, the Artin L-functions, can be analytically – as opposed to meromorphically – continued only in few cases, again by showing that they are equal to an $L(s,\pi)$. The classical abelian reciprocity laws and the Taniyama conjecture are both examples of this technique.

6. New and Different Problems

There is, in addition, a converse hypothesis: every standard automorphic L-function $L(s,\pi)$ defined by a representation π such that π_∞ is of Hodge type appears as a "factor" of a Hasse-Weil L-function.[8] This is not the time to attempt to give a precise meaning to the term "factor." For these factors, arising as they do from geometry and arithmetic, there is an entirely new complex of questions, all concerning their values at integers, especially negative integers, that transcend the simpler problem of analytic continuation and that are not related to the Riemann hypothesis.

There are, following Kato, three "phases" to the problem, to which the names of many, many people are attached: Tate, Birch, Swinnerton-Dyer, Shimura, Deligne, Beilinson, Iwasawa, Bloch, Kato, Fontaine, Perrin-Riou, ...

(1) When divided by appropriate factors – periods of integrals on the associated varieties as well as "regulators" – the values $L(n,\pi)$, $n\in\mathbb{Z}$, are algebraic. At integers n where the function has zeros or poles the order is predicted, and the appropriate power of $z-n$ is removed before calculating the *nonzero* value of the function thus obtained.
(2) These algebraic values lie in well-defined number fields and satisfy congruences that permit the introduction of p-adic L-functions.
(3) The values themselves, when normalized to be algebraic, or the values of the p-adic L-functions, are themselves very important, and are equal to numbers defined by diophantine data, thus implicitly by the solutions of diophantine equations.

[8] A third basic outstanding problem is to establish this for Maaß forms of Hodge type on $GL(2,\mathbb{A})$.

For the zeta function itself at even positive or odd negative integers phase (1) is classical and elementary. For example $\zeta(2) = \pi^2/6$ is an equality easily proved. Phase (2) is not so difficult, but phase (3) is very deep.

We are now faced with a number of questions, almost all embarrassing. First of all, to what extent can we establish that the factors of Hasse-Weil zeta-functions are indeed equal to standard automorphic L-functions. There is the classical case of the zero-dimensional varieties defined by abelian extensions of number fields, in which the methods are deep and, within their express limitations, general. There is also the Eichler-Shimura theory for quotients of the upper half-plane by subgroups of the modular group, but this is to some extent simply part of the theory of complex multiplication and thus of class fields. The theory for Shimura varieties, to which I shall return and which is a generalization of the theory of Eichler-Shimura, is also, in spite of the very many difficult and deep theorems that have been established by Kottwitz and many others, still largely a part of the theory of complex multiplication. Thus, as with the work of Arthur, still in progress, the work of Kottwitz and his collaborators may leave us, even when all clearly formulated questions asked at present are answered, with one more. Where do we go from here?

The issue is even more complex. Apparently we can only effect the analytic continuation of the factors of the Hasse-Weil zeta-function if we first exhibit them as automorphic L-functions, and it certainly only makes sense to inquire about particular properties of their values at negative integers after they have been analytically continued. On the other hand, analytic continuation is effected uniformly for automorphic L-functions $L(s,\pi)$, without inquiring whether π_∞ is of Hodge type. Thus in order to make sense of the values, we have to pass to a set of functions for which the values no longer are expected to have any special properties. This appears to me somewhat paradoxical!

On closer examination, matters are seen to be even more turbid. A Shimura variety S is at first a complex algebraic variety that can be represented as

$$G(\mathbb{Q})\backslash G(\mathbb{A})/K, \tag{E}$$

a set already introduced. The group G here is certainly not arbitrary, but there are several possibilities. It can for example be the symplectic group in any number of variables or a unitary group. The theory of complex multiplication then permits the definition of this complex variety as a variety over $\mathbb{Q}$ (or over a specific well-defined number field). The first goal of the theory is to express the Hasse-Weil zeta-function of what is now diophantine object as a product not of standard L-functions but of automorphic L-functions $L(s,\pi,\rho)$ associated to G itself or to one of its endoscopic groups H. (Then ${}^LH \subset {}^LG$ and ρ is a representation of ${}^LH$.) Of course functoriality predicts that $L(s,\pi,\rho)$ will be a standard L-function but it may be possible to achieve the first goal without proving this and even without being certain that $L(s,\pi,\rho)$ can be analytically continued.

What sometimes happens is that by one or the other of the special techniques listed (Hecke-Rankin-Selberg-Shimura) the functions $L(s,\pi,\rho)$ can be analytically continued, and then special values appear as integrals over a subset of (E) and can be interpreted as periods. It appears that accidental, conditional phenomena are being used to establish general principles, a philosophically disagreeable circumstance that may or may not be intrinsic to class-field theory and all its hypothetical generalizations. Even though at present these special techniques yield very little of what is foreseen, it is not entirely out of the question that they will remain an essential part of the final argument. It is useful to keep an open mind.

7. Shimura Varieties

The original purpose of expressing the Hasse-Weil zeta-functions of Shimura varieties as a product of automorphic L-functions was formed shortly after the introduction of these L-functions and for no other reason than to establish their utility and even necessity.

The techniques used to express the functions $L^{(i)}(s,S)$ as a product of functions $L(s,\pi,\rho)$ are similar to those used by Arthur, but there are additional ingredients. The product of functions $L^{(i)}(s,S)$ or rather its logarithm is expressed by spectral data. The trace formula converts the sum of spectral data into a sum of geometric data. On the other hand the Hasse-Weil zeta-function, or rather its logarithm, can also be expressed by geometric data, the number of points on the variety S over finite fields. Thus the primary problem is to relate the set of points on S with coordinates in a given finite field to conjugacy classes in $G(\mathbb{Q})$. This, it turns out, can be done, by no means easily, with the help of the theory of complex multiplication. There are two additional difficulties, not yet entirely overcome: combinatorial problems and the singularities of the completions of the variety S. By a stroke of luck, the combinatorial problems are the same as those of the fundamental lemma, and in so far as that lemma has been established they are solved. The difficulties arising from the singularities form a chapter, perhaps several chapters, in the study of intersection cohomology, and are being treated as such.

There is, so far as I know, no reason to believe that all motivic L-functions appear as factors of L-functions associated to Shimura varieties and the natural vector bundles on them, so that Shimura's reformulation (but I may misunderstand the somewhat obscure history) of the Taniyama conjecture pertains to an anomaly. Generalized in the most naive way (as no-one has ever suggested was appropriate!) the reformulation, sometimes known as the modular-curve conjecture, would ask that motivic L-functions appear as factors of the L-functions attached to Shimura varieties, whereas Taniyama's original conjecture generalized would only ask that they be identified as standard L-functions.

If the focus is on the L-functions that are attached to the Shimura varieties themselves, then the surprise, but not the disappointment, is that by and large the automorphic L-functions with which the motivic L-functions are identified are not those whose analytic continuation can be effected by the provisional methods at our disposition. For example, for the projective symplectic group $PSp(2n)$ that defines the Siegel varieties, which are the best known of the higher-dimensional Shimura varieties, the L-group is the spinor group attached to the orthogonal group in $2n+1$ dimensions and the principal factors of the zeta function are of the form $L(s,\pi,\sigma)$, where σ is the spinor representation. Except for $n=1$ and $n=2$, these are scarcely accessible at present.

8. Constructive Elements

In the most striking case so far in which it has been established that a factor of a Hasse-Weil zeta-function is a standard L-function, the proof by Wiles of the suggestion of Taniyama as modified by Shimura, the two aspects of $GL(2)$ – carrier of the standard L-functions and carrier of Shimura varieties – are entwined in very curious ways that I do not understand. In some sense the argument begins with an object given by the analytic theory, thus one obtained by examining $GL(2)$ in its first guise where no deformation is possible. Then it quickly abandons $GL(2)$ in this attire, and passes to a modular curve, thus to the Shimura variety attached to $GL(2)$, and begins a p-adic deformation argument.

These arguments reflect, I suppose, Wiles's experience with the conjectures of phase (3). For the layman the significance of those in phase (1) is sometimes easier to grasp. For an elliptic curve E over $\mathbb{Q}$ defined by an equation like $y^2 = ax^3 + bx^2 + cx + d$, Taniyama's suggestion, as I interpret it, is that $L^{(1)}(s, E)$ is a standard automorphic L-function for $GL(2)$. Shimura, again in my interpretation, goes further and states that there is a surjective map of a modular curve C of a certain level N onto E, and Weil allows us to predict the level. Apparently (although I have never found the time to carry out such computations) assured of all this – by the conjectures as theorems – we can with time and effort calculate the map $C \to E$ and thus the π for which $L^{(1)}(s, E) = L(s, \pi)$. The representation π in hand we can also calculate the value or the order of vanishing of $L(s, \pi)$ for any s. Then the conjectures of phase (1) at the integer $s = 1$, in the special case of an elliptic curve a part of the Birch-Swinnerton-Dyer conjecture, predict the rank of the group of $\mathbb{Q}$-rational points on E. What happens, however, so far as I can see is that, when this conjecture is proved, it is proved by exhibiting concretely points, the Heegner points, from which $\mathbb{Q}$-rational points can be explicitly constructed. Thus the proofs seem to give more information, even more appealing information, than the conjectures themselves.

Certainly phase (1) is, for me at least, fraught with more easily comprehended consequences than phases (2) and (3). Examined attentively, however, it has extremely problematic aspects, and I cannot resist posing a question to which some readers may have an answer. The Beilinson conjecture, a part of phase (1), includes the Tate conjecture explicitly. The Tate conjecture is certainly closely related to the Hodge conjecture and to its general forms. Does it imply the Hodge conjectures; or do the Hodge conjectures contain an analytic element? Can it be proved independently of the Hodge conjecture, or does phase (1) also contain an implicit analytic element? It is perhaps well to remind ourselves, when passing recklessly, gay of heart and light of foot, over phase (1) and on to phases (2) and (3) that we are abandoning questions whose very nature we have as yet failed to understand.

Even for a point over $\mathbb{Q}$, thus for the zeta function, phases (2) and (3) are extremely difficult. Their origins lie, I believe, in divisibility properties of class numbers of cyclotomic fields, thus in the work of Kummer, the founder of the theory of cyclotomic fields and therefore, in some sense, also the founder of class-field theory. If I understand correctly, phase (3) is for a point essentially the main conjecture of Iwasawa theory and was only solved a little more than ten years ago. It appears to be no accident that more than one of the principal contributors to its solution also contributed to the resolution of Fermat's theorem.

To deal with phase (2) or with phase (3) even for a point, it seems absolutely necessary to appeal to the theory of automorphic forms on $GL(2)$ – more precisely to the study of modular curves – thus, so to speak, to move from $GL(0)$ to $GL(2)$, and in particular (recall the remarks on the Riemann hypothesis) to exploit the theory of Eisenstein series not over $\mathbb{C}$ but over p-adic fields. One reason is that some of the arithmetic objects, whose existence is predicted by the conjectures of phase (3), have to be constructed explicitly, and this can only be done with the aid of the explicitly constructed diophantine objects already at hand, the modular curves.

For example, the value of the zeta function at $1-2m$, $m=1,2,\dots$, is expressed in terms of a Bernoulli number as $(-1)^m B_m/2m$. Let p be a prime number and μ_p a p-th root of unity. Let $\mathfrak{A}$ be the ideal class group of $\mathbb{Q}(\mu_p)$ and consider $\mathfrak{C}=\mathfrak{A}/\mathfrak{A}^p$. The Galois group $\mathrm{Gal}(\mathbb{Q}(\mu_p)/\mathbb{Q})$ acts on $\mathfrak{C}$ and $\mathfrak{C}$ decomposes as

$$\mathfrak{C}=\sum_{i \pmod{p-1}} \mathfrak{C}_i,$$

where

$$\mathfrak{C}_i=\{\mathfrak{c}\in\mathfrak{C}\mid \mathfrak{c}^\sigma=\mathfrak{c}^{\chi(\sigma)^i}\ \forall\sigma\in\mathrm{Gal}(\mathbb{Q}(\mu_p)/\mathbb{Q})\},$$

and

$$\mu_p^\sigma=\mu_p^{\chi(\sigma)}.$$

Suppose k is even and $2\le k\le p-3$. As what I suppose is one of the simplest expressions of phase (3), p divides $\zeta(1-k)$ if and only if the group $\mathfrak{C}_{1-k}$ is not trivial. Since this group is determined by the existence of abelian extensions of $\mathbb{Q}(\mu_p)$ with various properties, the implication

$$\mathfrak{C}_{1-k}\ne\{1\}\implies p\mid B_k,$$

proved by Herbrand entirely within the context of class-field theory and Kummer extensions, is a constraint on the existence of algebraic numbers with prescribed properties. On the other hand, the converse implication requires the construction of abelian extensions of $\mathbb{Q}(\mu_p)$ with prescribed properties, and this construction is not carried out with Kummer extensions directly. Rather[9] the construction utilizes the arithmetic theory of automorphic forms for $GL(2)$, exploiting among other things the appearance of zeta functions and Dirichlet L-functions as Fourier coefficients of Eisenstein series. Thus it is not so far in spirit from the techniques used by Shahidi for the study of automorphic L-functions. Difficult theorems for the smaller group are proved as consequences of easier theorems for the larger.

9. Conclusion

I have certainly not been able to digest the material pertinent to the questions broached in this review, and may never succeed in doing so. The question as to where we go from here, in so far as it applies to Hasse-Weil zeta-functions, and thus to Shimura varieties, especially in view of their connections with the problems evoked in Kato's three phases, is certainly not one I would venture to answer. The possibilities intimated by the recent intermingling of the analytic and the algebraic techniques are manifold but still, to me at least, obscure.

For functoriality too, it is best to be prudent. Once the distinction between the automorphic representations whose component at infinity is Hodge and the general class of all automorphic representations is made, there is one point to be observed. There is less to be established for the general class, and it will probably need to be established analytically and directly because the deformation provided by, say, a p-adic theory is unlikely to be available. On the other hand, the Artin L-functions, thus the factors of the Hasse-Weil zeta-functions of zero-dimensional varieties, are

[9]The relevant paper by Ribet appears in Inv. math., vol. 34, 1976.

difficult to distinguish from elements of this general class. To repeat myself in different words: the Hasse-Weil zeta-functions of zero-dimensional varieties will probably have to be treated directly, without the help of the auxiliary algebraic techniques such as deformation that may be available for the zeta functions of positive dimension.

In some sense the immediate future of functoriality is less bright than that of Shimura varieties. Beyond the work of Arthur and its possible extensions there is a clearly visible horizon that we have no idea how to cross. When discouraged by this reflection I recall, among other things, that the notion of endoscopy, which has nourished representation theory and harmonic analysis for almost two decades, arose not as an internal development of the analytic theory but from the study of the zeta functions of Shimura varieties.

There is certainly no lack of problems, large and small, any one of which may offer a clue.

School of Mathematics, Institute for Advanced Study, Princeton, NJ 08540, U.S.A.
E-mail address: `rpl@ias.edu`

Index

Selected Titles in This Series

(*Continued from the front of this publication*)

ISBN 0-8218-0609-2

9 780821 806098